Applied Mathematical Sciences
Volume 38

Editors
F. John   J.E. Marsden   L. Sirovich

Advisors
M. Ghil J.K. Hale J. Keller
K. Kirchgässner B.J. Matkowsky
J.T. Stuart A. Weinstein

Applied Mathematical Sciences

1. *John:* Partial Differential Equations, 4th ed.
2. *Sirovich:* Techniques of Asymptotic Analysis.
3. *Hale:* Theory of Functional Differential Equations, 2nd ed.
4. *Percus:* Combinatorial Methods.
5. *von Mises/Friedrichs:* Fluid Dynamics.
6. *Freiberger/Grenander:* A Short Course in Computational Probability and Statistics.
7. *Pipkin:* Lectures on Viscoelasticity Theory.
8. *Giacoglia:* Perturbation Methods in Non-linear Systems.
9. *Friedrichs:* Spectral Theory of Operators in Hilbert Space.
10. *Stroud:* Numerical Quadrature and Solution of Ordinary Differential Equations.
11. *Wolovich:* Linear Multivariable Systems.
12. *Berkovitz:* Optimal Control Theory.
13. *Bluman/Cole:* Similarity Methods for Differential Equations.
14. *Yoshizawa:* Stability Theory and the Existence of Periodic Solution and Almost Periodic Solutions.
15. *Braun:* Differential Equations and Their Applications, 3rd ed.
16. *Lefschetz:* Applications of Algebraic Topology.
17. *Collatz/Wetterling:* Optimization Problems.
18. *Grenander:* Pattern Synthesis: Lectures in Pattern Theory, Vol. I.
19. *Marsden/McCracken:* Hopf Bifurcation and Its Applications.
20. *Driver:* Ordinary and Delay Differential Equations.
21. *Courant/Friedrichs:* Supersonic Flow and Shock Waves.
22. *Rouche/Habets/Laloy:* Stability Theory by Liapunov's Direct Method.
23. *Lamperti:* Stochastic Processes: A Survey of the Mathematical Theory.
24. *Grenander:* Pattern Analysis: Lectures in Pattern Theory, Vol. II.
25. *Davies:* Integral Transforms and Their Applications, 2nd ed.
26. *Kushner/Clark:* Stochastic Approximation Methods for Constrained and Unconstrained Systems.
27. *de Boor:* A Practical Guide to Splines.
28. *Keilson:* Markov Chain Models—Rarity and Exponentiality.
29. *de Veubeke:* A Course in Elasticity.
30. *Shiatycki:* Geometric Quantization and Quantum Mechanics.
31. *Reid:* Sturmian Theory for Ordinary Differential Equations.
32. *Meis/Markowitz:* Numerical Solution of Partial Differential Equations.
33. *Grenander:* Regular Structures: Lectures in Pattern Theory, Vol. III.
34. *Kevorkian/Cole:* Perturbation Methods in Applied Mathematics.
35. *Carr:* Applications of Centre Manifold Theory.
36. *Bengtsson/Ghil/Källén:* Dynamic Meteorology: Data Assimilation Methods.
37. *Saperstone:* Semidynamical Systems in Infinite Dimensional Spaces.
38. *Lichtenberg/Lieberman:* Regular and Chaotic Dynamics, 2nd ed.
39. *Piccini/Stampacchia/Vidossich:* Ordinary Differential Equations in $\mathbf{R}^n$.
40. *Naylor/Sell:* Linear Operator Theory in Engineering and Science.
41. *Sparrow:* The Lorenz Equations: Bifurcations, Chaos, and Strange Attractors.
42. *Guckenheimer/Holmes:* Nonlinear Oscillations, Dynamical Systems and Bifurcations of Vector Fields.
43. *Ockendon/Taylor:* Inviscid Fluid Flows.
44. *Pazy:* Semigroups of Linear Operators and Applications to Partial Differential Equations.
45. *Glashoff/Gustafson:* Linear Operations and Approximation: An Introduction to the Theoretical Analysis and Numerical Treatment of Semi-Infinite Programs.
46. *Wilcox:* Scattering Theory for Diffraction Gratings.
47. *Hale et al:* An Introduction to Infinite Dimensional Dynamical Systems—Geometric Theory.
48. *Murray:* Asymptotic Analysis.
49. *Ladyzhenskaya:* The Boundary—Value Problems of Mathematical Physics.
50. *Wilcox:* Sound Propagation in Stratified Fluids.
51. *Golubitsky/Schaeffer:* Bifurcation and Groups in Bifurcation Theory, Vol. I.

(continued following index)

A.J. Lichtenberg
M.A. Lieberman

Regular and Chaotic Dynamics

Second Edition

With 225 Figures

Springer-Verlag
New York Berlin Heidelberg London Paris
Tokyo Hong Kong Barcelona Budapest

A.J. Lichtenberg
M.A. Lieberman
Department of Electrical Engineering
  and Computer Sciences
University of California
Berkeley, CA 94720
USA

Editors

F. John
Courant Institute of
  Mathematical Sciences
New York University
New York, NY 10012
USA

J.E. Marsden
Department of
  Mathematics
University of California
Berkeley, CA 94720
USA

L. Sirovich
Division of
  Applied Mathematics
Brown University
Providence, RI 02912
USA

Mathematics Subject Classification (1991): 70Kxx; 60Hxx; 34Cxx; 35Bxx

Library of Congress Cataloging in Publication Data
Lichtenberg, Allan J.
  Regular and chaotic dynamics/A.J. Lichtenberg, M.A. Lieberman.—
2nd ed.
    p. cm.—(Applied mathematical sciences; v. 38)
  Includes bibliographical references and index.
  ISBN 0-387-97745-7
  1. Nonlinear oscillations. 2. Stochastic processes.
  3. Hamiltonian systems. I. Lieberman, M.A. (Michael A.)
  II. Title. III. Series: Applied mathematical sciences (Springer-
Verlag New York Inc.); v. 38.
  QA1.A647   vol. 38 1992
  [QA867.5]
  510 s—dc20
  [531'.32]   91-39849

This book is the second revised and expanded edition of "Regular and Stochastic Motion," which was published in the Applied Mathematical Sciences series Vol. 38 by Springer-Verlag in 1983.

Printed on acid-free paper.

© 1983, 1992 Springer-Verlag New York, Inc.
All rights reserved. This work may not be translated or copied in whole or in part without the written permission of the publisher (Springer-Verlag New York, Inc., 175 Fifth Avenue, New York, NY 10010, USA), except for brief excerpts in connection with reviews or scholarly analysis. Use in connection with any form of information storage and retrieval, electronic adaptation, computer software, or by similar or dissimilar methodology now known or hereafter developed is forbidden.
The use of general descriptive names, trade names, trademarks, etc., in this publication, even if the former are not especially identified, is not to be taken as a sign that such names, as understood by the Trade Marks and Merchandise Marks Act, may accordingly be used freely by anyone.
Production managed by Francine Sikorski; manufacturing supervised by Robert Paella.
Typeset by Integral Typesetting, Norfolk, UK.
Printed and bound by R.R. Donnelley & Sons, Harrisonburg, VA.
Printed in the United States of America.

9 8 7 6 5 4 3 2

ISBN 0-387-97745-7 Springer-Verlag New York Heidelberg Berlin
ISBN 3-540-97745-7 Springer-Verlag Berlin Heidelberg New York

To Elizabeth and Marlene

Preface to the Second Edition

What's in a name? The original title of our book, *Regular and Stochastic Motion*, was chosen to emphasize Hamiltonian dynamics and the physical motion of bodies. The new edition is more evenhanded, with considerably more discussion of dissipative systems and dynamics not involving physical motion. To reflect this partial change of emphasis, we have substituted the more general terms in our title. The common usage of the new terms clarifies the emphasis of the book.

The main change in the book has been to expand the sections on dissipative dynamics, including discussion of renormalization, circle maps, intermittancy, crises, transient chaos, multifractals, reconstruction, and coupled mapping systems. These topics were either mainly in the mathematical literature or essentially unstudied when our first edition was written. The volume of work in these areas has surpassed that in Hamiltonian dynamics within the past few years.

We have also made changes in the Hamiltonian sections, adding many new topics such as more general transformation and stability theory, connected stochasticity in two-dimensional maps, converse KAM theory, new topics in diffusion theory, and an approach to equilibrium in many dimensions. Other sections such as mapping models have been revised to take into account new perspectives. We have also corrected a number of misprints and clarified various arguments with the help of colleagues and students, some of whom we acknowledge below. We have again chosen not to treat quantum chaos, partly due to our own lack of acquaintance with the subject.

Since writing the first edition, a number of monographs and texts have appeared that are complementary to ours, covering material with a different emphasis, or covering material not covered at all in our book. In writing the sections on dissipative dynamics, we benefited from monographs by Guckenheimer and Holmes (1983) and by Schuster (1988): The former is detailed and mathematically oriented with somewhat limited scope; the latter covers some of the topics we have added, at a similar level. New books on

Hamiltonian mechanics have recently appeared; one, a concise treatment but including quantum chaos and a detailed treatment of integrability by Tabor (1989), and another, a thorough account relating Hamiltonian chaos to plasma and wave problems by Sagdeev et al. (1988). There is also a recent general text by Jackson (1989) and an account of numerical methods by Chua and Parker (1989). The above list is by no means exhaustive.

There are many colleagues whose assistance was helpful in preparation of this new edition. We thank Per Bak, David Bruwiler, Jim Crutchfield, Celso Grebogi, and Jerrold Marsden for reading sections of the manuscript relevant to their work; Chris Goedde for supplying material not yet published from his thesis; and Jim Howard, Chris Silva, and M. Nambu for carefully reading portions of the original edition and pointing out a number of misprints and minor errors. Particular thanks go to Boris Chirikov and associates who translated the original edition into Russian. Besides bringing misprints to our attention, more substantive points were addressed. Once again, we gratefully acknowledge our research support in the dynamics area from the National Science Foundation, the Office of Naval Research, and the Department of Energy. This continuing support has been essential for our developing many of the topics that we treat in the new edition.

Preface to the First Edition

This book treats stochastic motion in nonlinear oscillator systems. It describes a rapidly growing field of nonlinear mechanics with applications to a number of areas in science and engineering, including astronomy, plasma physics, statistical mechanics, and hydrodynamics. The main emphasis is on intrinsic stochasticity in Hamiltonian systems, where the stochastic motion is generated by the dynamics itself and not by external noise. However, the effects of noise in modifying the intrinsic motion are also considered. A thorough introduction to chaotic motion in dissipative systems is given in the final chapter.

Although the roots of the field are old, dating back to the last century when Poincaré and others attempted to formulate a theory for nonlinear perturbations of planetary orbits, it was new mathematical results obtained in the 1960s, together with computational results obtained using high-speed computers, that facilitated our new treatment of the subject. Since the new methods partly originated in mathematical advances, there have been two or three mathematical monographs exposing these developments. However, these monographs employ methods and language that are not readily accessible to scientists and engineers, and also do not give explicit techniques for making practical calculations.

In our treatment of the material, we emphasize physical insight rather than mathematical rigor. We present practical methods for describing the motion, for determining the transition from regular to stochastic behavior, and for characterizing the stochasticity. We rely heavily on numerical computations to illustrate the methods and to validate them.

The book is intended to be a self-contained text for physical scientists and engineers who wish to enter the field, and a reference for those researchers already familiar with the methods. It may also be used as an advanced graduate textbook in mechanics. We assume that the reader has the usual undergraduate mathematics and physics background, including a mechanics course at the junior or senior level in which the basic elements of Hamiltonian theory have been covered. Some familiarity with Hamiltonian mechanics at the graduate level is desirable, but not necessary. An extensive review of the required background material is given in Sections 1.2 and 1.3.

The core ideas of the book, concerning intrinsic stochasticity in Hamiltonian systems, are introduced in Section 1.4. Our subsequent exposition in Chapters 2–6 proceeds from the regular to the stochastic. To guide the reader here, we have "starred" (*) the sections in which the basic material appears. These "starred" sections form the core of our treatment. Section 2.4a on secular perturbation theory is of central importance. The core material has been successfully presented as a 30-lecture-hour graduate course at Berkeley.

In addition to the core material, other major topics are treated. The effects of external noise in modifying the intrinsic motion are presented in Section 5.5 (using the results of Section 5.4d) for two degrees of freedom and in Section 6.3 for more than two degrees of freedom, and an application is given in Section 6.4. Our description of dissipative systems in Chapter 7 can be read more-or-less independently of our treatment of Hamiltonian systems. For studying the material of Chapter 7, the introduction in Section 1.5 and the material on surfaces of section in Section 1.2b and on Liapunov exponents in Sections 5.2b and 5.3 should be consulted. The topic of period doubling bifurcations is presented in Sections 7.2b, 7.3a, and Appendix B (see also Section 3.4d). Other specialized topics such as Lie perturbation methods (Section 2.5), superconvergent perturbation methods (Section 2.6), aspects of renormalization theory (Sections 4.3, 4.5), noncanonical methods (Section 2.3d), global removal of resonances (Section 2.4d and part of 2.4c), variational methods (Sections 2.6b and 4.6), and modulational diffusion (Section 6.2d) can generally be deferred until after the reader has obtained some familiarity with the core material.[1]

This book has been three and a half years in the writing. We have received encouragement from many friends and colleagues. We wish to acknowledge here those who reviewed major sections of the manuscript. The final draft has been greatly improved by their comments. Thanks go to H.D.I. Abarbanel, J.R. Cary, B.V. Chirikov, R.H. Cohen, D.F. Escande, J. Ford, J. Greene, R.H.G. Helleman, P.J. Holmes, J.E. Howard, O.E. Lanford, D.B. Lichtenberg, R. Littlejohn, B. McNamara, H. Motz, C. Sparrow, J.L. Tennyson, and A. Weinstein. Useful comments have also been received from G. Casati, A.N. Kaufman, I.C. Percival, and G.R. Smith. We are also pleased to acknowledge the considerable influence of the many published works in the field by B.V. Chirikov. Many of the ideas expressed herein were developed by the authors while working on grants and contracts supported by the National Science Foundation, the Department of Energy, and the Office of Naval Research. One of the authors (A.J.L.) acknowledges the hospitality of St. Catherine's College, Oxford, and one of the authors (M.A.L.) acknowledges the hospitality of Imperial College, London, where much of the manuscript was developed.

---

[1] The section numbers refer to the first edition. For the second edition, the reader should consult the Contents.

Contents

Preface to the Second Edition	vii
Preface to the First Edition	ix
List of Symbols	xvii

CHAPTER 1
Overview and Basic Concepts — 1

1.1. An Introductory Note	1
*1.2. Transformation Theory of Mechanics	7
*1.2a. Canonical Transformations	7
*1.2b. Motion in Phase Space	12
*1.2c. Action-Angle Variables	21
1.3. Integrable Systems	24
*1.3a. One Degree of Freedom	25
1.3b. Linear Differential Equations	29
1.3c. More than One Degree of Freedom	32
1.3d. Noncanonical Variables and Reduction	44
*1.4. Near-Integrable Systems	48
*1.4a. Two Degrees of Freedom	48
*1.4b. More than Two Degrees of Freedom	60
1.5. Dissipative Systems	62
1.5a. Strange Attractors	63
1.5b. The Lorenz System	65

CHAPTER 2
Canonical Perturbation Theory — 70

2.1. Introduction	70
2.1a. Power Series	73
2.1b. Asymptotic Series and Small Denominators	75
2.1c. The Effect of Resonances	77
*2.2. Classical Perturbation Theory	78
*2.2a. One Degree of Freedom	78
*2.2b. Two or More Degrees of Freedom	83

2.3. Adiabatic Invariance 93
　*2.3a. Introduction and Basic Concepts 93
　*2.3b. Canonical Adiabatic Theory 96
　*2.3c. Slowly Varying Harmonic Oscillator 100
　2.3d. Noncanonical Methods 102
2.4. Secular Perturbation Theory 109
　*2.4a. Removal of Resonances 110
　*2.4b. Higher-Order Resonances 117
　*2.4c. Resonant Wave-Particle Interaction 122
　2.4d. Global Removal of Resonances 128
2.5. Lie Transformation Methods 133
　2.5a. General Theory 135
　2.5b. Deprit Perturbation Series 136
　2.5c. Adiabatic Invariants 141
2.6. Superconvergent Methods 149
　2.6a. Kolmogorov's Technique 152
　2.6b. Singly Periodic Orbits 154

CHAPTER 3
Mappings and Linear Stability 162

*3.1. Hamiltonian Systems as Canonical Mappings 164
　*3.1a. Integrable Systems 164
　*3.1b. Near-Integrable Systems 167
　*3.1c. Hamiltonian Forms and Mappings 169
*3.2. Generic Behavior of Canonical Mappings 174
　*3.2a. Irrational Winding Numbers and KAM Stability 174
　*3.2b. Rational Winding Numbers and Their Structure 183
　*3.2c. Complete Description of a Nonlinear Mapping 188
　*3.2d. Numerical Examples 191
3.3. Linearized Motion 195
　*3.3a. Eigenvalues and Eigenvectors of a Matrix 196
　*3.3b. Two-Dimensional Mappings 200
　3.3c. Linear Stability in Higher Dimensions 208
　3.3d. General Stability Concepts 210
*3.4. Mapping Models 215
　*3.4a. Physical Problems and Models 216
　*3.4b. Numerical Results 220
　*3.4c. Fixed Points and Linear Stability 223
　*3.4d. Bifurcation Phenomena 226
　*3.4e. Hamiltonian Formulation 229
*3.5. The Separatrix Motion 231
　*3.5a. Driven One-Dimensional Pendulum 231
　*3.5b. The Separatrix Mapping 236
3.6. Mappings from Perturbed Linear Oscillators 237
　3.6a. Tiling of the Phase Space 238
　3.6b. Phase Space Structure 241

Contents   xiii

CHAPTER 4
Transition to Global Stochasticity   245

*4.1. Introduction   245
  *4.1a. Qualitative Description of Criteria   247
  *4.1b. The Standard Mapping   250
*4.2. Resonance Overlap   258
  *4.2a. Rationale for Criteria   258
  *4.2b. Calculation of Overlap Criteria   259
 4.3. Growth of Second-Order Islands   264
   4.3a. Elliptic Fixed Points   265
   4.3b. The Separatrix   268
*4.4. Stability of High-Order Fixed Points   271
  *4.4a. The Basic Elements of Greene's Method   271
  *4.4b. Numerical Evaluation   276
 4.5. Renormalization for Two Resonances   281
 4.6. Nonexistence of Invariant Curves   289
*4.7. Summary and Conclusions   291

CHAPTER 5
Stochastic Motion and Diffusion   293

 5.1. Introduction   293
 5.2. Definitions and Basic Concepts   294
  *5.2a. Ergodicity   294
  *5.2b. Liapunov Characteristic Exponents   296
   5.2c. Concepts of Stochasticity   302
   5.2d. Randomness and Numerical Errors   309
 5.3. Determination of Liapunov Exponents and KS Entropy   312
   5.3a. Analytical Estimates   314
  *5.3b. Numerical Methods   315
*5.4. Diffusion in Action Space   320
  *5.4a. The Fokker–Planck Equation   321
  *5.4b. Transport Coefficients   324
  *5.4c. Steady-State and Transient Solutions   326
 5.5. The Effects of Correlations   328
   5.5a. Higher-Order Transport Corrections   328
   5.5b. Diffusion near Critical K 339
   5.5c. Long-Time Correlations and Averaging   342
   5.5d. Quasiaccelerator Modes   345
 5.6. The Effect of Extrinsic Stochastic Forces   349
   5.6a. Introduction   349
   5.6b. Diffusion in the Presence of Resonances   353
   5.6c. Diffusion on Two Space and Time Scales   357
 5.7. Slow Separatrix Crossings   363

CHAPTER 6
Three or More Degrees of Freedom 373

*6.1. Resonance in Multidimensional Oscillations 373
    *6.1a. Geometric Relations 374
    *6.1b. Examples of Arnold Diffusion 380
6.2. Diffusion Rates Along Resonances 386
    *6.2a. Stochastic Pump Diffusion Calculation 387
    6.2b. Coupling Resonance Diffusion 393
    6.2c. Many Resonance Diffusion 397
    6.2d. Modulational Diffusion 399
6.3. Extrinsic Diffusion 406
    6.3a. Resonance Streaming 406
    6.3b. Diffusion of a Parameter 413
6.4. Diffusion in Toroidal Magnetic Fields 418
    6.4a. Magnetic Islands 418
    6.4b. Drift Surfaces and Diffusion in Static Fields 425
    6.4c. Time-Varying Fields 428
    6.4d. The Self-Consistent Problem 436
6.5. Many Degrees of Freedom 437
    6.5a. Strongly Nonlinear Coupled Systems 440
    6.5b. The Fermi–Pasta–Ulam Model and the Korteveg–de Vries Equation 443
    6.5c. Discretized Sine–Gordon Equation 450

CHAPTER 7
Bifurcation Phenomena and Transition to Chaos in Dissipative Systems 457

7.1. Simple and Strange Attractors 457
    7.1a. Basic Properties 458
    7.1b. Simple Attractors and Their Bifurcations 461
    7.1c. Examples of Strange Attractors 468
    7.1d. Geometric Properties of Strange Attractors 474
7.2. One-Dimensional Noninvertible Maps 478
    7.2a. Basic Properties 478
    7.2b. Periodic Behavior 482
    7.2c. Chaotic Motion 497
7.3. Bifurcation Phenomena in Two-Dimensional Maps 508
    7.3a. Dissipative Maps 509
    7.3b. Area-Preserving Maps 512
    7.3c. Transition Between Hamiltonian and Dissipative Chaos 519
7.4. Circle Maps 525
    7.4a. Basic Formalism 526
    7.4b. Critical Behavior and Universality 529
    7.4c. Supercritical Behavior 532
    7.4d. Physical Examples 534
7.5. Intermittency 537
    7.5a. Intermittency Mechanisms 538
    7.5b. Renormalization-Group Analysis 542

7.5c. Power Spectra and Experimental Results	545
7.6. Crises and Basin Boundaries	548
7.6a. Crises	549
7.6b. Fractal Basin Boundaries	557
7.7. Determination of Homoclinic Intersection	560

CHAPTER 8
Chaotic Motion in Dissipative Systems 570

8.1. Transient Chaos	570
8.1a. Weak Dissipation	571
8.1b. Transient Chaos Near a Crisis	579
8.2. Invariant Distributions on Strange Attractors	580
8.2a. Reduction to a One-Dimensional Map	581
8.2b. Direct Iteration of an Initial Distribution	583
8.3. Multifractals and Their Properties	588
8.3a. Generalized Dimensions	588
8.3b. Scaling Index Spectrum	592
8.4. Reconstructing Dynamics	597
8.4a. Delay Coordinates and Embedding Dimension	598
8.4b. Statistical Quantities	602
8.4c. Short-Time Prediction	604
8.5. Spatio-Temporal Dynamics	613
8.5a. Maps on a Lattice	613
8.5b. Self-Organized Criticality	620
8.6. The Fluid Limit	628
8.6a. Fourier Mode Expansions	630
8.6b. The Transition to Turbulence	634
8.6c. Hamiltonian Chaos in Fluids	638

APPENDIX A
Applications 644

A.1.	Planetary Motion	644
A.2.	Accelerators and Beams	646
A.3.	Charged Particle Confinement	648
A.4.	Charged Particle Heating	650
A.5.	Chemical Dynamics	651
A.6.	Quantum Systems	653

Bibliography	655
Author Index	671
Subject Index	679

List of Symbols

The major uses of symbols throughout the book are given first. Important special uses within a section are noted by section number. Minor uses do not appear. Scalars appear in italic type; vectors appear in boldface italic type; and tensors and matrices appear in boldface sans serif type.

a	coefficient; parameter; a_n, Fourier coefficient; a_k, Fourier or quasi-mode coefficient; a_1, a_2, continued fraction expansion
A	phase space area; A_n, Fourier coefficient (2.6); action integral (3.1)
A	vector potential
A	linear transformation matrix
$\mathcal{A}$	part of generating function (3.1); $\mathcal{A}_m$, Melnikov–Arnold integral
b	coefficient; constant; parameter; b_n, Fourier coefficient; b_{ij}, coefficient matrix (3.3); sub b, measured at a KAM barrier
B	magnetic field; friction coefficient; value of Jacobian (7.3); B_n, Fourier coefficient (2.6)
B	coefficient matrix (3.3)
c	velocity of light; constant; coefficient; parameter
c	diagonalized coefficient (2.6)
C	constant; parameter of one-dimensional map; Casimir; autocorrelation function
$\mathcal{C}$	correlation function
d	derivative; differential; parameter; embedding dimension; distance between neighboring trajectories (5.3); dimension
D	diffusion coefficient; dimension; total derivative (2.5); Melnikov distance (7.7); D^*, normalized diffusion coefficient; D_{QL}, quasilinear diffusion coefficient; D_∞, long time diffusion (5.5); $D_0, D_1, \ldots, D_{KY}$, fractal dimensions
e	electric charge; base of natural logarithm; parameter; e_n, error; $\hat{e}$, basis vector; e_{jr}, normal modes

E		energy
$\mathscr{E}$		complete elliptic integral of the second kind
f		function, forcing term; mean residue (4.4); invariant function (5.2); f(sub), fast; $f(\alpha)$, scaling index spectrum
$\boldsymbol{f}$		function; mapping function; impulsive force (3.4)
F		force coefficient in Hamiltonian; F_1, F_2, generating functions; ΔF, change in a generating function (5.8); F_i, Fibonacci numbers
$\mathscr{F}$		elliptic integral of the first kind
g		function; gravity; g^*, fixed point of renormalization; g_q, fractal generator
$\boldsymbol{g}$		function
G		nonlinearity coefficient in Hamiltonian; function
$\mathscr{G}$		part of generating function (3.1)
h		function; h_k, KS entropy; energy difference near a separatrix; h_λ, eigenfunction
H		Hamiltonian; H_0, value of Hamiltonian (energy); H_0, H_1, zero-order, first-order parts of H, etc.; $\bar{H}$, transformed H; $\hat{H}$, transformed to rotating coordinates; H_T, entropy (5.2)
$\mathscr{H}$		transformed H (1.3b)
i		summation index, $\sqrt{-1}$; i(sub), ith coordinate
I		invariant; transformed action; normalized momentum in standard map; I_1, I_2, involutions; inertia components (1.3); mutual information (8.4)
$\mathbf{I}$		identity matrix
$\mathscr{I}$		identity function on phase space
j		summation index
J		action; J_0, J_1, etc., zero, first-order parts; J_1, J_2, components; Jacobian; $\bar{J}$, transformed J; $\hat{J}$, transformed to rotating coordinates; J_{ij}, Poisson tensor elements
$\mathscr{J}$		$\mathscr{J}_n$, Bessel function of the first kind
k		integer; summation index; (integer) period of fixed point (3.4); amplitude or (subscripted) component of wave vector; perturbation amplitude; kT, thermal energy (6.4)
$\boldsymbol{k}$		wave vector
K		stochasticity parameter; transformed Hamiltonian (6.2); K_2, stochasticity parameter for separatrix map (4.3); K_α, nonlinearity parameter; circle map parameter
$\mathscr{K}$		complete elliptic integral of the first kind
l		integer; summation index; angular momentum (1.3)
L		Lagrangian; Lie operator; length; L_s, shear length (6.4); L_{g^*}, linearized period-doubling operator
m		integer; summation index; m_1, m_2, etc., Fourier path integers (5.5)
$\boldsymbol{m}$		integer vector

List of Symbols

M	phase advance parameter in Fermi map; mass; number of continuous derivatives (3.2); normalized potential amplitude; (4.5); M(sub) magnitude; number of phase space cells
M	Jacobian matrix
$\mathcal{M}$	mapping
n	normalized time units; integer; summation index; n(sub), step number
n	integer vector (2.4, 3.1)
N	degrees of freedom; N(sub), Nth coordinate
$\mathcal{N}$	noise power density
0	0(sub), unperturbed part; 0(sub), linearized value
$\mathcal{O}$	order of
p	momentum; integer; probability density; summation index (2.4); bifurcation family (4.1)
p	vector momentum
P	probability distribution; invariant distribution; transformed momentum; normalized perturbation amplitude (4.5)
q	coordinate; integer; Fourier transform variable (5.4d)
Q	Q_0, ratio of driving frequency to linearized oscillator frequency; inverse rotation number; Q_{jm}, Fourier amplitude (2.6)
r	integer; harmonic number; radius; parameter
r	vector position
R	axial ratio of phase space ellipse $(F/G)^{1/2}$; residue; radius; R(sub), resonance value; R_a, Rayleigh number (8.6); Reynolds number (8.6)
s	integer; harmonic number; s(sub), slow; s(sub), secondary; s(sub), stable; sx(sub), on separatrix
S	generating function F_2; generalized stochasticity parameter (4.5); scaling parameter (6.3b); shear parameter (6.4); S(sub), stochastic part (6.1); size (8.5); S_f, power spectrum
t	time; sub t, derivative with respect to time
T	period; mapping; Lie evolution operator (2.5); period doubling operator; T_∞, universal Hamiltonian map
$\mathcal{T}$	renormalization operator
u	normalized velocity; function of coordinates (1.2); periodic function (3.2)
U	potential
v	velocity; function of coordinates (1.2); periodic function (3.2)
v	Eulerian fluid velocity (8.6)
V	Hamiltonian perturbation amplitude; potential; V_1, V_2, potential amplitudes (4.5); V_k, Fourier harmonics of perturbation amplitude; V_p, p-dimensional volume (5.2)
V	vector field (7.1)
w	Lie generating function; deviation from separatrix energy
w	tangent vector $\Delta \mathbf{x}$ (5.2, 5.3)

W	W_t, transition probability; $W(|)$, transition probability; invariant manifold
x	coordinate; mapping variable; x^*, value of x at extremum; sub x, derivative with respect to x
$\boldsymbol{x}$	vector mapping variable; position
X	phase space variable in differential equations; guiding center coordinates (2.2); normalized amplitude (4.5); attractor (7.1); X_l, Fourier component (7.2)
$\mathbf{X}$	eigenvector matrix
y	coordinate; mapping variable; nice variable (2.3)
Y	phase space variable in differential equations; guiding center coordinates (2.2); normalized amplitude (4.5)
z	coordinate; nice coordinate (2.3); integer mapping; z_n, height difference; z_c, critical height difference (8.5)
$\boldsymbol{z}$	vector of generalized coordinates and momenta
Z	phase space variable in differential equations
α	rotation number (frequency ratio); α_l, invariant (1.2); α_l, golden mean (4.4); momentum variable (6.1); rescaling parameter; scaling index; α^2, coupling parameter (8.5)
β	parameter; momentum variable (6.1); Hamiltonian rescaling parameter (7.3)
γ	power law exponent; rescaling parameter (7.2)
Γ	Fourier coefficient (2.4); flow (5.5b); partition sum
$\boldsymbol{\Gamma}$	antisymmetric matrix (3.3)
δ	variation; $\delta\omega$ (or δJ), frequency (or action) distance between resonances; rescaling parameter; δ_H, Hamiltonian rescaling parameter; dissipation parameter; $\delta_1(\ )$, periodic delta function; $\delta(\ )$, Dirac delta function; δ_{ij}, Kronecker delta function
Δ	variation; $\Delta\omega$ (or ΔJ), small increment in frequency (or action) about a zero-order value; $\Delta\omega_{max}$ (or ΔJ_{max}) separatrix half-width in frequency (or action)
ϵ	perturbation strength; perturbation ordering parameter; small quantity
ζ	normalized coordinate; ζ_n, random variable (6.3b)
$\boldsymbol{\zeta}$	vector of delay coordinates (8.4)
η	power law exponent
θ	angle; angle conjugate to an action
$\boldsymbol{\theta}$	vector angle
Θ	2π or 1 (3.4); temperature variable (8.6); parameter (7.1)
$\mathbf{D}$	rotational transform $2\pi\alpha$
κ	elliptic integral argument, thermal conductivity (8.6)
λ	eigenvalue; unit step (4.5); modulation amplitude (6.2d)
$\boldsymbol{\lambda}$	matrix of eigenvalues
Λ	Fourier harmonic; perturbation amplitude; volume contraction rate

List of Symbols

μ	magnetic moment; parameter; phase space measure; coupling coefficient (6.2); extrinsically diffusing parameter (6.3b)
ν	collision frequency; eigenvalue (7.1); viscosity (8.6); decay rate
ξ	elliptic integral argument; normalized coordinate angle variable (3.2); random variable (6.3b)
π	pi
Π	product
ρ	gyroradius (ρ_L in 6.4)
σ	Liapunov exponent; phase of eigenvalue; parameter; standard deviation
Σ	summation; surface of section
τ	normalized time; time interval; phase space density (1.2); ϵt, slow time; (2.2, 2.3); τ_D, diffusion time (6.3); τ_c, collision time (6.4); rotation period; scaling index (8.3); time delay (8.4)
ϕ	pendulum angle; angle conjugate to an action; phase angle; toroidal angle (6.4)
Φ	potential; potential amplitude; transformed angle (1.3); Φ_0, potential amplitude
χ	constant angle; argument of Bessel function (4.3); angle in action space (6.3); variance (5.5)
ψ	phase angle; fixed angle parameter (3.1, 3.2); angle in action space (6.3); toroidal angle (6.4); stream function (8.6)
ω	radian frequency; ω_0, linear frequency; slowly varying frequency (2.5)
Ω	driving frequency; cyclotron frequency; rotation frequency; winding number; Ω_0, unperturbed winding number
A	scalar
$\boldsymbol{A}$	vector
$\mathbf{A}$	matrix or tensor
∂	partial derivative
$(\cdot)$	total derivative with respect to time
∇	gradient
$[,]$	Poisson bracket
$(\bar{\ })$	transformed to new coordinates; step of a mapping; average of quantity; renormalized quantity (4.5)
$(\ )'$	derivative with respect to argument; renormalized quantity
$(\ )^*$	fixed point of quantity; critical parameter
$(\ )_{c,cr}$	critical parameter
const	constant
$\langle\ \rangle$	average part over angles
$\{\ \}$	oscillating part
$\{,\}$	bracket operator
$(\tilde{\ })$	integral of oscillating part (2.3d)
$(\hat{\ })$	transformed to rotating coordinates; unit vector
$(\ )\wedge(\ )$	wedge operator (volume of parallelepiped)

⊥	perpendicular (sub)
∥	parallel (sub)
( )	matrix
| |	magnitude
∥ ∥	norm
sgn	sign of
det	determinant of matrix
Tr	trace of matrix
ln	natural logarithm
exp	exponential
div	divergence

CHAPTER 1
Overview and Basic Concepts

1.1. An Introductory Note

This volume grew out of developments in dynamics aimed at understanding the behavior of an oscillator for a slow change in parameters and at understanding the behavior of coupled oscillators when the coupling is weak. These two problems, first considered independently, were found to be intimately related for multiply periodic systems.

The understanding of slowly varying parameters was given its major impetus by Einstein at the Solvay Conference of 1911, when he suggested a physical significance for the action integral. He pointed out that its "adiabatic" constancy, first demonstrated by Liouville and Green three quarters of a century earlier, was directly related to the physical notion that the number of quanta should remain constant in a slowly varying system. The resulting WKB method (Wentzel, 1926; Kramers, 1926; Brillouin, 1927) become one of the cornerstones of the treatment of wave propagation in inhomogeneous media, and of wave mechanics. The underlying theory was developed by Bogoliubov and Mitropolsky (1961) and by Kruskal (1962) and is commonly known as the method of averaging.

The second development, that of treating coupled nonlinear oscillators, began with attempts to solve the three-body problem of celestial mechanics, which serves as a simplified model for the solar system. Early work dates back to the investigations of Hamilton and Liouville in the mid-nineteenth century, and stimulated the development of the Hamiltonian formalism that underlies most of our treatment of mechanics. Toward the end of the nineteenth century, many of the ideas concerning the stability of coupled nonlinear systems were investigated by Poincaré (1892) and applied to the problems of celestial mechanics. It was during this period that he, Von Zeipel

(1916), and others devised the perturbative methods that have proved so fruitful in describing the short-time behavior of these systems. The emergence of the quantum theory greatly stimulated these developments. Secular perturbation theory, which allows the local inclusion of resonant interaction between two degrees of freedom, was formalized in the early days of quantum mechanics (Born, 1927). In Chapter 2 we treat perturbation techniques using the earlier classical methods and the more modern Lie formalism (e.g., Deprit, 1969).

An important question that could not be answered by the early perturbative techniques was that of the long-time stability of the solar system. The prevailing belief was that planetary motion was "regular" (quasiperiodic) and might ultimately be "solved" by new mathematical methods. This belief was reinforced by the known solutions of the two-body problem and other simple mechanical systems and by the interpretation of the fossil record, which suggested the regularity of the earth's motion around the sun over hundreds of millions of years.

At the same time, a second, contradictory view of mechanical systems with many degrees of freedom was put forth by Boltzmann in an effort to understand the behavior of dilute gases. He argued that molecular motion should be considered random, with every molecule exploring the entire phase space energetically accessible to it. This point of view became known as the ergodic hypothesis, and was successful in explaining many of the observed properties of matter, becoming the foundation of classical statistical mechanics.

The first numerical attempt to test Boltzmann's ergodic hypothesis for a system with a modest number of degrees of freedom was made by Fermi *et al.* (1955), who used the model of mass points interacting nonlinearly with their nearest neighbors along a single line. They were surprised to find that their expectation of stochastic motion was not borne out numerically.

These deep contradictions between the existence of integrability, on the one hand, and the existence of ergodicity, on the other, were symptomatic of a fundamental unsolved problem of classical mechanics. Poincaré contributed to the understanding of these dilemmas by demonstrating the extremely intricate nature of the motion in the vicinity of unstable fixed points, a first hint that regular applied forces may generate stochastic motion in nonlinear oscillator systems. Subsequently, Birkhoff 1927) showed that both stable and unstable fixed points must exist whenever there is a rational frequency ratio (resonance) between two degrees of freedom. Successively higher-order resonances in the periodic motion change the topology of the phase trajectories leading to the formation of island chains on an increasingly fine scale. It was recognized that perturbative expansions did not include the effect of such resonances.

Although the work of Poincaré and Birkhoff had demonstrated the exceeding complexity of the phase space topology, the question of the ergodic hypothesis, whether a trajectory explores the entire region of the phase space that is energetically available to it, or whether it is constrained by the

1.1. An Introductory Note

existence of constants of the motion, was not definitively answered until quite recently. The KAM theorem, originally postulated by Kolmogorov (1954) and proved under different restrictions by Arnold (1963) and Moser (1962), states that for systems perturbed away from integrable ones, invariant surfaces continue to exist for most initial conditions. Thus, although the motion near the separatrix of each resonance is stochastic, the motion is constrained by nearby KAM curves and is not ergodic. In Chapter 3 we describe the KAM theorem and related topological results, which serve as the justification for many of the calculations in this book.

In addition to the new developments in perturbation theory and new mathematical results, the invention of the digital computer was a major impetus for the renewed interest in nonlinear mechanics. The early use of digital computation to integrate the equations of motion was quickly paired with Poincaré's surface-of-section method, such that the integration of N-dimensional flow equations was replaced by the iteration of $(N-1)$-dimensional mappings. As a result, the motion of a point in phase space could be followed over hundreds of thousands of oscillation periods. The remarkable structures that appeared on the finest spatial scales in these early computations quickly attracted the attention of both theoreticians and experimentalists. As a consequence, two central strands in the development of the material in this volume are a strong reliance on the results of numerical computation and, as described in Chapter 3, the close correspondence between N-dimensional flows and their $(N-1)$-dimensional Poincaré maps. Numerical computation lies at the heart of our description of mechanics and is generally considered the ultimate test of theoretical analyses. Numerical examples are found in every chapter to illustrate and to give insight.

From the detailed digital computations and accompanying theoretical developments, an extraordinary picture of the phase space for weakly perturbed systems has emerged. The invariant surfaces break their topology near resonances to form island chains. Within these islands the topology is again broken to form yet other chains of still finer islands. On ever finer scale, one sees islands within islands. But this structure is only part of the picture, for densely interwoven within these invariant structures are thin phase space layers in which the motion is stochastic. Much of Chapter 3 is devoted to explicating this structure.

For weakly perturbed systems with two degrees of freedom, the KAM surfaces isolate the thin layers of stochasticity from each other, and the stochastic excursions of the actions within any layer are exponentially small in the perturbation. As the perturbation strength increases, a transition can occur in which the isolating KAM surfaces disappear and the stochastic layers merge, resulting in globally stochastic motion that envelops the phase space. Often the phase space divides into three regions. One region contains primarily stochastic trajectories. It is connected to a second, mainly stochastic region containing large embedded islands. A third region consists of primarily regular trajectories and is isolated from the first two by KAM surfaces. A seminal numerical example illustrating the transition from mainly regular to mainly stochastic motion was formulated by Hénon and Heiles (1964) to simulate the behavior of the three-body problem. Numerical calculations and heuristic theories have evolved together over the past 20 years to elucidate the basic processes and determine the perturbation strength at which the transition occurs. These results are illustrated in Chapter 3 using the example of Fermi acceleration, originally proposed as a model for cosmic ray production, in which a ball bounces back and forth between a fixed and an oscillating wall. Then in Chapter 4 we determine the transition between locally contained and global stochasticity, using a number of calculational techniques (e.g. Chirikov, 1979; Greene, 1979a).

The methods discussed above have been applied to a wide range of problems, both in the initial development of the theory and in subsequent applications. Applications to the three-body problem and its many variants essentially initiated the field of study. New exciting astronomical applications are also emerging. For example, the structure of the rings of Saturn and of the asteroid belts is qualitatively understood in terms of resonances of the larger perturbing bodies. However, the detailed widths of rings and gaps and the recent satellite observations of complex structure in the Saturnian rings are yet to be understood.

Other important applications involve the multiperiodic motion of a charged particle moving in magnetic and electric fields. These studies began with the observation that the magnetic moment is an adiabatic invariant, associated with the gyromotion of a charged particle in a magnetic field (Alfven, 1950). The work was later extended to include the invariants of the other degrees of freedom. The problem also served as a springboard for the developments of asymptotic expansion and averaging techniques. Chirikov's (1960) investigations of the transition between ordered and stochastic motion, from which the first criterion for this transition (resonance overlap) was established, were also stimulated by the motion of charged particles in magnetic mirrors. This work was extended to include the effect of an r.f. field, which could resonate with gyromotion, leading to the calculation of a limit to r.f. heating due to the existence of KAM curves. A related problem of the motion of a particle moving in a magnetized plasma under the influence of a wave, illustrating many of these features, is used as an example of secular perturbation theory in Chapter 2 and of the determination of the transition from adiabatic to stochastic behavior in Chapter 4. Another interesting application of the theory is to particle motion in accelerators, in which some of the early explorations of the behavior of nonlinear multiply periodic systems were undertaken. In addition, Hamilton's equations can be used to describe other trajectories, such as the motion of magnetic field lines and of light rays in geometrical optics. In axisymmetrical toroidal geometry, the Hamiltonian describing the magnetic field line motion is integrable. There have been a number of studies of the breakup of toroidal magnetic surfaces due to perturbations, arising both from external currents and from self-

1.1. An Introductory Note

consistent currents of a confined plasma. Applications, such as these, are used as examples throughout the text; others are reviewed briefly in Appendix A.

The existence of motion that appears random in numerical simulations is well established. However, mathematical attempts to characterize the stochasticity as ergodic, mixing, etc., have not generally been successful. Sinai (1963) has proved these properties for the hard sphere gas. Other idealized models of Hamiltonian systems have been shown to have even stronger stochastic properties. These results, described in Chapter 5, give us confidence to attribute randomness to generic Hamiltonian systems when they have behavior characteristic of the idealized models.

The seemingly stochastic motion that often appears in these coupled systems is suggestive of a fundamental source for statistical mechanics, and thus has increasingly attracted researchers from that field. The complexity of the motion near unstable periodic solutions and the recognition that these unstable singular points are dense in the phase space give strong impetus to this point of view. More recently, considerable effort has gone into characterizing stochastic behavior of dynamical systems in terms of their Liapunov exponents, which measure the exponential rate of divergence of nearby trajectories. It is also of practical importance to calculate the diffusion in action, averaged over phases. Most early calculations started with an assumption of random phases. Clearly this assumption is incorrect if KAM curves exist, since these curves exclude some phases from being explored. Even if the phase space is essentially ergodic, such that the motion explores the entire energy surface, the time scale over which the phase is randomized must be determined. These questions have been explored numerically and analytically, and new insights into the decay of the phase correlations near KAM surfaces have emerged. We treat these topics in Chapter 5.

There is a marked difference between the stochasticity encountered in systems with two degrees of freedom and that encountered in systems with more than two degrees. Using topological arguments, Arnold (1963) showed that for systems with more than two degrees of freedom the stochastic layers are interconnected to form a web that is dense in the phase space. For initial conditions on the web, stochastic motion is driven along the layers, leading to a global diffusion not constrained by the KAM surfaces. This mechanism is commonly called Arnold diffusion and can be fast or slow, depending on the thickness of the stochastic layers. The diffusion exists (in principle) down to the limit of infinitesimal perturbations from integrable systems. Another interesting effect in multidimensional systems occurs when a slow modulation of one of the periodic motions is present. There can then be a transition to a strongly enhanced stochastic motion along the web known as modulational diffusion. This mechanism is counter to the intuitive notion that a slow modulation must lead to adiabatic behavior. Multidimensional systems having embedded resonances can also profoundly affect the diffusion due to external stochasticity (noise). For systems with two degrees of freedom, the noise generally acts equivalently to a third degree of freedom, allowing diffusion along resonance layers. The resonances can strongly enhance the diffusion rate. The above processes have been postulated as the mechanisms limiting particle lifetimes and beam intensities in accelerator storage rings. In Chapter 6 we treat diffusion processes arising from the higher dimensionality, both the intrinsic Arnold and modulational diffusion and the resonance-enhanced diffusion originating from external noise.

A new range of phenomena emerges for dissipative systems, for which the phase space volume is not constant, but rather contracts as the motion proceeds. The final state is motion on a subspace called an attractor, of lower dimensionality than the original phase space. The study of regular motion in such systems is an old endeavor, stretching back to Newton and the subsequent development of the theory of ordinary differential equations. It was recognized in these early studies that the motion can contract onto simple attractors such as points, closed orbits, or tori, representing a steady state that is, respectively, stationary, periodic, or quasiperiodic. In contrast, the recognition that intrinsically chaotic motion occurs in dissipative systems is quite recent, dating from the pioneering work of Lorenz (1963) who found such an attractor embedded in the motion described by a system of nonlinear ordinary differential equations. The term "strange attractor" was used by Ruelle and Takens (1971) to describe those attractors on which the motion appeared chaotic. The topology of strange attractors is quite remarkable, showing a geometric invariance in which the structure of the attractor repeats itself on ever finer spatial scales. Such structures, called fractals, have the curious property of fractional dimensionality, intermediate between that of a point and a line, a line and a plane, etc.

As the geometric structure of strange attractors has been clarified, a qualitative picture has emerged in which an important feature is the close correspondence between motion on a strange attractor and motion described by noninvertible one-dimensional maps. Such maps do not arise directly from dissipative flows, but are important examples of simple systems exhibiting chaotic behavior, which have found applications in such diverse fields as economics and ecology. Noninvertible maps undergo a sequence of bifurcations as a parameter is varied such that the period of an attracting periodic orbit repeatedly doubles, accumulating at a critical parameter value beyond which the behavior is chaotic. Feigenbaum (1978) has shown this process to have a universal nature.

It is in two-dimensional invertible maps (and their related three-dimensional flows), however, that strange attractors first appear. The period doubling bifurcation sequence is also present for these systems. The motion near separatrix layers and the calculation of invariant distributions are important elements in understanding their behavior. Despite the correspondence between one- and two-dimensional maps, our knowledge of the latter is incomplete; for example, there is no method presently known for finding the transition to a strange attractor in a multidimensional system.

The appearance of strange attractors in three-dimensional flows, such as the Lorenz system, suggests a possible route to fluid turbulence and has stimulated exceptionally precise experimental measurements near the transition from laminar to turbulent flow in real fluids. The Lorenz system was, in fact, obtained from the Rayleigh–Bénard convection problem of a fluid layer heated from below, by a Fourier expansion of the flow, followed by truncation to consider only three mode amplitudes. The chaotic motion present in the three-dimensional Lorenz system suggests a possible picture of turbulent behavior in some real fluid systems that is simpler than the previous picture of Landau (1941). We treat dynamical motion in dissipative systems in Chapters 7 and 8, considering one-dimensional noninvertible maps, two-dimensional invertible maps, and applications to the behavior of fluids.

*1.2. Transformation Theory of Mechanics

In this section we review briefly the concepts from Hamiltonian mechanics that are most pertinent to the phase space treatment of problems in dynamics. The treatment follows most closely that of Goldstein (1951, Chapters 8 and 9), but also relies heavily on Whittaker (1964). Most of the lengthy proofs are omitted here.

*1.2a. Canonical Transformations

Various equivalent forms of the equations of motion may be obtained from one another by coordinate transformations. One such form is obtained by introducing a Lagrangian function

$$L(\dot{\mathbf{q}}, \mathbf{q}, t) = T(\dot{\mathbf{q}}, \mathbf{q}) - U(\mathbf{q}, t), \qquad (1.2.1)$$

where the $\mathbf{q}$ and $\dot{\mathbf{q}}$ are the vector position and velocity over all the degrees of freedom, T is the kinetic energy, U is the potential energy, and any constraints are assumed to be time-independent. In terms of L the equations of motion are, for each coordinate q_i,

$$\frac{d}{dt}\frac{\partial L}{\partial \dot{q}_i} - \frac{\partial L}{\partial q_i} = 0. \qquad (1.2.2)$$

Equation (1.2.2) can be derived either from a variational principle ($\delta \int L\, dt = 0$) or by direct comparison with Newton's laws of motion. If we define the Hamiltonian by

$$H(\mathbf{p}, \mathbf{q}, t) \equiv \sum_i \dot{q}_i p_i - L(\dot{\mathbf{q}}, \mathbf{q}, t), \qquad (1.2.3)$$

in which $\dot{q}$ is considered to be a function of q and an as yet undefined variable p, and take the differential of H, we obtain

$$dH = \sum_i \frac{\partial H}{\partial q_i} dq_i + \sum_i \frac{\partial H}{\partial p_i} dp_i + \frac{\partial H}{\partial t} dt$$

$$= \sum_i \left(p_i - \frac{\partial L}{\partial \dot{q}_i} \right) d\dot{q}_i + \sum_i \dot{q}_i \, dp_i - \sum_i \left(\frac{d}{dt} \frac{\partial L}{\partial \dot{q}_i} \right) dq_i - \frac{\partial L}{\partial t} dt, \quad (1.2.4)$$

where we have substituted (1.2.2) into the third summation on the right. Equation (1.2.4) can be satisfied only if the p_i are defined by

$$p_i \equiv \frac{\partial L}{\partial \dot{q}_i}. \quad (1.2.5)$$

The first summation on the right vanishes identically, and equating the coefficients of the differentials, we obtain a form of the equations of motion involving first derivatives only:

$$\dot{p}_i = -\frac{\partial H}{\partial q_i}, \quad (1.2.6a)$$

$$\dot{q}_i = \frac{\partial H}{\partial p_i}, \quad (1.2.6b)$$

and

$$\frac{\partial L}{\partial t} = -\frac{\partial H}{\partial t}. \quad (1.2.7)$$

The set of p, q is known as *generalized momenta and coordinates*. Equations (1.2.6) and (1.2.7) are Hamilton's equations, and the character of their solutions forms the fundamental topic of this monograph. Any set of variables p, q whose time evolution is given by an equation of the form (1.2.6) is said to be *canonical*, with p_i and q_i said to be conjugate variables.

Mixed Variable Generating Functions. If we wish to transform from the canonical variables q, p to a new set $\bar{q}, \bar{p}$, we can relate them by a function of one old and one new variable, as follows: Since the Lagrangian is derived from a variational principle, by using (1.2.3), we have

$$\delta \left[\int_{t_1}^{t_2} \left(\sum_i p_i \dot{q}_i - H(p, q, t) \right) dt \right] = 0, \quad (1.2.8)$$

which holds in either the unbarred or barred coordinates. Thus the integrated form of (1.2.8) in the two sets of coordinates can differ by at most a complete

1.2. Transformation Theory of Mechanics

differential, which we express as

$$\sum_i p_i \dot{q}_i - H(\boldsymbol{p}, \boldsymbol{q}, t) = \sum_i \bar{p}_i \dot{\bar{q}}_i - \bar{H}(\bar{\boldsymbol{p}}, \bar{\boldsymbol{q}}, t) + \frac{d}{dt} F_1(\boldsymbol{q}, \bar{\boldsymbol{q}}, t), \qquad (1.2.9)$$

where we have arbitrarily chosen $F = F_1$, a function of $\boldsymbol{q}$ and $\bar{\boldsymbol{q}}$. Expanding the total derivative of F_1, we obtain

$$\frac{d}{dt} F_1(\boldsymbol{q}, \bar{\boldsymbol{q}}, t) = \sum_i \frac{\partial F_1}{\partial q_i} \dot{q}_i + \sum_i \frac{\partial F_1}{\partial \bar{q}_i} \dot{\bar{q}}_i + \frac{\partial F_1}{\partial t}. \qquad (1.2.10)$$

Taking the variables in (1.2.10) to be independent we find, on comparing terms in (1.2.9) and requiring the $\dot{q}_i$ and $\dot{\bar{q}}_i$ terms to vanish separately, that

$$p_i = \frac{\partial F_1}{\partial q_i}, \qquad (1.2.11a)$$

$$\bar{p}_i = -\frac{\partial F_1}{\partial \bar{q}_i}, \qquad (1.2.11b)$$

$$\bar{H}(\bar{\boldsymbol{p}}, \bar{\boldsymbol{q}}, t) = H(\boldsymbol{p}, \boldsymbol{q}, t) + \frac{\partial}{\partial t} F_1(\boldsymbol{q}, \bar{\boldsymbol{q}}, t). \qquad (1.2.11c)$$

We can also define generating functions in terms of other pairs of mixed (old and new) variables:

$$F_2(\boldsymbol{q}, \bar{\boldsymbol{p}}, t), \qquad F_3(\boldsymbol{p}, \bar{\boldsymbol{q}}, t), \qquad F_4(\boldsymbol{p}, \bar{\boldsymbol{p}}, t).$$

If, for example, we generate F_2 by means of a Legendre transformation

$$F_2(\boldsymbol{q}, \bar{\boldsymbol{p}}, t) = F_1(\boldsymbol{q}, \bar{\boldsymbol{q}}, t) + \sum_i \bar{q}_i \bar{p}_i, \qquad (1.2.12)$$

where $\bar{\boldsymbol{q}}$ is considered to be a function of $\boldsymbol{q}$ and $\bar{\boldsymbol{p}}$, then we obtain the *canonical transformation* equations

$$p_i = \frac{\partial F_2}{\partial q_i}, \qquad (1.2.13a)$$

$$\bar{q}_i = \frac{\partial F_2}{\partial \bar{p}_i}, \qquad (1.2.13b)$$

$$\bar{H}(\bar{\boldsymbol{p}}, \bar{\boldsymbol{q}}, t) = H(\boldsymbol{p}, \boldsymbol{q}, t) + \frac{\partial}{\partial t} F_2(\boldsymbol{q}, \bar{\boldsymbol{p}}, t). \qquad (1.2.13c)$$

F_3 and F_4 are defined by transformations similar to (1.2.12), leading to corresponding transformation equations.

With these transformations we can show, at least in a formal sense, how the equations of motion of a dynamical system can be solved. There are two cases of interest, one with a Hamiltonian explicitly a function of the time, and one called *autonomous*, with the Hamiltonian not an explicit function of time. In the first case, if we set $\bar{H} \equiv 0$, this is equivalent to obtaining new generalized coordinates whose time derivatives are zero from the canonical equations of motion. The new coordinates are constants that may be interpreted as the initial values of the untransformed coordinates. Thus the transformation equations are, in fact, the solution giving the position and momentum at any time in terms of the initial values. Substituting (1.2.13a) into (1.2.13c), with $\bar{H} = 0$, we obtain a partial differential equation for the transformation:

$$H\left(\frac{\partial F_2}{\partial \boldsymbol{q}}, \boldsymbol{q}, t\right) + \frac{\partial F_2}{\partial t} = 0, \tag{1.2.14}$$

where F_2, the solution to (1.2.14), is customarily called Hamilton's principal function. For the second case in which H is independent of time, we need only set $\bar{H}$ equal to a constant and the transformation (1.2.13c) becomes

$$H\left(\frac{\partial F_2}{\partial \boldsymbol{q}}, \boldsymbol{q}\right) = E, \tag{1.2.15}$$

the *Hamilton–Jacobi equation*. Here F_2 is known as Hamilton's characteristic function. Unless (1.2.14) or (1.2.15) are *separable* (see Section 1.2c), they usually prove to be as difficult to solve as the original canonical set. However, the Hamilton–Jacobi formalism is very useful in obtaining approximate series solutions for systems that are near-separable, or, as is more usually called, *near-integrable*.

Poisson Brackets. An important dynamical quantity is the *Poisson bracket* defined by

$$[u, v] = \sum_k \left(\frac{\partial u}{\partial q_k}\frac{\partial v}{\partial p_k} - \frac{\partial v}{\partial q_k}\frac{\partial u}{\partial p_k}\right), \tag{1.2.16}$$

where u and v are arbitrary functions of the generalized coordinates. The equations of motion can be written in Poisson bracket form by choosing u as a coordinate and v as the Hamiltonian, yielding

$$[q_i, H] = \sum_k \left(\frac{\partial q_i}{\partial q_k}\frac{\partial H}{\partial p_k} - \frac{\partial H}{\partial q_k}\frac{\partial q_i}{\partial p_k}\right) = \frac{\partial H}{\partial p_i}$$

Comparing with Hamilton's equations, we have

$$\dot{q}_i = [q_i, H] \tag{1.2.17}$$

and, similarly,

$$\dot{p}_i = [p_i, H]. \tag{1.2.18}$$

1.2. Transformation Theory of Mechanics

Introducing

$$z = (q_1, \ldots, q_n, p_1, \ldots, p_n),$$

we can write Hamilton's equations in the form

$$\dot{z} = [z, H]. \qquad (1.2.19)$$

Poisson brackets have the following properties, which can also be considered to define them more generally, as we discuss in Section 1.3d.

$$[u, v] = -[v, u], \qquad (1.2.20a)$$

$$[[u, v], w] + [[w, u], v] + [[v, w], u] = 0, \qquad (1.2.20b)$$

$$[u + av, w] = [u, w] + a[v, w], \qquad (1.2.20c)$$

$$[uv, w] = u[v, w] + [u, w]v. \qquad (1.2.20d)$$

Using Hamilton's equations for $\dot{p}$ and $\dot{q}$, the total time derivative of an arbitrary function $\chi = \chi(q, p, t)$ can be written in the form

$$\frac{d\chi}{dt} = \sum_i \left(\frac{\partial \chi}{\partial q_i} \frac{\partial H}{\partial p_i} - \frac{\partial \chi}{\partial p_i} \frac{\partial H}{\partial q_i} \right) + \frac{\partial \chi}{\partial t} = [\chi, H] + \frac{\partial \chi}{\partial t}. \qquad (1.2.21)$$

For χ not an explicit function of time, $\partial \chi/\partial t = 0$. If the Poisson bracket $[\chi, H]$ also vanishes, we say that χ commutes with the Hamiltonian and χ is a constant of the motion. Clearly if the Hamiltonian is not an explicit function of time, it is such a constant. Such Hamiltonians are called *autonomous*. If we choose the function χ to be one of the momentum variables p_i (not an explicit function of t) and if the Hamiltonian is independent of the conjugate coordinate (i.e., if $\partial H/\partial q_i = 0$), we obtain from (1.2.21) that $dp_i/dt = 0$, and thus p_i is a constant of the motion

$$p_i = \alpha_i = \text{const.} \qquad (1.2.22)$$

and

$$\dot{q}_i = \frac{\partial H}{\partial \alpha_i} = \omega_i. \qquad (1.2.23)$$

If $\partial H/\partial \alpha_i$ is a function only of α_i, then $\omega_i = $ const. and, on integration of (1.2.23),

$$q_i = \omega_i t + \beta_i, \qquad (1.2.24)$$

which gives a solution for the time dependence of the variables. If a canonical transformation can be found that transforms all the momenta to constants, (1.2.22) and (1.2.24) are a solution in the transformed coordinate system. The inverse transformation then gives the complete solution in the original coordinates. Solution of the Hamilton–Jacobi equation (1.2.15) supplies the required generating function for the transformation.

An alternative procedure uses a function $w(p, q, s)$ to generate a canonical transformation from old variables p, q to new variables $\bar{p}, \bar{q}$ by means of Hamilton's equations

$$\frac{dp_i}{ds} = -\frac{\partial w}{\partial q_i}, \qquad \frac{dq_i}{ds} = \frac{\partial w}{\partial p_i}, \qquad (1.2.25)$$

where s is any parameter, and the new variables are related to the old by

$$\bar{p} = p(s), \qquad \bar{q} = q(s). \qquad (1.2.26)$$

Equations (1.2.26) are a canonical transformation, obtained from the nonmixed variable *Lie generating function w*. Application of the Lie generating function to perturbation theory simplifies the calculations to higher orders. We discuss the method in Section 2.5.

Finally, we note that the usefulness of a transformation from old to new variables lies in the judicious choice of the transformation. A canonical transformation introduces no new physics, but may facilitate analysis or physical interpretation of the motion, whose underlying properties remain the same in the old or the new coordinates.

*1.2b. Motion in Phase Space

Consider the Hamilton equations of motion having most generally N degrees of freedom, given by (1.2.6) with i running from 1 to N.

Associated with these equations are $2N$ constants of motion that are the initial coordinates and momenta in the N degrees of freedom. These uniquely determine the subsequent motion, which may also be thought of as the motion of a point through a $2N$-dimensional space. Let us assume that we have solved (1.2.6) for p and q as a function of time. In the $2N$-dimensional space with coordinates p and q, we can follow the trajectories from an initial time t_1, corresponding to initial coordinates p_1 and q_1, to some later time t_2. We call this p–q space the *phase space* of the system. The trajectories of three representative initial conditions are shown in Fig. 1.1, where we represent the phase space in two dimensions with the N momenta p along the abscissa and the N positions q along the ordinate. There are three important properties of this phase space that we take note of at this time.

1. Trajectories in phase space do not intersect at a given instant of time. This is evident from the fact that the initial conditions uniquely determine the subsequent motion. Thus if two trajectories came together, they would have the same values of p and q at that time, and their subsequent motion would be identical. If the Hamiltonian is independent of time, the trajectories in phase space are independent of time, and cannot cross in the phase space. It is also apparent that a generalized phase space with an additional dimension of time will have noncrossing trajectories even if the Hamiltonian is a function of time.

1.2. Transformation Theory of Mechanics

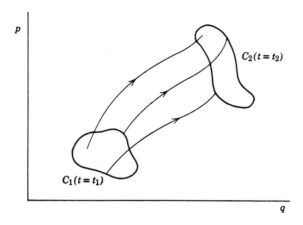

Figure 1.1. Trajectories in phase space.

2. A boundary in phase space C_1, which bounds a group of initial conditions at t_1, will transform into a boundary C_2 at t_2, which bounds the same group of initial conditions. This second property follows directly from the first, since any motion within the boundary, on approaching it, must then have the same initial conditions for subsequent motion as the boundary and thus move identically with the boundary. This second property has far-reaching consequences in that bounds can be placed on the motion of a large group of initial conditions by following a much smaller class of boundary initial conditions.
3. Consider an ensemble of initial conditions each representing a possible state of the system. We express the probability of a given ensemble or density distribution of system points in phase space as

$$\tau = \tau(\boldsymbol{p}, \boldsymbol{q}, t).$$

If we normalize τ such that[1]

$$\int_{\text{all space}} \tau \prod_i dp_i \, dq_i = 1, \qquad (1.2.27)$$

then $d\mathcal{N} = \tau \prod_i dp_i \, dq_i$ is the joint probability that at time t, the ensemble has an initial condition associated with the ith coordinates with position between q_i and $q_i + dq_i$ and momentum between p_i and $p_i + dp_i$. The rate of change of the number of phase points $d\mathcal{N}$, within the infinitesimal phase space volume

$$\prod_i dp_i \, dq_i, \qquad (1.2.28)$$

---

[1] This normalization cannot be performed if the phase space is infinite, but is not essential.

is obtained from the continuity equation

$$\frac{\partial d\mathcal{N}}{\partial t} + \sum_i \left(\frac{\partial}{\partial p_i}(d\mathcal{N}\dot{p}_i) + \frac{\partial}{\partial q_i}(d\mathcal{N}\dot{q}_i) \right) = 0. \qquad (1.2.29)$$

We divide by the volume to obtain the rate of change of density at a fixed position in phase space:

$$\frac{\partial \tau}{\partial t} + \sum_i \left(\dot{p}_i \frac{\partial \tau}{\partial p_i} + \tau \frac{\partial \dot{p}_i}{\partial p_i} + \dot{q}_i \frac{\partial \tau}{\partial q_i} + \tau \frac{\partial \dot{q}_i}{\partial q_i} \right) = 0.$$

From Hamilton's equation (1.2.6) for $\dot{p}_i$ and $\dot{q}_i$, the second and fourth terms of the summation cancel to obtain

$$\sum_i \left(\dot{p}_i \frac{\partial \tau}{\partial p_i} + \dot{q}_i \frac{\partial \tau}{\partial q_i} \right) + \frac{\partial \tau}{\partial t} = 0, \qquad (1.2.30)$$

which is a statement of the *incompressibility of the flow* in phase space. This result, known as Liouville's theorem, is a powerful tool for examining dynamical motion (see Lichtenberg, 1969, for detailed discussion and examples).

Integral Invariants. Because of the above properties, the phase space representation offers considerable simplifications in treating dynamical problems. In particular, it follows immediately from (1.2.30) that

$$\int \prod_i dp_i \, dq_i \qquad (1.2.31)$$

is an invariant of the motion, where the 2N-dimensional integral is constructed at a fixed time t. A hierarchy of such invariants, with the successive members of increasing dimensionality in the phase space, was first studied by Poincaré (1892) and called the integral invariants by him. A general derivation of these invariants can be found in Whittaker (1964). The integral invariants are fundamental to the theory of Hamiltonian flows and can be made the basis of the entire mechanics (Arnold, 1979). We examine the first member of that hierarchy [of which (1.2.31) is the Nth and final member]:

$$\int\int \sum_i dp_i \, dq_i = \text{const.}, \qquad (1.2.32)$$

where the integral is over a two-dimensional surface in phase space at a fixed time t.

If we apply Stokes' theorem to (1.2.32), we obtain the invariant

$$\oint \sum_i p_i \, dq_i = \text{const.}, \qquad (1.2.33)$$

where the integration is now over a closed path in phase space at a fixed

1.2. Transformation Theory of Mechanics

time t and is known as a *relative integral invariant* of the system. In general the application of Stokes' theorem reduces the order of the integration by one and transforms an integral invariant over an arbitrary domain into a relative invariant over a closed domain. The relative integral invariants are particularly important for oscillatory systems that we treat in the next subsection.

Extended Phase Space. We now consider a Hamiltonian H that is explicitly dependent on the time. The variational principle in (1.2.8), from which Hamilton's equations of motion are derived, clearly holds for integration with respect to any parameter that is independent of the variation implied by δ. Let us choose such a variable ζ and rewrite (1.2.8) as

$$\delta \int \left(\sum_{i=1}^{N} p_i \frac{dq_i}{d\zeta} - H \frac{dt}{d\zeta} \right) d\zeta = 0. \qquad (1.2.34)$$

Setting

$$\bar{p}_i = p_i, \qquad \bar{q}_i = q_i, \qquad i = 1, N,$$
$$\bar{p}_{N+1} = -H, \qquad \bar{q}_{N+1} = t, \qquad (1.2.35)$$

we have the new form of the variational equation

$$\delta \int \sum_{i=1}^{N+1} \bar{p}_i \frac{d\bar{q}_i}{d\zeta} d\zeta = 0,$$

where $-H$ and t are treated like any other momentum and coordinate, in a new $(2N + 2)$-dimensional *extended phase space*, and the flow is parametrized by the "time" ζ.

The new Hamiltonian $\bar{H}$ for the extended coordinates $(p, -H, q, t)$ can be obtained from the generating function

$$F_2 = \sum_{i=1}^{N} \bar{p}_i q_i + \bar{p}_{N+1} t. \qquad (1.2.36)$$

Using (1.2.13c), we find $\bar{H}(\bar{p}, \bar{q}) = H(p, q, t) - H$, where the canonical equations are now of the form

$$\frac{d\bar{p}_i}{d\zeta} = -\frac{\partial \bar{H}}{\partial \bar{q}_i}, \qquad \frac{d\bar{q}_i}{d\zeta} = \frac{\partial \bar{H}}{\partial \bar{p}_i}. \qquad (1.2.37)$$

The new Hamiltonian, generating the flow in the extended phase space, is explicitly independent of the "time" ζ. Further, (1.2.37) with $i = N + 1$ yields $t(\zeta) = \zeta$ and $\bar{H}$ = const. Thus the motion of a system with a time-dependent Hamiltonian is equivalent to that of a time-independent Hamiltonian with an additional degree of freedom.

A converse property also holds. Given a time-independent Hamiltonian $\bar{H}$ with N degrees of freedom and its $2N$-dimensional phase space, and choosing any generalized coordinate as the new "time" ζ, then the conjugate generalized coordinate represents a new "time" dependent Hamiltonian H with $N-1$ degrees of freedom, generating a dynamical motion in the $(2N-2)$-dimensional *reduced phase space*. For example, given

$$H(\boldsymbol{p}, \boldsymbol{q}) = H_0, \qquad (1.2.38)$$

we set

$$\bar{p}_i = p_i, \qquad \bar{q}_i = q_i, \qquad i = 1, N-1 \qquad (1.2.39)$$

and solve (1.2.38) for $p_N = p_N(\bar{\boldsymbol{p}}, \bar{\boldsymbol{q}}, q_N)$. Then putting $\bar{H} = p_N$ and $\zeta = q_N$, we obtain Hamilton's equations (1.2.37) in the reduced phase space $(\bar{\boldsymbol{p}}, \bar{\boldsymbol{q}})$, where the new Hamiltonian is an explicit function of the "time" ζ. The details can be found in Whittaker (1964, Section 141).

In this way, the theory developed for a time-independent Hamiltonian with N degrees of freedom also applies to a time-dependent Hamiltonian with $N-1$ degrees of freedom. In particular, a time-independent Hamiltonian with two degrees of freedom is dynamically equivalent to a time-dependent Hamiltonian with one degree of freedom.

The Action Integral. We now show the relationship between the relative integral invariant (1.2.33), which is an integral evaluated at a fixed time t, and the action integral for a one-dimensional oscillatory system described by a time-independent Hamiltonian. We define the action integral as

$$J = \frac{1}{2\pi} \oint p \, dq, \qquad (1.2.40)$$

where the integral is taken over one cycle of the oscillation in time.

In terms of the extended phase space, (1.2.33) for an oscillatory system with one degree of freedom can be rewritten as

$$\oint (p \, dq - H \, dt) = \text{const.},$$

where the path is now over $\zeta = \text{const.}$; but since ζ is arbitrarily chosen, the new path, which now includes variation of time, can be chosen so that part of it lies along the actual system trajectory in phase space. For the special case of $H = \text{const.}$, the second term vanishes around any closed path and we have

$$\oint p \, dq = \text{const.} \qquad (1.2.41)$$

Now if we take as the tube of trajectories around which we integrate the one shown in Fig. 1.2, we see that the path of integration consists of two

1.2. Transformation Theory of Mechanics

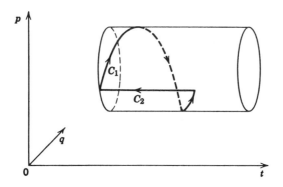

Figure 1.2. Path for action integral calculations.

parts given by

$$\oint p\, dq = \int_{C_1} p\, dq + \int_{C_2} p\, dq, \qquad (1.2.42)$$

where C_1 is the path along one complete cycle. For periodic systems, the end points of C_1 have identical values of q, and thus the path C_2 may be chosen such that $q = $ const. Therefore, from (1.2.42) and by comparison with (1.2.40),

$$\int_{C_1} p\, dq = \text{const.} = 2\pi J, \qquad (1.2.43)$$

which shows the equivalence of the action integral to the relative integral invariant in this case. The action integral is important, both in that it forms the canonical momentum in action-angle coordinates, as discussed in Section 1.2c, and that it is an *adiabatic* constant of the motion, that is, an approximate constant for a Hamiltonian that is changing slowly in time compared to the period of the oscillation described above. The adiabatic constancy of the action, which we treat in detail in Section 2.3, is of fundamental importance for the understanding of regular orbits in systems with time-dependent Hamiltonians or in systems with more than one degree of freedom.

Surface of Section. The definition of a Poincaré surface of section lies at the heart of the treatment of Hamiltonian flows. For an autonomous system with two degrees of freedom, the phase space is four dimensional. Referring to Fig. 1.3a, we choose a two-dimensional surface Σ_R in the phase space and label its two sides (say left and right). We then study the successive intersections of a trajectory with this surface. The intersections are generated each time the trajectory pierces the surface in a particular sense (say from left to right).

A particularly convenient choice for the surface of section Σ_R can be made.

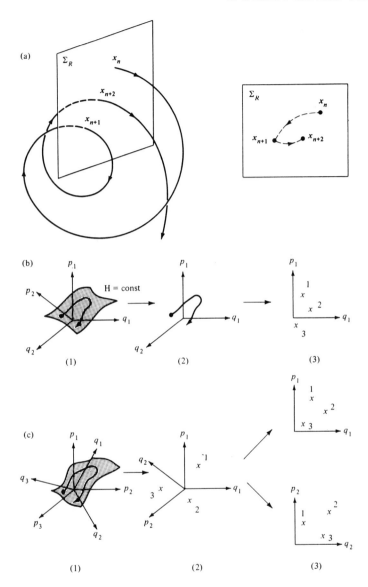

Figure 1.3. Motion in phase space and definition of the Poincaré surface of section. (a) Intersections of a trajectory with the surface of section. (b) Two degrees of freedom showing: (1) four-dimensional phase space with the trajectory on a three-dimensional energy surface; (2) projection of the trajectory onto the (p_1, q_1, q_2) volume; and (3) successive intersections of the trajectory with the two-dimensional surface of section $q_2 = $ const. (c) Three degrees of freedom showing: (1) six-dimensional phase space with trajectory on a five-dimensional energy surface; (2) three successive intersections of the trajectory with the four-dimensional surface of section $q_3 = $ const.; and (3) projections of these intersections of the surface of section on the (p_1, q_1) and (p_2, q_2) planes.

1.2. Transformation Theory of Mechanics

We first note that the trajectory lies on a three-dimensional energy surface $H(p_1, p_2, q_1, q_2) = H_0$ in the four-dimensional phase space, Fig. 1.3b(1). This equation determines any of the four variables, say p_2, in terms of the other three:

$$p_2 = p_2(p_1, q_1, q_2). \qquad (1.2.44)$$

We are therefore led to consider the projection of the trajectory onto a three-dimensional volume (p_1, q_1, q_2). Fig. 1.3b(2). If the motion is bounded, then the plane $q_2 = $ const. within this volume may be repeatedly crossed by the trajectory. This plane, consisting of a single coordinate q_1 and its canonical momentum p_1, is a convenient choice for the surface of section. If we plot the successive intersections of the motion with the surface of section, they will in general occur anywhere within a bounded area of the plane, If a constant of the motion

$$I(p_1, p_2, q_1, q_2) = \text{const.} \qquad (1.2.45)$$

exists in addition to H_0, then (1.2.44) and (1.2.45) can be combined to yield

$$p_1 = p_1(q_1, q_2). \qquad (1.2.46)$$

Thus the successive crossings of the motion with the surface of section must lie on a unique curve, given by (1.2.46) with $q_2 = $ const. In this way we can determine the existence of constants of the motion by examining the intersection of a trajectory with a surface of section. Once this existence is established, the smooth curves can be examined for local stability and other interesting features.

It will be noted that the particular surface of section (p_1, q_1) is just the *reduced phase space* of the original Hamiltonian system. Furthermore, successive crossings are obtained from one another by a canonical transformation generated by Hamilton's equations. Thus the area bounding a closed curve in the surface of section is conserved on successive crossings of the surface. This important property can be shown directly as follows: we write general differential relations

$$d\lambda = \frac{\partial \lambda}{\partial q} dq + \frac{\partial \lambda}{\partial \mu} d\mu,$$

$$dp = \frac{\partial p}{\partial q} dq + \frac{\partial p}{\partial \mu} d\mu, \qquad (1.2.47)$$

where the λ and μ may be thought of as the initial position and momentum taken in a surface of section. Substituting for the partial derivatives in terms of the generating function $F_2(\lambda, p)$, and solving for dp and dq in terms of $d\lambda$

and $d\mu$, we have

$$dq = \left(F_{\lambda p} - \frac{F_{pp}F_{\lambda\lambda}}{F_{\lambda p}}\right) d\lambda + \frac{F_{pp}}{F_{\lambda p}} d\mu,$$

$$dp = -\frac{F_{\lambda\lambda}}{F_{\lambda p}} d\lambda + \frac{1}{F_{\lambda p}} d\mu, \qquad (1.2.48)$$

where the notation $\partial F_2/\partial\lambda = F_\lambda$ has been used for compactness. The determinant of the coefficients of (1.2.48), which is equivalent to the Jacobian of the transformation from (λ, μ) to (q, p) variables, is seen to be equal to one, proving the area-preserving nature of the transformation.[2] The area-preserving property of a two-dimensional surface of section in a four-dimensional phase space will be very important in subsequent chapters, both for numerically looking for constants of the motion (as in Section 1.4) and for determining the stability of the linearized motion near a periodic solution (see Section 3.3).

The concept of a surface of section can be generalized to systems with $N > 2$ degrees of freedom. For a time-independent Hamiltonian with N degrees of freedom, the energy surface in phase space has dimensionality $2N - 1$, Fig. 1.3c(1). As previously done, we project out a single generalized coordinate, say p_N, and consider the successive intersection of the trajectory with the $(2N - 2)$-dimensional surface of section $q_N = $ const. with coordinates $p_1, \ldots, p_{N-1}, q_1, \ldots, q_{N-1}$ [Fig. 1.3c(2)]. As before, the surface of section is a reduced phase space in which the volume-preserving property holds. If one or more constants of the motion exist, then the intersections of the trajectory with the surface will all lie on a unique surface of dimensionality less than $2N - 2$; otherwise the intersections will fill a $(2N - 2)$-dimensional volume within the section.

For multidimensional systems, if the motion in each degree of freedom is approximately separable, the $N - 1$ projections of the surface of section onto the (p_i, q_i) planes, shown in Fig. 1.3c(3), are a useful way to visualize the motion of the trajectory point. For a regular trajectory, which is exactly separable in the (p_i, q_i) coordinates, the motion in each (p_i, q_i) plane is area-preserving, a constant of the motion exists in the ith degree of freedom, and the trajectory intersections lie on a smooth curve in the (p_i, q_i) plane. However, for a system with more than two degrees of freedom, even for regular trajectories, the trajectory intersections with the surface of section, projected onto an arbitrary (p_i, q_i) plane, do not generally lie on a smooth curve, but rather fill an annulus of finite area, whose thickness is related to the nearness to exact separability in the (p_i, q_i) coordinates. In this case, the intersections with the surface of section lie in an $(N - 1)$-dimensional surface, whose projection onto a (p_i, q_i) plane is a region of finite area. Examples embodying these features are described briefly in Section 1.4b and in detail in Chapter 6.

---

[2] For noncanonical coordinates some function of the coordinates, the *measure* is preserved, rather than the area.

*1.2c. Action-Angle Variables

For time-independent Hamiltonians with one degree of freedom, we have seen that a constant of the motion exists. For time-independent Hamiltonians with N degrees of freedom, N constants of the motion can still be found that decouple the N degrees of freedom if the Hamilton–Jacobi equation is completely separable in some coordinate system. Replacing the generating function F_2 by a common notation S,[3] we assume a separated solution

$$S = \sum_i S_i(q_i, \alpha_1 \ldots \alpha_N), \qquad (1.2.49)$$

where the α's are the new momenta associated with the N constants of the motion. If now the Hamiltonian can be written in the separated form[4]

$$H = \sum_i H_i\left(\frac{\partial S_i}{\partial q_i}, q_i\right),$$

then the Hamilton–Jacobi equation (1.2.15) splits into N equations:

$$H_i\left(\frac{\partial S_i}{\partial q_i}, q_i\right) = \alpha_i, \qquad (1.2.50)$$

because the q_i are independent. We can then solve for S_i in terms of q_i. The new momenta α_i are thus the separation constants of the Hamilton–Jacobi equation satisfying

$$\sum_i \alpha_i = H_0. \qquad (1.2.51)$$

The relation between the old and new generalized coordinates is given by (1.2.13). The new Hamiltonian $\bar{H}$ is a function of the momenta α_i only, and Hamilton's equations for the motion may be trivially solved.

The choice of the separation constants α_i as the new momenta is arbitrary. One could just as well choose as new momenta any N quantities J_i, which are independent functions of the α_i:

$$J_i = J_i(\boldsymbol{\alpha}). \qquad (1.2.52)$$

If these N equations are inverted,

$$\alpha_i = \alpha_i(\boldsymbol{J}) \qquad (1.2.53)$$

and inserted into (1.2.49), then the generating function for transformation

---

[3] In some mechanics texts S is reserved for time-dependent Hamiltonians, and W is employed in the above context.

[4] This is not the most general condition for separability.

to the new momenta J_i is found to be

$$\bar{S}(q, J) = S(q, \alpha(J)), \tag{1.2.54}$$

with the new Hamiltonian

$$\bar{H}(J) = \sum_i \alpha_i(J). \tag{1.2.55}$$

Again, Hamilton's equations may be trivially solved.

For completely separable, periodic systems, a special choice of the J's as functions of the α's is very useful. By periodic systems we mean those for which, in each degree of freedom, either (a) p_i and q_i are periodic functions of time with the same period, or (b) p_i is a periodic function of q_i. Case (a) is generally known as *libration*, case (b) as *rotation*. The periods of the motion in each degree of freedom need not be the same. If the periods are not in the ratio of a rational number, then the motion is called *conditionally periodic*. To define the action variables J_i as functions of the α's, we form the action integral as in (1.2.40). using (1.2.13a) for p_i

$$J_i = \frac{1}{2\pi} \oint p_i \, dq_i = \frac{1}{2\pi} \oint \frac{\partial S_i(q_i, \alpha)}{\partial q_i} \, dq, \tag{1.2.56}$$

where the $J_1 \ldots J_N$ are the new constant momenta. Inverting yields the new generating function $\bar{S}(q, J)$. From (1.2.24), the conjugate coordinate θ_i is given by

$$\theta_i = \omega_i t + \beta_i, \tag{1.2.57}$$

where ω_i and β_i are constants. Integrating θ_i over a complete oscillation period T, we have

$$\theta_i = \int_t^{t+T} d\theta_i = \omega_i T; \tag{1.2.58}$$

but from (1.2.13b),

$$d\theta_i = \frac{\partial}{\partial q_i} \frac{\partial \bar{S}}{\partial J_i} \, dq_i, \tag{1.2.59}$$

and substituting into (1.2.58), interchanging derivatives, and integrating over one cycle,

$$\Delta \theta_i = \frac{\partial}{\partial J_i} \oint \frac{\partial \bar{S}}{\partial q_i} \, dq_i = \frac{\partial}{\partial J_i} \oint p_i \, dq_i = 2\pi. \tag{1.2.60}$$

By comparing (1.2.58) with (1.2.60), we see that

$$\omega_i T = 2\pi, \tag{1.2.61}$$

that is, the constant ω_i is just the *radian frequency* of the oscillation. The

1.2. Transformation Theory of Mechanics

action-angle formulation thus provides a convenient way of obtaining the frequencies of oscillation without solving for the details of the motion. When describing the motion of near-integrable systems, one almost invariably makes a preparatory transformation to action-angle variables of the integrable part of the given system. The system is then prepared for further study using perturbation theory or other methods.

The Harmonic Oscillator. We illustrate the usefulness of the action-angle formalism with the harmonic (linear) oscillator, whose Hamiltonian is of the form

$$H = G\frac{p^2}{2} + F\frac{q^2}{2} = \alpha. \tag{1.2.62}$$

where G, F, and α are constants. Solving for $p(q, \alpha)$ and evaluating the action,

$$J = \frac{2}{\pi}\int_0^{q_{max}} \left(\frac{2\alpha}{G} - \frac{F}{G}q^2\right)^{1/2} dq, \tag{1.2.63}$$

where $q_{max} = (2\alpha/F)^{1/2}$, yields

$$J = \alpha(FG)^{-1/2}. \tag{1.2.64}$$

From (1.2.55), $\bar{H} = \alpha$, and rearranging (1.2.64),

$$\bar{H} = (FG)^{1/2}J, \tag{1.2.65}$$

independent of the angle variable. The angular frequency is given by $\omega_0 = \partial\bar{H}/\partial J = (FG)^{1/2}$. Substituting for α in (1.2.62) from (1.2.64), we obtain one transformation equation

$$p = p(q, J) = (2RJ - R^2q^2)^{1/2}, \tag{1.2.66}$$

where $R = (F/G)^{1/2}$. From (1.2.13a),

$$\bar{S} = \int_0^q (2RJ - R^2q^2)^{1/2} dq.$$

and from (1.2.13b), we obtain the second transformation equation

$$\theta = R\int_0^q (2RJ - R^2q^2)^{-1/2} dq, \tag{1.2.67}$$

which, on integration gives

$$q = (2J/R)^{1/2} \sin\theta. \tag{1.2.68a}$$

Combining with (1.2.66) yields

$$p = (2JR)^{1/2} \cos\theta. \tag{1.2.68b}$$

Equations (1.2.68) give the transformation from action-angle variables to the original variables p and q. This transformation is usually performed by means of a generating function of the type $F_1(q, \bar{q})$ given by

$$F_1 = \tfrac{1}{2}Rq^2 \cot \theta. \tag{1.2.69}$$

We also note that (1.2.62) is the equation of an ellipse, which can be canonically transformed to a circle by scaling the coordinates $p = \sqrt{R}p'$ and $q = q'/\sqrt{R}$. It is then obvious that a transformation from rectangular (p', q') coordinates to action-angle (J, θ) coordinates will represent the motion as the rotation of a radial vector of constant length J, yielding immediately the transformation equations (1.2.68). We also see that physically R is the ratio of the axes of the original ellipse.

We shall see in subsequent sections that the transformation to action-angle coordinates is useful when F and G are slowly varying functions, either of time or of a coordinate in some other degree of freedom, and J is then an *adiabatic invariant* of the motion; i.e., for F and G slowly varying, J does not vary much from its initial value even when ω_0 and $\bar{H}$ change by large amounts (see also Lichtenberg, 1969, Chapter 2). If the Hamiltonian is nonlinear, then (1.2.68) does not transform the system to action-angle form. Nevertheless (1.2.68) may be used in expansion techniques, as we shall see in Chapter 2. The general procedure developed in this section is still applicable to a nonlinear oscillator for obtaining the action-angle variables in terms of integrals, as we shall see in the following section.

1.3. Integrable Systems

We consider a Hamiltonian system with N degrees of freedom. If the Hamilton–Jacobi equation is separable into N independent equations, one for each degree of freedom, then we say the Hamiltonian, and the resulting motion, is *integrable* (*completely integrable* or *completely separable* are terms sometimes used). The separation constants α_i are known as *isolating integrals* or *global invariants of the motion*, since each invariant isolates a degree of freedom by the property that $\partial H/\partial p_i = f(q_i)$ in some canonical coordinate system. A Hamiltonian with N degrees of freedom is integrable if and only if N independent isolating integrals exist.

It is sufficient that the N independent integrals be *in involution*, i.e., their Poisson brackets with each other vanish, $[\alpha_i, \alpha_j] = 0$. This ensures that the α's are a complete set of new momenta in some transformed coordinate system. Any complete set of N functions of the α's, such as the action variables J_i, are a set of isolating integrals. The Poisson brackets of these functions automatically vanish. See Whittaker (1964, §147) for further details.

1.3. Integrable Systems

*1.3a. One Degree of Freedom

For one degree of freedom with H explicitly independent of time, we have seen from (1.2.21) that

$$H(p, q) = E, \qquad (1.3.1)$$

a constant of the motion. Thus all such Hamiltonians are integrable, with the momentum p determined as a function of q alone, independent of the time, by inverting (1.3.1):

$$p = p(q, E). \qquad (1.3.2)$$

The complete solution for p and q as a function of time can be obtained from the second of Hamilton's equations (1.2.6b), which yields

$$dt = \frac{dq}{\partial H/\partial p}, \qquad (1.3.3)$$

and on integration,

$$t = \int_{q_0}^{q} \frac{dq}{\partial H/\partial p}. \qquad (1.3.4)$$

Since $\partial H/\partial p$ is a function of p and q only, and p and q are related through (1.3.2), we have reduced the equation of motion to quadratures. The integral, however, often can only be evaluated numerically.

The Pendulum Hamiltonian. We illustrate the procedure with a simple example, that of a pendulum. This Hamiltonian has the basic form that arises in essentially all nonlinear resonance problems, and its solution lies at the core of our treatment of nonlinear motion in subsequence chapters. The equations of motion are

$$\dot{p} = -F \sin \phi,$$
$$\dot{\phi} = Gp, \qquad (1.3.5)$$

where $F = mgh$, $G = 1/(mh^2)$, mg is the gravitational force on the mass m, h the pendulum length, ϕ the angle from the vertical, and p the angular momentum conjugate to ϕ, satisfying Hamilton's equations. The Hamiltonian is the sum of kinetic energy $\frac{1}{2}Gp^2$ and potential energy $U = -F \cos \phi$:

$$H = \tfrac{1}{2}Gp^2 - F \cos \phi = E. \qquad (1.3.6)$$

Application of (1.3.4) reduces the problem to quadratures, with the solution obtained in terms of elliptic integrals. However, considerable information can be obtained by examining (1.3.6) for different values of the energy E, which we do in Fig. 1.4. The value of the Hamiltonian corresponds to the total energy, kinetic plus potential, of the system. If E is greater than the largest value of potential energy F, then p is always different from zero,

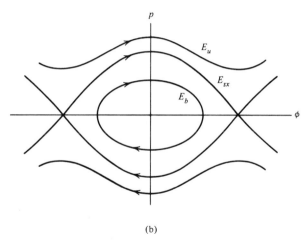

Figure 1.4. Correspondence between (a) an energy diagram and (b) a phase space diagram for the pendulum Hamiltonian.

resulting in unbounded motion in ϕ (rotation) corresponding to motion for $p > 0$ from left to right with energy E_u. For $E < F$, within the potential well, the value of initial energy corresponds to bounded motion (libration). For $E = F \equiv E_{sx}$, we have separatrix motion, in which the oscillation period becomes infinite. The motion has two singular points at $p = 0$: the origin at $\phi = 0$, which is a stable or *elliptic* singular point; and the joining of the two branches of the separatrix at $\phi = \pm\pi$, which is an unstable or *hyperbolic* singular point. A phase space trajectory near an elliptic singular point remains in its neighborhood, while a trajectory near a hyperbolic point diverges from it.

1.3. Integrable Systems

From (1.3.4), it is seen that the oscillation period is in general a function of the energy of the oscillator. Substituting from (1.3.6) for H into (1.3.4) evaluated over a complete period, we have

$$T = \frac{1}{(2G)^{1/2}} \oint \frac{d\phi}{(E + F \cos \phi)^{1/2}}, \qquad (1.3.7)$$

which can be evaluated in terms of elliptic integrals. In particular, on the separatrix we see from (1.3.5) that both the restoring force and the velocity go to zero at $\phi = \pi$, and thus T becomes infinite.

We can transform the Hamiltonian to action-angle variables (J, θ) using the evaluation of action as in (1.2.63) and the corresponding angle variable as obtained in the following discussion. We have

$$J(E) = \frac{2}{\pi} \int_0^{\phi_{max}} \left[\frac{2}{G}(E + F \cos \phi_1) \right]^{1/2} d\phi_1, \qquad (1.3.8)$$

$$\theta(\phi, E) = \left(G \frac{dJ}{dE} \right)^{-1} \int_0^{\phi} \frac{d\phi_1}{[(2/G)(E + F \cos \phi_1)]^{1/2}}, \qquad (1.3.9)$$

where $\phi_{max} = \pi/2$ for rotation $(E > F)$, and $\cos \phi_{max} = -H/F$ for libration $(E < F)$. The new Hamiltonian is obtained by putting $\bar{H} = E$ in (1.3.8) and inverting. By the use of the half-angle formulas, these expressions can be transformed into elliptic integrals, in terms of which (1.3.8) and (1.3.9) can be written (see Smith, 1977, Rechester and Stix, 1979)

$$J = R \frac{8}{\pi} \begin{cases} \mathscr{E}(\kappa) - (1 - \kappa^2)\mathscr{K}(\kappa), & \kappa < 1, \\ \frac{1}{2}\kappa\mathscr{E}(\kappa^{-1}), & \kappa > 1, \end{cases} \qquad (1.3.10)$$

$$\theta = \frac{\pi}{2} \begin{cases} [\mathscr{K}(\kappa)]^{-1}\mathscr{F}(\eta, \kappa), & \kappa < 1, \\ 2[\mathscr{K}(\kappa^{-1})]^{-1}\mathscr{F}(\tfrac{1}{2}\phi, \kappa^{-1}), & \kappa > 1. \end{cases} \qquad (1.3.11)$$

Here $R = (F/G)^{1/2}$ as before, $\mathscr{K}(\kappa)$ and $\mathscr{E}(\kappa)$ are the complete elliptic integrals of the first and second kind:

$$\begin{aligned}\mathscr{K}(\kappa) &= \int_0^{\pi/2} \frac{d\xi}{(1 - \kappa^2 \sin^2 \xi)^{1/2}}, \\ \mathscr{E}(\kappa) &= \int_0^{\pi/2} (1 - \kappa^2 \sin^2 \xi)^{1/2} \, d\xi,\end{aligned} \qquad (1.3.12)$$

where $\kappa \sin \eta = \sin \tfrac{1}{2}\phi$, $2\kappa^2 = 1 + E/F$, and $\mathscr{F}$ is the incomplete form of $\mathscr{K}$, i.e., $\mathscr{F}(\pi/2, \kappa) = \mathscr{K}(\kappa)$. The quantity κ is a measure of the normalized oscillator energy, with $\kappa = 1$ at $E/F = 1$, the separatrix energy, and $\kappa < 1$ for libration, $\kappa > 1$ for rotation.

From $dJ/dE = 1/\omega$ in (1.3.10), we obtain the normalized frequency

$$\frac{\omega(\kappa)}{\omega_0} = \frac{\pi}{2}\begin{cases}[\mathcal{K}(\kappa)]^{-1}, & \kappa < 1, \\ 2\kappa/\mathcal{K}(\kappa^{-1}), & \kappa > 1,\end{cases} \qquad (1.3.13)$$

where

$$\omega_0 = (FG)^{1/2} \qquad (1.3.14)$$

is the radian frequency for linearized oscillation about the elliptic singular point. The asymptotic value of $\mathcal{K}$ for κ near 1 yields a normalized frequency near the separatrix

$$\lim_{\kappa \to 1} \frac{\omega}{\omega_0} = \begin{cases} \dfrac{\pi}{2}\Big/ \ln\left[\dfrac{4}{(1-\kappa^2)^{1/2}}\right], & \kappa < 1, \\ \pi \Big/ \ln\left[\dfrac{4}{(\kappa^2-1)^{1/2}}\right], & \kappa > 1, \end{cases} \qquad (1.3.15)$$

which approaches zero logarithmically as κ approaches unity. The separatrix trajectory can be obtained using (1.3.6) and the separatrix condition $E = F$ to write

$$p_{sx} = \frac{2^{1/2}\omega_0}{G}(1 + \cos\phi_{sx})^{1/2}, \qquad (1.3.16)$$

where the subscript sx denotes the separatrix variables. Applying the half-angle formula to the square root, we have

$$p_{sx} = \pm\frac{2\omega_0}{G}\cos\frac{\phi_{sx}}{2}. \qquad (1.3.17)$$

The plus and minus signs give the upper and lower branches of the separatrix, respectively. From Hamilton's equations applied to (1.3.6), we have

$$\dot\phi_{sx} = Gp_{sx}, \qquad (1.3.18)$$

which, using (1.3.17), gives

$$\frac{d\phi_{sx}}{dt} = \pm 2\omega_0 \cos\frac{\phi_{sx}}{2}. \qquad (1.3.19)$$

Solving for dt and integrating with the initial condition $\phi = 0$ at $t = 0$,

$$\omega_0 t = \int_0^{\phi_{sx}} \frac{d\phi/2}{\cos(\phi/2)} = \ln\tan\left(\frac{\phi_{sx}}{4} + \frac{\pi}{4}\right), \qquad (1.3.20)$$

and, on inversion,

$$\phi_{sx} = 4\tan^{-1}[\exp(\omega_0 t)] - \pi. \qquad (1.3.21)$$

1.3. Integrable Systems

Orbits near the separatrix orbit are very similar to that on the separatrix itself, except that the oscillation period for a trajectory near the separatrix, from (1.3.13), tends to infinity as the initial conditions approach the separatrix.

Although the integrable Hamiltonian with one degree of freedom in (1.3.6) was derived for the pendulum, it has the basic form that arises in essentially all near-integrable systems with many degrees of freedom in which a resonance between degrees of freedom exists. In the neighborhood of the values of the action that give exact resonance, a Fourier expansion of the nonintegrable part of the Hamiltonian produces terms that have slow variation governed by (1.3.6). Suppose two angle variables ϕ and ψ in different degrees of freedom are near a resonance, such that the ratio of their frequencies ω_ϕ/ω_ψ is near a rational number r/l for some values of the actions. We then can make a canonical transformation to a new variable

$$\theta = l\phi - r\psi, \qquad (1.3.22)$$

such that θ is a slowly varying function of time, and one of the fast angle variables, say ϕ, is eliminated. The momentum canonical to θ is related to the deviation of an action, say, J_ϕ from its exact value at resonance J_0. If we average over the fast angle variable ψ, we obtain a one-degree-of-freedom Hamiltonian that is identical in form to the pendulum Hamiltonian (1.3.6). Since this Hamiltonian always emerges from Fourier analysis of the perturbation, followed by the employment of secular perturbation theory and the method of averaging, it has been called the "universal description of a nonlinear resonance" by Chirikov (1979). We refer to it as the *standard Hamiltonian*. It plays a fundamental role in the development of our subject. The perturbation theory from which the form arises is given in Section 2.4 and used extensively in succeeding chapters.

1.3b. Linear Differential Equations

Before proceeding to the examination of the nonlinear two-degrees-of-freedom problem, we consider the mathematically well-understood problem of a linear second-order differential equation with time-varying coefficients. For this problem solutions are known to exist and to correspond to regular motion. We examine the differential equation of the form

$$\ddot{x} + f(t)\dot{x} + g(t)x = 0; \qquad (1.3.23)$$

$f(t)$ and $g(t)$ are considered for the moment as arbitrary known functions of t. Since the equation is of second order and linear, a general solution can be constructed from a pair of linear independent solutions x_1 and x_2. An important property of this equation can be found by examining the

Wronskian determinant

$$W = \begin{vmatrix} x_1 & x_2 \\ \dot{x}_1 & \dot{x}_2 \end{vmatrix}. \qquad (1.3.24)$$

Differentiating both sides with respect to t, we obtain

$$\frac{dW}{dt} = \begin{vmatrix} x_1 & x_2 \\ \ddot{x}_1 & \ddot{x}_2 \end{vmatrix} + \begin{vmatrix} \dot{x}_1 & \dot{x}_2 \\ \dot{x}_1 & \dot{x}_2 \end{vmatrix}, \qquad (1.3.25)$$

where the second term is zero by virtue of having identical rows. Substitution for $\ddot{x}_1$ and $\ddot{x}_2$ from (1.3.23) and expanding, we have

$$\dot{W} = -x_1(f\dot{x}_2 + gx_2) + x_2(f\dot{x}_1 + gx_1).$$

The gx_1x_2 terms in each bracket cancel, leaving

$$\dot{W} = -fW.$$

This yields on integration

$$W(t) = W_0 \exp\left[-\int_{t_0}^{t} f(t)\,dt\right]. \qquad (1.3.26)$$

If $f(t) = 0$, then there is no dissipation and the resulting equation

$$\ddot{x} + g(t)x = 0 \qquad (1.3.27)$$

is clearly derivable from a Hamiltonian

$$H = \tfrac{1}{2}(p^2 + g(t)q^2), \qquad (1.3.28)$$

with $q = x$ and $p = \dot{x}$. For this case, (1.3.26) reduces to

$$W = \text{const.} \qquad (1.3.29)$$

Equation (1.3.29) holds whether $g(t)$ is periodic or not.

The solution to any second-order differential equation, whether periodic or not, is uniquely determined by the initial values of the function and its derivative. Thus for either independent solution we can write the transformation from an initial time $t = 0$ to any other time as

$$x_1(t) = m_{11}x_1(0) + m_{12}\dot{x}_1(0),$$
$$\dot{x}_1(t) = m_{21}x_1(0) + m_{22}\dot{x}_1(0),$$

where the coefficients m are dependent on the time but independent of the initial conditions. A consequence of (1.3.29) is that the determinant of the coefficients is equal to 1:

$$\det \mathbf{M} = \begin{vmatrix} m_{11} & m_{12} \\ m_{21} & m_{22} \end{vmatrix} = 1. \qquad (1.3.30)$$

1.3. Integrable Systems

This can be seen by writing the transformation matrix for the two solutions

$$\begin{bmatrix} x_1(t) & x_2(t) \\ \dot{x}_1(t) & \dot{x}_2(t) \end{bmatrix} = \begin{bmatrix} m_{11} & m_{12} \\ m_{21} & m_{22} \end{bmatrix} \begin{bmatrix} x_1(0) & x_2(0) \\ \dot{x}_1(0) & \dot{x}_2(0) \end{bmatrix}.$$

Taking the determinant of both sides and using the relationship, that the determinant of the product matrix is equal to the product of the two determinants, we obtain the Wronskian transformation

$$W(t) = \det \mathbf{M} \, W(0),$$

and since $W(t) = W(0)$, $\det \mathbf{M} = 1$, which is equivalent to the condition for area preservation. We consider only such conservative systems in the following analysis.

Periodic Coefficients. If $g(t)$ is periodic with period τ, then (1.3.27) has a pair of independent solutions of the form

$$x(t) = \operatorname{Re} w(t) \exp[i\psi(t)], \qquad (1.3.31)$$

such that w is periodic:

$$w(t) = w(t + \tau)$$

and

$$\exp\{i[\psi(t + \tau) - \psi(t)]\} = \exp(i\sigma),$$

where σ is independent of time. Thus $\dot{\psi}$ is also periodic. Equation (1.3.31) is the general Floquet form for an equation with periodic coefficients. If we differentiate (1.3.31) twice, and substitute the result into (1.3.27), $e^{i\psi}$ cancels, and equating real and imaginary parts we obtain

$$\ddot{w} - w\dot{\psi}^2 + g(t)w = 0 \qquad (1.3.32a)$$

and

$$2\dot{w}\dot{\psi} + w\ddot{\psi} = 0. \qquad (1.3.32b)$$

We rearrange the second of these equations to obtain

$$\frac{2\dot{w}}{w} + \frac{\ddot{\psi}}{\dot{\psi}} = 0,$$

which gives, on integration,

$$\dot{\psi} = \frac{1}{w^2}. \qquad (1.3.33)$$

Substituting (1.3.33) into (1.3.31), and writing the solution in sinusoidal form, we have

$$x(t) = w(t) \cos \psi(t), \qquad (1.3.34)$$

where now $\psi = \int_{t_0}^{t} (dt/w^2)$. An invariant can be found for this solution from the quantity

$$\cos^2 \psi + \sin^2 \psi = 1. \qquad (1.3.35)$$

The terms in (1.3.35) are found from (1.3.34) and its derivative to give the invariant

$$I(x, \dot{x}, t) = [w^{-2}x^2 + (w\dot{x} - \dot{w}x)^2]. \qquad (1.3.36)$$

Although $w(t)$ is not in general known explicitly, a solution is known to exist, and therefore the invariant I always exists for the Hamiltonian (1.3.28). Substituting (1.3.33) in (1.3.32a), w in (1.3.36) is seen to satisfy the differential equation

$$\ddot{w} + g(t)w - \frac{1}{w^3} = 0. \qquad (1.3.37)$$

Equation (1.3.37) is no easier to solve than the original equation (1.3.27). Lewis (1968) has pointed out, however, that *any* solution of (1.3.37) will, through (1.3.36), give the solution of (1.3.27) for all initial conditions. The invariant (1.3.36) was found useful for examining the motion in strong focused accelerators (Courant and Snyder, 1968), where $g(t)$ is piecewise constant, allowing an explicit determination of $w(t)$. The solutions are also explicitly known if $g(t) = a + b \cos t$, which gives the Matthieu equation. Lewis (1968) has considered general functions $g(t)$, showing that the invariant (1.3.36) could be derived using perturbation theory.

We point out that the above analysis applies to linear, nonautonomous, one-degree-of-freedom systems that correspond to a special class of two-degrees-of-freedom Hamiltonians. Since second-order, nonautonomous, linear systems, with periodic coefficients, are integrable, it is not surprising that an associated invariant can always be found. Attempts by Symon (1970) and Lewis and Leach (1980) to generalize these methods to nonlinear oscillators, while of mathematical interest, have not yet led to useful new results.

1.3c. More than One Degree of Freedom

For systems with more than one degree of freedom, (1.3.3) is generalized to

$$dt = \frac{dq_1}{\partial H/\partial p_1} = \frac{dq_2}{\partial H/\partial p_2} \cdots = \frac{dq_N}{\partial H/\partial p_N}. \qquad (1.3.38)$$

Only if $\partial H/\partial p_1 = f(q_1)$, a function of q_1 only, is the first equation reduced to quadratures (solved) and similarly for the succeeding equations. In general we must solve the entire set of differential equations simultaneously to obtain a complete solution. If other constants of the motion exist in addition to the Hamiltonian, however, then the number of simultaneous

1.3. Integrable Systems

equations may be reduced by one for each additional constant of the motion that is also an *isolating integral*; that is, an integral that in some transformed coordinate system makes $\partial H/\partial p_i = f(q_i)$. A transformation to action-angle form accomplishes this, since it ensures the more restrictive condition that $\partial H/\partial p_i = $ const. However, the transformation itself depends on the existence of the isolating integral, which may, however, be sufficiently imbedded in the dynamics that it is not easily recognizable. The isolating integrals can be associated with symmetries of the dynamical system. These symmetries may be obvious, in which case the proper coordinate transformation required to reduce the system to quadratures is usually straightforward. This is true, for example, of a particle in a central force field, as described below. When the symmetry is not obvious, as in the case of the Toda lattice, also treated below, uncovering an isolating integral is not at all a straightforward process. There is no procedure presently known for determining all the isolating integrals of a general Hamiltonian system, or even for finding their total number. It follows that there is no general test to determine integrability for a system with N degrees of freedom. Unless the symmetries are obvious, it is often numerical experiments that suggest the presence of a hidden isolating integral, and lead to its uncovering.

The Central Force Problem. We now illustrate the uncovering of an isolating integral in addition to the total energy and the reduction to quadratures for the simple example of a particle moving in a central field of force. This problem is well known to be integrable. Without loss of generality we reduce the problem to the two dimensions defined by the plane of the initial particle velocity and the origin. Motion in the third dimension is trivially separated from the other two by the isolating integral $p_z = 0$. In polar (r, θ) coordinates, the Hamiltonian is

$$H = \frac{1}{2m}\left(p_r^2 + \frac{p_\theta^2}{r^2}\right) + U(r), \tag{1.3.39}$$

where $p_r = m\dot{r}$, $p_\theta = mr^2\dot{\theta}$, m is the particle's mass, and U the potential corresponding to the central force $(F = -\partial U/\partial r)$. Since the system is conservative, $H = E$, a constant. In the form of (1.3.38) the equations of motion become

$$dt = \frac{d\theta}{\partial H/\partial p_\theta} = \frac{dr}{\partial H/\partial p_r}.$$

Taking the partial derivatives and eliminating p_r by the use of the Hamiltonian, we obtain

$$dt = \frac{d\theta}{p_\theta/mr^2} = \frac{dr}{[2m(E - U(r)) - p_\theta^2/r^2]^{1/2}/m}. \tag{1.3.40}$$

Now, unless p_θ is known as a function of θ and r, these equations cannot be

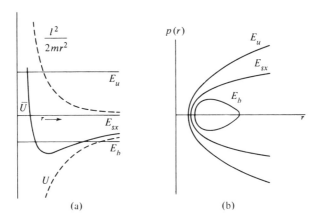

Figure 1.5(a, b). The equivalent one-dimensional potential and the corresponding phase space diagrams for an attractive central force.

solved. It is here that the existence of a second constant of the motion becomes essential. In the present case p_θ is a constant of the motion. This comes about because there is no θ force and θ therefore does not appear explicitly in the Hamiltonian; hence in (1.2.6a), $dp_\theta/dt = 0$, giving

$$p_\theta = l, \quad \text{a const.} \tag{1.3.41}$$

Substituting this in the second equation of (1.3.40), we reduce the r-motion to quadratures and the θ-motion follows by substitution. This can be seen directly from (1.3.39) by defining an equivalent potential $\bar{U}(r) = l^2/2mr^2 + U(r)$. For the problem with an attractive potential $U(r) = k/r^\beta$ with $2 < \beta < 0$ (this includes the Kepler problem with $\beta = 1$), the two terms of the potential are shown in Fig. 1.5a together with their sum or equivalent potential. We also show three values of the constant energy or Hamiltonian, E_b, E_{sx}, and E_u corresponding to bound, separatrix, and unbound motion. The corresponding phase–plane plots are shown in Fig. 1.5b. The motion here is seen to correspond to the one-dimensional case described in Fig. 1.4, except that the separatrix for the marginally bound case closes at infinity. The complete motion bound between radii r_1 and r_2 corresponding to energy E_b is shown in Fig. 1.6 in r, θ space for $\beta \neq 1$. The configuration space orbit is not closed because the ratio of the r and θ periods is not a whole number, an example of conditionally periodic motion. Nevertheless, the projections of the motion on the $r - p_r$ plane (or, alternatively, the values of r, p_r at each orbit crossing of a constant θ plane) form closed loops for this bound motion, as a consequence of the existence of the two isolating integrals of the motion $p_\theta = l$ and $H = E$. For $\beta = 1$ (the Kepler problem), the frequencies of the r and θ motion are the same, as we shall show, and the motion in configuration space is a closed curve (an ellipse).

We now show the transformation of the central force problem to action-

1.3. Integrable Systems

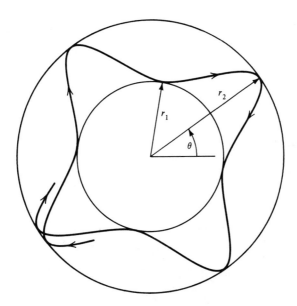

Figure 1.6. Illustration of the bounded orbits corresponding to the effective potential $\bar{U}$ of Fig. 1.5 for a non inverse square law force ($\beta \neq 1$).

angle variables. Introducing the characteristic functions in (1.3.39), we obtain the Hamilton–Jacobi equation as given in (1.2.15):

$$\frac{1}{2m}\left[\left(\frac{\partial S_r}{\partial r}\right)^2 + \frac{1}{r^2}\left(\frac{\partial S_\theta}{\partial \theta}\right)^2\right] + U(r) = E, \qquad (1.3.42)$$

where we have used the transformation $p_i = \partial S/\partial q_i$ and the fact that S is separable. Multiplying through by $2mr^2$, the Hamiltonian is also seen to be separable as in (1.2.50), leading to

$$\left(\frac{\partial S_\theta}{\partial \theta}\right)^2 = 2mr^2\left[E - \frac{1}{2m}\left(\frac{\partial S_r}{\partial r}\right)^2 - U(r)\right] = l^2. \qquad (1.3.43)$$

We have recovered the constant angular momentum, which, of course, follows directly from the fact that θ is a cyclic cordinate, that is, does not appear in the Hamiltonian. The second equation in (1.3.43) gives

$$\left(\frac{\partial S_r}{\partial r}\right)^2 = 2m(E - U(r)) - \frac{l^2}{r^2}. \qquad (1.3.44)$$

We form the action variables

$$2\pi J_\theta = \oint p_\theta \, d\theta = \oint \frac{\partial S_\theta}{\partial \theta} \, d\theta, \qquad (1.3.45)$$

and

$$2\pi J_r = \oint p_r \, dr = \oint \frac{\partial S_r}{\partial r} \, dr. \tag{1.3.46}$$

Substituting (1.3.43) and (1.3.44) into (1.3.45) and (1.3.46), respectively, we obtain

$$J_\theta = \frac{1}{2\pi} \int_0^{2\pi} l \, d\theta = l, \tag{1.3.47}$$

and

$$J_r = \frac{1}{2\pi} \int \left[2m(E - U(r)) - \frac{l^2}{r^2} \right]^{1/2} dr. \tag{1.3.48}$$

If, for example, $U(r) = -k/r$, the potential corresponding to an inverse square law attractive force, a simple integration procedure (see Goldstein, 1951, Section 9.7) leads to

$$J_r = -l + \frac{k}{2}\left(\frac{2m}{-E}\right)^{1/2}, \tag{1.3.49}$$

a constant. Rearrangement gives the new Hamiltonian

$$\bar{H} = E = -\frac{mk^2}{2(J_r + J_\theta)^2}, \tag{1.3.50}$$

where we have substituted J_θ for l. We note that the action variables appear only as a sum; hence there is an *intrinsic degeneracy* of the r and θ motion, with only one frequency of oscillation,

$$\omega = \frac{\partial \bar{H}}{\partial J_r} = \frac{\partial \bar{H}}{\partial J_\theta} = \frac{mk^2}{(J_r + J_\theta)^3}, \tag{1.3.51}$$

giving a closed orbit. If the central force has a different r dependence, the orbit would no longer be closed, as shown in Fig. 1.6.

The Toda Lattice. Our second example of an integrable Hamiltonian is that of the three-particle Toda lattice (Toda, 1970) whose Hamiltonian is given by

$$H = \tfrac{1}{2}(p_1^2 + p_2^2 + p_3^2) + \exp[-(\phi_1 - \phi_3)] + \exp[-(\phi_2 - \phi_1)] \\ + \exp[-(\phi_3 - \phi_2)] - 3, \tag{1.3.52}$$

which corresponds to three particles moving on a ring (Fig. 1.7), with exponentially decreasing repulsive forces between. In addition to the energy, there is a relatively obvious isolating integral, namely, the total momentum

$$P_3 = p_1 + p_2 + p_3 = \text{const.} \tag{1.3.53}$$

1.3. Integrable Systems

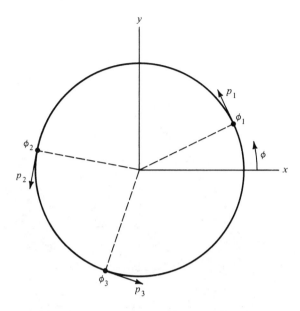

Figure 1.7. The three-particle Toda lattice.

This follows since the Hamiltonian is invariant to a rigid rotation $\phi_i \to \phi_i + \phi_0$, and may also be seen directly from Hamilton's equations. To exhibit this explicitly, we transform H to the new momenta $P_1 = p_1$, $P_2 = p_2$, with P_3 given by (1.3.53). Using the generating function

$$F_2 = P_1\phi_1 + P_2\phi_2 + (P_3 - P_1 - P_2)\phi_3 \qquad (1.3.54)$$

we find

$$\mathcal{H} = \tfrac{1}{2}[P_1^2 + P_2^2 + (P_3 - P_1 - P_2)^2]$$
$$+ \exp(-\Phi_1) + \exp[-(\Phi_2 - \Phi_1)] + \exp(\Phi_2) - 3, \qquad (1.3.55)$$

where the Φ's are canonical to the P's. Since $\mathcal{H}$ is independent of Φ_3, the constancy of P_3 is shown directly. Without loss of generality, we put $P_3 = 0$, corresponding to choice of a rotating system in which the total momentum is zero. There seems to be no obvious additional isolating integral. The Hamiltonian can be put in the form of a particle moving in a two-dimensional potential well by means of the generating function

$$F_2' = (4\sqrt{3})^{-1}[(p_x' - \sqrt{3}p_y')\Phi_1 + (p_x' + \sqrt{3}p_y')\Phi_2], \qquad (1.3.56)$$

together with a final noncanonical, but trivial transformation

$$p_x' = 8\sqrt{3}p_x, \quad x' = x; \quad p_y' = 8\sqrt{3}p_y, \quad y' = y; \quad \bar{H} = H'/\sqrt{3}$$

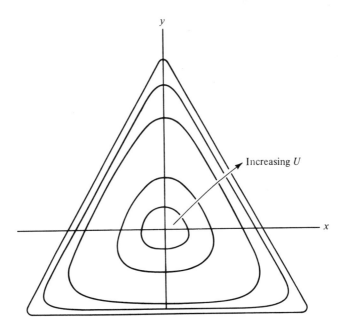

Figure 1.8. The potential well for the Toda Hamiltonian showing lines of constant potential U.

to yield the Toda Hamiltonian

$$\bar{H} = \tfrac{1}{2}(p_x^2 + p_y^2) + \tfrac{1}{24}[\exp(2y + 2\sqrt{3}x) \\ + \exp(2y - 2\sqrt{3}x) + \exp(-4y)] - \tfrac{1}{8}. \quad (1.3.57)$$

The potential curves, sketched in Fig. 1.8, vary smoothly outward from the origin in both x and y, displaying a threefold symmetry. If $\bar{H}$ is expanded to cubic terms in x and y, we obtain the Hénon and Heiles Hamiltonian

$$\bar{H}' = \tfrac{1}{2}(p_x^2 + p_y^2 + x^2 + y^2) + x^2 y - \tfrac{1}{3}y^3, \quad (1.3.58)$$

the motion of which is examined in the next section. This motion is known to be nonintegrable: Hénon and Heiles (1964), in their numerical experiments, found a transition with increasing energy $\bar{H}' = E$ from ordered to stochastic motion, with some stochasticity present for all energies, indicating that an isolating integral did not exist. The Toda Hamiltonian $\bar{H}$ was examined numerically by Ford et al. (1973) expecting the same result. They were surprised to discover the result that the trajectories were regular for arbitrary energy $\bar{H} = E$, with all intersections of a trajectory with the surface of section $x = 0$ falling on smooth invariant curves. In Fig. 1.9, these curves are shown for values of $E = 1$ and $E = 256$. These results are in dramatic contrast to those in the next section in which significant area filling trajectories are plainly visible down to energies as low as $E = \tfrac{1}{8}$. The resolution of this

1.3. Integrable Systems

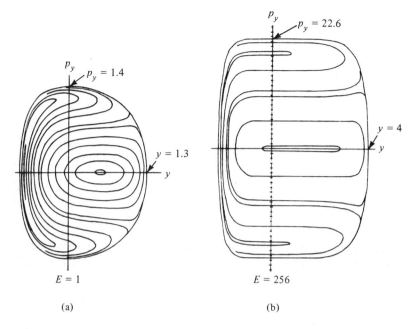

Figure 1.9. Surfaces of section for the Toda Hamiltonian with widely differing energies: (a) $E = 1$ and (b) $E = 256$ (after Ford et al., 1973).

difference is, of course, that a hidden symmetry exists for the Toda lattice, with its concomitant isolating integral. Hénon (1974), encouraged by Ford's numerical results, found the explicit, analytic form for the isolating integral

$$I = 8p_x(p_x^2 - 3p_y^2) + (p_x + \sqrt{3}p_y)\exp[(2y - 2\sqrt{3}x)]$$
$$- 2p_x \exp(-4y) + (p_x - \sqrt{3}p_y)\exp[(2y + 2\sqrt{3}x)] = \text{const.} \quad (1.3.59)$$

Setting $x = 0$, (1.3.59) can be used to eliminate p_x in (1.3.58), such that the invariant curves in Fig. 1.9 can be calculated directly. The existence of the three isolating integrals H, P_3, and I ensures integrability for the Toda Hamiltonian (1.3.52). Even in the original coordinates, I is related to no obvious conservation law or symmetry.

Finding Integrable Hamiltonians. Are there any general methods to test for the integrability of a given Hamiltonian? The answer, for the moment, is no. We can turn the question around, however, and ask if methods can be found to construct potentials that give rise to integrable Hamiltonians. The answer here is that a method exists, at least for a restricted class of problems, but the method becomes rapidly very tedious as the forms allowed for the integrals of the motion are expanded. The method was first applied by Whittaker (1964, §152) to particle motion in a two-dimensional potential with a Hamiltonian of the form

$$H = \tfrac{1}{2}p_1^2 + \tfrac{1}{2}p_2^2 + V(q_1, q_2). \tag{1.3.60}$$

Whittaker asked what potential functions V exist such that the system possess an integral up to quadratic order in the p's

$$I(p, q) = ap_1^2 + bp_2^2 + cp_1p_2 + ep_1 + fp_2 + g, \tag{1.3.61}$$

where the coefficients are functions of the q's. For I to be a constant of the motion, it must satisfy the Poisson bracket relation:

$$[I, H] = 0. \tag{1.3.62}$$

Substituting (1.3.60) and (1.3.61) in (1.3.62) and equating the coefficients of $p_1^m p_2^n$ separately, one finds a set of partial differential equations for the coefficients in terms of the potential V and its first derivatives. The terms linear in p are found to decouple from the other terms. From the terms linear in p, we find

$$\frac{\partial e}{\partial q_1} = 0, \quad \frac{\partial f}{\partial q_2} = 0, \quad \frac{\partial f}{\partial q_1} + \frac{\partial e}{\partial q_2} = 0, \tag{1.3.63}$$

and

$$e\frac{\partial V}{\partial q_1} + f\frac{\partial V}{\partial q_2} = 0; \tag{1.3.64}$$

from the terms quadratic in p, we find

$$\frac{\partial a}{\partial q_1} = 0, \quad \frac{\partial b}{\partial q_2} = 0, \quad \frac{\partial b}{\partial q_1} + \frac{\partial c}{\partial q_2} = 0, \quad \frac{\partial c}{\partial q_1} + \frac{\partial a}{\partial q_2} = 0, \tag{1.3.65}$$

and

$$\frac{\partial g}{\partial q_1} - 2a\frac{\partial V}{\partial q_1} - c\frac{\partial V}{\partial q_2} = 0, \quad \frac{\partial g}{\partial q_2} - 2b\frac{\partial V}{\partial q_2} - c\frac{\partial V}{\partial q_1} = 0. \tag{1.3.66}$$

Since (1.3.63) and (1.3.64) are independent of (1.3.65) and (1.3.66), the former can be solved for an invariant linear in the p's

$$I = q_1 p_2 - q_2 p_1,$$

which exists for the azimuthally symmetric potential

$$V = V(q_1^2 + q_2^2). \tag{1.3.67}$$

The invariant I is just the angular momentum, which we have seen to be constant for the central force problem. To look for solutions independent of

1.3. Integrable Systems

this symmetry, we set $e = f = 0$, and from (1.3.65) and (1.3.66) we obtain a second-order partial differential for V:

$$c\left(\frac{\partial^2 V}{\partial q_1^2} - \frac{\partial^2 V}{\partial q_2^2}\right) + 2(b - a)\frac{\partial^2 V}{\partial q_1 \partial q_2}$$
$$+ \left(\frac{\partial c}{\partial q_1} - 2\frac{\partial a}{\partial q_2}\right)\frac{\partial V}{\partial q_1} + \left(2\frac{\partial b}{\partial q_1} - \frac{\partial c}{\partial q_2}\right)\frac{\partial V}{\partial q_2} = 0. \quad (1.3.68)$$

where a, b, and c are found from the differential equations (1.3.65). Whittaker demonstrates that (1.3.68) has characteristics of the form

$$\frac{x^2}{\alpha^2} + \frac{y^2}{\alpha^2 - \gamma^2} = 1,$$

where x and y are related to q_1 and q_2 by a simple cordinate transformation, and α and γ are constants of integration. Taking new variables to be the parameters of these confocal ellipses and hyperboles,

$$x = \frac{\alpha\beta}{\gamma}, \quad y = \frac{1}{\gamma}[(\alpha^2 - \gamma^2)(\gamma^2 - \beta^2)]^{1/2},$$

the differential equation for V, (1.3.68), is satisfied by

$$V = \frac{\psi(\alpha) - \phi(\beta)}{\alpha^2 - \beta^2}, \quad (1.3.69)$$

where ψ and ϕ are arbitrary functions of their arguments.

The above result is interesting, but has not led to new solutions of physical problems. There has, however, been a recent rekindling of interest in constructing integrable Hamiltonians. Hall (1981) has applied the method to the motion of a particle in static electric and magnetic fields, explicitly introducing the vector potential into the problem. He found that Whittaker's solutions were not complete because the conservation of energy was not explicitly considered in the constraints. He also has examined other classes of invariants, not quadratic in the momenta. The hope here is to generate self-consistent currents for plasma confinement configurations whose invariants constrain the motion. Due to the sensitivity of integrability (witness the contrast of the Toda potential to the Hénon and Heiles truncation of that potential); there is little likelihood that truly integrable motion can be constructed, in this manner, from realizable potentials. The interesting question is how far can a potential be perturbed away from an integrable one such that most trajectories remain regular. This is the main subject of Chapters 2 and 4.

In another approach, Holt (1981) examined the Hamiltonian

$$H = H_0 + \epsilon V \quad (1.3.70)$$

and required that V be chosen such that in a perturbation expansion all terms higher than a given power of ϵ be identically zero. Using this procedure, he was able to uncover the invariant (1.3.59) of the Toda Hamiltonian. As Holt shows, it is also possible to obtain this invariant directly from the method of Whittaker by allowing the invariant to be cubic in the p's. The procedure is the same as described for the quadratic (in p) invariant, but is much more complicated. The method is unable, generally, to determine if an invariant exists for a given Hamiltonian of the form (1.3.70), because it is not possible to examine all invariants. On the other hand, if an invariant exists with a lower power of p (such as p^3 for the Toda lattice), it can be found, demonstrating the integrability of the original Hamiltonian. For more than two degrees of freedom, even for restricted classes of invariants, these techniques cannot be carried through.

A recent method for examining dynamical systems for integrability involves testing the system of equations for the Painlevé property. This property requires that a Laurent expansion of the solution, considering time as a complex variable, have only simple poles at all movable singularities. The movable singularities are those that depend on the initial conditions. Ablowitz et al., (1980) showed that there is a close connection between partial differential equations that have soliton solutions (are integrable) and the reduced ordinary differential equations which have the Painlevé property. Segur (1980) has reviewed this work and furthermore showed that the Lorenz model (see Section 1.5) for a dissipative system, generally having chaotic properties, is integrable for just those parameters for which the equations exhibit the Painlevé property. A number of well known examples of Hamiltonian systems were examined by Bountis et al. (1982) with the results again giving an exact correspondence between integrability and the Painlevé property. Although there is no rigorous proof of this correspondence, at least within the class of systems examined (systems with two degrees of freedom whose Hamiltonian is quadratic in the momentum) there is considerable evidence that a general correspondence exists. The method is, in principle, applicable to systems of higher dimensionality, but the correspondence has not yet been verified.

The method for finding particular integrable Hamiltonians involves choosing a Hamiltonian with a given form but with some arbitrary constants and requiring that these constants be chosen to satisfy the Painlevé property for the leading terms in the Laurent expansion. For example, the Hénon and Heiles Hamiltonian (1.3.58) is generalized to

$$H = \tfrac{1}{2}(\dot{x}^2 + \dot{y}^2 + Ax^2 + By^2) + x^2 y + \frac{\mu}{3} y^3$$

and the constants μ, A, and B are determined (Bountis et al., 1982). A similar procedure, using the constraints determined from Whittaker's procedure,

1.3. Integrable Systems

with invariants taken to fourth order, was applied to the generalized Hénon and Heiles Hamiltonian by Hall (1982). Both methods determine the integrability conditions

(a) $\mu = 1$, $A = B$,

(b) $\mu = 6$, any A and B,

(c) $\mu = 16$, $B = 16A$,

which have been verified by direct computation. It is not known, however, whether any fundamental connection exists between the Painlevé and the Whittaker procedure.

These ideas have been extended by various authors. Gramaticos et al. (1983), Hietarinta (1983), Yoshida (1984), and Dorizzi (1985) have studied various examples, applying the methods to new classes of problems. New mathematical results have been obtained by Ziglin (1984). Higher-degree-of-freedom systems have been analyzed by Inozentsev (1984). These and related results are summarized by Hietarinta (1986). A thorough exposition of the Painlevé method can be found in Tabor (1989).

Returning to the idea of approximate integrability, as introduced by Hall (1982), the procedure is to vary some parameter to produce an invariant to some order in an expansion. Hall shows how this could be done, following the direct methods described above. An alternative procedure is to approach integrability by minimizing some measure of nonintegrability. This technique has been developed by Cary (1982) and Cary and Hanson (1986) to increase the volume of good flux surfaces for confining plasmas in a toroidal magnetic field.

As a final comment on integrability, we have seen that nonlinear systems with one degree of freedom are integrable, while those with two degrees of freedom only rarely so. What happens as the number of degrees of freedom increases further? As just discussed, even the process of finding a sparse set of integrable potentials becomes difficult, if not impossible. However, in Section 6.5, numerical evidence indicates that the fraction of the phase space covered by regular trajectories may either increase or decrease with an increasing number of degrees of freedom. Remarkably, when we turn to systems governed by partial, rather than ordinary, differential equations, which in some sense correspond to the infinite-degree-of-freedom limit, we again find large classes of integrable systems. However, if we discretize these systems (reduce to finite but many dimensions), then the question of integrability returns in a more subtle form. We discuss these questions in Section 6.5. As described above, the partial differential equations that exhibit soliton (integrable) solutions can be reduced to ordinary differential equations with the Painlevé property. For further discussion of the methods of solution and of the relationship between partial and ordinary differential equations, the reader is referred to the literature (Lamb, 1980; Segur, 1980).

1.3d. Noncanonical Variables and Reduction

In Section 1.2a we defined the Poisson bracket $[u, v]$ in (1.2.16) and gave its properties in (1.2.20). Hamilton's equations could then be written as

$$\dot{z} = [z, H],$$

where $z = (q_1, \ldots, q_{N/2}, p_1, \ldots, p_{N/2})$. If we introduce the canonical Poisson tensor

$$\mathbf{J} = \begin{pmatrix} \mathbf{0} & \mathbf{1} \\ -\mathbf{1} & \mathbf{0} \end{pmatrix}, \qquad (1.3.71)$$

where $\mathbf{0}$ and $\mathbf{1}$ are $(N/2) \times (N/2)$ null and unit matrices, respectively, then, in terms of the J_{ij}'s we have

$$[u, v] = J_{ij} \frac{\partial u}{\partial z_i} \frac{\partial v}{\partial z_j}, \qquad (1.3.72)$$

and Hamilton's equations can be written as

$$\dot{z}_i = J_{ij} \frac{\partial H}{\partial z_j}. \qquad (1.3.73)$$

If we introduce a transformation $(z, H) \to (z', H')$ with $M_{ij} = \partial z_i'/\partial z_j$, the transformation matrix, then the Poisson bracket transforms to a new form $[u, v] \to \{u, v\}$, where

$$\{u, v\} = J_{ij}' \frac{\partial u}{\partial z_i'} \frac{\partial v}{\partial z_j'} \qquad (1.3.74)$$

and

$$J_{ij}' = M_{ik} M_{il} J_{kl} \qquad (1.3.75)$$

is the transformed bracket tensor. If the transformation is canonical, then we must have $J_{ij}' = J_{ij}$, and (1.3.74) still has the form of a Poisson bracket $[\cdot, \cdot]$. The new equations of motion are then in Hamiltonian form:

$$\dot{z}_i' = J_{ij} \frac{\partial H'}{\partial z_j'}. \qquad (1.3.76)$$

If the transformation is not canonical, then the new bracket $\{\cdot, \cdot\}$ is not in canonical form, although this noncanonical bracket still obeys the bracket properties (1.2.20).

For some purposes, the transformations to noncanonical coordinates are desirable; e.g., it may facilitate the application of perturbation theory for certain problems. For example, the Hamiltonian of a charged particle

1.3. Integrable Systems

gyrating in a static magnetic field is well known to be

$$H = \frac{1}{2m}\left|\boldsymbol{p} - \frac{e}{c}\boldsymbol{A}(\boldsymbol{q})\right|^2, \tag{1.3.77}$$

where $\boldsymbol{A}$ is the vector potential. Making the noncanonical transformation to new coordinates $\boldsymbol{r}' = \boldsymbol{q}$ and $\boldsymbol{v}' = (1/m)(\boldsymbol{p} - (e/c)\boldsymbol{A}(\boldsymbol{q}))$, we obtain the noncanonical bracket

$$\{u, v\} = \frac{1}{m}\left(\frac{\partial u}{\partial \boldsymbol{r}'} \cdot \frac{\partial v}{\partial \boldsymbol{v}'} - \frac{\partial u}{\partial \boldsymbol{v}'} \cdot \frac{\partial v}{\partial \boldsymbol{r}'}\right) + \frac{e}{m^2 c} \boldsymbol{B} \cdot \left(\frac{\partial u}{\partial \boldsymbol{v}'} \times \frac{\partial v}{\partial \boldsymbol{v}'}\right) \tag{1.3.78}$$

and the new Hamiltonian $H' = mv'^2/2$, where $\boldsymbol{B} = \nabla \times \boldsymbol{A}$ is the magnetic field. This bracket is in the form of (1.3.74) with

$$\boldsymbol{J}' = \begin{pmatrix} 0 & \frac{1}{m}\boldsymbol{1} \\ -\frac{1}{m}\boldsymbol{1} & \frac{1}{m}\boldsymbol{\Omega} \end{pmatrix}, \tag{1.3.79}$$

where $\Omega_{ij} = (e/mc)\epsilon_{ijk}B_k$ and ϵ_{ijk} is the third rank, antisymmetric unit pseudotensor.[5] This transformation is the starting point for guiding-center perturbation theory (Littlejohn, 1979, 1983).

Symmetry and Reduction. In section 1.3c we have seen, for the central force problem, that the existence of a symmetry in the potential allowed us to obtain a constant of the motion which reduced a two-degree-of-freedom problem to one degree of freedom. For this configuration, from Hamilton's equations, the constant was directly in the form of a canonical momentum, and thus its use left the Hamiltonian in canonical form. For the Toda lattice this canonical reduction to a lower dimensionality could be done for one of the three degrees of freedom but not for the second. Nevertheless, the existence of a second constant allowed the reduction to an integrable system, although it was not obtained explicitly.

The situation in which the symmetry leads directly to a reduced system in canonical form is the exception, rather than the rule. For a reduced system with a noncanonical bracket, Darboux's theorem (Littlejohn, 1979) shows how to construct new canonical coordinates for any nonsingular bracket; i.e., any bracket for which the bracket tensor J'_{ij} is nonsingular. However, the presence of constraints or symmetries in a Hamiltonian system sometimes gives rise to a reduced system having a singular noncanonical bracket. In this case, additional conserved quantities called *Casimirs* exist due to the degeneracy of the bracket. We describe this work following reviews by Marsden and Ratiu (1987), Finn and Sun (1987), and Boghosian (1987).

---

[5] ϵ_{ijk}: ± 1 if i, j, k is an even (odd) permutation of 1, 2, 3; $\epsilon_{ijk} = 0$ otherwise.

Consider the example of the pendulum equations of (1.3.5). In rectangular coordinates a quantity H, the sum of kinetic plus potential energy, can be expressed as

$$H = \frac{1}{2m}(p_x^2 + p_y^2) + mgy, \tag{1.3.80}$$

but now subject to the constraint

$$x^2 + y^2 = l^2. \tag{1.3.81}$$

Recognition that the constraint has polar symmetry allows the direct expression of the equations in polar coordinates with a single degree of freedom, with the resultant Hamiltonian (1.3.6) in canonical variables. However, the variables in (1.3.80) are not canonical, i.e., the motion does not satisfy Hamilton's equations, because of the additional constraint. The constraint can be imposed by a bracket tensor that requires $\dot{r} = 0$ and $\dot{p}_r = 0$:

$$\mathbf{J}'' = \begin{matrix} r \\ \theta \\ p_r \\ p_\theta \end{matrix} \begin{pmatrix} 0 & 0 & 0 & 0 \\ 0 & 0 & 0 & 1 \\ 0 & 0 & 0 & 0 \\ 0 & -1 & 0 & 0 \end{pmatrix}. \tag{1.3.82}$$

This matrix, and its associated bracket (1.3.74) are singular (det $\mathbf{J}'' = 0$), as it has rank $2 < 4$. In the original variables (x, y, p_x, p_y), the above matrix transforms to another $\mathbf{J}'$ still having rank 2, from which two constants of motion, namely, $x^2 + y^2 = r^2$ and $(xp_x + yp_y)/(x^2 + y^2)^{1/2} = p_r$ are obtained. These quantities are called Casimirs and are constants of motion because of the form of the bracket rather than the form of the Hamiltonian. In (θ, p_θ) space, of course, the bracket (as well as the bracket matrix) is canonical.

More precisely, if $K < N$ is the rank of the $N \times N$ tensor $\mathbf{J}'$ for the reduced system, then $\mathbf{J}'$ has a null space of dimension $M = N - K$ and there exist M linearly independent vectors V_i having $\mathbf{J} \cdot \mathbf{V} = \mathbf{0}$. It can be shown that there exist M functions $C_i(z')$, called the Casimirs of the degenerate bracket

$$\{u, v\} = J'_{ij} \frac{\partial u}{\partial z'_i} \frac{\partial v}{\partial z'_j},$$

such that $\{u, C_i\} = 0$ for any u. In the preceding example, $z' = (x, y, p_x, p_y)$, $N = 4$, $K = 2$, and $M = 2$.

As another example, we consider the free rigid body having as generalized coordinates the Euler angle θ, ϕ, and ψ with their canonical momenta p_θ, p_ϕ, and p_ψ. In the body frame in which the moment of inertia tensor $\mathbf{I}$ is

1.3. Integrable Systems

diagonal, the Hamiltonian is (Goldstein, 1980)

$$H = \frac{p_\psi^2}{2I_3} + \frac{1}{2I_2}[(p_\phi \csc\theta - p_\psi \cot\theta)\cos\psi - p_\theta \sin\psi]^2$$
$$+ \frac{1}{2I_1}[(p_\phi \csc\theta - p_\psi \cot\theta)\sin\psi + p_\theta \cos\psi]^2, \quad (1.3.83)$$

where I_1, I_2 and I_3 are the diagonal elements of $\mathbf{I}$. The phase space is six-dimensional, but the symmetry allows a reduction to the three components of the angular momentum vector in the body frame

$$m_1 = (p_\phi \csc\theta - p_\psi \cot\theta)\sin\psi + p_\theta \cos\psi,$$
$$m_2 = (p_\phi \csc\theta - p_\psi \cot\theta)\cos\psi - p_\theta \sin\psi, \quad (1.3.84)$$
$$m_3 = p_\psi,$$

such that we obtain the Euler equations for a rigid body,

$$\dot{m}_1 = \frac{I_2 - I_3}{I_2 I_3} m_2 m_3,$$
$$\dot{m}_2 = \frac{I_3 - I_1}{I_1 I_3} m_1 m_3, \quad (1.3.85)$$
$$\dot{m}_3 = \frac{I_1 - I_2}{I_1 I_2} m_1 m_2.$$

These equations can be written in the form

$$\dot{m}_i = \{m_i, H'\}, \quad (1.3.86)$$

where the noncanonical bracket in the reduced space is

$$\{u, v\} = -\mathbf{m} \cdot \left(\frac{\partial u}{\partial \mathbf{m}} \times \frac{\partial v}{\partial \mathbf{m}}\right) \quad (1.3.87)$$

and the new Hamiltonian is

$$H' = \sum_i \frac{m_i^2}{2I_i}. \quad (1.3.88)$$

The 3×3 bracket tensor

$$\mathbf{J}' = \begin{pmatrix} 0 & -m_3 & m_2 \\ m_3 & 0 & -m_1 \\ -m_2 & m_1 & 0 \end{pmatrix} \quad (1.3.89)$$

associated with the bracket (1.3.87) is singular, having rank $K = 2$. Hence there is a single Casimir,

$$C = m_1^2 + m_2^2 + m_3^2, \quad (1.3.90)$$

namely, the total angular momentum, which is conserved by the form of the bracket. Thus the motion of a free rigid body actually lives in a two-dimensional state space on the sphere $C = $ const., equivalent to a one-degree-of-freedom system, and therefore the motion is integrable.

From these examples it is apparent that the bracket method with Casimirs is a different way of looking at phase space transformations, and also a way of handling constraints and symmetries that are not easily incorporated within the canonical formalism. The new formalism has yielded new insights that have been useful. For example, it has allowed a new approach to the *nonlinear stability* of complicated systems such as hinged rigid bodies with attachments (Holmes and Marsden, 1983; Krishnaprasad and Marsden, 1987), and to infinite-dimensional systems such as fluids (Holm *et al.*, 1985). The stability concepts, first presented by Arnold (1966) and developed further by Holm *et al.* (1985) and by others, are outlined in Section 3.3d. The underlying theory, drawn from function theory, goes beyond the presentation in this book. Simpler accounts of the stability theory complementary to those presented here can be found in Marsden and Ratiu (1987) and in Finn and Sun (1987).

*1.4. Near-Integrable Systems

We now turn to a qualitative description of those generic Hamiltonian systems that we can treat as perturbations of integrable systems. We refer to these as *near-integrable* systems. We first review the simple case of an autonomous Hamiltonian with two degrees of freedom, or equivalently, a time-dependent Hamiltonian with one degree of freedom. As we saw in Section 1.2b, nonautonomous (time-dependent) systems can be included in the formalism by generalizing the coordinates to one higher degree of freedom. The distinguishing feature of near-integrable systems is the simultaneous presence of regular trajectories and regions of stochasticity, intimately mixed together, with regular trajectories separating regions of stochasticity. The stochastic trajectories are a self-generated consequence of the motion induced by Hamilton's equations, which are deterministic and do not contain additional *ad hoc* "stochastic" forces. We illustrate this by way of two examples much discussed in the literature: motion in the Hénon and Heiles potential and Fermi acceleration. For autonomous systems with greater than two degrees of freedom, regular trajectories no longer separate the stochastic regions, which unite into a "web" of stochasticity. This leads to the phenomenon of Arnold diffusion, described qualitatively to conclude this section.

*1.4a. Two Degrees of Freedom

We consider an autonomous periodic near-integrable system with two degrees of freedom, by which we mean a Hamiltonian of the form

$$H(J_1, J_2, \theta_1, \theta_2) = H_0(J_1, J_2) + \epsilon H_1(J_1, J_2, \theta_1, \theta_2), \quad (1.4.1)$$

1.4. Near-Integrable Systems

where J, θ are action-angle variables of the unperturbed motion, the perturbation ϵ is small, H_0 is a function of the actions alone, and H_1 is a periodic function of the θ's.

The generic character of motion for such systems is now fairly well understood. The trajectories lie in the three-dimensional surface $H = \text{const.}$ of the four-dimensional phase space. According to the KAM theorem reviewed in Chapter 3, a finite fraction of the phase space trajectories are regular, that is, are associated with first integrals of the motion. The remaining fraction exhibit stochastic or chaotic behavior. The stochastic and regular trajectories are intimately comingled, with a stochastic trajectory lying arbitrarily close to every point in the phase space in the same sense that any irrational can be approximated by a rational as closely as desired.

Regular Trajectories. Since the regular trajectories depend discontinuously on choice of initial conditions, their presence does not imply the existence of an isolating integral (global invariant) or symmetry of the system. However, regular trajectories, where they exist, represent exact invariants of the motion. These trajectories are either conditionally periodic in the angle variables, densely covering a toroidal surface of constant action in which the angle variables run around the two directions of the surface with incommensurable frequencies (the generic case); or are periodic closed curves, winding around the torus an integral number of turns (the concept of a torus will be discussed more fully in Chapter 3). Regular trajectories are most conveniently studied using perturbation theory and viewed within a surface of section in phase space, discussed in Section 1.2.

Various types of regular trajectories and their intersection with a surface of section $\theta_1 = \text{const.}$ are illustrated in Fig. 1.10, which shows in polar coordinates the J_2, θ_2 surface for two-degree-of-freedom Hamiltonians. Case (a) is a generic trajectory that covers the surface of the torus. The motion around the major axis of the torus is periodic in θ_1 with period 2π. The successive intersections of the trajectory with the surface of section at values of $\theta_2 = \theta_{21}, \theta_{22}, \theta_{23} \ldots$ lie on a closed *invariant curve* and densely cover the curve over long periods of time. Case (b) is an example of a resonance

$$k\omega_1(J) + l\omega_2(J) = 0, \qquad (1.4.2)$$

with $\omega_1 = \dot\theta_1$, $\omega_2 = \dot\theta_2$, k and l integers. The resonant trajectory is closed, periodic in θ_1 and θ_2. For $k = 5, l = 2$, shown in case (b), the successive interactions of this trajectory with the surface of section lie on five discrete points called *fixed points* or *periodic points* of the motion. We refer to this motion as a *primary resonance* since it is a closed periodic trajectory of the unperturbed Hamiltonian H_0. Resonance is a special case (b) of an invariant curve, for which the winding number k/l is rational. Resonances and their interaction play a crucial role in the appearance of stochastic motion in near-integrable systems.

Case (c) shows a surface of section of a generic trajectory in the

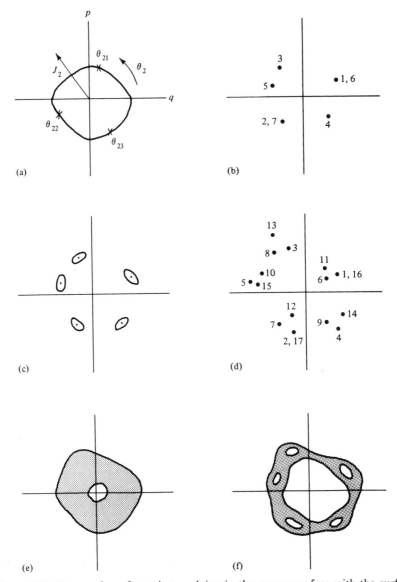

Figure 1.10. Intersection of a trajectory lying in the energy surface with the surface of section Σ defined by $\theta_1 = $ const. (a) Invariant curve generated by successive intersections of the trajectory with Σ. (b) Primary resonance $k = 5$, $l = 2$ showing the first seven intersections of the trajectory with Σ. (c) Primary islands around the $k = 5$, $l = 2$ fixed points generated by intersections of the trajectory with Σ. (d) Secondary resonance with three steps around the $k = 5$, $l = 2$ primary resonance. The first 17 intersections with Σ are shown. (e) Annular layer of stochasticity generated by intersections of a single trajectory with Σ lying between two invariant curves. (f) Stochasticity layer bounded by both primary and secondary invariant curves.

1.4. Near-Integrable Systems

neighborhood of the primary resonance of Fig. 1.10b. The successive intersections of the trajectory with the surface of section lie on a set of five smooth closed curves, called *primary islands*, encircling the fixed points of case (b). As a final illustration of the complexity of the possible motions, case (d) shows the surface of section for a closed periodic trajectory which winds three times about the $k = 5, l = 2$ primary resonance in 15 circuits of θ_1. This is an example of a *secondary resonance*, coupling the motion around a primary island to the unperturbed periodic motion. Secondary resonances are produced by the perturbation Hamiltonian H_1 and are in turn surrounded by secondary islands.

The very intricate structure of the regular trajectories should now be clear. Primary resonances give rise to primary islands, which give rise to secondary resonances and their islands, and so on *ad infinitum*. Calculation of regular trajectories (invariant curves and resonances) is treated in Chapters 2 and 3.

Regions of Stochasticity. The regions of stochastic trajectories are known to fill a finite portion of the energy surface in phase space. The successive intersections of a single stochastic trajectory with the surface of section fill a finite area. Two stochastic trajectories are illustrated in Fig. 1.10. Case (e) shows an annular layer of stochasticity filled by a single trajectory lying between two invariant curves similar to those shown in case (a). Periodic trajectories exist in this region, but the trajectories near them either do not move in stable islands around the fixed points (see Section 3.3), or the islands are too small to be visible. Case (f) shows a layer of stochasticity filled by a single trajectory near the islands of case (c).

Stochastic motion always occurs near separatrices separating invariant curves from their islands. Near a separatrix, the frequency of the island oscillation ω, given by an expression such as (1.3.15), approaches zero. The condition of resonance with the unperturbed oscillation at frequency ω_0

$$k\omega - \omega_0 = 0 \qquad (1.4.3)$$

then leads to a separation between the actions of neighboring (k and $k + 1$) resonances, which tends to zero as the separatrix is approached. One thus refers to the region of stochasticity that forms near a separatrix as a *resonance layer*. For a small perturbation ϵ, with two degrees of freedom, layers are thin and separated by invariant curves. The layers are isolated from each other, and motion from one layer to another is forbidden. As ϵ increases, the invariant curves separating neighboring chains of islands with their resonance layers are strongly perturbed and finally destroyed. The layers merge when the last invariant curve separating layers surrounding adjacent islands chains is destroyed. The merging of primary resonance layers leads to the appearance of *global* or *strong stochasticity* in the motion. The condition for the overlap of neighboring resonance layers and destruction of the last invariant curve is the topic of Chapter 4. The character of the chaotic motion within a resonance layer and within merged layers is the topic of Chapter 5

A hint of the reason for the chaotic behavior near separatrices is obtained from the concept of resonance overlap, which, as the overlap occurs to higher and higher orders (islands within islands) must make the motion exceedingly complicated. We can arrive at this same conclusion from a different point of view by observing the separatrix trajectory itself. Both from a theoretical point of view and from numerical calculations, the separatrix trajectory does not have the smooth nature to be found in the integrable problem of Fig. 1.4, but rather is exceedingly complicated. This motion, which we discuss in some detail in Section 3.2b, is limited to a region near the separatrix by the existence of nearby regular KAM surfaces (see Section 3.2a for discussion). However, with increasing perturbation strength the higher-order resonances move the KAM surfaces further from the separatrix, expanding the region over which the complicated separatrix motion affects the nearby trajectories.

Near each of the resonances of the system we have seen that the phase trajectories are strongly perturbed. At the resonance itself, there are certain values of the coordinates, the fixed points in Fig. 1.10b, d, for which periodic closed trajectories exist. The *stability of the linearized motion* obtained from an expansion about these periodic trajectories is also related to stochasticity. Linearly stable solutions lead to regular motion about the periodic trajectory that can only be disrupted on a long time scale for which the weak nonlinear terms become important. Linearly unstable solutions lead to exponential divergence of trajectories. The rate of divergence can be interpreted directly as a measure of the stochasticity (see Chapter 5). When all or nearly all of the periodic solutions are linearly unstable in a region of the phase space, we can expect the phase trajectory to be very chaotic within that region.

The Hénon and Heiles Problem. We illustrate many of these concepts with an example that has been much discussed in the literature, that of the motion in a two-dimensional potential well given by

$$U(x, y) = \tfrac{1}{2}(x^2 + y^2 + 2x^2y - \tfrac{2}{3}y^3), \qquad (1.4.4)$$

which is shown in Fig. 1.11. First treated numerically by Hénon and Heiles (1964) as a simple example of a nonlinear system of more than one degree of freedom, the potential (1.4.4) has been considered in great detail by many authors in an attempt to understand the underlying details of the system behavior. Unfortunately, this particular example presents a number of difficulties that are not generic to coupled nonlinear systems and therefore requires somewhat special treatment. Nevertheless, as a numerical example it illustrates very nicely the points we have been making. The well varies from a harmonic potential at small x and y to triangular equipotential lines on the edges of the well, as illustrated in Fig. 1.11. For a given value of total energy E, the Hamiltonian of a particle oscillating in the well is given in

1.4. Near-Integrable Systems

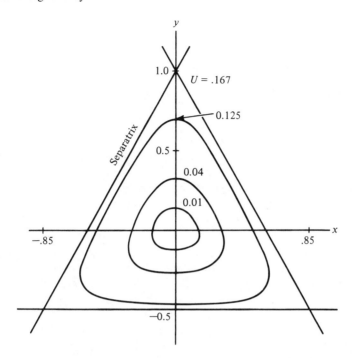

Figure 1.11. The potential well for the Hénon and Heiles Hamiltonian, showing lines of constant potential U, for closed equipotentials ($U \leq \frac{1}{6}$) only (after McNamara and Whiteman, 1966).

rectangular coordinates by $H = H_0 + \epsilon H_1 = E$, where

$$H_0 = \tfrac{1}{2}(p_x^2 + p_y^2) + \tfrac{1}{2}(x^2 + y^2), \tag{1.4.5a}$$

and

$$\epsilon H_1 = x^2 y - \tfrac{1}{3}y^3, \tag{1.4.5b}$$

where for convenience the particle mass is normalized to one and the small parameter ϵ is specified by choice of the energy, with $\epsilon \sim E$. If E is less than the limiting potential energy $U = \tfrac{1}{6}$, the particle will be trapped in the well. If we now take particles with various initial conditions and observe the crossings in the $p_y - y$ surface of section, we can see whether invariant curves exist. In the limit of small oscillations for a harmonic potential, we have the integrable Hamiltonian in action-angle variables (see Section 1.2c)

$$H_0 = \omega_1 J_1 + \omega_2 J_2, \tag{1.4.6}$$

where for the particular potential $\tfrac{1}{2}(x^2 + y^2)$ the frequencies of oscillation ω_1 and ω_2 are both equal to unity. If we now allow small perturbations due to the nonharmonic terms in the potential of (1.4.5b), we can write the Hamiltonian in the form

$$H = \omega_1 J_1 + \omega_2 J_2 + \epsilon H_1(J_1, J_2, \theta_1, \theta_2). \tag{1.4.7}$$

If the energy is sufficiently small, the crossings of the surface of section turn out to be closed curves for almost all initial conditions, except very near the resonances. These curves can be calculated either by numerically solving the coupled equations of motion or by computing the invariant to sufficient orders in ϵ using the methods of Chapter 2. If insufficient orders in the expansion are taken, closed curves will still be obtained but they will not be similar to the numerical solutions. On the other hand, if too many terms are kept, the answer will diverge from the correct solution because the series is asymptotic. McNamara and Whiteman (1966) compute J_2 to second and third order in ϵ for initial energy $E = 0.01$ and various initial conditions. They obtain surfaces of section as shown in Fig. 1.12a, b, respectively. We see that for this particular example the higher-order terms are very important, even changing the character of the solutions; that is, particles that are bounded in the surface near points A and B in the numerical calculation of Fig. 1.12c first appear in fourth order. The requirement of using higher-order perturbation theory is not generic to near-integrable systems, but is necessitated by the double difficulties for the Hénon and Heiles problem: (1) that the uncoupled oscillators are exactly linear and (2) that the uncoupled frequencies are identical. For close comparison with the numerical results of Hénon and Heiles, we use surfaces generated from eighth-order perturbation theory by Gustavson (1966), shown in Fig. 1.13. For surfaces at $E = 0.042$ ($\frac{1}{24}$) and $E = 0.083$ ($\frac{1}{12}$), invariant curves appear to exist everywhere, and the expansion procedure appears to converge to the proper limit, but neither of these appearances is strictly correct. We note that in addition to these invariant curves, very thin layers of stochasticity are densely distributed throughout the entire surface of section, associated with the resonances between the degrees of freedom. The thickness of the resonance layers is exponentially small in E^{-1}. For small values of E, the layers occupy a negligible fraction of the total area of the surface of section, and are not visible in these numerically generated pictures of the entire section.

For higher initial energy $E = 0.125$, we see three types of orbits: the simple invariant curve as with the lower energy; a multiple-loop orbit represented by the chain of five small islands, similar to Fig. 1.10c, in which the crossings jumps from loop to loop; and an apparently ergodic orbit similar to Fig. 1.10f, in which the random dots are successive crossings of the surface of section. For this orbit the action integral is no longer invariant and cannot be predicted from an expansion. Even at the dissociation energy, as in the final comparison, invariants exist in small isolated regions of the phase plane. The island orbits reflect the existence of an invariant formed near a primary resonance between the large amplitude, fundamental oscillator periods associated with the unperturbed x and y motion. Thus the five islands indicate that the trajectory returns to its initial neighborhood in the surface of section of the $p_y - y$ plane after intersecting that plane five times. The procedure for calculating the island invariants, including the amplitude of the islands, is given in Section 2.4, together with a detailed example.

1.4. Near-Integrable Systems

Figure 1.12. Surfaces of section computed using (a) the second and (b) the third order (in ϵ) invariants are compared in (c) with the numerical results; $E = 0.01$ (after McNamara and Whiteman, 1966).

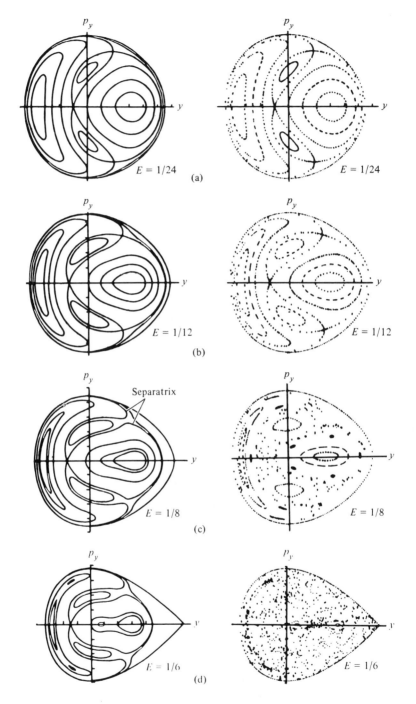

Figure 1.13(a–d). Comparison of surfaces of section computed from adiabatic theory with those computed numerically (after Gustavson, 1966).

1.4. Near-Integrable Systems

Although there is apparent correspondence between the surfaces calculated from perturbation theory and the numerically calculated surfaces, this does not ensure that the numerical surfaces are stable for all time, as predicted from the perturbation theory. Very small higher-order resonant terms may destroy the usual convergence. The KAM theory, however, comes to our rescue here by demonstrating that, at least for sufficiently small perturbation, invariant curves exist close to those calculated. This gives us confidence to assume that those curves that look smooth and stable, although possibly having unseen structure, are at least close to good invariant curves. Relying on this assumption, we develop heuristic criteria for destruction of invariants, related to criteria on the *overlapping of neighboring resonances* as proposed by Chirikov (1960). By comparing these criteria with the numerically determined surfaces of section, reasonably reliable quantitative estimates for destruction of invariants can be obtained. These calculations are the main subject of Chapter 4.

Fermi Acceleration. There is a very close connection between a Hamiltonian system with two degrees of freedom and an area-preserving mapping of a two-dimensional surface into itself. We have already noted this in Section 1.2, where we describe a surface of section of a system with two degrees of freedom. The transformation that describes the successive crossings of the surface of section is just such an area-preserving mapping. Conversely, a dynamical problem that can be expressed directly as a mapping also has a Hamiltonian representation obtainable by expansion of the transformation in a Fourier series.

We now introduce an example of a dynamical system that can be represented by an area-preserving mapping, and which illustrates the nature of stochastic trajectories in systems with two degrees of freedom. The mapping results from an idealization of the one-dimensional motion of a ball bouncing between a fixed and oscillating wall, originally examined by Fermi (1949) as a model for cosmic ray acceleration. Letting u_n be the ball velocity, normalized to the peak wall velocity, and ψ_n be the phase of the wall oscillation just before the nth collision of the ball with the oscillating wall, we have the mapping

$$u_{n+1} = |u_n + \sin \psi_n|, \qquad (1.4.8a)$$

$$\psi_{n+1} = \psi_n + \frac{2\pi M}{u_{n+1}}, \qquad (1.4.8b)$$

where M is the normalized distance between the walls, and the absolute magnitude of u_{n+1} is taken.

Transformations of this type can be examined numerically for many thousands of iterations, thus allowing both detailed knowledge of the structural behavior and statistical properties of the dynamical system to be determined. Figure 1.14 shows the $u - \psi$ surface with $M = 100$, for 623,000

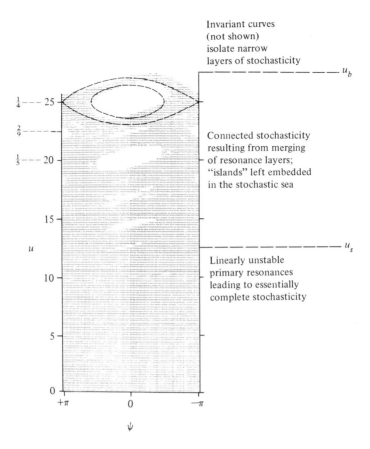

Figure 1.14. Surface of section for the Fermi problem, showing occupation of phase space cells for 623,000 iterations of a single initial condition. Dashed curves are calculated from secular perturbation theory (after Lieberman and Lichtenberg, 1972).

wall collisions of a single trajectory, with an initial condition at low velocity $u_0 \approx 1$. The surface has been divided into 200×100 cells, with a blank indicating no occupation of that cell. We find that the phase plane consists of three regions:

(1) a region for large u, in which invariant (KAM) curves predominate and isolate narrow layers of stochasticity near the separatrices of the various resonances;
(2) an interconnected stochastic region for intermediate values of u, in which invariant islands near linearly stable periodic solutions are embedded in a stochastic sea; and
(3) a predominantly stochastic region for small u, in which all primary periodic solutions (*primary fixed points* of the mapping) appear to be unstable.

1.4. Near-Integrable Systems

Both regions (2) and (3) exhibit *strong* or *global stochasticity* of the motion. In the latter region, although some correlation exists between successive iterations, over most of the region it is possible to approximate the dynamics by assuming a *random phase approximation* for the phase coordinate, thus describing the momentum coordinate by a diffusion equation. We explore this question more fully in Chapter 5.

Ergodic Systems. Just as the fully integrable systems are simpler than near-integrable systems, fully ergodic systems, with no regular trajectories are also simpler in some respects. Although the trajectories cannot be obtained, it is possible to deduce a number of general statistical properties. One example of such a system is that of a billiard ball in a two-dimensional periodic space undergoing collisions with a hard sphere in the space, as shown in Fig. 1.15a. The trajectory of the billiard ball is shown by the dark

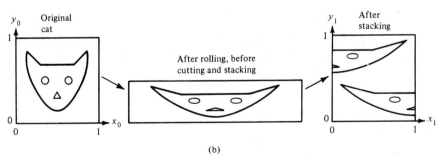

Figure 1.15. Two examples of systems that are ergodic and mixing. (a) Two-dimensional hard sphere gas, with $\alpha = \beta$; the motion is on a torus, with the top and bottom, and the two sides, identified. (b) The baker's transformation; after stretching and stacking, the cat becomes a mixed up cat (after Arnold and Avez, 1968).

line with arrows indicating the direction of motion. The reflections from the sphere follow the usual law that the angle of incidence α is equal to the angle of reflection β, as shown in the figure. The trajectory leaving one side of the enclosed space reenters from the opposite side with the same slope, as required by the periodicity. Sinai (1963) has shown this system to be both ergodic and mixing. An equivalent system is the motion of a ball in a lattice of hard spheres; the system is thus of interest in statistical mechanics (Ford, 1975).

Another well-known example of an ergodic system is the *baker's transformation*, which maps the unit square into itself, by the transformation

$$\begin{pmatrix} x_1 \\ y_1 \end{pmatrix} = \begin{cases} \begin{pmatrix} 2x_0 \\ y_0/2 \end{pmatrix}, & 0 \leq x_0 < \tfrac{1}{2}, \\ \begin{pmatrix} 2x_0 - 1 \\ (y_0 + 1)/2 \end{pmatrix}, & \tfrac{1}{2} \leq x_0 < 1. \end{cases} \quad (1.4.9)$$

Pictorially, using Arnold's well-known cat (Arnold and Avez, 1968), the map transforms as a baker rolling, then cutting and stacking his dough, as seen in Fig. 1.15b. This transformation, whose eigenvalues are all unstable, has also been shown to be mixing.

Thus we have good evidence for the statistical properties of transformations in the regions for which they exhibit unstable motion. The even stronger assumption of a random phase approximation, making the motion appear as a pure random walk in momentum space, depends also on having separate time scales for the spreading of the phase and momentum. For many dynamical problems of interest, the phase variable randomizes much more rapidly than the momentum variable, thus allowing this separation.

*1.4b. More than Two Degrees of Freedom

All the effects described previously for autonomous systems with two degrees of freedom are found in systems with more than two degrees of freedom. In the generic case, stochastic and regular trajectories are intimately comingled in the $2N$-dimensional phase space, and in the $(2N - 2)$-dimensional surface of section. Stochastic layers in phase space exist near resonances of the motion. The thickness of the layers expands with increasing perturbation, leading to primary resonance overlap, motion across the layers, and the appearance of *strong stochasticity* in the motion. In the limit of small perturbation, however, primary resonance overlap does not occur. A new physical behavior of the motion then makes its appearance: motion along the resonance layers—the so-called *Arnold diffusion*.

Arnold Diffusion. For systems with three or more degrees of freedom, resonance layers near the separatrices are not isolated from each other by

1.4. Near-Integrable Systems

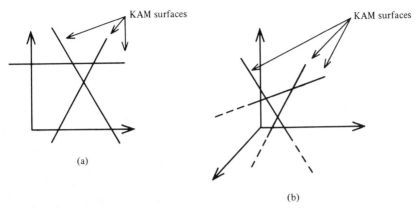

Figure 1.16. In (a) the two-dimensional energy surface (plane) is divided by lines into a set of bounded areas; in (b) the three-dimensional energy "surface" (volume) is not divided into bounded regions by lines (after Lieberman, 1980).

KAM surfaces. For two degrees of freedom, the two-dimensional KAM surfaces divide the three-dimensional energy "volume" in phase space into a set of closed volumes, each bounded by KAM surfaces, much as lines isolate regions of a plane, illustrated in Fig. 1.16a. For three degrees of freedom, the three-dimensional KAM "surfaces" do not divide the five-dimensional energy "volume" into a set of closed volumes, just as lines do not separate a three-dimensional volume into distinct regions, illustrated in Fig. 1.16b. For $N > 2$ degrees of freedom, the N-dimensional KAM surfaces do not divide the $(2N - 1)$-dimensional energy "volume" into distinct regions. Thus for $N > 2$, in the generic case, all stochastic layers of the energy surface in phase space are connected into a single complex network—the Arnold web. The web permeates the entire phase space, intersecting or lying infinitesimally close to every point. For an initial condition within the web, the subsequent stochastic motion will eventually intersect every finite region of the energy surface in phase space, even in the limit as the perturbation strength $\epsilon \to 0$. This motion is the Arnold diffusion.

The merging of stochastic trajectories into a single web was proved (Arnold, 1964) for a specific nonlinear Hamiltonian. A general proof of the existence of a single web has not been given, but many computational examples are known. From a practical point of view, there are two major questions concerning Arnold diffusion in a particular system: (1) what is the relative measure of stochastic trajectories in the phase space region of interest? and (2) for a given initial condition, how fast will the system diffuse along the thin threads of the Arnold web? The extent of the web in phase space can be estimated by means of resonance overlap conditions and will be considered in Chapter 6.

Overlap of resonances gives rise to a resonance layer thickness, with

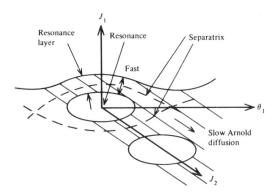

Figure 1.17. Illustrating the directions of the fast diffusion across a resonance layer and the slow diffusion along the resonance layer.

stochastic motion occurring *across* the layer as in systems with two degrees of freedom. The new feature of Arnold diffusion is the presence of stochastic motion *along* the resonance layer, produced as a result of the coupling of at least *three* resonances. We illustrate the motion along the resonance layer in Fig. 1.17. A projection of the motion onto the J_1, θ_1 plane is shown, illustrating a resonance with a stochastic layer. At right angles to this plane the action of the other coordinate J_2 is shown. If there are only two degrees of freedom in a conservative system, then the fact that the motion is constrained to lie on a constant energy surface restricts the change in J_2 for J_1 constrained to the stochastic layer. However, if there is another degree of freedom, or if the Hamiltonian is time-dependent, then this restriction is lifted, and motion along the stochastic layer in the J_2 direction can occur.

The diffusion rate along a layer has been calculated by Chirikov (1979) and by Tennyson *et al.* (1979) for the important case of three resonances, and is the main subject of Chapter 6. For coupling among many resonances, a rigorous upper bound on the diffusion rate has been obtained by Nekhoroshev (1977), but generally overestimates the rate by many orders of magnitude. A summary of numerical calculations of diffusion in the regime for which many resonances are important (Chirikov *et al.*, 1979) is given in Chapter 6.

1.5. Dissipative Systems

In contrast to Hamiltonian systems for which the conservation of the phase space volume is a fundamental constraint of the motion, a dissipative system is characterized by continued contraction of the phase space volume with increasing time. This leads to contraction onto a surface of lower dimensionality than the original phase space. The motion is no longer governed by

1.5. Dissipative Systems

canonical equations, but is generally described in the form of a set of first-order differential equations of the form

$$\frac{dx}{dt} = V(x), \qquad (1.5.1)$$

where if x and V have N components then the phase space of the system is N-dimensional. The trajectory of x is then called an N-dimensional flow. For regular motion the attractor of the flow represents a simple motion such as a fixed point (sink) or a singly periodic orbit (limit cycle). For flows in two dimensions these are, in fact, the only possibilities.

For three-dimensional flows, in addition to sinks and limit cycles, doubly periodic orbits may be possible. One might reason by analogy that these are the only possible attractors in a volume-contracting three-dimensional flow. This is not the case. It has been shown that attractors exist for dissipative flows in three or more dimensions that have very complicated geometric structures. These structures can be characterized as having a fractional dimension (see Section 7.1d), and are usually called *strange attractors*. The motion on strange attractors is chaotic.[6]

1.5a. Strange Attractors

Roughly speaking, a strange attractor is an attractor on which nearby orbits, though bounded, diverge exponentially. To visualize a strange attractor, we imagine a three-dimensional flow in the form of a layer containing infinitely many two-dimensional sheets. The layer expands along its width and folds over on itself, as shown in Fig. 1.18a. The two ends (AB and $A'B'$ in the figure) are smoothly joined together. Since $A'B'$, having two distinct sheets, joins to AB, having one sheet, there must be infinitely many sheets for the joining to be smooth. Otherwise, there would be a discontinuity yielding noninvertible flow. The infinitely leaved structure, when smoothly joined and embedded in a three-dimensional phase space, is shown in Fig. 1.18b.

From the construction, it is seen that orbits are bounded despite the fact that nearby orbits diverge exponentially. Furthermore, the structure of the leaves is such that on finer and finer scales the basic leaf pattern reappears. This similarity on finer scales is also characteristic of the island structure of Hamiltonian systems, and serves as one of the tools for analyzing both Hamiltonian and dissipative systems. The formation of a typical leaf structure, which can be characterized mathematically as a *Cantor set*, is discussed in Section 7.1.

---

[6] The word "chaotic" is commonly used to describe random motion in dissipative systems; the word "stochastic" is more often used for Hamiltonian systems. While we have tried to maintain this convention, the two words do not distinguish between "degrees of randomness" and we consider them to be essentially synonymous.

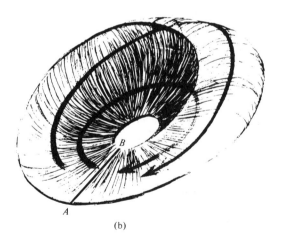

Figure 1.18. Qualitative illustration of a strange attractor. (a) An infinite layer of ribbons is stretched and folded. (b) The structure is joined head-to-tail and embedded in three-dimensional space (after Shaw, 1981).

1.5. Dissipative Systems

Equations (1.5.1), from which the strange attractor arises, are usually parameterized by some quantity (analogous to the perturbation strength in Hamiltonian systems) whose variation changes the character of the solutions. In Hamiltonian systems, we have seen for the Hénon and Heiles Hamiltonian and for the Fermi acceleration mapping that, as the perturbation strength changes, the trajectories in phase space change from mostly regular to mostly stochastic. Similarly, when a parameter is changed in a dissipative system, the system may change from periodic motion to the chaotic motion on a strange attractor. In many cases this change proceeds by successive doublings of the period of the singly periodic motion to some limit, beyond which the attractor changes character and becomes chaotic. Further change in the parameter can lead to an inverse process, or the appearance of simple attractor basins with other symmetries. Another very interesting characteristic of these systems is that a surface of section can usually be found in which the motion reduces, approximately, to a noninvertible mapping of a single variable; that is, a mapping that becomes multivalued on time reversal. This type of noninvertible mapping also has direct application to many physical problems, and will be studied in some detail in Section 7.2.

Not all of the above properties are easily seen from a single example. In the following section we shall examine the first example in which a strange attractor was studied, the Lorenz system (Lorenz, 1963). We shall return to this example again in Section 8.6, to study the physical system from which it arises. In Section 7.1 we also introduce other examples to look more fully at the various phenomena associated with dissipative systems exhibiting chaotic motion.

1.5b. The Lorenz System

A seminal example of chaotic flow arises from the hydrodynamic equations describing Rayleigh–Bénard convection. A fluid slab of finite thickness is heated from below, and a fixed temperature difference is maintained between the top cold surface and the bottom hot surface. Gravity acts downward. The fluid motion is described by the Navier–Stokes equation. Assuming only two-dimensional motion, the flow can be characterized by two variables, the stream function for the motion ψ and the departure Θ of the temperature profile from a one-dimensional profile that is linearly decreasing with the height.

The partial differential equations for the perturbed flow can be transformed to a set of ordinary differential equations by expanding ψ and Θ in a double Fourier series in x and z, with the Fourier coefficients functions of t alone. By truncating the series at a finite number of terms, we obtain motion in the finite-dimensional phase space of Fourier coefficients. The derivation of the equations for the Fourier coefficients starting from the Navier–Stokes equation is described in Section 8.6.

Lorenz (1963) studied a simplified system in which only the three "most important" coefficients were kept. In this simplified system the equations for the three coefficients become

$$\dot{X} = -\sigma X + \sigma Y,$$
$$\dot{Y} = -XZ + rX - Y, \qquad (1.5.2)$$
$$\dot{Z} = XY - bZ,$$

where X is the amplitude of the convection motion, Y the temperature difference between ascending and descending currents, Z the distortion of the vertical temperature profile from linearity, and σ, r, and b are dimensionless parameters whose physical meanings are discussed in Section 8.6.

The Lorenz system has been extensively studied with over 50 articles in print (see Helleman, 1980, for references). Usually σ and b are fixed at $\sigma = 10$, $b = \frac{8}{3}$, and the system is examined as r is varied. Some elementary properties of the Lorenz system are as follows (Lorenz, 1963; Lanford, 1976; Treve, 1978):

(i) The equations are invariant under the transformation $X \to -X$, $Y \to -Y, Z \to Z$.
(ii) The phase space volume contracts at a uniform rate given by (see Section 7.1a)

$$\Lambda = \frac{\partial \dot{X}}{\partial X} + \frac{\partial \dot{Y}}{\partial Y} + \frac{\partial \dot{Z}}{\partial Z} = -(\sigma + b + 1),$$

which is large for the parameters usually chosen: for $\sigma = 10$, $b = \frac{8}{3}$, then $\Lambda = -13\frac{2}{3}$. In one unit of time, the volume contracts by a factor of $e^\Lambda \approx 10^{-6}$.
(iii) All solutions are bounded for positive time, and large initial values of X, Y, Z are damped toward zero by the motion.

As r is increased from zero, the solutions have been found to change character as follows:

(i) For $0 < r < 1$, the origin $\mathbf{0} = (0, 0, 0)$ is the only fixed point, and it is an attracting point. This corresponds to steady state heat conduction in the Rayleigh–Bénard problem.
(ii) For $r > 1$, the point $\mathbf{0}$ loses stability and two new fixed points

$$X_{1,2} = (\pm[b(r-1)]^{1/2}, \pm[b(r-1)]^{1/2}, r-1)$$

are born. The points $X_{1,2}$ are attracting for $1 < r < r_2$, where

$$r_2 = \sigma(\sigma + b + 3)/(\sigma - b - 1) = 470/19 \approx 24.74.$$

This corresponds to a steady convection in the original Rayleigh–Bénard flow problem.
(iii) For $r > r_2$ all three points are unstable and a single point attractor does not exist.

1.5. Dissipative Systems

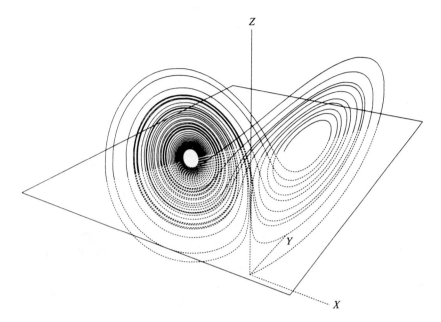

Figure 1.19. A chaotic trajectory for the Lorenz attractor at $r = 28$; the horizontal plane is at $Z = 27$ (after Lanford, 1977).

(iv) For $r > r_1 = 24.06$, there is a strange attractor on which chaotic motion takes place. Note that in a narrow range

$$24.06 < r < 24.74$$

three attractors coexist, two corresponding to steady convection and the third to chaotic motion. The system exhibits hysteresis; as r increases through 24.74, the convective motion becomes turbulent; as r decreases through 24.06, the turbulent motion becomes convective.

The value $r = 28$ was studied numerically by Lorenz. Figure 1.19 (after Lanford, 1976) shows in perspective a chaotic trajectory that starts at the origin **0** and its intersection with the surface of section $Z = 27$. The orbit first gets close to X_1 and then spirals away, finally being attracted close to X_2. It then spirals away from X_2 and is attracted back close to X_1. The period of rotation near $X_{1,2}$ is 0.62, and the spiral radius expands by about 6% with each rotation. The number of consecutive rotations about X_1 or X_2 varies widely in a manner that is unpredictable in practice; i.e., it depends sensitively on initial conditions.

The chaotic behavior of the attractor can be studied from the Poincaré map of the plane $Z = 27$ onto itself. The motion on the attractor has been proven to be mixing and ergodic (Bunimovich and Sinai, 1980). Figure 1.20 shows the power spectrum of $X(t)$; the continuous spectrum reflects the nonperiodic chaotic behavior of the motion on the attractor.

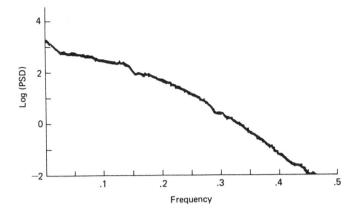

Figure 1.20. Power spectrum of $X(t)$ in the Lorenz attractor at $r = 28$. Here the lowest frequencies have the highest power. Such a "continuous" spectrum reflects the nonperiodic chaotic behaviour on the attractor (after Farmer et al., 1980).

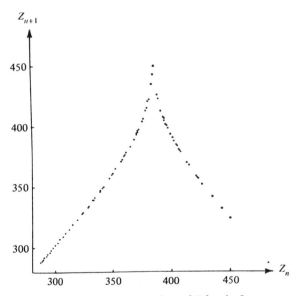

Figure 1.21. Maxima versus previous maxima of Z for the Lorenz attractor at $r = 28$, illustrating the embedded one-dimensional map (after Lorenz, 1963).

1.5. Dissipative Systems

Lorenz (1963) first observed that the plot of Z versus t appeared chaotic. With great insight he recorded successive maxima $Z_1, Z_2, \ldots$ and plotted Z_{n+1} versus Z_n, with the result shown in Fig. 1.21. To the left of the peak, the map represents successive orbits of either X_1 or X_2. The region to the right of the peak yields the transition between X_1 and X_2. The one-dimensional map is approximate with the infinitely many leaves of the attractor being melded into one, although the approximation is very good due to the high volume contraction rate.

The chaotic motion on the Lorenz attractor can be understood by iteration of the one-dimensional map shown in Fig. 1.21. The map has a slope whose magnitude is everywhere greater than unity. It is easy to show (see Section 7.2c) that nearby trajectories diverge exponentially. We explore the correspondence between strange attractors and one-dimensional maps more fully in Chapter 7. There are also other routes to chaos in nonlinear dissipative systems, which we consider in Chapter 7. In Chapter 8 the properties of dissipative chaotic systems are investigated.

CHAPTER 2

Canonical Perturbation Theory

2.1. Introduction

A basic method of solving nonlinear dynamical problems is by perturbation from known, integrable solutions. Given these, we seek "the integrable solutions" to a "nearby" system by expanding in the small parameter ϵ by which the two systems differ. For example, if the nearby system is slightly nonlinear, then the linearized motion may be obtained directly, and the nonlinear perturbation found as a series solution.

Implicit in the above scheme is the assumption that the sought-for integrable solutions of the nearby system actually exist. As we have seen (Chapter 1), this is not generally the case; most multidimensional nonlinear systems are not integrable. In such systems, chaotic trajectories associated with resonances among the degrees of freedom are densely distributed among the regular trajectories and have finite measure in the phase space. Perturbation theory fails to describe the complexity of this chaotic motion; formally, the series diverge.

Even for initial conditions for which the orbits are regular, there are difficulties in applying perturbation methods. Under perturbation, some solutions associated with particular resonances change their topology. These are the "islands" described in Section 1.4 and their measure in the phase space is also finite. The islands are microcosms of the original perturbed system, having their own chaotic and regular trajectories. The usual perturbation methods fail to describe the change in topology, and the special methods of *secular perturbation theory* have been devised to describe regular motion near a given resonance or a restricted set of resonances. There are no known methods that yield the regular solutions for the entire dense hierarchy of resonances.

2.1. Introduction

The change of topology leading to island formation, the chaotic regions, and the character of the chaotic solutions, form the fundamental topics of subsequent chapters in this volume. In this chapter, we examine the perturbation methods used to obtain solutions that in some sense "approximate" the actual motion in multidimensional nonlinear systems. If the actual trajectory is chaotic or involves significant change in topology, then the series solution may still approximate the motion in some coarse-grained sense, as when the chaotic motion is confined to a thin separatrix layer bounded by regular motion. On the other hand, perturbation methods fail to describe even qualitatively the chaotic motion in regions for which the main resonances overlap.

If the actual trajectory is regular, then one can hope to obtain series that uniformly converge to the true solution. However, the *classical* series described in this chapter, while extremely useful for theoretical calculations, are *not* convergent. In these methods, the amplitude and frequency of the perturbed motion are developed as power series in ϵ for fixed initial conditions. Since the resonances are dense in the frequency space, new resonances are continually encountered as the frequency is varied, leading to series expansions that are at best asymptotic.

The search for solutions that converge to the true solutions has led to the development of *superconvergent* expansion methods (Kolmogorov, 1954), along with a technique in which the frequency is held fixed and the initial conditions for which the expansion is performed are varied. These methods lead to the KAM result that for "sufficiently small" perturbations "sufficiently far" from resonance, convergent solutions can be found.

We begin the examination of perturbation methods with a brief description of some of the procedures using simple examples of dynamical systems and examining the motion directly from their defining differential equations. Even for nonlinear motion in a single degree of freedom, which is known to be completely integrable, a power series expansion in the amplitude of the oscillation alone leads to the appearance of *secular terms* (unbounded in time) and divergence of the series. By varying the amplitude and frequency together, Lindstedt (1882) and Poincaré (1892) overcame the problem of secularities to obtain convergent series solutions. Their technique is introduced in Section 2.1a and is presented in its general canonical form in Section 2.2a. It forms the basis for further developments of perturbation theory in subsequent sections.

For two or more degrees of freedom, real resonances between the fundamental frequencies and their harmonics give rise to further secularities, the so-called *small denominators*, and divergence of the classical series, which even for regular solutions do not converge fast enough to overcome the small denominators. Nevertheless, a number of techniques have been developed to suppress the divergence formally. The *method of averaging* leads directly to the calculation of *adiabatic invariants*, which are approximate integrals of the motion obtained by averaging over a fast angle variable in the system. Formally, the adiabatic invariants are calculated as asymptotic series in the perturbation strength ϵ. These ideas are introduced in Section 2.1b and are presented in canonical form in Section 2.3.

It is important to realize that the method of averaging yields the erroneous result that the perturbed system is everywhere integrable. The structure with intermingled chaotic motion and islands is wiped away in favor of everywhere integrable motion given by the adiabatic invariants. Whether this description is "valid" depends on the strength of the perturbation and the detail with which the actual motion is compared to that predicted by adiabatic theory. This point was emphasized in Section 1.4a, where, in the surface of section for the Hénon and Heiles problem (see Fig. 1.13 and discussion following), the actual motion and the adiabatic theory were contrasted. The formal divergence of the asymptotic series for the adiabatic invariant, for any finite ϵ, is another indication that the method of averaging does violence to the underlying phase space structure. Nevertheless, the theory has been highly successful in calculating the motion in nonlinear systems.

Near resonances among the degrees of freedom, the regular solutions are strongly perturbed and undergo a change in topology. A straightforward application of classical perturbation theory near a given resonance yields a resonant denominator and divergent series, as illustrated in Section 2.1c. By transforming to a rotating coordinate system at the resonance, the singularity is removed, and the usual averaging techniques may be employed to develop the adiabatic series. This *secular perturbation theory*, described in Section 2.4, forms the core of our use of perturbation theory to study features of the chaotic motion in subsequent chapters. The method is examined in depth in Sections 2.4a, b and a detailed example, the motion of a magnetically confined particle in the field of an electrostatic wave, is presented in Section 2.4c. An extension of this method, in which an entire set of resonant denominators is simultaneously removed, has been developed (Dunnett *et al.*, 1968) and is described in Section 2.4d.

For clarity of presentation in the above sections, the methods are developed only to first order in ϵ, and the canonical transformations are performed using mixed variable generating functions. These methods can be carried to higher order (Born, 1927), but the successive unraveling of the old and new variables becomes algebraically complex, and the corresponding series obscure. Nevertheless, higher-order calculations are often essential in calculating the motion, as in the Hénon and Heiles problem where first-order perturbation theory yields erroneous results even in the limit of very low energy. In Section 2.5, we introduce the methods of Lie transformation theory, which have superseded the older methods for development of the classical series to higher order in ϵ. Lie methods for proceeding to higher order are illustrated for one-dimensional problems and for calculating higher-order adiabatic invariants.

Section 2.6 is devoted to examination of superconvergent series methods.

2.1. Introduction

These methods are described briefly, and contrasted with classical methods in Section 2.6a. A special application to the calculation of singly periodic orbits in nonlinear perturbed systems is treated in Section 2.6b.

2.1a. Power Series

For illustrative purposes, we consider an integrable slightly nonlinear oscillator in one degree of freedom, such as small oscillations of the pendulum, shown in Fig. 2.1a and given by the Hamiltonian of (1.3.6). Expanding the first of Eqs. (1.3.5) to third order in ϕ, the differential equation for the motion can be written as

$$\ddot{x} + \omega_0^2 x = \tfrac{1}{6}\epsilon\omega_0^2 x^3, \tag{2.1.1}$$

where $x = \phi$ is the angle from the vertical, $\omega_0 = (FG)^{1/2}$ is the frequency of the linearized motion, and ϵ is a dimensionless measure of smallness introduced into the cubic term to identify it as the perturbation; we set $\epsilon = 1$ at the end of the calculation.

If we expand x in the most straightforward manner as

$$x = x_0 + \epsilon x_1 + \epsilon^2 x_2 + \cdots, \tag{2.1.2}$$

then, equating coefficients of ϵ, we obtain the zero-order harmonic oscillator solution

$$x_0 = A \cos \omega_0 t. \tag{2.1.3}$$

To first order in ϵ, we obtain the equation

$$\ddot{x}_1 + \omega_0^2 x_1 = \tfrac{1}{6}\omega_0^2 A^3 \cos^3 \omega_0 t. \tag{2.1.4}$$

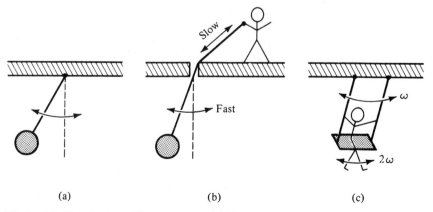

(a) (b) (c)

Figure 2.1. Nonlinear oscillator systems. (a) The pendulum in one degree of freedom; (b) adiabatic invariance for the pendulum; (c) an example of resonance.

Expanding the driving term as

$$\cos^3 \omega_0 t = \tfrac{1}{4}(\cos 3\omega_0 t + 3 \cos \omega_0 t),$$

the first term gives the well-behaved particular solution

$$x_{1a} = -\frac{A^3}{192} \cos 3\omega_0 t, \qquad (2.1.5)$$

but the second term is resonant with the driving term, producing a particular solution

$$x_{1b} = \frac{A^3}{64} (\omega_0 t \sin \omega_0 t + 2 \cos \omega_0 t). \qquad (2.1.6)$$

We see that the first part increases linearly with t, i.e., it is a *secular* term. This secularity arises, in this case, from an improper expansion, because the variation of frequency with amplitude was not taken into account.

The correct procedure, as developed by Lindstedt (1882), is to expand both amplitude and frequency in ϵ. We assume that $x = x(\omega t)$ is a periodic function of ωt with period 2π, and expand both x and ω in powers of ϵ, such that, along with (2.1.2), we have

$$\omega = \omega_0 + \epsilon \omega_1 + \epsilon^2 \omega_2 + \cdots. \qquad (2.1.7)$$

Substituting into the force equation gives

$$(\omega_0^2 + 2\epsilon \omega_0 \omega_1)(x_0'' + \epsilon x_1'') + \omega_0^2(x_0 + \epsilon x_1) - \tfrac{1}{6}\epsilon \omega_0^2(x_0 + \epsilon x_1)^3 = 0, \qquad (2.1.8)$$

where primes denote derivatives with respect to the argument ωt. In zero order, we have

$$\omega_0^2 x_0'' + \omega_0^2 x_0 = 0, \qquad (2.1.9)$$

which has the solution (2.1.3). To first order, we have

$$\omega_0^2 x_1'' + 2\omega_0 \omega_1 x_0'' + \omega_0^2 x_1 - \tfrac{1}{6}\omega_0^2 x_0^3 = 0, \qquad (2.1.10)$$

and by substituting the solution for x_0 into the above,

$$\omega_0^2 x_1'' + \omega_0^2 x_1 = \tfrac{1}{8}\omega_0^2 A^3 \cos \omega_0 t + 2 A \omega_0 \omega_1 \cos \omega_0 t$$
$$+ \tfrac{1}{24}\omega_0^2 A^3 \cos 3\omega_0 t. \qquad (2.1.11)$$

The periodicity of x requires that the driving term $\cos \omega_0 t$ be zero, for otherwise we obtain secular terms. We therefore choose ω_1 to eliminate the secularity

$$\omega_1 = -\tfrac{1}{16}A^2 \omega_0, \qquad (2.1.12)$$

2.1. Introduction

yielding a decrease in frequency with oscillation amplitude characteristic of the pendulum. With this choice of ω_1, the solution of (2.1.11) is

$$x_1 = A_1 \cos \omega_0 t + B_1 \sin \omega_0 t - \frac{A^3}{192} \cos 3\omega_0 t. \tag{2.1.13}$$

Introducing initial conditions

$$x_1(0) = \dot{x}_1(0) = 0, \tag{2.1.14}$$

we obtain

$$x_1 = \frac{A^3}{192}(\cos \omega_0 t - \cos 3\omega_0 t), \tag{2.1.15}$$

which gives the change in x due to the nonlinearity. The procedure can be carried to higher orders in ϵ. The essence of the method is the assumption of periodicity for x with a frequency ω that differs from the linear frequency ω_0. The additional freedom due to the expansion of ω allows the secularity to be eliminated at every order, thus achieving a uniformly convergent series solution. The canonical form of Lindstedt's method has been developed by Poincaré (1892) and Von Zeipel (1916) and is given in Section 2.2a.

2.1b. Asymptotic Series and Small Denominators

A problem related to the previous one, but with the fundamental difference that it is explicitly a function of time and thus represents a system with two degrees of freedom, is that of an oscillator with a slowly varying restoring force (or frequency). We write the equation of such an oscillator, shown in Fig. 2.1b, as

$$\ddot{x} + \omega^2(\epsilon t)x = 0, \tag{2.1.16}$$

where again ϵ is a dimensionless measure of smallness introduced to identify the perturbation and is set to unity at the end of the calculation. We may think of the small expansion parameter for this problem as

$$\left|\frac{1}{\omega}\frac{\dot{\omega}}{\omega}\right|, \tag{2.1.17}$$

which is the ratio of the oscillation period to the time scale over which the restoring force varies.

The technique involves expansion not directly in x, but in a subsidiary variable in which the higher-order terms measure progressively slower changes in time. First, with the change in independent variable $\tau = \epsilon t$, (2.1.16) becomes

$$\frac{d^2x}{d\tau^2} + \left(\frac{1}{\epsilon}\right)^2 \omega^2(\tau)x = 0. \tag{2.1.18}$$

We introduce the new dependent variable y by

$$x = \exp\left[\int^\tau y\, d\tau\right], \tag{2.1.19}$$

such that, writing $dx/d\tau$ as x',

$$x' = yx,$$

and

$$x'' = y^2 x + y' x,$$

which, when substituted into (2.1.18), gives the Ricatti equation

$$\epsilon^2(y^2 + y') + \omega^2 = 0. \tag{2.1.20}$$

We now expand y as a power series

$$y = \epsilon^{-1} y_0 + y_1 + \epsilon y_2 + \cdots \tag{2.1.21}$$

and substituting the expansion into (2.1.20), to lowest order in ϵ,

$$y_0 = \pm i\omega,$$

and to next order

$$2 y_0 y_1 + y_0' = 0. \tag{2.1.22}$$

Substituting (2.1.21) and (2.1.22) into (2.1.19),

$$x = \exp\left[\int \left(\epsilon^{-1} y_0 - \frac{1}{2}\frac{y_0'}{y_0}\right) d\tau\right],$$

which, on integration of the second term and substitution of $y_0 = \pm i\omega$, yields

$$x = \frac{A}{\omega^{1/2}} \exp\left[\pm i \int \omega\, dt\right]. \tag{2.1.23}$$

The first-order solution in ϵ, (2.1.23) has played an important role in quantum mechanics and is known as the WKB solution.

Let us consider a slow change in frequency from ω_1 for $t < t_1$ to ω_2 for $t > t_2$ and evaluate the action integral

$$J = \frac{1}{2\pi}\oint p\, dx \tag{2.1.24}$$

in the two regions $t < t_1$ and $t > t_2$ in which ω is assumed constant.

2.1. Introduction

Taking $p = m\dot{x}$ and the real form of (2.1.23),

$$x = A\omega^{-1/2}\cos(\omega t + \delta),$$

we find that $J = \frac{1}{2}mA^2$ in both regions, despite the fact that the frequency ω, and thus the energy $E = \frac{1}{2}m\dot{x}_{\max}^2 = \omega J$, have changed by arbitrarily large amounts. The action J is thus an adiabatic invariant of the motion, that is, a constant, within the approximation of the expansion that the parameters change slowly. The existence of the invariant allows a solution to be obtained, despite the fact that the Hamiltonian is not constant.

The action can be evaluated in the region $t_1 < t < t_2$ with slowly varying parameters, with the same result for x in lowest order. Furthermore, the expansion can be extended to all orders in the small parameter. This was done by Kulsrud (1957) for a linear oscillator and more generally by Kruskal (1962) and others (see Lichtenberg, 1969, and Section 2.3 for other references) for more general systems. There is, however, a fundamental difference between the expansion in this case and that of the previous subsection. Here the explicit time-dependent term is equivalent to motion with two degrees of freedom. In this case, the series solution does not generally converge to the exact solution, but rather is asymptotic to it, a concept that we define more precisely at the beginning of Section 2.3.

2.1c. The Effect of Resonances

If the nonlinear oscillator, treated in Section 2.1a, is perturbed by a periodic function of time, or if the slow variation of the oscillator, in Section 2.1b, is periodic, then resonances between the oscillator and the driving term destroy convergence of the expansions and modify or destroy the invariants. We illustrate this difficulty here with a simpler example; we treat the complete problem in subsequent sections.

We consider a driven, linear oscillator

$$\ddot{x} + \omega_0^2 x = g(t), \tag{2.1.25}$$

where the driving term $g(t)$ is periodic with period $2\pi/\Omega$. We know that the homogeneous solution is

$$x_h = A\cos\omega_0 t + B\sin\omega_0 t. \tag{2.1.26}$$

By Fourier analyzing $g(t)$, we obtain

$$g(t) = \frac{a_0}{2} + \sum_{n=1}^{\infty}(a_n\cos n\Omega t + b_n\sin n\Omega t). \tag{2.1.27}$$

We substitute this into (2.1.25) and, assuming that x has the same form as $g(t)$ and equating term by term, we have for the nth term of the particular solution,

$$x_{pn} = \frac{a_n \cos n\Omega t + b_n \sin n\Omega t}{\omega_0^2 - n^2\Omega^2}. \tag{2.1.28}$$

We see that whenever $\omega_0^2 = n^2\Omega^2$, there is a resonance leading to a blow-up of the oscillation. This is illustrated in Fig. 2.1c, which shows a child pumping a swing.

If the oscillator is nonlinear, then, as in Section 2.1a, all harmonics of ω_0 are present in the homogeneous solution. In this case, there is a resonance whenever ω_0/Ω is rational. We have seen that ω_0 is a function of the amplitude of the oscillation; there is thus a dense set of resonances at the rationals as the oscillation amplitude is varied. Furthermore, the resonances themselves change the amplitude and thus the oscillator frequency, breaking the exact resonance condition. These resonant denominators, which prevent convergence of the classical series in their neighborhood, are thus a manifestation of a physical phenomenon that locally changes the character of the phase space trajectories.

*2.2. Classical Perturbation Theory

Most multidimensional or driven systems are not integrable; that is, the solution to the Hamilton–Jacobi equation does not exist. However, for systems that do not differ by much from integrable ones, one can attempt to obtain solutions to a desired degree of accuracy by expanding the generating function in powers of a small parameter and then solving the Hamilton–Jacobi equation successively to each power. As we have discussed, the appearance of small denominators prevents the convergence of such series. Nevertheless, the results obtained in this manner may describe the system behavior quite well over some regions of the phase space. Furthermore, the method is very useful in obtaining approximate solutions to problems of one degree of freedom for which the closed form integrals are not easily evaluated, and for obtaining a preparatory transformation to action-angle variables in more than one degree of freedom.

*2.2a. One Degree of Freedom

We consider a Hamiltonian of the form

$$H = H_0(J) + \epsilon H_1(J, \theta) + \epsilon^2 H_2(J, \theta) + \cdots, \tag{2.2.1}$$

2.2. Classical Perturbation Theory

where H_0 is in action-angle form, such that its solution is

$$J = J_0, \tag{2.2.2a}$$

$$\theta = \omega t + \beta, \tag{2.2.2b}$$

$$\omega = \partial H_0/\partial J, \tag{2.2.2c}$$

with J_0, ω, β constants, independent of t. Following Poincaré (1892) and Von Zeipel (1916), we seek a transformation to new variables $\bar{J}$, $\bar{\theta}$ for which the new Hamiltonian $\bar{H}$ is a function of the action $\bar{J}$ alone. Using the generating function $S(\bar{J}, \theta)$, we expand S and $\bar{H}$ in power series in ϵ

$$S = \bar{J}\theta + \epsilon S_1 + \cdots, \tag{2.2.3}$$

$$\bar{H} = \bar{H}_0 + \epsilon \bar{H}_1 + \cdots, \tag{2.2.4}$$

where the lowest-order term in S has been chosen to generate the identity transformation $J = \bar{J}$ and $\bar{\theta} = \theta$. The old action and new angle are found from (1.2.13a) and (1.2.13b), respectively, as power series in ϵ:

$$J = \bar{J} + \epsilon \frac{\partial S_1(\bar{J}, \theta)}{\partial \theta} + \cdots, \tag{2.2.5a}$$

$$\bar{\theta} = \theta + \epsilon \frac{\partial S_1(\bar{J}, \theta)}{\partial \bar{J}} + \cdots. \tag{2.2.5b}$$

The new Hamiltonian is obtained from (1.2.13c). In order to find it, we must invert (2.2.5) to find the old variables in terms of the new. To order ϵ, this is easily done:

$$J = \bar{J} + \epsilon \frac{\partial S_1(\bar{J}, \bar{\theta})}{\partial \bar{\theta}} + \cdots, \tag{2.2.6a}$$

$$\theta = \bar{\theta} - \epsilon \frac{\partial S_1(\bar{J}, \bar{\theta})}{\partial \bar{J}} + \cdots. \tag{2.2.6b}$$

Then from (1.2.13c),

$$\bar{H}(\bar{J}, \bar{\theta}) = H(J(\bar{J}, \bar{\theta}), \theta(\bar{J}, \bar{\theta})). \tag{2.2.7}$$

Expanding the right-hand side of this equation in a power series in ϵ using (2.2.6),

$$H_0(J(\bar{J}, \bar{\theta})) = H_0(\bar{J}) + \epsilon \frac{\partial H_0}{\partial \bar{J}} \frac{\partial S_1}{\partial \bar{\theta}} + \cdots \tag{2.2.8a}$$

$$\epsilon H_1(J(\bar{J}, \bar{\theta}), \theta(\bar{J}, \bar{\theta})) = \epsilon H_1(\bar{J}, \bar{\theta}) + \cdots. \tag{2.2.8b}$$

Inserting these in (2.2.7) yields in zero order

$$\bar{H}_0 = H_0(\bar{J}), \tag{2.2.9}$$

and in first order

$$\bar{H}_1 = \omega(\bar{J})\frac{\partial S_1(\bar{J},\bar{\theta})}{\partial \bar{\theta}} + H_1(\bar{J},\bar{\theta}). \tag{2.2.10}$$

Since we are seeking a new Hamiltonian that is a function of $\bar{J}$ alone, we must choose S_1 in (2.2.10) to eliminate the $\bar{\theta}$-dependent part of H_1. Introducing the average part of H_1,

$$\langle H_1 \rangle = \frac{1}{2\pi}\int_0^{2\pi} d\bar{\theta}\, H_1(\bar{J},\bar{\theta}), \tag{2.2.11}$$

and the oscillating part

$$\{H_1\} = H_1 - \langle H_1 \rangle, \tag{2.2.12}$$

we have from (2.2.10) the two equations

$$\bar{H}_1 = \langle H_1 \rangle, \tag{2.2.13}$$

$$\omega \frac{\partial S_1}{\partial \bar{\theta}} = -\{H_1\}. \tag{2.2.14}$$

Combining (2.2.9) and (2.2.13), we have to first order the transformed Hamiltonian

$$\bar{H} = H_0(\bar{J}) + \epsilon \langle H_1(\bar{J},\bar{\theta}) \rangle + \cdots \tag{2.2.15}$$

with the new frequency $\bar{\omega} = \partial \bar{H}/\partial \bar{J}$, provided (2.2.14) can be solved for the S_1 required to eliminate $\{H_1\}$. We see that to first order the new Hamiltonian is the average of the old Hamiltonian over the phase.

To find S_1, we expand both $\{H_1\}$ and S_1 in Fourier series

$$\{H_1\} = \sum_{n \neq 0} H_{1n}(\bar{J})e^{in\bar{\theta}}, \tag{2.2.16a}$$

$$S_1 = \sum_n S_{1n}(\bar{J})e^{in\bar{\theta}}. \tag{2.2.16b}$$

From (2.2.14), it immediately follows that S_{10} = const. which can be set equal to zero and

$$S_{1n} = -\frac{H_{1n}}{in\omega}, \quad n \neq 0, \tag{2.2.17}$$

which yields a convergent Fourier series for S_1 provided $\omega(\bar{J}) \neq 0$. Substituting (2.2.17) into (2.2.16b), the coordinate transformation (2.2.6) is easily evaluated.

2.2. Classical Perturbation Theory

Higher-Order Expansions. It is sometimes necessary to carry the expansion to higher order in ϵ, either because the first-order correction is zero or because increased accuracy is desired. The Poincaré–Von Zeipel procedure can be carried to arbitrary order in ϵ, but the disentangling of the old and new variables, leading from (2.2.5) to (2.2.8), becomes algebraically tedious. If the complete inversion of the variables is not required, the procedure for calculating the new Hamiltonian, and therefore the perturbed frequency, can be relatively straightforward.

Although we defer our main consideration of higher-order calculations to Section 2.5, where modern methods using Lie transformations are introduced, we exhibit explicitly here the terms required for calculating the new Hamiltonian to second order in ϵ. The reader should consult Born (1927) and Giacaglia (1972) for a more detailed treatment of the higher-order expansions using the Poincaré–Von Zeipel series. With S and $\bar{H}$ written as power series in ϵ, as in (2.2.3) and (2.2.4), then (2.2.8a) and (2.2.8b) are written out explicitly (we assume only H_0 and H_1 are different from zero):

$$H_0(J(\bar{J},\theta)) = H_0(\bar{J}) + \sum_{m,n} \frac{1}{m!} \frac{\partial^m H_0}{\partial \bar{J}^m} \left(\epsilon^n \frac{\partial S_n}{\partial \theta}\right)^m, \qquad (2.2.18a)$$

$$\epsilon H_1(J(\bar{J},\theta),\theta) = \epsilon\left[H_1(\bar{J},\theta) + \sum_{m,n} \frac{1}{m!} \frac{\partial^m H_1}{\partial \bar{J}^m} \left(\epsilon^n \frac{\partial S_n}{\partial \theta}\right)^m\right]. \qquad (2.2.18b)$$

Assuming $\bar{H}$ a function of $\bar{J}$ alone, writing $\bar{H}$ as a power series in ϵ and equating $\bar{H} = H$ as given in (2.2.18), we have the zero and first orders in ϵ as in (2.2.9) and (2.2.10). Without inverting θ, we find to order ϵ^2

$$\bar{H}_2 = \frac{\partial H_0}{\partial \bar{J}} \frac{\partial S_2}{\partial \theta} + \frac{1}{2} \frac{\partial^2 H_0}{\partial \bar{J}^2}\left(\frac{\partial S_1}{\partial \theta}\right)^2 + \frac{\partial H_1}{\partial \bar{J}} \frac{\partial S_1}{\partial \theta}. \qquad (2.2.19a)$$

Since $\bar{H}_2$ is a function of $\bar{J}$ alone, we have

$$\bar{H}_2 = \left\langle \frac{1}{2} \frac{\partial^2 H_0}{\partial \bar{J}^2}\left(\frac{\partial S_1}{\partial \theta}\right)^2 + \frac{\partial H_1}{\partial \bar{J}} \frac{\partial S_1}{\partial \theta} \right\rangle_\theta, \qquad (2.2.19b)$$

with S_2, periodic in θ, given by

$$\frac{\partial H_0}{\partial \bar{J}} \frac{\partial S_2}{\partial \theta} = -\left\{\frac{1}{2} \frac{\partial^2 H_0}{\partial \bar{J}^2}\left(\frac{\partial S_1}{\partial \theta}\right)^2 + \frac{\partial H_1}{\partial \bar{J}} \frac{\partial S_1}{\partial \theta}\right\}_\theta, \qquad (2.2.19c)$$

where as previously $\langle\ \rangle$ and $\{\ \}$ refer to the average and oscillating parts. The frequency of oscillation is then obtained, as usual, from $\bar{\omega} = \partial \bar{H}/\partial \bar{J}$. Approximations to the frequency higher order in ϵ can be obtained in a similar way without inverting the transformation.

The Pendulum. To illustrate the expansion procedure, we calculate to first order the nonlinear libration of a pendulum. The Hamiltonian, given in

(1.3.6), is of the form

$$H_p = \tfrac{1}{2}Gp^2 - F\cos\phi = E. \tag{2.2.20}$$

We choose for H_{0p} the quadratic terms, which generate a linear oscillation, and consider the remaining terms as the perturbation. Expanding H_p in a Taylor series and dropping the constant term,

$$H_p = \tfrac{1}{2}Gp^2 + \tfrac{1}{2}F\phi^2 - \frac{\epsilon}{4!}F\phi^4 + \frac{\epsilon^2}{6!}F\phi^6 - \cdots,$$

where the first two terms comprise H_{0p}. The small parameter ϵ has been inserted to identify conveniently the perturbation terms, and for future reference, the order ϵ^2 term has been kept. The actual expansion parameter is the ratio of libration to separatrix energy, so we may set $\epsilon = 1$ at the end of the calculation.

To prepare the system for perturbation theory, we transform to action-angle variables of the unperturbed system using (1.2.69), to obtain the new Hamiltonian, $H = E + F$,

$$H = \omega_0 J - \frac{\epsilon}{6}GJ^2\sin^4\theta + \frac{\epsilon^2}{90}\frac{G^2 J^3}{\omega_0}\sin^6\theta - \cdots, \tag{2.2.21}$$

where $\omega_0 = (FG)^{1/2}$ is the unperturbed libration frequency. Expanding the powers of $\sin\theta$ in a Fourier series yields

$$H_0 = \omega_0 J, \tag{2.2.22a}$$

$$H_1 = -\frac{GJ^2}{48}(3 - 4\cos 2\theta + \cos 4\theta), \tag{2.2.22b}$$

$$H_2 = \frac{G^2 J^3}{2880\omega_0}(10 - 15\cos 2\theta + 6\cos 4\theta - \cos 6\theta). \tag{2.2.22c}$$

Applying the expansion technique, and averaging (2.2.22b) over θ, we obtain immediately from (2.2.15) the new Hamiltonian to first order:

$$\bar{H} = \omega_0 \bar{J} - \frac{\epsilon}{16}G\bar{J}^2, \tag{2.2.23}$$

with the new frequency

$$\bar{\omega} = \frac{\partial \bar{H}}{\partial \bar{J}} = \omega_0 - \frac{\epsilon}{8}G\bar{J}, \tag{2.2.24}$$

which shows the reduction in frequency with libration amplitude, in agreement with the expansion of (1.3.13) to first order in κ. Eliminating $\bar{J}$ to first order between (2.2.23) and 2.2.24), we obtain $\bar{\omega}(\bar{H})$, which we plot as the

2.2. Classical Perturbation Theory

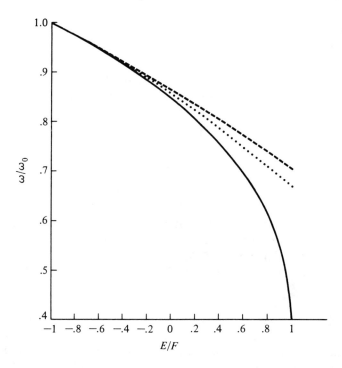

Figure 2.2. Frequency versus energy for the pendulum, with Hamiltonian $H = \frac{1}{2}Gp^2 - F\cos\phi = E$, and $\omega_0 = (FG)^{1/2}$. The solid line is the exact result from (1.3.13). The dashed line is obtained using first-order perturbation theory. The dotted line is the result from second-order perturbation theory (see Section 2.5b).

dashed curve in Fig. 2.2, which we compare with the exact result computed from (1.3.13).

The generating function is obtained by integrating (2.2.14), yielding

$$S_1 = -\frac{GJ^2}{192\omega_0}(8\sin 2\theta - \sin 4\theta). \tag{2.2.25}$$

Using this expression, the transformation (2.2.6) from new to old variables is easily evaluated.

*2.2b. Two or More Degrees of Freedom

If the perturbed Hamiltonian is autonomous with more than one degree of freedom or is explicitly time dependent with one or more degrees of freedom, then the previous expansion procedure does not converge. To see this, we generalize the Poincaré–Von Zeipel technique to an autonomous Hamiltonian with N degrees of freedom. Explicit time dependence may be treated by introducing an extended phase space. We write

$$H(\boldsymbol{J}, \boldsymbol{\theta}) = H_0(\boldsymbol{J}) + \epsilon H_1(\boldsymbol{J}, \boldsymbol{\theta}), \tag{2.2.26}$$

where $\boldsymbol{J}$ and $\boldsymbol{\theta}$ are the N-dimensional actions and angles of H_0, and H_1 is a multiply periodic function of the angles:

$$H_1 = \sum_m H_{1m}(\boldsymbol{J}) e^{i\boldsymbol{m}\cdot\boldsymbol{\theta}}, \tag{2.2.27}$$

where

$$\boldsymbol{m}\cdot\boldsymbol{\theta} = m_1\theta_1 + \cdots + m_N\theta_N \tag{2.2.28}$$

with the m_i being integers, and the sum being N-fold over the m_i. We again seek a transformation to new vectors $\bar{\boldsymbol{J}}, \bar{\boldsymbol{\theta}}$, for which the new Hamiltonian $\bar{H}$ is a function of $\bar{\boldsymbol{J}}$ alone. We introduce the near-identity generating function whose first-order term is also an N-fold sum, multiply periodic in $\boldsymbol{\theta}$:

$$S = \bar{\boldsymbol{J}}\cdot\boldsymbol{\theta} + \epsilon \sum_m S_{1m}(\bar{\boldsymbol{J}}) e^{i\boldsymbol{m}\cdot\boldsymbol{\theta}} + \cdots. \tag{2.2.29}$$

As in one dimension, we write the old variables in terms of the new using the generating function, and insert these into (1.2.13c). Equating like powers of ϵ, we have to zero order

$$\bar{H}_0(\bar{\boldsymbol{J}}) = H_0(\bar{\boldsymbol{J}}), \tag{2.2.30}$$

and to first order

$$\bar{H}_1 = \boldsymbol{\omega}(\bar{\boldsymbol{J}})\cdot\frac{\partial S_1(\bar{\boldsymbol{J}}, \bar{\boldsymbol{\theta}})}{\partial \bar{\boldsymbol{\theta}}} + H_1(\bar{\boldsymbol{J}}, \bar{\boldsymbol{\theta}}), \tag{2.2.31}$$

where

$$\boldsymbol{\omega}(\bar{\boldsymbol{J}}) = \frac{\partial H_0(\bar{\boldsymbol{J}})}{\partial \bar{\boldsymbol{J}}} \tag{2.2.32}$$

is the frequency vector for the unperturbed motion.

Proceeding as in one dimension, we average (2.2.31) over *all* the angle variables to find

$$\bar{H} = H_0(\bar{\boldsymbol{J}}) + \epsilon \langle H_1(\bar{\boldsymbol{J}}, \bar{\boldsymbol{\theta}})\rangle \tag{2.2.33}$$

and

$$\boldsymbol{\omega}\cdot\frac{\partial S_1}{\partial \bar{\boldsymbol{\theta}}} = -\{H_1\}. \tag{2.2.34}$$

The solution for S_1 involves an integral over the zero-order orbits of the

2.2. Classical Perturbation Theory

system since

$$\frac{dS_1}{dt} = \frac{\partial S_1}{\partial t} + \frac{\partial S_1}{\partial \bar{\theta}} \cdot \frac{d\bar{\theta}}{dt} + \frac{\partial S_1}{\partial \bar{J}} \cdot \frac{d\bar{J}}{dt} \qquad (2.2.35)$$

and, to zero order, the first and third terms vanish, yielding the left-hand side of (2.2.34). Thus

$$S_1 = -\int^t dt' \{H_1(\bar{J}, \bar{\theta}(t'))\}. \qquad (2.2.36)$$

Alternatively, we solve for S_1 by integrating the Fourier series for H_1 term by term to obtain

$$S(\bar{J}, \bar{\theta}) = \bar{J} \cdot \bar{\theta} + \epsilon i \sum_{m \neq 0} \frac{H_{1m}(\bar{J})}{m \cdot \omega(\bar{J})} e^{im \cdot \bar{\theta}} + \cdots. \qquad (2.2.37)$$

We are at once faced with the problem of small denominators, since for any $\bar{J}$ an m can be found such that $m \cdot \omega$ is arbitrarily close to zero. This clearly prevents the series from converging. We point out again that this phenomenon represents a physical as well as a mathematical difficulty. It arises from real resonances that, as we shall see in Section 2.4, change the topology of the phase space trajectories. Nevertheless, considerable effort has gone into developing expansion procedures that at least stave off the singularity to higher order in the expansion. In defense of these seemingly hopeless procedures, they yield solutions that converge in certain regions of phase space to the actual solutions for motion taken over finite but long periods of time. Furthermore, the solutions can in some cases closely approximate the motion within a prescribed *coarse graining* of the phase space for arbitrarily long times. This latter result is due to the actual convergence of certain (KAM) series solutions for *some* values of $\bar{J}$. As we shall see in the next two chapters, for two degrees of freedom these solutions closely bound the resonant ones, thus constraining the more complicated resonant trajectories to approximate the nonresonant ones.

Explicit Time Dependence. We now develop the first-order equations for a system in one degree of freedom and the time, which will be useful to us in later sections. We start with the Hamiltonian

$$H = H_0(J) + \epsilon H_1(J, \theta, t), \qquad (2.2.38)$$

where the perturbation is periodic in θ with period 2π and in the time with period $2\pi/\Omega$:

$$H_1 = \sum_{l,m} H_{1lm}(J) e^{i(l\theta + m\Omega t)}. \qquad (2.2.39)$$

As in Section 2.2a, we take S of the form

$$S = \bar{J}\theta + \epsilon S_1(\bar{J}, \theta, t), \qquad (2.2.40)$$

giving the old variables in terms of the new as in (2.2.6). Due to the explicit time dependence of H_1, (2.2.7) is modified to

$$\bar{H}(\bar{J}, \bar{\theta}, t) = H(J, \theta, t) + \epsilon \frac{\partial S_1(\bar{J}, \theta, t)}{\partial t}. \tag{2.2.41}$$

Expanding in ϵ,

$$\bar{H}_0 = H_0(\bar{J}), \tag{2.2.42}$$

$$\bar{H}_1 = \frac{\partial S_1}{\partial t} + \omega \frac{\partial S_1}{\partial \bar{\theta}} + H_1. \tag{2.2.43}$$

Again choosing S_1 to eliminate the oscillating part of H_1, we have

$$\bar{H} = H_0 + \epsilon \langle H_1 \rangle, \tag{2.2.44}$$

where the average is taken over both the θ and t oscillations, and

$$\frac{\partial S_1}{\partial t} + \omega \frac{\partial S_1}{\partial \bar{\theta}} = -\{H_1\}. \tag{2.2.45}$$

To determine S_1, we expand in a Fourier series:

$$S_1 = i \sum_{l, m \neq 0} \frac{H_{1lm}(\bar{J})}{l\omega(\bar{J}) + m\Omega} \exp[i(l\bar{\theta} + m\Omega t)]. \tag{2.2.46}$$

We again have the small denominators, preventing convergence of the series.

Although resonances appear explicitly in lowest order, classical canonical perturbation theory can be quite useful in determining the invariants for values of the action sufficiently far from the primary resonances. To illustrate, we choose a simple form for H_1 in which the high harmonics in θ are absent,

$$H_1 = U(J, t) + V(J, t) \cos \theta. \tag{2.2.47}$$

To determine S_1, explicitly, we expand both H_1 and S_1 in Fourier series

$$H_1(\bar{J}, \bar{\theta}, t) = \sum b_{lm}(\bar{J}) \cos(l\bar{\theta} - m\Omega t), \tag{2.2.48}$$

$$S_1(\bar{J}, \bar{\theta}, t) = \sum a_{lm}(\bar{J}) \sin(l\bar{\theta} - m\Omega t), \tag{2.2.49}$$

where the sum is over $l = 0, 1$, and over all m. Substituting (2.2.49) in (2.2.45), we solve for the coefficients a_{lm} for $l, m \neq 0$:

$$a_{lm} = \frac{b_{lm}}{m\Omega - l\omega(\bar{J})}. \tag{2.2.50}$$

Provided we choose $\bar{J}$ such that the denominators are sufficiently far from zero, the a_{lm} are well behaved, and to order ϵ we have the new Hamiltonian in action-angle form, as in the one-dimensional case, in (2.2.44):

$$\bar{H} = H_0(\bar{J}) + \epsilon b_{00}(\bar{J}). \tag{2.2.51}$$

2.2. Classical Perturbation Theory

In terms of the old variables, the new invariant is

$$\bar{J}(J, \theta, t) = J - \epsilon \frac{\partial S_1(J, \theta, t)}{\partial \theta}. \tag{2.2.52}$$

Any function $I(J)$ is also an invariant. In Section 2.4d, we exploit this fact to construct invariants valid globally near the primary resonances.

Wave-Particle Interaction. We illustrate the procedures and limitations of canonical perturbation theory in more than one dimension with a practical example, that of a charged particle gyrating in a uniform magnetic field and interacting with an electrostatic wave. The problem has been considered for a wave propagating at an angle to the field by Smith and Kaufman (1975), and for propagation perpendicular to the field by Karney and Bers (1977) and Fukuyama et al. (1976). The system is illustrated in Fig. 2.3.

We first introduce action-angle variables for the unperturbed system. The unperturbed Hamiltonian is

$$H_{0p} = \frac{1}{2M} \left| \mathbf{p} - \frac{e}{c} \mathbf{A} \right|^2, \tag{2.2.53}$$

where M is the mass, e is the charge, c the velocity of light,

$$\mathbf{A}(\mathbf{x}) = -B_0 y \hat{\mathbf{x}} \tag{2.2.54}$$

is the vector potential for a uniform field B_0, and

$$\mathbf{p} = M\mathbf{v} + \frac{e}{c} \mathbf{A} \tag{2.2.55}$$

Figure 2.3. A charge moves in the field of a uniform static magnetic field B_0 in the z-direction, and is perturbed by an electrostatic wave with amplitude Φ_0, frequency ω, and wavevector $\mathbf{k}$, with $\mathbf{k}$ lying in the y–z plane. In the absence of the wave, the charge gyrates perpendicular to B_0 and moves uniformly along B_0.

is the canonical momentum conjugate to x. We transform to *guiding center variables* using the generating function

$$F_1 = M\Omega[\tfrac{1}{2}(y - Y)^2 \cot \phi - xY]. \tag{2.2.56}$$

Using (1.2.11), we obtain the new variables

$$\tan \phi = \frac{v_x}{v_y}, \tag{2.2.57a}$$

$$P_\phi = \frac{Mc}{e}\mu = \frac{1}{2}\frac{Mv_\perp^2}{\Omega} = \frac{1}{2}M\Omega\rho^2, \tag{2.2.57b}$$

$$Y = y + \rho \sin \phi, \tag{2.2.57c}$$

$$X = x - \rho \cos \phi, \tag{2.2.57d}$$

where

$$\Omega = \frac{eB_0}{Mc} \tag{2.2.58}$$

is the gyration frequency, μ is the magnetic moment, $v_\perp^2 = v_x^2 + v_y^2$, and $\rho = v_\perp/\Omega$ is the gyration radius. X and Y are the guiding center position, P_ϕ the angular momentum, and ϕ the gyration angle. The new momenta are P_ϕ, $M\Omega X$, and the linear momentum P_z, with their corresponding coordinates ϕ, Y, and z, respectively. The transformed Hamiltonian is found to be

$$H'_0 = \frac{P_z^2}{2M} + P_\phi\Omega. \tag{2.2.59}$$

Assuming a perturbation in the form of an electrostatic wave

$$\Phi = \Phi_0 \sin(k_z z + k_\perp y - \omega t) \tag{2.2.60}$$

with an electric field $\boldsymbol{E} = -\nabla\Phi$, we have in the guiding center coordinates

$$H'_1 = e\Phi_0 \sin(k_z z + k_\perp Y - k_\perp \rho) \sin(\phi - \omega t), \tag{2.2.61}$$

where

$$\rho(P_\phi) = \left(\frac{2P_\phi}{M\Omega}\right)^{1/2}. \tag{2.2.62}$$

Since

$$H' = H'_0 + \epsilon H'_1 \tag{2.2.63}$$

is independent of the momentum $M\Omega X$, we have the corresponding coordinate $Y = $ const., and we shift z or t by a constant to eliminate the constant phase $k_\perp Y$ in (2.2.61). The nonlinearity in H' arises from the dependence of the phase on $\sin \phi$ and ρ. Because z and t appear only in the linear

2.2. Classical Perturbation Theory

combination $k_z z - \omega t$, time may be eliminated by transforming to the wave frame using the generating function

$$F_2 = (k_z z - \omega t)P_\psi + P_\phi \phi. \tag{2.2.64}$$

We obtain new variables P_ψ and ψ and new Hamiltonian H, using relations (1.2.13),

$$P_z = \frac{\partial F_2}{\partial z} = k_z P_\psi, \tag{2.2.65a}$$

$$\psi = \frac{\partial F_2}{\partial P_\psi} = k_z z - \omega t, \tag{2.2.65b}$$

and

$$H = k_z^2 P_\psi^2/2M - P_\psi \omega + P_\phi \Omega + \epsilon e \Phi_0 \sin(\psi - k_\perp \rho \sin \phi)$$
$$= E, \quad \text{a const.} \tag{2.2.66}$$

Here, as previously, ϵ is an arbitrary ordering parameter to be set equal to one at the end of the calculation. The harmonic resonances in the nonlinear forcing function can be exposed by expansion in a Bessel series to obtain

$$H = k_z^2 P_\psi^2/2M - P_\psi \omega + P_\phi \Omega + \epsilon e \Phi_0 \sum_m \mathcal{J}_m(k_\perp \rho) \sin(\psi - m\phi). \tag{2.2.67}$$

We have seen that it is necessary to stay sufficiently far from the unperturbed resonances that the Fourier amplitudes decrease faster than the near-resonant denominators. The Bessel coefficients $\mathcal{J}_m(k_\perp \rho)$ give the fall-off of the Fourier amplitudes. The unperturbed frequencies are obtained from (2.2.67) as

$$\omega_\phi = \frac{\partial H_0}{\partial P_\phi} = \Omega, \tag{2.2.68a}$$

$$\omega_\psi = \frac{\partial H_0}{\partial P_\psi} = \frac{k_z^2}{M} P_\psi - \omega = k_z v_z - \omega. \tag{2.2.68b}$$

In first order the perturbation excites only resonances between the frequency ω_ψ and the various harmonics of ω_ϕ, so the resonance condition is

$$\omega_\psi - m\Omega = 0. \tag{2.2.69}$$

For $k_z = 0$, substituting for ω_ψ from (2.2.68b), we find that (2.2.69) becomes

$$\omega + m\Omega = 0. \tag{2.2.70}$$

For $k_z \neq 0$, solving (2.2.69) for P_ψ, we have the condition

$$P_\psi = \frac{M}{k_z^2}(\omega + m\Omega) \tag{2.2.71}$$

at resonance. We treat these resonant cases by secular perturbation theory in Section 2.4c.

Because P_ψ is an action variable of the system, the resonances satisfying (2.2.71) are inherently imbedded in the motion for wave propagation at an angle to the field. We therefore treat the case of perpendicular wave propagation ($k_z \equiv 0$), where the resonance condition (2.2.70) is not met. In this case, as will be shown later, there are no primary resonances for a sufficiently small perturbation. Applying (2.2.34) to (2.2.67), we have

$$-\omega \frac{\partial S_1}{\partial \psi} + \Omega \frac{\partial S_1}{\partial \phi} = -e\Phi_0 \sum_m \mathcal{J}_m(k_\perp \bar{\rho}) \sin(\psi - m\phi), \qquad (2.2.72)$$

where $\bar{\rho} = \rho(\bar{J}_\phi)$. Equation (2.2.72) has the solution

$$S_1 = -e\Phi_0 \sum_m \mathcal{J}_m(k_\perp \bar{\rho}) \frac{\cos(\psi - m\phi)}{\omega + m\Omega}, \qquad (2.2.73)$$

from which we obtain the old actions P_ψ and P_ϕ in terms of the new:

$$P_\psi = \bar{P}_\psi + \epsilon \frac{\partial S_1}{\partial \psi}$$

$$= \bar{P}_\psi + \epsilon e\Phi_0 \sum_m \mathcal{J}_m(k_\perp \bar{\rho}) \frac{\sin(\psi - m\phi)}{\omega + m\Omega}. \qquad (2.2.74)$$

Inverting, we have to first order

$$\bar{P}_\psi = P_\psi - \epsilon e\Phi_0 \sum_m \mathcal{J}_m(k_\perp \rho) \frac{\sin(\psi - m\phi)}{\omega + m\Omega}$$

$$= \text{const.} \qquad (2.2.75)$$

Similarly, we obtain

$$\bar{P}_\phi = P_\phi + \epsilon e\Phi_0 \sum_m m\mathcal{J}_m(k_\perp \rho) \frac{\sin(\psi - m\phi)}{\omega + m\Omega}$$

$$= \text{const.}, \qquad (2.2.76)$$

with $\rho(P_\phi)$ given by (2.2.62). Now, for example, using (2.2.76) and setting one of the phase variables equal to a constant, P_ϕ may be plotted as a function of the other phase variable as a series of level curves for a set of values of the invariant $\bar{P}_\phi$. These level curves are equivalent to a surface of section plot. Karney (1977) has plotted level curves of $k_\perp \rho(P_\phi)$ versus ψ with $\phi = \pi$, which we reproduce in Fig. 2.4a, b. The ordinate and abscissa are not canonically conjugate variables, but still reveal the general character of the phase space. These results are compared in Figs. 2.5a and 2.5b with numerical trajectories computed by Karney. Since $\omega = 30.11\Omega$ for the unperturbed system, at small perturbation amplitude, the level curves lie sufficiently far

2.2. Classical Perturbation Theory

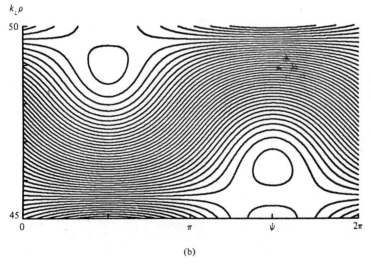

Figure 2.4. The nonresonant invariant curves of $k_\perp \rho$ versus ψ in a surface of section $\phi = \pi$ for off-resonant interaction of a gyrating particle interacting with a perpendicular propagating wave. The ratio of the applied frequency to the gyrofrequency is $\omega/\Omega = 30.11$. (a) Low wave amplitude—no trapping; (b) higher amplitude—with trapping (after Karney, 1977).

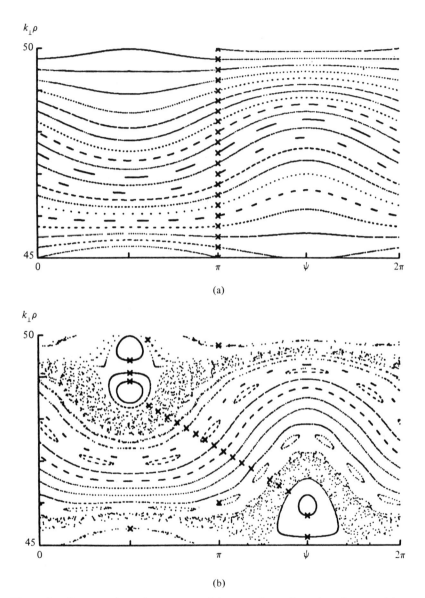

Figure 2.5. Exact trajectories computed in a surface of section $\phi = \pi$, with parameters as in Fig. 2.4. The x's mark the initial conditions (after Karney, 1977).

from the primary resonance that analytic and numerical results are quite similar. At higher perturbation strength, the perturbed frequency may exhibit exact resonance. The first-order generating function, having poles at the unperturbed resonance, remains finite and reproduces the gross features of the resonant behavior. The higher-order island structure is not described by this procedure, but requires the calculation of second-order islands from secular perturbation theory (Section 2.4). The stochastic regions are not directly calculable by the perturbation methods of this chapter, but their size may be estimated, as described in Chapter 4.

2.3. Adiabatic Invariance

*2.3a. Introduction and Basic Concepts

As we saw in Section 2.1b, for an oscillator with a slowly varying aperiodic parameter, an expansion can be constructed to obtain an *adiabatic* constant of the motion. The expansion parameter was the ratio of the fast oscillation period to the time scale for the slowly varying parameter. For multiply periodic systems, this same general procedure can be used to obtain series that do not explicitly contain small denominators. The method, first developed by Poincaré (1892), was placed on a more rigorous basis by Birkhoff (1927).

A more systematic form of the expansion procedure by Bogoliubov and associates (see Krylov and Bogoliubov, 1936; Bogoliubov and Zubarev, 1955; Bogoliubov and Mitropolsky, 1961) received the name of the *multiple scale method of averaging*. A slightly different form better adapted to canonical representation was introduced by Kruskal (1962) and is presented in Section 2.3d. Kruskal showed that the adiabatic invariants can be generated to all orders in the expansion parameter, and that the resulting series are asymptotic. Bogoliubov's and Kruskal's works deal directly with the differential equations, eschewing Hamiltonian transformation theory. The methods are applicable to general sets of differential equations, not necessarily in Hamiltonian form. McNamara and Whiteman (1967), Whiteman and McNamara (1968), and Stern (1970a, b) introduced useful canonical methods of calculation, but in higher order these have been superseded by Lie techniques (McNamara, 1978) as described in Section 2.5. The relationships between the various methods have been reviewed by McNamara and Whiteman (1967) and by Giacaglia (1972).

Asymptotic Series. Birkhoff (1927) first showed that an oscillator with slowly varying frequency, e.g., of the type given in (2.1.16),

$$\ddot{x} + \omega^2(\epsilon t)x = 0,$$

has an *asymptotic* series expansion. That is, a sequence of approximates

$$X_n(t, \epsilon) = \sum_{i=0}^{n} \epsilon^i x_i(t) \quad (2.3.1)$$

to the actual solution x can be constructed such that for any fixed n and t

$$\lim_{\epsilon \to 0} \epsilon^{-n}[x(t, \epsilon) - X_n(t, \epsilon)] = 0. \quad (2.3.2)$$

We recall several important properties of asymptotic series. First, two different functions may have the same asymptotic series, and thus construction of an asymptotic series does not define a unique function. Suppose that two functions x_1 and x_2 differ by the quantity, exponentially small in the limit $\epsilon \to 0$,

$$\Delta x = a(t) \exp\left[\frac{-b(t)}{\epsilon}\right]. \quad (2.3.3)$$

Since

$$\lim_{\epsilon \to 0} \epsilon^{-n} \exp\left(\frac{-b}{\epsilon}\right) = 0, \quad (2.3.4)$$

for $n > 0$, Δx has the asymptotic expansion $X_n = 0$ for all n (see Fig. 2.6a). Thus x_1 and x_2 have the same asymptotic series. In fact, direct calculations of adiabatic invariants for oscillatory systems with slowly varying parameters are found to have this exponential variation (see, e.g., Hertweck and Schlüter, 1957; Howard 1970; Cohen et al., 1978). For multiply periodic systems, these exponentially small changes in the invariant are a consequence of the resonant terms that modify the phase space, and ultimately destroy the modified invariants for ϵ not sufficiently small. Thus an "adiabatic" invariant constructed from an asymptotic series is, in the limit $\epsilon \to 0$, an approximation to order $\exp(-b/\epsilon)$ of an exact motion, which may lie (a) on a smooth KAM surface, (b) on a chain of KAM islands, or (c) within a thin stochastic layer.

A second property of asymptotic series is their formal divergence, that is, for any fixed ϵ and t,

$$\lim_{n \to \infty} [x(t, \epsilon) - X_n(t, \epsilon)] \to \infty. \quad (2.3.5)$$

The general behavior of the X_n as n increases for fixed ϵ, illustrated in Fig. 2.6b, is at first to better approximate x, but for n greater than some $n_{\max}(\epsilon)$, the succeeding X_n become increasingly poor approximates of x, ultimately diverging as $n \to \infty$. Thus we need find only the first $n_{\max}$ terms in the expansion, but their sum will err from the actual solution by roughly the magnitude of the last $(n_{\max})$ term.

A final point is that the asymptotic expansion is not valid for times long compared to the "slow" time scale, i.e., an adiabatic invariant may not be

2.3. Adiabatic Invariance

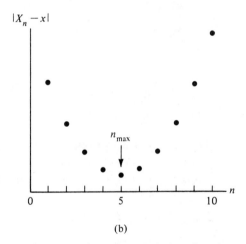

Figure 2.6. Properties of asymptotic series. (a) A function $\Delta x = \exp(-1/\epsilon)$ whose asymptotic expansion is zero; (b) divergence of asymptotic series for a fixed ϵ. The nth approximate X_n to the actual solution x first approaches x and then diverges from it. The "best" approximate is for $n = n_{\max}$.

even approximately conserved as $t \to \infty$. Nonconservation of these invariants—known as *Arnold diffusion*—arises in systems with *three* or more degrees of freedom and is the main topic of Chapter 6.

Slow Perturbations. We now describe the difference in the ordering of the expansion of a near-integrable Hamiltonian for small perturbations and for slow (or "adiabatic") ones. For *small* perturbations, the Hamiltonian has the general form

$$H = H_0(J, t) + \epsilon H_1(J, \theta, t) + \cdots,$$

where H_0 describes completely integrable motion, and ϵ is a small parameter characterizing the magnitude of the nonintegrable part of H. For small perturbations, the derivatives of H_0 and H_1 are assumed to be of the same order as H_0 and H_1 themselves; e.g.,

$$\left|\frac{\partial H_0}{\partial t}\right| \sim |H_0|, \quad \left|\frac{\partial H_1}{\partial J}\right| \sim |H_1|, \quad \text{etc.}$$

For slow perturbations, the terms produced by differentiation are assumed to be smaller by order ϵ than the terms from which they are derived, e.g., for slow time variation,

$$\left|\frac{\partial H_0}{\partial t}\right| \sim \epsilon |H_0|, \quad \text{etc.}$$

To keep track of this ordering, we often insert the small parameter ϵ, writing

$$H_0 = H_0(\epsilon t),$$

such that

$$\frac{\partial H_0}{\partial t} = \epsilon H'_0,$$

where prime denotes differentiation with respect to the argument $\tau = \epsilon t$.

In this section we are interested in systems for which the variation in all but one of the degrees of freedom, and in the time, is slow. Accordingly, we write the Hamiltonian in the form

$$H = H_0(J, \epsilon y, \epsilon t) + \epsilon H_1(J, \theta, \epsilon y, \epsilon t) + \cdots, \qquad (2.3.6)$$

where J and θ are the action-angle variables for the unperturbed ($\epsilon \equiv 0$) motion in the single fast degree of freedom, and the $y = (p, q)$ are the "slow" canonical variables, not necessarily in action-angle form, in the remaining degrees of freedom. Since the system is effectively one-dimensional when $\epsilon = 0$, it is integrable and J and θ can always be found. The small parameter ϵ in (2.3.6) will "automatically" keep track of the ordering when we differentiate H to construct the perturbation series, and may be set to unity at the end of the calculation.

*2.3b. Canonical Adiabatic Theory

We now construct to first order the classical adiabatic invariant for the Hamiltonian (2.3.6). In zero order, the invariant is the action J associated with the fast degree of freedom. To calculate the effect of the perturbation

2.3. Adiabatic Invariance

ϵH_1, we proceed as in Section 2.2b to find a transformation from J, θ, y to $\bar{J}, \bar{\theta}, \bar{y}$ such that the new Hamiltonian

$$\bar{H} = \bar{H}_0 + \epsilon \bar{H}_1 + \cdots \tag{2.3.7}$$

is independent of the "fast" phase variable $\bar{\theta}$. Introducing the near-identity generating function

$$S = \bar{J}\theta + \bar{p}\cdot q + \epsilon S_1(\bar{J}, \theta, \bar{p}, q, t) + \cdots, \tag{2.3.8}$$

we have, to first order, the transformations

$$J = \bar{J} + \epsilon \frac{\partial S_1}{\partial \theta}, \tag{2.3.9a}$$

$$\bar{\theta} = \theta - \epsilon \frac{\partial S_1}{\partial \bar{J}}, \tag{2.3.9b}$$

$$p = \bar{p} + \epsilon \frac{\partial S_1}{\partial q}, \tag{2.3.9c}$$

$$\bar{q} = q - \epsilon \frac{\partial S_1}{\partial \bar{p}}. \tag{2.3.9d}$$

Inserting these into H_0 and expanding to first order in ϵ,

$$H_0(J, \epsilon y, \epsilon t) = H_0(\bar{J}, \epsilon \bar{y}, \epsilon t) + \epsilon \omega \frac{\partial S_1}{\partial \theta}, \tag{2.3.10}$$

where $\omega = \partial H_0/\partial \bar{J}$ is the fast frequency. Note that the terms in

$$-\frac{\partial H_0}{\partial \bar{q}} \cdot \frac{\partial S_1}{\partial \bar{p}}, \quad \frac{\partial H_0}{\partial \bar{p}} \cdot \frac{\partial S_1}{\partial \bar{q}}, \tag{2.3.11}$$

are second order in ϵ and have been dropped. Now from (1.2.13c),

$$\bar{H}(\bar{J}, \bar{\theta}, \epsilon \bar{y}, \epsilon t) = H(J, \theta, \epsilon y, \epsilon t)$$

$$+ \epsilon \frac{\partial S(\bar{J}, \theta, \epsilon \bar{p}, \epsilon q, \epsilon t)}{\partial (\epsilon t)}. \tag{2.3.12}$$

Expanding $\bar{H}, H$, and S, using (2.3.9), and equating like powers of ϵ, we have to zero order

$$\bar{H}_0(\bar{J}, \epsilon \bar{y}, \epsilon t) = H_0(\bar{J}, \epsilon \bar{y}, \epsilon t), \tag{2.3.13}$$

and to first order

$$\bar{H}_1(\bar{J}, \bar{\theta}, \epsilon \bar{y}, \epsilon t) = \omega \frac{\partial S_1}{\partial \bar{\theta}} + H_1(\bar{J}, \bar{\theta}, \epsilon \bar{y}, \epsilon t), \tag{2.3.14}$$

with $S_1 = S_1(\bar{J}, \bar{\theta}, \epsilon \bar{y}, \epsilon t)$. Again the term $\partial S_1/\partial t$ in (2.3.12) is second order and has been dropped from (2.3.14).

Since we require $\bar{H}_1$ to be independent of $\bar{\theta}$, we choose S_1 to eliminate the oscillating part (in $\bar{\theta}$) of H_1. Holding the slow angle variables fixed, we define the average over $\bar{\theta}$ alone as

$$\langle H_1 \rangle_{\bar{\theta}} = \frac{1}{2\pi} \int_0^{2\pi} H_1 \, d\bar{\theta}, \tag{2.3.15}$$

and the oscillating part over $\bar{\theta}$ as

$$\{H_1\}_{\bar{\theta}} = H_1 - \langle H_1 \rangle_{\bar{\theta}}. \tag{2.3.16}$$

Separating (2.3.14) into its average and oscillating part yields for $\bar{H}$ to first order

$$\bar{H}(\bar{J}, \epsilon \bar{y}, \epsilon t) = H_0 + \epsilon \langle H_1 \rangle_{\bar{\theta}}, \tag{2.3.17}$$

and for S_1 the equation

$$\omega \frac{\partial S_1}{\partial \bar{\theta}} = -\{H_1\}_{\bar{\theta}}, \tag{2.3.18}$$

which is easily integrated. To zero order, the adiabatic invariant is J. In first order, the new invariant is $\bar{J}$, which is given in terms of the old variables from (2.3.9a) by

$$\bar{J}(J, \theta, \epsilon y, \epsilon t) = J - \epsilon \frac{\partial S_1}{\partial \theta}. \tag{2.3.19}$$

Substituting (2.3.18) into (2.3.19) and writing θ for the dummy variable $\bar{\theta}$, we have

$$\bar{J} = J + \frac{\epsilon \{H_1\}_\theta}{\omega}. \tag{2.3.20}$$

Actually, any function of $\bar{J}$ may be chosen as the adiabatic invariant.

The Small Denominators. Where are the singularities that prevented convergence of the series in Section 2.2b? They are most easily seen if we consider the y's to be in action-angle form to first order in ϵ: $y = (J_y, \theta_y)$. In this case, and not dropping the terms (2.3.11) and $\partial S_1/\partial t$, (2.3.18) is replaced by

$$\omega \frac{\partial S_1}{\partial \bar{\theta}} + \epsilon \omega_y \cdot \frac{\partial S_1}{\partial (\epsilon \bar{\theta}_y)} + \epsilon \frac{\partial S_1}{\partial (\epsilon t)} = -\{H_1\}_{\bar{\theta}}. \tag{2.3.21}$$

Since S_1 and $\{H_1\}_{\bar{\theta}}$ are periodic in the θ's and Ωt, we expand in Fourier series to find

$$S_1 = i \sum_{\substack{n=(k,l,m) \\ k \neq 0}} \frac{H_{1n}(\bar{J}, \bar{J}_y)}{k\omega + \epsilon \boldsymbol{m} \cdot \boldsymbol{\omega}_y + \epsilon l \Omega} \exp[i(k\bar{\theta} + \boldsymbol{m} \cdot \epsilon \bar{\boldsymbol{\theta}}_y + l\epsilon t)]. \tag{2.3.22}$$

2.3. Adiabatic Invariance

We see that high-order resonances (m, l large) between the slow variables and the fast oscillation of $\bar\theta$ give rise to resonant denominators. Sufficiently near these resonances, neglect of the order ϵ terms in (2.3.21) is incorrect. We should not be surprised to find that the adiabatic series, which neglects these resonant effects, is asymptotic, therefore formally divergent and only valid over times less than or of the order of the slow time scales.

The adiabatic expansion given above can be carried to higher order. In each order, an equation for S_n of the type (2.3.18) for S_1 must be solved. Resonant denominators never appear, their effects continually being pushed into higher order as the expansion is continued. Higher-order expressions for the adiabatic invariant will be presented in Section 2.5.

A Hierarchy of Invariants. Construction of an adiabatic invariant, if it actually exists, effectively reduces the system from N to $N - 1$ degrees of freedom within the limit of the adiabatic approximation. This follows because the transformed Hamiltonian, given as an asymptotic series in ϵ,

$$\bar H = \bar H(\bar J, \epsilon \bar y, \epsilon t, \epsilon),$$

is independent of $\bar\theta$; hence, $\bar J$ is a constant. If one of the remaining degrees of freedom undergoes an oscillation that is fast compared to the other degrees of freedom, then we can introduce a second small parameter ϵ_2, transform to action-angle variables of the fast motion for the unperturbed ($\epsilon_2 \equiv 0$) system and find a second adiabatic invariant. This process may be continued to obtain a hierarchy of adiabatic invariants, until the system is reduced to one degree of freedom, which can be integrated to obtain the final invariant. The process is well known in plasma physics where, for a charged particle gyrating in a magnetic mirror field, we first find the magnetic moment invariant μ associated with the fast gyration, then find the longitudinal invariant $J_\|$ associated with the slower bounce motion, and finally find the flux invariant Φ associated with the drift motion. The three degrees of freedom are shown in Fig. 2.7 (see Hastie et al., 1967, for a more complete discussion). The three small parameters in this case are ϵ, the ratio of bounce frequency to gyration frequency; ϵ_2, the ratio of guiding center drift frequency to bounce frequency; and ϵ_3, the ratio of the frequency of the time-varying magnetic field to the drift frequency.

The hierarchy of invariants is subject to the limitations of the adiabatic theory, in which resonances may modify or destroy the invariants. The modification and destruction of the adiabatic invariants for a gyrating particle in a static magnetic mirror field has been studied by Chirikov (1960, 1979). Similar studies for a mirror-confined particle interacting with an oscillatory electric field were performed by Jaeger et al. (1972) and Lieberman and Lichtenberg (1972). Even in the absence of overlap of the primary resonances, for more than two degrees of freedom, particles can diffuse by Arnold diffusion, which is the main topic of Chapter 6. However, it is shown there that, in the limit $\epsilon \to 0$, both the Arnold diffusion rate and the fractional

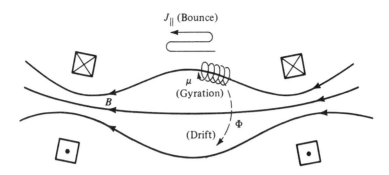

Figure 2.7. A hierarchy of adiabatic invariants for the charged particle gyrating in a nonaxisymmetric, magnetic mirror field. The three adiabatic invariants are the magnetic moment μ, the longitudinal invariant $J_\parallel$, and the guiding center flux invariant Φ. The latter is superfluous in a static magnetic field, since the energy is strictly conserved.

volume of the Arnold web in phase space tend to zero. Then for practical applications, the hierarchy of adiabatic invariants may represent the actual motion to a very good approximation. The adiabatic theory has been and will continue to be a fruitful approach to understanding the motion in dynamical systems.

*2.3c. Slowly Varying Harmonic Oscillator

To illustrate the general method and the effect of resonances, we now calculate, to first order, the adiabatic invariant for the slowly varying linear oscillator

$$H_l = \tfrac{1}{2}G(\tau)p^2 + \tfrac{1}{2}F(\tau)q^2, \tag{2.3.23}$$

where the small parameter has been inserted using $\tau = \epsilon t$ to order the perturbation series. To prepare the system, we transform to action-angle variables J, θ of $H_{l0} = H_l$ ($\epsilon = 0$). In treating the harmonic oscillator, we use the generating function $F_1(q, \theta, \tau)$ given by

$$F_1 = \tfrac{1}{2}Rq^2 \cot\theta, \tag{2.3.24}$$

where $R(\tau) = (F/G)^{1/2}$. From (1.2.11), we obtain the transformation equations (1.2.68) and the transformed Hamiltonian

$$H = \omega_0 J + \epsilon \frac{1}{2}\frac{R'}{R} J \sin 2\theta, \tag{2.3.25}$$

where $\omega_0(\tau) = (FG)^{1/2}$, and the prime denotes differentiation with respect to τ. The system is now in the form (2.3.6) and is prepared for the Poincaré-Von

2.3. Adiabatic Invariance

Zeipel expansion. To zero order, the adiabatic invariant is just

$$J = \frac{H_0}{\omega_0} = \text{const.}, \tag{2.3.26}$$

i.e., the number of quanta $\hbar\omega_0$ are conserved as the frequency of oscillation slowly varies (Einstein, 1911). To find the first-order invariant, we apply (2.3.20) to (2.3.25) to obtain at once

$$\bar{J} = J(1 + \epsilon P \sin 2\theta) = \text{const.}, \tag{2.3.27}$$

with $P(\epsilon t) = (R'/2\omega_0 R)$, showing that to first order, J has a small component oscillating with twice the frequency of the fast variable. We can verify the constancy of $\bar{J}$ by taking the time derivative (dot $\equiv d/dt$) of (2.3.27),

$$\dot{\bar{J}} = \dot{J} + \epsilon \dot{P} J \sin 2\theta + 2\epsilon P J \cos 2\theta + \mathcal{O}(\epsilon^2). \tag{2.3.28}$$

From Hamilton's equations applied to (2.3.25), the first and third terms on the right cancel, leaving to first order in ϵ,

$$\dot{\bar{J}} = \epsilon \dot{P} J \sin 2\theta. \tag{2.3.29}$$

If the standard slow perturbation ordering is used, i.e.,

$$\dot{P} \sim \epsilon P,$$

then $\dot{\bar{J}}$ is of order ϵ^2; hence, $\bar{J}$ is a first-order invariant.

Passage Through Resonance. Now consider the possibility of secular changes in the adiabatic invariant due to resonances between the oscillation and slow periodic changes in the oscillator parameters. Expanding $\dot{P}$ in a Fourier series, we obtain

$$\dot{P} = \epsilon \sum_{n \neq 0} a_n \exp(in\omega_1 \epsilon t), \tag{2.3.30}$$

where $\epsilon\omega_1$ is the frequency of the slow oscillation and ω_1/ω_0 is of order unity. Using $\dot{P}$ in (2.3.29), we obtain

$$\dot{\bar{J}} = \frac{\epsilon^2}{2i} \bar{J} \sum_{n \neq 0} a_n \{\exp[i(n\omega_1 \epsilon t + 2\theta)] - \exp[i(n\omega_1 \epsilon t - 2\theta)]\}, \tag{2.3.31}$$

where we have substituted $\bar{J}$ for J to first order in ϵ. Integrating (2.3.31) over a period of the slow oscillation yields $\Delta \bar{J}/\bar{J} \sim \epsilon^2$, unless there is a commensurability between the oscillations in t and θ:

$$\frac{\omega_0}{\epsilon\omega_1} = \frac{s}{2}, \tag{2.3.32}$$

where s is an integer of order ϵ^{-1}. In this case, the $n = \pm s$ terms in the sum will be constants in t, and the integral over the slow period then yields

$$\frac{\Delta \bar{J}}{\bar{J}} \sim \epsilon^2 |a_s| \cdot \frac{2\pi\epsilon^{-1}}{\omega_1}, \qquad (2.3.33)$$

or $\Delta \bar{J}/\bar{J} \sim \epsilon$. Thus if the resonance is maintained for times $\sim 2\pi\epsilon^{-1}/\omega_0$, the first-order invariant is destroyed implying a strong violation of adiabaticity. This does not, however, imply that an invariant of the motion does not exist, only that it does not take the form described above. In fact, we have seen in Section 1.3b that a linear oscillator (2.3.23) with periodic time-varying coefficients is integrable, implying the existence of an invariant. If the oscillator is nonlinear, on the other hand, the invariant may exist, may be topologically modified, or may be completely destroyed. Although the nonlinear oscillator is of more interest, the linear one has been used here to illustrate the expansion techniques and their limitations.

2.3d. Noncanonical Methods

There are certain types of problems that cannot be analyzed using the classical methods of the previous subsections. Foremost among these is the "guiding center problem," alluded to in the previous subsection, for the motion of a charged particle in a magnetic field (see Fig. 2.7). The Hamiltonian is given by

$$H = \frac{1}{2m}|\mathbf{p} - \frac{e}{c} \mathbf{A}(\mathbf{q}, t)|^2, \qquad (2.3.34)$$

where

$$\mathbf{p} = m\mathbf{v} + \frac{e}{c} \mathbf{A}(\mathbf{q}, t) \qquad (2.3.35)$$

is the canonical momentum and $\mathbf{A}$ is the vector potential related to the field by $\mathbf{B} = \nabla \times \mathbf{A}$. The motion of a charged particle in a uniform magnetic field $\mathbf{B}_0$ has the well-known integrable solution consisting of gyration perpendicular to $\mathbf{B}_0$ and uniform drift along $\mathbf{B}_0$. A small nonuniformity in $\mathbf{B}$ can markedly change the motion, introducing both "bounce" oscillations along the field and drifts across the field, as described in Fig. 2.7. Because the gyrofrequency is clearly fast compared to bounce or drift frequency, for a small perturbation from a uniform field, a small parameter is introduced into the equations by taking $m/e = \mathcal{O}(\epsilon)$, which ensures that the gyrofrequency $\Omega = (e/mc)B = \mathcal{O}(\epsilon^{-1})$ is large. However, in the Hamiltonian formulation given by (2.3.34), the leading terms in both $\mathbf{p}$ and $\mathbf{A}$ are $(e/mc)B = \mathcal{O}(\epsilon^{-1})$ and nearly cancel, thus not clearly giving a proper ordering in ϵ. This difficulty is not inherent to the guiding center approach, but only to its

2.3. Adiabatic Invariance

Hamiltonian description. In fact, it was this and similar problems that motivated Krylov and Bogoliubov (1936), Bogoliubov and Zubarev (1955), Bogoliubov and Mitropolsky (1961), and Kruskal (1962) to formulate an adiabatic perturbation theory for a system of differential equations, not necessarily derivable from a Hamiltonian or, if so derivable, then not necessarily written in terms of canonical variables. As in the canonical case, the methods are based on the existence of a fast variable in a single degree of freedom, with slow variables in the remaining degrees of freedom, and involve averaging over the fast time scale. They are often called *multiple-scale averaging* methods. While effective and very general, they are inherently messy to use, especially as we proceed to higher order. In a canonical formulation, the differential equations are derivable from a scalar function H, and coordinate transformations are similarly specified by a scalar, namely, a generating function. The multiple scale averaging methods lack both of these simplifying features.

In this subsection, we first outline Kruskal's method, which we then illustrate by calculating the first-order invariant of the slowly varying harmonic oscillator. The work of Kruskal, rather than Krylov and Bogoliubov, is chosen because Kruskal shows how the noncanonical perturbation solutions relate to action integrals and adiabatic invariants when the system of differential equations can be derived from a Hamiltonian.

We draw attention to a perturbation method due to Littlejohn (1979, 1983) in which noncanonical coordinates are used, but where the differential equations and the coordinate transformations are, as in the Poincaré–Von Zeipel method, specified by scalars. Littlejohn's method bridges the gap between the classical canonical theory and the multiple-scale averaging methods of Krylov and Kruskal. The reader should consult his paper for further details.

Kruskal's Averaging Method. Following Kruskal, we start with the autonomous system of differential equations

$$\dot{x} = f(x, \epsilon) \tag{2.3.36}$$

having the property that, for $\epsilon = 0$ ($f = f_0(x)$), the trajectory $x(t)$ traces out a closed curve, or loop, as t progresses. To prepare the system for perturbation theory, we introduce a transformation to separate the slow and fast motion. If x is an N-component vector, then there must be $(N-1)$ "slow" variables $y = y(x)$ describing the motion of the loop that satisfy

$$(f_0 \cdot \nabla) y = 0, \tag{2.3.37}$$

with the remaining "fast" variable $\theta = \theta(x)$, which is periodic in zero order and measures the position along the loop. In terms of the new variables, equations (2.3.36) have the form

$$\dot{y} = \epsilon g(y, \theta), \qquad \dot{\theta} = \omega(y, \theta), \tag{2.3.38}$$

where θ represents the rapidly varying phase of a single fast oscillator and is associated with one of the y_i, which in canonical form would be the associated action. As in canonical perturbation methods, the preparatory transformation may itself be found as an expansion in terms of a small parameter.

The multiple time-scale averaging procedure is to find a new set of "nice" variables z and ϕ such that the time evolution of the nice variables is not dependent on ϕ:

$$\dot{z} = \epsilon h(z), \qquad \dot{\phi} = \Omega(z), \qquad (2.3.39)$$

and where z, ϕ, h, and Ω can be found independently to each order ϵ. Thus we need four equations that relate the nth-order quantities to the same quantities of $(n-1)$ order; since the zero-order quantities can be determined, the complete solution is therefore available by induction. In order to obtain these relations, we express the total derivatives in (2.3.39) in terms of the variables y and θ by use of (2.3.38) and the transformation equations

$$\nabla_y z \cdot \epsilon g + \left(\frac{\partial z}{\partial \theta}\right)\omega = \epsilon h(z),$$

$$\nabla_y \phi \cdot \epsilon g + \left(\frac{\partial \phi}{\partial \theta}\right)\omega = \Omega(z). \qquad (2.3.40)$$

In addition we require z and ϕ to be periodic, that is,

$$z(y, \theta + 2\pi) = z(y, \theta),$$
$$\phi(y, \theta + 2\pi) = \phi(y, \theta) + 2\pi, \qquad (2.3.41)$$

where ϕ is an angle-like variable. In order to have sufficient conditions to solve for z and ϕ in terms of y and θ, we introduce somewhat arbitrary initial conditions

$$z(y, 0) = y, \qquad \phi(y, 0) = 0. \qquad (2.3.42)$$

Other choices of initial conditions are also possible, but those given here simplify the transformations. We now obtain expressions from which z, ϕ, h, and Ω can be calculated to any order in ϵ. We assume that we have been able to calculate y, θ, g, and ω exactly, although these quantities can also be obtained by expansion. By dividing (2.3.40) by ω, integrating and determining the constants from the initial conditions, we obtain

$$z = y + \epsilon \int_0^\theta (h(z) - \nabla_y z \cdot g) \frac{d\theta}{\omega}, \qquad (2.3.43a)$$

$$\phi = \int_0^\theta (\Omega(z) - \epsilon \nabla_y \phi \cdot g) \frac{d\theta}{\omega}, \qquad (2.3.43b)$$

2.3. Adiabatic Invariance

and from the periodicity conditions,

$$\int_0^{2\pi} (h(z) - \nabla_y z \cdot g) \frac{d\theta}{\omega} = 0, \qquad (2.3.44a)$$

$$\int_0^{2\pi} (\Omega(z) - \epsilon \nabla_y \phi \cdot g) \frac{d\theta}{\omega} = 2\pi. \qquad (2.3.44b)$$

Equations (2.3.43) and (2.3.44) are four equations that can be used to solve for the unknown variables z, ϕ, h, and Ω to any given order in ϵ. They correspond, in the canonical theory, to the separation of Eq. (2.3.14) for the first-order generating function S_1 into average and oscillating parts (2.3.17) and (2.3.18), respectively.

The next step is to expand z and ϕ in powers of ϵ, e.g.,

$$z = \sum_n \epsilon^n z_n, \qquad (2.3.45)$$

and h and Ω are also expanded, taking the form, e.g.,

$$\epsilon h(z) = \epsilon h_1(z_0) + \epsilon^2 \left[h_2(z_0) + z_1 \cdot \frac{\partial}{\partial z} h_1(z_0) \right] + \cdots . \qquad (2.3.46)$$

To zero order in ϵ, we have $z_0 = y$. The first-order variables are obtained from (2.3.43) and (2.3.44) with $z = y$ and $h = h_1$ as follows. From (2.3.43a),

$$z_1 = \int_0^\theta (h_1(z) - \nabla_y y \cdot g) \frac{d\theta}{\omega}, \qquad (2.3.47)$$

subject to periodicity condition (2.3.44a):

$$\int_0^{2\pi} (h_1(z) - \nabla_y y \cdot g) \frac{d\theta}{\omega} = 0. \qquad (2.3.48)$$

Using $\nabla_y y \cdot g = g$, and defining the average part of g by

$$\langle g \rangle = \int_0^{2\pi} g \frac{d\theta}{\omega} \bigg/ \int_0^{2\pi} \frac{d\theta}{\omega} \qquad (2.3.49)$$

and the integral of the oscillating part of g by

$$\tilde{g} = \int_0^\theta (g - \langle g \rangle) \frac{d\theta}{\omega}, \qquad (2.3.50)$$

then from (2.3.48),

$$\langle h_1(z) \rangle = \langle g \rangle, \qquad (2.3.51)$$

and from (2.3.47),

$$z_1 = -\tilde{g}. \qquad (2.3.52)$$

A similar calculation determines ϕ to first order by use of (2.3.43b) and (2.3.44b). We note that the physical significance of the transformation is that at each order we obtain a variable whose average part has been subtracted, so that an integration over the angle variable does not produce secular terms. The periodic variables, obtained through the transformation, may be very general in nature. For an oscillator of one degree of freedom with a slowly varying restoring force (slowly varying frequency), y may be the system Hamiltonian. For a time-independent Hamiltonian with more than one degree of freedom, y may represent the vector action. However, the formal results by Kruskal require the construction of the invariant in the usual way:

$$J = \int_0^{2\pi} \sum_{i=1}^{N} p_i \frac{dq_i}{d\phi} d\phi, \qquad (2.3.53)$$

where the p_i and q_i can be the components of z or, more generally, functions of z. Kruskal demonstrates the canonical nature of J and ϕ by showing that the Poisson bracket

$$[\phi, J] = 1. \qquad (2.3.54)$$

The actual proof with J given by (2.3.53) is rather involved, but we should already suspect from the form of (2.3.53) that J is the proper action variable.

Slowly Varying Oscillator. We illustrate Kruskal's method with the time-varying harmonic oscillator. Although the method was developed for application to multidimensional systems, the one-dimensional oscillator illustrates its basic aspects. We have

$$\frac{dp}{dt} = -F(\epsilon t)q, \qquad \frac{dq}{dt} = Gp, \qquad (2.3.55)$$

where for convenience we have taken $G = $ const. To put this in the standard form of an autonomous set of first-order equations, we introduce a new variable $\tau = \epsilon t$, giving (dot $\equiv d/dt$)

$$\dot{p} = -F(\tau)q, \qquad \dot{q} = Gp, \qquad \dot{\tau} = \epsilon, \qquad (2.3.56)$$

so that to lowest order in ϵ, we have $\tau = $ const. and thus $F = $ const. We choose for the vector variable y [note that (,) denotes the components of a two-vector]:

$$y = (H, \tau), \qquad (2.3.57)$$

where H is the Hamiltonian

$$H = \frac{F}{2} q^2 + \frac{G}{2} p^2, \qquad (2.3.58)$$

2.3. Adiabatic Invariance

which is constant to zero order. H and τ are, to within a sign, the canonical coordinates in an extended phase space. The zero-order solution is just the usual harmonic oscillator

$$q = q_0 \sin(\omega_0 t + \beta),$$
$$p = Rq_0 \cos(\omega_0 t + \beta),$$
(2.3.59)

where $R = (F/G)^{1/2}$ and $\omega_0 = (FG)^{1/2}$ as before. Defining the angle variable $\theta = (\omega_0 t + \beta)$, we obtain from (2.3.59)

$$\theta = \tan^{-1}\left(\frac{Rq}{p}\right). \quad (2.3.60)$$

We wish to find a transformation to a new set of "nice" variables, independent of the fast phase angle. We therefore must apply the transformations, order by order, as given by (2.3.43) and (2.3.44). In order to do this, we see that it is necessary obtain expressions for the derivatives $\dot{H}$ and $\dot{\theta}$. By differentiating (2.3.58) and substituting from (2.3.56), we obtain

$$\dot{H} = \frac{\epsilon}{2} F' q^2, \quad (2.3.61)$$

where F' is the derivative with respect to τ. From (2.3.58) and (2.3.60), we can solve for q and p in terms of the variables H and θ, obtaining

$$q^2 = \left(\frac{2H}{F}\right)\sin^2\theta, \quad p^2 = \left(\frac{2H}{G}\right)\cos^2\theta; \quad (2.3.62)$$

substituting for q^2 in the expression for $\dot{H}$, we find

$$\dot{H} = 2\epsilon\left(\frac{R'}{R}\right) H \sin^2\theta, \quad (2.3.63)$$

where we have used the relation $R'/R = F'/2F$. In terms of the definitions of the previous section, we have

$$\dot{y} \equiv (\dot{H}, \dot{\tau}) = \epsilon g(H, \tau, \theta). \quad (2.3.64)$$

Similarly, we obtain the derivative of the angle variable by differentiating (2.3.60) and substituting from (2.3.56):

$$\dot{\theta} = \frac{1}{1 + (Rq/p)^2}\left\{\epsilon\frac{R'}{R}\frac{Rq}{p} + \omega_0\left[1 + \left(\frac{Rq}{p}\right)^2\right]\right\}. \quad (2.3.65)$$

Substituting for Rq/p from (2.3.60) with the proper trigonometric identities,

$$\dot{\theta} = \omega_0 + \epsilon(R'/2R)\sin 2\theta \equiv \omega(H, \tau, \theta). \quad (2.3.66)$$

Equations (2.3.63) and (2.3.66) can now be used to obtain the "nice" variables z and ϕ. Using (2.3.63), (2.3.64), and (2.3.66) in (2.3.49), we obtain to first order

$$\langle g \rangle = \left(\frac{R'H}{R}, 1 \right) \tag{2.3.67}$$

where $h_1 = \langle g \rangle$ from (2.3.51). Setting $z = (\bar{H}, \tau)$, we find by evaluating the integral in (2.3.50) and using (2.3.51),

$$\bar{H} = H \left[1 + \epsilon \left(\frac{R'}{2\omega_0 R} \right) \sin 2\theta \right]. \tag{2.3.68}$$

A similar calculation yields trivially $\bar{\tau} = \tau$, which with (2.3.68) gives the first-order nice variables so that $\dot{z}$ is independent of ϕ to first order, as required. Similarly we can construct the transformation to obtain the angle variable ϕ.

Once we have obtained the nice variables, we must express the constant of the motion in terms of these variables. We have

$$\bar{J} = \frac{1}{2\pi} \oint p \, dq = \frac{1}{2\pi} \int_0^{2\pi} p \frac{dq}{d\phi} d\phi, \tag{2.3.69}$$

where p and q are functions of $\bar{H}$ and ϕ. To first order in ϵ, it is sufficient here to carry out the integration in terms of the variable θ rather than ϕ, thereby simplifying the transformations. Northrop et al. (1966) have used this simplification in their work. Substituting for p and q from (2.3.62) and taking the derivative of q, we obtain

$$\bar{J} = \frac{1}{2\pi} \int_0^{2\pi} \frac{2H}{\omega_0} \cos^2 \theta \, d\theta,$$

and substituting for $\bar{H}$ from (2.3.68), we have to order ϵ,

$$\bar{J} = \frac{1}{2\pi} \int_0^{2\pi} \frac{2\bar{H}}{\omega_0} \left(1 - \frac{\epsilon}{2\omega_0} \frac{R'}{R} \sin 2\theta \right) \cos^2 \theta \, d\theta.$$

Performing the integration, the second term integrates to zero, and we have

$$\bar{J} = \frac{\bar{H}}{\omega_0} \tag{2.3.70}$$

as the adiabatic constant. This expression is identical with (2.3.27), derived from canonical perturbation theory. For computational purposes, we reintroduce H through the transformation in (2.3.68). If H is evaluated in a region in which the parameters are not varying, $R' = 0$ and we obtain the usual adiabatic result that

$$\frac{H}{\omega_0} = \text{const.} \tag{2.3.71}$$

We see that (2.3.71) is only correct to zero order, in a region in which the parameters are varying with time.

2.4. Secular Perturbation Theory

Near a resonance in the unperturbed Hamiltonian, a resonant denominator appears in the first-order adiabatic invariant calculated by the standard methods of Section 2.3. The resonant variables can be eliminated from the unperturbed Hamiltonian by a canonical transformation to a frame of reference that rotates with the resonant frequency. The new coordinates then measure the slow oscillation of the variables about their values at resonance, which is an elliptic fixed point of the new phase plane. This technique was early applied to the nonlinear theory of accelerator dynamics (see Laslett, 1967, for treatment and additional references) and to electron cyclotron resonance in a mirror field (Seidl, 1964). The procedure is also similar to that used by Chirikov (1960). The techniques of Section 2.3 are then used to average over the rapidly rotating phase after the resonance has been removed. In the following, we assume an autonomous Hamiltonian in two degrees of freedom. The extension to nonautonomous systems is easily accomplished by introducing an extended phase space (see Section 1.2).

If the perturbation is large enough, then secondary resonances appear that may in turn modify or destroy the modified invariant calculated in Section 2.4a. The secularity of secondary resonances may be removed by a procedure analogous to the removal of the secularity of the original resonance; this removal is described in Section 2.4b. The way that secondary resonances destroy the modified invariant, mirroring the way that the original resonances destroy the "unmodified" invariant, forms the basis of methods of calculating the transition from regular to chaotic behavior in Hamiltonian systems, as will be shown in Chapter 4.

The above ideas are illustrated in Section 2.4c using the example of wave-particle resonance that was introduced in Section 2.2b. Near resonance between the particle gyration and the doppler-shifted wave frequency, the modified invariant is calculated, and the effect of secondary resonances is shown, along with numerical examples illustrating the fundamental effects.

Transformation to a rotating coordinate system is not the only way to effect a topological modification of an adiabatic invariant near resonances. There is a certain freedom in choosing the form of the invariant, since if J is an invariant of the unperturbed system, then $I(J)$ is also. By choosing

$dI/dJ = 0$ near resonant values of J, a change of topology can be accommodated when the system is perturbed. This method, developed by Dunnett et al. (1968) is described in Section 2.4d and again illustrated using the example of wave-particle resonance.

*2.4a. Removal of Resonances

We assume a Hamiltonian of the form

$$H = H_0(\boldsymbol{J}) + \epsilon H_1(\boldsymbol{J}, \boldsymbol{\theta}), \qquad (2.4.1)$$

where H_0 is solvable in action-angle variables and H_1 is periodic in $\boldsymbol{\theta}$:

$$H_1 = \sum_{l,m} H_{l,m}(\boldsymbol{J}) \exp(i\boldsymbol{n} \cdot \boldsymbol{\theta}), \qquad (2.4.2)$$

where $\boldsymbol{n} = (l, m)$ is an integer vector. If a resonance exists between the unperturbed frequencies

$$\frac{\omega_2}{\omega_1} = \frac{r}{s}, \quad r, s \text{ integers}, \qquad (2.4.3)$$

where

$$\omega_1(\boldsymbol{J}) = \frac{\partial H_0}{\partial J_1}, \quad \omega_2(\boldsymbol{J}) = \frac{\partial H_0}{\partial J_2}, \qquad (2.4.4)$$

then an attempt to solve the motion by the perturbation theory of Sections 2.2 and 2.3 leads to secular growth of the solution. We shall take (2.4.3) to represent either a primary resonance in the system or a secondary resonance created by harmonic frequencies of an island oscillation generated by the primary resonance. In either case, the secularity can be removed by applying a transformation that eliminates one of the original actions J_1 or J_2. We choose the generating function

$$F_2 = (r\theta_1 - s\theta_2)\hat{J}_1 + \theta_2 \hat{J}_2, \qquad (2.4.5)$$

which defines a canonical transformation from $\boldsymbol{J}, \boldsymbol{\theta}$ to $\hat{\boldsymbol{J}}, \hat{\boldsymbol{\theta}}$, such that

$$J_1 = \frac{\partial F_2}{\partial \theta_1} = r\hat{J}_1, \qquad (2.4.6a)$$

$$J_2 = \frac{\partial F_2}{\partial \theta_2} = \hat{J}_2 - s\hat{J}_1, \qquad (2.4.6b)$$

$$\hat{\theta}_1 = \frac{\partial F_2}{\partial \hat{J}_1} = r\theta_1 - s\theta_2, \qquad (2.4.6c)$$

2.4. Secular Perturbation Theory

$$\hat{\theta}_2 = \frac{\partial F_2}{\partial \hat{J}_2} = \theta_2. \tag{2.4.6d}$$

These coordinates put the observer in a rotating frame in which the rate of change of the new variable

$$\dot{\hat{\theta}}_1 = r\dot{\theta}_1 - s\dot{\theta}_2 \tag{2.4.7}$$

measures the slow deviation from resonance.

In removing a resonance by use of a generating function as in (2.4.5), there is an arbitrary choice of which of the original phase variables to leave unchanged. We assume here that $\dot{\theta}_2$ is the slower of the two frequencies, and we therefore leave $\hat{\theta}_2 = \theta_2$ unchanged, such that in averaging the Hamiltonian over the fast phase after the transformation, the average is taken over the slower of the original variables. This choice is convenient if higher-order resonances are to be removed, in order to retain the lowest harmonic of the second-order interaction.[1]

Applying (2.4.6) to the Hamiltonian (2.4.1), we obtain from (1.2.13c)

$$\hat{H} = \hat{H}_0(\hat{J}) + \epsilon \hat{H}_1(\hat{J}, \hat{\theta}), \tag{2.4.8}$$

where

$$\hat{H}_1 = \sum_{l,m} H_{l,m}(\hat{J}) \exp\left\{\frac{i}{r}[l\hat{\theta}_1 + (ls + mr)\hat{\theta}_2]\right\}. \tag{2.4.9}$$

Now proceeding as in Section 2.3, we can "average" over $\hat{\theta}_2$, as in (2.3.17), to obtain the transformed Hamiltonian to first-order

$$\bar{H} = \bar{H}_0(\hat{J}) + \epsilon \bar{H}_1(\hat{J}, \hat{\theta}_1), \tag{2.4.10}$$

where

$$\bar{H}_0 = \hat{H}_0(\hat{J}) \tag{2.4.11}$$

and

$$\bar{H}_1 = \langle \hat{H}_1(\hat{J}, \hat{\theta}) \rangle_{\hat{\theta}_2} = \sum_{p=-\infty}^{\infty} H_{-pr,ps}(\hat{J}) \exp(-ip\hat{\theta}_1). \tag{2.4.12}$$

The averaging is valid near the resonance, where $\dot{\hat{\theta}}_2 \gg \dot{\hat{\theta}}_1$. Since $\bar{H}$ is independent of $\hat{\theta}_2$, we have the result that

$$\hat{J}_2 = \hat{J}_{20} = \text{const.} \tag{2.4.13}$$

This is the first term of the series for the adiabatic invariant of Hamiltonian (2.4.8). We see from (2.4.6b) that $\hat{J}_2$ represents a combined invariant for the

---

[1] The final result does not depend on the choice of $\hat{\theta}_2$. One can take any combination of θ_1 and θ_2 that is linearly independent of $\hat{\theta}_1$.

system, namely,

$$\hat{J}_2 = J_2 + \frac{s}{r} J_1 = \text{const.} \quad (2.4.14)$$

The effect of the rotating coordinates is to exhibit explicitly the modified invariant of the system near resonance. However, for a high-order resonance in which $s \gg r$, the modified invariant $\hat{J}_2$ is just a constant multiple of the unmodified invariant J_1. Hence, the only resonances of importance for modifying the invariants are those with low harmonic numbers s.

With $\hat{J}_2$ a constant, the $\hat{J}_1 - \hat{\theta}_1$ motion given by Hamiltonian (2.4.10) is effectively the motion in a single degree of freedom, and therefore integrable. A stationary point or set of points $\hat{J}_{10}, \hat{\theta}_{10}$ exists in the $\hat{J}_1 - \hat{\theta}_1$ phase plane at

$$\frac{\partial \bar{H}}{\partial \hat{J}_1} = 0, \quad \frac{\partial \bar{H}}{\partial \hat{\theta}_1} = 0, \quad (2.4.15)$$

which represent periodic solutions for the perturbed Hamiltonian. The periodic solutions for the unperturbed Hamiltonian, found at a particular J given by (2.4.3), are degenerate in θ, existing for all θ. Under perturbation, the degeneracy is removed, leaving only the periodic solutions given by (2.4.15) (see also discussion of the Poincaré–Birkhoff theorem in Section 3.2b).

The Fourier amplitudes $H_{-pr, ps}$ generally fall off rapidly as p increases. Thus, to a good approximation, we can describe the integrable $\hat{J}_1 - \hat{\theta}_1$ motion using only the $p = 0, \pm 1$ terms:

$$\bar{H} = \hat{H}_0(\hat{J}) + \epsilon H_{0,0}(\hat{J}) + 2\epsilon H_{r, -s}(\hat{J}) \cos \hat{\theta}_1, \quad (2.4.16)$$

where without loss of generality we have taken $H_{-r,s} = H_{r,-s}$, which can always be accomplished by a trivial change of variables $\hat{\theta}_1 \to \hat{\theta}_1 + \text{const.}$

Applying (2.4.15) to (2.4.16), we find the location of the fixed points:

$$\frac{\partial \hat{H}_0}{\partial \hat{J}_{10}} + \epsilon \frac{\partial H_{0,0}}{\partial \hat{J}_{10}} + 2\epsilon \frac{\partial H_{r,-s}}{\partial \hat{J}_{10}} \cos \hat{\theta}_{10} = 0, \quad (2.4.17)$$

$$-2\epsilon H_{r,-s} \sin \hat{\theta}_{10} = 0. \quad (2.4.18)$$

From (2.4.18) we have two fixed points located at $\hat{\theta}_{10} = 0, \pi$. From (2.4.17) and for the special case of exact resonance $\hat{J}_{10}$ is obtained, to zero order in ϵ, from

$$\frac{\partial \hat{H}_0}{\partial \hat{J}_{10}} = r \frac{\partial H_0}{\partial J_1} - s \frac{\partial H_0}{\partial J_2} = r\omega_1 - s\omega_2 = 0, \quad (2.4.19)$$

To first order $\hat{J}_{10}$ is obtained from

$$\frac{\partial^2 \hat{H}_0}{\partial \hat{J}_{10}} + \epsilon \frac{\partial H_{0,0}}{\partial \hat{J}_{10}} \pm 2\epsilon \frac{\partial H_{r,-s}}{\partial \hat{J}_{10}} = 0, \quad (2.4.20)$$

2.4. Secular Perturbation Theory

where the positive sign corresponds to $\hat{\theta}_{10} = 0$ and the negative sign to $\hat{\theta}_{10} = \pm \pi$.

We now consider two cases:

(1) If the resonance in the unperturbed Hamiltonian is satisfied only for particular values of J_1 and J_2, then the Hamiltonian H_0 is *accidentally degenerate*. This is the generic case, for which the unperturbed Hamiltonian, transformed to the rotating system, is a function of both transformed actions:

$$\hat{H}_0 = \hat{H}_0(\hat{J}_1, \hat{J}_2). \tag{2.4.21}$$

(2) If the resonance condition is met for all values of J_1 and J_2, then the Hamiltonian H_0 is *intrinsically degenerate*. Evidently

$$H_0 = H_0(sJ_1 + rJ_2) \tag{2.4.22}$$

for intrinsic degeneracy, such that (2.4.3) is satisfied for all J_1 and J_2. Transforming to the rotating system using (2.4.6a) and (2.4.6b), we find

$$\hat{H}_0 = \hat{H}_0(\hat{J}_2); \tag{2.4.23}$$

i.e., H_0 is independent of $\hat{J}_1$ for intrinsic degeneracy. This case often arises as a primary resonance in physical systems of interest. While primary resonances may be accidental or intrinsic, secondary resonances are nearly always accidental due to the complicated way in which the frequencies depend on the actions. The special properties of intrinsically degenerate systems have been considered by Jaeger and Lichtenberg (1972) and Izrailev (1980).

Accidental Degeneracy. From (2.4.16) and using (2.4.21), the excursions in $\hat{J}_1$ and $\hat{\theta}_1$ are

$$\dot{\hat{J}}_1 = \mathcal{O}(\epsilon H_{r,-s}), \qquad \dot{\hat{\theta}}_1 = \mathcal{O}(1), \tag{2.4.24}$$

so we can expand (2.4.16) about the stationary point in $\hat{J}_1$, but not in $\hat{\theta}_1$. Introducing

$$\Delta \hat{J}_1 = \hat{J}_1 - \hat{J}_{10}, \tag{2.4.25}$$

we have

$$\bar{H}_0(\hat{J}) = \hat{H}_0(\hat{J}_0) + \frac{\partial \hat{H}_0}{\partial \hat{J}_{10}} \Delta \hat{J}_1 + \frac{1}{2} \frac{\partial^2 \hat{H}_0}{\partial \hat{J}_{10}^2} (\Delta \hat{J}_1)^2 + \cdots. \tag{2.4.26}$$

We drop the second term on the right in (2.4.26) due to the condition (2.4.17). Ignoring the first (constant) term, inserting (2.4.26) into (2.4.16), and keeping only terms of lowest order in ϵ and in $\Delta \hat{J}_1$, we obtain the Hamiltonian describing the motion near resonance

$$\Delta \bar{H} = \tfrac{1}{2} G (\Delta \hat{J}_1)^2 - F \cos \hat{\theta}_1. \tag{2.4.27}$$

Here, G is the *nonlinearity parameter*

$$G(\hat{J}_0) = \frac{\partial^2 \hat{H}_0}{\partial \hat{J}_{10}^2}, \quad (2.4.28)$$

and F is the product of the perturbation strength and the Fourier mode amplitude:

$$F(\hat{J}_0) = -2\epsilon H_{r,-s}(\hat{J}_0). \quad (2.4.29)$$

This is a remarkable result, which suggests that *the motion near a typical resonance is like that of a pendulum*, complete with libration, separatrix, and rotation motion. The result (2.4.27) has been used by Chirikov (1979) and others to describe the generic motion of Hamiltonian systems near resonance, and it forms the basis for our treatment of chaotic motion near the separatrices associated with these resonances. In some sense (2.4.27) provides a universal description of the motion near all resonances; in recognition of this, we sometimes refer to $\Delta \bar{H}$ as the *standard Hamiltonian*.

The transformation of the original motion under perturbation and near resonance is illustrated in Fig. 2.8a. For $GF > 0$, the stable fixed point is at $\hat{\theta}_1 = 0$, the unstable fixed point at $\pm \pi$. The frequency for the $\hat{J}_1 - \hat{\theta}_1$ libration near the stable point is slow:

$$\hat{\omega}_1 = (FG)^{1/2} = \mathcal{O}[(\epsilon H_{r,-s})^{1/2}], \quad (2.4.30)$$

and this frequency decreases to zero as the separatrix is approached, being always much smaller than the frequency of the $\hat{J}_2 - \hat{\theta}_2$ oscillation, which is of order unity. The maximum excursion $\Delta \hat{J}_{1\,\text{max}}$ is small, occurs on the separatrix, and is given by half the separatrix width (at $\hat{\theta}_1 = 0$)

$$\Delta \hat{J}_{1\,\text{max}} = 2\left(\frac{F}{G}\right)^{1/2} = \mathcal{O}[(\epsilon H_{r,-s})^{1/2}]. \quad (2.4.31)$$

Near the stable fixed point, the phase plane orbits are elliptic, with the ratio of lengths of the semiaxes of the ellipse given by

$$\frac{\Delta \hat{J}_1}{\Delta \hat{\theta}_1} = \left(\frac{F}{G}\right)^{1/2} = \mathcal{O}[(\epsilon H_{r,-s})^{1/2}]. \quad (2.4.32)$$

To complete the formal solution, we should transform to action-angle variables for the slow libration (or rotation) of $\Delta \hat{J}_1$ and $\hat{\theta}_1$. The transformation is generally unnecessary unless a secondary resonance is to be removed; we therefore postpone this step until Section 2.4b.

Intrinsic Degeneracy. Proceeding as for the accidental degeneracy, but noting from (2.4.23) that $\hat{H}_0$ is independent of $\hat{J}_1$, we replace (2.4.24) by

$$\dot{\hat{J}}_1 = \mathcal{O}(\epsilon H_{r,-s}), \quad \dot{\hat{\theta}}_1 = \mathcal{O}(\epsilon H_{0,0}, \epsilon H_{r,-s}), \quad (2.4.33)$$

2.4. Secular Perturbation Theory

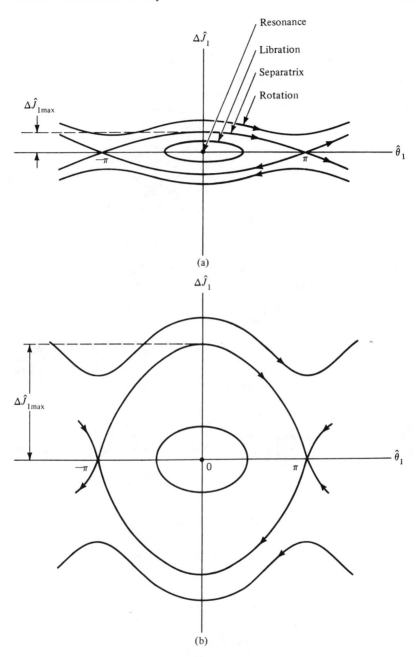

Figure 2.8. Adiabatic motion near an isolated resonance. (a) Accidental degeneracy. The motion is shown in the rotating (hat) coordinate system. The maximum excursion in the action $\hat{J}_1$ and the libration frequency $\hat{\omega}_1$ are both small, of order $\epsilon^{1/2}$. (b) Intrinsic degeneracy. The maximum excursion in action is large, of order unity, while the libration frequency is very slow, of order ϵ.

which shows that the excursions in $\hat{J}_1$ and $\hat{\theta}_1$ are of the same order. We cannot linearize (2.4.16) in $\hat{J}_1$ alone, as we did for the accidentally degenerate case. In the general case, we solve (2.4.16) for the $\hat{J}_1 - \hat{\theta}_1$ motion by transforming to action-angle variables, as described in Section 1.3a. Rather than do this, we examine the general character of the solution by linearizing about the elliptic fixed point $\hat{\theta}_{10} = 0$, writing $\Delta\hat{\theta}_1 = \hat{\theta}_1$, and expanding (2.4.16) in powers of $\Delta\hat{\theta}_1$ and $\hat{J}_1$ to quadratic terms:

$$H_{0,0}(\hat{J}) = H_{0,0}(\hat{J}_0) + \frac{\partial H_{0,0}}{\partial \hat{J}_{10}}(\Delta\hat{J}_1) + \frac{1}{2}\frac{\partial^2 H_{0,0}}{\partial \hat{J}_{10}^2}(\Delta\hat{J}_1)^2 + \cdots, \quad (2.4.34)$$

$$H_{r,-s}(\hat{J}) = H_{r,-s}(\hat{J}_0) + \frac{\partial H_{r,-s}}{\partial \hat{J}_{10}}(\Delta\hat{J}_1) + \frac{1}{2}\frac{\partial^2 H_{r,-s}}{\partial \hat{J}_{10}^2}(\Delta\hat{J}_1)^2 + \cdots, \quad (2.4.35)$$

$$\cos\hat{\theta}_1 = 1 - \tfrac{1}{2}(\Delta\hat{\theta}_1)^2 + \cdots. \quad (2.4.36)$$

The terms linear in $\Delta\hat{J}_1$ drop out by virtue of (2.4.17). Ignoring the constant terms, we have the harmonic oscillator Hamiltonian,

$$\Delta\bar{H} = \tfrac{1}{2}G(\Delta\hat{J}_1)^2 + \tfrac{1}{2}F(\Delta\hat{\theta}_1)^2, \quad (2.4.37)$$

where

$$G = \frac{\partial^2 \hat{H}_0}{\partial \hat{J}_{10}^2} + \epsilon\frac{\partial^2 H_{0,0}}{\partial \hat{J}_{10}^2} + 2\epsilon\frac{\partial^2 H_{r,-s}}{\partial \hat{J}_{10}^2}, \quad (2.4.38)$$

$$F = -2\epsilon H_{r,-s}. \quad (2.4.39)$$

For intrinsic degeneracy, the first term in (2.4.38) is zero. In this case, G and F are both of order ϵ. The frequency of the $\hat{J}_1 - \hat{\theta}_1$ oscillation near the elliptic point is

$$\hat{\omega}_1 = (GF)^{1/2} = \mathcal{O}(\epsilon), \quad (2.4.40)$$

and the ratio of the semiaxes of the ellipse is

$$\frac{\Delta\hat{J}_1}{\Delta\hat{\theta}_1} = \left(\frac{F}{G}\right)^{1/2} = \mathcal{O}(1). \quad (2.4.41)$$

The transition from accidental to intrinsic degeneracy near the elliptic fixed point is governed by (2.4.38) with the first term passing to the limit of zero.

A similar linearization near the fixed point $\hat{\theta}_1 = \pm\pi$ yields hyperbolic orbits whose asymptotes are inclined at angles, $\pm\chi$ to the $\hat{\theta}_1$ axis, with

$$\tan\chi = \left(\frac{F}{G}\right)^{1/2}. \quad (2.4.42)$$

Thus, provided $G \neq 0$, the intrinsic and accidental cases have roughly similar behavior. The intrinsic case is illustrated in Fig. 2.8b. Further consideration

2.4. Secular Perturbation Theory

of the distinction between accidental and intrinsic degeneracy, with application to mappings and the KAM theorem, is given in Section 3.2a. A brief discussion of the exceptional case $G = 0$ is also given there. Generally, we find that the weak nonlinearity present for intrinsic degeneracy leads to complicated behavior; accidental degeneracy is usually simpler to understand.

*2.4b. Higher-Order Resonances

If ϵ is not sufficiently small, secondary resonances that are present in Hamiltonian (2.4.9) contribute secular terms that modify or destroy the adiabatic invariant $\hat{J}_2$. These resonances are between harmonics of the $\hat{J}_1 - \hat{\theta}_1$ phase oscillation derived in Section 2.4a and the fundamental frequency ω_2. They give rise, in the adiabatic limit, to island chains, as shown in Fig. 2.9a. They can be removed in a manner analogous to that used in Section 2.4a, although the results have some additional features, which we shall see by explicitly carrying through some of the steps. We start with the average Hamiltonian (2.4.10), which we must first express in action-angle variables of the $\hat{J}_1 - \hat{\theta}_1$ motion. Rather than solving the Hamilton–Jacobi equation (1.2.50) directly, we proceed as in Section 2.2a using perturbation theory for the motion near the elliptic singular point. We let the transformed Hamiltonian be K_0 and the action-angle variables be I_1 and ϕ_1. If (2.4.16) is a good approximation to (2.4.10), then we have immediately from (2.2.23) with I_2 written for $\hat{J}_2$ for notational consistency

$$K_0(I_1, I_2) = \hat{H}_0(\hat{J}_{10}, I_2) + \hat{\omega}_1 I_1 - \frac{\epsilon}{16} G I_1^2 + \cdots, \qquad (2.4.43)$$

where G and $\hat{\omega}_1$ are functions of I_2 given by (2.4.38) and (2.4.40). Equation (2.4.43) is the formal solution in the case that the average over $\hat{\theta}_2$ is valid. It is independent of angles so that $I_2 = \hat{J}_2$ and I_1 are the two constants of the motion. The transformation to action-angle variables is shown in Fig. 2.9b.

To take into account the effect of a secondary resonance in modifying this solution, we reintroduce the terms $\hat{H}'_1$ ignored in the simple average over $\hat{\theta}_2$:

$$\hat{H}'_1(\hat{\boldsymbol{J}}, \hat{\boldsymbol{\theta}}) = \hat{H}_1(\hat{\boldsymbol{J}}, \hat{\boldsymbol{\theta}}) - \bar{H}_1(\hat{\boldsymbol{J}}, \hat{\theta}_1), \qquad (2.4.44)$$

with Fourier expansion

$$\hat{H}'_1 = \sum_{l,m}{}' H_{lm}(\hat{\boldsymbol{J}}) \exp\left[i\frac{l}{r}\hat{\theta}_1 + i\left(l\frac{s}{r} + m\right)\hat{\theta}_2\right], \qquad (2.4.45)$$

where the prime denotes that terms with $ls + mr = 0$ are excluded from the

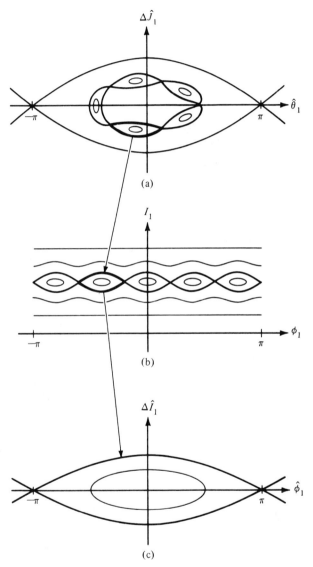

Figure 2.9. Motion near a secondary resonance. (a) The secondary islands in the $\hat{J}_1 - \hat{\theta}_1$ coordinates, which appear when $5\hat{\omega}_1 = \hat{\omega}_2$. (b) Transformation to action-angle variables $I_1 - \phi_1$ of the unperturbed $\hat{J}_1 - \hat{\theta}_1$ libration. The resonant value of the action is obtained from $5\hat{\omega}_1(I_1) = \hat{\omega}_2$. (c) Transformation to a rotating coordinate system associated with the $\hat{I}_1 - \hat{\phi}_1$ secondary librations, obtaining again the phase plane of a pendulum.

2.4. Secular Perturbation Theory

sum. Expanding about the elliptic singularity $\hat{\theta}_{10} = 0$ gives

$$\hat{H}'_1 = \sum_{l,m}{}' H_{lm}(\hat{J}_{10} + \Delta\hat{J}_1, \hat{J}_2) \exp\left[i\frac{l}{r}\Delta\hat{\theta}_1 + i\left(l\frac{s}{r} + m\right)\hat{\theta}_2\right]. \quad (2.4.46)$$

Transforming this perturbation Hamiltonian to lowest-order action-angle variables for the $\Delta\hat{J}_1 - \Delta\hat{\theta}_1$ libration using (1.2.68), and writing ϕ_2 for $\hat{\theta}_2$, we obtain

$$K_1 = \sum_{l,m}{}' H_{lm}(\hat{J}_{10}, I_2) \exp\left[i\left(l\frac{s}{r} + m\right)\phi_2\right] \exp\left[\frac{il}{r}\left(\frac{2I_1}{R}\right)^{1/2} \sin\phi_1\right], \quad (2.4.47)$$

where $R = (F/G)^{1/2}$. We have considered the case of accidental degeneracy, and thus neglected the excursion in $\Delta\hat{J}_1$, assumed to be small by virtue of (2.4.32). Expanding the second exponential

$$K_1 = \sum_{l,m,n}{}' \Gamma_{lmn}(\hat{J}_{10}, I) \exp\left[in\phi_1 + i\left(l\frac{s}{r} + m\right)\phi_2\right], \quad (2.4.48)$$

where

$$\Gamma_{lmn} = \hat{H}_{lm}(\hat{J}_{10}, I_2) \mathcal{J}_n\left[\frac{l}{r}\left(\frac{2I_1}{R}\right)^{1/2}\right], \quad (2.4.49)$$

and $\mathcal{J}_n$ is the Bessel function of order n. From (2.4.48) it is evident that there can be higher-order resonances between ϕ_2 and ϕ_1, such that the average over ϕ_2 is not zero. To obtain the new invariant, we write the total Hamiltonian using (2.4.43) and (2.4.48) as

$$K = K_0(I_1, I_2) + \epsilon_2 K_1(I_1, I_2, \phi_1, \phi_2), \quad (2.4.50)$$

which has a form similar to (2.4.1) with ϵ_2 a new ordering parameter, so that we can use the method of Section 2.4a to remove the secondary resonance. Assuming a resonance

$$\frac{\hat{\omega}_2}{\hat{\omega}_1} = \frac{p}{q}, \quad (2.4.51)$$

where

$$\hat{\omega}_2 = \frac{\partial K_0}{\partial I_2} = \omega_2 = \mathcal{O}(1) \quad (2.4.52)$$

as given by (2.4.4), and

$$\hat{\omega}_1 = \frac{\partial K_0}{\partial I_1} = \mathcal{O}(\epsilon^{1/2}) \quad (2.4.53)$$

as given by (2.4.30), we transform to new variables $\hat{I}_1$, $\hat{I}_2$, $\hat{\phi}_1$, and $\hat{\phi}_2$ as in

(2.4.6), where

$$\hat{\phi}_1 = p\phi_1 - q\phi_2 \qquad (2.4.54)$$

is the slow variable derived from the generating function

$$F_2 = (p\phi_1 - q\phi_2)\hat{I}_1 + \phi_2\hat{I}_2. \qquad (2.4.55)$$

Averaging over the fast phase $\hat{\phi}_2 = \phi_2$ we obtain from (2.4.48) and (2.4.51)

$$nq = -\left(l\frac{s}{r} + m\right)p,$$

where nq, $l(s/r)p$, mp, are integers. This is equivalent to keeping only the terms in the transformed (2.4.45) with

$$n = -jp, \qquad l = kr, \qquad m = jq - ks, \qquad (2.4.56)$$

where j and k are integers. For $j = \pm 1$, the pth harmonic of the island oscillation frequency resonates with the qth harmonic of the oscillation in $\phi_2 = \hat{\theta}_2$, with the higher harmonics given by $|j| > 1$. Performing the average over the Hamiltonian $\hat{K}$ in the rotating system,

$$\bar{K} = \bar{K}_0(\hat{I}_1, \hat{I}_2) + \epsilon_2 \bar{K}_1(\hat{I}_1, \hat{I}_2, \hat{\phi}_1) \qquad (2.4.57)$$

with

$$\bar{K}_0(\hat{I}) = K_0(I_1(\hat{I}), I_2(\hat{I})) \qquad (2.4.58)$$

and

$$\bar{K}_1 = \sum_j K_{-jp, jq} \exp(-ij\hat{\phi}_1), \qquad (2.4.59)$$

where

$$K_{-jp, jq} = \sum_k \Gamma_{kr, jq-ks, -jp} \qquad (2.4.60)$$

is the Fourier coefficient for the jth harmonic of the $\hat{\phi}_1$ oscillation. Since $\bar{K}$ is independent of $\hat{\phi}_2$, we have immediately that

$$\hat{I}_2 = I_2 + \left(\frac{q}{p}\right)I_1 = \text{const.}, \qquad (2.4.61)$$

which is the adiabatic invariant for the island oscillation. We also see that the $\hat{I}_1 - \hat{\phi}_1$ motion is integrable. Comparing (2.4.57) and (2.4.59) with (2.4.10) and (2.4.12), we see that all the results of Section 2.4a apply to the higher-order resonance of this subsection. The transformed motion in the rotating system is shown in Fig. 2.9c.

Island Oscillation Amplitude. To estimate the strength of the secondary resonances, we take the largest term in (2.4.60), which is $|j| = 1$ and $|k| = 1$,

2.4. Secular Perturbation Theory

and we also let $q = 1$, corresponding to resonance with the fundamental of the $\phi_2 = \hat{\theta}_2$ oscillation. From (2.4.49) we see that the largest term in (2.4.60) is proportional to

$$\mathscr{J}_p\left[\left(\frac{2I_1}{R}\right)^{1/2}\right],$$

where from (2.4.51) p is an integer of order $\epsilon^{-1/2}$. Since $I_{1\,\mathrm{max}}$ is of order R, the maximum value of the argument $(2I_1/R)^{1/2}$ is of order unity. Then for p large (ϵ small), we can approximate the Bessel function by the first term in the expansion for small argument as

$$\mathscr{J}_p\left(\sqrt{\frac{2I_1}{R}}\right) \sim \frac{(I_1/2R)^{p/2}}{p!} < \mathcal{O}\left[\frac{1}{(\epsilon^{-1/2})!}\right]. \tag{2.4.62}$$

From (2.4.62) we also see that the amplitude of the interaction term is proportional to $I_1^{p/2}$ such that the island oscillations decrease in size rapidly with decreasing I_1. Replacing $H_{-r,s}$ with $K_{-p,q}$ in (2.4.30) and (2.4.31) and using (2.4.62), we see that the amplitude and frequency of the oscillation in $\hat{I}_1$ will be at least a factor of $\mathcal{O}[1/(\epsilon^{-1/2})!]^{1/2}$ smaller than the amplitude and frequency of the oscillation in $\hat{J}_1$. We call the new $\hat{I}_1$ oscillation an island oscillation because it appears as a chain of islands in the $\hat{J}_1 - \hat{\theta}_1$ phase plane. We explicitly exhibit the second-order island Hamiltonian in the example of Section 2.4c.

Although the procedure for exhibiting the invariant curves of the island oscillation, arising from the second-order resonance, is the same as that used for obtaining the invariant curves of the primary resonance, the results have a somewhat different character. The amplitude of the island resonance depends strongly on ϵ as in (2.4.62). The strength of the primary resonance is related weakly to ϵ, as $\epsilon^{1/2}$. Thus, for relatively small ϵ, island oscillations away from the separatrix rapidly become of negligible importance. For relatively large ϵ, on the other hand, the island oscillation may be as important as the primary resonance in determining the limits of adiabatic invariance.

The rapid reduction in size of higher-order island chains near elliptic singularities manifests itself in the stability of these islands at modest levels of perturbation amplitudes. Thus, even with rather large perturbations, as in the case of the coupled oscillator of Fig. 1.13c, not only do invariants persist in the neighborhood of primary periodic solutions, but also in the neighborhood of second-order periodic solutions, as can be seen from the stable second-order island chain. The consequences of this stability will be explored more fully in subsequent chapters.

This procedure for removing secondary resonances near an elliptic fixed point (Jaeger and Lichtenberg, 1972) is a type of *renormalization method*, in which we develop a transformation from an nth to an $(n + 1)$st order resonance that preserves the form of the Hamiltonian, but where the parameters are transformed in proceeding to higher order. This idea forms the basis of several methods of calculating the transition from regular to chaotic motion. These methods will be developed in Chapter 4.

*2.4c. Resonant Wave-Particle Interaction

We return to the example of Section 2.2b, that of the effect of an electrostatic wave on a particle gyrating in a magnetic field. Here we consider the case of resonance between the doppler-shifted wave frequency and the unperturbed gyromotion. We were unable to find invariants in this case by the standard perturbation techniques because of the appearance of resonant denominators in the transformation equations. Secular perturbation theory has shown us how to remove the resonance locally. We now apply this technique, following Lichtenberg (1979a), to the two cases discussed previously, that of oblique wave propagation $k_z \neq 0$, which corresponds to accidental degeneracy, and that of propagation perpendicular to the magnetic field $k_z = 0$, which corresponds to intrinsic degeneracy.

Accidental Degeneracy. For $k_z \neq 0$, an accidental degeneracy occurs whenever (2.2.71) holds, which is satisfied for a series of m values for particles with different z momentum. Choosing a particular resonance $m = l$, we can transform to the rotating frame using the generating function of (2.4.5), which for the variables in (2.2.67) is

$$F_2 = (\psi - l\phi)\hat{P}_\psi + \phi \hat{P}_\phi. \qquad (2.4.63)$$

Applying F_2 to (2.2.67) gives the new Hamiltonian

$$\hat{H} = \frac{k_z^2}{2M} \hat{P}_\psi^2 + \Omega(\hat{P}_\phi - l\hat{P}_\psi) - \hat{P}_\psi \omega$$
$$+ \epsilon e \Phi_0 \sum_m \mathscr{J}_m(k_\perp \rho) \sin[\hat{\psi} - (m-l)\hat{\phi}], \qquad (2.4.64)$$

where ρ is implicitly a function of the actions. Sufficiently close to a resonance, $\hat{\psi}$ is slowly varying, and we can average over the fast phase $\hat{\phi}$. All of the perturbation terms average to zero except for $m = l$, giving for the Hamiltonian in (2.4.64)

$$\bar{H} = \frac{k_z^2}{2M} \hat{P}_\psi^2 + \Omega(\hat{P}_\phi - l\hat{P}_\psi) - \hat{P}_\psi \omega + \epsilon e \Phi_0 \mathscr{J}_l(k_\perp \rho) \sin \hat{\psi}. \qquad (2.4.65)$$

We trivially shift $\hat{\psi} \to \hat{\psi} + \frac{1}{2}\pi$ such that $\sin \hat{\psi} \to \cos \hat{\psi}$, yielding a Hamiltonian $\bar{H}$ of the form (2.4.16). Since $\bar{H}$ is independent of $\hat{\phi}$,

$$\hat{P}_\phi = P_\phi + lP_\psi = \hat{P}_{\phi 0}. \qquad (2.4.66)$$

2.4. Secular Perturbation Theory

The fixed points are, from (2.4.18) and (2.4.20), located at

$$\psi_0 = 0, \pi, \tag{2.4.67a}$$

$$\frac{k_z^2}{M}\hat{P}_\psi - l\Omega - \omega = \mp \epsilon e \Phi_0 \frac{\partial \mathcal{J}_l(k_\perp \rho)}{\partial \hat{P}_\psi}, \tag{2.4.67b}$$

where

$$\rho = \left(\frac{2}{M\Omega}\right)^{1/2} (\hat{P}_\phi - l\hat{P}_\psi)^{1/2}. \tag{2.4.68}$$

Equation (2.4.67b) implicitly determines $\hat{P}_{\psi 0}$. Linearizing in $\hat{P}_\psi$ but not in $\hat{\psi}$, we obtain the pendulum Hamiltonian (2.4.27), with

$$G = \frac{k_z^2}{M} \tag{2.4.69}$$

and

$$F = -\epsilon e \Phi_0 \mathcal{J}_l(k_\perp \rho_0) \tag{2.4.70}$$

given by (2.4.28) and (2.4.29), respectively. We have, from (2.4.30), the frequency near the elliptic singular point for the perturbed motion

$$\hat{\omega}_\psi = \left|\frac{\epsilon e \Phi_0 \mathcal{J}_l k_z^2}{M}\right|^{1/2}. \tag{2.4.71}$$

The peak amplitude at the separatrix, as obtained from (2.4.31), is

$$\Delta \hat{P}_{\psi\,\text{max}} = \frac{2\hat{\omega}_\psi}{G}. \tag{2.4.72}$$

Both $\hat{\omega}_\psi$ and $\Delta \hat{P}_\psi$ are proportional to the square root of the small perturbation. The separation of adjacent resonances is given from (2.2.71) as

$$\delta \hat{P}_\psi = \frac{M\Omega}{k_z^2}. \tag{2.4.73}$$

The ratio of the momentum oscillation to momentum separation is then

$$\frac{2\Delta \hat{P}_{\psi\,\text{max}}}{\delta \hat{P}_\psi} = \frac{4\hat{\omega}_\psi}{\Omega}. \tag{2.4.74}$$

Intrinsic Degeneracy. We compare the above results with the intrinsically degenerate system for which $k_z \equiv 0$ in (2.4.65). Expanding about the singular point in both $\Delta \hat{P}_\psi$ and $\Delta \hat{\psi}$, we obtain the harmonic oscillator Hamiltonian with

$$G = \epsilon e \Phi_0 \frac{\partial^2 \mathcal{J}_l(k_\perp \rho_0)}{\partial \hat{P}_{\psi 0}^2}, \tag{2.4.75}$$

$$F = -\epsilon e \Phi_0 \mathcal{J}_l(k_\perp \rho_0). \tag{2.4.76}$$

The frequency and momentum excursion are

$$\hat{\omega}_\psi = \epsilon e \Phi_0 \left| \mathcal{J}_l \frac{\partial^2 \mathcal{J}_l}{\partial \hat{P}_{\psi 0}^2} \right|^{1/2}, \qquad (2.4.77)$$

$$\Delta \hat{P}_{\psi \max} = \frac{2\hat{\omega}_\psi}{G}. \qquad (2.4.78)$$

Comparing these with (2.4.71) and (2.4.72), we observe that, for intrinsic degeneracy, the frequency of the beat oscillation is of order ϵ, which is $\epsilon^{1/2}$ slower than for accidental degeneracy, while the excursion in momentum is of order unity, $\epsilon^{-1/2}$ larger than for accidental degeneracy.

Unlike the situation for wave propagation at an angle to the magnetic field, the driving frequency is fixed at ω, and thus there can be resonance only at a single harmonic of Ω, although this resonance may appear at a succession of values of $\hat{P}_\psi$. This can be seen by setting $k_z = 0$ in (2.4.67b) to obtain

$$\omega + l\Omega \pm \epsilon e \Phi_0 \frac{\partial \mathcal{J}_l(k_\perp \rho)}{\partial \hat{P}_\psi} = 0, \qquad (2.4.79)$$

which gives the values of $\hat{P}_\psi$ at the fixed points of the motion. We note that these roots can occur over a range of values of $k_\perp \rho$. In particular, for $\omega + l\Omega = 0$, they occur for

$$\mathcal{J}'_l(k_\perp \rho) = 0. \qquad (2.4.80)$$

If $\omega + l\Omega = \delta\omega$, then, from (2.4.79), we see that

$$\left| \epsilon e \Phi_0 \frac{\partial \mathcal{J}_l(k_\perp \rho)}{\partial \hat{P}_\psi} \right| > |\delta\omega| \qquad (2.4.81)$$

for resonance to occur.

There is also a transition from accidental to intrinsic degeneracy as the wave direction approaches a normal to the magnetic field. The transition can be found by keeping the $\mathcal{O}(\epsilon)$ term in (2.4.38) for G:

$$G = \frac{k_z^2}{M} + \epsilon e \Phi_0 \frac{\partial^2 \mathcal{J}_l(k_\perp \rho)}{\partial \hat{P}_\psi^2}, \qquad (2.4.82)$$

such that intrinsic degeneracy occurs for

$$\frac{k_z^2}{M} \lesssim \epsilon e \Phi_0 \left| \frac{\partial^2 \mathcal{J}_l(k_\perp \rho)}{\partial \hat{P}_\psi^2} \right|. \qquad (2.4.83)$$

Second-Order Resonance. To obtain the Hamiltonian for secondary resonance, as in Section 2.4b, we transform to action-angle variables I, θ for the

2.4. Secular Perturbation Theory

$\Delta\hat{P}_\psi - \hat{\psi}$ oscillation, which yields for $k_z \neq 0$,

$$K_0(I, \hat{P}_\phi) = \hat{\omega}_\psi I - \frac{\epsilon}{16} G I^2 + \cdots, \quad (2.4.84)$$

where G and $\hat{\omega}_\psi$ are given by (2.4.69) and (2.4.71), and

$$\hat{\psi} = \hat{\psi}_0 + \left(\frac{2I}{R}\right)^{1/2} \sin\theta + \mathcal{O}(\epsilon) \quad (2.4.85)$$

with $R = (F/G)^{1/2}$. The perturbation Hamiltonian is expressed to lowest order in action-angle variables as in (2.4.47):

$$K_1 = e\Phi_0 \sum_{m \neq l} \mathcal{J}_m(k_\perp \rho) \sin\left[\hat{\psi}_0 + \left(\frac{2I}{R}\right)^{1/2}\sin\theta - (m-l)\hat{\phi}\right]$$

$$= e\Phi_0 \sum_{n, m \neq l} \mathcal{J}_m(k_\perp \rho) \mathcal{J}_n\left[\left(\frac{2I}{R}\right)^{1/2}\right] \sin[\hat{\psi}_0 + n\theta - (m-l)\hat{\phi}]. \quad (2.4.86)$$

Taking the most important term $m = l + 1$, then for resonance near a given n,

$$K_{1,l+1} = K_n \sin(\hat{\psi}_0 + n\theta - \hat{\phi}) \quad (2.4.87)$$

with

$$K_n = e\Phi_0 \mathcal{J}_{l+1}(k_\perp \rho) \mathcal{J}_n\left[\left(\frac{2I}{R}\right)^{1/2}\right].$$

Transforming to a new locally slow variable, we obtain

$$\hat{\theta} = n\theta - \hat{\phi} + \hat{\psi}_0 - \tfrac{1}{2}\pi, \quad (2.4.88a)$$

$$I = n\hat{I}. \quad (2.4.88b)$$

Averaging (2.4.86) over the fast phase $\hat{\phi}$, we obtain the Hamiltonian for the secondary islands

$$\Delta \bar{K} = \tfrac{1}{2} G_s (\Delta \hat{I})^2 - F_s \cos\hat{\theta}, \quad (2.4.89)$$

where from (2.4.84), with $\epsilon \equiv 1$,

$$G_s = -\frac{G}{8}, \quad (2.4.90a)$$

and from (2.4.87),

$$F_s = -K_n. \quad (2.4.90b)$$

The frequency near the singularity and half-width momentum excursion of the second-order islands are then given, as previously, by

$$\hat{\omega}_s = (F_s G_s)^{1/2} \quad (2.4.91)$$

and
$$\Delta \hat{I}_{\max} = \frac{2\hat{\omega}_s}{G_s}. \tag{2.4.92}$$

We calculate the distance between the nth and $(n + 1)$st secondary resonance using
$$\begin{aligned} n\hat{\omega}_\psi - \Omega &= 0, \\ (n+1)(\hat{\omega}_\psi + G_s \delta I) - \Omega &= 0, \end{aligned} \tag{2.4.93}$$

which, with (2.4.88b), yields
$$\delta \hat{I} = \frac{\hat{\omega}_\psi}{G_s}. \tag{2.4.94}$$

Substituting for $\Delta I_{\max}$ from (2.4.92), we obtain
$$\frac{2\Delta \hat{I}_{\max}}{\delta \hat{I}} = \frac{4\hat{\omega}_s}{\hat{\omega}_\psi}, \tag{2.4.95}$$

which is identical in form to that obtained for the primary resonances in (2.4.74). By induction, higher-order resonances have the same form, i.e., the relation is universal. Note that the secondary and higher-order resonances are accidentally degenerate.

Numerical Results. Detailed numerical calculations have been made for both accidental degeneracy (oblique wave propagation) by Smith and Kaufman (1975) and Smith (1977), and for intrinsic degeneracy (perpendicular wave propagation) by Karney and Bers (1977), Karney (1978), and Fukuyama *et al.* (1976).

For oblique incidence, Smith and Kaufman examined the resonances near $l = k_z v_z / \Omega = -1, 0, 1$ in the wave frame ($\omega \equiv 0$), choosing $k_\perp (2E/M)^{1/2}/\Omega = 1.48$, where E is given by (2.2.66) with $\omega \equiv 0$, $\epsilon \equiv 1$ and $k_z = k_\perp$. This choice lies near a maximum in the Bessel function $\mathcal{J}_l(k_\perp \rho)$. Their results for $v_z \propto P_\psi$ versus ψ are shown for $k_z^2 e \Phi_0 / M\Omega^2 = 0.025$ and 0.1, in Fig. 2.10a, b, respectively. In Fig. 2.10a the primary islands are seen. The relative island frequency near the fixed point is numerically found to be $\hat{\omega}_\psi/\Omega = 1/10$, in agreement with that calculated from (2.4.71). In this case the second-order islands are too small to be seen. In Fig. 2.10b we see a five-island chain appearing around the singular points, indicating $\hat{\omega}_\psi/\Omega = \frac{1}{5}$, again in agreement with (2.4.71). The approximate second-order island size can be calculated from (2.4.92). The intermingling of points originating from the regions near the separatrices of the $l = -1$ and $l = 0$ resonance is the stochastic behavior discussed in Section 1.4, which is suppressed by the averaging procedures used in this chapter. Note also the appearance of an island chain with two periodic points, which has grown to visible size,

2.4. Secular Perturbation Theory

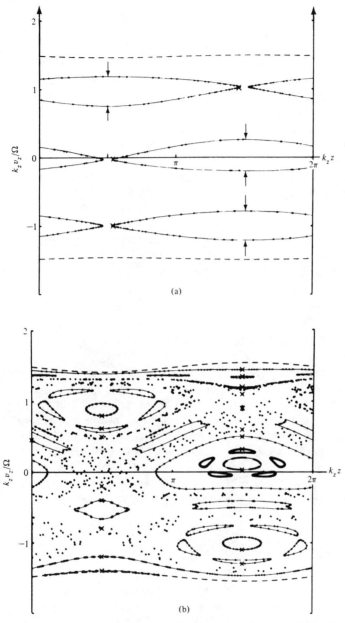

Figure 2.10. Surface of section plots $k_z v_z/\Omega \propto P_\psi$ versus $k_z z = \psi$ for oblique angle ($k_z \neq 0$), wave-particle resonance. Three resonances are shown, calculated by integration of Hamilton's equations. (a) Weak perturbation. The initial conditions, indicated by the x's, were chosen to yield trajectories very close to the separatrices. The points representing the trajectories have been connected with hand-drawn curves. The wave amplitude is given by $k_z^2 e\Phi_0/M\Omega^2 = 0.025$. (b) Strong perturbation. The wave amplitude is increased to 0.1 (after Smith and Kaufman, 1975).

between the two major ones. These phenomena are the subject matter of the subsequent chapters.

We contrast this result with the case of the perpendicular propagation, at resonance, as computed by Karney (1977). In Fig. 2.11a, b, their numerical results are given at the surface of section $\phi = \pi$ for $l = 30$, and the value of perturbation parameter chosen so that the fundamental rotation number for the primary islands is $\alpha = \hat{\omega}_s/\Omega = \frac{1}{9}$ for the smaller perturbation and $\alpha = \frac{1}{5}$ for the larger perturbation. For the smaller perturbation, the invariant curves are almost the same as obtained from the averaged Hamiltonian (2.4.65) with $k_z = 0$. For the larger perturbation, the five-island chain appears, as expected, and also the other features that we have become accustomed to. The primary island sizes are roughly the same in both cases, because, as we have seen in the case of intrinsic degeneracy, the island size is not a function of perturbation amplitude. As with oblique propagation, the phase space was examined for values of $k_\perp \rho$ near the maximum of $\mathcal{J}_l(k_\perp \rho)$; otherwise the island frequency can be considerably smaller for a given value of $e\Phi_0$. Similar observations for a different problem were made by Ford and Lunsford (1970).

2.4d. Global Removal of Resonances

We now describe a method which in certain cases can be applied to remove an entire set of resonances. The method was first applied to the motion of a charged particle in a spatially modulated magnetic field (Dunnett et al., 1968), and later to the wave-particle resonance problem of Section 2.4c; the latter will be summarized here. The extension to higher order in the expansion parameter has been given explicitly by McNamara (1978) and will be described in the next section.

The method of Dunnett and co-workers was developed to treat autonomous systems in two degrees of freedom with H_0 in the special action-angle form:

$$H_0(\boldsymbol{J}) = a(J_1) + \omega_2 J_2, \qquad (2.4.96)$$

with ω_2 constant, such that both ω_1 and ω_2 are independent of one of the actions, here chosen to be J_2. This form is not uncommon, and is in fact generic to nonautonomous systems in one degree of freedom in an extended phase space (see Section 1.2b).

The method exploits a certain latitude in constructing invariants. In the surface of section $\theta_2 = $ const., the unperturbed $J_1 - \theta_1$ invariant curves are the straight lines $J_1 = $ const. But any arbitrary function $I_0(J_1)$ also generates these lines:

$$I_0(J_1) = \text{const.} \qquad (2.4.97)$$

2.4. Secular Perturbation Theory

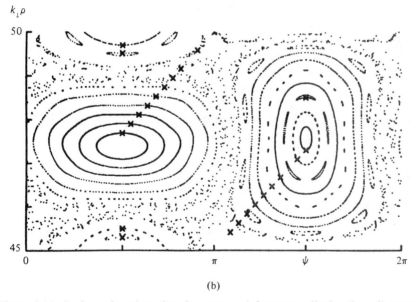

Figure 2.11. Surface of section plots $k_\perp \rho$ versus ψ for perpendicular ($k_z = 0$) wave-particle resonance, $\omega/\Omega = 30$, calculated by integration of Hamilton's equations. (a) For small perturbation strength; (b) for larger perturbation strength. The x's mark the initial conditions (after Karney, 1977).

Thus we may consider I_0 to be the invariant for the unperturbed system, choosing its specific form as desired.

The failure of canonical perturbation theory near a resonance has been alluded to in Section 2.2b, and a specific example exhibiting the resonant denominators was given there and in Section 2.4c. This failure has a simple physical interpretation; near a resonance, the topology of the true invariant curves, $I = $ const., differs from that of the unperturbed curves $I_0 = $ const. Generally, if ϵ is small, the contours of $I_0 + \epsilon I_1 = $ const. can be topologically different from the contours of $I_0 = $ const. only if I_1 is large. Thus the appearance of large I_1 is simply the response of perturbation theory to the change in topology. This suggests that a valid perturbation expansion can be obtained if the zero-order invariant is chosen so that a small I_1 can make the topology of the $I_0 + \epsilon I_1$ curves differ from that of the I_0 curves. This is the case if dI_0/dJ_1 vanishes, so that the level curves of I_0, which are straight lines, display a maximum or minimum with respect to the unperturbed action J_1 in the surface of section.

Perturbation Theory. To generate the series for the new invariant I, we note from (1.2.21) that any constant of the motion satisfies

$$[I, H] = 0. \tag{2.4.98}$$

Expanding H and I, we have

$$H(\boldsymbol{J}, \boldsymbol{\theta}) = H_0(\boldsymbol{J}) + \epsilon H_1(\boldsymbol{J}, \boldsymbol{\theta}), \tag{2.4.99}$$

$$I(\boldsymbol{J}, \boldsymbol{\theta}) = I_0(J_1) + \epsilon J_1(\boldsymbol{J}, \boldsymbol{\theta}) + \cdots, \tag{2.4.100}$$

where we choose I_0 to be a function of J_1 alone. Equating like powers of ϵ to zero order,

$$[I_0, H_0] = 0, \tag{2.4.101}$$

which is satisfied by construction since I_0 and H_0 are functions of the actions alone. In first order,

$$[I_1, H_0] + [I_0, H_1] = 0 \tag{2.4.102}$$

or

$$\omega_1 \frac{\partial I_1}{\partial \theta_1} + \omega_2 \frac{\partial I_1}{\partial \theta_2} = \frac{dI_0}{dJ_1} \frac{\partial H_1}{\partial \theta_1}. \tag{2.4.103}$$

Fourier analyzing H_1 and I_1:

$$H_1 = \sum_{l,m} H_{lm}(\boldsymbol{J}) e^{i(l\theta_1 + m\theta_2)}, \tag{2.4.104a}$$

$$I_1 = \sum_{l,m} I_{lm}(\boldsymbol{J}) e^{i(l\theta_1 + m\theta_2)}. \tag{2.4.104b}$$

2.4. Secular Perturbation Theory

From (2.4.103),

$$(\omega_1 l + \omega_2 m)I_{lm} = l \frac{dI_0}{dJ_1} H_{lm}. \tag{2.4.105}$$

To find the invariant, we put

$$\frac{dI_0}{dJ_1} = \prod_{l',m'} C_{l'm'}(\omega_1 l' + \omega_2 m'), \tag{2.4.106}$$

where the product is over l' and m' corresponding to the l's and m's with nonzero amplitudes H_{lm}, and the C's are suitably chosen coefficients. By construction, dI_0/dJ_1 vanishes at each resonance, and I_{lm} obtained from (2.4.105) is then finite. The invariant is then

$$I = I_0 + \epsilon I_1 = \text{const.}, \tag{2.4.107}$$

with I_0 from (2.4.106), and I_1 from (2.4.105) and (2.4.104b).

Wave-Particle Resonance. To illustrate the method, we consider the Hamiltonian of (2.2.67) for the case of accidental degeneracy with wave propagation at 45° (Taylor and Laing, 1975). Looking in the wave frame ($\omega \equiv 0$), and introducing dimensionless variables: $k_\perp = k_z = 1$, $\Omega = 1$, $M = 1$, $\rho = (2J_\phi)^{1/2}$, $e\Phi_0 = 1$,

$$H = \tfrac{1}{2}P_\psi^2 + P_\phi + \epsilon \sum_{m=0}^{\infty} \mathscr{J}_m(\rho) \sin(\psi - m\phi). \tag{2.4.108}$$

Then from (2.4.106), we choose C's such that

$$\frac{dI_0}{dP_\psi} = \prod_{m=0}^{\infty} C_m(P_\psi - m)$$

$$= P_\psi \prod_{m=1}^{\infty} \left(1 - \frac{P_\psi}{m}\right)$$

$$= \sin \pi P_\psi. \tag{2.4.109}$$

From (2.4.105),

$$I_m = \sin \pi P_\psi \frac{\mathscr{J}_m(\rho)}{P_\psi - m}, \tag{2.4.110}$$

and the first-order invariant is

$$I = \frac{1}{\pi} \cos \pi P_\psi - \epsilon \sin \pi P_\psi \sum_m \mathscr{J}_m(\rho) \frac{\sin(\psi - m\phi)}{P_\psi - m}, \tag{2.4.111}$$

in which the second term is small even at resonances.

Two plots of the invariant $I = \text{const.}$ in the surface of section $\phi = \pi$ are shown in Fig. 2.12 for the same parameters as the numerical orbit

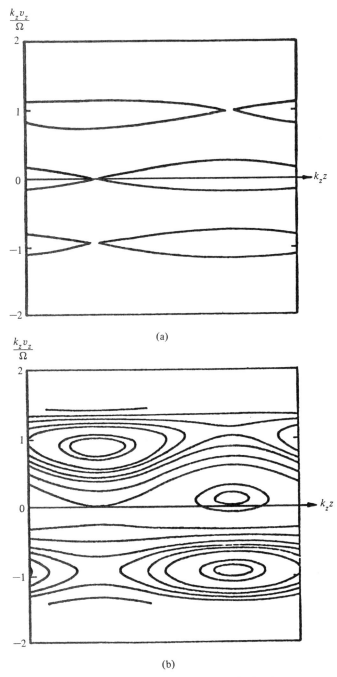

Figure 2.12. First-order invariant after global removal of primary resonances, for oblique angle wave-particle resonance problem. The parameters are the same as given in Fig. 2.10 (after Taylor and Laing, 1975).

calculations of Fig. 2.10. The agreement between these figures is good; of course the chaotic regions that have developed for the strong perturbation in Fig. 2.10b are not found from the invariant curves.

A Single Resonance. The relation of the invariant I to that calculated from secular perturbation theory for a single resonance may be seen as follows: for the Hamiltonian

$$H = a(J_1) + \omega_2 J_2 + \epsilon \Lambda \sin(l\theta_1 - m\theta_2), \qquad (2.4.112)$$

the transformation to rotating coordinates (2.4.6) yields

$$\hat{H} = a(l\hat{J}_1) + \omega_2(\hat{J}_2 - m\hat{J}_1) + \epsilon \Lambda \sin \hat{\theta}_1. \qquad (2.4.113)$$

Since $\hat{H}$ is independent of $\hat{\theta}_2$, $\hat{J}_2 = $ const. Coupled with $\hat{H} = $ const., we have an invariant for the $\hat{J}_1 - \hat{\theta}_1$ motion:

$$\hat{I} = a(l\hat{J}_1) - m\omega_2 \hat{J}_1 + \epsilon \Lambda \sin \hat{\theta}_1. \qquad (2.4.114)$$

Applying the method of this subsection, (2.4.106) yields

$$\frac{dI_0}{dJ_1} = a'(J_1) - \frac{m}{l}\omega_2, \qquad (2.4.115)$$

which leads to the invariant I, which is identical to $\hat{I}$.

The extension of the Dunnett, Laing, and Taylor method to more general Hamiltonians, not limited to the form (2.4.96), has not been made. It is also not clear how to choose I_0 when all harmonics H_{lm} are nonzero, since I_0' must then be zero at all the rationals. Nevertheless, the method is quite successful in accomplishing a global removal of resonances within these limitations.

2.5. Lie Transformation Methods

In calculating the series for the invariants of the preceding sections, it is often necessary to carry the expansion to higher than first order. We saw an example of this in the results of a problem in two degrees of freedom, with the Hénon and Heiles potential, as described in Section 1.4a. Because of the intrinsic degeneracy, the asymptotic expansion had to be carried to at least second order to represent reasonably the proper phase space trajectories even at low energy. There are other cases, for example the calculation of the ponderomotive force to be presented in Section 2.5c, for which the first-order calculation gives a null result, and the first nonzero term is of second or higher order.

Carrying the expansion procedure to higher than first order in ϵ becomes increasingly tedious when using the classical technique of Poincaré and Von

Zeipel. The classical approach proceeds by the use of a mixed variable generating function, e.g., $S(\boldsymbol{\theta}, \bar{\boldsymbol{J}}, t)$, to perform a transformation from old variables $\boldsymbol{J}, \boldsymbol{\theta}$ to new variables $\bar{\boldsymbol{J}}, \bar{\boldsymbol{\theta}}$. As a consequence, the transformation itself appears in mixed form:

$$\bar{\boldsymbol{\theta}}(\boldsymbol{\theta}, \bar{\boldsymbol{J}}, t) = \frac{\partial S(\boldsymbol{\theta}, \bar{\boldsymbol{J}}, t)}{\partial \bar{\boldsymbol{J}}}, \qquad (2.5.1)$$

whereas what is desired is $\bar{\boldsymbol{\theta}}(\boldsymbol{\theta}, \boldsymbol{J}, t)$, namely, the new variables in terms of the old, or *vice versa*. Again, the relation between the new and old Hamiltonians is given in mixed form:

$$\bar{H}(\bar{\boldsymbol{\theta}}, \bar{\boldsymbol{J}}, t) = H(\boldsymbol{\theta}, \boldsymbol{J}, t) + \frac{\partial S(\boldsymbol{\theta}, \bar{\boldsymbol{J}}, t)}{\partial t}, \qquad (2.5.2)$$

where $\boldsymbol{J}$ and $\boldsymbol{\theta}$ are considered known functions of $\bar{\boldsymbol{J}}$ and $\bar{\boldsymbol{\theta}}$ determined by the generating function S. If the old and new Hamiltonians and the generating function are expressible as a power series in ϵ, then it is reasonably simple to obtain perturbation expansions to any order (Born, 1927). However, even in order ϵ^2, these expressions are lengthy [see (2.2.19)] and no pattern is manifested. To order ϵ^3 or higher, the amount of algebra is disheartening, and any relationships that reside in the physics may lie hidden.

An important breakthrough in Hamiltonian perturbation theory was the introduction of Lie transforms by Hori (1966) and Garrido (1968). Using the Lie methods, no functions of mixed variables appear, and all the terms in the series are repeating Poisson brackets, making the theory canonically invariant. Deprit (1969) improved the method to obtain expressions for the nth-order term of an expansion of the transformation in a power series in ϵ. Dewar (1976) developed a transformation theory in closed form, which can be used to study systems where the transformation is not expandable in a power series. The Lie method has been used to study magnetic moment invariance in a dipole field by Dragt and Finn (1976a). Howland (1977) has applied Lie methods to study Kolmogorov's superconvergent perturbation theory. Contributions have also been made by Kaufman and co-workers (see Johnston and Kaufman, 1978; Cary and Kaufman, 1977; Kaufman et al., 1978; Kaufman, 1978) and by McNamara (1978). A more recent expansion method based on a technique of Dragt and Finn (1976b) is described by Cary (1981) and is particularly effective for high-order calculations. Presentations of the Lie formalism can be found in Nayfeh (1973), Giacaglia (1972), and in tutorial articles by Cary (1978, 1981) and Littlejohn (1978). In this section we draw heavily on the latter two sources.

In Section 2.5a we develop the basic Lie formalism, which we then apply in Section 2.5b to obtain the Deprit power series expansions. These expansions are used to find the second-order correction to the pendulum Hamiltonian as an illustration of the method. In Section 2.5c we modify the Lie method to obtain the series for adiabatic invariants, illustrating the

2.5. Lie Transformation Methods

method with a second-order calculation of the invariant for the time-varying harmonic oscillator and with a calculation of the ponderomotive force for a particle in the field of an electrostatic wave. We conclude by describing briefly the higher-order calculation of modified adiabatic invariants, due to McNamara (1978), using the example of wave-particle resonance.

2.5a. General Theory

We now consider autonomous systems and give the generalization to time-dependent Hamiltonians later. Let $x = (p, q)$ be the vector of generalized momenta and coordinates representing the system point in phase space. We consider a function $w(\bar{x}, \epsilon)$, called a *Lie generating function*, which satisfies the equation

$$\frac{d\bar{x}}{d\epsilon} = [\bar{x}, w]. \tag{2.5.3}$$

The components of (2.5.3) are just Hamilton's equations in Poisson bracket notation applied to the "Hamiltonian" w, with the parameter ϵ the "time." Equation (2.5.3) therefore generates a canonical transformation for any ϵ; i.e., the solution for any initial system point x,

$$\bar{x} = \bar{x}(x, \epsilon), \tag{2.5.4}$$

represents a transformation from x to $\bar{x}$, which satisfies the Poisson bracket conditions

$$[\bar{q}_i, \bar{q}_j] = [\bar{p}_i, \bar{p}_j] = 0, \qquad [\bar{q}_i, \bar{p}_j] = \delta_{ij}. \tag{2.5.5}$$

Corresponding to this transformation, we introduce the *evolution operator* T, which evaluates any function g at the transformed point $\bar{x}(x, \epsilon)$, producing a new function f at the original point; i.e., for

$$f = Tg, \tag{2.5.6}$$

then

$$f(x) = g[\bar{x}(x, \epsilon)]. \tag{2.5.7}$$

If g is the identity function $\mathscr{I}(x) = x$, then it follows from (2.5.4) that

$$\bar{x} = Tx. \tag{2.5.8}$$

To find the transformation T explicitly, we introduce the *Lie operator* L:

$$L = [w, \ ] \tag{2.5.9}$$

and note from (2.5.3) and (2.5.8) that

$$\frac{dT}{d\epsilon} = -TL \tag{2.5.10}$$

with the formal solution

$$T = \exp\left[-\int^{\epsilon} L(\epsilon')d\epsilon'\right]. \qquad (2.5.11)$$

For any canonical transformation and, in particular that generated by w, the new Hamiltonian $\bar{H}$ is related to the old Hamiltonian H by

$$\bar{H}(\bar{x}(x,\epsilon)) = H(x); \qquad (2.5.12)$$

i.e., the new Hamiltonian at the new position is equal to the old Hamiltonian at the old position. Comparing this to (2.5.7) and applying (2.5.6), we have

$$\bar{H} = T^{-1}H. \qquad (2.5.13)$$

For nonautonomous systems, w, L, and T are explicit functions of t, which is held fixed in deriving the transformation T, and thus the analysis leading to (2.5.11) is still correct. However, the old and new energies are not equal, and thus (2.5.13) is incorrect. The correct form for time-dependent systems (Cary, 1978) is

$$\bar{H} = T^{-1}H + T^{-1}\int_0^{\epsilon} d\epsilon' T(\epsilon') \frac{\partial w(\epsilon')}{\partial t}, \qquad (2.5.14)$$

which should be compared to (1.2.11c) or (1.2.13c). This expression was first obtained by Dewar (1976), and, together with (2.5.11) and (2.5.9) for T and L, it provides a complete description of canonical transformations using Lie generating functions.

2.5b. Deprit Perturbation Series

To obtain the perturbation series, we expand w, L, T, H, and $\bar{H}$ as power series in ϵ (Deprit, 1969; Cary, 1978):

$$w = \sum_{n=0}^{\infty} \epsilon^n w_{n+1}, \qquad (2.5.15)$$

$$L = \sum_{n=0}^{\infty} \epsilon^n L_{n+1}, \qquad (2.5.16)$$

$$T = \sum_{n=0}^{\infty} \epsilon^n T_n, \qquad (2.5.17)$$

$$H = \sum_{n=0}^{\infty} \epsilon^n H_n, \qquad (2.5.18)$$

$$\bar{H} = \sum_{n=0}^{\infty} \epsilon^n \bar{H}_n, \qquad (2.5.19)$$

2.5. Lie Transformation Methods

where from (2.5.9),

$$L_n = [w_n, \ ]. \tag{2.5.20}$$

Inserting (2.5.16) and (2.5.17) into (2.5.10) and equating like powers of ϵ, we obtain a recursion relation for the T_n ($n > 0$):

$$T_n = -\frac{1}{n} \sum_{m=0}^{n-1} T_m L_{n-m}, \tag{2.5.21}$$

which with $T_0 = 1$, yields T_n in terms of the L_n in all orders. We also point out that the T's and L's do not generally commute; i.e., $L_i L_j \neq L_j L_i$, etc.

We also need the inverse operators T_n^{-1}. Differentiating the equation

$$TT^{-1} = 1$$

with respect to ϵ yields

$$\frac{dT^{-1}}{d\epsilon} = LT^{-1}, \tag{2.5.22}$$

from which the recursion relation for the T_n^{-1} is found ($n > 0$):

$$T_n^{-1} = \frac{1}{n} \sum_{m=0}^{n-1} L_{n-m} T_m^{-1}, \tag{2.5.23}$$

where again $T_0^{-1} = 1$. To third order

$$T_1 = -L_1, \tag{2.5.24a}$$

$$T_2 = -\tfrac{1}{2} L_2 + \tfrac{1}{2} L_1^2, \tag{2.5.24b}$$

$$T_3 = -\tfrac{1}{3} L_3 + \tfrac{1}{6} L_2 L_1 + \tfrac{1}{3} L_1 L_2 - \tfrac{1}{6} L_1^3, \tag{2.5.24c}$$

$$T_1^{-1} = L_1, \tag{2.5.25a}$$

$$T_2^{-1} = \tfrac{1}{2} L_2 + \tfrac{1}{2} L_1^2, \tag{2.5.25b}$$

$$T_3^{-1} = \tfrac{1}{3} L_3 + \tfrac{1}{6} L_1 L_2 + \tfrac{1}{3} L_2 L_1 + \tfrac{1}{6} L_1^3. \tag{2.5.25c}$$

The general rule for finding T_n^{-1}, given T_n, is as follows: replace L_n by $-L_n$ and invert the order of all noncommuting L operators.

To obtain the nth-order perturbation equation, we premultiply (2.5.14) by T and differentiate with respect to ϵ:

$$\frac{\partial T}{\partial \epsilon} \bar{H} + T \frac{\partial \bar{H}}{\partial \epsilon} = \frac{\partial H}{\partial \epsilon} + T \frac{\partial w}{\partial t}. \tag{2.5.26}$$

Using (2.5.10) to eliminate $\partial T/\partial \epsilon$ (with $dT/d\epsilon \to \partial T/\partial \epsilon$, since here T also depends explicitly on t) and premultiplying by T^{-1},

$$\frac{\partial w}{\partial t} = \frac{\partial \bar{H}}{\partial \epsilon} - L\bar{H} - T^{-1} \frac{\partial H}{\partial \epsilon}. \tag{2.5.27}$$

Inserting the series expansions and equating like powers of ϵ, we obtain in nth order ($n > 0$)

$$\frac{\partial w_n}{\partial t} = n\bar{H}_n - \sum_{m=0}^{n-1} L_{n-m}\bar{H}_m - \sum_{m=1}^{n} mT_{n-m}^{-1}H_m. \qquad (2.5.28)$$

By writing out the first term in the first sum,

$$L_n \bar{H}_0 = L_n H_0 = [w_n, H_0],$$

and the last term in the last sum, we get for $n > 0$ the final result

$$D_0 w_n = n(\bar{H}_n - H_n) - \sum_{m=1}^{n-1} (L_{n-m}\bar{H}_m + mT_{n-m}^{-1}H_m), \qquad (2.5.29)$$

where

$$D_0 \equiv \frac{\partial}{\partial t} + [\ , H_0] \qquad (2.5.30)$$

is the total time derivative taken along the unperturbed orbits, and $\bar{H}_0 = H_0$ in zero order. To third order, (2.5.29) yields

$$D_0 w_1 = \bar{H}_1 - H_1, \qquad (2.5.31a)$$

$$D_0 w_2 = 2(\bar{H}_2 - H_2) - L_1(\bar{H}_1 + H_1), \qquad (2.5.31b)$$

$$D_0 w_3 = 3(\bar{H}_3 - H_3) - L_1(\bar{H}_2 + 2H_2)$$
$$- L_2(\bar{H}_1 + \tfrac{1}{2}H_1) - \tfrac{1}{2}L_1^2 H_1. \qquad (2.5.31c)$$

Equation (2.5.31a) should be compared to the equivalent first-order equation (2.2.10), which is obtained using the Poincaré–Von Zeipel expansion. In both cases, given H_1, we choose $\bar{H}_1$ in some way, usually to eliminate secularities in the generating function w_1 or S_1, and then solve for the generating function. Proceeding to second order, one uses w_1 on the right-hand side of (2.5.31b), chooses $\bar{H}_2$ to eliminate secularities in w_2, and solves for w_2. In this way we proceed systematically to as high an order as desired.

Although the set of equations, of which (2.5.31) give the first three members, are formally correct for any number of degrees of freedom, for more than one degree of freedom resonant denominators appear, as they did in the mixed variable perturbation technique. The formal method of eliminating resonances to each order is given in Section 2.5c. The resonances prevent convergence of the series as described in previous sections.

The Pendulum. To illustrate the use of the Deprit series, we continue the example in Section 2.2a to obtain the nonlinear libration motion of the pendulum to second order. The old Hamiltonian was given as an expression in terms of action-angle variables of the unperturbed (linear) system $J - \theta$

2.5. Lie Transformation Methods

in (2.2.22). In zero order, using (2.2.22a),
$$\bar{H}_0(J) = \omega_0 J. \tag{2.5.32}$$

In first order (2.5.31a) reads
$$\omega_0 \frac{\partial w_1}{\partial \theta} = \bar{H}_1 - H_1, \tag{2.5.33}$$

where H_1 is given from (2.2.22b) as
$$H_1 = -\frac{GJ^2}{48}(3 - 4\cos 2\theta + \cos 4\theta).$$

Choosing $\bar{H}_1$ to eliminate the average part of H_1, which would give rise to a secularity, we have
$$\bar{H}_1(J) = \langle H_1 \rangle = -\tfrac{1}{16} GJ^2, \tag{2.5.34}$$

where $\langle \ \rangle$ denotes the average over θ. Then (2.5.33) becomes
$$\omega_0 \frac{\partial w_1}{\partial \theta} = -\{H_1\}, \tag{2.5.35}$$

where $\{\ \}$ denotes the oscillating part. Solving for w_1,
$$w_1 = \frac{1}{192} \frac{GJ^2}{\omega_0} (\sin 4\theta - 8 \sin 2\theta). \tag{2.5.36}$$

We note that in first order $\bar{H}_1$ in (2.5.34) and w_1 in (2.5.36) are identical to our previous results using mixed variables (2.2.23) and (2.2.25) with J replacing $\bar{J}$. Proceeding to second order, (2.5.31b) reads
$$\omega_0 \frac{\partial w_2}{\partial \theta} = 2(\bar{H}_2 - H_2) - L_1(\bar{H}_1 + H_1), \tag{2.5.37}$$

which, in terms of average and oscillating parts of H_1 and H_2, yields
$$\omega_0 \frac{\partial w_2}{\partial \theta} = 2(\bar{H}_2 - \langle H_2 \rangle - \{H_2\}) - 2[w_1, \langle H_1 \rangle] - [w_1, \{H_1\}]. \tag{2.5.38}$$

We again choose $\bar{H}_2$ to make the average of the right-hand side equal to zero. The first Poisson bracket has no average part, since w_1 is oscillatory and $\langle H_1 \rangle$ is averaged. The second Poisson bracket involves the product of two oscillatory terms and therefore has an average part. Thus
$$\bar{H}_2 = \langle H_2 \rangle + \tfrac{1}{2} \langle [w_1, \{H_1\}] \rangle. \tag{2.5.39}$$

Working out the second term using (2.2.22b) and (2.5.36),
$$[w_1, \{H_1\}] = -\frac{1}{1152} \frac{G^2 J^3}{\omega_0} (17 - 9 \cos 2\theta + \cos 6\theta). \tag{2.5.40}$$

Then using (2.2.22c) and the above,

$$\bar{H}_2 = -\frac{1}{256} \frac{G^2 J^3}{\omega_0}.$$

Note that $\bar{H}_2$ is not simply the average of H_2, but involves an additional term quadratic in the first-order quantities. In order to find $\bar{H}_2$, we need w_1 but not w_2. In general, to find $\bar{H}_n$ we need the w's only to order $n-1$. To find w_2 we solve (2.5.38), which, using (2.5.39), now reads

$$\omega_0 \frac{\partial w_2}{\partial \theta} = -2\{H_2\} - 2[w_1, \langle H_1 \rangle] - \{[w_1, \{H_1\}]\}. \qquad (2.5.41)$$

The solution is not very illuminating and will not be given here. Using w_2, the transformtion from old to new variables could be evaluated to second order. Adding the results for $\bar{H}_0$, $\bar{H}_1$, and $\bar{H}_2$, and writing $\bar{J}$ for the variable J, we have to order ϵ^2

$$\bar{H} = \omega_0 \bar{J} - \frac{1}{16} G \bar{J}^2 - \frac{1}{256} \frac{G^2 \bar{J}^3}{\omega_0}, \qquad (2.5.42)$$

where we have set $\epsilon = 1$. To understand the substitution of $\bar{J}$ for J, we need to realize that the Lie technique involves operations on functions, rather than on variables. This was seen in the original derivation for which the transformation (2.5.8) was interpreted in terms of the transformation of functions as in (2.5.6), with f and g suitably defined. The result is that the arguments of the functions are really dummy variables. Thus the result we have found for the function $\bar{H}$ should be interpreted as $\bar{H}(\bar{J})$, a function of the transformed action $\bar{J}$.

Once we have $\bar{H}$ as a function of $\bar{J}$ only, the new frequency follows at once, to second order, as

$$\bar{\omega} = \frac{\partial \bar{H}}{\partial \bar{J}} = \omega_0 \left[1 - \frac{1}{8} \frac{G \bar{J}}{\omega_0} - \frac{3}{256} \frac{G^2 \bar{J}^2}{\omega_0^2} \right], \qquad (2.5.43a)$$

which can be compared to the first-order result (2.2.24). Eliminating the action $\bar{J}$ from (2.5.43a) by use of the equation $\bar{H}(\bar{J}) = E + F$, where $\bar{H}(\bar{J})$ is given by (2.5.42) with J replaced by $\bar{J}$, and using $\omega_0 = \sqrt{FG}$, we obtain to second order the frequency as a function of energy

$$\bar{\omega} = \omega_0 \left[1 - \frac{1}{8} \left(\frac{E+F}{F} \right) - \frac{5}{256} \left(\frac{E+F}{F} \right)^2 \right], \qquad (2.5.43b)$$

which is shown plotted as the dotted line in Fig. 2.2.

2.5c. Adiabatic Invariants

Despite the effects of resonances in locally modifying or destroying an adiabatic invariant, even in first order in ϵ, it is often necessary to carry the asymptotic expansion to higher than first order. This becomes increasingly tedious when using the averaging procedures of Kruskal or Bogoliubov. The reason is that the inverse transformations become increasingly complicated when both angle and action variables must be considered together. Explicit transformations to order ϵ^2 have been given by McNamara and Whiteman (1967), and a specific example, using some shortcuts, has been worked out by Northrop et al. (1966).

An alternative procedure, the Poisson bracket method, introduced by Whittaker (1944) and developed by McNamara and Whiteman (1967) and by Giacaglia (1972), offers easier computational methods to higher order. The two procedures have been shown to be essentially equivalent to second order, by McNamara and Whiteman, and to all orders, in a restricted class of problems, by Stern (1970a). More recently, McNamara (1978) has shown that the Poisson bracket technique is a special case of the method using Lie transforms. The Lie method has essentially superseded the older Poisson bracket techniques and forms the basis of the description of higher-order adiabatic invariants given here. For more details regarding the older techniques, the reader is referred to the papers cited above.

The Adiabatic Ordering. For *slow* perturbations, the Hamiltonian has the form

$$H = H_0(J, \epsilon y, \epsilon t) + \epsilon H_1(J, \theta, \epsilon y, \epsilon t) + \cdots, \quad (2.5.44)$$

as described in Section 2.3a, where J, θ are action-angle variables for the single fast degree of freedom, and $y = (p, q)$ are the generalized coordinates of the remaining slow degrees of freedom. Derivatives with respect to the y's and t are smaller by order ϵ than the terms from which they are derived. The series developed in Section 2.5b do not embody this ordering and must therefore be used with care. For slow perturbations, the Lie operator has the form

$$L = L_f + \epsilon L_s, \quad (2.5.45)$$

where the fast part is

$$L_f = \left(\frac{\partial w}{\partial \theta} \frac{\partial}{\partial J} - \frac{\partial w}{\partial J} \frac{\partial}{\partial \theta} \right), \quad (2.5.46a)$$

and the slow part is

$$L_s = \sum_i \left[\frac{\partial w}{\partial (\epsilon q_i)} \frac{\partial}{\partial (\epsilon p_i)} - \frac{\partial w}{\partial (\epsilon p_i)} \frac{\partial}{\partial (\epsilon q_i)} \right], \quad (2.5.46b)$$

with $w = w(J, \theta, \epsilon p, \epsilon q, \epsilon t)$. Using (2.5.23), we see also that T_n^{-1} is given in terms of the coefficients of the power series expansions of L_f and L_s as an nth-order polynomial in ϵ. Finally, the term $\partial w_n/\partial t$ in the nth-order perturbation equation (2.5.29) is itself of order ϵ: $\partial w_n/\partial t \to \epsilon \partial w_n/\partial(\epsilon t)$. One procedure for solving this equation is to expand w_n and $\bar{H}_n$ as power series in ϵ

$$w_n = \sum_{k=0}^{\infty} \epsilon^k w_{nk}, \quad \text{etc.,} \tag{2.5.47}$$

and equate like powers of ϵ in (2.5.29). We obtain a chain of equations that can be solved successively for $w_{n0}, w_{n1}, \ldots$. At each step in the chain, we choose the corresponding $\bar{H}_{nk}$ to eliminate the secularity in the fast phase variable θ. In practice, when doing nth-order perturbation theory, we need to find the w_{mk} only for $m + k \leq n$. The procedure used here is equivalent, in any order, to the method of averaging as described in Section 2.3. It is systematic in that it automatically separates the fast and slow variables by order ϵ, thus allowing an average over the fast variable in any order to eliminate the secular terms. However, the same limitations apply here as described in Section 2.3.

The first equation in the chain ($k = 0$) is

$$\omega_\theta \frac{\partial w_{n0}}{\partial \theta} = n(\bar{H}_{n0} - H_n)$$

$$- \sum_{m=1}^{n-1} (L_{f,n-m} \bar{H}_{m0} + m T_{f,n-m}^{-1} H_m), \tag{2.5.48}$$

where the T_f^{-1}'s are obtained from (2.5.25) with the L's replaced by L_f's. Every equation in the chain has this same form

$$\omega_\theta \frac{\partial w_{nk}}{\partial \theta} = n\bar{H}_{nk} + \Phi, \tag{2.5.49}$$

where Φ is a known function of the generalized coordinates and time. Therefore resonant denominators never arise when solving the chain, since the left-hand side of (2.5.49) contains the derivative of the fast phase alone. Actually, as we know, resonances between the fast phase and high harmonics of the slow variables cause the resulting solution for w_n to be only asymptotic.

The procedure for obtaining the invariants in higher order is best illustrated by example. We first continue the calculation of Section 2.3c to obtain the adiabatic invariant for the slowly varying harmonic oscillator to second order. As a second application, we obtain the "ponderomotive force" acting on a particle moving in the presence of a fast-oscillating wave.

Slowly Varying Harmonic Oscillator. From (2.3.25), writing ω for ω_0 for ease

2.5. Lie Transformation Methods

of notation, the Hamiltonian is

$$H = \omega J + \epsilon \frac{\omega'}{2\omega} J \sin 2\theta, \qquad (2.5.50)$$

where

$$\omega = \omega(\epsilon t) \qquad (2.5.51)$$

and the prime is the derivative with respect to the argument ϵt. We have for convenience taken $G = 1$, $F = \omega^2(\epsilon t)$ in (2.3.23). In zero order,

$$\bar{H}_0 = H_0.$$

The first-order equation (2.5.31a) is

$$\omega \frac{\partial w_1}{\partial \theta} + \epsilon \frac{\partial w_1}{\partial (\epsilon t)} = \bar{H}_1 - H_1. \qquad (2.5.52)$$

We need the solution for w_1 to first order in ϵ, so we put

$$w_1 = w_{10} + \epsilon w_{11}, \qquad \bar{H}_1 = \bar{H}_{10} + \epsilon \bar{H}_{11}, \qquad (2.5.53)$$

to find

$$\omega \frac{\partial w_{10}}{\partial \theta} = \bar{H}_{10} - \frac{1}{2} \frac{\omega'}{\omega} J \sin 2\theta, \qquad (2.5.54a)$$

$$\omega \frac{\partial w_{11}}{\partial \theta} = \bar{H}_{11} - \frac{\partial w_{10}}{\partial (\epsilon t)}. \qquad (2.5.54b)$$

Setting $\bar{H}_{10} = 0$ to keep w_{10} from becoming secular, we can integrate (2.5.54a). Similarly, setting $\bar{H}_{11} = 0$ to keep w_{11} from becoming secular, we have

$$\bar{H}_1 = \bar{H}_{10} + \epsilon \bar{H}_{11} = 0. \qquad (2.5.55)$$

Substituting w_{10}, obtained from the integration, into (2.5.54b) and integrating, we obtain

$$w_1 = w_{10} + \epsilon w_{11}$$

$$= \frac{1}{4} \frac{\omega'}{\omega^2} J \cos 2\theta - \frac{\epsilon}{8\omega} \left(\frac{\omega'}{\omega^2}\right)' J \sin 2\theta. \qquad (2.5.56)$$

In second order, (2.5.31b) is

$$\omega \frac{\partial w_2}{\partial \theta} + \epsilon \frac{\partial w_2}{\partial (\epsilon t)} = 2(\bar{H}_2 - H_2) - L_1(\bar{H}_1 + H_1). \qquad (2.5.57)$$

Solving this to zero order in ϵ, we drop the second term on the left and to zero order put $L_1 = L_{f1} = [w_{10}, \ ]$ in the second term on the right to obtain

$$\omega \frac{\partial w_{20}}{\partial \theta} = 2\bar{H}_{20} - [w_{10}, H_1], \qquad (2.5.58)$$

where from (2.5.50). $H_2 \equiv 0$. The Poisson bracket is

$$[w_{10}, H_1] = -\frac{1}{4}\frac{\omega'^2}{\omega^3} J \qquad (2.5.59)$$

having only an average part. Thus to keep w_{20} from becoming secular, we choose

$$\bar{H}_{20} = -\frac{1}{8}\frac{\omega'^2}{\omega^3} J \qquad (2.5.60)$$

and find $w_{20} = 0$.

To second order, the new Hamiltonian is thus

$$\bar{H} = \omega \bar{J} - \frac{1}{8}\frac{\omega'^2}{\omega^3} \bar{J}, \qquad (2.5.61)$$

where, as previously, we interpret J as $\bar{J}$. Differentiating with respect to $\bar{J}$,

$$\bar{\omega} = \omega\left[1 - \frac{1}{8}\left(\frac{\omega'}{\omega^2}\right)^2\right], \qquad (2.5.62)$$

the new frequency.

The new action $\bar{J}$ is found in terms of the old variables from (2.5.8) as

$$\bar{J} = TJ. \qquad (2.5.63)$$

Using (2.5.24a) and (2.5.24b), we have to second order

$$\bar{J} = J - \epsilon[w_1, J] + \tfrac{1}{2}\epsilon^2[w_{10}, [w_{10}, J]], \qquad (2.5.64)$$

where we need w_1 to first order in the second term on the right. Using (2.5.56),

$$\bar{J} = J + \frac{\epsilon}{2}\frac{\omega'}{\omega^2} J \sin 2\theta + \frac{\epsilon^2}{8}\left(\frac{\omega'}{\omega^2}\right)^2 J$$

$$+ \frac{\epsilon^2}{4\omega}\left(\frac{\omega'}{\omega^2}\right)' J \cos 2\theta, \qquad (2.5.65)$$

where $\bar{J} = $ const. from (2.5.61). Thus (2.5.65) produces phase space trajectories of $J - \theta$ along the level lines of $\bar{J} = $ const. This result agrees with calculations by Littlewood (1963), Whiteman and McNamara (1968), and Stern (1970a), but with much reduced complexity of calculation compared to the pre-Lie methods.

The Ponderomotive Force. As a second example of an adiabatic calculation, we find the average force seen by a charged particle in the field of an

2.5. Lie Transformation Methods

electrostatic wave with a slowly varying amplitude. It is well known that the average force is quadratic in the wave amplitude, thus requiring a second-order calculation. We take the Hamiltonian of the form

$$H_w(p, x, t) = \frac{1}{2m} p^2 + \epsilon e \Phi(\epsilon kx) \cos(kx - \omega t), \qquad (2.5.66)$$

where ϵ is an ordering parameter to be set equal to one at the end of the calculation. In an extended phase space, the Hamiltonian becomes

$$H_e(p, E, x, t) = \frac{1}{2m} p^2 + E + \epsilon e \Phi(\epsilon kx) \cos(kx - \omega t), \qquad (2.5.67)$$

with $E = -H_w$ the momentum conjugate to t. Using the generating function

$$F_2 = (-kx + \omega t) J_\theta + kx J_\phi, \qquad (2.5.68)$$

we have

$$E = \frac{\partial F_2}{\partial t} = \omega J_\theta, \qquad \phi = \frac{\partial F_2}{\partial J_\phi} = kx,$$

$$p = \frac{\partial F_2}{\partial x} = k(J_\phi - J_\theta), \qquad \theta = \frac{\partial F_2}{\partial J_\theta} = \omega t - kx. \qquad (2.5.69)$$

The new Hamiltonian H is now in a form that exhibits the fast and slow phases θ and $\epsilon\phi$, respectively,

$$H = H_0 + \epsilon e \Phi(\epsilon\phi) \cos\theta, \qquad (2.5.70)$$

where the unperturbed Hamiltonian

$$H_0 = \frac{k^2}{2m} (J_\phi - J_\theta)^2 + \omega J_\theta \qquad (2.5.71)$$

has the frequencies

$$\omega_\phi = \frac{k^2}{m} (J_\phi - J_\theta) = kv_x, \qquad (2.5.72)$$

$$\omega_\theta = \omega - \frac{k^2}{m} (J_\phi - J_\theta) = \omega - kv_x. \qquad (2.5.73)$$

For the ordering we have chosen, we see from the above that the particle must not be near resonance with the wave, which requires that

$$\left| \frac{\omega}{k} - v_x \right| \gg v_x. \qquad (2.5.74)$$

We now introduce the Lie transformations. In zero order, $\bar{H}_0 = H_0$. In first order,

$$\omega_\theta \frac{\partial w_1}{\partial \theta} + \epsilon \omega_\phi \frac{\partial w_1}{\partial (\epsilon \phi)} = \bar{H}_1 - e\Phi \cos \theta. \tag{2.5.75}$$

We need w_1 to first order in ϵ, so we put

$$w_1 = w_{10} + \epsilon w_{11}, \quad \bar{H}_1 = \bar{H}_{10} + \epsilon \bar{H}_{11} \tag{2.5.76}$$

to find

$$\omega_\theta \frac{\partial w_{10}}{\partial \theta} = \bar{H}_{10} - e\Phi \cos \theta, \tag{2.5.77a}$$

$$\omega_\theta \frac{\partial w_{11}}{\partial \theta} = \bar{H}_{11} - \omega_\phi \frac{\partial w_{10}}{\partial (\epsilon \phi)}. \tag{2.5.77b}$$

Since H_1 has zero average value, we choose $\bar{H}_{10} = 0$, and on integrating (2.5.77a),

$$w_{10} = -\frac{e\Phi}{\omega_\theta} \sin \theta. \tag{2.5.78a}$$

Using this in the second equation, there is no secularity if we set $\bar{H}_{11} = 0$ and integrating,

$$w_{11} = -\frac{\omega_\phi}{\omega_\theta} \frac{e\Phi'}{\omega_\theta} \cos \theta. \tag{2.5.78b}$$

Proceeding to second order, we need w_2 to zero order in ϵ so we drop higher-order terms in (2.5.31b) to get

$$\omega_\theta \frac{\partial w_{20}}{\partial \theta} = 2\bar{H}_{20} + \frac{\partial w_{10}}{\partial J_\theta} \frac{\partial H_1}{\partial \theta}, \tag{2.5.79}$$

since $H_2 = 0$, $\bar{H}_1 = 0$, and $\partial H_1/\partial J_\theta = 0$. Choosing $\bar{H}_{20}$ to eliminate the secularity in w_{20},

$$\bar{H}_{20} = -\frac{1}{2} \left\langle \frac{\partial w_{10}}{\partial J_\theta} \frac{\partial H_1}{\partial \theta} \right\rangle_\theta, \tag{2.5.80}$$

and using (2.5.78a),

$$\bar{H}_{20} = \frac{1}{4} \frac{e^2 \Phi^2(\epsilon \phi)}{\omega_\theta^2} \frac{\partial \omega_\theta}{\partial J_\theta}, \tag{2.5.81}$$

which shows the quadratic dependence on wave amplitude. Transforming $\bar{H} = \bar{H}_0 + \epsilon \bar{H}_1 + \epsilon^2 \bar{H}_2$ back to the original coordinates (with $\epsilon \equiv 1$),

$$H_w^p = \frac{p^2}{2m} + \frac{k^2}{4m} \frac{e^2 \Phi^2(kx)}{(\omega - kv_x)^2} = \text{const.}, \tag{2.5.82}$$

2.5. Lie Transformation Methods

showing that H_w^p, the ponderomotive Hamiltonian, is the sum of the kinetic energy and a "ponderomotive potential energy" $U(x)$. The ponderomotive force is

$$F^p = -\frac{\partial U}{\partial x} = -\frac{e^2 k^3}{2m} \frac{\Phi \Phi'}{(\omega - k v_x)^2}, \qquad (2.5.83)$$

which is zero near a maximum or a minimum in the wave amplitude; the latter is the stable equilibrium position of the charge.

The ponderomotive force gets very large near resonance, but the adiabatic ordering $\omega_\theta \gg \omega_\phi$ ceases to be valid there. This case must be treated using secular perturbation theory, or, in higher order, with the method given below.

Removal of Resonances. As was shown in Section 2.4, adiabatic invariants are topologically modified in the neighborhood of resonances. For a single resonance, the transformation (2.4.6) to a rotating coordinate system (the "hat" variables) accounts for the topological change and forms the basis of secular perturbation theory, which in first order was presented in Section 2.4a. In the rotating system, for two degrees of freedom, the separation into a fast and a slow phase variation has been accomplished. Therefore, the methods of this section may be directly applied to calculate the modified invariants in the neighborhood of a resonance to higher than first order.

McNamara (1978) has applied the Lie technique to the method of Dunnett et al. (1968), which was described in Section 2.4d. Recall that using this method, we construct a first-order invariant for a certain class of problems in which an entire set of resonances is removed at once. The method is based on the observation that if J is an invariant of the unperturbed system, then any function $I_0(J)$ is also an invariant. By choosing I_0' to vanish at the resonant values of J, the change of topology for the invariant $\bar{I}$ of the perturbed system can be accommodated. The Lie technique naturally allows the function I_0 to replace the identity function $\mathcal{I}(J) = J$ as follows: by the method in this subsection, the evolution operator T is calculated to the desired order n. Then instead of writing the invariant

$$\bar{J} = TJ = \text{const.} \qquad (2.5.84)$$

as in (2.5.63), we use the functional form as in (2.5.13) to obtain

$$\bar{I} = T^{-1} I_0 = \text{const.}, \qquad (2.5.85)$$

where $I_0(J)$ is chosen to eliminate the poles from the function $\bar{I}(J)$. McNamara shows that I_0 need not be chosen until the nth order has been reached and gives rules for choosing it.

Wave-Particle Resonance. We illustrate the results of this technique for the Hamiltonian (2.4.108), corresponding to wave propagation at 45° for the problem described in Section 2.2b. In first order from (2.5.31a),

$$P_\psi \frac{\partial w_1}{\partial \psi} + \frac{\partial w_1}{\partial \phi} = -\sum_{m=0}^{\infty} \mathscr{I}_m(\rho) \sin(\psi - m\phi), \qquad (2.5.86)$$

with the solution

$$w_1 = \sum_{m=0}^{\infty} \mathscr{I}_m(\rho) \frac{\cos(\psi - m\phi)}{P_\psi - m}. \qquad (2.5.87)$$

The invariant is then, using (2.5.25a),

$$\bar{I} = I_0(P_\psi) + \epsilon[w_1, I_0] \qquad (2.5.88)$$

or

$$\bar{I} = I_0 + \epsilon \frac{\partial w_1}{\partial \psi} \frac{dI_0}{dP_\psi}. \qquad (2.5.89)$$

With the choice of dI_0/dP_ψ given in (2.4.109), we remove the resonance in (2.5.87) and obtain $\bar{I} = I$ as given in (2.4.111). The primary resonances at $P_\psi = +1, 0, -1$ are well described by the invariant, as discussed in Section 2.4d.

Proceeding to second order, one solves (2.5.31b) to obtain w_2. The invariant is from (2.5.25):

$$\bar{I} = I_0 + \epsilon[w_1, I_0]$$
$$+ \frac{\epsilon^2}{2}([w_2, I_0] + [w_1, [w_1, I_0]]), \qquad (2.5.90)$$

which has second-order poles for $P_\psi =$ integers and first-order poles for $P_\psi =$ half integers. Choosing

$$\frac{dI_0}{dP_\psi} = \sin^2 \pi P_\psi \sin 2\pi P_\psi \qquad (2.5.91)$$

eliminates these poles from $\bar{I}$. A plot of the second-order invariant $\bar{I}$ is given in Fig. 2.13. This should be compared with the plot of the first-order invariant shown in Fig. 2.12b and with the exact numerical orbit calculations shown in Fig. 2.10b. The half-integer resonances are well represented in Fig. 2.13, and the drifting orbits are likewise similar to those seen in Fig. 2.10b. However, the primary (integer) resonances are spuriously broken up, although for a smaller perturbation strength the agreement is much better (McNamara, 1978). There are many additional questions to be answered concerning the behavior of these invariants.

2.6. Superconvergent Methods

Figure 2.13. Resonance invariant correct to $\mathcal{O}(\epsilon^2)$ at $\epsilon = 0.1$. The second harmonics are well represented but the breakup of the primary resonances is quite wrong. This figure should be compared with the orbit computations in Fig. 2.10b, and the resonant invariant correct to $\mathcal{O}(\epsilon)$ in Fig. 2.12b (after McNamara, 1978).

2.6. Superconvergent Methods

We now describe a class of perturbation methods originally due to Kolmogorov (1954), which, as described in Chapter 3, play a fundamental role in the proof of the KAM theorem. The key feature is their extremely rapid convergence when proceeding to higher order. In the methods described earlier in this chapter, the Hamiltonian

$$H = H_0 + \epsilon H_1$$

is transformed by successive canonical transformations that are chosen to increase the order of perturbation by one power of ϵ in every step. Letting H_n be the untransformed part of the Hamiltonian after the nth transformation,

$$\epsilon H_1 \to \epsilon^2 H_2 \to \epsilon^3 H_3 \to \cdots \epsilon^n H_n.$$

This is clearly seen from the Lie series (2.5.31), where choosing the Lie generating function w_1 eliminates ϵH_1 but leaves $\epsilon^2 H_2$, choosing w_2 eliminates $\epsilon^2 H_2$ but leaves $\epsilon^3 H_3$, etc. Kolmogorov showed that successive canonical transformations may be chosen such that the order of the perturbation is increased by the square of the preceding one for each step:

$$\epsilon H_1 \to \epsilon^2 H_2 \to \epsilon^4 H_3 \to \cdots \epsilon^{2^{n-1}} H_n.$$

Thus after the nth transformation, one has solved the problem to order $\epsilon^{2^{n-1}}$. This phenomenon is called *superconvergence* or, sometimes, quadratic convergence.

An analogy between the two methods, which is also of practical use, arises when finding the root of a function. Following Moser (1973) and Berry (1978), we illustrate this procedure. We wish to find the value of x that satisfies the equation

$$f(x) = 0, \qquad (2.6.1)$$

where we know an x_0 that satisfies an unperturbed problem that may be just an educated guess as to the root. Starting at x_0, we obtain the next approximation x_1 from a Taylor's expansion about x_0,

$$f(x_0) + f'(x_0)(x_1 - x_0) = 0. \qquad (2.6.2)$$

If we carried the expansion to higher order in the small parameter $\epsilon = (x - x_0)$, we would then generate a series in powers of ϵ. At the nth step, we would solve the finite polynomial equation

$$\sum_{m=0}^{n} \frac{1}{m!} f^{(m)}(x_0)(x_n - x_0)^m = 0 \qquad (2.6.3)$$

for the nth approximation x_n. By subtracting (2.6.3) from the Taylor series expansion of $f(x)$ about x_0, we obtain the error $e_n = x - x_n$ after the nth step

$$e_n \sim \frac{1}{(n+1)!} \frac{f^{(n+1)}(x_0)}{f'(x_0)} \epsilon^{n+1}, \qquad (2.6.4)$$

which is of order ϵ^{n+1}. This linear convergence is characteristic of the standard perturbation methods.

To exhibit quadratic convergence, we resort to Newton's root-finding method, with the first step given by (2.6.2). The error e_1 for the first step is found by expanding $f(x)$ about x_0 to quadratic terms and using (2.6.2),

$$e_1 \sim \alpha(x_0)(x - x_0)^2 = \alpha \epsilon^2, \qquad (2.6.5)$$

where

$$\alpha(x_0) = -\frac{1}{2} \frac{f''(x_0)}{f'(x_0)}. \qquad (2.6.6)$$

Thus the error is of order ϵ^2. For the next step, we repeat the expansion of

2.6. Superconvergent Methods

f about x_1 (not x_0) to find

$$x_2 - x_1 = -\frac{f(x_1)}{f'(x_1)} \tag{2.6.7}$$

with, using (2.6.5), an error

$$e_2 \sim \alpha(x_1)(x - x_1)^2 \sim \alpha^2(x_0)\alpha(x_1)\epsilon^4, \tag{2.6.8}$$

which is the order of ϵ^4. The rapid convergence occurs because the value about which the expansion is made is moved closer to the true value at each iteration. By induction, the error after n iterations is

$$(x - x_n) \sim \prod_{i=1}^{n} \alpha^{2^{n-i}}(x_{i-1})\epsilon^{2^n}. \tag{2.6.9}$$

The normal assumption is that $f''/f' = \mathcal{O}(1)$, such that the error is quadratically reduced with every step. However, even if f has a resonance very near the sought-for root, $f \sim (x - x_r)^{-1}$, such that the usual ordering breaks down with $x_r \sim \mathcal{O}(x_0)$ and thus,

$$\frac{f''}{f'} = \mathcal{O}(\epsilon^{-1}), \tag{2.6.10}$$

the quadratic convergence can still overcome this resonant behavior.[2] The analogous superconvergent expansion in the KAM procedure also overcomes the effect of nearby zeros in the denominators, due to resonances, which generate terms of the order of (2.6.10).

The superconvergent transformation method has not been widely applied to practical perturbation calculations. As we shall see, the method is generally too cumbersome to use in this way. Chirikov (1979) illustrates the superconvergent method by treating the one-dimensional pendulum using mixed variable generating functions. However, it is more natural to examine superconvergence using Lie transformations; this was first done by Howland (1977) for autonomous systems. In Section 2.6a, we show how superconvergence arises in Lie perturbation theory, following closely the analysis of Cary (1978), which applies as well to time-dependent systems. McNamara (1978) sketches how to use these superconvergent Lie methods in finding the modified invariants of the previous subsection; the reader should consult his paper for further details.

Related to the use of perturbation theory to obtain approximate solutions for near-integrable systems is the problem of finding the *singly periodic* solutions in such systems. These solutions are important because they are dense in the set of all solutions, and, once found, nearby solutions can be examined using linear theory (see Section 3.3). Techniques developed

---

[2] Actually, one requires $f''/f' = \mathcal{O}(\epsilon^{-\tau})$ with $\tau < 1$ for convergence; a similar restriction on τ is required for proof of the KAM theorem. See Section 3.2a for further details.

by Helleman, Eminhizer, and other collaborators (see Eminhizer et al., (1976)) are superconvergent and provide highly efficient means for calculating the periodic solutions. These methods are described in Section 2.6b, and illustrated using the Hénon and Heiles potential.

2.6a. Kolmogorov's Technique

Following Cary (1978), we start with the perturbed Hamiltonian

$$H = \sum_{n=0}^{\infty} \epsilon^n H_n(x, t), \tag{2.6.11}$$

where H_0 is integrable and $x = (J, \theta)$ are the action-angle variables of H_0. We shall introduce successive Lie transforms to successive new Hamiltonians, which we shall distinguish with a superscript. Following the general Lie method of Section 2.5b, we begin with the Deprit equations (2.5.31) for the first Lie generating function $w^{(1)}$:

$$\bar{H}_0^{(1)} = H_0, \tag{2.6.12a}$$

$$D_0 w_1^{(1)} = \bar{H}_1^{(1)} - H_1, \tag{2.6.12b}$$

$$D_0 w_2^{(1)} = 2(\bar{H}_2^{(1)} - H_2) - [w_1^{(1)}, (\bar{H}_1^{(1)} + H_1)]. \tag{2.6.12c}$$

In the usual Lie perturbation technique, we solve these equations order by order. In each order, we choose the new Hamiltonian $\bar{H}_i^{(1)}$ to eliminate secularities and then integrate the equation to find $w_i^{(1)}$. This is substituted into the next equation and the process is repeated. Solving each equation corresponds to a single "step," and after n steps, we have the new Hamiltonian $\bar{H}$, a function of the action $\bar{J}$ alone, to order ϵ^n.

In Kolmogorov's technique, we choose the $\bar{H}$'s and w's differently. The general rule is that we calculate the w's simultaneously to as many orders as we can, by choosing the $\bar{H}$'s as before to eliminate secularities, and integrating to find the w's, but without coupling w's of different orders together in the same equation. In the remaining orders, we set the w's equal to zero and use the equations to find the $\bar{H}$'s, which are then determined trivially. We regard this entire procedure as a single "step."

For the first step, we can "simultaneously" solve only one equation, namely, (2.6.12b). That is, setting $\bar{H}_1^{(1)} = \langle H_1 \rangle$, $w_1^{(1)}$ is determined from

$$D_0 w_1^{(1)} = -\{H_1\}, \tag{2.6.13}$$

$w_i^{(1)} = 0$ for $i > 1$, $\bar{H}_2^{(1)}$ is given by (2.6.12c) with $w_2^{(1)} = 0$, etc. We cannot solve simultaneously for $w_2^{(1)}$ when obtaining (2.6.13) because (2.6.12c) requires that $w_1^{(1)}$ is already known.

2.6. Superconvergent Methods

For the second step, we take

$$\bar{H}^{(1)} = \sum_{n=0}^{\infty} \epsilon^n \bar{H}_n^{(1)} \tag{2.6.14}$$

to be the new "old Hamiltonian," which we write in the form

$$H_0^{(1)} = \bar{H}_0^{(1)} + \epsilon \bar{H}_1^{(1)}, \tag{2.6.15a}$$

$$H_1^{(1)} = 0, \tag{2.6.15b}$$

and

$$H_i^{(1)} = \bar{H}_i^{(1)} \tag{2.6.15c}$$

for $i > 1$, i.e., we absorb all solvable parts into the unperturbed Hamiltonian. Applying a second transformation to $H^{(1)}$, we have $\bar{H}_0^{(2)} = H_0^{(1)}$, $w_1^{(2)} = 0$, and $\bar{H}_1^{(2)} = 0$, since there is no first-order perturbation. The succeeding Deprit equations are

$$D_0^{(1)} w_2^{(2)} = 2(\bar{H}_2^{(2)} - H_2^{(1)}), \tag{2.6.16a}$$

$$D_0^{(1)} w_3^{(2)} = 3(\bar{H}_3^{(2)} - H_3^{(1)}), \tag{2.6.16b}$$

$$D_0^{(1)} w_4^{(2)} = 4(\bar{H}_4^{(2)} - H_4^{(1)}) - [w_2^{(2)}, \bar{H}_2^{(2)} + H_2^{(1)}], \tag{2.6.16c}$$

etc., where

$$D_0^{(1)} \equiv \frac{\partial}{\partial t} + [\ , H_0^{(1)}] \tag{2.6.17}$$

is the total time derivative taken along the *first-order* orbits of the original system. We can now simultaneously solve the first *two* equations (2.6.16a) and (2.6.16b) choosing $\bar{H}_2^{(2)}$ and $\bar{H}_3^{(2)}$ to eliminate secularities in $w_2^{(2)}$ and $w_3^{(2)}$, respectively. In the remaining equations we set the $w^{(2)}$'s to zero and solve for the $\bar{H}^{(2)}$'s. We now have a new Hamiltonian of the form

$$H^{(2)} = H_0^{(2)} + \epsilon^4 H_4^{(2)} + \epsilon^5 H_5^{(2)} + \cdots, \tag{2.6.18}$$

where

$$H_0^{(2)} = \bar{H}_0^{(2)} + \epsilon^2 \bar{H}_2^{(2)} + \epsilon^3 \bar{H}_3^{(2)} \tag{2.6.19}$$

is a function of action alone, and

$$H_i^{(2)} = \bar{H}_i^{(2)} \tag{2.6.20}$$

for $i > 3$ are the perturbations. Thus in the second step, terms of both $\mathcal{O}(\epsilon^2)$ and $\mathcal{O}(\epsilon^3)$ have been eliminated.

Proceeding to the third step, we could simultaneously eliminate terms of orders four through seven. This is seen by induction from the general Deprit equation (2.5.29): suppose the perturbation terms in $H^{(n)}$ are $\mathcal{O}(\epsilon^{2^n})$ or greater. Applying the transformation generated by $w^{(n+1)}$, we must have $w_i^{(n+1)} = 0$, $L_i^{(n+1)} = 0$, $(T^{-1})_i^{(n+1)} = 0$, and $\bar{H}_i^{(n+1)} = 0$ for $i < 2^n$. From

(2.5.29), this implies

$$D_0^{(n)} w_i^{(n+1)} = i(\bar{H}_i^{(n+1)} - H_i^{(n)}) \qquad (2.6.21)$$

for $u < 2^{n+1}$. Thus the equations from $i = 2^n$ through $i = 2^{n+1} - 1$ are simultaneously solvable.

We note that each "step" in Kolmogorov's technique is much more complicated than the corresponding step using the ordinary technique and, moreover, that each step involves a different integration over the orbits of the system, perhaps expressing the general conservation law that one may climb to a given height using many small steps or a few large ones! We conclude that it may not be the best way of doing practical calculations. On the other hand, Kolmogorov's technique increases the singularity from a resonant denominator d by just one power for each "large" step; i.e., when doing perturbation theory to order 2^n, resonance denominators of order $1/d^n$ appear in Kolmogorov's technique, but these are of order $1/d^{(2^{n-1})}$ using the standard technique. This fact, together with the quadratic convergence, is crucial in the KAM demonstration that *convergent* series expansions can be found sufficiently far from resonances.

2.6b. Singly Periodic Orbits

For the class of singly periodic orbits, which close on themselves, it is possible to construct convergent series solutions for systems with more than one degree of freedom. This is possible because the exact periodicity condition allows a representation in a single Fourier series, thus avoiding the resonant denominators encountered in Section 2.2b. However, in order to ensure exact resonance, within a perturbation scheme, it is necessary to vary the initial conditions, while holding the frequency fixed, at each successive approximation. This is in contrast to the usual perturbation methods in which, for a given set of initial conditions, the successive approximations improve the accuracy of determining the frequency.

The technique for obtaining periodic solutions by holding the frequency and phase fixed, while varying the initial displacement and momentum, was developed by Helleman, Eminhizer, and co-workers and applied to a variety of problems. These have included the driven anharmonic oscillator (Eminhizer *et al.*, 1976), the two-degree-of-freedom oscillator with the Hénon and Heiles potential (Helleman and Bountis, 1979; see Section 1.4a), and Hénon's quadratic twist mapping (Helleman, 1977; see Section 3.2d). They named the expansion procedure the "backward scheme" as it proceeds backwards from the frequency to the initial conditions, rather than in the more usual direction.

Although the class of periodic solutions has zero measure in the phase space, they densely fill the space. This corresponds to the number-theoretic concept that the rationals have zero measure on the unit line, while at the

2.6. Superconvergent Methods

same time closely approximating the irrationals. Thus at first sight we might think that determination of the periodic solutions might be equivalent to determination of all solutions. This is, in fact, not the case, as the periodic solutions do not in general bound the aperiodic ones. Thus, although the solutions may initially be close together, they can evolve to positions arbitrarily far apart in the phase space at a later time. For two degrees of freedom, KAM solutions can bound nearby solutions and are thus much more useful in characterizing the behavior of the entire set of initial conditions. Nevertheless, the periodic solutions can be quite useful in two ways. They can indicate certain properties of the general structure of the solutions, such as the appearance of second- or higher-order islands of a given rotation number, and they can be used as a base about which linear stability can be investigated. This latter property can be used in determining global stability, as we shall see in Section 4.4.

Variational Theory. We consider closed orbits in autonomous systems with N degrees of freedom. The motion is then simply periodic; i.e., there exist $N - 1$ resonance conditions of the form

$$l_k \cdot \omega = 0, \tag{2.6.22}$$

where the l's are linearly independent, N-dimensional integer vectors. We can then write all the ω's in terms of integer multiples of a basic recurrence frequency ω_r,

$$\omega_j = m_j \omega_r, \tag{2.6.23}$$

and thus the coordinates q_j are given by a single Fourier series

$$q_j = \sum_{m=-\infty}^{\infty} Q_{jm} \exp(im\omega_r t), \tag{2.6.24a}$$

with their velocities

$$\dot{q}_j = \sum_{m=-\infty}^{\infty} im\omega_r Q_{jm} \exp(im\omega_r t). \tag{2.6.24b}$$

The variational equation for closed orbits is most conveniently given in terms of the Lagrangian (1.2.1) and is found from Hamilton's principle

$$\delta \oint L(\dot{q}_j, q_j) dt = \delta \langle L \rangle = 0, \tag{2.6.25}$$

where $\langle L \rangle$ is the integral of L over a recurrence time $2\pi/\omega_r$. We consider $\dot{q}_j$ and q_j to be functions of the Fourier mode coefficients Q_{jm} given in (2.6.24), and perform the variation with respect to the Q's. Thus,

$$\delta \langle L \rangle \equiv \frac{\partial \langle L \rangle}{\partial Q_{jm}} dQ_{jm}, \tag{2.6.26}$$

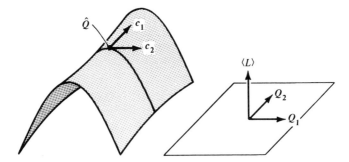

Figure 2.14. A two-dimensional saddle. The stationary point $\hat{Q}$, where the time-averaged Lagrangian $\langle L \rangle$ is stationary, is generally a saddlepoint, neither a maximum nor a minimum. Also shown are the new coordinates c_1, c_2 for which the curvature matrix is diagonal.

and from the calculus of variations,

$$\frac{\partial \langle L \rangle}{\partial Q_{jm}} = 0 \tag{2.6.27}$$

if

$$\frac{d}{dt}\frac{\partial L}{\partial \dot{q}_j} - \frac{\partial L}{\partial q_j} = 0, \tag{2.6.28}$$

which are Lagrange's equations. Note that $\langle L \rangle$ is a purely algebraic function of the Q's (and ω_r). Given $\langle L \rangle$, the Q's are determined by finding the roots of the set of equations (2.6.27), i.e., the stationary point of $\langle L \rangle$. The difficulty in finding these roots is that the stationary point $\hat{Q}$ is generally not a minimum or a maximum of $\langle L \rangle$ but a saddlepoint (see Fig. 2.14) whose type changes discontinuously as the mode coefficients Q_{jm} (and thus the initial $\dot{q}_j$ and q_j) are varied. Suppose we try a simple Newton root-finding method. Setting $f = \partial \langle L \rangle / \partial Q$, then from (2.6.27),

$$f(Q_0) + \Delta Q \cdot \frac{\partial f}{\partial Q_0} = 0, \tag{2.6.29}$$

where Q is the vector of the Q_{jm}'s, Q_0 is the guess, and $\Delta Q = Q_1 - Q_0$ is the first correction. Substituting for f, we have

$$\Delta Q \cdot \frac{\partial^2 \langle L \rangle}{\partial Q_0^2} = -\frac{\partial \langle L \rangle}{\partial Q_0}. \tag{2.6.30}$$

Diagonalizing the curvature matrix and transforming to the diagonalized

2.6. Superconvergent Methods

system,

$$\frac{\partial^2 \langle L \rangle}{\partial Q_0^2} \to \lambda(c_0) \quad \text{(diagonal)},$$

$$\Delta Q \to \Delta c, \quad (2.6.31)$$

$$\frac{\partial \langle L \rangle}{\partial Q_0} \to \langle L \rangle'(c_0),$$

(see Fig. 2.14), then (2.6.30) becomes

$$\lambda_n(c_0)\Delta c_n = -\langle L \rangle'_n(c_0), \quad (2.6.32)$$

where n labels the components of the vector of mode coefficients c in the diagonalized system. The problem now is that some of the principal curvatures $\lambda_n(\hat{c})$ are negative, some positive, and others near zero. Let us order them with increasing n from negative to positive, with one curvature $\lambda_m(\hat{c}) = 0$ (see Helleman, 1978). When computed with an initial guess using Newton's method, this ordering is not preserved in the iteration. This can be corrected by inserting a convergence factor into Newton's method

$$(\lambda_n - \bar{\lambda})\Delta c_n = -\langle L \rangle'_n, \quad (2.6.33)$$

where $\bar{\lambda}$ is chosen to make $\lambda_n - \bar{\lambda}$ always have the same sign as one iterates. This relocated Newton's method converges (linearly) to $\hat{c}_n$. Furthermore, when the iterated solution gets sufficiently near $\hat{c}_n$, we can reduce $\bar{\lambda}$ to zero, thus reverting to the original Newton's method with its quadratic convergence. A convenient choice for $\bar{\lambda}$ is $\lambda_m(c)$, since for $n \gtrless m$, $\lambda_n(c) - \lambda_m(c) \gtrless 0$, as required for convergence in (2.6.33), and $\lambda_m \to 0$ as $c \to \hat{c}$. We thus specify the value of $n = m$ at which the curvature is zero (and specify ω_r) instead of the usual initial values $\dot{q}_j(0)$ and $q_j(0)$. The usual constraints on the Q's

$$q_j(0) = \sum_m Q_{jm}, \quad (2.6.34a)$$

$$\dot{q}_j(0) = \sum_m im\omega_r Q_{jm}, \quad (2.6.34b)$$

are then replaced by a new constraint for each vector c

$$\langle L \rangle'_m(c) = 0, \quad (2.6.35)$$

which ensures $\lambda_m = 0$ to first order in Δc_m.

In practice, we work in the original, not the diagonalized coordinates (the Q's, not the c's), and moreover we truncate the Fourier series at a finite number of terms. Various other devices are also introduced to reduce the computational labor or modify the convergence rate; the reader should refer to the original literature for further details.

The Hénon and Heiles Problem. Here we follow most closely the work of Bountis (1978) and Helleman and Bountis (1979) as an example of the general method for systems with two degrees of freedom. We have the resonance condition

$$s\omega_1 - r\omega_2 = 0, \qquad (2.6.36)$$

where

$$\alpha = \frac{r}{s} \qquad (2.6.37)$$

is a specified constant, r and s being relatively prime integers. The frequencies ω_1 and ω_2 are no longer independent, but related to the common frequency

$$\omega_r = \frac{\omega_1}{r} = \frac{\omega_2}{s} \qquad (2.6.38)$$

with the recurrence time being $\tau_r = 2\pi/\omega_r$. Taking the spatial coordinates as x and y, corresponding to frequencies ω_1 and ω_2, respectively, the solutions can then be written in terms of a single Fourier series (2.6.24a) for each degree of freedom:

$$x(t) = \sum_{n=-\infty}^{\infty} A_n \exp(in\omega_r t), \quad y(t) = \sum_{n=-\infty}^{\infty} B_n \exp(in\omega_r t). \quad (2.6.39)$$

In the backward method, we could specify ω_r and two of the four initial conditions x, y, $\dot{x}$, and $\dot{y}$, solving recursively for the mode coefficients A_n and B_n and the remaining two initial conditions. Alternatively, since only r and s need be specified to ensure a periodic solution, we specify three initial conditions, with ω_r and the fourth condition found from the series solution. The primary frequencies associated with the Fourier series of (2.6.39) are, of course, the original frequencies of (2.6.36), and thus we expect the main terms in the Fourier expansion to be $A_{\pm r}$ and $B_{\pm s}$. The recognition of these primary terms can help in obtaining rapid convergence.

To illustrate the procedure, following Bountis (1978), we consider the Hénon and Heiles potential given by (1.4.4) and Fig. 1.11, which give Lagrange's equations, equivalent to (2.6.28),

$$\ddot{x} = -x - 2xy, \qquad (2.6.40a)$$

$$\ddot{y} = -y + y^2 - x^2. \qquad (2.6.40b)$$

Recognizing that the primary frequencies are those given in (2.6.36), the above equations are rewritten as

$$\ddot{x} + \omega_1^2 x = \epsilon[(\omega_1^2 - 1)x - 2xy], \qquad (2.6.41a)$$

$$\ddot{y} + \omega_2^2 y = \epsilon[(\omega_2^2 - 1)y + y^2 - x^2], \qquad (2.6.41b)$$

2.6. Superconvergent Methods

where the right-hand sides are in some sense small and are therefore multiplied by a small parameter ϵ in order to solve recursively. Equations (2.6.41) are equivalent to the "relocated" Newton equation (2.6.33) (Helleman and Bountis, 1978). Now the Fourier coefficients in (2.6.39) are expanded in powers of ϵ:

$$x(t) = \sum_{j=0}^{\infty} \epsilon^j \sum_{n=-\infty}^{\infty} A_{nj} \exp(in\omega_r t),$$
$$y(t) = \sum_{j=0}^{\infty} \epsilon^j \sum_{n=-\infty}^{\infty} B_{nj} \exp(in\omega_r t). \qquad (2.6.42)$$

Substituting these in (2.6.41) and equating powers of ϵ for each Fourier mode, we get a set of recursive relations for the Fourier coefficients:

$$(r^2 - n^2)\omega_r^2 A_{n, j+1} = (r^2\omega_r^2 - 1)A_{nj} - 2 \sum_{k=-\infty}^{\infty} A_{kj} B_{n-k, j}, \qquad (2.6.43a)$$

$$(s^2 - n^2)\omega_r^2 B_{n, j+1} = (s^2\omega_r^2 - 1)B_{nj} + \sum_{k=-\infty}^{\infty} B_{kj} B_{n-k, j} - \sum_{k=-\infty}^{\infty} A_{kj} A_{n-k, j}. \qquad (2.6.43b)$$

These equations can be used to solve for all but the main coefficients $n = r, s$, for which the left-hand sides (l.h.s.) vanish. These coefficients are most easily obtained by specifying, in addition to r and s, three initial conditions, rather than two conditions and the fundamental frequency ω_r. For simplicity, one can choose $x = x_0$ at $t = 0$, and take the Fourier coefficients to be real, $A_n = A_{-n}$, $B_n = B_{-n}$, such that $\dot{x}(0) = \dot{y}(0) = 0$. With these convenient choices, A_r is now computed at each iteration, directly from (2.6.42) as

$$A_{r, j+1} = \tfrac{1}{2}x_0 - \tfrac{1}{2}A_{0j} - \sum_{\substack{n=1 \\ n \neq r}}^{\infty} A_{nj}, \qquad (2.6.44)$$

and the frequency from (2.6.43a), with the l.h.s. $= 0$, is also found recursively

$$\omega_{r, j+1}^2 = \frac{1}{r^2 A_{rj}} \left[A_{rj} + 2\sum_{k} A_{kj} B_{r-k, j} \right]. \qquad (2.6.45)$$

Rather than find B_0 from (2.6.43b) with $n = 0$ and B_s from the same equation with $n = s$, the procedure was reversed to achieve faster convergence. If we estimate the lowest order $B_{0,0}$ from (2.6.43b) with the l.h.s. $= 0$ and $B_{\pm s, 0}$ and $A_{\pm s, 0}$ the only nonzero coefficients, then we find

$$(s^2\omega_r^2 - 1) + 2B_{0,0} = 0. \qquad (2.6.46)$$

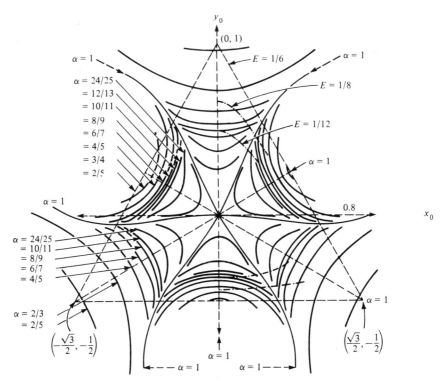

Figure 2.15. Converged initial displacements x_0, y_0 of the periodic solutions of the Hénon–Heiles system [with $\dot{x}(0) = 0 = \dot{y}(0)$] at different values of $\alpha \equiv r/s$; only the "primary" solutions are used; i.e., r and s are relatively prime (after Helleman and Bountis, 1979).

Then B_s can be obtained recursively from (2.6.43b) with $n = 0$,

$$2B_{s,j+1}^2 = B_{0j} - \sum_{k \neq \pm s} B_{kj}^2 + \sum_k A_{kj}^2, \qquad (2.6.47)$$

where B_{0j} is obtained, in each order, from the same equation with $n = s$,

$$B_{0,j+1} = \frac{1}{2B_{sj}}\left[(1 - s^2\omega_r^2)B_{sj} - \sum_{k \neq 0,s} B_{kj}B_{s-k,j} + \sum_k A_{kj}A_{s-k,j}\right]. \qquad (2.6.48)$$

Equation (2.6.47) is the equivalent to the constraint equation (2.6.35). All of the other A_n and B_n are obtained recursively from (2.6.43) with n truncated at some convenient value. At any order the initial y displacement is then obtained directly from (2.6.39), with $\epsilon \equiv 1$.

Using this expansion procedure, Bountis (1978) plots curves of initial displacement x_0, y_0 at fixed α and continuously varying ω_r. Successive values of $\alpha = r/s$ are chosen to better approximate $\alpha = 1$ (the relation between the linearized frequencies). This is reproduced in Fig. 2.15. Each constant α curve

2.6. Superconvergent Methods

represents a continuous range of ω_r values, but the solutions are discrete in α, a small variation in α leading to very different values of ω_r, r, and s.

Although we have followed the approach of the originators of the method in presenting the expansions in terms of the Lagrange equations of motion, the development could equally well have been carried out in canonical form by use of the substitution $\dot{x} = p$. However, for the periodic solutions, there is no apparent advantage in presenting the solutions in canonical form. In fact, one prefers a form in which the coordinates appear as simple powers (squares, cubes, fourth powers) in the Lagrangian. Otherwise, the method becomes cumbersome as large numbers of Fourier expansions have to be multiplied together, leading to high-order multiple sums in the recursion equations for the Fourier coefficients.

CHAPTER 3

Mappings and Linear Stability

Certain topological considerations can facilitate our visualization and consequent understanding of multiply periodic motion. They lead naturally to a set of difference equations, i.e., a *mapping* of the dynamical trajectory onto a subspace of the system phase space. These mappings allow easy numerical visualization of the motion for problems of two degrees of freedom. Moreover, mathematical proofs concerned with the existence of various types of orbits and theoretical and numerical calculations of stochastic behavior can usually be approached most conveniently from the equations of a mapping. On the other hand, as we have seen in Chapter 2, regular motion is often conveniently described in terms of differential equations. Conversion of (Hamilton's) differential equations into mappings, and *vice versa*, are common devices for calculating the motion of most nonlinear dynamical systems.

We consider three ways in which mappings can arise:

(a) directly from the physical statement of a dynamical problem, such as in Fermi acceleration discussed in Chapter 1 and treated in Section 3.4;
(b) from the successive intersections of a dynamical trajectory with a surface of section, as in the Hénon–Heiles problem of Chapter 1; and
(c) as an approximation to the motion, valid over some limited time scale or range of phase space, as for the motion near the separatrix of a periodically perturbed pendulum, treated in Section 3.5.

In this chapter we concern ourselves with canonical mappings related to the trajectories of nonlinear coupled oscillators with N degrees of freedom, but with special emphasis on systems with two degrees of freedom.

3. Mappings and Linear Stability

In Section 3.1 we make the connection between Hamiltonian systems and canonical mappings by means of a surface of section in phase space. We introduce the perturbed twist mappings, show how to calculate the mapping in terms of the given Hamiltonian, and how, given a mapping, to recover a Hamiltonian description. These results apply generally to systems with two or more degrees of freedom.

In Section 3.2 we discuss the mathematical results that lead to important rigorous estimates of a system's behavior, including the KAM theorem that proves orbit stability for sufficiently incommensurable orbits under sufficiently small perturbation. We examine the condition of "moderate" non-linearity that is necessary for the theoretical treatment of regular motion. For commensurate orbits, the Poincaré–Birkhoff fixed point theorem is described, and the structure near these orbits is examined, including elliptic and hyperbolic fixed points, resonance layers and stochastic motion. We present the results of various mathematical investigations, but make no attempt at mathematical rigor. For the more mathematically oriented reader, the reviews by Arnold and Avez (1968) and Moser (1973), in which many of the mathematical proofs are outlined, can supply much of this information. For the most part, we confine our attention to systems with two degrees of freedom.

In Section 3.3 we consider linearized motion and stability of motion near fixed points. The calculation is done explicitly for two- and four-dimensional mappings relating the eigenvalues of the linearized motion to properties of the defining matrices. In Section 3.3d, more general concepts of stability are considered.

In Section 3.4 we introduce various mapping models that approximate physical systems. Particular attention is given to the class of problems known as "Fermi acceleration." The fixed points (periodic solutions) and the stability of the linearized motion in their neighborhood are studied analytically and compared with numerical results. The Hamiltonian form of the mapping is also derived.

Section 3.5 treats the problem of motion near the separatrix of a periodically perturbed pendulum. The conversion from Hamiltonian form to a mapping and the character of the linearized motion are described. The formulation has been used by Chirikov (1979) to calculate the transitions between regular and stochastic motion, and we review the procedure in Chapter 4. It is also useful for the calculation of Arnold diffusion rates presented in Chapter 6.

In Section 3.6 we consider mappings that are obtained from perturbed linear oscillators. For continuous systems these perturbations were analyzed in Section 2.4 using secular perturbation theory for the intrinsically degenerate case. The result was that small perturbations lead to large changes in the phase space topology. This is also found for the corresponding mappings, leading to very interesting tilings of the phase space.

*3.1. Hamiltonian Systems as Canonical Mappings

*3.1a. Integrable Systems

Consider an oscillator with two degrees of freedom with a time-independent Hamiltonian. Since H is assumed integrable, it may be expressed in action-angle variables (see Section 1.2c) as

$$H(J_1, J_2) = E, \tag{3.1.1}$$

where E is the constant energy of the system, and J_1 and J_2 are constants of the motion. The constant energy allows the phase space motion to be reduced from four to three dimensions. The constancy of either action allows the further reduction to a two-dimensional surface in the three-dimensional constant energy space. On the two-dimensional surface, the angular motion is parametrized by the frequencies associated with each degree of freedom:

$$\theta_1 = \omega_1 t + \theta_{10}, \qquad \theta_2 = \omega_2 t + \theta_{20}, \tag{3.1.2}$$

where each angle variable is periodic with period 2π.

Motion on a Torus. A convenient way of describing the motion, which can be generalized to more than two degrees of freedom, is that of motion on a torus in phase space. Assuming a given energy E and examining one of the two degrees of freedom (say 1), then J_1 parametrizes the surface as radii of concentric circles with θ_1 the angle around each circle. The complete surface is specified by the addition of θ_2, at right angles to θ_1, forming the torus, as shown in Fig. 3.1a.

Choosing a given E and fixing J_1 also fixes J_2. Since $\omega_1 = \omega_1(J)$ and $\omega_2 = \omega_2(J)$, the ratio

$$\alpha = \frac{\omega_1}{\omega_2} \tag{3.1.3}$$

is also fixed. For $\alpha = r/s$ with r, s relatively prime integers, the two frequencies are commensurate, and the motion degenerates into the periodic trajectory of a one-dimensional curve on the torus, which closes on itself after r revolutions in θ_1 and s revolutions in θ_2. Generally, α will be irrational in which case a trajectory maps onto the entire surface. Because r and s can both be very large, the periodic orbits are arbitrarily close to one another in action space.

The concept of motion on a torus is particularly useful in that it can be generalized to systems with more than two degrees of freedom. Each constant action reduces the dimensionality of the phase space of the trajectory by one, such that for an N-degree of freedom system with N constant actions the motion in the $2N$-dimensional phase space is reduced to that on an N-dimensional surface or manifold on which the N angle variables run. The

3.1. Hamiltonian Systems as Canonical Mappings

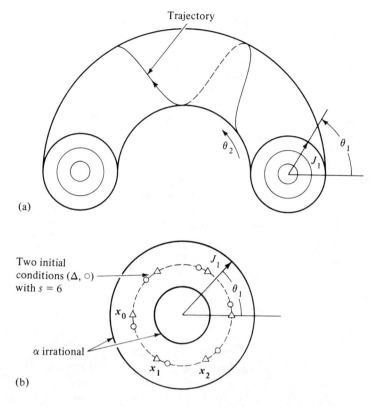

Figure 3.1. Motion of a phase space point for an integrable system with two degrees of freedom. (a) The motion lies on a torus $J_1 = $ const., $J_2 = $ const. (b) Illustrating trajectory intersections with a surface of section $\theta_2 = $ const. after a large number of such intersections.

topological properties of the surface are those of an *N-dimensional torus*, that is, in analogy to Fig. 3.1a, the N phase variables are orthogonal to one another and periodic with period 2π, and the surface is parametrized by the action variables.

An important consequence of the above discussion is that integrable motion in any canonical space

$$p = p(J, \theta), \qquad q = q(J, \theta), \tag{3.1.4}$$

is decomposable as

$$p(t) = \sum_m p_m(J) \exp[i(m \cdot \omega t + m \cdot \beta)], \tag{3.1.5a}$$

$$q(t) = \sum_m q_m(J) \exp[i(m \cdot \omega t + m \cdot \beta)], \tag{3.1.5b}$$

where m is an N-dimensional vector with integer components, and $\boldsymbol{\beta}$ is a constant vector. Equations (3.1.5) follow by Fourier expansion of the angle variables and in general yield quasiperiodic motion for p, q.

To obtain the periodic solutions, we set

$$\boldsymbol{m} \cdot \boldsymbol{\omega} = 0. \tag{3.1.6}$$

Since $\boldsymbol{m}$ has integer components, this also implies integer relations among the components of $\boldsymbol{\omega}$, i.e., $\boldsymbol{\omega} = \boldsymbol{n}\omega_0$, where the n_i have no common factor. Closure occurs after a period

$$T = \frac{2\pi}{\omega_0} = \frac{2\pi n_i}{\omega_i}, \tag{3.1.7}$$

where n_i is the number of circuits required for the ith degree of freedom with angular frequency ω_i.

Twist Mapping. A convenient way to study phase space trajectories, particularly in problems with two degrees of freedom, is by means of a surface of section, which was described in detail in Section 1.2b. For the Hamiltonian (3.1.1), there are two useful choices for a surface of section: the $J_1 - \theta_1$ plane (θ_2 = const.) and the $J_2 - \theta_2$ plane (θ_1 = const.). Choosing the former, we examine the intersections of the trajectory (Fig. 3.1a) with the $J_1 - \theta_1$ surface of section. Successive intersections have J_1 = const. and are separated by a time $\Delta t = 2\pi/\omega_2$. During this interval, θ_1 advances by $\omega_1 \Delta t = 2\pi\alpha(J_1)$, where α is the rotation number. Since $J_2 = J_2(J_1, E)$, for a given E, α can be considered a function of J_1 alone. Dropping the subscript 1 for notational reasons, we have the equations that describe the motion from the nth to the $(n + 1)$st intersections:

$$J_{n+1} = J_n, \tag{3.1.8a}$$

$$\theta_{n+1} = \theta_n + 2\pi\alpha(J_{n+1}), \tag{3.1.8b}$$

where for convenience we write α a function of J_{n+1} rather than J_n. The mapping given by (3.1.8) is called a *twist mapping* in which circles map into circles but with the rotation number in general dependent on the radius of the circle. For α irrational, any initial condition on a circle uniformly fills the circle as $n \to \infty$. For $\alpha = r/s$ rational, r and s relatively prime, we have the *fixed points* of the mapping, for which any initial condition recurs after exactly s intersections. Two such initial conditions are shown in Fig. 3.1b for $s = 6$.

A twist mapping need not be written in action-angle form. For example, the equations

$$x_{n+1} = x_n \cos \psi - y_n \sin \psi, \tag{3.1.9a}$$

$$y_{n+1} = x_n \sin \psi + y_n \cos \psi, \tag{3.1.9b}$$

3.1. Hamiltonian Systems as Canonical Mappings

with ψ a fixed parameter, describe a linear twist mapping. A perturbation of this mapping will be described in Section 3.2d.

As pointed out in Section 1.2b, a two-dimensional twist mapping must be area-preserving

$$\frac{\partial(J_{n+1}, \theta_{n+1})}{\partial(J_n, \theta_n)} \equiv [\theta_{n+1}, J_{n+1}] = 1 \tag{3.1.10}$$

with a similar requirement for (3.1.9).

These results generalize for integrable systems with N degrees of freedom and $H = $ const. Choosing, say $\theta_N = $ const. as a surface of section, we have the twist mapping for the $N - 1$ remaining pairs of action-angle coordinates

$$J_{n+1} = J_n, \tag{3.1.11a}$$

$$\theta_{n+1} = \theta_n + 2\pi\alpha(J_{n+1}), \tag{3.1.11b}$$

where $\alpha_i = \omega_i/\omega_N$ is the rotation number of the ith pair. The Poisson bracket conditions, equivalent to area preservation for a two-dimensional mapping, obviously hold.

*3.1b. Near-Integrable Systems

Area-Preserving Mappings. We now consider a two-dimensional integrable system that is perturbed slightly such that the Hamiltonian is a function of the angles

$$H(\boldsymbol{J}, \boldsymbol{\theta}) = H_0(\boldsymbol{J}) + \epsilon H_1(\boldsymbol{J}, \boldsymbol{\theta}). \tag{3.1.12}$$

At the $J_1 - \theta_1$ surface of section $\theta_2 = $ const. (mod 2π), we expect the twist mapping (3.1.8) to change to the *perturbed twist mapping*

$$J_{n+1} = J_n + \epsilon f(J_{n+1}, \theta_n), \tag{3.1.13a}$$

$$\theta_{n+1} = \theta_n + 2\pi\alpha(J_{n+1}) + \epsilon g(J_{n+1}, \theta_n), \tag{3.1.13b}$$

where f and g are periodic in θ. Since the transformation from n to $n+1$ is generated by Hamilton's equations, this mapping must be area-preserving. We have chosen f and g to be functions of J_{n+1}, rather than J_n, so that the area-preserving property takes a particularly simple form. Indeed, the generating function

$$F_2 = J_{n+1}\theta_n + 2\pi\mathscr{A}(J_{n+1}) + \epsilon\mathscr{G}(J_{n+1}, \theta_n) \tag{3.1.14}$$

generates (3.1.13) with

$$\alpha = \frac{d\mathcal{A}}{dJ_{n+1}}, \tag{3.1.15a}$$

$$f = -\frac{\partial \mathcal{G}}{\partial \theta_n}, \tag{3.1.15b}$$

$$g = \frac{\partial \mathcal{G}}{\partial J_{n+1}}, \tag{3.1.15c}$$

and thus

$$\frac{\partial f}{\partial J_{n+1}} + \frac{\partial g}{\partial \theta_n} = 0 \tag{3.1.16}$$

for area preservation. Maps with J_{n+1} and θ_{n+1} given explicitly in terms of J_n and θ_n can be written using Lie transform techniques (Dragt and Finn, 1976b; see also Section 2.5), which directly give such transformations, but these are not as convenient for our purposes.

If f is a function of J_{n+1}, then (3.1.13a) defines J_{n+1} implicitly in terms of J_n. Computationally, J_{n+1} is easily determined by Newton's method, or by successive approximations, in which the new approximation $J_{n+1}^{(i)}$ is found by using the previous approximation $J_{n+1}^{(i-1)}$ in f. Both methods are rapidly convergent (see Section 2.6).

For many interesting mappings, f is independent of J, and $g \equiv 0$. Then (3.1.13) takes the form of a *radial twist mapping*:

$$J_{n+1} = J_n + \epsilon f(\theta_n), \tag{3.1.17a}$$

$$\theta_{n+1} = \theta_n + 2\pi\alpha(J_{n+1}). \tag{3.1.17b}$$

In this form, successive applications of (3.1.17a) and (3.1.17b) generate the mapping explicitly. We later consider two such mappings in detail, the simplified Fermi mapping in Section 3.4 and the separatrix mapping in Section 3.5.

If we linearize (3.1.17b) about a period 1 fixed point $J_{n+1} = J_n = J_0$, for which $\alpha(J_0)$ is an integer, then for a nearby action

$$J_n = J_0 + \Delta J_n, \tag{3.1.18}$$

the substitution of a new action

$$I_n = 2\pi\alpha' \Delta J_n \tag{3.1.19}$$

converts (3.1.17) to the *generalized standard mapping*

$$I_{n+1} = I_n + Kf^*(\theta_n), \tag{3.1.20a}$$

$$\theta_{n+1} = \theta_n + I_{n+1}, \mod 2\pi. \tag{3.1.20b}$$

Here

$$K = 2\pi\alpha'\epsilon f_{\max} \tag{3.1.21}$$

3.1. Hamiltonian Systems as Canonical Mappings

is the *stochasticity parameter*, and $f^* = f/f_{\max}$ is the jump in the action, normalized to a maximum value of unity. Thus the generalized standard mapping is locally equivalent (in J) to any radial twist mapping. For $f^* = \sin\theta_n$, (3.1.20) becomes the *standard mapping* (also known as the Chirikov–Taylor mapping)

$$I_{n+1} = I_n + K\sin\theta_n, \qquad (3.1.22a)$$

$$\theta_{n+1} = \theta_n + I_{n+1}, \qquad (3.1.22b)$$

which has been used by Chirikov (1979) and Greene (1979a) to estimate the transition from regular to chaotic motion. We discuss the standard mapping and apply it to calculate the transition to chaotic motion in Chapter 4.

For sufficient symmetry, mappings may be factored into the product of simpler mappings. The factoring of the radial twist mapping into the product of two *involutions*, and its use in determining the fixed points, are described in Sections 3.3b and 3.4d.

The generalization of the perturbed twist mapping to more than two degrees of freedom with $H = \text{const.}$ is obtained from a generating function

$$F_2 = \mathbf{J}_{n+1}\cdot\boldsymbol{\theta}_n + 2\pi\mathscr{A}(\mathbf{J}_{n+1}) + \epsilon\mathscr{G}(\mathbf{J}_{n+1}, \boldsymbol{\theta}_n) \qquad (3.1.23)$$

for the $N-1$ pairs of action-angle variables in the surface of section leading to

$$\mathbf{J}_{n+1} = \mathbf{J}_n + \epsilon f(\mathbf{J}_{n+1}, \boldsymbol{\theta}_n), \qquad (3.1.24a)$$

$$\boldsymbol{\theta}_{n+1} = \boldsymbol{\theta}_n + 2\pi\boldsymbol{\alpha}(\mathbf{J}_{n+1}) + \epsilon g(\mathbf{J}_{n+1}, \boldsymbol{\theta}_n), \qquad (3.1.24b)$$

which are canonical by construction from F_2. If $\mathscr{G}$ is not a function of $\mathbf{J}_{n+1}$, then we have a $(2N-2)$-dimensional radial twist mapping. The difference equations for the billiards problem, mentioned in Section 1.4b and described in detail in Chapter 6, are one such mapping.

We represent all these mappings symbolically by introducing $x = (\mathbf{J}, \boldsymbol{\theta})$ and writing

$$x_{n+1} = Tx_n, \qquad (3.1.25)$$

where T represents the mapping.

*3.1c. Hamiltonian Forms and Mappings

As mentioned previously, conversion between Hamiltonian forms and mappings has wide application to the calculation of motion in dynamical systems. As will be seen in the next section, the generic behavior of Hamiltonian systems is usually described by mathematicians in terms of mappings. Numerical calculations over millions of periods of the nonlinear oscillations can readily be performed from a mapping. The analytical determination of diffusion equations for chaotic motion is usually approached from the mapping equations. On the other hand, the regular characteristics of mappings are often best approached from a Hamiltonian description. As we shall see, mappings have their own Hamiltonian representation that can be used to connect their analysis to the general body of Hamiltonian theory. We first show how a Hamiltonian description can be converted to a mapping, and then consider the inverse problem of writing certain classes of mappings in Hamiltonian form.

Conversion to a Mapping. In two dimensions, the functions f and g in the perturbed twist mappings (3.1.13) are determined to order ϵ as follows: from Hamilton's equations,

$$\frac{dJ_1}{dt} = -\epsilon \frac{\partial H_1}{\partial \theta_1}. \tag{3.1.26}$$

We integrate (3.1.26) over one period of the θ_2 motion around the torus to get

$$\Delta J_1 = -\epsilon \int_0^{T_2} dt \, \frac{\partial H_1}{\partial \theta_1} (J_{n+1}, J_2, \theta_n + \omega_1 t, \theta_{20} + \omega_2 t), \tag{3.1.27}$$

where (recall) J_2, ω_1, and ω_2 are all functions of J_{n+1}. To order ϵ, we use the unperturbed values of J, θ, i.e., integrate $\partial H_1/\partial \theta_1$ over the unperturbed orbits. The jump in the action J is then

$$\epsilon f(J_{n+1}, \theta_n) = \Delta J_1(J_{n+1}, \theta_n). \tag{3.1.28}$$

The corresponding jump in the phase is most conveniently found from the area-preserving condition (3.1.16), which when applied to the perturbed twist mapping yields

$$g(J, \theta) = -\int^\theta \frac{\partial f}{\partial J} d\theta'. \tag{3.1.29}$$

For the radial twist mapping, $g \equiv 0$.

When calculating the jump in action, it is not necessary to express the unperturbed motion of the remaining degree of freedom in action-angle form. From the Hamiltonian

$$H = H_0(J_1, p_2, q_2) + \epsilon H_1(J_1, \theta_1, p_2, q_2), \tag{3.1.30}$$

we again have (3.1.26), from which, to order ϵ

$$\Delta J_1 = -\epsilon \int_0^{T_2} dt \, \frac{\partial H_1}{\partial \theta_1} (J_{n+1}, \theta_n + \omega_1 t, p_2(t), q_2(t)). \tag{3.1.31}$$

We shall return to this expression when we consider the separatrix mapping in Section 3.5.

The calculation of the jump in action can be generalized to N degrees of

3.1. Hamiltonian Systems as Canonical Mappings

freedom to obtain the jumps in the $N - 1$ surface of section actions

$$\Delta J = -\epsilon \int_0^{T_N} dt \, \frac{\partial H_1}{\partial \boldsymbol{\theta}} (J_{n+1}, J_N, \boldsymbol{\theta}_n + \boldsymbol{\omega}t, \theta_N + \omega_N t), \quad (3.1.32)$$

with $\epsilon f = \Delta J$. Then as in (3.1.15), $\mathcal{G}$ in (3.1.23) is found by integration of f over the θ_n's, and $g = \partial \mathcal{G}/\partial J_{n+1}$ yielding the mapping (3.1.24).

Conversion to a Hamiltonian Form. We consider the class of radial twist mappings (3.1.17). We often want to represent these equations in Hamiltonian form, where the iteration number n plays the role of the time. We can do this by introducing the periodic delta function into (3.1.17a):

$$\delta_1(n) = \sum_{m=-\infty}^{\infty} \delta(n-m) = 1 + 2 \sum_{q=1}^{\infty} \cos 2\pi q n, \quad (3.1.33)$$

where the last form is the Fourier expansion. Then (3.1.17) takes the form

$$\frac{dJ}{dn} = \epsilon f(\theta) \delta_1(n), \quad (3.1.34a)$$

$$\frac{d\theta}{dn} = 2\pi \alpha(J), \quad (3.1.34b)$$

where J_n and θ_n are $J(n)$ and $\theta(n)$ just before "time" n. Equations (3.1.34) are in Hamiltonian form with

$$H(J, \theta, n) = 2\pi \int^J \alpha(J') \, dJ' - \epsilon \delta_1(n) \int^\theta f(\theta') \, d\theta'. \quad (3.1.35)$$

Note that H is nonautonomous with one degree of freedom. The above calculation easily generalizes to the N degree of freedom radial twist map. We shall apply (3.1.35) to the Fermi mapping considered in Section 3.4.

Generating Functions and Phase Space Areas. A useful formulation of phase space areas for a mapping can be developed in terms of generating functions. Considering some interval $\theta_1 < \theta < \theta_2$, the phase space area (or action) for any smooth single-valued curve $J(\theta)$ is defined by the integral

$$A = \int_{\theta_1}^{\theta_2} J(\theta) \, d\theta. \quad (3.1.36)$$

Consider now the transformation of A under a mapping step. A mapping step will then transform the curve into a new curve spanning a new interval. Providing the curve remains continuous in the transformation, it will define a new area. The change in area may be related to a generating function of the mapping, as follows. Let $F_1(\theta_{n+1}, \theta_n)$ be a generating function for the mapping such that

$$J_n = -\frac{\partial F_1}{\partial \theta_n}, \qquad J_{n+1} = \frac{\partial F_1}{\partial \theta_{n+1}}. \tag{3.1.37}$$

Now if we define an arbitrary variable λ, which parametrizes the curve, then we can formally write

$$\frac{dF_1}{d\lambda} = \frac{\partial F_1}{\partial \theta_{n+1}} \frac{d\theta_{n+1}}{d\lambda} + \frac{\partial F_1}{\partial \theta_n} \frac{d\theta_n}{d\lambda}, \tag{3.1.38}$$

and substituting from (3.1.37) we obtain

$$\frac{dF_1}{d\lambda} = J_{n+1} \frac{d\theta_{n+1}}{d\lambda} - J_n \frac{d\theta_n}{d\lambda}. \tag{3.1.39}$$

Equation (3.1.39) may be integrated over the appropriate ranges of the phase variable to obtain

$$F_1(\theta_2'(\theta_2), \theta_2) - F_1(\theta_1'(\theta_1), \theta_1) = \int_{\theta_1'}^{\theta_2'} J_{n+1}\, d\theta_{n+1} - \int_{\theta_1}^{\theta_2} J_n\, d\theta_n, \tag{3.1.40a}$$

or

$$\Delta F_1(\theta_2, \theta_1) = A_{n+1} - A_n, \tag{3.1.40b}$$

where θ_1' and θ_2' are the interval ends after the mapping step. We thus have a rather remarkable result that the change in the area under an arbitrarily chosen curve in a mapping step is equal to the difference in the generating function for the mapping, defined at its end points, independent of the choice of the intervening curve. This result, obtained by MacKay et al. (1984) and Bensimon and Kadenoff (1984), and also described by Mather (1984), has interesting consequences which we will return to in Section 5.5b. The quantity ΔF_1 is a change in action, which is denoted by Mather as ΔW. We note that a phase-spanning phase space trajectory transforms into itself, and thus must have $A_{n+1} = A_n = 0$ for that special case.

Symplectic Integration. The conversion of a Hamiltonian to a canonical mapping can be viewed as a scheme to integrate Hamilton's equations while preserving the canonical form of the solution. Such numerical schemes are called symplectic integration algorithms (de Vogelaere, 1958; Ruth, 1983; Channel and Scoval, 1988). They are important for numerically maintaining numerical accuracy of essential properties of the solution in problems for which millions of fundamental oscillation periods are required, e.g., in celestial mechanics, particle storage rings, etc. An example is the N-degree-of-freedom Hamiltonian

$$H = \frac{|\mathbf{p}|^2}{2} + V(\boldsymbol{\theta}), \tag{3.1.41}$$

3.1. Hamiltonian Systems as Canonical Mappings

which can be integrated using the first-order mapping

$$p_{n+1} = p_n - \Delta t \nabla V(\theta_n),$$
$$\theta_{n+1} = \theta_n + \Delta t p_{n+1}. \tag{3.1.42}$$

The usual method of obtaining a symplectic map, given the Hamiltonian of the continuous system is by means of a generating function. Thus, using the general form (3.1.23), a generating function can be constructed from the Hamiltonian (3.1.41),

$$F_2 = p_{n+1} \cdot \theta_n + \Delta t \left(\tfrac{1}{2}|p_{n+1}|^2 + V(\theta_n)\right), \tag{3.1.43}$$

which generates the mapping (3.1.42) insuring its symplectic nature. For example, applying this scheme to the pendulum Hamiltonian $H = p^2/2 + F \cos \theta$, we obtain, using (3.1.43), the standard map (3.1.22) with $I_n = p_n \Delta t$ and $K = F(\Delta t)^2$. Although symplectic algorithms preserve the canonical form of the solution, e.g., the existence of KAM tori (see Section 3.2), they only approximate the actual solution. For the pendulum, the solutions are integrable and are symmetric about $p = 0$. The solutions of the standard map are neither integrable nor symmetric (see Section 4.1). Nevertheless, the orbits are essentially reproduced, due to the existence of the KAM curves that tightly bound regions of stochasticity in the mapping. As we shall see in Section 6.5, for coupled oscillator systems with many dimensions, the technique can be vital to exploring the long time system behavior.

Higher-order symplectic algorithms have been developed. For the Hamiltonian (3.1.41), introducing the maps $T_1(\tau)$, given by

$$\bar{\theta} = \theta + \tau p,$$
$$\bar{p} = p, \tag{3.1.44}$$

and $T_2(\tau)$, given by

$$\bar{\theta} = \theta,$$
$$\bar{p} = p - \tau \nabla V(\theta), \tag{3.1.45}$$

we note that (3.1.44) gives the solution for the Hamiltonian $|p|^2/2$ and (3.1.45) gives the solution for the Hamiltonian $V(\theta)$. Expanding the solution generated by H to first order in Δt and comparing this to successive applications of T_1 and T_2, we obtain the first-order algorithm (3.1.42) given by $T = T_1(\Delta t) \circ T_2(\Delta t) + O((\Delta t)^2)$, where $\circ$ denotes the *composition* of T_2 and T_1. Similarly, expanding the solution to fourth order in Δt, Forest and Berz (1989) give an explicit fourth-order algorithm, the Baker–Campbell–Hausdorf formula:

$$T = T_1(s_1) \circ T_2(d_1) \circ T_1(s_2) \circ T_2(d_2) \circ T_1(s_2) \circ T_2(d_1) \circ T_1(s_1) + O((\Delta t)^2), \tag{3.1.46}$$

where the constants are $s_1 = \Delta t/2(2 - \beta)$, $s_2 = \Delta t(1 - \beta)/2(2 - \beta)$, $d_1 = \Delta t/(2 - \beta)$, and $d_2 = -\beta \Delta t/(2 - \beta)$, with $\beta = \sqrt[3]{2}$. This symplectic algorithm is as efficient as the fourth-order Runge–Kutta algorithm commonly used to integrate general sets of ordinary differential equations, but the latter algorithm does not preserve the canonical form of the solutions, while the former does.

*3.2. Generic Behavior of Canonical Mappings

We consider now the general behavior and phase space structure for systems of nonlinear coupled oscillators. The major result, the KAM theorem guaranteeing the existence of invariant tori, applies to systems with two or more degrees of freedom. However, certain implications of the theorem are profoundly different for two degrees and for more than two degrees of freedom, namely, the existence of Arnold diffusion in the latter case. We defer discussion of this question and detailed examples to Chapter 6. Here, we focus attention on two-dimensional (area-preserving) mappings to illustrate the KAM theorem and generic structure of phase space. The appropriate generalizations to more than two degrees of freedom are given where applicable.

*3.2a. Irrational Winding Numbers and KAM Stability

If an integrable system is perturbed slightly such that the Hamiltonian is a function of the angles as given by (3.1.12), then, as we have seen in Chapter 2, resonances between the degrees of freedom may destroy the convergence of the various power series expansions about the unperturbed system. Nevertheless, it is possible to prove a theorem (the KAM theorem) that, provided certain conditions are satisfied (to be enumerated below), there exists an invariant torus (J, θ) parametrized by ξ, satisfying the relations

$$J = J_0 + \mathbf{v}(\xi, \epsilon) \qquad (3.2.1a)$$

and

$$\theta = \xi + \mathbf{u}(\xi, \epsilon). \qquad (3.2.1b)$$

Here u and v are periodic in ξ and vanish for $\epsilon = 0$, and $\dot{\xi} = \omega$, the unperturbed frequencies on the torus. The conditions to be satisfied are

(1) the linear independence of the frequencies

$$\sum_i m_i \omega_i(J) \neq 0 \qquad (3.2.2)$$

over some domain of J (sufficient nonlinearity), where the ω_i are the components of $\omega = \partial H_0/\partial J$ and the m_i are the components of the integer vector m;
(2) a smoothness condition on the perturbation (sufficient number of continuous derivatives of H_1);
(3) initial conditions sufficiently far from resonance to satisfy

$$|m \cdot \omega| \geq \gamma |m|^{-\tau} \qquad (3.2.3)$$

for all m, where τ is dependent on the number of degrees of freedom and

3.2. Generic Behavior of Canonical Mappings

the smoothness of H_1, and γ is dependent on ϵ, on the magnitude of the perturbation Hamiltonian H_1, and on the nonlinearity G of the unperturbed Hamiltonian H_0.

Since (3.2.3) cannot be met for γ too large, and γ increases with ϵ, $|H_1|$, and $1/G$, there is a condition of sufficiently small perturbation for KAM tori to exist. Conditions (1) and (2) also imply a condition of *moderate nonlinearity*. If the conditions of the theorem are met, then the circle of a twist mapping perturbs to a near-circle, as shown in Fig. 3.2a, without change of topology. This is the intersection of a KAM torus with a surface of section.

The theorem was proved by Arnold (1961, 1962, 1962b) for analytic H_1 (all derivatives existing), following a conjecture by Kolmogorov (1954), and by Moser (1962) for a sufficient number of continuous derivatives. It provides the basis for the existence of invariants in nonlinear coupled systems. The theorem is generally called the KAM theorem in recognition of their work. We discuss the meaning of the theorem, the general method of proof, and the significance of the applicability conditions.

We illustrate the difficulty in proving the theorem by considering a perturbed twist mapping with two degrees of freedom. The successive intersections of the perturbed trajectory with the surface of section (see Fig. 3.1a) are described by a pair of difference equations. These give the new values of the surface of section variables J_1 and θ_1 in terms of their values at the previous intersection. In terms of the assumed invariant curve (3.2.1a), the J_1 difference equation has the form, equivalent to (3.1.13a) of the perturbed twist mapping,

$$J_1(\theta_1 + 2\pi\alpha) = J_1(\theta_1) + v(\theta_1), \tag{3.2.4}$$

which we try to solve by taking Fourier components in θ_1

$$J_1(\theta_1) = \sum a_k \exp(ik\theta_1), \qquad v(\theta_1) = \sum b_k \exp(ik\theta_1). \tag{3.2.5}$$

Since we wish to find the a_k in terms of the b_k,

$$J_1(\theta_1 + 2\pi\alpha) - J(\theta_1) = -\sum a_k[1 - \exp(ik2\pi\alpha)] \exp(ik\theta_1) \tag{3.2.6}$$

such that the coefficients a_k are

$$a_k = \frac{-b_k}{1 - \exp(ik2\pi\alpha)}. \tag{3.2.7}$$

The a_k do not approach zero as fast as the b_k and are undefined at α rational. This is just the problem of zero denominators that prevents the convergence of perturbation theory. If α is a function of J_1, then the value of J_1 must be chosen such that the denominator is never resonant. This requires both an appropriately modified expansion procedure and sufficiently rapidly decreasing b_k. The latter restriction, which is of practical importance for our study, will be considered later in this section. We now discuss the methodology of the various proofs, the actual proofs being too complicated to present here. Basically, the method is to vary the initial conditions at each step in the

Figure 3.2. Conditions for the KAM theorem to hold. (a) Illustration of the sufficient nonlinearity condition discussed in the text; the perturbed curve lies near the unperturbed curve. (b) Illustration of the smoothness condition; the region between primary resonances is examined for secondary resonances. (c) The action region between two primary resonances converted to a frequency scale using $\Delta\omega = G\Delta J$. Secondary resonances are shown hatched. For sufficient smoothness, the secondary resonances are isolated.

3.2. Generic Behavior of Canonical Mappings

expansion procedure to remain sufficiently far from all resonances so that the expansion can proceed to the next stage.

Linear Independence or Sufficient Nonlinearity. As we have seen, if a resonance exists between two degrees of freedom in the unperturbed system, the phase space trajectories induced by the perturbation in the neighborhood of that resonance are distorted. If the unperturbed frequencies are functions of the actions, then the resulting change in actions usually perturbs the system off resonance and limits the excursion in actions. If the maximum excursion in the perturbed action J is much less than the unperturbed action J_0, then KAM-invariant curves may exist that are "near" the unperturbed invariant $J = J_0$. This is the meaning of the linear independence requirement $\mathbf{m} \cdot \boldsymbol{\omega} \neq 0$ given in (3.2.2), which in (3.2.1a) guarantees $\nu(\xi, \epsilon) \to 0$ as $\epsilon \to 0$.

To explore this in more depth, let us calculate the condition for linear dependence of the frequencies. For simplicity, consider two degrees of freedom and assume the frequencies $\omega_1(J_1, J_2)$ and $\omega_2(J_1, J_2)$ are related by

$$f(\omega_1, \omega_2) = 0.$$

Differentiating, we have

$$df = \frac{\partial f}{\partial \omega_1}\left(\frac{\partial \omega_1}{\partial J_1} dJ_1 + \frac{\partial \omega_1}{\partial J_2} dJ_2\right) + \frac{\partial f}{\partial \omega_2}\left(\frac{\partial \omega_2}{\partial J_1} dJ_1 + \frac{\partial \omega_2}{\partial J_2} dJ_2\right) = 0$$

valid for arbitrary dJ_1 and dJ_2, yielding

$$\boldsymbol{\omega}_J \cdot \mathbf{f}_\omega = \begin{pmatrix} \dfrac{\partial \omega_1}{\partial J_1} & \dfrac{\partial \omega_2}{\partial J_1} \\ \dfrac{\partial \omega_1}{\partial J_2} & \dfrac{\partial \omega_2}{\partial J_2} \end{pmatrix} \begin{pmatrix} \dfrac{\partial f}{\partial \omega_1} \\ \dfrac{\partial f}{\partial \omega_2} \end{pmatrix} = 0. \quad (3.2.8)$$

For $\det \boldsymbol{\omega}_J \neq 0$, the only solution is $\mathbf{f}_\omega = 0$. Then no relation valid for all $\mathbf{J}$ of the form

$$f = m_1 \omega_1 + m_2 \omega_2 = 0 \quad (3.2.9)$$

exists. A necessary condition for *linear independence* (nonlinear dependence of the frequencies) is then

$$\det \boldsymbol{\omega}_J \neq 0, \quad (3.2.10)$$

which is the usually quoted condition.

For a given resonance, a less-restrictive condition can be used, requiring only that the frequency not be constant along the direction of the actual change in $\mathbf{J}$. If we consider two degrees of freedom, we have the Hamiltonian of (2.4.1)

$$H = H_0(J_1, J_2) + \epsilon \sum_{l,m} H_{lm}(J_1, J_2) \exp[i(l\theta_1 - m\theta_2)]. \quad (3.2.11)$$

Choosing a particular resonance $l = r$, $m = s$, and $\omega_2/\omega_1 = r/s$, Hamilton's equations yield $dJ_1/dJ_2 = -r/s$. Inserting this into (3.2.8) with $\partial f/\partial \omega_1 = r$, $\partial f/\partial \omega_2 = -s$ as in (3.2.9), we find the nonlinearity condition

$$r^2 \frac{\partial^2 H_0}{\partial J_1^2} - 2rs \frac{\partial^2 H_0}{\partial J_1 \, \partial J_2} + s^2 \frac{\partial^2 H_0}{\partial J_2^2} \neq 0, \tag{3.2.12}$$

which is also a sufficient condition for nonlinearity as used in the KAM proof. A more general treatment of this result, for an arbitrary number of degrees of freedom, can be found in Chirikov (1979; Section 3.3) giving

$$m_i \frac{\partial^2 H_0}{\partial J_i \, \partial J_j} m_j \neq 0. \tag{3.2.13}$$

It is instructive to derive (3.2.12) using the treatment of secular perturbation theory given in Section 2.4. Applying the generating function, as in (2.4.5),

$$F_2 = (r\theta_1 - s\theta_2)\hat{J}_1 + \theta_2 \hat{J}_2, \tag{3.2.14}$$

to Hamiltonian (3.2.11) we obtain the new variables as in (2.4.6). Expanding, as we did there, about the resonant value of the action, and averaging over the fast variables, we obtain to lowest order in ϵ

$$\Delta \hat{H} = \frac{\partial^2 \hat{H}_0}{\partial \hat{J}_1^2} \frac{\Delta \hat{J}_1^2}{2} + 2\epsilon H_{rs} \cos \hat{\theta}_1, \tag{3.2.15}$$

where for convenience we have chosen H_{rs} to be real. If $\partial^2 \hat{H}_0/\partial \hat{J}_1^2 = 0$, then the nonlinearity only appears in the next higher order and the separatrix width is not constrained to be of order $\epsilon^{1/2}$. Thus we obtain the nonlinearity condition equivalent to (3.2.12)

$$G = \frac{\partial^2 \hat{H}_0}{\partial \hat{J}_1^2} \neq 0. \tag{3.2.16}$$

This condition separates *accidentally degenerate* ($\partial^2 \hat{H}_0/\partial \hat{J}_1^2 \neq 0$) or *strongly nonlinear* systems from *intrinsically degenerate* ($\partial^2 \hat{H}_0/\partial \hat{J}_1^2 = 0$) or *weakly nonlinear* systems. For accidental degeneracy, the nonlinearity condition required by KAM theory is obeyed. To show the equivalence of (3.2.12) and (3.2.16), we expand

$$\frac{\partial^2 \hat{H}_0}{\partial \hat{J}_1^2} = \frac{\partial}{\partial \hat{J}_1} \left(\frac{\partial H_0}{\partial J_1} \frac{\partial J_1}{\partial \hat{J}_1} + \frac{\partial H_0}{\partial J_2} \frac{\partial J_2}{\partial \hat{J}_1} \right) \neq 0, \tag{3.2.17}$$

and remembering that $\partial J_1/\partial \hat{J}_1 = r$ and $\partial J_2/\partial \hat{J}_1 = -s$, we obtain (3.2.12).

For a fixed value of ϵ, the required nonlinearity in G can be estimated from the condition illustrated in Fig. 3.2a, that the excursion in the action ΔJ_1 be much less than the unperturbed action J_0. The full separatrix width

3.2. Generic Behavior of Canonical Mappings

$\Delta J_1 = r \Delta \hat{J}_1$, which for (3.2.15) yields the condition

$$4r \left(\frac{2\epsilon H_{rs}}{G} \right)^{1/2} \ll J_0$$

or

$$G \gg \frac{32 r^2 \epsilon H_{rs}}{J_0^2}. \tag{3.2.18}$$

Intrinsic degeneracy, for which the linear independence (3.2.10) condition is not satisfied, occurs in a great many systems of physical interest. The question arises whether KAM curves exist in these systems. The answer seems to be that the structure of KAM theory (invariant curves, islands, etc.) is usually, but not always, present in intrinsically degenerate systems. We have seen two examples of KAM behavior for intrinsic degeneracy in the Hénon and Heiles system of Section 1.4 and the problem of perpendicular wave propagation treated in Section 2.4. Another example is the problem of beam–beam interaction in particle storage rings (Tennyson, 1979). For these problems, the zero-order Hamiltonian is of the form

$$H_0 = \boldsymbol{\omega}_0 \cdot \boldsymbol{J} + \epsilon H_{10}(\boldsymbol{J}) \tag{3.2.19}$$

in which the $m = 0$ term of the perturbation itself induces the nonlinearity. Although G is now of order ϵ, the nonlinearity condition (3.2.18) can still be met. Thus, KAM curves may still exist in intrinsically degenerate systems. If one proceeds to examine the second- and higher-order resonances of such systems, as in Section 2.4, one almost always finds accidental degeneracy, which satisfies the KAM nonlinearity condition. Thus the KAM structures of intrinsically and accidentally degenerate systems are generally similar.

We now mention that there are exceptional cases of intrinsic degeneracy for which KAM curves do not exist. A notable example is the system that has been studied by Lunsford and Ford (1972), and by Contopoulos (1978). Following Contopoulos, we start with the Hamiltonian

$$H = \omega_1 I_1 + \omega_2 I_2 + \omega_3 I_3 + \epsilon [\alpha \cos(m_1 \theta_1 + m_2 \theta_2 + m_3 \theta_3) \\ + \beta \cos(n_1 \theta_1 + n_2 \theta_2 + n_3 \theta_3)], \tag{3.2.20}$$

where α and β are higher than linear functions of the actions I_i and the m_i and n_i are integers. Since the angles only appear in two linear combinations, a transformation as in (2.4.5) transforms (3.2.20) to a Hamiltonian possessing only two angles:

$$\hat{H} = \hat{\omega}_1 \hat{I}_1 + \hat{\omega}_2 \hat{I}_2 + \hat{\omega}_3 \hat{I}_3 + \epsilon [\alpha \cos \hat{\theta}_1 + \beta \cos \hat{\theta}_2], \tag{3.2.21}$$

where $\hat{\theta}_1 = m_1 \theta_1 + m_2 \theta_2 + m_3 \theta_3$, $\hat{\theta}_2 = n_1 \theta_1 + n_2 \theta_2 + n_3 \theta_3$, $\hat{\omega}_1 = m_1 \omega_1 + m_2 \omega_2 + m_3 \omega_3$, etc. Since θ_3 does not appear in the Hamiltonian, $\hat{I}_3$ is a constant of the motion, and a reduced Hamiltonian can be written with only two degrees of freedom. However, if we choose the frequencies such that

$\hat{\omega}_1 = \hat{\omega}_2 = 0$, as done by Lunsford and Ford, then the reduced Hamiltonian becomes

$$\hat{H} = \epsilon[\alpha \cos \hat{\theta}_1 + \beta \cos \hat{\theta}_2]. \tag{3.2.22}$$

Unlike the case of particle motion in a perpendicular electrostatic wave (see Section 2.4c), no separation into fast and slow variables can be made; hence, the methods of secular perturbation theory do not apply. In effect, the reduced system no longer has a small parameter, that is, it is not close to an integrable system. In this case, we do not expect KAM curves to exist, even as $\epsilon \to 0$. This was found to be the case, by Lunsford and Ford, from numerical integration of the equations of motion.

Smoothness Condition. A necessary condition on the smoothness of the driving term (number of continuous derivatives) can be obtained from the observation that KAM curves can only exist away from all island perturbations. If the islands between two lower-order resonances fill up the entire phase space between them, we can reasonably conclude that a KAM curve cannot be found. Again looking at the simplest case of two degrees of freedom, we use the Hamiltonian of (3.2.11). For clarity we choose $\omega_1/\omega_2 = s$, with H_0 depending linearly on J_2, such that the neighboring primary resonances as J_1 is varied are separated by $\delta\omega_1 = \omega_2$, where ω_2 is a constant, independent of J_1 and J_2, and illustrated in Fig. 3.2b. Between these primary resonances are sums of secondary resonances at values of $\omega_1/\omega_2 = s + p/q$ (p, q integer, $p < q$). Choosing $l = q$ in (3.2.11), m runs through the integer values for secondary resonances,

$$m(p, q) = p + sq. \tag{3.2.23}$$

Using the values from (2.4.31) for the separatrix width $\Delta \hat{J}_1$ of each resonance

$$\Delta \hat{J}_1 = 4 \left(\frac{2\epsilon H_{qm}}{G} \right)^{1/2}$$

and the definitions $\Delta J_1 = q\Delta \hat{J}_1$ and $G = q^2 \partial^2 H_0/\partial J_1^2 = q^2 \bar{G}$,

$$\sum \Delta J_1 = 4 \left(\frac{2\epsilon}{\bar{G}} \right)^{1/2} \sum_{p,q} H_{qm}^{1/2}, \tag{3.2.24}$$

where we sum over all of the secondary resonances. In terms of frequency,

$$\Delta \omega_1 = \frac{\partial \omega_1}{\partial J_1} \Delta J_1 = \bar{G} \Delta J_1, \tag{3.2.25}$$

the ratio of the sum of the secondary island widths, shown hatched in Fig.

3.2. Generic Behavior of Canonical Mappings

3.2c, to the distance between primary islands is

$$\frac{\Sigma \Delta \omega_1}{\delta \omega_1} = \frac{4(2\epsilon \bar{G})^{1/2}}{\omega_2} \sum_{p,q} H_{qm}^{1/2}. \tag{3.2.26}$$

If we now specify a smoothness to the perturbation Hamiltonian, say M continuous derivatives, then since m is linear in q, the Fourier amplitudes fall off for large q as

$$H_{qm} \sim \frac{\Lambda_0}{q^{M+2}}, \tag{3.2.27}$$

where Λ_0 is a constant. Substituting (3.2.27) in (3.2.26) and noting that

$$\sum_p H_{qm}^{1/2} \sim q H_{qm}^{1/2}, \tag{3.2.28}$$

we have

$$\frac{\Sigma \Delta \omega_1}{\delta \omega_1} \sim \frac{4(2\epsilon \bar{G} \Lambda_0)^{1/2}}{\omega_2} \sum_{q=1}^{\infty} q^{-M/2}. \tag{3.2.29}$$

The summation in (3.2.29) converges to a positive number σ for $M > 2$. Thus, independent of the coefficient in front of the summation, we have the important condition for existence of a KAM surface that the number of continuous derivatives M satisfies

$$M > 2. \tag{3.2.30}$$

We can compare this result with the actual KAM condition be rewriting (3.2.3) in two dimensions as

$$\left| \frac{\omega_1}{\omega_2} - \frac{r}{s} \right| > \gamma' s^{-(\tau+1)}. \tag{3.2.31}$$

We can think of the left-hand side as the width of a single secondary resonance that we are excluding and sum over all r and s. Taking the total width to be examined as that between two lowest-order resonances, i.e., the unit interval, there are at most s values of r in the interval. The length $\mathcal{M}$ (Lebesgue measure, see, e.g., Siegel and Moser, 1971) excluded by the sum is obtained by multiplying (3.2.31) by s and summing over s. We find

$$\mathcal{M} < \gamma' \sum_{s=1}^{\infty} s^{-\tau}. \tag{3.2.32}$$

Comparing (3.2.32) with (3.2.29), we can identify τ with $\frac{1}{2}M$ and γ' with $\mathcal{O}(\omega_1/\omega_2)$, $\omega_1 = (\epsilon \bar{G} \Lambda_0)^{1/2}$. As with (3.2.29), the sum in (3.2.32) is finite for $\tau > 1$, which Moser (see Siegel and Moser, 1971, p. 242) has shown to be sufficient for the existence of a KAM surface (torus or curve). Using the fact that the Hamiltonian of (3.2.11) is the integral of the mapping [see (3.1.27)], this would imply that for a two-dimensional mapping two continuous derivatives of the mapping and three of the corresponding Hamiltonian should be sufficient for the existence of KAM surfaces. Moser (1973, p. 53) asserts that $M > 4$ is sufficient to prove the existence of a KAM surface and conjectures that in fact $M > 3$. In a numerical example in Section 3.4b, we find that for $M \geqslant 2$ an invariant torus exists, a result also found numerically by Chirikov (1978). The same example shows that for $M = 1$ no KAM curves exist. However, in a counterexample Takens (1971) finds a case for which the invariant does not exist for $M = 2$. Thus we can conjecture with Moser (1973) that $M > 3$ is always sufficient, speculate further that $M = 2$ is necessary, and depending on the example, invariant curves may or may not exist with two continuous derivatives of the perturbation Hamiltonian.

The calculation leading to (3.2.30), for two degrees of freedom, has been performed in a somewhat more abstract manner by Chirikov (1979) for N degrees of freedom, leading to the *necessary* condition

$$M \geqslant 2N - 2. \tag{3.2.33}$$

Moser (1966) has also determined a rigorous *sufficient* condition

$$M \geqslant 2N + 2 \tag{3.2.34}$$

(note that this is a weaker condition than implied by Moser's two-dimensional conjecture $M > 3$), provided γ, which tends to zero with ϵ, is also chosen sufficiently small.

Sufficient Irrationality and Moderate Nonlinearity. Assuming that the sum in (3.2.29) converges to σ, we see that KAM surfaces may not exist if $\alpha = \omega_1/\omega_2$ falls within one of the cross-hatched regions in Fig. 3.2c. Since the width of these regions is proportional to $(\epsilon G)^{1/2}$ and decreases with increasing q, there is a condition that α lies sufficiently far from a rational p/q. For small ϵ, this condition is easily met, but as ϵ increases, only those irrationals that are hardest to approximate by rationals will yield KAM surfaces. The most irrational number in this sense is the golden mean $\frac{1}{2}(\sqrt{5} - 1)$. Greene (1979a) has given a very sharp criterion for the onset of strong stochasticity, as ϵ is increased, in terms of the disappearance of the KAM surface associated with the golden mean. We describe his method and results in Chapter 4.

Setting the left-hand side of (3.2.29) equal to unity gives the requirement

$$4(2\epsilon \bar{G} \Lambda_0)^{1/2} \lesssim \frac{\omega_2}{\sigma}, \tag{3.2.35}$$

which is the condition of small ϵ mentioned, and also a condition of sufficiently small nonlinearity for fixed ϵ. Combining (3.2.18) and (3.2.35)

3.2. Generic Behavior of Canonical Mappings

yields the condition of *moderate nonlinearity*

$$\frac{32\epsilon\Lambda_0}{J_0^2} < \bar{G} < \frac{\omega_2^2}{32\epsilon\Lambda_0\sigma^2}, \tag{3.2.36}$$

where we have used $(r^2/q^2)H_{rs} \sim \Lambda_0$ in (3.2.18). A similar criterion has been given by Chirikov (1979).

The general proof of the KAM theorem (Moser, 1962) requires an exeedingly small value of ϵ to be chosen to ensure that, for a given resonance, the perturbation amplitude ϵH_{lm} is sufficiently small. Chirikov (1959) recognized that for actual systems, the largest perturbation parameter could be estimated from a criterion for the overlap of the adjacent primary islands whose widths are shown in Fig. 3.2b. Numerical calculations indicated that this criterion gave a reasonably good value for the size of the perturbation parameter required to destroy the last KAM surface between these islands. Using a combination of analytical and numerical results and including the half and third integer resonances $q = 2$ and $q = 3$, Chirikov (1979) has refined the criterion to obtain quite accurate predictions of the transition. The use of overlap and related criteria to determine the value of ϵ at this transition, for a standard class of perturbations, is the subject of Chapter 4.

*3.2b. Rational Winding Numbers and Their Structure

We continue our examination of the perturbed twist map, given by (3.1.13). We have seen that if we examine irrational surfaces sufficiently far from the rational $\alpha = r/s$, the KAM theorem tells us that the surfaces retain their topology and are only slightly deformed from the unperturbed circles. However, on the rational surface $\alpha = r/s$, and in a neighborhood about it, the KAM theorem fails. For this region we can fall back on an earlier theorem for some clue to the structure of the mapping near rational α.

Poincaré–Birkhoff Theorem. For the unperturbed twist mapping (3.1.8), we have seen that *any* point on the circle $\alpha(J) = r/s$ is a fixed point of the mapping with period s (see Fig. 3.1b). The theorem states that for some even multiple of s, i.e., $2ks$ ($k = 1, 2, \ldots$), fixed points remain after the perturbation. The theorem is easy to prove and the proof can be outlined, using Fig. 3.3, as follows.

If we assume for definiteness that $\alpha(J)$ increases outward (strong spring), then there is a KAM curve outside the rational surface, which for s iterations of the mapping maps counterclockwise (outside arrows) $\alpha > r/s$, and one inside the rational surface, which maps clockwise (inside arrows) $\alpha < r/s$. Therefore between these two there must be a curve (solid line, not a KAM curve) whose angular coordinate θ is unchanged after s iterations of the mapping. These points are then radially mapped from the solid curve to

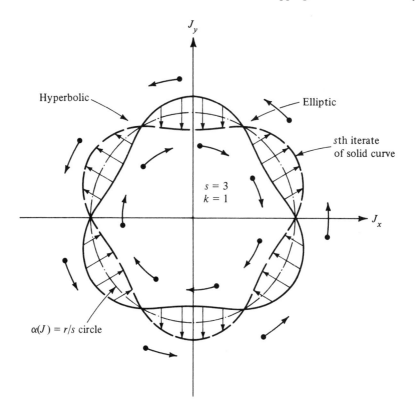

Figure 3.3. Illustrating the Poincaré–Birkhoff theorem that some fixed points are preserved in a small perturbation. The intersections of the heavy solid and dashed curves are the preserved fixed points.

some dashed curve (not a KAM curve), as shown in the figure. Due to the area-preserving property of the transformation, the solid and dashed curves must enclose an equal area. This is only possible if the two curves cross each other an even number of times. Each crossing when iterated s times returns to its initial position, so each of the s iterates is itself a fixed point. Thus, for an even number of crossings, there must be $2ks$ such points, which are the Poincaré–Birkhoff fixed points.

The theorem makes no claim about the value of the integer k, although generally $k = 1$. If we further examine the mapping in the neighborhood of the fixed points, we notice in Fig. 3.3 that there are two distinct types of behavior. Near the fixed point labeled *elliptic*, points with $\alpha \neq r/s$, tend to connect with the $\alpha = r/s$ radial transformation, and thus to circle about the fixed point. Near the fixed point labeled *hyperbolic*, on the other hand, successive transformations take the points further from the neighborhood of the fixed point. This behavior has already been noted for the phase space motion of a simple pendulum or nonlinear spring in Section 1.3. We found

3.2. Generic Behavior of Canonical Mappings

chains of alternating elliptic and hyperbolic singular points, with regular phase space trajectories encircling the elliptic fixed points and a separatrix trajectory connecting the hyperbolic points. For small perturbation amplitudes, the alternation of elliptic and hyperbolic singular points about the resonance curve is a generic property of the system.

Elliptic Points. By transforming to a coordinate system fixed on an elliptic singular point, we were able, in Section 2.4, to investigate systematically successively smaller regions of regular motion in the phase plane. Indeed, the detailed examination of an elliptic orbit in the neighborhood of the fixed point reveals a set of higher-order resonances, or fixed points, which have their own motions similar to that just analyzed above but on a finer scale. We also saw in Section 2.4 [see Eq. (2.4.62) and following discussion] that the higher-order perturbations become extremely small in s, i.e., proportional to $1/s$! If the original perturbation from an integrable system is small, most of the space is filled with KAM curves that are topologically similar to the integrable curves. In the remaining space of topologically modified regions surrounding the elliptic fixed points, new KAM curves exist, and these fill an even larger portion of the remaining space. But is the entire space then, filled with KAM curves of increasing complexity on a finer and finer scale, until suddenly with increasing perturbation the entire picture breaks down leading to stochasticity? The answer to this question is no; stochastic regions *generically* exist in the neighborhood of the separatrices (hyperbolic singular points) and these stochastic regions grow as the perturbation amplitude increases.

Hyperbolic Points. To understand the behavior of a mapping in the neighborhood of a hyperbolic singular point, we again rely on results from the theory of mappings. We have already seen that an oscillator with one degree of freedom, such as the pendulum, has a separatrix that smoothly joins the hyperbolic orbits emanating from and entering the singular point(s). For the pendulum there is one hyperbolic point (mod 2π), but in general we have a chain of ks hyperbolic points, which, in an integrable system, has a smooth separatrix that joins a hyperbolic point to its nearest neighbors. For a near-integrable system with two or more degrees of freedom or the equivalent mapping of (3.1.13), the situation is much more complicated. We present here a qualitative account of this behavior, following Berry (1978), based on the work of Poincaré (1899), Birkhoff (1935), and Smale (1965) and as reviewed by Arnold and Avez (1968) and Moser (1973).

At the hyperbolic singularity four curves join, corresponding to the two incoming trajectories of the separatrix Hamiltonian H^+, and the two outgoing trajectories of the separatrix Hamiltonian H^-. A point x lies on H^+ if the repeated transformation $T^n x$ as n approaches infinity brings x to the singularity, while it lies on H^- if the inverse transformation $\lim_{n \to \infty} T^{-n} x$ brings x to the singular point. Since the period is infinite on the separatrix trajectory, the movement of x toward the singularity becomes increasingly slow as the hyperbolic fixed point is approached. Now consider the separatrix joining adjacent hyperbolic singular points of the same set. Rather than joining smoothly, as with the one-dimensional oscillator, the H^- curve leaving one hyperbolic point generically intersects the H^+ curve arriving at the neighboring (shifted by $2\pi/ks$) hyperbolic point. This intersection is called a *homoclinic point*, as it connects outgoing and incoming trajectories of the topologically same hyperbolic point. (Intersections between trajectories from neighboring resonances are designated *heteroclinic points*.) If a single intersection exists, then it is easy to show that there are infinitely many intersections, all of which are also homoclinic points.

The extraordinarily complicated motion may be visualized a little more clearly with a picture of the type first drawn by Melnikov (1962). Consider the hyperbolic point and its $2\pi/ks$ iterate as shown in Fig. 3.4a. The intersection of the H^+ and H^- trajectories at the homoclinic point x implies the crossing at x' and then again at x'', with x'' closer to x' than x' is to x. Since the areas enclosed by the intersections (cross-hatched areas) are mappings of one another, they are conserved, and thus H^- oscillates more strongly between x' and x'' than between x and x'. Successive crossings are closer together and oscillate more wildly. Although we have not drawn it, H^+ must have emanated from the singular point on the left, oscillating just as wildly as the H^- we have drawn, and so also for the other two branches. Thus in the neighborhood of the singular point on the right, H^+ from below also oscillates wildly and intersects H^- from above an infinite number of times. This is illustrated in Fig. 3.4b, from a numerical example by Dragt and Finn (1976a). Both the oscillating H^- approaching the singular point from its image on the left and the oscillating H^+ leaving the neighborhood of the singular point (not shown in Fig. 3.4a) are shown in Fig. 3.4b.

To calculate numerically these curves, say H^- emanating from the left, we choose a point on H^- near the singularity and draw a short straight line segment between the point and its image under T. This line segment is then repeatedly mapped forward, generating the oscillating H^- curve approaching the singularity on the right.

But the homoclinic points by themselves do not tell the whole story of this most complex region near a separatrix. Since the local rotation number around the elliptic singular point goes to infinity at the separatrix, then in the neighborhood of the separatrix there are an infinite number of second-order resonances corresponding to high rotation numbers. Each has its own set of alternating elliptic and hyperbolic singular points, with its own wild motion, repeatedly intersecting both each other and the first-order separatrix trajectory in heteroclinic points. All of these trajectories appear to fill densely the space available to them. The intersections of the orbits in homoclinic points, in fact, demonstrate that a KAM torus cannot exist at such a point as the corresponding trajectory changes its topology. The detailed argument has been presented by Dragt and Finn (1976a). For small perturbations,

3.2. Generic Behavior of Canonical Mappings

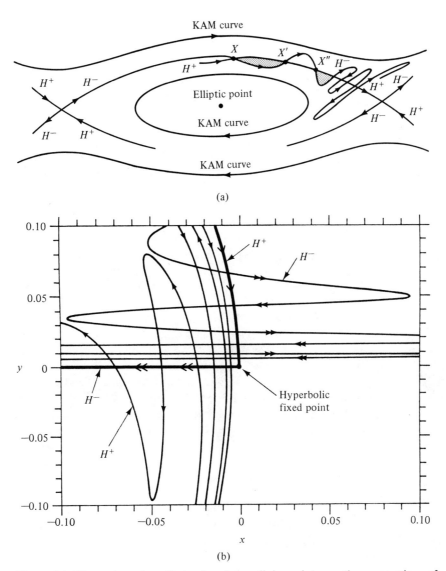

Figure 3.4. Illustrating the effect of a homoclinic point on the generation of stochasticity near a separatrix. (a) The stable (H^+) and unstable (H^-) branches of the separatrix intersect infinitely many times. (b) Detail of the intersections near the hyperbolic fixed points (after Dragt and Finn, 1976a).

however, all of this exceedingly complicated behavior occurs in a region of phase space bounded between KAM curves (shown as heavy lines in Fig. 3.4a).

*3.2c. Complete Description of a Nonlinear Mapping

We are now in a position to describe the complete behavior of a mapping, e.g., the surface of section of a coupled nonlinear oscillator with two degrees of freedom. The topology of the mapping has been given in Arnold and Avez (1968, p. 91), but in a rather more schematic presentation than necessary. We consider a modest size coupling term such that many of the constant J surfaces, as visualized in Fig. 3.1b without coupling, remain as KAM surfaces with the coupling term present. To be specific, we take the oscillator in the surface of section to be of the weak spring type $d\omega_1/dJ_1 < 0$, and also, for definiteness, let $\alpha = \omega_{01}/\omega_{02} = 1/\pi$ (irrational) at $J_1 = 0$. As J_1 increases outward, ω_1 decreases until the first important resonance is reached at $\omega_1 = \omega_2/4$, giving a chain of four islands in the surface of section. In other words, the trajectory labeled as $x_0, x_1, \ldots$ in the surface of section of Fig. 3.1b takes four traversals around the torus to return to its initial position provided that the orbit's initial phase corresponds to that of a fixed point of the mapping. For other nearby phases, $dJ_1/dt \neq 0$ and the mapping is that of the island trajectories around the stable fixed points. We have already shown how to calculate these island trajectories in Section 2.4, and will give further examples in Sections 3.4 and 3.5, and in subsequent chapters. From our previous discussion, we found that the elliptic and hyperbolic fixed points alternate around the constant J surface, which was also our conclusion from the expansion of Section 2.4. Near the hyperbolic fixed points, the presence of the homoclinic and heteroclinic points assures a region of chaotic motion bounded for larger and smaller J_1 by KAM surfaces. Continuing to larger J_1, we hit the next important resonance at $\omega_1 = \omega_2/5$, which gives a chain of five islands. At still larger J_1, there appear the six-island resonance, the seven-island resonance, etc. In all this we must keep in mind that there are an infinity of intermediate resonances, between each major resonance, at all of the rationals, but that these resonances, as we have seen in Section 2.4, have rapidly decreasing amplitudes. For example, the resonance between $\alpha = \frac{1}{4}$ and $\alpha = \frac{1}{5}$, at $\alpha = \frac{2}{9}$, i.e., $r = 2$ and $s = 9$, has an amplitude that is related to the $\alpha = \frac{1}{5}$ resonance by the square root of their Fourier coefficients. That is, we have from (2.4.31)

$$\Delta \hat{J}_1 \sim \left| \frac{\epsilon H_{r,-s}}{\partial^2 \hat{H}_0/\partial \hat{J}_1^2} \right|^{1/2}, \qquad (3.2.37)$$

where $H_{r,-s}$ is the Fourier coefficient of the approximate resonant harmonic.

3.2. Generic Behavior of Canonical Mappings

The maximum value of the Fourier coefficient, as found in Section 2.4b, is

$$H_{r,-s}(\max) \approx \mathscr{J}_s(\pi),$$

where $\mathscr{J}$ is the ordinary Bessel function, and the ratio of island amplitudes at $s = 5$ and $s = 9$ is roughly

$$\frac{\Delta J(s/r = 9/2)}{\Delta J(s/r = 5)} \approx \left[\frac{\mathscr{J}_9(\pi)}{\mathscr{J}_5(\pi)}\right]^{1/2} \approx 0.1. \tag{3.2.38}$$

Thus at a magnification in which the main resonances are of a modest size, the ones associated with the next higher harmonic of r are quite small.

With the above discussion in mind, we qualitatively sketch the mapping in Fig. 3.5a. The solid lines represent KAM surfaces. The ones surrounding the origin ($J_1 = 0$) are distorted from the original circles, and thus correspond to a new invariant $\bar{J}_1 = \text{const.}$, on which J_1 varies in a regular, single-valued way with phase. Approximate values of this invariant are calculated by the *method of averaging* as in Section 2.3. The solid lines surrounding the elliptic singular points cannot be calculated in this way, but are calculated, approximately, by the *method of secular perturbations* in Section 2.4. For the trajectories near the separatrices, an invariant does not exist, and the trajectories are area filling. Several such trajectories are shown as the shaded region surrounding the various island chains and limited in the J_1 excursion by KAM curves. Between the $\alpha = \frac{1}{4}$ and $\alpha = \frac{1}{5}$ island chains lies the $\alpha = \frac{2}{9}$ chain of smaller, but still visible, islands. There are, of course, all the other primary fractional island chains, $\alpha = r/s$, not shown and either generally of an amplitude too small to be seen on the scale of the figure, or giving rise to purely chaotic motions through being embedded in the shaded regions of the figure.

Now, what happens when we expand the scale of a single island in one of the island chains? We do this in Fig. 3.5b, first transforming to the new variables $\hat{J}_1, \hat{\theta}_1$, as in Section 2.4, and introducing the new action

$$I_1 = \frac{1}{2\pi}\oint \hat{J}_1 \, d\hat{\theta}_1, \tag{3.2.39}$$

corresponding to the closed lines around a given α elliptic singular point. This transforms the island chain into a single set of concentric circles. However, resonances between the local transform around the fixed point and the fundamental frequencies now create *second-order island chains*, which again modify these new circles. We have already calculated these second-order islands in Section 2.4b. As we saw there, the amplitude and frequency of the islands are both exponentially small in the square root of the inverse rotation number $(1/\alpha)^{1/2}$. By an argument similar to that leading to (3.2.37), we can calculate the widths of the second-order islands. We leave the details of a similar calculation to Chapter 4; here we qualitatively illustrate the results in Fig. 3.5b. The ratios of second-order island amplitudes to the

Figure 3.5. Regular and stochastic trajectories for a Hamiltonian with relatively large perturbation (a) near the primary fixed point; (b) expanded (and circularized) scale near a second-order fixed point.

3.2. Generic Behavior of Canonical Mappings

second-order island separations, are considerably smaller than were found for the first-order island chains. We have chosen an example of a rather large perturbation. For significantly smaller first-order islands, the second-order island chains would be vanishingly small even after the enlargement, that is, on a scale for which adjacent island resonances are simultaneously observed, the island width would not be visible.

We have already used the KAM result, given in (3.2.1), to show that, for sufficiently small perturbation from an integrable system, most of the space is occupied by invariant curves. The present picture, with exponentially small, higher-order resonances occupying increasingly small percentages of the phase space, reinforces this picture. The following may be a helpful way of looking at this ordering: place a point down at any place on the mapping; magnify the region around that point until the structure becomes visible. For reasonably small perturbations from an integrable system, the probability is high that the point will appear to lie on a smooth KAM curve, although around it will lie stochastic regions near the separatrices of island chains. Now repeat the procedure of placing a point at random on the expanded section of the map and repeat the expansion until the next order of detail is visible in its neighborhood. After this expansion much less of the space will appear to be stochastic, and thus the probability of the point appearing to lie on an invariant curve will be much higher. On each repetition of this process, the stochastic region becomes an exponentially smaller portion of the space, and the probability of the point falling on an invariant curve tends exponentially to unity. Nevertheless, on any such trial, the point has a finite probability of falling into a stochastic region. Once this happens, it is possible to expand the region of stochasticity until it fills the entire field of observation, and subsequently all future expansions would yield only stochastic trajectories.

*3.2d. Numerical Examples

The numerical study of a surface of section mapping obtained by integration of Hamilton's equations is very time consuming because the intersections of the trajectory with the surface of section can only be determined by solving the equations of motion over time scales much shorter than the mapping period T_2. Direct iteration of an area-preserving mapping exhibits essentially the same behavior and can be easily studied over hundreds of thousands of mapping periods. These mappings have been extensively used to uncover the structure of nonlinear coupled oscillator systems. We discuss one such mapping, the Fermi acceleration mechanism (Fermi, 1949), in Section 3.4. Here we illustrate some of the phenomena previously discussed in this section with two examples, the first being the quadratic twist mapping studied by Hénon (1969):

$$\begin{pmatrix} x_1 \\ y_1 \end{pmatrix} = T \begin{pmatrix} x_0 \\ y_0 \end{pmatrix} = \begin{bmatrix} x_0 \cos \psi - (y_0 - x_0^2) \sin \psi \\ x_0 \sin \psi + (y_0 - x_0^2) \cos \psi \end{bmatrix}, \quad (3.2.40)$$

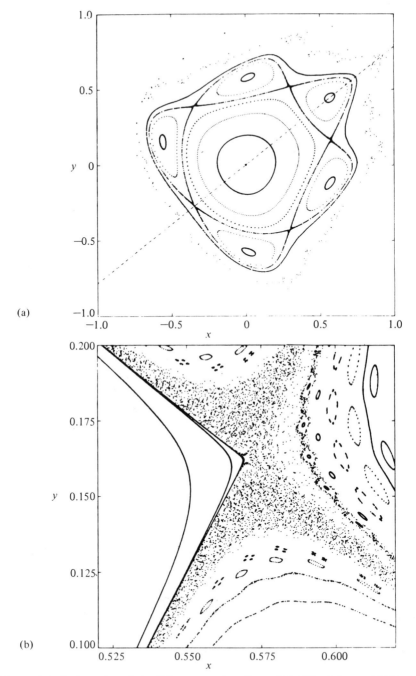

Figure 3.6. Trajectories of the Hénon mapping [Eqs. (3.2.40)] with $\alpha = 0.2114$. (a) Mapping including the origin and the first island chain; (b) expanded mapping near a separatrix of the first island chain. Each island chain, etc., is generated from a single (x_0, y_0) (after Hénon, 1969).

3.2. Generic Behavior of Canonical Mappings

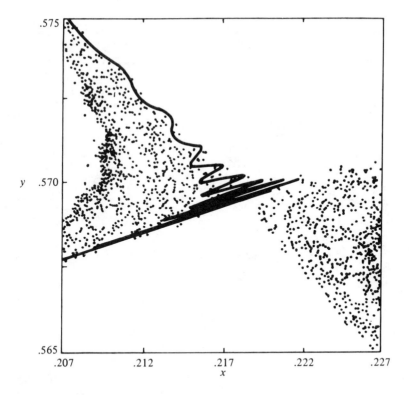

Figure 3.7. Transformation of a small line segment of initial conditions on the separatrix (solid curve) through sufficient mapping iterations to develop the behavior of the separatrix mapping near a homoclinic point (after Berry, 1978).

where $\psi = 2\pi\alpha_0$, and α_0 is the rotation number of the linear twist mapping (3.1.9). The perturbation from the linear mapping, x_0^2, is small near the origin, which is an elliptic fixed point of the mapping. In Fig. 3.6a,b, we reproduce Hénon's results for $\psi = 76.11°$. In Fig. 3.6a, we see the first major resonance with the mapping period at a rotation number $\alpha = \frac{1}{5}$, corresponding to a 72° rotation. The nonlinear term is thus of the "weak spring" type, slowing down the rotation until the first major resonance occurs. Examination of the mapping near a hyperbolic fixed point of this resonance yields the remarkable picture seen in Fig. 3.6b. Chains of second- and third-order islands are seen, together with an area-exploring trajectory (50,000 iterations) very close to the hyperbolic fixed point. In order to explore the oscillation of the separatrix trajectory, due to the presence of homoclinic points, Cuthill (unpublished) mapped a short line segment emanating from one of the hyperbolic points, as shown in Fig. 3.7. After 146 iterations, the line has stretched into the long oscillatory curve in the figure, characteristic of the oscillation of the H^- trajectory as shown in Fig. 3.4.

As sketched in Fig. 3.5, and observed qualitatively in Fig. 3.6, Hamiltonian

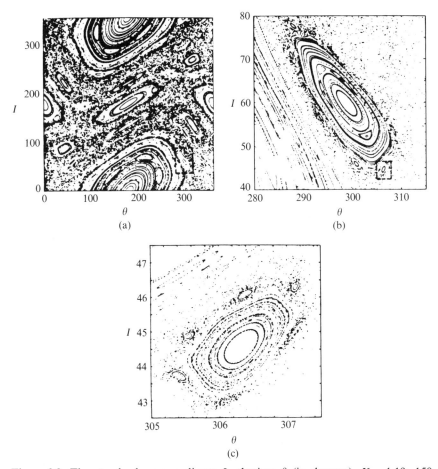

Figure 3.8. The standard map: ordinate I, abscissa θ (in degrees). $K = 1.19$; 150 random initial conditions for each graph; number of iterations increasing from (a) to (c) as $N = 500$, 5000, and 10,000. In (b) the boxed region in (a) is magnified; in (c), the boxed region in (b) is further magnified (after Lichtenberg, 1985).

systems have structure on all scales. This property will be used in Chapter 4 in various renormalization approaches to determine the parameter values at which the last KAM surface is destroyed. To see this complexity we look at the standard mapping (3.1.22) for $K = 1.19$. In Fig. 3.8a, a number of initial conditions are shown, exploring both regular (KAM) orbits and stochastic orbits. Second-order island chains are seen to exist near the large island structures. In Fig. 3.8b, one island from a chain (inside the dashed box in Fig. 3.8a) is magnified, revealing a third-order structure which is then magnified by the same factor to obtain Fig. 3.8c. This self-similarity of island structures is a generic feature of a divided phase space, one containing both regular and stochastic regions, and has been used to examine various

3.3. Linearized Motion

properties of the mappings (Lichtenberg, 1985; Meiss, 1986). Around each island torus there also exist "cantori," the remnants of the destroyed KAM tori. We will see in Section 5.5b that the diffusion of phase points is related to the phase space flux passing through gaps in the cantori.

Many of the regular features of the mappings, shown in Figs. 3.6 and 3.8, can be calculated analytically. Rather than do this here, we return to these calculations for other examples in subsequent sections. These calculations will follow more naturally after an investigation of the stability of fixed points, which we present in the next section.

3.3. Linearized Motion

It is in principle straightforward to find the period k fixed points of a mapping by requiring that, for the kth iterate

$$T^k x_0 = x_0, \qquad (3.3.1)$$

although for k large the actual calculation may be cumbersome. The fixed points can be obtained directly from the Hamiltonian: the equivalent equations to be solved are for *periodic orbits*, which in a surface of section $x = (p, q)$ can be written

$$\frac{\partial H}{\partial x_i} = 0 \qquad (3.3.2)$$

and can generally only be solved by series (Eminhizer et al., 1976). Once the fixed points of the mapping have been found, we can examine the linear stability about a given fixed point.

The procedure for examining linear stability is to expand x as $x = x_0 + \Delta x$. Keeping only the linear terms in Δx, we obtain an equation of the form

$$\Delta x_{n+k} = \mathbf{A} \cdot \Delta x_n, \qquad (3.3.3)$$

where $\mathbf{A}$ is a transformation independent of Δx.

As we have already seen in the previous section, a small perturbation of an integrable system creates a set of alternating elliptic (stable) and hyperbolic (unstable) fixed points. We have seen this behavior numerically in the example of Section 3.2d. For large perturbations, the topological argument no longer applies, and these fixed points may all be unstable. In Hamiltonian systems which are not integrable (generically with two or more degrees of freedom), linear stability appears to be both necessary and sufficient for nonlinear stability, in the sense that linear stability guarantees the existence of KAM surfaces arbitrarily close to the fixed point.[1]

---

[1] Rotation numbers with period 3 or 4 are special cases.

*3.3a. Eigenvalues and Eigenvectors of a Matrix

From (3.3.3), and putting $x = (p, q) = \Delta x_n$ and $\bar{x} = (\bar{p}, \bar{q}) = \Delta x_{n+k}$ for ease of notation, the linearized equations for an M-dimensional mapping are written in matrix form as

$$\bar{x} = \mathbf{A} \cdot x, \tag{3.3.4}$$

where $\mathbf{A}$ is an $M \times M$ matrix independent of x. We assume the rank of $\mathbf{A}$ is M; hence, $\det \mathbf{A} \neq 0$. The eigenvalues of (3.3.4) are the constants that satisfy the equation

$$\mathbf{A} \cdot x = \lambda x, \tag{3.3.5}$$

thus leaving the vector x unchanged except for a constant. Clearly (3.3.5) requires that λ satisfies

$$\det(\mathbf{A} - \lambda \mathbf{I}) = 0, \tag{3.3.6}$$

where $\mathbf{I}$ is the identity matrix. This is an Mth-order equation for λ having M characteristic roots. Each eigenvalue corresponds to a normal mode of the system, that is, a possible independent mode of behavior. From (3.3.5) it is also seen that each value of λ must satisfy $\lim_{n \to \infty} \lambda^n$ bounded, in order to have stable motion for repeated applications of $\mathbf{A}$.[2] The set of vectors x_k that correspond to the values of λ_k satisfying (3.3.6) are the eigenvectors or normal modes of the motion. They can be found, once the λ_k are known, by solving the set of equations

$$\mathbf{B} \cdot x_k = \sum_j b_{ij} x_{jk} = 0 \tag{3.3.7}$$

with $\mathbf{B}(\lambda_k) = \mathbf{A} - \lambda_k \mathbf{I}$. If all of the eigenvalues of $\mathbf{A}$ are different, this can easily be done by moving a column l of (3.3.7) to the right-hand side and omitting the corresponding row l. Setting the lth component $x_k = c_k B_{ll}$, where $B_{ll} \neq 0$ is the cofactor of b_{ll} and c_k is an arbitrary constant, the resulting set of $M - 1$ equations can be solved by Cramer's rule for the remaining components of $x_k = c_k x_k'$:

$$x_{jk} = c_k B_{lj}(\lambda_k); \quad j = 1, \ldots, M. \tag{3.3.8}$$

Here B_{lj} is the cofactor of b_{lj}. Since the rank of $\mathbf{B}$ is $M - 1$, there exists at least one $B_{ll} \neq 0$. If the eigenvalues are not all distinct, then some of the x_k will depend on two or more arbitrary constants, but a similar technique can be used.

---

[2] For repeated eigenvalues with $|\lambda| = 1$ the linear system can be unstable; see Section 3.3d.

3.3. Linearized Motion

If we form columns of the matrix **X** from the M independent eigenvectors x_k, then from (3.3.5),

$$\mathbf{A} \cdot \mathbf{X} = \mathbf{X} \cdot \mathbf{\Lambda}, \tag{3.3.9}$$

where, if the eigenvalues are all distinct, then $\mathbf{\Lambda}$ is a diagonal matrix whose elements $\Lambda_{ii} = \lambda_i$.[3] From (3.3.9),

$$\mathbf{\Lambda} = \mathbf{X}^{-1} \cdot \mathbf{A} \cdot \mathbf{X}, \tag{3.3.10}$$

which shows that for distinct eigenvalues, **X** diagonalizes **A**. Introducing the new eigenvectors through the transformation

$$u_k = \mathbf{X}^{-1} \cdot x_k, \qquad x_k = \mathbf{X} \cdot u_k, \tag{3.3.11}$$

we have from (3.3.5)

$$\mathbf{\Lambda} \cdot u_k = \lambda_k u_k \tag{3.3.12}$$

so that in the transformed space the u_k can be chosen as a set of M orthonormal unit vectors e_k.

Symmetry of the Eigenvalues. If the transformation generated by **A** is canonical, M is an even integer $2N$, and the Poisson bracket conditions hold

$$[\bar{q}_i, \bar{q}_j] = [\bar{p}_i, \bar{p}_j] = 0,$$
$$[\bar{q}_i, \bar{p}_j] = \delta_{ij}, \qquad i,j = 1, \ldots, N. \tag{3.3.13}$$

We introduce the $2N$-dimensional antisymmetric matrix

$$\mathbf{\Gamma} = \begin{pmatrix} 0 & -\mathbf{I} \\ \mathbf{I} & 0 \end{pmatrix}, \tag{3.3.14}$$

where each entry in $\mathbf{\Gamma}$ is an $N \times N$ block, $\mathbf{\Gamma}^T = \mathbf{\Gamma}^{-1} = -\mathbf{\Gamma}$ (T denotes the transpose), and $\det \mathbf{\Gamma} = 1$. We may then write (3.3.13) in the form

$$[\bar{x}_i, \bar{x}_j] = \sum_{k,l} A_{ik} \Gamma_{kl} A_{jl} = \Gamma_{ij}, \tag{3.3.15}$$

where

$$A_{ik} = \frac{\partial \bar{x}_i}{\partial x_k}, \tag{3.3.16}$$

showing that the components of **A** are not all independent. Expressing this in matrix form, we have

$$\mathbf{A} \cdot \mathbf{\Gamma} \cdot \mathbf{A}^T = \mathbf{\Gamma} \tag{3.3.17a}$$

with its inverse

$$\mathbf{A}^T \cdot \mathbf{\Gamma} \cdot \mathbf{A} = \mathbf{\Gamma}. \tag{3.3.17b}$$

A matrix that satisfies this condition is said to be *symplectic*.

---

[3] If the eigenvalues are not all distinct, see Arnold (1979) or any text on linear algebra for details.

We now show that if λ is an eigenvalue of $\mathbf{A}$, then $1/\lambda$ is also an eigenvalue. The eigenvalues of any matrix and its transpose are identical, so from (3.3.5),

$$\mathbf{A}^T \cdot \mathbf{y} = \lambda \mathbf{y} \tag{3.3.18}$$

or

$$(\mathbf{A}^T)^{-1} \cdot \mathbf{y} = (1/\lambda)\mathbf{y}. \tag{3.3.19}$$

From (3.3.17a),

$$\mathbf{A} \cdot (\mathbf{\Gamma} \cdot \mathbf{y}) = (1/\lambda)(\mathbf{\Gamma} \cdot \mathbf{y}), \tag{3.3.20}$$

which shows that $1/\lambda$ is an eigenvalue of $\mathbf{A}$, with eigenvector $\mathbf{x} = \mathbf{\Gamma} \cdot \mathbf{y}$. Thus we write

$$\lambda_{i+N} = \lambda_i^{-1}, \qquad i = 1, \ldots, N. \tag{3.3.21}$$

Since $\mathbf{A}$ is real, complex roots must come in complex conjugate pairs. For λ complex with $|\lambda| \neq 1$ and $\operatorname{Im} \lambda \neq 0$, the roots must then appear as 4-tuples

$$\lambda, \lambda^*, 1/\lambda, 1/\lambda^*$$

symmetrical with respect to the real axis and the unit circle (Fig. 3.9). For $\operatorname{Im} \lambda = 0$, we have pairs $\lambda, 1/\lambda$ lying on the real axis. For either of these cases the motion is unstable. For $|\lambda| = 1$, we have pairs $\lambda, \lambda^* = 1/\lambda$ lying on the unit circle, and the motion is stable.

One consequence of the symmetry of the eigenvalues, easily shown, is that the characteristic equation (3.3.6) can be written in the form

$$\lambda^N + a_1 \lambda^{N-1} + \cdots + a_{2N-1}\lambda + 1 = 0, \tag{3.3.22}$$

where the coefficients are symmetric

$$a_1 = a_{2N-1}, \qquad a_2 = a_{2N-2}, \ldots. \tag{3.3.23}$$

For stability, then, with distinct eigenvalues, we require all roots to lie on the unit circle. For multiple roots, the question of stability is more complicated (see Arnold, 1979, p. 227); one generally has marginal stability.

We now show that if $\mathbf{A}$ is symplectic, then $\mathbf{X}$ can also be put in symplectic form by properly choosing the arbitrary multiplicative factors c_k in the columns of $\mathbf{X}$. We consider the antisymmetric matrix

$$\mathbf{S} = \mathbf{X} \cdot \mathbf{\Gamma} \cdot \mathbf{X}^T \tag{3.3.24}$$

with components

$$S_{ij} = \mathbf{x}_i \cdot \mathbf{\Gamma} \cdot \mathbf{x}_j^T. \tag{3.3.25}$$

From (3.3.9) and (3.3.24), we obtain the result

$$\mathbf{\Lambda} \cdot \mathbf{S} \cdot \mathbf{\Lambda}^T = \mathbf{S}, \tag{3.3.26a}$$

3.3. Linearized Motion

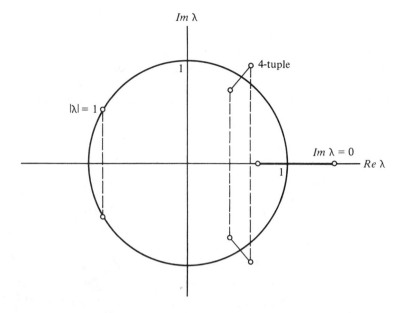

Figure 3.9. Distribution of eigenvalues of a symplectic matrix.

which in component form is

$$\lambda_i \lambda_j S_{ij} = S_{ij}, \quad (3.3.26b)$$

which shows that $S_{ij} = 0$ unless $\lambda_i \lambda_j = 1$. From (3.3.21), the nonzero elements of S are then

$$S_{i,i+N} = -S_{i+N,i}, \quad i = 1, \ldots, N. \quad (3.3.27)$$

Choosing $S_{i,i+N} = -1$, and using (3.3.8) in (3.3.25), we find

$$c_i c_{i+N} \mathbf{x}_i^{\prime T} \cdot \mathbf{\Gamma} \cdot \mathbf{x}_{i+N}^{\prime} = -1. \quad (3.3.28)$$

This equation determines, say c_{i+N}, given c_i. Thus by construction, $\mathbf{S} = \mathbf{\Gamma}$, and by comparing (3.3.24) with (3.3.17a), we see that $\mathbf{X}$ is symplectic. The construction, and therefore $\mathbf{X}$, are not unique because the first N values of c_i can be chosen arbitrarily.

The skew-symmetric product $\mathbf{x}_1^T \cdot \mathbf{\Gamma} \cdot \mathbf{x}_2$ is invariant to a symplectic transformation $\bar{\mathbf{x}} = \mathbf{A} \cdot \mathbf{x}$, i.e.,

$$\mathbf{x}_1^T \cdot \mathbf{\Gamma} \cdot \mathbf{x}_2 = \bar{\mathbf{x}}_1^T \cdot \mathbf{\Gamma} \cdot \bar{\mathbf{x}}_2, \quad (3.3.29)$$

which can be seen using (3.3.17b). Equation (3.3.29) is often taken as the definition of a symplectic transformation. Letting $\mathbf{x}_1 = \mathbf{x}$, $\mathbf{x}_2 = \mathbf{A} \cdot \mathbf{x}$, we see that the quadratic form

$$Q = \mathbf{x}^T \cdot \mathbf{\Gamma} \cdot \mathbf{A} \cdot \mathbf{x} \quad (3.3.30)$$

is an invariant of the mapping $\mathbf{A}$.

Only in two dimensions, for which the equation for λ is a quadratic, are the eigenvalues and eigenvectors readily found analytically. This, however, is the case that applies to the two-dimensional mappings that are of central importance to our study of coupled nonlinear systems. For higher dimensionality, solutions can sometimes be found without resorting to numerical techniques, but each problem must be treated as an individual case.

*3.3b. Two-Dimensional Mappings

A two-dimensional mapping is described by a set of two first-order difference equations. In action-angle variables in the surface of section, we have the perturbed twist mapping of (3.1.13), repeated with $\epsilon \equiv 1$ for ease of notation:

$$J_{n+1} = J_n + f(J_{n+1}, \theta_n), \quad (3.1.13a)$$

$$\theta_{n+1} = \theta_n + 2\pi\alpha(J_{n+1}) + g(J_{n+1}, \theta_n), \quad (3.1.13b)$$

where f and g are periodic in θ with period 2π (or sometimes, for convenience, unity), and the rotation number $\alpha = \omega_1/\omega_2$ describes the advancing of the phase θ_1 (written as θ here) for a 2π advance of θ_2. The quantities α, f, and g are written in terms of J_{n+1}, rather than J_n, so that the symplectic property takes on the simple form (3.1.16) of area-preservation. A mapping need not be written in action-angle form; Hénon's quadratic twist mapping is an example.

Fixed Points. Equations (3.1.13) possess a fixed point of period k at $x_0 = (J_0, \theta_0)$ when

$$x_0 = T^k(x_0) \quad (3.3.31)$$

and x_0 is not a fixed point at any period less than k, i.e., an initial condition located exactly at x_0 will first reappear after exactly k iterations. For every positive integer k, there is a denumerably infinite set of fixed points. Fixed points of period k occur in families $\{x_{01}, x_{02}, \ldots, x_{0k}\}$ of exactly k members each, where $x_{0i} = T^i x_{01}$. These families may be organized into a hierarchy (Greene, 1968).

To obtain all the period k fixed points, we solve the $2k + 2$ algebraic equations

$$\begin{aligned} J_{i+1} &= J_i + f(J_{i+1}, \theta_i) \quad i = 1, \ldots, k, \\ \theta_{i+1} &= \theta_i + 2\pi\alpha(J_{i+1}) + g(J_{i+1}, \theta_i), \\ J_{k+1} &= J_1, \\ \theta_{k+1} &= \theta_1 \pm 2\pi m, \quad m = 0, 1, \ldots, \end{aligned} \quad (3.3.32)$$

where m is an integer relatively prime to k. Except for $k = 1$, and possibly $k = 2$, these equations are very difficult to solve analytically or even

3.3. Linearized Motion

numerically. For k large, simple root-finding methods, such as Newton's method, fail, due essentially to the closeness of nearby solutions in the $(2k + 2)$-dimensional space. The special series and variational methods (Eminhizer et al., 1976; Helleman, 1978; Helleman and Bountis, 1979) devised to deal with these problems are described in Section 2.6.

For the radial twist map, with f not a function of J and $g = 0$, successive substitution of the equation for x_i into the equation for x_{i+1} yields a two-parameter set of equations of the form

$$f_1(J_1, \theta_1) = 0,$$
$$f_2(J_1, \theta_1) = 0, \qquad (3.3.33)$$

but again these are very difficult to solve for large k due to the extremely convoluted nature of the functions f_1 and f_2.

Involution Products. There is an important class of mappings for which the fixed points can be found from a one-parameter equation. If a mapping, or its underlying Hamiltonian, has sufficient symmetry, then it may be factored into the product of two involutions (de Vogeleare, 1958; Greene, 1979b):

$$T = I_2 I_1. \qquad (3.3.34)$$

Two successive iterations of an involution mapping automatically reproduce the initial conditions, i.e.,

$$I_1^2 = I, \qquad I_2^2 = I, \qquad (3.3.35)$$

where I is the unit mapping. The radial twist mapping is an involution product provided f is antisymmetric about some angle (chosen for convenience to be zero). Then I_1 is given by

$$\bar{J} = J_n + f(\theta_n), \qquad (3.3.36a)$$
$$\bar{\theta} = -\theta_n, \qquad (3.3.36b)$$

and I_2 by

$$J_{n+1} = \bar{J}, \qquad (3.3.37a)$$
$$\theta_{n+1} = -\bar{\theta} + 2\pi\alpha(\bar{J}). \qquad (3.3.37b)$$

The factorization into involutions greatly aids in determining the fixed points. Since all initial conditions automatically repeat themselves after two iterations, involutions have only period 1 fixed points, and these are easy to find. For example, I_1 has lines of fixed points x_1 given by

$$\theta_1 = 0, \pi, \quad \text{for any } J_1, \qquad (3.3.38)$$

and I_2 has lines of fixed points x_2 for

$$2\theta_2 = 2\pi\alpha(J_2) - 2\pi m, \qquad (3.3.39)$$

where m is an integer. We can show, using (3.3.34) and (3.3.35), that if both x and $T^n x$ are fixed points of I_1 (or I_2), then x is a fixed point of T^{2n}. We apply this method in Section 3.4 to obtain the stable period 2 fixed points for the Fermi mapping, which bifurcate out of a period 1 fixed point when the latter goes unstable. Greene has used these methods to obtain the high-period fixed points for the standard mapping (Greene, 1979a) and the Hénon and Heiles problem (Greene, 1979b), which can both be cast in involution product form. Any odd power of T can be written as an involution product, and its fixed points determined in the same manner, for example,

$$T^3 = (I_2 I_1 I_2)(I_1 I_2 I_1) \tag{3.3.40}$$

is the product of two involutions.

A mapping that is a product of two involutions is *reversible*, i.e., the dynamical system is invariant under the transformation $t \to -t$. Quispel and Roberts (1988) and Roberts et al. (1991) have studied such systems. They have been able to find reversible systems that are not area preserving by constructing one of the involutions to be nonarea preserving. Such mappings have lines of symmetry with the Jacobian determinant changing from $J < 1$ to $J > 1$ on crossing the symmetry line. The mapping can have an elliptic fixed point with surrounding KAM curves on the symmetry line, and symmetrically placed attractors and repellers outside the stable region. This allows for both Hamiltonian chaos (stochastic regions) near the separatrices of stable island chains that are bounded by KAM curves and transient chaos (see Section 8.1) in regions of the phase space that are connected to the attractors.

Linearized Mapping. We now expand the mapping around the k successive fixed points in a family

$$x_{01} \to x_{02} \to \cdots \to x_{0k} \to x_{01}$$

to find the linearized equation for motion near the first member in the family x_{01}

$$\Delta x_{n+k} = \mathbf{M}(x_{0k}) \cdot \Delta x_{n+k-1}$$
$$= \mathbf{M}(x_{0k}) \cdot \mathbf{M}(x_{0,k-1}) \cdot \Delta x_{n+k-2}, \tag{3.3.41}$$

etc., or

$$\Delta x_{n+k} = \mathbf{A}_1 \cdot \Delta x_n. \tag{3.3.42}$$

Here

$$\mathbf{A}_1 = \mathbf{M}(x_{0k}) \cdot \mathbf{M}(x_{0,k-1}) \cdots \mathbf{M}(x_{01}) \tag{3.3.43}$$

is the ordered product of k matrices $\mathbf{M}_i = \mathbf{M}(x_{0i})$, evaluated at the successive

3.3. Linearized Motion

fixed points, with

$$\mathbf{M}(x) = \begin{pmatrix} \dfrac{\partial J_{n+1}}{\partial J_n} & \dfrac{\partial J_{n+1}}{\partial \theta_n} \\ \dfrac{\partial \theta_{n+1}}{\partial J_n} & \dfrac{\partial \theta_{n+1}}{\partial \theta_n} \end{pmatrix} \qquad (3.3.44)$$

the Jacobian matrix of the mapping. For motion near x_{0i}, the ith member of the family, a similar calculation yields

$$\mathbf{A}_i = \mathbf{M}_{i-1} \cdot \mathbf{M}_{i-2} \cdots \mathbf{M}_i \qquad (3.3.45)$$

which is a cyclic permutation of $\mathbf{A}_1$.

For the perturbed twist map, $\mathbf{M}$ is obtained after substituting J_{n+1} from (3.1.13a) into (3.1.13b). Setting $\epsilon = 1$ for simplicity, we obtain

$$M_{11} = \left(1 - \frac{\partial f}{\partial \bar{J}}\right)^{-1}, \qquad M_{12} = M_{11} \frac{\partial f}{\partial \theta},$$

$$M_{21} = M_{11}\left(2\pi \frac{d\alpha}{dJ} + \frac{\partial g}{\partial \bar{J}}\right), \qquad M_{22} = 1 + M_{21}\frac{\partial f}{\partial \theta} + \frac{\partial g}{\partial \theta}. \qquad (3.3.46)$$

Here $\mathbf{M}$ is the Jacobian matrix, whose determinant gives the transformation of a differential area:

$$\det \mathbf{M} = \left(1 - \frac{\partial f}{\partial \bar{J}}\right)^{-1}\left(1 + \frac{\partial g}{\partial \theta}\right). \qquad (3.3.47)$$

Setting $\det \mathbf{M} = 1$, we obtain the area-preserving condition (3.1.16).

For a period 1 fixed point, $\mathbf{A}_1 = \mathbf{M}_1$, which determines the stability and motion near the fixed point. For a family of fixed points of period greater than one, the eigenvalues and eigenvectors for the first member are found from

$$\mathbf{A}_1 \cdot x = \lambda x, \qquad (3.3.48)$$

and for the second member, from

$$\mathbf{A}_2 \cdot y = \lambda y. \qquad (3.3.49)$$

But putting $\mathbf{A}_2 = \mathbf{M}_1 \cdot \mathbf{A}_1 \cdot \mathbf{M}_1^{-1}$ in (3.3.49) reduces it to (3.3.48), with $y = \mathbf{M}_1 \cdot x$. By induction, $\mathbf{A}_i$ and $\mathbf{A}_{i-1}$ are likewise related. It follows that the eigenvalues, and the stability of the fixed points, are the same for all members of the family. The eigenvectors for the ith member are given in terms of the first by

$$x_i = \mathbf{M}_{i-1} \cdot \mathbf{M}_{i-2} \cdots \mathbf{M}_1 \cdot x_1. \qquad (3.3.50)$$

Linear Stability and Invariant Curves. The eigenvalues of the linearized mapping are the roots of the equation (3.3.35):

$$\begin{vmatrix} A_{11} - \lambda & A_{12} \\ A_{21} & A_{22} - \lambda \end{vmatrix} = 0$$

or

$$\lambda^2 - \lambda \operatorname{Tr} \mathbf{A} + 1 = 0, \tag{3.3.51}$$

where we have set $\det \mathbf{A} = 1$ in (3.3.51) because of the area-preserving property. $\operatorname{Tr} \mathbf{A}$ is the trace $A_{11} + A_{22}$ of the matrix. The solutions for λ are

$$\lambda_{1,2} = \frac{\operatorname{Tr} \mathbf{A}}{2} \pm i \left[1 - \left(\frac{\operatorname{Tr} \mathbf{A}}{2} \right)^2 \right]^{1/2}. \tag{3.3.52}$$

We see that $\lambda_1 \lambda_2 = 1$ and $\lambda_1 + \lambda_2 =$ a real number, with solutions of three forms:

(1) They may be complex conjugates on the unit circle

$$\lambda_{1,2} = e^{\pm i\sigma}, \tag{3.3.53}$$

representing stable solutions with

$$\operatorname{Tr} \mathbf{A} = 2 \cos \sigma \tag{3.3.54}$$

and

$$|\operatorname{Tr} \mathbf{A}| < 2. \tag{3.3.55}$$

(2) They may be real reciprocals

$$\lambda_2 = \lambda_1^{-1} \tag{3.3.56}$$

such that

$$|\lambda_{1,2}| = e^{\pm \sigma} \tag{3.3.57}$$

and

$$|\operatorname{Tr} \mathbf{A}| = |2 \cosh \sigma| > 2, \tag{3.3.58}$$

representing growing and decaying solutions. The unstable case can also be divided into two possibilities: $\operatorname{Tr} \mathbf{A} > 2$ ($\lambda_1 > 1$) and $\operatorname{Tr} \mathbf{A} < -2$ ($\lambda_1 < -1$).

(3) The roots may be equal such that

$$\lambda_1 = \lambda_2 = \pm 1. \tag{3.3.59}$$

These cases also follow directly from the symmetry of the eigenvalues. We now describe the physical meaning of these three solutions.

3.3. Linearized Motion

Elliptic Orbits. For case (1), we combine λ_1 and λ_2 to obtain transformations of the form

$$\Delta J_n = a \cos n\sigma + b \sin n\sigma, \qquad \Delta\theta_n = c \cos n\sigma + d \sin n\sigma, \qquad (3.3.60)$$

which represent elliptical motion around the fixed point. The ellipse is found from the invariance of the quadratic form (3.3.30). For two dimensions, using $p = \Delta J$ and $q = \Delta\theta$ to emphasize that $\Delta J - \Delta\theta$ are not action-angle variables,

$$A_{12}q^2 + (A_{11} - A_{22})qp - A_{21}p^2 = Q. \qquad (3.3.61)$$

We diagonalize the matrix **A** as in (3.3.10). This rotates the coordinates by the angle χ to the principal axis system, in which the cross term vanishes (see Fig. 3.10a), yielding

$$A'_{12}q'^2 - A'_{21}p'^2 = Q, \qquad (3.3.62)$$

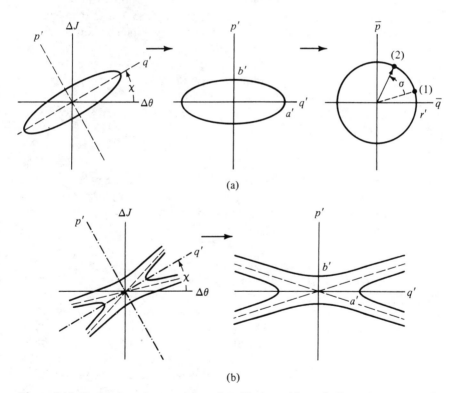

Figure 3.10. Reduction to normal form for elliptic and hyperbolic orbits. (a) Elliptic orbits, showing first a rotation by χ to the principal axes (p', q'), followed by a change of scale to the normal coordinates $(\bar{p}, \bar{q})$; $a' = |Q/A'_{12}|^{1/2}$; $b' = |Q/A'_{21}|^{1/2}$, $r' = (a'b')^{1/2}$. (b) Hyperbolic orbits, showing a rotation by χ to the principal axes (p', q'). The asymptotes are given by $p' = \pm (A'_{12}/A'_{21})q'$.

where

$$\tan 2\chi = \frac{A_{11} - A_{22}}{A_{12} + A_{21}}, \tag{3.3.63}$$

$$A'_{11} = A'_{22} = \tfrac{1}{2} \operatorname{Tr} \mathbf{A}, \tag{3.3.64a}$$

$$A'_{12} - A'_{21} = A_{12} - A_{21}, \tag{3.3.64b}$$

$$A'_{12} A'_{21} = (\tfrac{1}{2} \operatorname{Tr} \mathbf{A})^2 - 1. \tag{3.3.64c}$$

The latter two equations can be solved for the off-diagonal terms of $\mathbf{A}'$. For case (1), A'_{12} and A'_{21} have opposite signs, and (3.3.62) is the equation of an ellipse with the ratio

$$R = \frac{p'_{\max}}{q'_{\max}} = \left| \frac{A'_{12}}{A'_{21}} \right|^{1/2}. \tag{3.3.65}$$

A final transformation, $p' = R^{1/2} \bar{p}$, $q' = R^{-1/2} \bar{q}$, reduces the orbit to a circle, for which the mapping is a simple rotation by an angle σ, and therefore the linear twist mapping of an integrable system. If we include the nonlinear terms, proportional to the higher powers of Δx, then we have a perturbed twist map, having its own KAM structure, thus demonstrating the hierarchy of structure near elliptic fixed points illustrated in Fig. 3.5.

Hyperbolic Orbits. For case (2), we again combine λ_1 and λ_2 to obtain transformations of the form

$$\Delta J_n = a \lambda_1^n + b \lambda_1^{-n}, \qquad \Delta \theta_n = c \lambda_1^n + d \lambda_1^{-n}, \tag{3.3.66}$$

which represent motion on one or both branches of a hyperbola.

Rotating to the principal axis system (see Fig. 3.10b), we again have (3.3.62). Now, however, A'_{12} and A'_{21} have the same sign, giving the equation of a set of hyperbolas with asymptotes

$$p' = \pm R q'. \tag{3.3.67}$$

Writing (3.3.66) in the rotated system, it is seen that the trajectory traces a single hyperbolic branch for λ_1 positive. If λ_1 is negative, however, then on each successive iteration p' and q' reflect about the origin and the trajectory traces two opposite branches. This is shown in Fig. 3.11.

The usual case in which elliptic and hyperbolic singular points alternate have hyperbolic trajectories in which points stay on a single branch of the hyperbola. An example of this behavior, already treated in the introduction, is the pendulum with the Hamiltonian given in (1.3.6), which we write for simplicity in the form

$$H = \tfrac{1}{2} p^2 - \cos \psi = \text{const}. \tag{3.3.68}$$

The corresponding phase space trajectories are given in Fig. 1.4. Linearizing

3.3. Linearized Motion

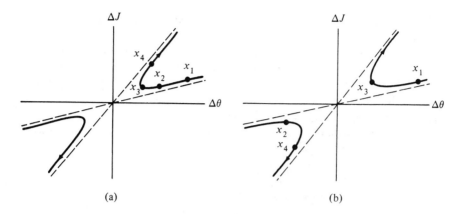

Figure 3.11. A succession of four points mapped near (a) an ordinary hyperbolic point $\lambda_1 > 1$; and (b) a reflection hyperbolic point $\lambda_1 < -1$.

around the fixed points at $p = 0$ and $\psi = 2\pi n$, we obtain

$$p^2 + (\Delta\psi)^2 = \text{const}. \tag{3.3.69}$$

Similarly, the linearization around $p = 0$ and $\psi = \pi(n + 1)$ yields

$$p^2 - (\Delta\psi)^2 = \text{const}. \tag{3.3.70}$$

Clearly (3.3.69) parametrizes ellipses, while (3.3.70) parametrizes hyperbolas. These correspond, respectively, to the $\psi = 0$ and $\psi = \pi$ fixed points of the complete Hamiltonian, as shown in Fig. 1.4.

The second type of hyperbolic fixed point, that of reflection, is also important for our analysis. As we shall see, it occurs for perturbations of nonlinear mappings that are sufficiently large that a given period k elliptic fixed point is transformed into a reflection hyperbolic one. The entire set of fixed points corresponding to a given T^k is then hyperbolic. The phase plane in the neighborhood of such a resonance is then strongly stochastic. We shall see this explicitly and calculate the condition for it to occur in Section 3.4.

Equal Eigenvalues. The third case for which $\lambda_1 = \lambda_2 = \pm 1$ is special, in that it corresponds to the exact transition between the stable and unstable cases. For $\lambda_1 = \lambda_2 = 1$, and rotating by the angle ψ to the principal axis system, the mapping becomes

$$\begin{pmatrix} p'_{n+k} \\ q'_{n+k} \end{pmatrix} = \begin{pmatrix} 1 & 0 \\ A_{21} & 1 \end{pmatrix} \begin{pmatrix} p'_n \\ q'_n \end{pmatrix}. \tag{3.3.71}$$

From (3.3.71) we see that the invariant curves describe a sheared flow along a straight trajectory with $p' = \text{const}$.[4] For $p'_0 = 0$, there is a line of fixed

---

[4] Or one may have $A'_{21} = 0$ and $A'_{12} = A_{12}$ giving a trajectory with $q' = \text{const}$.

points. For $\lambda_1 = \lambda_2 = -1$, there is a similar sheared flow along two straight lines $p' = \pm \text{const}$. For $p'_0 = 0$, there is a half-line of period 2 fixed points $q'_0 > 0$ with $q'_k = -q'_0$, etc. Keeping the nonlinear terms that perturb this system, the Poincaré–Birkhoff theorem tells us that some of those points survive the perturbation, producing the bifurcation phenomena illustrated in Section 3.4d. We also shall find, in Section 4.4, that this special case $\lambda_1 = \lambda_2 = -1$ is of particular significance in determining the transition between confined and unconfined stochastic motion, using the method of Greene (1979a).

3.3c. Linear Stability in Higher Dimensions

For two-dimensional area-preserving maps the eigenvalues were determined from the characteristic equation

$$P(\lambda) = \lambda^2 - (\text{Tr }\mathbf{A})\lambda + 1 = 0 \tag{3.3.72}$$

with simple explicit solutions. For an N-dimensional mapping, $P(\lambda)$ is in the symmetric (reflexive) form given by (3.3.22) and (3.3.23),

$$P(\lambda) = \lambda^N + a_1 \lambda^{N-1} + a_2 \lambda^{N-2} + \cdots + a_2 \lambda^2 + a_1 \lambda + 1. \tag{3.3.73}$$

Dividing (3.3.73) by $\lambda^{N/2}$, we obtain the reduced characteristic polynomial,

$$Q(\rho) = \rho^{N/2} + b_1 \rho^{N/2-1} + \cdots + b_{N/2}, \tag{3.3.74}$$

where

$$\rho = \lambda + \lambda^{-1} \tag{3.3.75}$$

is called the stability index. Setting $Q(\rho) = 0$ we obtain $N/2$ such indices.

Solving (3.3.75) for λ, we obtain

$$\lambda = [\rho \pm (\rho^2 - 4)^{1/2}]/2. \tag{3.3.76}$$

For each value of ρ, we obtain two values of λ, yielding the N eigenvalues of the tangent map $\mathbf{A}$.

For many-dimensional symplectic mappings, we must distinguish two notions of linear stability:

(1) A periodic orbit is *linearly stable* if, given an $\epsilon > 0$, there is a $\delta(\epsilon) > 0$ such that all orbits of the tangent map $\mathbf{A}$ initially within δ of 0 remain within ϵ of 0 for all time.
(2) A periodic orbit is *spectrally stable* if all eigenvalues of the tangent map have magnitude equal to one.

Linear stability implies spectral stability, but the converse is not true; a repeated root having magnitude one may yield secular (nonexponential) growth. For example, the case of two equal eigenvalues $\lambda_1 = \lambda_2 = 1$ given in (3.3.71) is spectrally stable, but the solution $p_n = p_0$ and $q_n = q_0 + A_{21} n p_0$

3.3. Linearized Motion

is linearly unstable for $A_{21} \neq 0$. Other examples are given in the next subsection. Although repeated roots are nongeneric in Hamiltonian systems, they are important because they determine the linear stability boundaries. It can be shown (Howard and MacKay, 1987) that the boundaries for spectral and linear stability are identical for symplectic maps. Hence we can use the concept of spectral stability to describe linear stability limits without excluding the case of multiple eigenvalues. Spectral stability $|\lambda| = 1$ implies, using (3.3.76), that all stability indices ρ are real with $|\rho| \leq 2$.

Stability boundaries can be calculated for arbitrary dimensional symplectic maps and the calculation has been carried out in four, six, and eight dimensions by Howard and MacKay (1987). We summarize here the results for the four-dimensional case, previously obtained by Broucke (1969). The characteristic equation can be put in the reflexive form

$$P(\lambda) = \lambda^4 - A\lambda^3 + B\lambda^2 - A\lambda + 1 = 0, \qquad (3.3.77)$$

where $A = \text{Tr }\mathbf{A}$ and $2B = (\text{Tr }\mathbf{A})^2 - \text{Tr}(\mathbf{A})^2$. The reduced characteristic polynomial is

$$Q(\rho) = \rho^2 - A\rho + B - 2 = 0, \qquad (3.3.78)$$

with solution

$$\rho = \tfrac{1}{2}A \pm (\tfrac{1}{4}A^2 - B + 2)^{1/2}, \qquad (3.3.79)$$

and, from (3.3.76) we obtain the four values of λ.

The boundaries of stability are explored by examining the three ways in which the λ's can leave the unit circle. These are indicated in Fig. 3.12.

(1) *Tangent bifurcation.* A complex conjugate pair of roots moves to $(1, 0)$ and separates along the positive real axis when

$$B < 2A - 2. \qquad (3.3.80)$$

(2) *Period-doubling bifurcation.* A complex conjugate pair of roots coalesces at $(-1, 0)$ and separates along the negative real axis when

$$B < -2A - 2. \qquad (3.3.81)$$

(3) *Krein collision.* A nonconjugate pair of roots merges and moves off the unit circle into the complex plane when

$$B > 2 + \tfrac{1}{4}A^2. \qquad (3.3.82)$$

As shown in Fig. 3.12, all four roots necessarily destabilize together in a Krein collision.

The region of stability is shown in Fig. 3.13. Not all maps reaching the parabola [equality in (3.3.82)] can destabilize on further change of a parameter. An additional condition must be met for the roots to leave the unit circle, given by Krein's theorem [see Howard and MacKay (1987) for details]. The generic case is that roots touching the parabola will in fact destabilize.

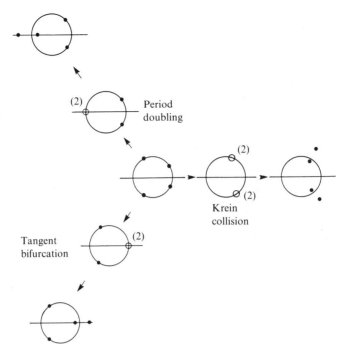

Figure 3.12. The three destabilization routes for symplectic maps (after Howard and MacKay, 1987).

Using the procedure described above, the stability boundaries can also be found for six- and eight-dimensional maps, but the cubic and quartic reduced polynomials to be solved require increasingly complicated representations. For details the reader should consult Howard and MacKay (1987). In four dimensions, linear stability theory has been applied to a number of problems such as the restricted three-body problem (Broucke, 1969), two frequency cyclotron resonance heating (Howard et al., 1986), and in somewhat different form for accelerators (e.g., Chao, 1979). An analogous calculation for Hamiltonian equilibria has been given by Howard and MacKay (1987a).

3.3d. General Stability Concepts

The distinction between spectral and linear stability for periodic orbits of a map also holds for general equilibria in dynamical systems. Following Finn and Sun (1987), if we linearize a Hamiltonian system

$$\dot{z}_i = [z_i, H] = F_i(z), \qquad i = 1, \ldots, N, \qquad (3.3.83)$$

3.3. Linearized Motion

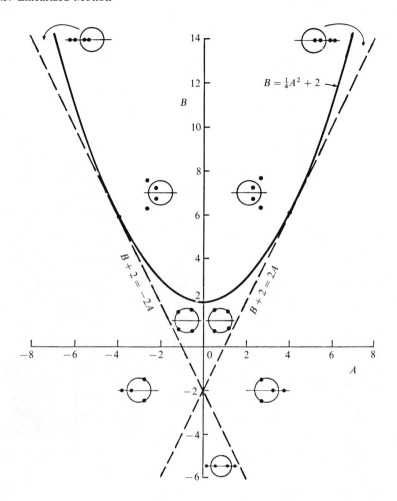

Figure 3.13. Stability diagram for a general fourth-order symplectic matrix. Corresponding eigenvalue configurations are shown as inserts. Only the inner arrowhead is stable (after Howard et al., 1986).

about an equilibrium state z_e given by $F_i(z_e) = 0$ or, in canonical form, by

$$\left(\frac{\partial H}{\partial z_i}\right)_{z=z_e} = 0, \qquad (3.3.84)$$

then we obtain

$$\dot{\zeta}_i = A_{ij}(z_e)\zeta_j, \qquad (3.3.85)$$

with $\zeta_i = z_i - z_{ei}$ and

$$A_{ij} = \left(\frac{\partial F_i}{\partial z_j}\right)_{z=z_e}. \qquad (3.3.86)$$

In general, the solutions of (3.3.85) are linear combinations of $e^{\sigma_i t}$, where σ_i are the characteristic exponents and $\lambda_i = e^{\sigma_i}$ are the N eigenvalues of **A**. However, if there is a repeated eigenvalue, then the solutions may involve linear combinations of $e^{\sigma_i t}$, $te^{\sigma_i t}$, $t^2 e^{\sigma_i t}$, ... up to the multiplicity of the root, and thus exhibit secular growth. As an example, the Hamiltonian

$$H = \tfrac{1}{2}q^2 + \tfrac{1}{2}p^2 + pq \tag{3.3.87}$$

has $\sigma_1 = \sigma_2 = 0$ and hence is spectrally stable, but has the solution

$$q(t) = At + q_0, \tag{3.3.88}$$

$$p(t) = -At + p_0, \tag{3.3.89}$$

and is thus linearly unstable. As another example, the four-degree-of-freedom Hamiltonian

$$H = q_2 p_1 - q_1 p_2 + q_4 p_3 - q_3 p_4 + q_2 q_3 \tag{3.3.90}$$

has four roots at $\sigma = i$ and four at $-i$, and thus is spectrally stable. However, Hamilton's equations yield harmonic solutions for the q's of the form $q = A \cos t + B \sin t$ which then drive the p's at their natural frequencies, obtaining p's of the form

$$p = A \cos t + Bt \sin t \tag{3.3.91}$$

which exhibits secular growth.

More general criteria are required to describe the stability of nonlinear systems. We say that a Hamiltonian system is *formally stable* if the second variation $\delta^2 H$ is positive or negative definite. This means that the N eigenvalues of the symmetric matrix

$$h_{ij} = \left(\frac{\partial^2 H}{\partial z_i \, \partial z_j}\right)_{z=z_e} \tag{3.3.92}$$

must all be positive or all be negative. Physically, the level sets of H must be a set of nested ellipsoids about the equilibrium point z_e. It can be shown that formal stability implies linear stability, but the converse is not true. For example, the Hamiltonian $H = p(q^2 + p^2)$ with equilibrium at $(0, 0)$ has a linear solution $p = p_0$, $q = q_0$, and is linearly stable. However, the second variation of H is 0 (the eigenvalues of h_{ij} are both 0), and hence the system is not formally stable. To see this, we transform H to action-angle form,

$$\bar{H} = -(2J)^{3/2} \sin \theta, \tag{3.3.93}$$

where $J = (p^2 + q^2)/2$ and $\sin \theta = -p/(2J)^{1/2}$. From (3.3.93), we obtain

$$\dot{J} = [(2J)^3 - \bar{H}^2]^{1/2}, \tag{3.3.94}$$

which yields a divergent solution for all initial $J > 0$, $\theta \neq \pi$. In this case, the linear forces are always zero, so the nonlinear forces dominate.

3.3. Linearized Motion

As a more generic example, consider the Hamiltonian

$$H = \tfrac{1}{2}(p_1^2 + q_1^2) - (p_2^2 + q_2^2) + \epsilon[\tfrac{1}{2}p_2(p_1^2 - q_1^2) - q_1 q_2 p_1]. \quad (3.3.95)$$

A transformation to action-angle variables of the linear system as in (1.2.68) yields

$$\bar{H} = J_1 - 2J_2 + \sqrt{2J_1} J_2^{1/2} \sin(2\theta_1 + \theta_2). \quad (3.3.96)$$

The generating function

$$F_2 = I_1(2\theta_1 + \theta_2) + I_2(2\theta_2 - \theta_1) \quad (3.3.97)$$

then transforms (3.3.96) to the form

$$\bar{\bar{H}} = -5I_2 + \sqrt{2}(2I_1 - I_2)(I_1 + 2I_2)^{1/2} \sin \phi_1. \quad (3.3.98)$$

Since $\bar{\bar{H}}$ is independent of ϕ_2, $I_2 = $ const. (set equal to 0); then

$$\bar{\bar{H}} = (2I_1)^{3/2} \sin \phi_1 \quad (3.3.99)$$

as in (3.3.93). Thus the system is not formally stable, although it is linearly stable. For this example, the linear forces are not zero in general; however, there is a direction in phase space; i.e., $J_1 - 2J_2 = $ const., where the nonlinear forces can dominate the system behavior.

The most general stability criterion is nonlinear stability. We say a system is *nonlinearly stable* if given an $\epsilon > 0$, there exists a $\delta > 0$ such that for $\|z(0) - z_e\| < \delta$, $\|z(t) - z_e\| < \epsilon$ for all times $t > 0$. In finite dimensions, formal stability implies nonlinear stability; hence, we treat the two as equivalent.

Energy–Casimir Method. For noncanonical coordinates, the equilibrium (3.3.84) and the formal stability condition based on (3.3.92) must be generalized. The resulting formalism is known as the energy–Casimir method (e.g., see Holm et al., 1985), or the energy-momentum method (see Marsden and Simo, 1990). In presenting the method we follow the (relatively) nonmathematical reviews of Finn and Sun (1987) and Marsden and Ratiu (1987).

Using the notation of Section 1.3d, for the equations of motion

$$\dot{z}_i = \{z_i, H\},$$

with the noncanonical bracket $\{\cdot, \cdot\}$, the equilibria are given by $\{z_i, H\} = 0$; i.e.,

$$J'_{ik} \frac{\partial H}{\partial z_k} = 0. \quad (3.3.100)$$

Because of the degeneracy of the bracket, this does not imply $\partial H / \partial z_i = 0$,

Figure 3.14. Illustrating rotation axes of symmetry of a rigid body.

but rather

$$\frac{\partial H}{\partial z_i} = \frac{\partial}{\partial z_i} \sum_{j=1}^{M} a_j C_j, \quad (3.3.101)$$

where M is the dimension of the null space of $\bm{J}'$. Each Casimir $C = \sum a_j C_j$ corresponds to a distinct equilibrium $(\partial/\partial z_i)(H - C) = 0$. However, it may turn out that there are equilibria that require another invariant of the motion I related to a symmetry rather than to a degeneracy of the bracket. That is, a more general class of equilibria may be described by

$$\frac{\partial}{\partial z_i}(H - C - bI) = 0, \quad (3.3.102)$$

where b is a constant multiplier.

As an example of the use of Casimirs in determining stability we consider a free rigid body that is described by the Euler equations (1.3.85) with the bracket (1.3.87) and the Casimir (1.3.90). Assuming that the diagonal elements of the moment of inertia tensor are ordered as $I_1 > I_2 > I_3$ as the axes of rotation go from longest to shortest, as illustrated in Fig. 3.14, we inquire as to the stability of these rotations. Consider equilibrium as in Fig. 3.14a with

$$m_2 = m_3 = 0, \quad m_1 = m_{10}. \quad (3.3.103)$$

Taking the Casimir to be of the simplest form

$$C = a(m_1^2 + m_2^2 + m_3^2),$$

then the equilibrium condition gives

$$\frac{\partial}{\partial m_1}(H - C) = \frac{m_1}{I_1} - 2am_1 = 0 \quad (3.3.104)$$

or

$$a = \frac{1}{2I_1}. \quad (3.3.105)$$

3.4. Mapping Models

We then examine the second derivatives for positive (or negative) definiteness

$$\frac{\partial^2(H-C)}{\partial m_i^2} = \frac{\partial^2}{\partial m_i^2}\left[\frac{1}{2}\left(\frac{m_1^2}{I_1} + \frac{m_2^2}{I_2} + \frac{m_3^2}{I_3}\right) - \frac{1}{2I_1}(m_1^2 + m_2^2 + m_3^2)\right]. \quad (3.3.106)$$

Carrying out the derivatives, the value for the m_1 derivative is zero due to (3.3.104), while the other two derivatives with respect to m_2 and m_3 yield, respectively,

$$\frac{1}{I_2} - \frac{1}{I_1} > 0,$$
$$\frac{1}{I_3} - \frac{1}{I_2} > 0, \quad (3.3.107)$$

which satisfies the formal stability condition of (3.3.92). Carrying out the same procedure with respect to the equilibrium

$$m_1 = m_3 = 0, \quad m_2 = m_{20}, \quad (3.3.108)$$

we obtain

$$a = \frac{1}{2I_2} \quad (3.3.109)$$

and the stability is then determined from

$$\frac{\partial^2(H-C)}{\partial m_1^2} = \frac{1}{I_1} - \frac{1}{I_2} < 0,$$
$$\frac{\partial^2(H-C)}{\partial m_2^2} = 0, \quad (3.3.110)$$
$$\frac{\partial^2(H-C)}{\partial m_3^2} = \frac{1}{I_3} - \frac{1}{I_2} > 0,$$

which is not positive or negative definite and therefore is not formally stable. As would be expected by comparing (3.3.107) with (3.3.109), the third degree of freedom yields a negative definite result and is therefore stable. We thus obtain the nonintuitive result that rotations about the shortest and longest axes are stable, but rotation about the middle axis is unstable. There are various difficulties, both practical and technical, in employing the energy–Casimir method, and the reader in such difficulty is advised to turn to the referenced papers.

3.4. Mapping Models

In this section we consider several physical systems that result in two-dimensional mappings. We then examine in some detail one of the simpler mappings, but one that is sufficiently generic to be illustrative of a large class.

One of the earliest dynamical mappings was that proposed by Fermi (1949) to model the acceleration of cosmic rays by momentum transfer from magnetic field structures. The model, consisting of a ball bouncing between a fixed and a sinusoidally oscillating wall was examined, numerically, by Ulam and associates (Ulam, 1961). The particle motion appeared to be stochastic, but did not increase its energy on the average. Ulam's result was explained by a combination of analytical and numerical work of Zaslavskii and Chirikov (1965), Brahic (1971), and Lieberman and Lichtenberg (1972). There are, in fact, various versions of this model, all characterized by a phase advance equation which is inversely proportional to the velocity. We shall designate the general class as Fermi acceleration mappings, and choose the simplest form for detailed examination.

*3.4a. Physical Problems and Models

The problem of a particle bouncing between a fixed and an oscillating wall is illustrated in Fig. 3.15a. It has been treated using an exact dynamical mapping for a sawtooth wall velocity by Zaslavskii and Chirikov (1965), and for arbitrary wall velocity by Lichtenberg et al. (1980). A "simplified" mapping, in which the oscillating wall imparts momentum to the particle but occupies a fixed position, was introduced by Lieberman and Lichtenberg (1972) and studied for arbitrary wall velocities.

To find the exact mapping for this system, we introduce a fixed surface of section at some $x = $ const. Defining $u_n = v_n/2\omega a$ to be the normalized velocity, $\theta_n = \omega t_n$ to be the phase of the moving wall at the nth collision with the fixed surface $x = 0$, then a difference equation for the motion of the

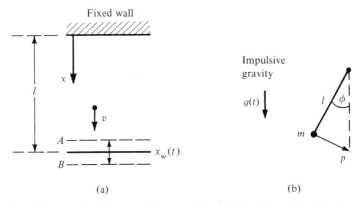

Figure 3.15. Systems having mapping models. (a) Fermi acceleration in which a particle bounces between an oscillating wall and a fixed wall. (b) Kicked rotor (pendulum driven by periodic impulsive force).

3.4. Mapping Models

particle can be determined in terms of a wall motion $x_w(t) = aF(\psi)$, where F is an even periodic function of the phase $\psi = \omega t$, with period 2π and with $F_{max} = -F_{min} = 1$. We obtain, in implicit form, the equations of motion

$$u_{n+1} = u_n + f(\psi_c), \tag{3.4.1a}$$

$$\theta_{n+1} = \psi_c + \frac{[2\pi M - \tfrac{1}{2}F(\psi_c)]}{u_{n+1}}, \tag{3.4.1b}$$

$$\psi_c = \theta_n - \frac{1}{2}\frac{F(\psi_c)}{u_n}. \tag{3.4.1c}$$

Here ψ_c is the phase at the next collision with the moving wall, after the nth collision with the fixed surface $x = 0$, $M = l/2\pi a$, with l the distance between the walls, and $f = dF/d\psi$ is the velocity impulse given to the ball. In this form it is easy to see that measuring the distance from the fixed wall as x, conjugate to v, then the phase θ is a time-like variable conjugate to the energy $E = u^2$. That is, in the extended phase space $(v, x, -E, t)$, the choice of a surface $x = 0$ gives an area-preserving mapping for the remaining pair $(-E, \theta)$. This can be confirmed by direct computation of the Jacobian yielding

$$J \equiv \frac{\partial(E_{n+1}, \theta_{n+1})}{\partial(E_n, \theta_n)} = 1.$$

As we show in Section 5.2a, this implies that a stochastic orbit has a uniform invariant distribution over the accessible (E, θ) phase space. Hence, assuming all phases are accessible, the normalized energy E has a uniform invariant distribution.

Because of its implicit form (3.4.1) is not convenient for numerical or analytical study. Substituting $E = u^2$ in (3.4.1) and expanding to first order in f (and F), we obtain

$$E_{n+1} = E_n + 2\sqrt{E_n}f(\theta_n) \tag{3.4.2a}$$

and, from (3.1.29), we obtain the first-order phase advance equation

$$\theta_{n+1} = \theta_n + \frac{2\pi M}{\sqrt{E_{n+1}}} + \frac{F(\theta_n)}{\sqrt{E_{n+1}}}. \tag{3.4.2b}$$

We refer to (3.4.2) as the "moving wall" Fermi mapping. A still simpler form can be constructed if the oscillating wall imparts momentum to the ball, according to the wall velocity, without the wall changing its position in space. The problem defined in this manner has many of the features of the more physical problem and can be analytically treated with various wall-forcing functions. In this simplified form the mapping is

$$u_{n+1} = u_n + f(\psi_n), \qquad (3.4.3a)$$

$$\psi_{n+1} = \psi_n + \frac{2\pi M}{u_{n+1}}. \qquad (3.4.3b)$$

A fundamental difference between the exact and simplified mappings, however, is that the canonical variables are different, leading to different variables in which the invariant distribution is uniform. For the simplified problem, a proper canonical set of variables are the ball velocity and phase just before the nth impact with the moving wall. The normalized velocity u then has a uniform invariant distribution, as will be seen in Section 5.4c.

Keeping these considerations in mind, a very useful map for investigating dynamics of physical systems is the simplified Fermi map with a sinusoidal forcing function

$$u_{n+1} = |u_n + \sin \psi_n|, \qquad (3.4.4a)$$

$$\psi_{n+1} = \psi_n + \frac{2\pi M}{u_{n+1}} \quad (\text{mod } 2\pi). \qquad (3.4.4b)$$

The mapping in (3.4.4) serves as an approximation (with suitably defined variables) to many physical systems in which the transit time between kicks is inversely proportional to a velocity. The absolute-value signs in (3.4.4) correspond to the velocity reversal, at low velocities, $u < 1$, which appears in the exact equations (3.4.1). The absolute value has no effect on the region $u > 1$, which is the primary region of interest.

The simplified map is in the form of a radial twist mapping (3.1.17), which we discussed in Section 3.1b. Many physical problems lead to the more general form of an invertible twist mapping as in (3.1.24). For example, the cyclotron resonance of a charged particle gyrating in a magnetic mirror having quadratic potential and resonating with an r.f. field near the turning points in the mirror leads to a mapping of the form (Lichtenberg et al., 1986)

$$\begin{aligned} E_{n+1} &= E_n + A E_{n+1}^{1/3} \, \text{Ai}(CE_{n+1}) \sin \psi_n, \\ \psi_{n+1} &= \psi_n + \frac{2\pi M}{E_{n+1}^{1/2}} + g(E_{n+1}, \psi_n), \end{aligned} \qquad (3.4.5)$$

where E is the perpendicular energy, ψ is the phase angle between the particle velocity and the accelerating field, Ai(x) is the Airy function of argument x, A, C, and M are constants depending on the parameters, and g is obtained as described in Section 3.1b,

$$g = -\frac{\partial}{\partial E_{n+1}} \int^{\psi_n} E_{n+1}^{1/3} \, \text{Ai}(CE_{n+1}) \sin \psi' \, d\psi'.$$

3.4. Mapping Models

Although care must be used, many of the features of this physical problem can be captured by use of the simpler mapping (3.4.4).

A mapping that has a quite different phase advance equation can be found by having gravity act impulsively on the pendulum ("kicked rotor"), as in Fig. 3.15b, and also describes a ball returning to an oscillating wall under he action of gravity (Pustylnikov, 1978). The mapping, introduced by Taylor and analyzed extensively by Chirikov (1979) is now known as the Chirikov–Taylor mapping or *standard mapping*. This can be obtained from the pendulum equations of (1.3.5) by changing the momentum in discrete steps, to obtain

$$p_{n+1} - p_n = -F \sin \phi_n,$$
$$\phi_{n+1} - \phi_n = G p_{n+1} \Delta t.$$

Here the momentum is changed at time t_n corresponding to phase ϕ_n by applying gravity as a periodic δ-function. The phase then shifts by an amount proportional to the new constant momentum over the time Δt between kicks. A change in scaling of the momentum variable $I = G p \Delta t$ puts this into the standard form

$$\begin{aligned} I_{n+1} &= I_n + K \sin \theta_n, \\ \theta_{n+1} &= \theta_n + I_{n+1} \quad (\mathrm{mod}\ 2\pi), \end{aligned} \qquad (3.4.6)$$

where $K = FG\Delta t$ and $\theta = \phi + \pi$. In contrast to the Fermi map, the standard map is 2π periodic in the action I, as well as the angle. We have already shown in Section 3.1b that (3.4.6) is a local approximation to more general mappings. It has played an important role in understanding the transition from regular to stochastic motion, and we treat it in this context in Chapter 4.

Another mapping that has somewhat different properties over local portions of the action space is one for which the sign in the phase advance equation changes at some value of the action. Such mappings naturally occur in the dynamics of particles in circular accelerators where the sign transition occurs with increasingly relativistic motion (Symon and Sessler, 1956). It has also been observed in two frequency cyclotron heating (Howard et al., 1956) and is an underlying cause of the period 3 "catastrophe" of the standard mapping, which we consider in connection with accelerator modes in Section 5.5d (see van der Weele et al., 1988). A simple example of such a mapping, studied by Howard and Hohs (1984), is

$$\begin{aligned} x' &= x - K \sin \theta, \\ \theta' &= \theta + x - \alpha x^3, \end{aligned}$$

where $\alpha > 0$. As we shall see in studying the stability of fixed points in Section 3.4c, when $x - \alpha x^3$ changes sign from positive to negative with increasing x, the stable (unstable) fixed point switches from $\theta = 0$ ($\theta = \pi$) to $\theta = \pi$ ($\theta = 0$). As the parameter α is changed, the phase space structure between these fixed points undergoes a change in topology called reconnection, in analogy to similar changes in the topology of magnetic field lines (Taylor, 1974). The interaction of neighboring island chains by the reconnection process is different from the interaction of neighboring island chains which leads to the destruction of KAM curves (considered in Chapter 4), and does not necessarily lead to global stochasticity connecting the separatrices of neighboring island chains (see Howard and Hohs for a numerical study of the reconnection process).

*3.4b. Numerical Results

The difference equations for these two-dimensional mappings are readily solvable on a high-speed computer for millions of wall collisions. For example, to explore the entire phase space for the simplified Fermi map (3.4.4), one divides the phase interval $(0, 2\pi)$ into, say, 100 increments and the velocity interval $(0, u_{\max})$ into 200 increments. Keeping track of the number of times a particle is found within any of the 20,000 cells of the phase space, the results of the calculations with $M = 100$, for 623,000 collisions of a single particle, are shown in Fig. 1.14. The symbol in each cell represents the number of cell occupations (nor readily visible here), with a blank denoting zero occupations. The particle is initially given a phase and low velocity chosen randomly. Subsequent collisions allow it to explore stochastically the available phase space. The final phase-plane plot is independent of the initial conditions. The unoccupied islands are bounded by KAM curves, and therefore are inaccessible from outside. The centers of the islands are elliptic singularities in the phase plane. We show in the next subsection that for $u < (\pi M/2)^{1/2}$, the linearized motion about all the principal singularities is unstable, as is readily verified from the numerical phase plot. Elliptic singular points indicated in the figure, at velocities $u/M = \frac{1}{4}, \frac{1}{5}, \frac{1}{6}, \ldots$, correspond to resonances for which the wall oscillates, respectively, four, five, and six times, ..., for one bounce of the ball.

Between the above period 1 fixed points ($k = 1$), we find island patterns from the higher k values. For example, a $k = 2$ resonance appears near $u = 22.2$ ($u/M = \frac{2}{9}$) between the central resonances at $u = 25$ ($u/M = \frac{1}{4}$) and $u = 20$ ($u/M = \frac{1}{5}$), corresponding to nine oscillations of the wall for every two bounces of the ball.

As already discussed in Section 1.4a, Fig. 1.14 shows that, for smooth forcing functions f in (3.4.3), the phase plane divides up into three distinct regions with increasing ball velocity:

(1) a low-velocity region in which all $k = 1$ fixed points are unstable, leading to stochastic motion over almost the entire region;
(2) an intermediate velocity region in which islands of stability, surrounding elliptic fixed points, are embedded in a stochastic sea; and
(3) a higher-velocity region in which bands of stochastic motion, near the

3.4. Mapping Models

separatrices of the island trajectories joining the hyperbolic fixed points, are isolated from each other by regular orbits.

The existence of region (3), in which invariant curves span the entire range of phase, bounds the energy gain of the particle. If the forcing function is not sufficiently smooth, then region (3) does not exist, in agreement with the KAM theorem.

An example of a nonsmooth forcing function is the simplified Fermi map with sawtooth wall velocity,

$$u_{n+1} = |u_n + \psi_n/2\pi - \tfrac{1}{2}|, \tag{3.4.7a}$$

$$\psi_{n+1} = \psi_n + \frac{2\pi M}{u_{n+1}}, \quad \text{mod } 2\pi. \tag{3.4.7b}$$

The results of the calculations with $M = 10$, for ten particles, are given in Fig. 3.16, after 163,840 wall collisions per particle. The density distribution $P(u)$, integrated over phases and over all collisions, is given to the right of the phase space. For $u < \tfrac{1}{2}M^{1/2}$, the linearized motion about all the principal singularities is unstable, as shown by the numerical phase plot. The elliptic singular point of the main island at $u/M = 1$ corresponds to one-to-one resonance between the particle oscillation and the wall oscillation. The successive central resonances at lower velocities $u/M = \tfrac{1}{2}, \tfrac{1}{3}, \tfrac{1}{4}, \ldots$, correspond to resonances for which the wall oscillates, respectively, twice, three times, four times, ..., for one bounce of the ball.

Again, between the above period 1 fixed points ($k = 1$), we find island patterns from the higher k values. For example, a $k = 2$ resonance appears near $u = 6.67$ ($u/M = \tfrac{2}{3}$) between the central resonances at $u = 10$ ($u/M = 1$) and $u = 5$ ($u/M = \tfrac{1}{2}$), corresponding to three oscillations of the wall for every two bounces of the ball. Now, however, this island chain is in turn separated from the main island by higher k islands whose centers are the elliptic singularities. As the island chains converge on the edges of the central islands, they indicate approximate rotation numbers of these islands. Because the island trajectories themselves correspond to nearly linear equations (due to the choice of a linear velocity law), these trajectories are not appreciably disturbed from ellipses and do not exhibit second-order island structures.

In contrast to the sine map, phase-spanning orbits for the sawtooth map encounter discontinuities in the forcing function that destroy the phase spanning KAM curves. We also find numerically an important result for a cubic forcing function which does exhibit phase-spanning KAM curves. Since $df/d\psi$ is continuous, but $d^2f/d\psi^2$ is not, then the corresponding Hamiltonian for this example

$$H \sim \int f \, d\psi$$

has $l = 2$ continuous derivatives necessary and sufficient for the existence of a KAM curve. This is one less than the best estimate (sufficient) given by

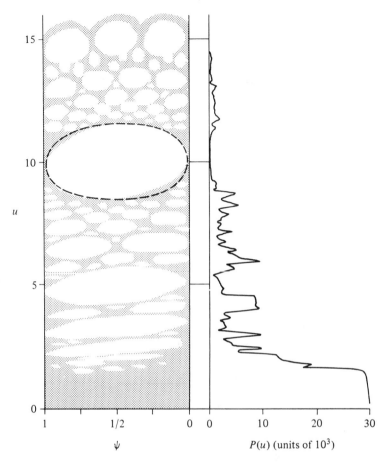

Figure 3.16. Phase space $u - \psi$ and the velocity distribution $P(u)$ for (3.4.7), the sawtooth wall velocity with $M = 10$. There are ten initial conditions used, each having 163,840 collisions with the moving wall. The dashed curve is calculated from secular perturbation theory (after Lieberman and Lichtenberg, 1972).

Moser (1963) for any mapping. The dashed curves in Figs. 3.16 and 1.14 show the separatrices for two island oscillations as calculated from Hamiltonian theory (see Section 3.4e). With a linear force (Fig. 3.16), the separatrix is approximately an ellipse. With a sinusoidal force (Fig 1.14), the trajectories near the separatrix are unstable because of second-order island formation. For this case a Hamiltonian trajectory is also given, which corresponds to a maximum phase excursion near the stability limit. The slight skewing of the islands in the numerically calculated plots arises from a term neglected in the Hamiltonian approximation, corresponding to a rotation of the principal axes, as shown in Fig. 3.10a.

*3.4c. Fixed Points and Linear Stability

Fixed Points. First we consider the velocity and phase equations for the simplified Fermi mappings, which have the form

$$u_{j+1} = |u_j + f(\psi_j)|, \qquad (3.4.8a)$$

$$\psi_{j+1} = \psi_j + \frac{2\pi M}{u_{j+1}}. \qquad (3.4.8b)$$

A few simple properties of these equations can be shown. Summing over the k velocity equations, and assuming $u_j > f(\psi_j)$ for all j, we obtain a relation among the phases for each family of fixed points:

$$\sum_{j=1}^{k} f(\psi_j) = 0. \qquad (3.4.9)$$

Summing over all k phase equations, we obtain the "average" velocity $\bar{u}_{km}$ of each family m of the period k fixed points:

$$\bar{u}_{km} = \frac{kM}{m}, \qquad (3.4.10)$$

where m is an integer relatively prime to k, and where

$$\bar{u}_{km}^{-1} = k^{-1} \sum_{j=1}^{k} u_j^{-1}. \qquad (3.4.11)$$

For each k, the integer m is used to order the families of fixed points. The k members of each (k, m) family are all found within a velocity spread $\Delta u_{\max} = (k-1)|f|_{\max}$.

As can be seen from Figs. 1.14 and 3.16, the fixed points with large islands are those for $u \gg 1$, for which the quantity $\epsilon = |f|_{\max}/\bar{u}_{km}$ is small. These are primary fixed points, i.e., they exist even in the limit $\epsilon \to 0$. For $\epsilon \equiv 0$, the k members of each (k, m) family are located at

$$(\psi_j, u_j) = (\psi_0 + 2\pi j m/k, kM/m), \qquad j = 1, \ldots, k,$$

(m, k) relatively prime, where ψ_0 is arbitrary. The effect of a small but finite ϵ is to determine the possible values for ψ_0.

Linear Stability. We now consider the stability analysis. For the simplified sine wall-velocity problem, we linearize (3.4.4) at the period 1 fixed points $u_0 = M/m$, $\psi_0 = 0, \pi$ to obtain

$$\Delta u_{n+1} = \Delta u_n \pm \Delta \psi_n, \qquad (3.4.12)$$

$$\Delta \psi_{n+1} = \Delta \psi_n - \frac{2\pi M}{u_0^2}(\Delta u_n \pm \Delta \psi_n), \qquad (3.4.13)$$

where $+$ and $-$ belong to 0 and π, respectively. Here the transformation matrix, directly from (3.4.12) and (3.4.13), is

$$\mathbf{A} = \begin{bmatrix} 1 & \pm 1 \\ -\dfrac{2\pi M}{u_0^2} & 1 \mp \dfrac{2\pi M}{u_0^2} \end{bmatrix} \qquad (3.4.14)$$

and det $\mathbf{A} = 1$, as required for area preservation. From (3.3.55) we have the condition for stability

$$|\text{Tr } \mathbf{A}| = \left|2 \pm \dfrac{2\pi M}{u_0^2}\right| < 2. \qquad (3.4.15)$$

The fixed point at $\psi = \pi$ has Tr $\mathbf{A} = 2 + 2\pi m^2/M$ so that the orbits are the regular hyperbolic type for all m. These alternate with the fixed point at $\psi = 0$, which has Tr $\mathbf{A} = 2 - 2\pi m^2/M$ so that these points are stable provided $m < (2M/\pi)^{1/2}$, i.e., provided

$$u_0 > u_1 = (\tfrac{1}{2}\pi M)^{1/2}. \qquad (3.4.16)$$

For $u < u_1$ the period 1 fixed points alternate between the two hyperbolic types, and the neighboring region of the phase plane is stochastic. For $u_0 < (\pi M/2)^{1/2}$, we note that Tr $\mathbf{A} < -2$, which gives a hyperbolic orbit of the reflection type. This type of singularity is just the one that occurs when the elliptic singular points of a given order are transformed into hyperbolic singular points, thus destroying local stability. The physical meaning of this transition becomes clearer if we calculate the rotation angle about the singular point, given by (3.3.54),

$$\cos \sigma = \tfrac{1}{2} \text{Tr } \mathbf{A}. \qquad (3.4.17)$$

At the stability boundary, $\cos \sigma = -1$ and thus the phase shift $\sigma = \pi$ per mapping period. This π resonance is a well-known property of all periodically excited systems. That is, when the period of excitation is in resonance with the half-period of the oscillation, a *stop band* is reached and the periodic oscillation is destroyed. This transition from stable to unstable oscillations has been thoroughly investigated in such diverse examples. as wave motion in periodic systems (Brillouin, 1953) and particle motion in accelerators (Courant and Snyder, 1958).

For the sawtooth velocity (3.4.7), a similar calculation gives $u_1 = M^{1/2}/2$. For period 2 and $u \gg 1$ we find Tr $\mathbf{A} \simeq 2 - m^2/M$ and thus stable orbits for $u_{01} \simeq u_{02} > M^{1/2}$; and for the sinusoidal force, we find stability provided $u_{01} = u_{02} > (\pi M)^{1/2}$, where the subscripts 01 and 02 refer to the two u-values of the periodic orbit. In both cases we note that, for fixed M, as u decreases the period 2 primary fixed points become unstable before the period 1 fixed points.

The stability analysis for the $k = 3, 4, 5$, etc., primary fixed points becomes progressively more difficult. However, by an expansion procedure, Lieberman and Lichtenberg (1972) obtained an expression for the stability of

3.4. Mapping Models

primary fixed points for the simplified Fermi problem for large k. They showed that the larger the value of k, the higher is the associated stability boundary u. The period 1 stability boundary therefore represents an important transition velocity for a particle. In the cases studied, no primary period k islands exist below this velocity. We denote this transition value as $u = u_s$ (see Fig. 1.14). Except for secondary islands embedded in the phase space below $u = u_s$, which we consider in the next subsection, and which generally occur over a narrow parameter "window" in M, all phase space positions are accessible and have access to all other positions. Thus for the class of problems considered, it is sufficient to calculate the stability boundary of the $k = 1$ fixed points. A comparison of the calculated stability boundaries for $k = 1$ with the computational result for u_s is shown in Fig. 3.17, for the simplified mapping with sinusoidal wall oscillation and for a range of M values. The value of u at the first KAM curve, which bounds the acceleration from below, $u = u_b$, is also shown. As will be shown in Chapter 4, u_b is related to, but somewhat larger than, the value of u at the fixed points for the transition to globally stochastic motion, because it includes the island width.

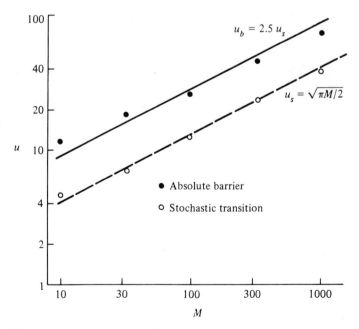

Figure 3.17. Absolute barrier u_b (●) and stochastic transition velocity u_s (○) as a function of M for the sinusoidal wall velocity (3.4.4). The circles and dots are found from numerical iteration of the difference equations (after Lieberman and Lichtenberg, 1972).

*3.4d. Bifurcation Phenomena

We now examine the character of the motion near a period 1 fixed point at $\psi = 0$ as it passes from stable to unstable behavior. Figure 3.18 shows 100,000 iterations of a stochastic orbit for the simplified Fermi mapping (3.4.4) with $M = 14$. For the fixed point at $u_0 = \frac{14}{3}$, $\psi_0 = 0$, and for this value of M, $u_s = 4.68947 > u_0$, and we therefore expect this fixed point to be unstable. The fixed point is reflection hyperbolic unstable as calculated, but nearby orbits do not connect to the surrounding stochastic sea. Outside the

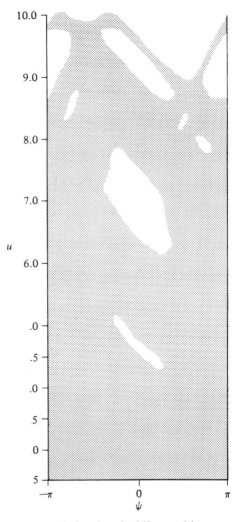

Figure 3.18. Phase space $u - \psi$ showing the bifurcated lowest order fixed point for u_0 slightly less than u_s. The mapping equation is (3.4.4) with $M = 14$.

3.4. Mapping Models

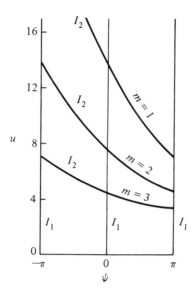

Figure 3.19. Sketch of the lines of fixed points for the involutions of the Fermi mapping; $M = 14$.

stochastic separatrix layer, a dumbbell-shaped KAM surface surrounds the unstable fixed point and two period 2 fixed points exist within the separatrix orbit that do not exist at values of M for which $u_s < u_0$. The topology of this finer scale structure is similar, but not identical, to the original topology in which some KAM orbits enclose single fixed points while others isolate separatrices.

To calculate the location of the period 2 fixed points, we write the mapping as the product of two involutions I_1 and I_2 as described in Section 3.3b. The lines of fixed points of I_1 and I_2, given by (3.3.38) and (3.3.39), are shown in Fig. 3.19. The bifurcated fixed points move on the I_2 line

$$u = \frac{\pi M}{(\psi + \pi m)} \tag{3.4.18}$$

away from its intersection with the I_1 line $\psi = 0$. These $k = 2$ fixed points can be calculated directly from (3.3.32), as has been done by Lichtenberg *et al.* (1980). Rather than do this, we appeal to the theorem of Section 3.3b relating $k = 2$ fixed points to the $k = 1$ fixed points of the product involutions. Letting

$$x = \left(\frac{\pi M}{\psi + \pi m}, \psi\right), \tag{3.4.19}$$

then $Tx = (u_1, \psi_1)$, with

$$u_1 = \frac{\pi M}{(\psi + \pi m)} + \sin \psi, \quad \psi_1 = \psi + \frac{2\pi M}{u_1} - 2\pi m. \tag{3.4.20}$$

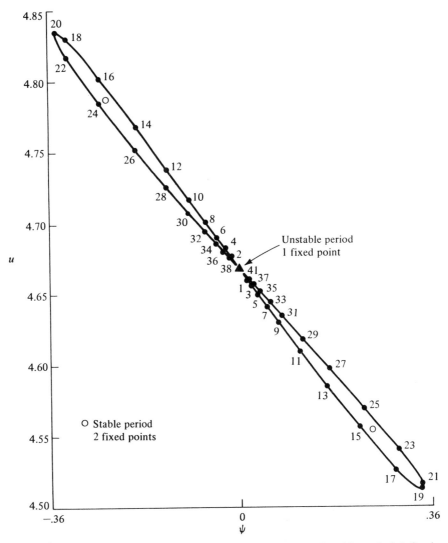

Figure 3.20. Detail of a stable orbit surrounding the pair of stable period 2 fixed points (○), inside the separatrix of the unstable period 1 fixed point (▲).

If we set (u_1, ψ_1) to be a fixed point I_2, we find

$$\frac{2\psi}{(\pi m)^2 - \psi^2} = \frac{1}{\pi M} \sin \psi. \qquad (3.4.21)$$

For $M = 14$, $m = 3$, and putting $\sin \psi \approx \psi - \psi^3/6$, we find $\psi = \pm 0.23351$, with u given by (3.4.18). These two fixed points are shown in Fig. 3.20, together with an orbit corresponding to an initial condition near the unstable

$k = 1$ fixed point, which now forms a figure 8 separatrix layer encircling the $k = 2$ fixed points.

Now what happens as we continue to decrease M, thus driving u_0 further below u_s? To understand this we note that near the I_2 line we can use (3.4.18) in (3.4.4a) to write a *one-dimensional* mapping for u

$$u_{n+1} = u_n + \sin \pi \left(\frac{M}{u_n} - m \right). \tag{3.4.22}$$

These maps arise naturally in the study of dissipative systems, and are the main topic of Section 7.2. We show in Section 7.2 that there is a cascade of successive bifurcations to stable high-period orbits $2 \to 4 \to 8 \to 16$, etc., as M is decreased to a limiting value, below which all these orbits are unstable. The detailed behavior of the $k = 2$ fixed points with varying M, including the bifurcation to $k = 4$, is given in Lichtenberg et al. (1980). The general treatment of Hamiltonian bifurcation theory is given in Section 7.3.

*3.4e. Hamiltonian Formulation

Just as the complete motion of coupled or driven oscillators can be written as difference equations in a surface of section, conversely, a physical problem formulated in terms of difference equations can be written in Hamiltonian form. This approximation allows the use of the method of averaging and secular perturbation theory, which we have developed in Chapter 2. As described in Section 3.1c, the difference equations can be transformed to differential equations by introducing the periodic delta function (3.1.33) in the force equation. For the mapping (3.4.4),

$$\frac{du}{dn} = \sum_{l=-\infty}^{\infty} \exp(i2\pi l n) \sin \psi, \tag{3.4.23}$$

and

$$\frac{d\psi}{dn} = \frac{2\pi M}{u}, \tag{3.4.24}$$

where the time variable is n. Equations (3.4.23) and (3.4.24) have the Hamiltonian form

$$H = 2\pi M \ln u + \sum \exp(i2\pi l n) \cos \psi, \tag{3.4.25}$$

with u and ψ the canonical coordinates.

Averaged Equations. For large u ($u \gg M$), the bounce frequency 2π is large compared to $\dot\psi$, i.e.,

$$(\psi_{n+1} - \psi_n)/2\pi \ll 1. \tag{3.4.26}$$

As described in Section 2.4a, (3.4.25) can be averaged over n to obtain a first integral of the motion $2\pi M \ln u + \cos \psi = C$. However, velocities sufficiently large that (3.4.26) holds are not of major interest. For the velocity range of interest $1 \ll u \leqslant M$, (3.4.26) is not satisfied. We can, however, introduce a change in variable, as in secular perturbation theory (Section 2.4),

$$\Delta \hat{u} = u - \frac{M}{m}, \qquad \hat{\phi} = \psi - 2\pi mn, \qquad m \text{ an integer}, \qquad (3.4.27)$$

so as to transform to a coordinate system around a period 1 fixed point at $u_0 = M/m$. In the hat variables, with $\Delta \hat{u} \ll u_0$, (3.4.23) and (3.4.24) take the form

$$\frac{d(\Delta \hat{u})}{dn} = \sum_l \exp(i2\pi ln) \sin \hat{\phi}, \qquad (3.4.28a)$$

$$\frac{d\hat{\phi}}{dn} = -\frac{2\pi M}{u_0^2} \Delta \hat{u}. \qquad (3.4.28b)$$

Equations (3.4.28) can be derived from the Hamiltonian

$$\Delta \hat{H} = -\frac{2\pi M}{u_0^2} \frac{(\Delta \hat{u})^2}{2} + \sum_l \exp(i2\pi ln) \cos \hat{\phi}. \qquad (3.4.29)$$

This is just the island oscillation Hamiltonian of Section 2.4a, including the rapidly varying terms. If the motion in the $\Delta \hat{u} - \hat{\phi}$ phase plane is assumed to be slow on the time scale n, i.e.,

$$(\hat{\phi}_{n+1} - \hat{\phi}_n) \ll 2\pi, \qquad (3.4.30)$$

(3.4.29) can be averaged over n to give the averaged Hamiltonian

$$\Delta \bar{H} = -\frac{2\pi M}{u_0^2} \frac{(\Delta \hat{u})^2}{2} + \cos \hat{\phi} = C, \qquad (3.4.31)$$

which describes the trajectories near the period 1 fixed points at $\hat{\phi} = 0, \pi$, and $\Delta \hat{u} = 0$. From the elliptic singular point at $\hat{\phi} = 0$, the Hamiltonian curves of $\Delta \bar{H}$ consist of encircling orbits out to the separatrix, beyond which there are passing orbits. The maximum excursion of $\Delta \hat{u}$ occurs for the separatrix trajectory, $C = +1$, giving, as in (2.4.31),

$$(\Delta \hat{u})_{\max} = 2u_0(2\pi M)^{-1/2}. \qquad (3.4.32)$$

In Fig. 1.14 the Hamiltonian curves from (3.4.31) (shown as dashed lines) are compared with the results from the numerical calculations. The linearized frequency about the elliptic fixed point, as calculated from (2.4.30), is

$$\hat{\omega}_0 = \frac{(2\pi M)^{1/2}}{u_0}, \qquad (3.4.33)$$

which is to be compared with the bounce frequency of $2\pi M/u_0$ in order to determine the number of second-order islands. We defer this calculation, which bears on the transition from regular to stochastic motion, to Chapter 4.

*3.5. The Separatrix Motion

We have seen in Section 2.4 that by applying secular perturbation theory to a given resonance in a near-integrable system and averaging over the fast phases, the Hamiltonian describing the motion near the resonance is transformed to the standard form

$$H = H_0(\mathbf{J}) + \tfrac{1}{2}Gp^2 - \epsilon F \cos\phi, \qquad (3.5.1)$$

yielding for the fast motion

$$\mathbf{J} = \text{const.}, \qquad (3.5.2)$$

$$\boldsymbol{\theta} = \boldsymbol{\omega}_\theta(\mathbf{J})t + \boldsymbol{\theta}_0, \qquad (3.5.3)$$

and for the slow variables the smooth integrable motion of a pendulum (see Section 1.3a). We also know that this picture is incomplete; resonances between the fast phases and the slow ϕ motion lead to secondary island formation and regions of stochasticity. In particular, the motion near the separatrix of (3.5.1) has a chaotic component, as shown in Section 3.2b, even in the limit $\epsilon \to 0$.

For two degrees of freedom, and reintroducing the resonant terms that were dropped in averaging the resonance $\omega_2/\omega_1 = r/s$ over the fast phase, (3.5.1) has the form, using (2.4.9)

$$H = H_0(\mathbf{J}) + \tfrac{1}{2}Gp^2 - \epsilon F \cos\phi + \epsilon \sum_{\substack{p \geq 1 \\ q \neq 0}} \Lambda_{pq} \cos\left(\frac{p}{r}\phi - \frac{q}{r}\theta + \chi_{pq}\right), \qquad (3.5.4)$$

where G, F, Λ, and χ are functions of J, the action conjugate to the fast phase θ. In the $p - \phi$ surface of section $\theta = \text{const.}$, the smooth motion of (3.5.1) is perturbed, producing a thin stochastic layer surrounding the separatrix. This combination of separatrix motion with superimposed stochastic component is shown as the shaded area in Fig. 3.21a.

We can also view the motion in the $J - \theta$ surface of section $\phi = \text{const.}$ In this space, the unperturbed motion is a constant J line, which perturbs to form the stochastic layer, shown shaded in Fig. 3.21b. In the $J - \theta$ plane the chaotic motion is clearly seen separated from the (trivial) integrable motion. We therefore develop the surface of section mapping in terms of the $J - \theta$ variables, choosing for convenience the surface of section $\phi \approx \pm\pi$.

*3.5a. Driven One-Dimensional Pendulum

Before finding the jump in J, we adopt a simpler alternate problem. The perturbation Hamiltonian of (3.5.4) has terms explicit in the phase, which drive not only the $p - \phi$ motion but also the action J. A rigorous formulation of the motion in the neighborhood of the separatrix then requires keeping

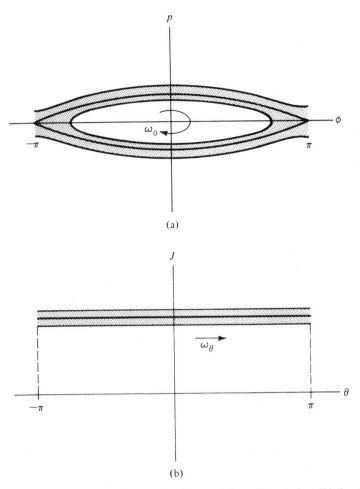

Figure 3.21. Motion near the separatrix of a pendulum. The dark solid line is the unperturbed orbit; the shaded region shows the perturbed stochastic trajectory. (a) The p, ϕ surface of section, $\theta = $ const.; (b) the $J - \theta$ surface of section, $\phi = $ const.

track of the change in frequency $\omega_\theta(J)$. A simpler problem, which still retains the features we are interested in, is to consider ω_θ as a fixed driving frequency. This simplification decouples the Hamiltonian of the perturbed pendulum from the complete problem. The phase θ may now be thought of as a function proportional to the time such that (3.5.4) is now a driven one-dimensional pendulum with Hamiltonian

$$H' = \tfrac{1}{2}Gp^2 - \epsilon F \cos\phi + \epsilon \sum \Lambda_{pq} \cos\left(\frac{p}{r}\phi - \frac{q}{r}\Omega t + \chi_{pq}\right), \quad (3.5.5)$$

where G, F, Λ, Ω, and χ are constants. There being no "action" J, $H_0(J)$ has

3.5. The Separatrix Motion

been dropped. H' and H are entirely equivalent to order ϵ, as can be seen by introducing the extended phase space (see Section 1.2b) for (3.5.5), with the new canonical variables $J = -H'/\Omega$ and $\theta = \Omega t$, and where in the extended phase space

$$H = H' + \Omega J. \tag{3.5.6}$$

Comparing dJ/dt for (3.5.4) with dH'/dt for (3.5.5), we see that the jump in J is now proportional to the jump in H' itself

$$\Delta J = \frac{-\Delta H'}{\Omega}. \tag{3.5.7}$$

Calculation of the Jump in Action. We integrate dH'/dt, or equivalently, dJ/dt, over one period of the unperturbed separatrix motion of the pendulum, as in (3.1.31). As will be seen below, the jump is exponentially small in the frequency ratio

$$\frac{q}{r}\frac{\Omega}{\omega_0}.$$

We assume

$$\omega_0 = (\epsilon FG)^{1/2} \ll \Omega, \tag{3.5.8}$$

where ω_0 is the frequency for small librations of the pendulum. The Fourier coefficients Λ_{pq} fall off with increasing p and $|q|$. It thus suffices to keep only the dominant term $p = q = 1$. From (3.1.31), with $\Lambda \equiv \Lambda_{11}$

$$\Delta J = -\frac{\epsilon \Lambda}{r} \int_{-\infty}^{\infty} dt \sin\left[\frac{1}{r}\phi(\omega_0 t) - \frac{1}{r}(\Omega t + \theta_n)\right], \tag{3.5.9}$$

where the unperturbed separatrix motion (1.3.21) for ϕ (see Fig. 3.22a)

$$\phi(s) = 4\tan^{-1} e^s - \pi \tag{3.5.10}$$

and the rotation for θ

$$\theta = \Omega t + \theta_n \tag{3.5.11}$$

have been used, and χ has been lumped into θ_n, which measures the phase of θ at the nth crossing of the surface of section $\phi \approx \pm\pi$. Changing variables to $s = \omega_0 t$ and expanding the sine, only the symmetric part contributes to the integral:

$$\Delta J = \epsilon \frac{\Lambda}{\omega_0 r} \mathcal{A}_2(Q_0) \sin \theta_n, \tag{3.5.12}$$

where

$$\mathcal{A}_m(Q_0) = \int_{-\infty}^{\infty} ds \cos\left[\frac{m}{2}\phi(s) - Q_0 s\right] \tag{3.5.13}$$

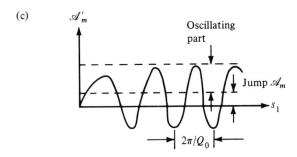

Figure 3.22. Illustrating the definition of the Melnikov–Arnold integral. (a) The phase $\phi(s)$ and (b) the frequency $\phi'(s)$ for motion along the separatrix of a pendulum Hamiltonian; (c) the definition of $\mathcal{A}'_m(s_1)$ showing the oscillating part, and the jump $\mathcal{A}_m$, which is the Melnikov–Arnold integral.

is the Melnikov–Arnold integral and

$$Q_0 = \frac{1}{r}\frac{\Omega}{\omega_0} \qquad (3.5.14)$$

is the frequency ratio.

3.5. The Separatrix Motion

Melnikov–Arnold Integral. The integral in (3.5.13) is actually improper; no limit exists. Thus in the form

$$\mathscr{A}'_m(Q_0, s_1) = 2 \int_0^{s_1} ds \cos\left[\frac{m}{2}\phi(s) - Q_0 s\right] \quad (3.5.15)$$

we note that for $s_1 \to \infty$, as shown in Fig. 3.22c, the integral is the sum of a rapidly oscillating part and a jump. The oscillating part may be large compared with the jump, but averages to zero on the time scale of the separatrix motion. The jump arises from the region $s \lesssim 1/Q_0$ where the instantaneous frequency of the separatrix phase (see Fig. 3.22b) has its maximum $2\omega_0$, corresponding to the closest approach of the separatrix frequency to the driving frequency, and leading to the qualitative behavior shown in Fig. 3.22c. The jump in (3.5.13) has been evaluated by Melnikov (see Chirikov, 1979). For m integer, the method of residues can be used to find the jump exactly. For $Q_0 < 0$, we find

$$\mathscr{A}_m(Q_0) = (-1)^m \mathscr{A}_m(-Q_0) \exp(\pi Q_0), \quad (3.5.16)$$

which is generally very small. For $Q_0 > 0$,

$$\mathscr{A}_1 = \frac{2\pi \exp(\pi Q_0/2)}{\sinh(\pi Q_0)}, \quad (3.5.17)$$

$$\mathscr{A}_2 = 2Q_0 \mathscr{A}_1, \quad (3.5.18)$$

with the recurrence relation for $m > 2$

$$\mathscr{A}_m = \frac{2Q_0 \mathscr{A}_{m-1}}{(m-1)} - \mathscr{A}_{m-2}. \quad (3.5.19)$$

For $Q_0 \gg m$, the asymptotic expansion is

$$\mathscr{A}_m = \frac{4\pi (2Q_0)^{m-1} \exp(-\pi Q_0/2)}{(m-1)!}. \quad (3.5.20)$$

The detailed evaluation of these integrals can be found in Chirikov (1979, Appendix A).

Returning to (3.5.12) and using the asymptotic expansion for $\mathscr{A}_2$ in view of $Q_0 \propto \epsilon^{-1/2}$, we obtain the mapping function f for the perturbed twist map (3.1.13a):

$$f = f_0 \sin \theta_n, \quad (3.5.21)$$

where

$$f_0 = 8\pi \Lambda \Omega^{-1} Q_0^2 \exp\left(\frac{-\pi Q_0}{2}\right). \quad (3.5.22)$$

*3.5b. The Separatrix Mapping

The change in phase θ is determined by the time required for the pendulum in near-separatrix motion to swing from $\phi = -\pi$ to π. From (1.3.15), the half-period for libration near the separatrix is

$$T = \omega_0^{-1} \ln \frac{32}{|w|}, \qquad (3.5.23)$$

where

$$w(J) = -\frac{\epsilon F + \Omega J}{\epsilon F} \qquad (3.5.24)$$

is the relative deviation of the energy $-\Omega J$ from the separatrix energy ϵF. The change in θ is just ΩT, which leads to the rotation number for the twist map (3.1.13b)

$$2\pi\alpha = \frac{\Omega}{\omega_0} \ln \frac{32}{|w|}. \qquad (3.5.25)$$

Since f is independent of J, from (3.1.16) there is no additional phase shift arising from the perturbation; $g \equiv 0$ in (3.1.13b).

It is more convenient to change variables from J to w given by (3.5.24). In the w and θ variables, the mapping is then

$$w_{n+1} = w_n - w_0 \sin \theta_n, \qquad (3.5.26a)$$

$$\theta_{n+1} = \theta_n + Q_0 r \ln \left|\frac{32}{w_{n+1}}\right|, \qquad (3.5.26b)$$

where

$$w_0 = \frac{\Omega J_0}{F} = 8\pi \left(\frac{\Lambda}{F}\right) Q_0^2 \exp\left(\frac{-\pi Q_0}{2}\right). \qquad (3.5.27)$$

This is the separatrix mapping, called by Chirikov (1979) the *whisker mapping* following Arnold's use of the term.

Fixed Points and Stability. We do not consider the separatrix mapping in the detail of the previous section for the simplified Fermi mapping, but summarize some salient features. The Fermi and separatrix mappings are very similar. Both are radial twist mappings, the product of two involutions (see Section 3.1b). They have the same type of action nonlinearity in the phase equation, leading to increasing phase shifts and consequent stochastic behavior as the "action" (w or u) is decreased toward zero.

The period 1 fixed points are found from (3.5.26) to be

$$Q_0 r \ln \left|\frac{32}{w_1}\right| = 2\pi m$$

or

$$w_1 = \pm 32 \exp\left(\frac{-2\pi m}{Q_0 r}\right), \quad m \text{ integer}, \tag{3.5.28a}$$

$$\theta_1 = 0, \pi. \tag{3.5.28b}$$

Evaluating the stability from (3.3.55), we have

$$Tr\, \mathbf{A} = 2 + \frac{w_0}{w_1} Q_0 r \cos \theta_1, \tag{3.5.29}$$

and thus for $w_1 > 0$ the $\theta_1 = 0$ fixed points are all unstable. For $\theta_1 = \pi$, there is stability for

$$w_1 > w_s = \frac{w_0 Q_0 r}{4} \tag{3.5.30}$$

or

$$w_1 > 2\pi r \left(\frac{\Lambda}{F}\right) Q_0^3 \exp\left(\frac{-\pi Q_0}{2}\right). \tag{3.5.31}$$

As in the case of the Fermi mapping, we expect w_s to be an important transition value for the action, with mostly stochastic behavior for $w < w_s$.

The separatrix mapping is of great importance in understanding the chaotic behavior in near-integrable systems. We have seen that separatrices always form surrounding the resonances in these systems. The separatrix mapping, which describes the motion near a separatrix, clearly exhibits chaotic behavior as $w \to 0$. Because of its universality in dynamical problems, the separatrix mapping has been extensively studied (Chirikov, 1979) to determine its stochastic barrier w_b, where the first KAM curve spans the space in θ, and to understand the statistical properties of the chaotic motion within this barrier. We summarize these studies in Chapters 4 and 5. Furthermore, as will be shown in Chapter 6, the separatrix motion is at the root of the treatment of Arnold diffusion, which occurs generically in systems with three or more degrees of freedom.

3.6. Mappings from Perturbed Linear Oscillators

In Section 2.4c we described the breakdown of perturbation theory using the example of resonant wave-particle interaction, for which an electromagnetic wave interacts with a charged particle gyrating in a static magnetic field. There we saw that there were two distinct cases: perturbation of a nonlinear oscillator, for an obliquely propagating wave, and perturbation of a linear oscillator, for a wave propagating perpendicular to the field. By comparing Figs. 2.10 and 2.11 we saw, in contrast to the nonlinear oscillator, that the topology of the phase space of the linear oscillator was dramatically changed at small perturbation amplitude, although the regions of stochastic phase space increased in similar fashion for linear and nonlinear oscillators as the perturbation was increased. The same situation holds for δ-function perturbations of integrable systems, from which we obtain mappings. In contrast to the examples of Section 3.4, in which the mapping equations could be considered as perturbations of nonlinear oscillators, quite different phase space topologies can arise from mappings obtained by perturbing a linear oscillator. This problem has been considered in a series of papers by Zaslavskii and associates (Zaslavskii et al., 1986; Chernikov et al., 1987) in which the structure of thin stochastic webs in phase space, generated by such mappings, has been explored.

In Section 3.6a, we obtain this global web structure. In Section 3.6b the finer structure within the stochastic web is explored, by use of a separatrix mapping, as in Section 3.5. The global diffusion through such a web will be calculated, using these results, in Section 5.6, following Lichtenberg and Wood (1989).

3.6a. Tiling of the Phase Space

We consider a charged particle gyrating in a z-directed magnetic field $\hat{z}B_0$, as in Section 2.2b, and interacting with an electric field such that the electric force is a periodic δ-function kick in the x-directed velocity

$$\Delta \dot{x} = \frac{e}{M} E_0 T \sin kx \sum_{n=-\infty}^{\infty} \delta(t - nT). \quad (3.6.1)$$

The field can be thought of as arising from an infinite set of waves of wave number k and frequencies $\omega_m = 2\pi m/T$. The unperturbed orbit is governed by the equations

$$\dot{x} = \Omega y, \qquad \dot{y} = -\Omega x, \quad (3.6.2)$$

where $\Omega = eB_0/M$, the gyrofrequency. The motion can then be represented by a mapping consisting of a kick in $\dot{x}$, followed by a rotation. In terms of normalized variables $u = (k/\Omega)\dot{x}$, $v = (k/\Omega)\dot{y}$, the mapping equations can be written

$$u_{n+1} = (u_n + K_\alpha \sin v_n) \cos \alpha + v_n \sin \alpha,$$
$$v_{n+1} = -(u_n + K_\alpha \sin v_n) \sin \alpha + v_n \cos \alpha, \quad (3.6.3)$$

where $\alpha = \Omega T$ is the rotation angle between kicks, and $K_\alpha = (e/m)E_0 kT^2/\alpha$ is the stochasticity coefficient. The mapping is composed of a product of two involutions and is therefore measure-preserving. At a resonance we have $\alpha = 2\pi p/q$. Taking $p = 1$ for simplicity, we see that q is the number of

3.6. Mappings from Perturbed Linear Oscillators

Figure 3.23. A portion of the separatrix web for mapping (3.6.3) with $\alpha = \pi/2$ and $K_\alpha = 0.6$. One thousand initial conditions within the web were iterated 10^3 times to generate the web (after Lichtenberg and Wood, 1989).

δ-function kicks per gyroperiod. There is an approximate q-fold rotational symmetry which becomes exact in the limit of $K_\alpha \to 0$. Furthermore, for $q = 2, 3, 4,$ and 6, there is also an exact translational symmetry in this limit. The combination of these two symmetries ensures that the phase space is tiled by the separatrix joining the unstable periodic points of the mapping. For other values of q, the separatrices are much more complicated, but for small K_α tend to produce structures that have a skeleton approximated by a q-fold rotation of a set of parallel lines. With finite K_α, higher-order resonances produce stochastic layers of finite thickness.

We illustrate these types of behavior with numerical results for $q = 4$ and $q = 5$. In Fig. 3.23 the separatrix web is shown for a rather large value of $K_\alpha = 0.6$, in order to be able to see the web thickness. One thousand initial values of u and v were chosen within the stochastic region at the center of

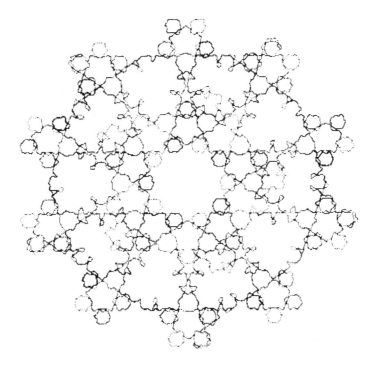

Figure 3.24. A portion of the separatrix web with $\alpha = 2\pi/5$ and $K_\alpha = 0.8$.

the figure and were followed for 10^3 mapping iterations. The edge of the stochastic region is bounded by KAM curves within each mesh. Because of the translational symmetry there is no obvious origin, with the web unbounded and apparently uniform for arbitrarily large values of u and v. In contrast, in Fig. 3.24 a portion of the web is shown for $q = 5$. In this case, without translational symmetry, there is an obvious center to the figure in u–v space. The five-sided symmetry in the separatrix web shows up both on the small and the large, reminiscent both of quasicrystals and Penrose tilings. These similarities have been pointed out and discussed in some detail in the papers by Zaslavskii and associates. The separatrix has also been found numerically to become thinner, and therefore more difficult to penetrate by numerical iteration of the mapping, as u and v extend away from the origin. It is not known whether the motion is bounded.

The mapping of (3.6.3) is not structurally stable in that small perturbations can dramatically change the resulting dynamics. We have already seen in Section 2.2b, in considering the interaction of a gyrating particle with a single wave, that wave propagation not exactly perpendicular to the magnetic field introduces nonlinear terms that can change the topology. Similarly, a relativistic correction to the gyromotion introduces a nonlinearity that limits the extent of the web even for the linear, translationally symmetric resonances (Longcope and Sudan, 1987). By propagating the wave at the correct small

3.6. Mappings from Perturbed Linear Oscillators

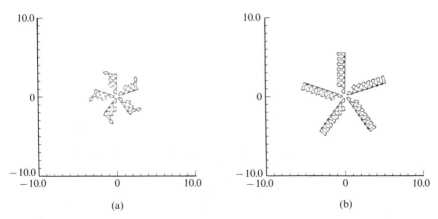

Figure 3.25. Integration of the exact equations of motion with $\omega_1/\Omega = 5$. (a) Five harmonics; (b) ten harmonics (after Murakami et al., 1988).

angle with respect to the magnetic field the linear properties are recovered for the relativistic case (Karimabadi and Angelopoulos, 1989) but not in a structurally stable manner. Also, as a practical matter, if we consider interaction with a finite number of waves, rather than with a periodic δ-function electric field, the structure of the phase space is also modified. In this situation the equations of motion must be numerically integrated to obtain the surface of section mapping for the separatrix phase space. This was done by Murakami et al. (1988) with $q = 5$ to obtain the interesting results shown in Fig. 3.25. The easily accessible part of the separatrix layer produced lobes radiating outward, with the number of lobes increasing with the number of waves, as seen in the figure for $N = 5$ and 10. Thinner connected portions of the separatrix layer probably exist but have not been explored over the finite integration time. The regions between the "spokes" in Fig. 3.25 do get filled in at higher perturbation amplitude, such that the separatrix layer looks more like that shown in Fig. 3.24.

3.6b. Phase Space Structure

A particular value of q for which a simple Hamiltonian form can be obtained, in lowest order, is $q = 4$. In this case (3.6.3) reduces to

$$u_{n+1} = v_n,$$
$$v_{n+1} = -(u_n + K_\alpha \sin v_n). \tag{3.6.4}$$

The twist can then be removed, to order K_α, by iterating the map four times, keeping only the lowest order terms, to obtain

$$v_{n+4} = v_n - 2K_\alpha \sin u_n,$$
$$u_{n+4} = u_n + 2K_\alpha \sin v_{n+4}.$$
(3.6.5)

The map (3.6.5) is again a product of two involutions and therefore measure-preserving. Although (3.6.5) was derived from (3.6.4) a subtle difference has arisen which becomes very important for the calculation of the separatrix width; that is, the map (3.6.4) has one kick per step or four kicks per gyroperiod while the map (3.6.5) has two kicks per step or two kicks per gyroperiod, each kick being twice as large. The first-order structure is the same in both cases, but not the thickness of the separatrix layer, which depends exponentially on the ratio of the δ-function period to the period going around a cell. In order to recover the original separatrix width, we must expand (3.6.4) to order K_α^2.

As in Section 3.4c, we can construct a Hamiltonian for the mapping of (3.6.5),

$$H_4 = -2K_\alpha(\cos v + \cos u) - 2K_\alpha \cos u \sum_{\substack{j=-\infty \\ j \neq 0}}^{\infty} \cos\left(2\pi j \frac{(n+\epsilon)}{4}\right)$$
$$- 2K_\alpha \cos v \sum_{\substack{j=-\infty \\ j \neq 0}}^{\infty} \cos\left(2\pi j \frac{(n-\epsilon)}{4}\right),$$
(3.6.6)

where the summation is the Fourier representation of the period δ function and the limit $\epsilon \to 0^+$ is to be taken. Since the lowest-order oscillation about the fixed point has the frequency $\Omega_4 = 2K_\alpha$, for small K_α this frequency is very slow compared to the wave frequency. This allows averaging over the rapidly varying terms to obtain the averaged Hamiltonian

$$\bar{H}_4 = -2K_\alpha(\cos v + \cos u).$$
(3.6.7)

The averaged Hamiltonian is clearly 2π periodic. From derivatives of (3.6.7) the fixed points are found to lie at $u = 0, \pi$ and $v = 0, \pi$. The separatrices are obtained from $\cos v + \cos u = 0$ to be the straight lines

$$v = \pm(u + \pi) + 2\pi m, \quad m \text{ integer}.$$
(3.6.8)

When the rapidly varying terms are included, a separatrix layer of stochasticity appears, as in the nonlinear case, which increases with increasing K_α. This was shown in Fig. 3.23 for $K_\alpha = 0.6$. Both the distortion of the separatrix from the straight-line 45° grid, and the web thickness are clearly seen for this value of K_α.

Near the separatrix of an oscillator that is perturbed by higher frequencies,

3.6. Mappings from Perturbed Linear Oscillators

Figure 3.26. Structure of the phase space in the neighborhood of an x point of the stochastic web for $K_\alpha = 0.4$ (after Lichtenberg and Wood, 1989).

the motion is approximated by a separatrix mapping as described in Section 3.5:

$$\Delta w = w_0 \sin \theta, \qquad (3.6.9)$$

$$\Delta \theta = 2\pi\tau(w)/q, \qquad (3.6.10)$$

where w is the energy (or action) deviation from the separatrix, τ is the full period (in mapping steps) to traverse the web, and w_0 is an exponentially small function of the ratio Q_0 of the mapping period to the linearized mesh frequency. In Section 3.5 we expressed w_0 in terms of an Melnikov–Arnold integral in the form

$$w_0 = 20\, Q_0^2 \exp\left[-\frac{\pi Q_0}{2}\right]; \qquad (3.6.11)$$

here the constant 20 was obtained numerically and is slightly different from the analytic value of 16 obtained in Section 3.5.

The rotation period within the separatrix layer can also be found, approximately, from the averaged Hamiltonian in (3.6.7), as we have done in Section 1.3a, to obtain

$$\tau = \frac{2}{K_\alpha} \int_0^{\cos^{-1}(w-1)} du\, \frac{1}{[1 - (w - \cos u)^2]^{1/2}}, \qquad (3.6.12)$$

where, as previously, w is the deviation in energy (or action) units from the value of H on the separatrix ($H_{sx} = 0$). The elliptic integral (3.6.12) is logarithmic near the separatrix, and has been found by numerical integration to be

$$\tau(w) \cong \frac{2}{K_\alpha}(2 - \ln w). \qquad (3.6.13)$$

In Fig. 3.26a we show a portion of the separatrix region near an unstable fixed point, which is blown up in Fig. 3.26b. The same type of local structure is seen within a separatrix layer as has been seen in the Fermi acceleration mapping for small M (see also Figs. 3.6 and 3.7). Finding the border of stochasticity, i.e., the first KAM surface that separates the separatrix layer from the central meshes having regular motion, is the subject of Chapter 4.

The remarkable uniformity of the stochastic web over the entire phase space allows us to calculate a global rate of diffusion, despite the complex structure within the web itself. This is in contrast to the standard mapping in which calculations of diffusion become increasingly difficult when significant regular motion coexists with the stochastic regions. We return to these questions in Sections 5.5 and 5.6 where we calculate the diffusion for various mapping models.

CHAPTER 4

Transition to Global Stochasticity

*4.1. Introduction

For near-integrable systems with two degrees of freedom, we have seen in Chapter 3 the existence of chaotic regions surrounding the separatrices associated with resonances. These regions persist for any nonzero perturbation strength ϵ, although their area tends to zero as $\epsilon \to 0$. Therefore, there is no abrupt "transition to stochasticity" at some critical ϵ, and one must define carefully the meaning of any such criterion.

One possible characterization is to give some measure of the fraction of phase space area that is chaotic. We could then find the minimum ϵ for which this fraction exceeds some arbitrarily chosen value, say $\frac{1}{10}$ or $\frac{1}{2}$. This characterization has led to examination of divergence rates of neighboring trajectories. Calculations of Lyapunov exponents (Benettin et al., 1979) and the related KS entropy (Chirikov, 1979) distinguish between chaotic and regular motion and furthermore distinguish the strength of the stochasticity within the chaotic regions. These methods are treated in Section 5.3.

A more natural characterization of the "transition to stochasticity" stems from the observation, seen in various examples having two degrees of freedom, that a sharp transition occurs, with increasing perturbation strength between regions of the phase space in which stochastic motion of the action is closely bounded by KAM surfaces and regions for which the stochastic motion is interconnected over large portions of the space. In the former regions (see Fig. 4.1), the excursion of the action in the chaotic region is limited to the order of the separatrix width, with $\Delta J/J \propto \sqrt{\epsilon}$ as described in Section 2.4a. In the latter regions, the action may chaotically wander over large portions of the phase plane, with $\Delta J/J \sim O(1)$. Clearly, a quantitative

245

4. Transition to Global Stochasticity

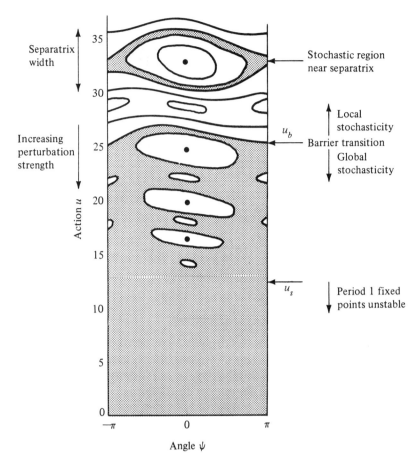

Figure 4.1. An illustration of the transition from local to global stochasticity as the perturbation strength is increased.

determination of the parameters for which this transition takes place yields important information concerning the system's behavior. We designate these two regions as the *local* (or *isolated*, or *weak*) and the *global* (or *connected*, or *strong*) stochastic regions, respectively, and denote the transition as the *barrier transition* or the *transition to global stochasticity*. It is this transition that is considered in detail in this chapter.

We have already seen in the Fermi acceleration problem, treated in Section 3.4, that a stochastic barrier velocity u_b occurs,[1] separating the stochastic sea at low velocities, from the stochastic areas near separatrices of the resonances at high velocities (see Fig. 4.1 as well as Figs. 1.14 and 3.17). This barrier is different from the velocity u_s, below which all period 1 fixed points

---

[1] Recall, for the Fermi problem, that the effective perturbation strength is proportional to $1/u^2$.

4.1. Introduction

are unstable. In fact, $u_b > u_s$, which implies that the disappearance of stable period 1 fixed points is sufficient but not necessary for connected stochasticity and is thus too strong a condition. We look for a sharper criterion, i.e., one that is both necessary and sufficient. Unfortunately, there is no strictly analytic treatment to determine such a condition. To overcome this difficulty, plausibility arguments reinforced by numerical calculations are used to establish procedures for determining the transition to global stochasticity. In this chapter we consider five procedures, introduced qualitatively in Section 4.1a, each contributing different insights into the process. As a model to test the procedures, we use the *standard mapping*, the properties of which are discussed in Section 4.1b.

*4.1a. Qualitative Description of Criteria

The earliest procedure to determine the transition to global stochasticity, advanced by Chirikov (1960) and later refined by him (Chirikov, 1979), is now known as the *overlap criterion*. In its simplest form, it postulates that the last KAM surface between two lowest-order resonances is destroyed when the sum of the half-widths of the two island separatrices formed by the resonances, but calculated independently of one another, just equals the distance between the resonances. The distances are measured either in action or frequency space, whichever is more convenient. This criterion has an intuitive appeal, since we know that regions near the separatrix are, in fact, stochastic such that the touching of two stochastic regions cannot leave a good surface between them. Rigorously, however, the overlap criterion is neither necessary nor sufficient. One can imagine the last KAM surface breaking up well before the islands overlap, due to the interaction of the slowly varying terms outside of the two separatrices. Alternatively, the solution of the complete problem may significantly modify the island widths so that they do not actually overlap when the single resonance calculations predict they do. In actual fact, numerical results indicate that the overlap criterion is too severe a condition for the transition to connected stochasticity. It can, however, be made sharper by considering both the width of the stochastic region near the separatrix and several of the more important secondary resonances lying between the main ones. This procedure, as developed by Chirikov (1979), is the main subject of Section 4.2. Without the added complexity of these modifications, the simplest form of the criterion still serves as a rough estimate of the transition and has been used in a wide variety of problems by Chirikov (1959, 1960, 1971, 1978, 1979a), Ford and co-workers (e.g., Walker and Ford, 1969), Rosenbluth et al. (1966), Jensen (1984), and by many others (see Chirikov, 1979, for a more complete bibliography). It should be pointed out that the loss of linear stability of the main resonances, which gave the stochastic transition u_s in the example of

Section 3.4, can also be used as a rough criterion for the disappearance of the KAM barrier.

In a related procedure, Jaeger and Lichtenberg (1972) calculate the amplitude of second-order islands that results from resonance between harmonics of the libration motion near the major resonances and the fundamental of one of the original periodic motions. They show that at resonance overlap, the ratio of second-order island size to spacing has grown to a size comparable to the ratio of primary island size to spacing, and, by induction, higher-order islands have similar ratios. At this point the local winding number of the primary island about its fixed point is $\alpha = \frac{1}{4}$, indicating four second-order islands. The construction makes obvious the fact that a simple overlap criterion is too severe. Numerically it was found that an island amplitude two-thirds the size, corresponding to a $k = 6$ island chain, is sufficient to destroy the last KAM curve between two first-order islands. The technique has been applied to a number of problems, including, for example, Fermi acceleration (Lieberman and Lichtenberg, 1972) and cyclotron heating (Jaeger *et al.*, 1972; Lieberman and Lichtenberg, 1973), giving good agreement with numerical computations. This technique is particularly useful in determining the width of the stochastic layer surrounding a separatrix, for problems that do not have the structure of the standard mapping. For example, for a perturbed linear oscillator, we have seen in Section 2.4c and Section 3.6 that the primary islands tend to fill the phase space, leaving layers of stochasticity. This method was used to determine the layer thickness by Fukuyama *et al.* (1976) and for the phase space tilings, described in Section 3.6, by Lichtenberg and Wood (1989). The technique has also been employed in problems in which a single island dominates the phase space (Lichtenberg, 1984).

We point out that the modified overlap criterion of Chirikov also makes use of second-order islands to determine the stochastic width near the separatrix, but there the expansion is performed near the separatrix rather than near the singular point. We consider the second-order island method in Section 4.3. The expansion technique, using successive applications of secular perturbation theory, has already been described in Section 2.4.

The third method of determining the stochastic barrier returns to the examination of linear stability. The idea is that, although loss of stability of the lowest-period islands ($k = 1$) is too strong a condition, the linear stability transition of the high k number islands close to a KAM surface may give a sharper criterion for the destruction of that surface. This proposition has been investigated numerically by Greene (1968, 1979a) and found to be correct. More specifically, if a surface with an irrational winding number α is approximated more and more closely by a ratio of integers, then the asymptotic stability of the motion about the fixed points is directly correlated to the existence of a KAM surface at the given α. In particular, between two lowest harmonic ($k = 1$), equal amplitude primary resonances, the irrational number that is furthest away from neighboring rationals (see Section 4.4)

4.1. Introduction

can be shown to be the golden mean[2] ($\sqrt{5} - 1)/2$. Thus one expects that, with increasing perturbation parameter, the last KAM surface to disappear would be this one. Again, numerical calculations by Greene show this to be the case. Therefore a sharp estimate for the transition that destroys the last KAM barrier to stochastic wandering between $k = 1$ islands is found by determining the stability of the rational iterates of the golden mean, asymptotically. In order to make this process numerically tractable, rapid convergence for successively higher iterates is required. Rapid convergence is facilitated by use of the fact that near an irrational surface that is about to disappear, the high-order islands have local rotation number $Q_0 = \frac{1}{6}$, which occurs just before local stability of the island chain is lost. This fact correlates with the observation in Section 4.2 that the final barrier to global stochasticity disappears when the local rotation number of the *primary* island is $Q_0 = \frac{1}{6}$, and therefore correlates directly with the growth to large amplitude of second-order island chains. This method has, in fact, the characteristics of a *renormalization* procedure, which we consider next.

The fourth method that we describe relies directly on renormalization analysis. It was introduced by Escande and Doveil (1981) to determine the amplitudes of two nearest-neighbor island chains that are required to destroy the last phase spanning KAM curve that separates their stochastic separatrix layers. The method proceeds, largely analytically, by *renormalization* of the original Hamiltonian to a series of new Hamiltonians of the same form, which describe successively higher-order resonances in the system. The procedure yields a transition that is typically within 3–10% of the actual value found numerically. Results have been obtained for the general case where the two primary resonances have arbitrary separation, amplitudes, and frequency ratio. In this case, the "last KAM curve" is not necessarily found at the rotation number equal to the golden mean. The method and the results are discussed in Section 4.5. Another renormalization procedure, applied directly to the KAM curve, but requiring numerical calculation of the renormalization equation, has been introduced by Shenker and Kadanoff (1982) and developed further by MacKay (1983). Accurate values of the destruction of the golden mean KAM curve have been found by this method. For details the reader should consult the cited literature.

It is also possible to determine the existence of a KAM torus from a variational approach. The method was formulated by Percival (1974) in Hamiltonian form and by Percival (1979a) and Klein and Li (1979) in Lagrangian form. In the latter form, the method closely parallels the work described in Section 2.6b on simply periodic orbits and yields sharp bounds on the destruction of invariant tori, similar to Greene's estimate (Percival, 1979b). The transition is obtained when the analysis fails to converge to a KAM curve. Percival and his students have applied the method to several

---

[2] Actually, any irrational that is the sum of an integer plus the golden mean is "furthest from the rationals" in this sense.

nonlinear problems in molecular dynamics (Percival and Pomphrey, 1976, 1978; Percival, 1977). For further details the reader should consult the cited literature.

A rigorous upper bound on the value of K for the existence of a phase spanning KAM curve in the standard map can also be obtained. A basic calculation, without mathematical details was given by Lieberman and Lichtenberg (1972). A rigorous and sharper upper bound was found by Mather (1984). MacKay and Percival (1985) showed that Mather's technique could be improved using a computer-aided procedure. We discuss the basic method and the improvements in Section 4.6. The proofs can be found in the original papers.

In Section 4.7 we conclude with a quantitative summary of the various criteria for the transition to global stochasticity, along with some comments regarding their application to practical calculations. The simple overlap criterion is adequate for order-of-magnitude estimates and easy to apply. For sharper criteria, the results of Greene's calculation for the standard mapping, or, in the more general case, the calculation of Escande and Doveil, can sometimes be applied to the problem at hand without the need to carry out their procedures. These criteria combine into an estimate known as the "two-thirds rule," which is reasonably sharp and quite easy to apply. Other perspectives regarding the various criteria for the barrier transition and comprehensive bibliographies may be found in reviews by Chirikov (1979), Tabor (1981), and Escande (1985).

If a system has more than two degrees of freedom, then an abrupt barrier transition no longer exists. This is due to the connection throughout phase space of all bands of stochasticity associated with resonance separatrices. The resulting Arnold diffusion is generally weak compared to diffusion in the globally stochastic regime. Thus the transition to global stochasticity can be a meaningful concept for short time diffusion in systems with more than two degrees of freedom.

*4.1b. The Standard Mapping

For calculating the transition between orbits bounded by KAM surfaces and orbits that can wander stochastically between primary resonances of equal amplitude and frequency, we use the model of the standard mapping (3.1.22). This allows us to explore the transition in terms of the strength of the *stochasticity parameter* K as used both by Chirikov and Greene. We saw in Section 3.1b that this mapping locally approximated more general nonlinear mappings, and we shall now show that these include the Fermi and the separatrix mapping. The parameter K can generally be related to values of the action (or energy) in differing parts of the phase plane for coupled systems with two degrees of freedom.

4.1. Introduction

Taking the Fermi mapping of (3.4.4), repeated below,

$$u_{n+1} = |u_n + \sin \psi_n|, \tag{4.1.1a}$$

$$\psi_{n+1} = \psi_n + 2\pi M / u_{n+1}, \tag{4.1.1b}$$

we obtain the standard mapping by linearization in action space near a given period 1 fixed point. These are located at

$$\frac{2\pi M}{u_1} = 2\pi m, \quad m \text{ integer}. \tag{4.1.2}$$

Putting $u_n = u_1 + \Delta u_n$, and shifting the angle

$$\theta_n = \psi_n - \pi, \quad -\pi < \theta_n \leq \pi,$$

the mapping equations then take the standard mapping form

$$I_{n+1} = I_n + K \sin \theta_n, \tag{4.1.3a}$$

$$\theta_{n+1} = \theta_n + I_{n+1}, \tag{4.1.3b}$$

where

$$I_n = \frac{-2\pi M \Delta u_n}{u_1^2} \tag{4.1.4}$$

is the new action and

$$K = \frac{2\pi M}{u_1^2} \tag{4.1.5}$$

is the stochasticity parameter. We thus have related K to the old action u_1. The conversion from Fermi to standard mapping is illustrated in Fig. 4.2 for two different values of u_1 leading to two different values of K.

We proceed in like fashion for the separatrix mapping (3.5.26), repeated here with $r = 1$ for future application:

$$w_{n+1} = w_n - w_0 \sin \theta_n, \tag{4.1.6a}$$

$$\theta_{n+1} = \theta_n + Q_0 \ln \left| \frac{32}{w_{n+1}} \right|, \tag{4.1.6b}$$

with period 1 fixed points at

$$Q_0 \ln \left| \frac{32}{w_1} \right| = 2\pi m, \quad m \text{ integer}. \tag{4.1.7}$$

Putting $w_n = w_1 + \Delta w_n$ and linearizing in w, we again obtain the standard mapping (4.1.3) with

$$I_n = \frac{-Q_0 \Delta w_n}{w_1} \tag{4.1.8}$$

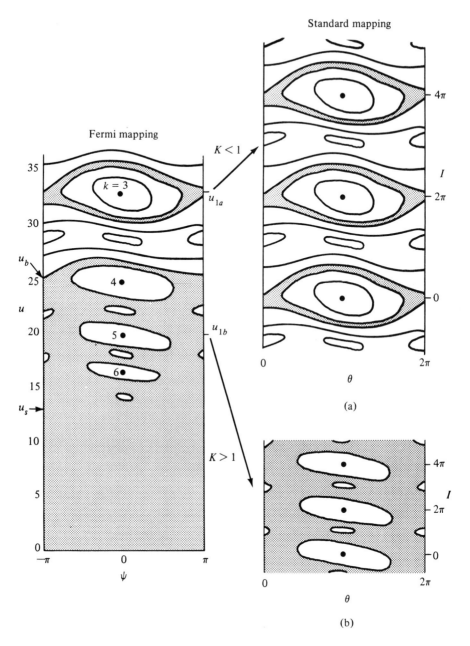

Figure 4.2. Local approximation of the Fermi mapping by the standard mapping. (a) Linearization about u_{1a} leading to K small and local stochasticity; (b) linearization about u_{1b} leading to K large and global stochasticity.

4.1. Introduction

and

$$K = \frac{Q_0 w_0}{w_1}. \quad (4.1.9)$$

Period 1 Fixed Points. Period 1 fixed points of the standard mapping are obtained by requiring that the phase (mod 2π) and the action are stationary, yielding

$$I_1 = 2\pi m, \quad m \text{ integer},$$
$$\theta_1 = 0, \pi. \quad (4.1.10)$$

There are two fixed points for each integer m. Expanding about a fixed point, we obtain as in Section 3.3c the linearized matrix of the transformation

$$\mathbf{A} = \begin{bmatrix} 1 & \pm K \\ 1 & 1 \pm K \end{bmatrix}, \quad (4.1.11)$$

where the upper sign corresponds to $\theta_1 = 0$ and the lower to $\theta_1 = \pi$. Det $\mathbf{A} = 1$ as required for area-preserving mappings. From (3.3.55), the stability condition depends on the trace of $\mathbf{A}$, yielding

$$|2 \pm K| < 2. \quad (4.1.12)$$

Thus the point at $\theta_1 = 0$ is always unstable, while for

$$K > 4 \quad (4.1.13)$$

the elliptic fixed point at $\theta_1 = \pi$ changes to reflection hyperbolic. There is no stable motion about period 1 fixed points for $K > 4$.

For a general mapping, θ is always modulo 2π but I is not. However, for the standard mapping, (4.1.3b) implies that I can also be considered modulo 2π. This peculiarity gives rise to a second type of period 1 fixed point. If we consider that the phase and action (both mod 2π) are stationary, then we replace (4.1.10) by

$$I_{1l} = 2\pi m, \quad K \sin \theta_{1l} = 2\pi l, \quad m, l \text{ integers}. \quad (4.1.14)$$

For $l \neq 0$, we have the *accelerator modes* (Chirikov, 1979), so-called because the action at the fixed point increases by $2\pi l$ for every iteration. The stability condition (4.1.12) is replaced by

$$|2 \pm K \cos \theta_{1l}| < 2, \quad (4.1.15)$$

which implies that stability windows for period 1 fixed points exist for successively higher values of K as l increases ($\cos \theta_{1l}$ decreases), in contrast to the stability limit $K < 4$ for $l = 0$. This is a curiosity of the standard mapping, and it is clearly inappropriate to consider accelerator modes for mappings which only locally approximate the standard mapping.

Higher-Period Fixed Points. Chirikov (1979), Greene (1979a), Lichtenberg et al. (1980), and Schmidt (1980) have examined several of the families of fixed points with period $k > 1$. Two types of fixed points may be distinguished: the *primary* families, so-called because they exist down to the limit $K \to 0$, and the *bifurcation* families, which exist only above certain threshold values of K.

We consider in detail only period 2 fixed points. Each family has two members (I_1, θ_1) and (I_2, θ_2). Iterating the mapping twice, we obtain

$$I_2 = I_1 + K \sin \theta_1, \tag{4.1.16a}$$

$$\theta_2 = \theta_1 + I_2 - 2\pi m_1, \tag{4.1.16b}$$

$$I_1 = I_2 + K \sin \theta_2, \tag{4.1.16c}$$

$$\theta_1 = \theta_2 + I_1 - 2\pi m_2, \tag{4.1.16d}$$

where m_1 and m_2 are integers. The stability follows from the trace of the matrix

$$\mathbf{A} = \begin{bmatrix} 1 & K \cos \theta_2 \\ 1 & 1 + K \cos \theta_2 \end{bmatrix} \begin{bmatrix} 1 & K \cos \theta_1 \\ 1 & 1 + K \cos \theta_1 \end{bmatrix}, \tag{4.1.17}$$

which for stablity leads to

$$-4 < 2K(\cos \theta_1 + \cos \theta_2) + K^2 \cos \theta_1 \cos \theta_2 < 0. \tag{4.1.18}$$

From the sum of (4.1.16a) and (4.1.16c), we see there are two cases (a) $\theta_2 = -\theta_1$ and (b) $\theta_2 = \theta_1 - \pi$, with $0 < \theta_1 \leq \pi$. For case (a), from (4.1.16),

$$2\pi p - 4\theta_1 = K \sin \theta_1 \tag{4.1.19}$$

determines θ_1, with $p = m_1 - m_2 = 1$ giving the primary family, and $p > 1$ the bifurcation families. From (4.1.18) the stability condition is

$$-4 < K \cos \theta_1 < 0. \tag{4.1.20}$$

The primary family is found at $I_{1,2} = 2\pi(m_2 + \frac{1}{2})$, $\theta_{1,2} = \pm \frac{1}{2}\pi$ for $K \ll 1$, and is unstable for any value of K since $\theta_1 \leq \frac{1}{2}\pi$. The first bifurcation family $p = 2$ appears at $K = 4$, where the period 1 fixed point at $I_1 = 2\pi m_2$, $\theta_1 = \pi$ goes unstable, and remains stable for $4 < K < 2\pi$. There are windows of stability for arbitrarily high K's as p increases.

For case (b) from (4.1.16),

$$2\pi(p - 1) = K \sin \theta_1 \tag{4.1.21}$$

determines θ_1, with the stability condition

$$K^2 \cos^2 \theta_1 < 4. \tag{4.1.22}$$

The primary family is found at $I_{1,2} = 2\pi(m_2 + \frac{1}{2})$, $\theta_1 = \pi$, $\theta_2 = 0$, and is stable for $K < 2$. The first bifurcation family $p = 2$ appears at $K = 2\pi$, where the $p = 2$ bifurcation family of case (a) goes unstable. The case (b) family is

4.1. Introduction

stable for $6.28 < K < 6.59$. As for case (a), there are windows of stability for arbitrarily high K

$$(2\pi)^2(p-1)^2 < K^2 < (2\pi)^2(p-1)^2 + 4, \quad (4.1.23)$$

referred to by Chirikov (1979) as the *islets of stability*. Other mappings may not have this special property (see Lichtenberg et al., 1980, for details). Higher-period fixed points have been examined by Schmidt (1980) and Schmidt and Bialek (1981).

The Barrier Transition. Equation (4.1.3) for the standard mapping, with I and θ taken modulo 2π, can be iterated easily for hundreds of thousands of steps, thus examining numerically the $I - \theta$ phase plane for various values of K. Figure 4.3 shows a sequence of phase planes for K increasing from small to large values. At low K, the stable periods 1 and 2 primary fixed points and the unstable points with their chaotic separatrix layers are clearly seen. The transition from local to global stochasticity takes place between $K = 0.95$ and 1.00. More detailed numerical calculations give $K \approx 0.9716$ at the barrier transition (see Section 4.4). As K increases, the primary period 2 and then period 1 points destabilize, but, as seen from the figure, islets of stability can be found for large values of K. The standard mapping is thus typical of generic Hamiltonian systems, with neither integrability nor complete chaos being found for any nonzero value of K.

A detailed numerical plot of the standard mapping, with a value of $K = 0.97$, just below the barrier transition value, is shown for four orbits in Fig. 4.4. As in the numerical calculations of Fermi acceleration, in Section 3.4, the shaded regions are explored by the stochastic wandering of a mapping trajectory, with the blank regions forbidden by the existence of KAM trajectories surrounding the main-island elliptic fixed points. Here KAM orbits still exist, isolating the period 2 separatrix trajectory from the period 1 trajectory, but the period 4 trajectory has been engulfed by the period 1 separatrix stochasticity.

Hamiltonian Form. We can construct a Hamiltonian for the standard mapping by introducing a periodic δ-function force into the equations of motion and then expanding in a Fourier series as described in Section 3.1b and carried out in Section 3.4e for the Fermi problem. We obtain

$$H = \frac{I^2}{2} + K \cos \theta \sum_{m=-\infty}^{\infty} \exp(i2\pi mn), \quad (4.1.24)$$

where the iteration number n is a time variable. We assume here that θ is a slow variable

$$\frac{d\theta}{dn} \ll 2\pi, \quad (4.1.25)$$

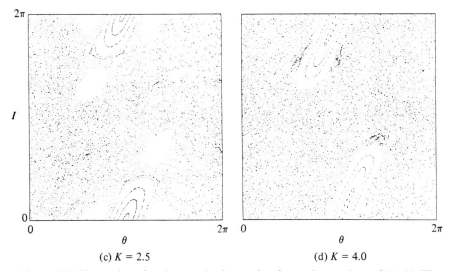

Figure 4.3. Phase plane for the standard mapping for various values of K. (a) The primary period 1 and 2 families are seen clearly; the stochasticity is local, occurring only near separatrices that are isolated. (b) The last KAM curve between the period 1 and period 2 islands is destroyed, leading to global stochasticity. (c) The primary period 2 fixed point is unstable; a period 4 second-order island chain has been emitted from the primary fixed point. (d) The primary period 1 fixed point is just unstable; the surrounding KAM surfaces have elongated to accommodate the period 2 stable fixed points being born by bifurcation (after Berman, 1980).

4.1. Introduction

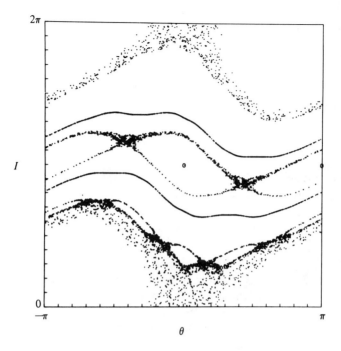

Figure 4.4. Numerical plot of four orbits of the standard mapping for $K = 0.97$. The phase scale is shifted by π from Fig. 4.3 (after Greene, 1979a).

Anticipating that the only term with time variation that will contribute significantly is one with a slowly varying phase, we keep only terms with $m = 0$ and $m = \pm 1$ to obtain

$$H = \frac{I^2}{2} + K \cos \theta + 2K \cos \theta \cos 2\pi n. \tag{4.1.26}$$

Assuming the third term on the right is a perturbation of the motion, which tends to average to zero, we have an unperturbed Hamiltonian

$$H_0 = \frac{I^2}{2} + K \cos \theta, \tag{4.1.27}$$

which is just that of the pendulum (see Section 1.3a). Its phase space trajectories, shown in Fig. 1.4, are ellipses around the fixed point $I = 0, \theta = \pi$, changing to a separatrix trajectory through $\theta = 0, 2\pi$. The motion is that of libration out to the separatrix, on which the period becomes infinite and that of rotation beyond. For the simple form of the pendulum Hamiltonian in (4.1.27), the libration frequency near the elliptic point $\theta = \pi$ is

$$\omega_0 = K^{1/2}, \tag{4.1.28}$$

and the excursion of I is found from (4.1.27) by taking $\cos \theta = 1$ at $I = 0$

such that the maximum excursion at $\cos\theta = -1$ is

$$\Delta I_{\max} = 2K^{1/2}. \tag{4.1.29}$$

Since the distance between primary resonances δI is, in the standard mapping, just equal to the periodicity 2π, the ratio of the full separatrix width to the distance between resonances is

$$\frac{2\Delta I_{\max}}{\delta I} = \frac{4K^{1/2}}{2\pi}. \tag{4.1.30}$$

This can be related to the central angular frequency and therefore the local rotation number $\alpha_0 = \omega_0/2\pi$ through (4.1.28):

$$\frac{2\Delta I_{\max}}{\delta I} = 4\alpha_0 = \frac{4}{Q_0}, \tag{4.1.31}$$

where $Q_0 = 1/\alpha_0$ is the number of iterations per libration period. Equation (4.1.31) is a universal relation for all neighboring island chains of any order, relating the relative island size to its rotation number. For example, the numerical observation that the transition to global stochasticity occurs near the appearance of a set of six second-order islands implies, from (4.1.31), the two-thirds rule $2\Delta I_{\max}/\delta I \approx 2/3$.

*4.2. Resonance Overlap

*4.2a. Rationale for Criteria

In this section we follow the work of Chirikov (1979) to construct a sharp quantitative criterion for the transition to global stochasticity, i.e., the value of the perturbation required to destroy the last KAM surface between two $k = 1$ primary resonances. We use the Hamiltonian form of the standard mapping to calculate the conditions for the period 1 separatrices to just touch, finding the simple overlap criterion $K = 2.47$. Next we take the period 2 separatrix width into account, finding the improved value $K = 1.46$. This is closer to, but still larger than, the experimentally determined $K \approx 0.99$ found for the standard mapping by Chirikov (1979). Allowing for some inaccuracy in the experimentally determined value we must add the thickness of the stochastic region to the maximum island width calculated on the separatrix to account for this difference. (Chirikov found that period 3 resonance was unimportant.) To do this, we calculate the overlap of the second-order resonances (second level in Chirikov's terminology) near the period 1 separatrix. This can be done either (1) by transforming the separatrix mapping in Section 3.5 to a new standard mapping near the separatrix, as done here, with a subsequent recomputation of the second-order island

4.2. Resonance Overlap

Hamiltonian, or (2) by directly computing the second-order island amplitude near the separatrix using the elliptic integral representation of the standard Hamiltonian, which we discuss in Section 4.3b. However, neither calculation can be expected to give the correct stochastic thickness, because the overlap condition in second order requires the same corrections as the first-order calculation, i.e., the inclusion of the two iteration resonance and the thickness of the second-order separatrices by including the overlap of the third-order islands. The process might be expected to converge to the correct answer. Rather than undertaking such a tedious task, Chirikov short-circuits the process by introducing a factor into the second-order mapping. This "weighting factor" in second order then brings the analytical results, including both first- and second-order islands, into agreement with the numerical computations (see Chirikov, 1979, for details).

*4.2b. Calculation of Overlap Criteria

Simple Overlap. The simplest criterion, illustrated in Fig. 4.5a, is to take twice $\Delta I_{\max}$, from (4.1.29), equal to the distance between $k = 1$ resonances, $\delta I = 2\pi$, giving

$$4K^{1/2} = 2\pi \tag{4.2.1}$$

or

$$K = \left(\frac{\pi}{2}\right)^2 = 2.47. \tag{4.2.2}$$

This K is too large, so we improve on it by calculating the $k = 2$ island chain occurring at a value of action midway between primary resonances.

Overlap of First and Second Harmonics. As illustrated in Fig. 4.5b, we define the overlap condition as the touching of the first and second harmonic separatrices

$$\Delta I_{1\max} + \Delta I_{2\max} = \delta I_{12} = \pi, \tag{4.2.3}$$

where the subscripts 1 and 2 stand for the period 1 and period 2 resonances, respectively.

To calculate the ΔI_2 resonance width, we need the second harmonic Fourier component, as described in Section 2.4. For the standard mapping this component appears only in second order in the expansion, as $K \sin \theta$ has only a first harmonic component of angle to lowest order in K. Since the region of phase space in which the second harmonic resonance is found is far from primary resonances, standard perturbation theory may be used to expand to that order.

We need the transformed Hamiltonian $\bar{H}$ to second order, so we use the Lie methods of Section 2.5b. The Hamiltonian (4.1.24) can be written as

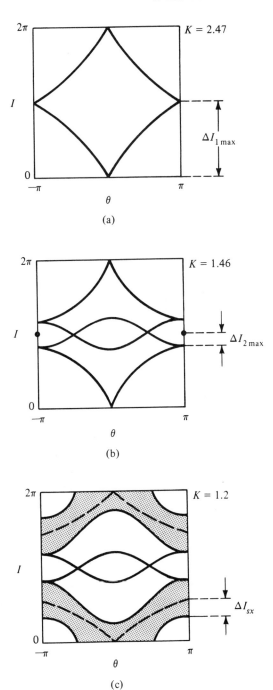

Figure 4.5. Overlap criteria. (a) Primary resonance overlap. (b) Overlap of primary and secondary resonances. (c) Improved overlap criterion, including separatrix width of primary resonance.

4.2. Resonance Overlap

$H = H_0 + \epsilon H_1$ or

$$H = \tfrac{1}{2}I^2 + \epsilon K \sum_{m=-\infty}^{\infty} \cos(\theta - 2\pi mn), \qquad (4.2.4)$$

where ϵ has been inserted to order the perturbation, and will be set to unity at the end of the calculation. We first calculate the Lie generating function w_1 from the first-order Deprit equation (2.5.31a)

$$\left(\frac{\partial}{\partial n} + I\frac{\partial}{\partial \theta}\right)w_1 = \bar{H}_1 - H_1. \qquad (4.2.5)$$

Since $\langle H_1 \rangle = 0$, where $\langle\ \rangle$ denotes the average over zero-order orbits, we choose $\bar{H}_1 = 0$ and solve (4.2.5) for w_1 as

$$w_1 = K \sum_m \frac{\sin(\theta - 2\pi mn)}{2\pi m - I}. \qquad (4.2.6)$$

We expect to find the second-order islands for I midway between primary resonances

$$I = (2p + 1)\pi, \qquad p \text{ integer}, \qquad (4.2.7)$$

so w_1 is nonsingular. The transformed Hamiltonian $\bar{H}$ is now calculated from the second-order Deprit equation

$$\left(\frac{\partial}{\partial n} + I\frac{\partial}{\partial \theta}\right)w_2 = 2(\bar{H}_2 - H_2) - [w_1, \bar{H}_1 + H_1]. \qquad (4.2.8)$$

H_2 and $\bar{H}_1$ are both zero. We choose $\bar{H}_2$ to eliminate the average of the right-hand side

$$\bar{H}_2 = \tfrac{1}{2}\langle[w_1, H_1]\rangle. \qquad (4.2.9)$$

Evaluating the Poisson bracket

$$\bar{H}_2 = -\frac{1}{2}\left\langle \frac{\partial w_1}{\partial I}\frac{\partial H_1}{\partial \theta}\right\rangle, \qquad (4.2.10)$$

and inserting the expansions for w_1 and H_1

$$\begin{aligned}\bar{H}_2 &= \frac{K^2}{2}\left\langle \sum_m \frac{\sin(\theta - 2\pi mn)}{(2\pi m - I)^2}\sum_{m'}\sin(\theta - 2\pi m'n)\right\rangle \\ &= \frac{K^2}{4}\left\langle \sum_{m,m'}\frac{\cos[2\pi(m'-m)n] - \cos[2\theta - 2\pi(m'+m)n]}{(2\pi m - I)^2}\right\rangle.\end{aligned} \qquad (4.2.11)$$

We average over the zero-order orbit $\theta = In$. The first cosine sum vanishes unless $m' = m$, leading to a constant term in $\bar{H}_2$, which we ignore. The second cosine sum vanishes unless

$$2I - 2\pi(m' + m) = 0, \qquad (4.2.12)$$

or, using (4.2.7),
$$m + m' = 2p + 1, \tag{4.2.13}$$

which yields from (4.2.11)
$$\bar{H}_2 = -\frac{K^2}{4} \sum_m \frac{\cos[2\theta - 2\pi(2p+1)n]}{(2\pi)^2 (m - p - \tfrac{1}{2})^2}. \tag{4.2.14}$$

The sum over m is independent of p
$$\sum_m \frac{1}{(m - p - \tfrac{1}{2})^2} = \pi^2 \tag{4.2.15}$$

so that to second order,
$$\bar{H} = \tfrac{1}{2} I^2 - \left(\frac{K}{4}\right)^2 \cos 2\hat{\theta}. \tag{4.2.16}$$

Here
$$\hat{\theta} = \theta - \pi(2p+1)n \tag{4.2.17}$$

is the slowly varying phase. From (4.2.16), the maximum excursion of I is
$$\Delta I_{2\max} = \tfrac{1}{2} K. \tag{4.2.18}$$

The value of K at overlap including both primary and secondary resonances is then obtained by substituting from (4.1.29) and (4.2.18) into (4.2.3)
$$2K^{1/2} + \tfrac{1}{2} K = \pi, \tag{4.2.19}$$

which yields $K = 1.46$ as found by Chirikov. Equation (4.2.19) still overestimates K. We can further improve the estimate by (1) taking account of higher-period ($k > 2$) resonances and (2) taking account of the finite width of the stochastic layer near the separatrix. Chirikov made both of these calculations, finding the dominant effect to be the stochastic layer width, which we now calculate.

Thickness of the Separatrix Layer. We start with the mapping describing the motion in the vicinity of the separatrix of the primary resonances derived in Section 3.4 and repeated in (4.1.6) as
$$w_{n+1} = w_n - w_0 \sin \phi_n, \tag{4.1.6a}$$
$$\phi_{n+1} = \phi_n + Q_0 \ln \left|\frac{32}{w_{n+1}}\right|. \tag{4.1.6b}$$

Here w is the relative deviation of energy from the separatrix energy K
$$w = \frac{\Delta H}{K}, \tag{4.2.20}$$

4.2. Resonance Overlap

ϕ is the phase of the driving term [$\phi = 2\pi n$ in (4.1.26)] as the primary resonance phase θ passes the hyperbolic fixed point, and Q_0 is the ratio of driving frequency to primary island frequency at the elliptic fixed point

$$Q_0 = \frac{2\pi}{K^{1/2}}. \tag{4.2.21}$$

As we observed in Section 4.1b, linearization in w about a (second-order) period 1 fixed point w_1 near the separatrix transforms (4.1.6) into the standard mapping. Substituting for w_0 from (3.5.27) with $\Lambda/F \equiv 1$ and the stochasticity parameter from (4.1.9), we obtain

$$K_2 = \frac{8\pi Q_0^3}{w_1} \exp\left(\frac{-\pi Q_0}{2}\right), \tag{4.2.22}$$

where K_2 is that K associated with the second-order islands.[3] Substituting (4.2.21) into (4.2.22) and requiring for self-consistency of an overlap condition for first- and second-order resonances that $K_2 = K$, we obtain the half-width of the primary stochastic layer as

$$w_1 = \frac{4(2\pi)^4}{K^{5/2}} \exp\left(\frac{-\pi^2}{K^{1/2}}\right), \tag{4.2.23}$$

where we have reintroduced K from (4.2.21).

We must relate w_1 to the separatrix half-width ΔI_{sx} at the phase $\theta = \pi$ of the elliptic fixed point. For the pendulum

$$H = \tfrac{1}{2}I^2 + K \cos \theta, \tag{4.2.24}$$

with separatrix energy $H = K$, the action at $\theta = \pi$ is, from (4.1.29), $I_0 = 2K^{1/2}$. From (4.2.20), at energy

$$K + \Delta H = K(1 + w_1), \tag{4.2.25}$$

the action at $\theta = \pi$ is, using (4.2.24),

$$I^2 = 2K(2 + w_1), \tag{4.2.26}$$

or, for $w \ll 1$,

$$I = 2K^{1/2}\left(1 + \frac{w_1}{4}\right). \tag{4.2.27}$$

Thus,

$$\Delta I_{sx} = I - I_0 = \tfrac{1}{2}w_1 K^{1/2}. \tag{4.2.28}$$

---

[3] This result is a factor of two larger than that obtained by Chirikov. Our perturbation strength in (4.1.26) is $2K$; Chirikov uses K here, which does not correspond to the standard mapping.

The improved overlap criterion, illustrated in Fig. 4.5c, is then

$$\left(1 + \frac{w_1}{4}\right)2K^{1/2} + \frac{K}{2} = \pi, \tag{4.2.29}$$

that is, the sum of the widths of the primary island and the stochastic region outside of its separatrix (the stochastic layers outside and inside the separatrix are taken to be symmetric), added to the width of the second harmonic island just equals the distance between the two primary island centers. Equations (4.2.23) and (4.2.29) must then be solved self-consistently for K to determine the onset of global stochasticity. Performing the calculation, we obtain $K = 1.2$. By including additional heuristic corrections, Chirikov (1979) finds $K = 1.06$. Extensive numerical iteration of the standard mapping (Chirikov, 1979) sets a close upper bound at $K \approx 0.99$. In Section 4.4 we shall see that a sensitive criterion for stochasticity, for the standard mapping, gives the transition between local and global stochasticity at $K = 0.9716$.

Comparison with Fermi Acceleration. In the Fermi acceleration problem, linearization about a period 1 fixed point at u_1 gives the standard mapping with, from (4.1.5), $K = 2\pi M/u_1^2$. For the fully nonlinear mapping, we found the barrier to global stochasticity, numerically (see Fig. 3.17), to be at $u_b = 2.8\sqrt{M}$. Setting $u_b = u_1$, this gives a value of $K = 0.8$. The difference from $K \approx 1.0$ in the standard mapping can be readily explained by a difference in definitions in the two results. Referring to Fig. 1.14, we see that u_b occurs at $2.8\sqrt{M} = 28$, as given in Fig. 3.17, but that this is the upper limit of the stochastic motion, corresponding to the last KAM surface. However, the proper analogy to the standard mapping relates K to the value of u_1 at the center of the primary island related to the last KAM surface. From Fig. 1.14 we see that this occurs at $u_1 = 25 = 2.5\sqrt{M}$. The corresponding value of $K = 1.0$ ($\alpha = \frac{1}{6}$ for the primary island) agrees with the results for the standard mapping. The nonlinearity in action associated with Fermi acceleration, which decreases the perturbation strength K with increasing u, creates sufficient asymmetry to destroy the last KAM surface at $u < u_1$ but retains the one for $u > u_1$. Since the value of u at the fixed point just below u_B does not necessarily correspond to $K = 1$, the position of u_B can vary as much as the separation between neighboring islands.

4.3. Growth of Second-Order Islands

We consider here the amplitude of the motion near second-order resonances for the driven pendulum with Hamiltonian (4.1.26). Because of the nonlinearity

4.3. Growth of Second-Order Islands

of the libration of the undriven pendulum, the motion contains harmonic components of the slow fundamental frequency ω_0. These components can resonate with the fast driving motion in n, having frequency 2π, to produce local distortions in the phase plane or *second-order islands*. As the complete representation is rather cumbersome, we separate the problem into expansions valid near the elliptic fixed point and near the separatrix.

4.3a. Elliptic Fixed Points

The features of the second-order islands that are found near the elliptic singular point have been treated in detail in Section 2.4b. Transforming to action-angle variables $J - \phi$ for the pendulum Hamiltonian of the form $H_0 = \frac{1}{2}I^2 + \omega_0^2 \cos\theta$ and using the methods of Section 2.2a, we have from first-order perturbation theory for libration the new Hamiltonian

$$\bar{H}^0(J) = \omega_0 J - \tfrac{1}{16}J^2, \tag{4.3.1}$$

as in (2.4.43), with the corresponding frequency

$$\omega = \omega_0 - \tfrac{1}{8}J. \tag{4.3.2}$$

We consider a simple generalization of Hamiltonian (4.1.26), including an arbitrary amplitude V and phase shift θ_0 in the perturbation

$$H_1 = V\cos(\theta + \theta_0)\cos\Omega t. \tag{4.3.3}$$

Transforming this term to $J - \phi$ coordinates,

$$\bar{H} = \bar{H}_0 + \bar{H}_1,$$

where

$$\bar{H}_1 = V\cos\left[\left(\frac{2J}{\omega_0}\right)^{1/2}\sin\phi + \theta_0\right]\cos[\Omega t]. \tag{4.3.4}$$

This can be expanded in a Fourier series

$$\bar{H}_1 = V\mathcal{J}_0(\chi)\cos\theta_0 \cos\Omega t$$

$$+ V\cos\theta_0 \sum_{\substack{l>0 \\ l\text{ even}}} \mathcal{J}_l(\chi)[\cos(l\phi - \Omega t) + \cos(l\phi + \Omega t)]$$

$$- V\sin\theta_0 \sum_{\substack{l>0 \\ l\text{ odd}}} \mathcal{J}_l(\chi)[\sin(l\phi - \Omega t) + \sin(l\phi + \Omega t)], \tag{4.3.5}$$

where

$$\chi(J) = \left(\frac{2J}{\omega_0}\right)^{1/2} \tag{4.3.6}$$

and the $\mathcal{J}_l$ are Bessel functions of the first kind. For the standard mapping (4.1.26), $V = 2K$, $\Omega t = 2\pi n$, $\theta_0 = 0$, and only even harmonics are present.

The terms in $\bar{H}_1$ will average to zero over t except in the neighborhood of the action J for which

$$l\omega(J) = \Omega, \qquad (4.3.7)$$

where ω is the nonlinear pendulum frequency. For this l, one term in (4.3.5) is slowly varying. We then use the transformation of secular perturbation theory, as described in Section 2.4b, to a locally slow phase variable

$$\hat{\phi} = \begin{cases} l\phi - \Omega t, & l \text{ even}, \\ l\phi - \Omega t - \dfrac{\pi}{2}, & l \text{ odd}, \end{cases} \qquad (4.3.8)$$

with corresponding action

$$\hat{J} = \frac{J}{l}. \qquad (4.3.9)$$

Averaging over t, as previously, all terms except the lth are approximately zero. In the remaining term, we expand the action about the fixed point at $\partial \bar{H}_0 / \partial \hat{J}_0 = 0$ to obtain, locally, the Hamiltonian

$$\Delta \bar{H} = \frac{\partial^2 \bar{H}_0}{\partial \hat{J}_0^2} \frac{(\Delta \hat{J})^2}{2} + \Lambda_l \cos \hat{\phi}, \qquad (4.3.10)$$

where

$$\Lambda_l = V \mathcal{J}_l(\chi_0) \times \begin{cases} \cos \theta_0, & l \text{ even}, \\ -\sin \theta_0, & l \text{ odd}, \end{cases} \qquad (4.3.11)$$

and from (4.3.1) and (4.3.9),

$$\frac{\partial^2 \bar{H}_0}{\partial \hat{J}_0^2} = -\frac{l^2}{8}. \qquad (4.3.12)$$

The Hamiltonian (4.3.10), describing the second-order islands, has the same form as the Hamiltonian (4.1.27), describing the primary islands. The frequency separation of second-order resonances is

$$\delta\omega = \omega(J_l) - \omega(J_{l+1}) = \frac{\Omega}{l} - \frac{\Omega}{l+1} \simeq \frac{1}{l}\omega. \qquad (4.3.13)$$

If only even or odd harmonics exist, $\delta\omega = 2\omega/l$. From (4.3.2) and (4.3.9),

$$\delta\omega = \tfrac{1}{8} l \delta \hat{J} \qquad (4.3.14)$$

leading to the separation in action

$$\delta \hat{J} = \frac{8m_s \omega}{l^2}, \qquad (4.3.15)$$

4.3. Growth of Second-Order Islands

where $m_s = 1$ for no symmetry and $m_s = 2$ for even or odd symmetry. Using the same procedure as with the first-order islands, the maximum excursion in a second-order resonance occurs on its separatrix,

$$\Delta \hat{J}_{\max} = (2/l)(8\Lambda_l)^{1/2}, \tag{4.3.16}$$

and the second-order linearized libration frequency is

$$\hat{\omega}_l = l\left(\frac{\Lambda_l}{8}\right)^{1/2}. \tag{4.3.17}$$

Equations (4.3.14)–(4.3.17) can be combined to give, as in (4.1.31) for the primary islands,

$$\frac{2\Delta \hat{J}_{\max}}{\delta \hat{J}} = \frac{4}{m_s} \frac{\hat{\omega}_l}{\omega}. \tag{4.3.18}$$

We now show that for the standard mapping, the second-order island amplitude is small compared to the distance between second-order islands, except when the first-order island size-to-spacing is also large. Evaluating (4.3.18) using $m_s = 2$, $V = 2K$, and

$$\omega \approx \omega_0 = K^{1/2},$$

we obtain a measure of the interaction

$$\frac{2\Delta \hat{J}_{\max}}{\delta \hat{J}} = [l^2 \mathscr{J}_l(\chi)]^{1/2}. \tag{4.3.19}$$

The argument of the Bessel function has its largest allowable value at $\chi = \pi$ (the separatrix trajectory). Thus for $l \gg \pi$ the Bessel function is exponentially small, and the second-order islands are unimportant. By a similar argument, islands arising from higher-order iterations of the mapping can also be neglected.

We are now in a position to estimate for what value of the rotation number the second-order island perturbation can be as important as that of the primary islands. Setting the right-hand sides of (4.3.19) and (4.1.31) equal for the same rotation number ($Q_0 = l$), and for the maximum second-order island size at $\chi = \pi$ (corresponding to the island chain being close to the separatrix), we obtain

$$l^4 \mathscr{J}_l(\pi) = 16. \tag{4.3.20}$$

The left-hand side is about as large as the right-hand side for $l = 4$ and $l = 6$ (the odd harmonics are absent). We therefore expect the interaction of second-order islands to be as important as the interaction of the primary islands when the stochasticity parameter has increased to sufficient size that the $l = 4$ and $l = 6$ island chains have appeared surrounding the elliptic singular point. By induction, the same result exists between second- and third-order islands, and consequently to all orders. Thus we obtain the remarkable result: at a critical value of primary rotation number of $\alpha \approx \frac{1}{5}$, all higher-order islands simultaneously have about the same rotation number and the same ratio of separatrix width to spacing as the primary resonance. The result of an abrupt transition to the destruction of the last KAM surface between islands is thus physically very plausible. From (4.1.28), the corresponding perturbation parameter is $K = 1.2$.

Detailed numerical calculations of the second-order island structure have been made for the standard mapping and for a number of Hamiltonians with two degrees of freedom in which an appropriate portion of the phase plane in a surface of section is similar to that of the standard mapping. A close correspondence to the problem is of a particle gyrating in a magnetic field and resonating with an obliquely propagating wave (Smith and Kaufman, 1975; Smith, 1977), which was treated in Section 2.2b. Their Hamiltonian, in the wave frame, can be written in the form [see (2.2.67) and preceding discussion]

$$H = \frac{k_z^2 P_\psi^2}{2M} - P_\psi \omega + P_\phi \Omega + \epsilon e \Phi_0$$

$$\times \sum_{m=-\infty}^{\infty} \mathcal{J}_m[k_\perp \rho(P_\phi)] \sin(\psi - m\phi). \quad (4.3.21)$$

Choosing P_ϕ such that $\mathcal{J}_m$ is of order of unity, a few neighboring harmonics were analyzed near resonance. They chose $m = -1, 0, 1$, and with the appropriate choice of perturbation amplitude $e\Phi_0$, the result of Fig. 2.10b is obtained. Chains of five second-order islands are seen around each primary resonance. (In this example, odd harmonics are present.) The last KAM surface between resonances is also seen to have disappeared, as the dots represent a single set of initial conditions that wanders freely between the primary resonances.

4.3b. The Separatrix

If we wish to determine the second-order island structure near the separatrix, we must either determine the action from perturbation theory using the separatrix trajectory as the unperturbed motion, or calculate the action directly in the separatrix region from the exact pendulum solution of Section 1.3a. While both methods involve rather complicated calculations, determination of the action near the separatrix has been carried out by a number of authors, and we can use their results without exhibiting all of the details of the calculation. We follow here the work of Smith (1977) and Smith and Pereira (1978) who have analyzed the problem directly from the exact solution.

The pendulum Hamiltonian (4.1.27) can be written in action-angle form, with the action J and angle ϕ given by (1.3.10) and (1.3.11). For libration

4.3. Growth of Second-Order Islands

(see Smith, 1977)

$$J = \omega_0(\tfrac{8}{\pi})[\mathscr{E}(\kappa) - (1 - \kappa^2)\mathscr{K}(\kappa)], \qquad \kappa < 1, \tag{4.3.22}$$

where $\mathscr{E}$ and $\mathscr{K}$ are the complete elliptic integrals of the first and second kind, and

$$2\kappa^2 = 1 + \frac{H}{\omega_0^2}, \tag{4.3.23}$$

where H is the energy and $\kappa = 1$ at $H = \omega_0^2 = K$, the separatrix energy. The pendulum frequency is, from (1.3.13),

$$\omega(\kappa) = \frac{\tfrac{1}{2}\pi\omega_0}{\mathscr{K}(\kappa)} \tag{4.3.24}$$

with an asymptotic value for κ near unity

$$\omega(\kappa) = \frac{\tfrac{1}{2}\pi\omega_0}{\ln[4/(1 - \kappa^2)^{1/2}]}. \tag{4.3.25}$$

A general overlap condition for second-order islands in the standard mapping can be written from (4.3.18), with $\hat{\omega}$ found from (4.3.10), as

$$\frac{2\Delta\hat{J}_{max}}{\delta\hat{J}} = 2\left|\frac{\Lambda_l\, d\omega}{dJ}\right|^{1/2} \frac{l}{\omega}, \tag{4.3.26}$$

where the quantities are now evaluated at the action for which

$$l\omega(J) = 2\pi \tag{4.3.27}$$

and only even harmonics are present. From (4.3.22) and (4.3.23),

$$\frac{d\omega}{dJ} = \frac{d\omega}{d\kappa}\frac{d\kappa}{dJ}, \tag{4.3.28}$$

obtaining, after using the various elliptic integral expressions,

$$\frac{d\omega}{dJ} = -\frac{1}{16}\frac{\omega^3 J}{\omega_0^4 \kappa^2(1 - \kappa^2)}. \tag{4.3.29}$$

We find Λ_l from a Fourier analysis of the third term in (4.1.26), generalized to an arbitrary perturbation amplitude by replacing $2K$ with V, where $\theta = \theta(J, \phi)$ from (1.3.11). The calculation is rather complicated, involving complex integration to evaluate an improper integral. Smith and Pereira (1978, see their Appendix A) find

$$\Lambda_l = V\frac{(\pi/\mathscr{K})^2 l q^{l/2}}{1 - (-q)^l}, \qquad l \text{ even}, \tag{4.3.30a}$$

where

$$q = \exp\left\{\frac{-\pi \mathcal{K}[(1-\kappa^2)^{1/2}]}{\mathcal{K}(\kappa)}\right\}. \tag{4.3.30b}$$

Substituting (4.3.29) and (4.3.30) into (4.3.26), we can plot the ratio $2\Delta\hat{J}_{max}/\delta\hat{J}$ for any perturbation amplitude V.

Near the separatrix $\kappa^2 \to 1$ and we find

$$q^{1/2} = \exp\left(-\frac{\pi}{2}Q_0\right),$$

where $Q_0 = 2\pi/\omega_0$ is the ratio of driving frequency to linearized pendulum frequency. Using (4.3.30a) with $V = 2K = 2\omega_0^2$, and dropping $(-q)^l$ which is small,

$$\Lambda_l = 16\pi\omega \exp\left(-\frac{\pi}{2}Q_0\right).$$

Inserting this expression together with (4.3.29) into (4.3.26),

$$\left(\frac{2\Delta\hat{J}_{max}}{\delta\hat{J}}\right)^2 = \frac{16}{\pi}Q_0^3 \exp\left(-\frac{\pi}{2}Q_0\right)\frac{1}{1-\kappa^2}. \tag{4.3.31}$$

From the definitions (4.2.20) for w and (4.3.23) for κ, we obtain the correspondence

$$\tfrac{1}{2}w_1 = 1 - \kappa^2. \tag{4.3.32}$$

For consistency with (4.2.1) for primary resonance overlap,

$$\left(\frac{2\Delta\hat{J}_{max}}{\delta\hat{J}}\right)^2 = \left(\frac{4K_2^{1/2}}{2\pi}\right)^2. \tag{4.3.33}$$

We equate the right-hand sides of (4.3.31) and (4.3.33) to obtain the condition

$$K_2 = \frac{8\pi Q_0^3}{w_1}\exp\left(-\frac{\pi}{2}Q_0\right), \tag{4.3.34}$$

which is identical to that found in (4.2.22) using the separatrix mapping.

Fukuyama *et al* (1976) have looked at the simple "overlap" of second-order islands for wave-particle resonance using the general elliptic integrals with a few minor simplifications to obtain the result in Fig. 4.6. They consider the fraction r of the action space satisfying the simple overlap criterion [(4.3.26) equal to one] and plot this fraction against the rotation number

$$\alpha = \frac{\omega_0}{2\pi} = \frac{K^{1/2}}{2\pi}.$$

The rapid onset of stochasticity is seen to occur around $\alpha \approx 0.2$, i.e., when a five-island chain appears nearest the primary fixed point. This is consistent with our previous results.

4.4. Stability of High-Order Fixed Points

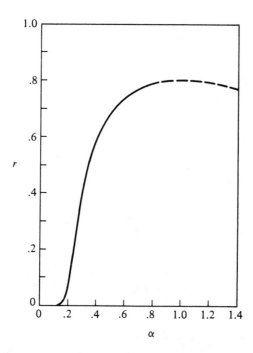

Figure 4.6. Fraction r of the action space satisfying the simple overlap criterion versus rotation number $\alpha = \omega_0/2\pi$ (after Fukuyama et al., 1976).

*4.4. Stability of High-Order Fixed Points

*4.4a. The Basic Elements of Greene's Method

We now describe a method for finding the exact transition to global stochasticity developed by Greene (1968, 1979a). The method postulates a correspondence, illustrated in Fig. 4.7, between two properties of the system: (1) the disappearance of a KAM surface with an irrational winding number α; and (2) the destabilization of the elliptic singular points of the high-harmonic rational iterates ($\alpha \cong r/s$, r, s relatively prime integers), which approach the irrational α in the limit $s \to \infty$.

The Mean Residue. The residue of a tangent mapping at a period s fixed point is defined as

$$R = \tfrac{1}{4}(2 - \text{Tr } \mathbf{A}), \tag{4.4.1}$$

where $\mathbf{A}$ is the ordered product of the s Jacobian matrices of the mapping evaluated at the s fixed points (see Section 3.3b). Comparison with (3.3.54)

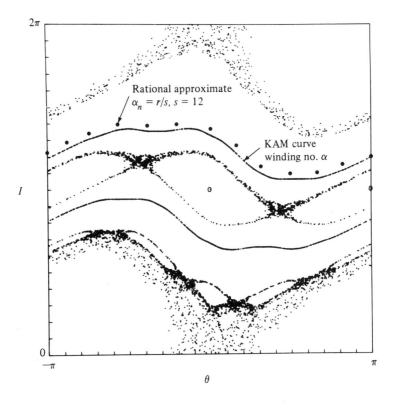

Figure 4.7. The basic correspondence of Greene's method; the value of K at which the large period rational approximate $\alpha_n = r/s$ goes linearly unstable signifies the loss of stability of the nearby KAM curve with irrational winding number $\alpha \approx \alpha_n$.

shows that (for $0 \leq R \leq 1$)

$$R = \sin^2\left(\frac{\sigma}{2}\right), \qquad (4.4.2)$$

where σ is the phase shift per iteration of the mapping. There is therefore stable motion about the family of fixed points for

$$0 < R < 1. \qquad (4.4.3)$$

Over a full cycle of the periodic orbit, the matrix of the linearized standard mapping of (4.1.3) as obtained from (3.3.43) and (3.3.44) is, for rational winding number $\alpha = r/s$,

$$\mathbf{A} = \prod_{i=1}^{s} \begin{bmatrix} 1 & K\cos\theta_i \\ 1 & 1 + K\cos\theta_i \end{bmatrix}, \qquad (4.4.4)$$

from which Tr $\mathbf{A}$ can be determined. A more convenient form for calculating

4.4. Stability of High-Order Fixed Points

the residue is found from a result of Helleman and Bountis (see Greene, 1979a):

$$R = -\tfrac{1}{4} \det \begin{bmatrix} 2+K\cos\theta_1 & -1 & 0 \cdots & -1 \\ -1 & 2+K\cos\theta_2 & & \\ 0 & & & \\ \vdots & & & \\ -1 & & & 2+K\cos\theta_s \end{bmatrix}, \quad (4.4.5)$$

where the $s \times s$ matrix is tridiagonal except for an additional -1 in two corners.

When K is large, we expect that $R \propto K^s$. Greene has also shown this to be true for small K and postulates it to be true for all K. This implies that the residue is exponential with the orbit length, which is proportional to s. For orbits with $R > 0$, there is then a transition, for large s, from values of R approaching zero for stable orbits to very large values of R for unstable orbits. It is therefore natural to investigate the behavior of a quantity proportional to the sth root of T defined by

$$f = \left(\frac{|R|}{\beta}\right)^{1/s}, \quad (4.4.6)$$

which Greene calls the mean residue. The constant β is chosen for rapid convergence in the numerical iteration process, i.e., to obtain reliable answers for relatively small s, and can always be set equal to one. The test for stability is then $f < 1$.

Continued Fraction Approximates. It can be shown (see, e.g., Khinchin, 1964) that the best way to approximate irrationals by rationals is with a continued fraction expansion. For an irrational between 0 and 1, this expansion is represented in the form

$$\alpha = \cfrac{1}{a_1 + \cfrac{1}{a_2 + \cfrac{1}{a_3 + \cdots}}} \quad (4.4.7)$$

written more conveniently as $\alpha = [a_1, a_2, a_3, \ldots]$, where the a_n's are positive integers. That is, a_1 is determined by reciprocating α and taking the integer part, a_2 is determined by reciprocating the remainder and again taking the integer part, etc. The expansion is unique, and the successive iterates of this continued fraction r_n/s_n, where a_n is the last term taken, best approximate α in the sense that no other r/s is closer to α for $s \leq s_n$. From the continued fraction form it follows that the r_n/s_n alternate about α. Furthermore, it can be shown that the convergence is given by

$$\left| \alpha - \frac{r_n}{s_n} \right| \leq \frac{1}{s_n s_{n-1}}, \tag{4.4.8}$$

which is quadratic in s for s large. As an example, we take $\alpha = \pi - 3 \approx 0.14159$, the fractional part of π. The continued fraction expansion is

$$[7, 15, 1, 293, 1, \ldots]. \tag{4.4.9}$$

A rapid increase in the size of the a_n's indicates rapid convergence of the series. A set of successive continued fraction iterates of a given irrational is the most rapidly converging set of fixed points, approximating that irrational, that can be explored for stability.

Now, if we wish to explore the *last* KAM surface to be destroyed, we expect that surface to be the one furthest from rationals, which implies the value of α obtained by using the lowest values of the a_n's. Clearly that value of winding number is

$$\alpha_I = [1, 1, 1, \ldots,] = \frac{\sqrt{5} - 1}{2}. \tag{4.4.10}$$

α_I has long been known to be of special significance, and has been given the name of the *golden mean*. We therefore examine the stability of the approximates α_{I_n} to determine the transition to global stochasticity. The golden mean has been found to be, within numerical error, the winding number associated with the last KAM curve to be destroyed in the standard mapping, thus verifying Greene's intuition. Any mapping that can be locally approximated by the standard mapping will also have this property. However, the primary resonances for the standard mapping have equal periods, amplitudes, and phases. In more general systems of primary resonances, the golden mean is *not* found to be associated with the last KAM curve and the transition to global stochasticity (Escande and Doveil, 1981). We shall consider this situation in Section 4.5.

For numerical work the convergence is improved if low-order approximates to R are close to the limiting value of R as $s \to \infty$. In the case of the golden mean, Greene has shown numerically that, for the critical value of $K = K_c$, as $s \to \infty$, the residue converges to the limit

$$R(\alpha_I, K_c) = 0.25. \tag{4.4.11}$$

Thus taking $\beta = \frac{1}{4}$ in (4.4.6) makes $f = 1$ in this limit such that we might expect rapid convergence with the number of iterates n. This indeed proves to be the case.

The above also suggests the procedure to find the disappearance of the last KAM surface in any localized region of the action space. We choose a particular irrational surface by requiring that its partial fraction expansion

4.4. Stability of High-Order Fixed Points

converges as slowly as possible, that is,

$$\alpha = [a_1 \ldots a_n, 1 \ldots 1], \tag{4.4.12}$$

where $a_1 \ldots a_n$ are chosen to place the winding number in the appropriate portion of the space. Again, numerical results indicate that the residues associated with the approximates of the α of (4.4.12) have a limit given by (4.4.11). This also implies that $\beta = \frac{1}{4}$ is a good convergence factor for looking at the stability properties of any portion of the plane.

It is efficient, but not necessary, to use continued fraction expansions in employing the general technique of examining high-iterate fixed points. For example, Lunsford and Ford (1972) found that a choice of $\alpha^{-1} = k \pm 1/n$, where k is a set of harmonics chosen over the range of interest, $k = 4, 5, 6, 7, 8$, etc., and n is allowed to run over a set of integers $1 < n \leqslant n_0$, proved a convenient (but not precise) method of determining the transition. Another choice, based on fractal diagrams, has been advanced by Schmidt and Bialek (1981). We contrast their results with Greene's in the next subsection.

Other Observations. Greene (1979a) has pointed out a number of other interesting observations that have come out of the numerical analysis. In our somewhat more restricted context, we mention a few of them. For a more complete discussion the reader should consult the original paper.

First, we note from (4.4.2) that $R = \frac{1}{4}$ corresponds to $\sigma = \pi/3$. That is, the disappearance of the last KAM surface in a region of the phase space corresponds to a change of topology of a large s family of elliptic fixed points, which lie close to the KAM surface. This topological change results in each fixed point being surrounded by six higher-order islands. But this is also the condition for six second-order islands to appear surrounding the primary island in the standard mapping, i.e., $\omega_0/2\pi = \frac{1}{6}$. Thus, periodic trajectories of shortest ($s = 1$) and limiting infinite length ($s \to \infty$), which are rational iterates of the golden mean, both have the same second-order rotation number. In fact, Greene's numerical results suggest that all continued fraction iterates of the golden mean have this property.

A second observation of Greene is of practical importance for numerical work. He shows (see Section 5.6a for details) that in the tangent space of (linearized motion near) an orbit lying on a KAM curve, a circle of numerical error will transform into an ellipse in such a way that the error rapidly increases along the KAM curve but not across it. The result is that numerical errors spread the trajectory parallel to a KAM surface, rather than across it; thus computations can be carried out over very large numbers of iterations, without introducing spread in KAM surfaces.

Finally, we point out that the method is a sensitive numerical test for integrability. If $R \equiv 0$ over the entire phase space, the eigenvalues are $\lambda_1 = \lambda_2 = 1$, corresponding to the situation described in (3.3.71). This sheared flow implies that periodic orbits exist continuously on certain phase space surfaces, i.e., the invariant surfaces do not break up into island structures at any level. Such a structure should, in principle, be completely integrable.

*4.4b. Numerical Evaluation

The Numerical Procedure. We briefly review here procedures to perform the numerical calculations in this section. The details of the procedures used by Greene are described more fully in his papers (Greene, 1968, 1979a, 1979b).

(1) Consider that the problem has been represented as a mapping, such as the standard mapping or the Fermi acceleration mapping. For a general two-degree-of-freedom problem, this requirement may present some difficulties. Given a Hamiltonian, the mapping can be found by perturbation methods (see Section 3.1c) or constructed by numerically integrating the equations of motion to find, for all initial conditions, the succeeding intersection of the trajectory with a chosen surface of section. Greene (1979b) has used natural symmetries to simplify this task for the Hénon and Heiles potential.
(2) Find the primary ($k = 1$) elliptic fixed points to high accuracy, analytically if possible.
(3) On a symmetry line of the phases, numerically compute the rotation number α as a function of the distance in action away from the fixed point. This must be done as a long-time average. However, the high-harmonic fixed points $\alpha = r/s$, which have the same symmetry as the primary fixed points, will be accurately determined after s mappings.
(4) Choose a set of iterates $\alpha_n = r_n/s_n$ near some irrational α where the test for stochasticity is to be made. If one is looking for the transition to global stochasticity in systems with primary resonances having equal amplitude, period, and phase, then the set of iterates are those found from the continued fraction expansion of the golden mean.
(5) Find the linearized mapping **A** in the neighborhood of the singular points, defined by

$$x^s - x_0 = \mathbf{A}(x - x_0),$$

where x^s is the value of x (near x_0) after s iterations of the mapping, and x_0 is the coordinates of the particular fixed point under investigation. This can normally be done numerically from the second partial derivatives of the Hamiltonian evaluated at the fixed point (see Greene, 1979b, for details).

Numerical Results. For the standard mapping of Section 4.1b, Greene (1979a) used the method to explore the properties previously discussed. He changed variables to transform the mapping to be periodic in the unit square, rather than in 2π, but this does not alter any results. Figure 4.4 gives his

4.4. Stability of High-Order Fixed Points

Table 4.1. The Residues f and R Versus r/s

	$K = 0.9$		$K = 0.9716$	
$r_n s_n$	$f^{(+)}$	r_n/s_n	$f^{(+)}$	R
2/3	0.93896	1/1	0.971635	0.24291
3/5	0.919959	1/2	0.971635	0.23602
5/8	0.92775	2/3	1.014042	0.26068
⋮				
34/55	0.92427	3/5	0.993528	0.24201
		⋮		
55/89	0.92409	377/610	0.99999965	0.24995
89/144	0.92406	610/987	1.00000009	0.25002
144/233	0.92401[a]	987/1597	0.9999970	0.24988

[a] $R = 2.5 \times 10^{-9}$.

numerical results for four orbits for $K = 0.97$, probably just before destruction of the last KAM curve. Because of symmetry about the value of the action corresponding to two islands with rotation number $\alpha = \frac{1}{2}$, there are two last KAM curves. Also, the $k = 1$ separatrix trajectory is probably diffusing very slowly beyond the limits shown, which could correspond to a just-destroyed KAM curve. That is, due to computational limits placed on the number of iterations, the fate of the separatrix orbit is not certain from this picture. Other slightly less stable KAM surfaces, corresponding to other irrational winding numbers may still exist. However, at $K = 0.9716$, there is a clearly different behavior of the KAM surface at the golden mean, and at $K = 0.975$ the invariant is certainly destroyed because trajectories are found to diffuse through it (although very slowly).

Greene has calculated the values of f at the rational continued fraction iterates of the golden mean for $K = 0.9716$ and for $K = 0.9$, which we can compare. These results are shown in Table 4.1. We see a clear transition from $f < 1$ at $K = 0.9$ to $f \approx 1$, asymptotically, at $K = 0.9716$. We also note the dramatic change in R for long orbit lengths (large s), from the asymptotically small value for stable orbits to the value near $R = 0.25$ at the transition.

The physical picture of this breakup of the KAM surface at the golden mean can be explored by plotting the physical positions of the fixed points at successive pairs of continued fraction iterates. This is done in Fig. 4.8, comparing $K = 0.95$ for which $f(\lim_{n \to \infty} \alpha_n) \cong 0.977$ with the case of $K = 0.9716$ for which, as we have seen, $f(\lim_{n \to \infty} \alpha_n) \cong 1.000$. Because each successive golden mean iterate has approximately $1/\alpha$ more fixed points than the preceding one, and because two successive iterates are used to bracket the actual α, the horizontal scale is expanded by $(1/\alpha)^2$ to keep the same

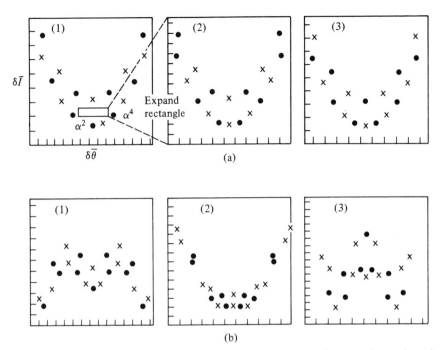

Figure 4.8. Fixed point positions (● stable, × unstable) of successive pairs of continued fraction iterates. (a) $K = 0.95$; (b) $K = 0.9716$ (after Greene, 1979a).

number of points in successive frames [labeled (1), (2), and (3)]. To preserve the aspect ratio where the fixed points lie roughly on a parabola, the vertical scale is expanded by approximately $(1/\alpha)^4$. We see in Fig. 4.8a for $K = 0.95$ that successive iterates appear to converge uniformly to the KAM surface. This is contrasted with the results for $K = 0.9716$, in Fig. 4.8b, where the successive sets of fixed points appear to have an underlying variability that has structure at every scale. A reasonable explanation for this is that the island structure on the previous scale is always sufficiently large to disrupt the next finer scale. It seems reasonable that an invariant curve will not exist under these circumstances.

To summarize from Greene's work, we have qualitatively the picture in Fig. 4.9, where we sketch the local rotation number $\hat{\alpha}$ for the island motion versus the order s of the iterate. For $\hat{\alpha} > \frac{1}{2}$, we have unstable fixed points. For $K \ll 1$, $\hat{\alpha}$ starts at a value well below $\frac{1}{6}$ at $s = 1$ (the main island where $\hat{\alpha} = \alpha$) and decays exponentially with increasing s. For K near but less than 0.9716, $\hat{\alpha}$ still decreases with increasing s. At $K = 0.9716$, the results in Table 4.1 show that $\hat{\alpha} = \frac{1}{6}$ ($R = 0.25$) for all iterates; all iterates have stable fixed points, but the ratio of island size to island separation is approximately equal for all iterates. For K near but greater than 0.9716, $\hat{\alpha}$ increases with increasing s, and thus for sufficiently large s, $\hat{\alpha} > \frac{1}{2}$, yielding unstable fixed points. We therefore see the correspondence between the two pictures of higher-order

4.4. Stability of High-Order Fixed Points

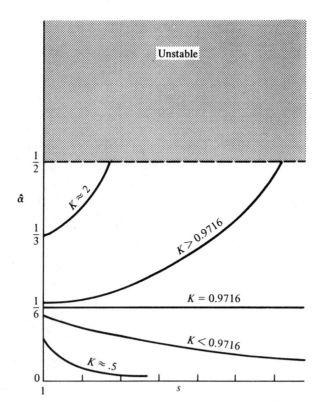

Figure 4.9. Qualitative picture of rotation number $\hat{\alpha}$ for the island motion versus the order s of the rational iterate.

island amplitude growth and destabilization of the higher-order iterates, i.e., the flat line at $\hat{\alpha} = \frac{1}{6}$ occurs at the limiting K, above which some higher iterates must be unstable. This picture is strengthened by results of Escande and Doveil (1981) whose methods are described in the next section. They show that when a KAM curve with a given α is destroyed, then all fixed points with iterates large enough are destabilized. To complete the picture for $K \gg 1$, using (4.4.5), we have seen that $\hat{\alpha}$ increases exponentially with increasing s, clearly destabilizing the higher-order iterates.

Another viewpoint that connects the destabilization of fixed points with the disappearance of KAM curves orders the marginally stable fixed points in a *fractal* diagram (Schmidt and Bialek, 1981). The basic idea is to order the fixed points by rotation number α, with the first two orders given by

$$\alpha_1 = \frac{1}{n}, \quad \alpha_2 = \frac{1}{n \pm 1/m}, \quad (4.4.13)$$

respectively, where the n and m values are the positive integers. For m large the fixed points approach the island separatrix of the associated n, while for

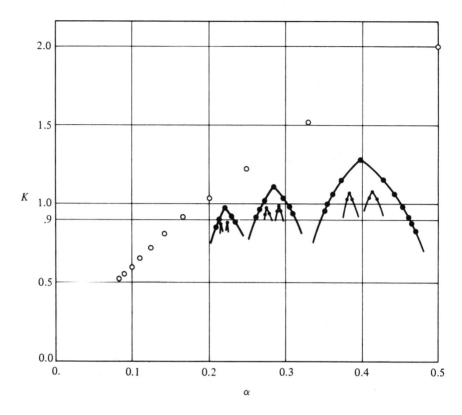

Figure 4.10. Stochasticity parameter K versus rotation number α for marginally stable families of fixed points for the standard mapping. The number of islands in the chain is $1/\alpha$. The fractal nature of the diagram is evident; the pattern of values of critical K at which families of fixed points bifurcate repeats itself on finer and finer scales (after Schmidt and Bialek, 1981).

$m = 1$, $\alpha_2 = \alpha_1$ of the neighboring island chains. For the first three such orderings, the value of K at which the fixed points become unstable is plotted, for the standard mapping, in Fig. 4.10. Excluding the main island ($n = 1$), the values of K for which the $n = 2, 3, 4, \ldots$ fixed points (open circles) become unstable are seen to fall on a smooth curve of descending values of K. (The curve is symmetric about $\alpha = 0.5$, with the other half not shown.) Between each pair of n values, the values of K for which the second-order fixed points become unstable are shown (solid circles). The curve formed by interconnecting these points looks like that of the first-order curve with the scales changed. The third-order fixed points repeat the picture on a still finer scale. This scaling is characteristic of fractals, as we discuss in Section 7.1d.

Schmidt and Bialek postulate that the values of K at the peaks of the curves, in each order l, follow a geometric series

$$\frac{K_l - K_{l+1}}{K_{l-1} - K_l} = \Delta_l, \tag{4.4.14}$$

where $\Delta_l =$ constant for large l. From the fractal diagram, the disappearance of the last KAM curve between any rationals can then be predicted from the first few island orders that have peaks in that region. The use of (4.4.14) is given credence by the similar properties of bifurcation trees, both in dissipative and Hamiltonian systems (see Sections 7.2 and 7.3). Schmidt and Bialek have tested the predictions of Fig. 4.10, for the existence of a few of the KAM curves, against numerical computations, obtaining good agreement.

Finally, we discuss briefly the applicability of these fixed point destabilization techniques to other problems. Lunsford and Ford (1972) earlier applied a simpler form of the method to the Hénon and Heiles coupled oscillator, described in Section 1.4a, of a particle moving in the potential

$$U(x, y) = \tfrac{1}{2}(x^2 + y^2 + 2x^2y - \tfrac{2}{3}y^3).$$

They examined the surface of section mappings for particle energies given by $E = \tfrac{1}{12}$ and $E = \tfrac{1}{8}$, shown in Figs. 1.13b and 1.13c, respectively. The symmetry line $p_y = 0$, across the main island, is explored. The criterion for stability using a slightly different definition of the mean residue $f = |R|^{2/Q}$ is that $f < 1$. The iterates are chosen by taking the island number $Q_n = s_n/r_n =$ integer $\pm 1/n$, where n runs through integer values greater than 1. Since the s_n/r_n are not chosen as convergents on an irrational surface, f varies widely, peaking at values larger than 1 near the primary resonances where s_n/r_n is integer. It falls to minimum values close to the irrationals. Examining the region near the $Q_n = 5$ (five-island) resonance, f is found to dip below unity for $E = \tfrac{1}{12}$. In contrast, for $E = \tfrac{1}{8}$, f climbs and stays above unity near the $Q_n = 5$ resonance, which implies that the KAM curves associated with the central island in that neighborhood have been destroyed. This is confirmed in Fig. 1.13c, in which we see the usual picture of a five-island chain, with a trajectory within this chain being able to communicate directly with the stochastic sea outside the chain. Greene (1979b) has given the estimate $E = 0.118$ for the breakup of a KAM surface near the five-island trajectory. We conclude that problems having the inherent analytical difficulties characteristic of the Hénon and Heiles potential can still yield to this numerical technique.

4.5. Renormalization for Two Resonances

We now describe a method (Escande and Doveil, 1981) for determining the destruction of a KAM torus that lies between two primary resonances. The method, in the spirit of the comments in Section 4.2, examines the structure near the KAM torus on finer and finer scales. With proper selection of the torus that is examined, the transition to strong (or "connected") stochasticity is determined.

Basically, the method proceeds by a renormalization of the original Hamiltonian to a series of new Hamiltonians of the same form, which describes successively higher-order resonances in the original system. At each step in the renormalization, the amplitudes of the two most important resonances, bounding the sought-for KAM torus, are examined. If the amplitudes tend to zero as the renormalization proceeds, then the KAM torus is shown to exist.

Following Escande and Doveil, we consider the Hamiltonian for a particle interacting with two waves

$$H_w = \frac{p^2}{2m} - V_1 \cos(k_1 x - \omega_1 t) - V_2 \cos(k_2 x - \omega_2 t) \tag{4.5.1}$$

and introduce the change of variables

$$\psi = k_1 x - \omega_1 t,$$

$$I = \frac{p/m}{v_2 - v_1}, \tag{4.5.2}$$

$$\Omega = k_1(v_2 - v_1),$$

where $v_1 = \omega_1/k_1$ and $v_2 = \omega_2/k_2$ are the phase velocities of the waves. This change of variables transforms (4.5.1) to

$$H_I = \tfrac{1}{2}I^2 - M \cos \psi - P \cos k(\psi - \Omega t), \tag{4.5.3}$$

where

$$k = \frac{k_2}{k_1} \tag{4.5.4}$$

is the wavenumber ratio,

$$M = \frac{V_1}{m(v_2 - v_1)^2} \tag{4.5.5}$$

is the amplitude of the "main" resonance, and

$$P = \frac{V_2}{m(v_2 - v_1)^2} \tag{4.5.6}$$

is the amplitude of the perturbing resonance. The Hamiltonian (4.5.3) for the motion of a pendulum perturbed by a wave with frequency Ω is the basic form that will be preserved after renormalization. For rotations in the unperturbed ($P = 0$) system, transforming to action-angle variables, we have the Hamiltonian $H_0(J)$, with J, θ, and the rotation frequency $\omega(J)$ given by

4.5. Renormalization for Two Resonances

(1.3.10), (1.3.11), and (1.3.13) for $\kappa > 1$. In the limit of uniform rotation ($M = 0$),

$$\omega(J) = J = I.$$

We consider now the existence of a KAM torus with inverse rotation number

$$Q(J) = \frac{k\Omega}{\omega(J)} \tag{4.5.7}$$

with the perturbation present. The Hamiltonian is

$$H = H_0(J) - P \cos k[\psi(J, \theta) - \Omega t], \tag{4.5.8}$$

where for rotation

$$\psi = \theta + \chi(J, \theta),$$

with χ, given in terms of elliptic functions, being 2π periodic in θ. This allows H to be written as a Fourier series

$$H = H_0(J) - P \sum_{n=-\infty}^{\infty} U_n(J) \cos[(k+n)\theta - k\Omega t]. \tag{4.5.9}$$

We see secondary resonances between harmonics of the rotation frequency $\omega(J_n)$ and the fundamental of the driving frequency $k\Omega$:

$$(k + n)\omega(J_n) - k\Omega = 0, \tag{4.5.10}$$

with

$$Q_n = \frac{k\Omega}{\omega(J_n)} + k + n \tag{4.5.11}$$

the inverse rotation number of the nth resonance at the action J_n. Figure 4.11 shows an example ($k = 2$) of these resonances in the original $I - \psi$ and transformed $J - \theta$ coordinates. The M resonance is located at $J_m = 0$ with the rotation modes above the separatrix value $J_s = 4\sqrt{M/\pi}$, and the P resonance ($n = 0$) at $J_p \approx 1$ (actually $\omega(J_p) = \Omega$). A few of the secondary resonances given by (4.5.10) for $n \geq 1$ are labelled in the figure, there being Q_n islands at resonance n.

Now consider a possible KAM torus at $J = J_0$, lying between the M and P resonances, with inverse rotation (island) number

$$Q(J_0) = \frac{k\Omega}{\omega(J_0)}. \tag{4.5.12}$$

Without loss of generality we can take[4]

$$Q > k + 1; \tag{4.5.13}$$

---

[4] If $Q < k + 1$ we exchange the V_1 and V_2 resonances in (4.5.1); this leads to $Q > k + 1$ for the exchanged system.

284 4. Transition to Global Stochasticity

Figure 4.11. Surfaces of section for a system with two primary resonances. (a) Original $(I - \psi)$ coordinates. (b) Transformed $(J - \theta)$ coordinates. Parameters are $X/Y = 1$, $S = 0.4$, and $k = 2$ (after Escande and Doveil, 1981).

4.5. Renormalization for Two Resonances

hence, the torus lies between the secondary resonances n and $n + 1$ (see Fig. 4.11), where $n > 1$ is the integer part of

$$z = Q - k = n + \delta k \tag{4.5.14}$$

and δk is the fractional part of z. It will be convenient to specify n and δk rather than Q.

To obtain the renormalized Hamiltonian associated with Q, (4.5.9) is averaged over all secondary resonances except n and $n + 1$. Since the system is accidentally degenerate, the excursion in action is small (see Section 2.4a) so we also expand H_0 to quadratic order in $\Delta J = J - J_0$ to obtain

$$H \approx H_0(J_0) + \omega(J_0)\Delta J + \tfrac{1}{2}G(J_0)(\Delta J)^2$$
$$- PU_n(J_0) \cos[(k + n)\theta - k\Omega t]$$
$$- PU_{n+1}(J_0) \cos[(k + n + 1)\theta - k\Omega t], \tag{4.5.15}$$

where G as usual is the nonlinearity parameter

$$G = \frac{\partial^2 H_0}{\partial J_0^2}. \tag{4.5.16}$$

Choosing the secondary resonance nearer to Q as $\bar{M}$ resonance, with the remaining secondary resonance the $\bar{P}$ resonance, and transforming away the constant and linear terms in ΔJ in the usual way, we get the renormalized Hamiltonian

$$\bar{H} = \tfrac{1}{2}\bar{I}^2 - \bar{M} \cos \bar{\psi} - \bar{P} \cos \bar{k}(\bar{\psi} - \bar{\Omega}t). \tag{4.5.17}$$

Here

$$\bar{\psi} = (k + n + \lambda)\theta - k\Omega t, \tag{4.5.18}$$

$$\bar{\Omega} = \frac{(2\lambda - 1)k\Omega}{(k + n + 1 - \lambda)}, \tag{4.5.19}$$

$$\bar{z} \equiv \bar{n} + \delta\bar{k} = \frac{(1 - \lambda - \delta k)}{(\delta k - \lambda)}, \tag{4.5.20a}$$

$$\bar{k} = \frac{(k + n + 1 - \lambda)}{(k + n + \lambda)}, \tag{4.5.20b}$$

$$\bar{M} = PU_{n+\lambda}\beta^2 G, \tag{4.5.20c}$$

$$\bar{P} = PU_{n+1-\lambda}\beta^2 G, \tag{4.5.20d}$$

where again, $\bar{n}$ and $\delta\bar{k}$ are the integer and fractional part of

$$\bar{z} = \bar{Q} - \bar{k} \tag{4.5.21}$$

and, in the above, λ is a unit step function of δk alone

$$\lambda(\delta k) = \begin{cases} 0, & \delta k < \tfrac{1}{2}, \\ 1, & \delta k \geq \tfrac{1}{2}, \end{cases} \quad (4.5.22)$$

and

$$\beta = \frac{(k+n)(k+n+1)}{k}.$$

Equations (4.5.20) define a renormalization transformation $\mathcal{T}$

$$\mathcal{T}: (\delta k, n, k, M, P) \to (\bar{\delta k}, \bar{n}, \bar{k}, \bar{M}, \bar{P}),$$

which can be iterated. This five-dimensional (non-Hamiltonian) mapping is very difficult to treat analytically. Escande and Doveil note that $\bar{\delta k}$ and $\bar{n}$ are functions of δk alone, so they study $\mathcal{T}$ at period 1 fixed points of the one-dimensional mapping, from (4.5.20a),

$$\bar{\delta k} = \begin{cases} (1 - \lambda - \delta k) \\ (\delta k - \lambda) \end{cases}_f, \quad (4.5.23)$$

where $\{\ \}_f$ denotes the fractional part. The fixed points δk_n^λ are found at

$$\delta k_n^0 = \tfrac{1}{2}[(n^2 + 2n + 5)^{1/2} - n - 1]$$
$$= [n+1, n+1, n+1 \ldots], \quad (4.5.24a)$$
$$\delta k_n^1 = \tfrac{1}{2}[(n^2 + 4n)^{1/2} - n]$$
$$= [1, n, 1, n \ldots], \quad (4.5.24b)$$

where the latter forms are the continued fraction expansions. At these fixed points, where the island number is

$$Q = Q_n^\lambda = k + n + \delta k_n^\lambda, \quad (4.5.25)$$

we have a three-dimensional mapping

$$\mathcal{M}_n^\lambda : (k, X, Y) \to (\bar{k}, \bar{X}, \bar{Y}),$$

where $X = 2\sqrt{M}$ and $Y = 2\sqrt{P}$, the unperturbed widths of the M and P resonances, are used for convenience instead of M and P. The mapping $\mathcal{M}_n^\lambda$ has two attracting fixed points, the first at $(k_n^\lambda, 0, 0)$ and the second at $(k_n^\lambda, \infty, \infty)$ (Escande and Doveil, 1981), where k_n^λ is the stable period 1 fixed point of the one-dimensional mapping (4.5.20b). Setting $\bar{k} = k = k_n^\lambda$ in (4.5.20b) and using (4.5.24) for the stable (positive) root,

$$k_n^\lambda = \delta k_n^\lambda + 1 - \lambda. \quad (4.5.26)$$

If the iterates of $\mathcal{M}_n^\lambda$ for the initial (k, X, Y) and the chosen n (and its corresponding λ) converge toward the first attracting point, then the continually renormalized values of X and Y tend to zero and a KAM torus must exist. If the iterates converge to the second attracting point, then there is a strong numerical indication that the torus is destroyed.

4.5. Renormalization for Two Resonances

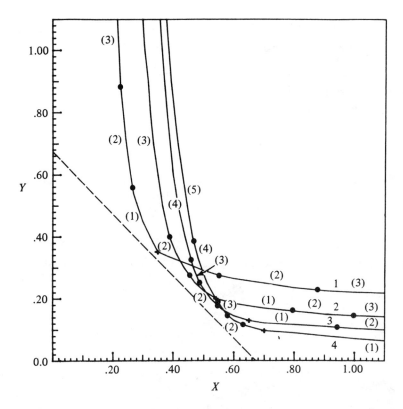

Figure 4.12. Y versus X for values of $k = 1, 2, 3$, and 4. The numbers in parentheses refer to the n values. In all cases, $\lambda = 1$. The dashed line is the result for $X + Y = 2/3$ (after Escande and Doveil, 1981).

Figure 4.12 shows, on a graph of Y versus X for fixed k, the upper envelope of the boundary between these two attracting regions. This is defined for a given X/Y ratio, as the maximum (with respect to n and λ, and thus Q_n^λ) value of $S = X + Y$ for which a KAM torus exists. The envelopes are plotted as curves of Y versus X with $k = 1, 2, 3$, and 4 as a parameter, the various segments of the curve being labelled with their n values (in all cases, $\lambda = 1$). To the left of the plus on each curve, the island number Q is found using (4.5.25); to the right of the plus, M and P have exchanged their roles so that in place of (4.5.25),

$$Q_n^\lambda = k + (n + \delta k_n^\lambda)^{-1}.$$

The envelopes vary continuously with k so that, for example, the $k = \tfrac{3}{2}$ envelope lies roughly midway between $k = 1$ and $k = 2$. It can also be shown that (k, X, Y) and $(1/k, Y, X)$ define the same system, so Fig. 4.12 gives the KAM threshold for $k = \tfrac{1}{2}, \tfrac{1}{3}$, and $\tfrac{1}{4}$ also. For comparison, the two-thirds rule $S = \tfrac{2}{3}$ is plotted as the dashed line in the figure.

Escande and Doveil (1981) have compared the theory to the stochasticity threshold found by direct numerical integration of the equations of motion for (4.5.3). Varying X/Y over the range 1–5 and k over the range 1–4, the predicted threshold was 3–10% lower than that calculated numerically. They have also compared the theory with Greene's calculation for the standard mapping, which has the Hamiltonian ($k = 1$, $\Omega = 1$)

$$H = \frac{I^2}{2} - \frac{S^2}{16} \sum_{m=-\infty}^{\infty} \cos(\psi - mt), \qquad (4.5.27)$$

where in (4.5.27) the primary resonances are separated by $\delta I = 1$ rather than 2π as in (4.1.24). Greene's results give $S = 0.6275$, corresponding to

$$K = (\pi S/2)^2 = 0.9716,$$

which, as mentioned in the preceding section, agrees very well with direct numerical iteration of the mapping. For the system with just two resonances,

$$H = \frac{I^2}{2} - \frac{S^2}{16} \cos \psi - \frac{S^2}{16} \cos(\psi - t), \qquad (4.5.28)$$

Fig. 4.12 gives ($X = Y$ and $k = 1$ here) $S = 0.6995$, or $K = 1.21$, with

$$Q = Q_1^1 = 2 + \frac{\sqrt{5} - 1}{2}$$

as the inverse rotation number of the envelope. This gives $\alpha = \frac{1}{2}(3 - \sqrt{5})$, which is one of the two golden mean KAM curves found by Greene. As expected for only two resonances, the limiting value of S (or K) is larger than for the mapping that has an infinite number of resonances. Numerical integration of the equations of motion for (4.5.28) yielded $S = 0.74$, consistent with Fig. 4.12, given the reduced accuracy of the numerical value as obtained from a Hamiltonian rather than from a mapping.

It should be apparent from the discussion leading to the renormalized Hamiltonian (4.5.17) that the method, though yielding a value close to the actual stochasticity threshold, is approximate. The major source of error is the consideration, at each stage in the renormalization, of just two of the infinite number of secondary resonances. This effect is seen clearly when comparing the infinite resonance system (4.5.27) with the two resonance system (4.5.28). For this extreme case, comparison of the numerical results indicates that the effect of the additional resonances is to reduce considerably the stochasticity threshold (from $S = 0.74$ to $S = 0.63$). Other sources of error, such as the quadratic expansion leading to (4.5.15), and the consideration of only the Q_n^λ's (and not the entire range of Q's) are less important than the neglect of all but two resonances.

The extension of the method to autonomous systems with two degrees of freedom, where M and P are functions of V, and the extension to treat the

4.6. Nonexistence of Invariant Curves

libration modes of the main resonance are relatively straighforward (Escande and Doveil, 1981). From the latter, one may hope to study bifurcation phenomena near fixed points of the system.

4.6. Nonexistence of Invariant Curves

In Section 3.2a we discussed the KAM theorem for the existence of invariant curves of Hamiltonians that are perturbed away from integrability. For the standard mapping, improvements on the sufficient condition for the existence of a phase-spanning KAM curve have determined a lower bound on the perturbation strength K (assumed positive) required for the existence of a phase-spanning KAM curve, of $K = \frac{1}{34}$ (Herman, 1985). This rigorous result, although much better than the earlier KAM estimates is still clearly well below the value of $0.9716\ldots$, obtained by various nonrigorous techniques, as discussed in the previous sections.

It is also possible to obtain a rigorous upper bound to the perturbation strength for the nonexistence of a phase-spanning KAM curve which is much closer to the actual value. A calculation, without mathematical details, was given by Lieberman and Lichtenberg (1972), giving nonintegrability for $K \geqslant 2$. A rigorous proof, using a theorem by Birkhoff (1920) was given by Mather (1984), who also improved the nonintegrability condition to $K = \frac{4}{3}$. Using Mather's technique, MacKay and Percival (1985) showed that the upper bound could be sharpened, using a constructive computer proof, obtaining $K > \frac{63}{64} \approx 0.9844$, which is quite close to the best nonrigorous result. This upper bound could presumably be improved by using more computer time, as we will describe, briefly, below. MacKay and Percival called the method *converse KAM*, as it was an upper rather than a lower bound on the perturbation strength that was being determined. Below, we first present the simplest estimate and then we discuss more recent work. We consider the particular case of the standard map (4.1.3), although the result in Lieberman and Lichtenberg (1972) was applicable to the general case of radial twist maps, such as the Fermi map (4.1.1).

For the standard map

$$I_{n+1} = I_n + f(\theta_n), \qquad (4.6.1\text{a})$$

$$\theta_{n+1} = \theta_n + I_{n+1}, \qquad (4.6.1\text{b})$$

with $f = K \sin \theta$, if we assume an invariant curve in θ, for all n, then we can parametrize the invariant curve by the action $I(\theta)$ and the rotation angle $\Omega(\theta)$. Substituting these quantities in (4.6.1), we obtain

$$\Omega(\theta) = I(\theta) + f(\theta). \qquad (4.6.2)$$

Equation (4.6.2) can also be written from (4.6.1b), in the form

$$\Omega(\theta) = I(\theta + \Omega(\theta)). \tag{4.6.3}$$

Without proof, Lieberman and Lichtenberg (1972) assumed that a necessary condition for the existence of phase spanning KAM curves is that Ω and I be continuous single-valued functions of θ. Differentiating (4.6.2) and (4.6.3) we obtain

$$\Omega_\theta = I_\theta + f_\theta, \tag{4.6.4}$$

$$\Omega_\theta = I_T(1 + \Omega_\theta), \tag{4.6.5}$$

where the subscript θ denotes differentiation with respect to θ, and I_T is I_θ evaluated at $\theta + \Omega(\theta)$. Writing (4.6.5) as

$$I_T = \frac{\Omega_\theta}{1 + \Omega_\theta} \quad \text{or} \quad \Omega_\theta = \frac{I_T}{1 - I_T} \tag{4.6.6}$$

we see that I can become double valued for $\Omega_\theta < -1$. Similarly, Ω can become double valued for $I_T > 1$, or, equivalently, $I_\theta > 1$. Solving for I_θ from (4.6.4), and substituting in the smallest value of $\Omega_\theta = -1$ we have

$$-(1 + f_\theta) < I_\theta < 1. \tag{4.6.7}$$

Removing I_θ from the inequality and rearranging, we have

$$f_\theta < -2 \tag{4.6.8}$$

as a *sufficient* condition for the *nonexistence* of a KAM curve and therefore an upper bound on the perturbation strength. For the standard map (4.6.1), $f = K \sin \theta$, and therefore the upper bound is

$$K = 2. \tag{4.6.9}$$

The rigorous proof of this result requires conditions on the deformations of an invariant curve that were supplied by a theorem of Birkhoff (1920) and used in the formal proof by Mather (1984). Birkhoff's work contained an additional concept, which allowed Mather to sharpen the result of (4.6.9), that of a *Lipshitz cone*. That is, the slope of a phase spanning invariant curve (called an *invariant circle* in the mathematical literature, because it encircles a phase space cylinder modulo 2π in the phase) is bounded at any point by a pair of slopes. The previous calculation had an implicit assumption that the invariant circle $I(\theta)$ whose slope was transformed to the vertical originated from a section with a slope of zero, i.e., a Lipshitz cone of angle zero. In fact, this is more stringent a requirement than necessary. Considering the minimum angle of the cone, Mather (1984) sharpened the criterion to $K = \frac{4}{3}$. MacKay and Percival realized that this result could be further sharpened by taking more iterations of the mapping. The result can no longer be expressed analytically, but by rigorously bounding the slope an upper bound could be obtained at each order of iteration. Iterating an orbit interval to a length of 196, they obtained a rigorous upper bound of $K = \frac{63}{64} \approx 0.9844$. This is quite close to the result of the most accurate estimate of the transition of $K \approx 0.9716$ obtained either from Greene's residue method (Section 4.4) or a renormalization calculation, and actually below the best numerical calculation of direct penetration of the golden mean orbit made by Chirikov of $K \approx 0.99$.

*4.7. Summary and Conclusions

In Table 4.2 we list the values of K, in the standard mapping, for which various phenomena occur that are related to the transition from local to global stochasticity. The actual transition at which the last KAM curve between two primary islands is lost has been fairly well established from various renormalization techniques to be $K = 0.9716\ldots$, which corresponds closely to the primary island rotation number at the fixed point $Q_0 = 2\pi/\omega_0 \cong \frac{1}{6}$. The other phenomena, such as the various "overlap" criteria, add insight and are useful estimates of transitions for various coupled oscillator problems. The rigorous upper bound of $K = 0.9844$ is very close to the estimates from renormalization techniques and probably closer than the best estimate ($K \cong 0.99$) from numerically iterating the mapping. Loss of linear stability at the primary resonance, which occurs at the considerably larger value of $K = 4$ gives a crude estimate for the transition; but more importantly, it gives an estimate of the parameter value for which the phase space is mostly stochastic.

Table 4.2. Phenomena Related to Transition to Global Stochasticity for the Standard Mapping[a]

Physical phenomena	K	Q_0[b]
Loss of linear stability of primary resonance	4	2
Overlap of primary resonances from perturbation analysis	$\left(\frac{\pi}{2}\right)^2$	3–4
Improved "overlap" criterion	1.2[c]	5
Second-order islands important	1.2	5
Renormalization of two resonances near golden mean	1.2	5–6
Rigorous upper bound	$\frac{63}{64}$	6
Renormalization techniques including loss of stability at rational iterates of golden mean	$0.9716\ldots$	6

[a] $I_{n+1} = I_n + K \sin \theta_n$; $\theta_{n+1} = \theta_n + I_{n+1}$.
[b] $Q_0 = 2\pi/\omega_0$ is the island number, where, from the tangent mapping, $\cos \omega_0 = 1 - K/2$.
[c] 1.06 with heuristic corrections (Chirikov, 1979).

For the standard mapping, from (4.1.31), the criterion of $Q_0 = \frac{1}{6}$, for the destruction of the last KAM surface, corresponds to a ratio of the full separatrix width to the distance between resonances of

$$\frac{2\Delta I_{max}}{\delta I} \approx \frac{2}{3}. \tag{4.7.1}$$

For mappings in which the island amplitudes are somewhat unequal, such as the Fermi acceleration mapping, the "two-thirds rule" of (4.7.1) can be generalized to

$$\frac{(\Delta I_{max})_1 + (\Delta I_{max})_2}{\delta I_{12}} \geq \frac{2}{3}. \tag{4.7.2}$$

For the Fermi acceleration mapping, the numerical position of the barrier given in Fig. 3.17 is consistent with this result, as this surface lies between the two lowest (in u) island chains for which (4.7.2) is not satisfied. When neighboring primary resonances have greatly different amplitudes, the standard mapping results do not apply. If only two resonances are present, a quite sharp estimate can be obtained graphically from Fig. 4.12, where phase differences are ignored, but large variations in period and amplitude are allowed. The figure also gives the rotation numbers, not necessarily the "golden mean", of the last KAM curve, thus locating the value of the unperturbed action at the barrier transition.

A model with two independent driving frequencies of arbitrary amplitude in the Fermi acceleration mapping (Howard *et al.*, 1982) indicates that the two-thirds rule works surprisingly well even for quite disparate island amplitudes. The two-resonance criterion in Fig. 4.12 yields equally good results. However, for the systems that have many resonances with differing amplitudes and frequencies, the two most important interacting resonances in the region of interest must be carefully chosen.

CHAPTER 5
Stochastic Motion and Diffusion

5.1. Introduction

In the globally stochastic region of the phase space for a system with two degrees of freedom, in which KAM curves spanning the phase coordinate do not exist, a complete description of the motion is generally impractical. We can then seek to treat the motion in a statistical sense. That is, the evolution of certain average quantities can be determined, rather than the trajectory corresponding to a given set of initial conditions (e.g., Chandrasekhar, 1943; Wang and Uhlenbeck, 1945). Such a formulation in terms of average quantities is also the basis for statistical mechanics (see, e.g., Penrose, 1970).

The procedures for obtaining a statistical description, which are discussed in this chapter, depend on a number of underlying mathematical results that take us beyond the scope of this monograph. However, some understanding of the main features of the results is useful. For the mathematically oriented, many of these results are presented in Arnold and Avez (1968). In Section 5.2, following their treatment, we describe these results in a less mathematically rigorous way, introducing the concepts of ergodicity, mixing, K- and C-systems. The existence of "randomness in deterministic systems," and the effects of roundoff error in numerical studies of chaotic behavior are also examined briefly. For other physically oriented, shorter expositions of this subject matter, the reviews of Ford (1975), Berry (1978), and Helleman (1980) are recommended.

Some of the basic questions concerned with the analysis of a stochastic region of a dynamical system are the following: How can the nonexistence of isolating integrals be unambiguously determined? What quantity or quantities are useful in describing the motion in the stochastic region? In what sense do numerical computations describe the real system behavior? What are the conditions under which the phase space description can be simplified such that a diffusion equation in action space alone can describe the important properties of the system behavior? What are the effects of extrinsically stochastic forces on the diffusion in the system? A related question, which we leave for the following chapter, is how do these properties change as the number of degrees of freedom is increased?

A number of investigators have addressed these questions, primarily during the past decade. Among these are Hénon and Heiles, Zaslavskii, Chirikov and associates, Froeschlé, Ford and associates, and Galgani and associates. Hénon and Heiles (1964), in their pioneering numerical study to determine the stochastic region for two coupled oscillators, used the observation that "Experience has shown previously that ... in a region occupied by curves the distance (between orbits) increases only slowly, about linearly, but ... in the ergodic region the distance increases rapidly, roughly exponentially." Subsequent developments have placed this observation on a more rigorous footing, relating the exponentiation to the positive Kolmogorov entropy and the Liapunov characteristic exponents. These exponents and the related Kolmogorov entropy have been computed for a number of specific examples. The technique, which we describe in detail in Sections 5.2 and 5.3, has become standard in testing for stochasticity.

In many problems, the density distribution in action space is the important quantity. Its dynamics is simplified by an average over phases to find the dynamical friction and diffusion coefficients for the action, which can then be used for determining its time evolution. If the phases are decorrelated after each mapping step, then a *random phase approximation* can be used which greatly simplifies the calculation. This procedure is described in Section 5.4.

Phase correlations, which always exist in a phase space with both regular and stochastic regions, complicate the calculation procedures. Close to the borders between stochastic and regular regions, the correlations become pronounced, requiring entirely different procedures for determining the diffusion. The existence of *accelerator modes* can also lead to nondiffusive behavior. These topics are treated in Section 5.5.

Other phenomena of interest for diffusion calculations are the effect of noise and the effect of slow changes of system parameters. These topics are treated in Sections 5.6 and 5.7.

5.2. Definitions and Basic Concepts

*5.2a. Ergodicity

For a mapping T, the time mean of an observable function in phase space $f(x)$ is defined by

5.2. Definitions and Basic Concepts

$$f^*(x) = \lim_{N \to \infty} \frac{1}{N} \sum_{n=0}^{N-1} f(T^n x). \tag{5.2.1}$$

It can be shown that for almost all x: (a) $f^*(x)$ exists; (b) f^* is an *invariant function*; i.e., is independent of initial conditions on a given orbit

$$f^*(T^n x) = f^*(x);$$

and (c) the space-mean value of $f^*(x)$ is equal to the space-mean value of $f(x)$. The space mean is defined as

$$\bar{f} = \int_{\mathcal{M}} f(x)\, d\mu, \tag{5.2.2}$$

where $\mathcal{M}$ having dimension M is the phase space of the system and μ is the invariant measure, i.e., $d\mu = P(x)\, d^M x$, where $P(x)$ is the invariant distribution. For a Hamiltonian system with x the canonical variables, $P = 1$; the invariant distribution for dissipative systems is obtained as described in Section 8.2. A dynamical system is defined as ergodic if

$$f^*(x) = \bar{f} \tag{5.2.3}$$

for almost all x. From this definition it is clear that the time mean of an ergodic system cannot depend on x. Since the initial f can be arbitrarily chosen, a system can only be ergodic if the system trajectory is accessible to all parts of the phase space, i.e., passes arbitrarily close to any point in phase space infinitely many times (the converse is not true). This excludes, for example, a system that includes invariant (KAM) curves, which is then called *decomposable*. Thus it is usually said that a stochastic region in a coupled two-degree-of-freedom system of the type we have been considering is not ergodic because the complete system also includes the invariant curves and is therefore decomposable.

Whether a system is considered ergodic depends on the subspace over which ergodicity is defined. An autonomous Hamiltonian cannot give rise to an ergodic flow on the entire phase space, because the energy is an exact invariant of the motion and thus the system point cannot explore all energies. However, the flow may be ergodic on a constant energy surface. If other invariants besides the energy do exist, then the system can be ergodic only on a subspace of the phase space that conserves all these invariants. In a sense, ergodicity is universal, and the central question is to define the subspace over which it exists.

We explore these ideas using the example of translations on a torus given by the mapping

$$T(\theta_1, \theta_2) = (\theta_1 + \omega_1, \theta_2 + \omega_2) \pmod{2\pi}, \tag{5.2.4}$$

where $\alpha = \omega_1/\omega_2$ is irrational. The orbit is everywhere dense on the torus and therefore its density can be given by an invariant function f^*. We can see that this implies ergodicity by comparing its Fourier coefficients through a mapping. Before the mapping the Fourier coefficients are

$$a_k = \int_M \exp(-i\mathbf{k} \cdot \mathbf{x}) f^*(\mathbf{x}) \, d\mu,$$

and after the mapping the Fourier coefficients of $f^*(T\mathbf{x})$ are

$$b_k = \int_M \exp[-i\mathbf{k} \cdot (\mathbf{x} - \boldsymbol{\omega})] f^*(\mathbf{x}) \, d\mu = \exp(i\mathbf{k} \cdot \boldsymbol{\omega}) a_k.$$

From the invariant property of f^* this implies that $a_k = b_k = 0$ except when $\mathbf{k} \cdot \boldsymbol{\omega} = 0$ for any $\mathbf{k} \neq 0$. Thus we concude that for α irrational ($\mathbf{k} \cdot \boldsymbol{\omega} \neq 0$) only a_0 remains, and f^* is a constant, as required for ergodicity.

Physically, the mapping (5.2.4) corresponds to the twist mapping in a surface of section $\theta_2 = $ const. (omitting the subscript 1):

$$J_{n+1} = J_n, \qquad \theta_{n+1} = \theta_n + 2\pi\alpha, \tag{5.2.5}$$

which is illustrated in Fig. 5.1a for $\alpha = \frac{1}{5}$ (rational) for two initial conditions (circles and crosses). Choosing $f(\theta)$ as shown in Fig. 5.1b, we see that $f^* \to 0$ for the circles and $f^* \to f_0$ for the crosses; thus the motion is not ergodic. For α irrational, the entire circle is covered by the orbit, and $f^*(\mathbf{x}) \to \bar{f}$, yielding ergodic behavior on the circle.

Now, the mapping (5.2.4) corresponds to the motion of an integrable Hamiltonian system on a given KAM torus, where θ_1 and θ_2 are the angle variables. Thus in a restricted sense we can speak of the motion as being ergodic on the invariant torus, while not ergodic within the entire phase space. From the above example of regular quasi-periodic motion, it is clear that ergodicity does not imply stochasticity. Our definition of ergodicity over a restricted portion of the complete phase space does admit the supposition that a stochastic region, such as a separatrix layer, defined by the invariant curves that bound it, is also ergodic. However, this may be of somewhat limited usefulness as it excludes adiabatic islands that are of finite measure and may be quite numerous in portions of the stochastic sea.

*5.2b. Liapunov Characteristic Exponents

Liapunov exponents play an important role in the theory of both Hamiltonian and dissipative dynamical systems. They provide a computable, quantitative measure of the degree of stochasticity for a trajectory. In addition, there is a close link between the Liapunov exponents and other measures of randomness such as the Kolmogorov entropy (see Section 5.2c) and the capacity (see Section 7.1d).

Roughly speaking, the Liapunov exponents of a given trajectory characterize the mean exponential rate of divergence of trajectories surrounding it.

5.2. Definitions and Basic Concepts

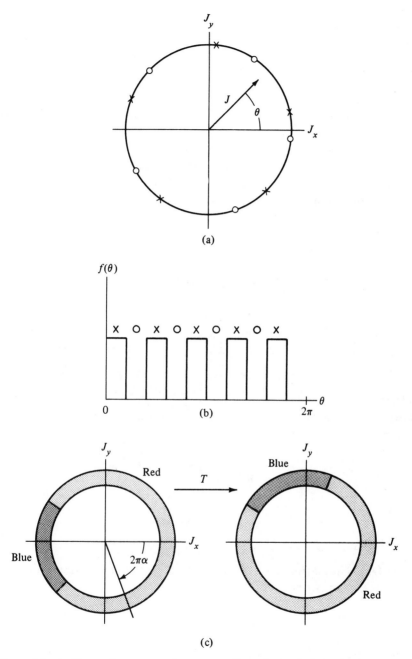

Figure 5.1. An illustration contrasting nonergodic and ergodic behavior for the twist mapping; in neither case is the motion mixing. (a) Two initial conditions ($\times$ and $\bigcirc$) on a rational surface; (b) initial choice of $f(\theta)$ showing that two initial conditions do not have the same time average; (c) initial conditions on an irrational surface; although the motion is ergodic the red and blue regions never mix.

Characterization of the stochasticity of a phase space trajectory in terms of the divergence of nearby trajectories was introduced by Hénon and Heiles (1964) and further studied by Zaslavskii and Chirikov (1972), Froeschlé and Scheidecker (1973), and Ford (1975).

The theory of Liapunov exponents (Liapunov, 1907) was applied to characterize stochastic trajectories by Oseledec (1968). The connection between Liapunov exponents and exponential divergence was given by Benettin et al. (1976) and by Pesin (1977), who also established the connection to Kolmogorov entropy. The procedure for computing the Liapunov exponents was developed by Benettin et al. (1980) and is described in Section 5.3. Here, following mostly Benettin et al. (1976, 1980) and Benettin and Galgani (1979), we give the properties of the Liapunov exponents. We refer the reader to the literature for the mathematical proofs.

We define the Liapunov exponents for the flow $x(t)$ generated by an autonomous first-order system

$$\frac{dx_i}{dt} = V_i(x), \quad i = 1, \ldots, M. \tag{5.2.6}$$

Consider a trajectory in M-dimensional phase space and a nearby trajectory with initial conditions x_0 and $x_0 + \Delta x_0$, respectively, as shown in Fig. 5.2a. These evolve with time yielding the tangent vector $\Delta x(x_0, t)$, with its Euclidean norm

$$d(x_0, t) = \|\Delta x(x_0, t)\|.$$

Writing for convenience $w = \Delta x$, the time evolution for w is found by linearizing (5.2.6) to obtain

$$\frac{dw}{dt} = \mathbf{M}(x(t)) \cdot w, \tag{5.2.7a}$$

where

$$\mathbf{M} = \frac{\partial V}{\partial x} \tag{5.2.7b}$$

is the Jacobian matrix of V. We now introduce the mean exponential rate of divergence of two initially close trajectories

$$\sigma(x_0, w) = \lim_{\substack{t \to \infty \\ d(0) \to 0}} \left(\frac{1}{t}\right) \ln \frac{d(x_0, t)}{d(x_0, 0)}. \tag{5.2.8}$$

It can be shown that σ exists and is finite. Furthermore, there is an M-dimensional basis $\{\hat{e}_i\}$ of w such that for any w, σ takes on one of the M (possibly nondistinct) values

$$\sigma_i(x_0) = \sigma(x_0, \hat{e}_i), \tag{5.2.9}$$

5.2. Definitions and Basic Concepts

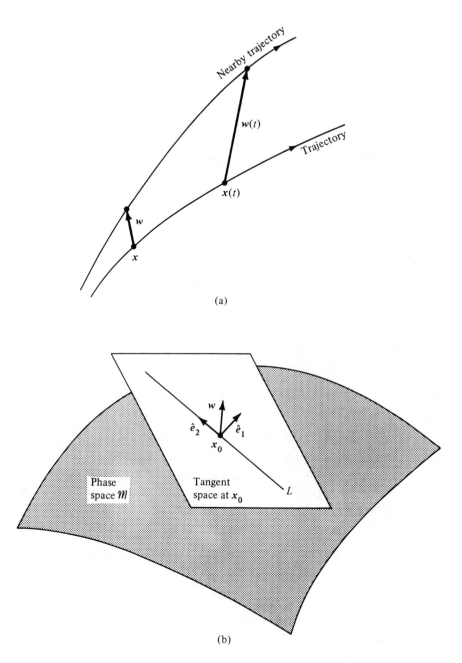

Figure 5.2. Definition of Liapunov characteristic exponents; $w(t) = \Delta x(t)$ is a tangent vector. (a) Two nearby initial conditions that separate as time evolves. (b) Tangent spaces of the Liapunov exponents σ_1 and σ_2 for a two-dimensional flow. For any tangent vector w that does not lie along the line L, $\sigma(w) = \sigma_1$. For w lying along L, $\sigma(w) = \sigma_2 < \sigma_1$ (after Benettin and Galgani, 1979).

which are the Liapunov characteristic exponents. These can be ordered by size,

$$\sigma_1 \geq \sigma_2 \geq \cdots \geq \sigma_M.$$

The Liapunov exponents are independent of the choice of metric for the phase space (Oseledec, 1968).

These properties are easy to understand for the special case of a periodic orbit with period τ. In this case the trajectory returns to x_0 after a time τ, and (5.2.7) defines a linear mapping $w(x_0, n\tau) \to w(x_0, (n+1)\tau)$ that can be written (suppressing the x dependence) as

$$w_{n+1} = \mathbf{A} \cdot w_n. \tag{5.2.10}$$

As described in Section 3.3, $\mathbf{A}$ has M (possibly complex) eigenvalues whose magnitudes can be ordered

$$|\lambda_1| \geq |\lambda_2| \geq \cdots \geq |\lambda_M|.$$

Let $\hat{e}_i$ denote the corresponding eigenvectors. Then for $w_0 = \hat{e}_i$, (5.2.10) implies

$$w_n = \lambda_i^n \hat{e}_i, \tag{5.2.11}$$

and so one concludes from (5.2.8) that

$$\sigma(\hat{e}_i) = \frac{1}{\tau} \ln|\lambda_i| = \sigma_i. \tag{5.2.12}$$

Furthermore, for

$$w_0 = c_1 \hat{e}_1 + \cdots + c_M \hat{e}_M,$$

it follows from (5.2.11) that the first nonvanishing coefficient c_i dominates in the subsequent evolution of w_n. We see from (5.2.8) that if $c_1 \neq 0$, then $\sigma(w_0) = \sigma_1$; if $c_1 = 0$ but $c_2 \neq 0$, then $\sigma(w_0) = \sigma_2$; and so on. In descending order, each Liapunov exponent "lives" in a space of dimensionality one less than that of the preceding exponent. This situation is illustrated in Fig. 5.2b for a two-dimensional flow. We see that for almost all w, $\sigma = \sigma_1$.

For a nonperiodic orbit, we cannot define eigenvalues and eigenvectors. Nevertheless, Oseledec (1968) has proven the existence of the basis vectors $\hat{e}_1, \ldots, \hat{e}_M$ and the Liapunov exponents for nonperiodic orbits. This result is not surprising since the periodic orbits are dense in the phase space. Thus a periodic orbit of arbitrarily large period can always be found arbitrarily close to any nonperiodic orbit.

For any flow generated by (5.2.6), at least one of the Liapunov exponents must vanish. This follows because in the direction along the flow, w grows only linearly with time. For a Hamiltonian flow, $x = (p, q)$ and

$$V = \left(\frac{-\partial H}{\partial q}, \frac{\partial H}{\partial p}\right).$$

5.2. Definitions and Basic Concepts

In this case the Liapunov exponents have a particular symmetry

$$\sigma_i = -\sigma_{2N-i+1}, \tag{5.2.13}$$

where $2N = M$ with N the number of degrees of freedom. For a periodic orbit, this can be seen by using the symmetry of the eigenvalues (3.3.21) in (5.2.12). It follows that there are at least two Liapunov exponents that vanish in the case of a Hamiltonian flow. On the constant energy surface, $M = 2N - 1$ and one of the vanishing exponents is absent.

Higher-Order Exponents. We have considered the Liapunov exponents of the vectors w, which are also called Liapunov exponents of order 1. Oseledec (1968) has generalized the concept to describe the mean rate of exponential growth of a p-dimensional volume in the tangent space, where $p \leq M$. We use the wedge operator notation

$$V_p = w_1 \wedge w_2 \wedge \cdots \wedge w_p$$

for the volume V_p of a p-dimensional parallelepiped whose edges are the vectors $w_1, w_2, \ldots, w_p$. Then,

$$\sigma^{(p)}(x_0, V_p) = \lim_{t \to \infty} \frac{1}{t} \ln \frac{\|V_p(x_0, t)\|}{\|V_p(x_0, 0)\|} \tag{5.2.14}$$

defines a Liapunov exponent of order p. Oseledec (1968) and Benettin et al. (1980) show that $\sigma^{(p)}$ is given as the sum of p Liapunov exponents of order 1. Just as we have $\sigma(x_0, w) = \sigma_1(x_0)$ for almost all w's, so we find here that $\sigma^{(p)}$ is the sum of the p largest Liapunov exponents

$$\sigma^{(p)} = \sigma_1^{(p)} = \sigma_1 + \sigma_2 + \cdots + \sigma_p \tag{5.2.15}$$

for almost all initial V_p's. Equation (5.2.15) is used to compute numerically the Liapunov exponents (see Section 5.3). For $p = M$, we obtain the mean exponential rate of growth of the phase space volume as

$$\sigma^{(M)} = \sum_{i=1}^{M} \sigma_i(x_0). \tag{5.2.16}$$

For a measure-preserving flow (Hamiltonian systems are a special case), we see that

$$\sum_{i=1}^{M} \sigma_i(x_0) = 0,$$

while for a dissipative system, this sum must be negative (see Section 7.1).

Relation Between Maps and Flows. Liapunov exponents are defined for maps as well as flows. For the M-dimensional map

$$x_{n+1} = F(x_n), \tag{5.2.17}$$

we can define σ^{map} through (5.2.8) with t replaced by n, leading to the M Liapunov exponents σ_i^{map}. Equivalently, one can introduce the eigenvalues $\lambda_i(n)$ of the matrix

$$\mathbf{A}_n = [\mathbf{M}(x_n) \cdot \mathbf{M}(x_{n-1}) \ldots \mathbf{M}(x_1)]^{1/n}, \tag{5.2.18}$$

where

$$\mathbf{M} = \frac{\partial \mathbf{F}}{\partial x}$$

is the Jacobian matrix of $\mathbf{F}$. The Liapunov exponents are then given by

$$\sigma_i^{\text{map}} = \lim_{n \to \infty} \ln|\lambda_i(n)|. \tag{5.2.19}$$

If a flow of dimension M generates a Poincaré map of dimension $M - 1$, then the Liapunov exponents for the map are proportional to the Liapunov exponents for the flow

$$\sigma_i^{\text{map}}(x) = \bar{\tau}\sigma_i(x), \tag{5.2.20}$$

where the zero exponent for the flow has been deleted. The constant of proportionality $\bar{\tau}$ is the mean time between successive intersections of the trajectory with the surface of section. (For an autonomous Hamiltonian system, two zero exponents are deleted in obtaining the exponents for the Poincaré section.)

5.2c. Concepts of Stochasticity

Mixing. The concept of mixing is very simple. Following Arnold, we take a shaker that consists of 20% rum and 80% cola, representing the initial distribution of the "incompressible fluid" in phase space. If we then stir or shake the liquid repeatedly, we expect that after the fluid has been stirred sufficiently often (n stirrings with $n \to \infty$) every part of the shaker, however small, will contain "approximately" 20% rum. When stated rigorously, this defines a *mixing system*. Mixing implies a *coarse-graining* of the phase space. A small but finite part of the phase space volume must be examined.

It is easy to show (see Arnold and Avez, 1968, p. 20) that mixing implies ergodicity. However, the converse is not true; ergodicity does not imply mixing. This can be seen at once from a counterexample of the translations on the torus, described by (5.2.4). If we paint one bundle of trajectories red and a neighboring bundle blue, clearly they will maintain their relationship to one another as they ergodically cover the torus, and are therefore not mixing. This property is illustrated in Fig. 5.1c for the twist mapping, where the blue and red portions of the circle do not mix even when α is irrational.

A mixing system approaches an equilibrium; $f(x) \to \bar{f}$, as $t \to \infty$. We

5.2. Definitions and Basic Concepts

illustrate this property with the baker's transformation

$$\begin{pmatrix} x_{n+1} \\ y_{n+1} \end{pmatrix} = \begin{cases} \begin{pmatrix} 2x_n \\ \frac{1}{2}y_n \end{pmatrix}, & 0 \leq x_n \leq \frac{1}{2}, \\ \begin{pmatrix} 2x_n - 1 \\ \frac{1}{2}(y_n + 1) \end{pmatrix}, & \frac{1}{2} \leq x_n \leq 1, \end{cases} \qquad (5.2.21)$$

which we have discussed qualitatively in Section 1.4. Following Ford (1975) and Berry (1978), we show that a "coarse-grained" observable function approaches an equilibrium state that is uniform, independent of initial conditions. From (5.2.21) the evolution equation for f is

$$f_{n+1}(x, y) = \begin{cases} f_n\left(\frac{x}{2}, 2y\right), & 0 \leq y \leq \frac{1}{2}, \\ f_n\left(\frac{x+1}{2}, 2y - 1\right), & \frac{1}{2} \leq y \leq 1. \end{cases} \qquad (5.2.22)$$

Since f is rapidly filamenting in y, we coarse-grain f_n by averaging over this fine-scale variation:

$$g_n(x) = \int_0^1 f_n(x, y) \, dy.$$

We shall see in Section 5.4 that this is equivalent to integrating over the rapidly varying phase. Performing this average on (5.2.22), we obtain

$$\begin{aligned} g_{n+1}(x) &= \int_0^{1/2} f_n\left(\frac{x}{2}, 2y\right) dy + \int_{1/2}^1 f_n\left(\frac{x+1}{2}, 2y - 1\right) dy \\ &= \frac{1}{2}\left[g_n\left(\frac{x}{2}\right) + g_n\left(\frac{x+1}{2}\right) \right]. \end{aligned} \qquad (5.2.23)$$

Thus the mapping of the coarse-grained f continually averages f in each half of the mapping. Such a procedure must lead to a final f that is uniform in x, as required for mixing.

We intuitively expect that the stochastic orbits that we have encountered in previous sections are mixing over the bounded portion of the phase space for which they exist. It is hard to prove the mixing property of dynamical systems rigorously. However, Sinai has proven this for the completely ergodic problem (see Section 1.4a) of the hard sphere gas (angle of incidence equals angle of reflection for circles on a two-torus). The proof depends on the continual defocussing of a bundle of trajectories as they encounter the negative curvature of the circle or sphere in the toroidal space (see Fig. 1.15a). Although the proof for the hard sphere gas does not verify our supposition of the mixing properties for general near-integrable systems, it gives us confidence to extend the rigorous result, empirically, by way of numerical examples.

K-Systems. K-systems, so called after Kolmogorov (1959), who studied them, are systems that have positive KS (Kolmogorov) entropy. The KS entropy (Krylov, 1950; Kolmogorov, 1959; Sinai, 1962, 1966) is defined by introducing a partition of phase space. Dividing the phase space Φ at time $t = 0$ into a set $\{A_i(0)\}$ of small cells, each having a finite measure, we let each cell evolve backwards under the flow for a unit time, obtaining a new set of cells $\{A_i(-1)\}$. Each element of the intersection of these two sets $B(-1) = \{A_i(0) \cap A_j(-1)\}$ typically has a smaller measure than that of an element $A_i(0)$. Continuing in this way, we can generate the elements of the set

$$B(-2) = \{A_i(0) \cap A_j(-1) \cap A_k(-2)\},$$

etc. Under what conditions does the measure of an element of $B(-t)$ decrease exponentially as $t \to \infty$? It can be shown that

$$h(\{A_i(0)\}) = -\lim_{t \to \infty} \frac{1}{t} \sum_{i=1}^{R_t} \mu[B_i(-t)] \ln \mu[B_i(-t)] > 0$$

for this to occur, where R_t is the number of elements of $B(-t)$, and $\mu[B_i(-t)]$ is the measure of each element. The quantity h has the significance of a mean exponential rate. The KS entropy h_k is the maximum of h over all measurable initial partitions of phase space, i.e., letting the initial cell size tend to zero.

Since the KS entropy is positive only when there is an exponential decrease in the average measure of an element of B (going backward in time), we should not be surprised to learn that the KS entropy is related to the average rate of exponential divergence of nearby trajectories (going forward in time), i.e., to the Liapunov exponents. Pesin (1977) has obtained the relation that can be written as

$$h_k = \int_{\mathcal{M}} \left[\sum_{\sigma_i(x) > 0} \sigma_i(x) \right] d\mu, \qquad (5.2.24)$$

where the sum is over all positive Liapunov exponents and the integral is over a specified region of phase space. The KS entropy is generally understood to be a measure applied to a single region of connected stochasticity, excluding regular regions, embedded islands, etc. In this case the σ's are independent of x and the integral over $\mathcal{M}$ is unity,

$$h_k = \sum_{\sigma_i > 0} \sigma_i. \qquad (5.2.25)$$

For a two-degree-of-freedom autonomous Hamiltonian, only σ_1 is greater than zero and

$$h_k = \sigma_1. \qquad (5.2.26)$$

Another interpretation is to apply (5.2.24) to the entire (compact) region

5.2. Definitions and Basic Concepts

of phase space (Benettin et al., 1976). In this case and for two degrees of freedom, h_k is the average over all phase space of the mean exponential rate of divergence of nearby trajectories. We know that $\sigma_1(x)$ vanishes in the regular regions. Suppose for simplicity that σ_1 is equal to a positive constant $\bar{\sigma}_1$ in the chaotic regions of a system. Then (5.2.24) implies

$$h_k = \mu_s \bar{\sigma}_1, \qquad (5.2.27)$$

where μ_s is the fraction of the phase space that is stochastic. As will be seen in Section 5.3, a numerical computation of $\sigma_1(x)$ can be used to test for stochasticity of a trajectory, and (5.2.26) or (5.2.27) can be used in conjunction with these computations to determine the KS entropy. We should note that σ_1 is not usually the same constant for all regions; each isolated region of stochasticity generally has a different value of σ_1. Thus if (5.2.24) is applied to the entire region of phase space, then the proper average over the values of σ_1 should be taken.

C-Systems. C-systems, studied by Anosov (1962, 1963) and sometimes called Anosov-systems after him, are characterized by a linear tangent space that is decomposable into three disjoint parts: (1) a space along the trajectory that has a vanishing Liapunov exponent; (2) a divergent space in which the trajectories diverge exponentially at a rate that is locally bounded from below for all initial conditions and all times; and (3) a convergent space in which the trajectories converge exponentially at a rate that is locally bounded from above for all initial conditions and all times. This is a very strong condition, which near-integrable Hamiltonian systems never satisfy. In fact, it has been shown that C-systems are structurally stable; i.e., a perturbed C-system remains a C-system.

Let us therefore investigate the properties of a C-system by means of a simple, if unphysical, example, of the type favored by mathematicians. Following Arnold, we consider the mapping T given by

$$\begin{pmatrix} x_{n+1} \\ y_{n+1} \end{pmatrix} = \begin{pmatrix} 1 & 1 \\ 1 & 2 \end{pmatrix} \begin{pmatrix} x_n \\ y_n \end{pmatrix}, \qquad \text{mod } 1. \qquad (5.2.28)$$

Within the unit square T is also the tangent map **A**, with det $\mathbf{A} = 1$, and the mapping is therefore area-preserving. In Fig. 5.3 (after Arnold and Avez, 1968, p. 6), we illustrate the distortion of the plane, together with the picture of Arnold's famous cat, carrying the mapping to T^2. It is clear that the procedure of mapping back onto the unit square continually filaments the cat (the phase space) and, furthermore, mixes the filaments. Using the results of Section 3.3c to calculate the eigenvalues and eigenvectors, we solve the characteristic equation

$$\begin{vmatrix} 1 - \lambda & 1 \\ 1 & 2 - \lambda \end{vmatrix} = 0$$

Figure 5.3. Arnold's cat mapping, showing the cat A transformed to TA and then to $T^2 A$. This is a C-system (after Arnold and Avez, 1968).

to obtain the eigenvalues

$$\lambda_{1,2} = \frac{3 \pm \sqrt{5}}{2} \tag{5.2.29}$$

and the eigenvectors from (3.3.7) as

$$\begin{aligned}\zeta_1 &= \left(\frac{2}{5-\sqrt{5}}\right)^{1/2}\left(\frac{\sqrt{5}-1}{2}\hat{x} + \hat{y}\right), \\ \zeta_2 &= \left(\frac{2}{5+\sqrt{5}}\right)^{1/2}\left(-\frac{\sqrt{5}+1}{2}\hat{x} + \hat{y}\right).\end{aligned} \tag{5.2.30}$$

From (5.2.30) we see that the ζ_1 direction is always expanding while the orthogonal direction ζ_2 is contracting, and, of course, $\lambda_1 = 1/\lambda_2$ and $0 < \lambda_2 < 1 < \lambda_1$, which follows from the area-preserving property. Since the expansion rate of two orbits is proportional to the distance apart of the orbits, we can emphasize the exponential divergence of the orbits by writing the eigenvalues as

$$\lambda_{1,2} = e^{\pm \sigma} \tag{5.2.31}$$

with $\sigma = \ln[(3 + \sqrt{5})/2]$, the same positive constant for all initial conditions. (Recall that, more generally, σ is a function of phase space coordinates, so that we must average over the trajectory to obtain a characteristic rate of growth, $\sigma_1(x)$, for a given trajectory.) From (5.2.31) we see that for the cat mapping, $\sigma_1(x) = \sigma > 0$ for all initial conditions, and thus the cat mapping is a K-system. It has been shown (Arnold and Avez, 1968, p. 75) that C-systems imply K-systems. For the cat mapping of (5.2.28), we have the particularly simple result

$$h_k = \sigma_1 = \ln[\tfrac{1}{2}(3 + \sqrt{5})].$$

Bernoulli Shifts. As a final example of chaotic behavior, we describe systems known as Bernoulli shifts. Let us consider a phase space partitioned into M cells. With probability p_i, we label each cell with a distinct color a_i. Assuming the system state changes every second, then the dynamical motion is described by a random sequence of colors. The "mapping" in this case is a shift of the sequence one place to the right. For example, coin tossing involves two colors: "0" for heads and "1" for tails, each with probability $\tfrac{1}{2}$, leading to motion described by a random binary sequence such as

$$0111011\ldots.$$

The usual expression for the entropy H_T for a sequence of length n generated by a Bernoulli shift having M colors is

$$H_T = -n \sum_{i=1}^{M} p_i \ln p_i,$$

which is maximized for the Bernoulli partition $p_i = 1/M$ yielding

$$H_T = n \ln M.$$

We can generate such a shift by means of the simple map

$$x_{n+1} = Mx_n, \quad \text{mod } 1,$$

so that we obtain a new member of the Bernoulli sequence after each iteration. The KS entropy for this linear map is easily found to be $h_k = \ln M$,

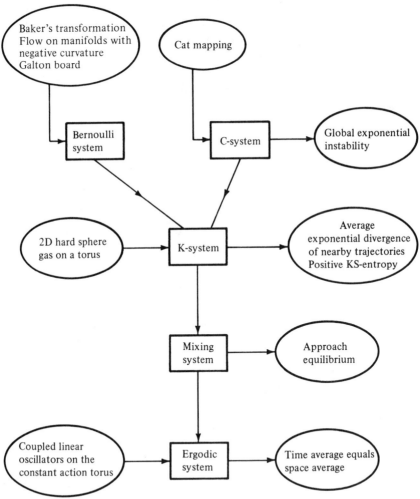

Figure 5.4. Systems exhibiting random behavior and their relationships. Near-integrable systems with their intermingled regions of regular and chaotic behavior do not fall into any of these categories.

5.2. Definitions and Basic Concepts

such that $H_T = h_k n$. Thus the KS entropy represents the rate of growth of H_T with "time" n. The entropy H_T increases without bound as $n \to \infty$.

We summarize the relationships among the various categories of random behavior in Fig. 5.4. The boxes show these, with the arrows to neighboring boxes giving the directions of implication, i.e., K-systems are mixing and ergodic, etc. The strongest notions of randomness are at the top, the weakest at the bottom. The distinguishing characteristic of each category is given to the right of the box. Some examples of systems having these properties are shown entering at the left of the box.

Near-integrable systems such as the Hénon-Heiles or Fermi problems, with their intermingled regions of regular and chaotic behavior, do not fall into any of these categories. However, in the chaotic regions, the motion near homoclinic points is locally equivalent to the baker's transformation, thus yielding the random behavior of a Bernoulli shift. Furthermore, as will be seen in Section 5.3, there is strong numerical evidence that each connected stochastic region, such as a separatrix layer, has positive KS entropy, and thus exhibits K-system behavior within its domain.

5.2d. Randomness and Numerical Errors

Random Sequences. We now ask in what sense random behavior can arise in deterministic dynamical systems. There has been considerable research in defining the meaning of randomness (Kolmogorov, 1965; Solomonoff, 1964; see Chaitin, 1975, for a simple exposition). An intuitive notion is that a random sequence must not display a pattern. We can devise tests for the frequency of a sequence of bits, pairs of bits, etc., which a random sequence must satisfy. If a sequence satisfies all tests for the absence of patterns, we could define it as "random."

The modern definition of randomness is based on the observation that the information embodied in a random sequence of bits cannot be "compressed," or reduced to a more compact form. Thus we cannot give a rule for calculating a random sequence of N bits that is substantially shorter than the trivial rule that merely reads back the sequence. These ideas can be put on a formal basis by first defining the minimal program size K_C of sequence S_N of N bits: K_C is the minimum number of bits required to instruct a computer C to generate the N bits of the sequence. This definition depends on C, but it can be decomposed into a machine-independent and a machine-dependent part

$$K_C \geq K_N(S) - C_C,$$

where $C_C \geq 0$ depends on C but is independent of S_N; the *complexity* $K_N(S)$ depends on S_N but is independent of C. Clearly the complexity is bounded between some constant (any computer must be given one command) and some maximum

$$K_N^{\max}(S) = N,$$

for which the computer copies back the N bits of the sequence.

It can be shown that there are sequences of N bits having maximum complexity. Furthermore, consider any computable test for randomness that is necessary but not sufficient (e.g., the ratio of 1's to 0's in the sequence must approach 1 as $N \to \infty$). Then the following theorem has been proven: A sequence can pass all computable tests for randomness if and only if

$$\lim_{N \to \infty} \left(\frac{K_N(S)}{N} \right) > 0.$$

If one defines a measure for all possible sequences of bits, by associating each sequence with a point on the real line $[0, 1]$, such that the measure of a set of sequences is the measure of the real line interval associated with the set of sequences, then one can further prove the theorem that almost all sequences are random, i.e., the set of nonrandom sequences has zero measure.

We can now see how a random sequence arises from a deterministic equation. Consider, for example, the one-dimensional mapping defined over the interval $[0, 1]$

$$x_{n+1} = 10x_n, \quad \text{mod } 1, \tag{5.2.32}$$

with initial condition

$$0 \leqslant x_0 < 1.$$

We divide the "phase space" $0 \leqslant x < 1$ into ten equal intervals having as labels a_i the integer part of $10x$. Then for example the initial condition $x_0 = 0.157643\ldots$ through the mapping (5.2.32) generates the sequence

$$\{a_1\} = 1, 5, 7, 6, 4, 3, \ldots.$$

Is this sequence random? We see that the answer hinges on whether the initial condition x_0 is random. However, by the previous theorem, "almost all" x_0's are random, and thus the sequence is random. It is conjectured that the motion near homoclinic points within a separatrix layer is random in this sense. Conversely, the regular motion on a KAM surface is nonrandom according to the above definition.

Roundoff Errors. In many problems of interest, computers have been used to study Hamiltonian mappings, with their intermingled structure of chaos and regular motion, by iterating the equations over millions of mapping periods. In what sense do computations using finite precision arithmetic, with their resulting roundoff and other "computer noise" errors, represent the dynamics? It is easy to verify the essential correctness of numerically

5.2. Definitions and Basic Concepts

generated phase plane plots, measures of stochasticity in the chaotic regions, etc., by varying the numerical precision, and such checks are generally a part of any numerical study. Greene (1979a), for example, has studied the effects of roundoff error on the determination of the transition to global stochasticity and found them to be negligible (see Section 4.4a for further details). Benettin *et al.* (1978) have studied the effects of computer errors on the calculation of time averages such as the Liapunov characteristic exponents, showing that for C-systems these effects are negligible. However, C-systems are structurally stable, and the effect of computer errors on other systems is still an open question.

On the other hand, certain features of the motion are drastically altered by finite precision arithmetic. As shown later in this section, the initial conditions are generally lost as the mapping is iterated, with the error in the initial condition increasing exponentially for the chaotic orbits and linearly for the regular orbits. For arithmetic and initial conditions with N-bit precision, the initial condition is lost completely after $n_{max} \sim \mathcal{O}(N)$ iterations for a chaotic orbit, and after $n_{max} \sim \mathcal{O}(2^N)$ iterations for a regular orbit. In both cases, the system will be far from its initial conditions if run forward n_{max} times and then backwards n_{max} times.

A more serious theoretical problem is the blurring of the distinction between chaotic and regular orbits when using finite bit precision, with a resulting coarse-graining of the dynamics. Suppose the dynamical state for a two-dimensional Hamiltonian mapping is described by two, N-bit numbers. Then there are only a finite number $M = (2^{2N})$ of system states, and, after $n \leqslant M$ iterations, the system must return to one of the system states previously visited. Thus if the dynamics (but not the phase space) is coarse-grained, then although the dynamics is described exactly, *all* trajectories are periodic orbits or cycles (this is true even for dissipative systems). In what sense can we characterize the motion in such systems as random?

These questions have been explored by Rannou (1974) using the integer mapping

$$a_{i+1} = a_i + b_i + \left[\frac{m}{2\pi} \left(1 - \cos \frac{2\pi}{m} b_i \right) \right], \quad \text{mod } m,$$

$$b_{i+1} = b_i - \left[\frac{Km}{2\pi} \left(\sin \frac{2\pi}{m} a_{i+1} + 1 - \cos \frac{2\pi}{m} a_{i+1} \right) \right], \quad \text{mod } m,$$

(5.2.33)

where $[X]$ represents the integer nearest to X, K is a stochasticity parameter, and a and b are integers with range

$$-m/2 \leqslant a, b \leqslant m/2.$$

The mapping (5.2.33) is a one-to-one mapping of the $M = m^2$ points in (a, b) space onto themselves. There is no roundoff or computer error of any kind. For the cases studied numerically, with m a multiple of 4, there are two period 1 fixed points $(0, 0)$ and $(-m/4, 0)$ for $K < 4$; the former is stable, the latter is unstable for small K. For $K = 1.3$, $m = 400$, we find 48 elliptically shaped invariant cycles with appreciable thickness that closely encircle the stable fixed point $(0, 0)$, and 9 cycles that have extremely long periods that fill rather uniformly the region surrounding the elliptic cycles and consequently appear to represent chaotic motion. For $K = 10$, there are only 12 (long) cycles and thus only chaotic motion appears to be present.

Rannou defines randomness for these periodic orbits as follows: there are $M!$ possible one-to-one mappings of M points onto themselves. The *random mapping* is defined by attributing the same probability $1/M!$ to each of these $M!$ mappings. For the random mapping it can be shown that

(1) The probability of obtaining a cycle of length n from a given point (a, b) is $1/M$, independent of n.
(2) The average length of the cycle originating at (a, b) is $\frac{1}{2}(M + 1)$.
(3) The average number of cycles of all lengths is nearly equal to $\ln M + \gamma$, where γ is Euler's constant.

Good numerical agreement is obtained (see Rannou, 1974) between these properties of the random mapping and the apparently chaotic motion seen for the cases $K = 1.3$ and $K = 10$ for mapping (5.2.33), and with $300 \leqslant m \leqslant 800$. Thus we conclude that chaotic motion is intrinsic to systems described by exact dynamics, independent of computer noise or roundoff errors. This strongly suggests that the chaotic motion observed in generic Hamiltonian systems is intrinsic to the dynamics and is not produced by discretization effects associated with finite precision arithmetic. On the contrary, as we have seen, discretization leads to periodic behavior. Computational experiments in which the precision is changed also tend to confirm this conclusion.

5.3. Determination of Liapunov Exponents and KS Entropy

We recall (see Fig. 5.2) the definition of the Liapunov characteristic exponent σ of a trajectory x and nearby trajectory $x + w$, where w is a tangent vector:

$$\sigma(x, w) = \lim_{\substack{t \to \infty \\ d(0) \to 0}} \left(\frac{1}{t}\right) \ln \frac{d(t)}{d(0)}. \tag{5.2.8}$$

Analytical and numerical calculations of the σ's (especially the maximum value of $\sigma = \sigma_1$ with respect to variations of w) are widely used as measures of the degree of stochasticity in a near-integrable system. We note first (Casati et al., 1980) that integrable Hamiltonian systems have all the σ's equal to zero. We illustrate this using the twist mapping, which for continuous time

5.3. Determination of Liapunov Exponents and KS Entropy

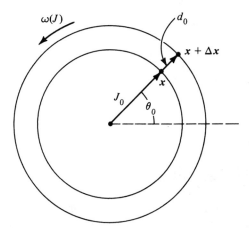

Figure 5.5. Illustrating the linear increase in the separation $d(t)$ of neighboring trajectories for the integrable twist mapping.

variation can be written as

$$J(t) = J_0, \tag{5.3.1a}$$
$$\theta(t) = \theta_0 + 2\pi\omega(J)t. \tag{5.3.1b}$$

As shown in Fig. 5.5, the orbits are circles, and the maximum separation of nearby trajectories is found when Δx lies along the radial direction. In this case two nearby trajectories have

$$\theta_1(t) = \theta_0 + 2\pi\omega_0 t, \tag{5.3.2a}$$
$$\theta_2(t) = \theta_0 + 2\pi\omega_0 t + 2\pi\omega_0' d_0 t, \tag{5.3.2b}$$

where ω_0' is the derivative with respect to J. The separation of neighboring trajectories is found from

$$\begin{aligned} d^2(t) &= J_0^2(\theta_2 - \theta_1)^2 + d_0^2 \\ &= d_0^2[(2\pi\omega_0'J_0 t)^2 + 1], \end{aligned} \tag{5.3.3}$$

which yields for large t a linear rate of separation. Then from (5.2.8)

$$\begin{aligned} \sigma &= \lim_{t \to \infty} \frac{1}{2t} \ln[(2\pi\omega_0'J_0 t)^2 + 1] \\ &= 0. \end{aligned} \tag{5.3.4}$$

For trajectories that separate at an exponential rate, characteristic of behavior in the chaotic regions of near-integrable systems,

$$d(t) \sim d_0 \exp[\sigma_1(x)t], \qquad \sigma_1 > 0, \tag{5.3.5}$$

and

$$\sigma_{\max} = \lim_{t \to \infty} \left(\frac{1}{t}\right) \ln \exp(\sigma_1 t) = \sigma_1. \tag{5.3.6}$$

In particular, for C-systems having the two eigenvalues

$$\lambda_{1,2} = e^{\pm \bar{\sigma}}, \qquad \bar{\sigma} > 0,$$

we have $\sigma_1 = \bar{\sigma}$.

5.3a. Analytical Estimates

For a reasonably simple dynamical problem, for which the adiabatic regions form a negligible part of the phase space, it is possible to determine analytically the maximal Liapunov exponent and the related KS entropy. Chirikov (1979) has done this for the standard mapping (see Section 4.1) for large K. If we linearize the standard mapping around any point, not necessarily a fixed point, then the characteristic equation gives the maximum (growing) eigenvalue, which, for large K, is

$$\lambda^+ = K |\cos \theta|. \tag{5.3.7}$$

We find σ_1 (and h_k, since the phase space is assumed essentially all stochastic) by averaging $\ln \lambda^+$ over the phase space. For this example, it is just an average over θ yielding

$$\sigma_1 = \frac{1}{2\pi} \int_0^{2\pi} \ln(|K \cos \theta|) \, d\theta = \ln \frac{K}{2}. \tag{5.3.8}$$

This analytical result was then compared with a numerical determination σ_n for initial conditions taken in the stochastic sea. At $K \cong 6$, which is about the smallest value that one could expect close agreement, Chirikov found $\sigma_n/\sigma_1 = 1.02$.

As we observed in Section 4.1, islets of stability can persist for arbitrarily large K for the standard mapping. These islets tend to become small with increasing K, but since they may exist for fixed points with arbitrarily large periods, the question arises as to whether they may occupy a significant fraction of the phase space and thus distort the calculation of entropy, even for large K. Chirikov investigated this question for the standard mapping in two ways. In the first, he subdivided the phase space into 100×100 bins and calculated the fraction μ_s of bins crossed by a trajectory with initial values in the stochastic sea. Clearly this "coarse-graining" procedure only gives useful results for relatively low K for which the islands are relatively large. In the second procedure, he chose arbitrary initial conditions for (say) $N = 100$ particles and computed the fraction μ_r whose value of KS entropy tended to zero. We expect $\mu_s + \mu_r$ to equal approximately unity. The numerical results are shown in Table 5.1. Chirikov concluded that the sum

5.3. Determination of Liapunov Exponents and KS Entropy

Table 5.1. Numerically Determined Fraction of the Phase Space that is Stochastic (μ_s) and Regular (μ_r) for the Standard Mapping[a]

K	8.888	6.59	6.21	5	4	3	2	1	0.5
μ_s	—	—	—	0.98	0.92	0.89	0.79	0.44	—
μ_r	$<10^{-4}$	0.0099	0.004	0.014	0.08	0.11	0.19	0.52	0.96

[a] After Chirikov (1979, Table 5.3).

of the stable region tends to zero area as K increases, so that stochastic concepts can be employed for large K.

*5.3b. Numerical Methods

For practical determination of the stochasticity of a given region of the phase plane, a numerical computation of the Liapunov exponents can be employed. The computation of the maximum exponent σ_1 has been used extensively as a test for stochasticity (e.g. Froeschlé and Scheidecker, 1973; Chirikov et al., 1973; Ford, 1975; Benettin et al., 1976; Shimada and Nagashima, 1979).

Calculation of σ_1. We have shown [see discussion following (5.2.12)] that $\sigma(x, w) = \sigma_1(x)$ for almost all initial tangent vectors w. To compute σ, we choose an initial w_0 and integrate the equation for w (5.2.7a) along an orbit $x(t)$, obtaining numerically the quantity

$$d(t) = \|w(t)\|,$$

where for convenience the initial norm d_0 is chosen to be unity. The difficulty is that if the norm of w increases exponentially with t, this leads to overflow and other computation errors. To circumvent this problem, we choose a small fixed time interval τ as shown in Fig. 5.6, and we renormalize w to a norm of unity every τ seconds. Thus we iteratively compute the values

$$d_k = \|w_{k-1}(\tau)\|, \tag{5.3.9a}$$

$$w_k(0) = \frac{w_{k-1}(\tau)}{d_k}, \tag{9.3.9b}$$

where $w_k(\tau)$ is obtained by integrating (5.2.7a), with the initial value $w_k(0)$, along the trajectory from $x(k\tau)$ to $x((k+1)\tau)$. From (5.2.8). we define the quantity

$$\sigma_n = \frac{1}{n\tau} \sum_{i=1}^{n} \ln d_i. \tag{5.3.10}$$

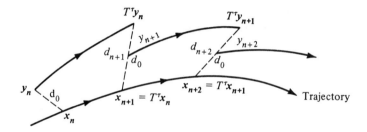

Figure 5.6. Numerical calculation of the maximal Liapunov characteristic exponent. Here $y = x + w$ and τ is a finite interval of time (after Benettin et al., 1976).

For τ not too large, it can be shown that

$$\sigma_\infty = \lim_{n \to \infty} \sigma_n = \sigma_1 \tag{5.3.11}$$

exists and is independent of τ. In the regular regions of near-integrable systems $\sigma_\infty = 0$, while in a connected stochastic region, σ_∞ is always positive and independent of the initial value of x. Since τ is arbitrary, (5.3.10) can be applied to calculate σ for maps as well as flows.

Following Benettin et al. (1976) we illustrate the method as applied to the Hénon and Heiles problem, which we have discussed in Section 1.4a. In Fig. 5.7, we give values of σ_n versus $n\tau$ for energy $E = 0.125$ (see Fig. 1.13) calculated for six orbits. Three of the orbits are taken with initial conditions that appear to lie on adiabatic trajectories and the other three orbits have initial conditions that appear to be in the large stochastic sea of Fig. 1.13. The expected result is substantiated, i.e., that the former three initial conditions have σ_n's that appear to approach zero with increasing n, while the latter three appear to approach a single positive value $\sigma_\infty = \sigma_1$.

From calculations such as those shown in Fig. 5.7, Benettin et al. (1976) numerically determined σ_∞ in the stochastic and regular regions as a function of energy. Their results are shown in Fig. 5.8, with both the nonvanishing values of σ_∞ and the vanishing values (on the abscissa) shown. The solid curve is an exponential fit to the nonvanishing values of σ_∞. Using Hénon and Heiles' (1964) estimate for $1 - \mu_s$ as a function of energy E, they obtain from (5.2.27) an approximate value of the KS entropy $h_k(E)$, shown dashed in Fig. 5.8. They speculate that $h_k(E) > 0$ for all $E > 0$, which seems reasonable from our knowledge of the generic behavior, although this result cannot be seen in Fig. 5.8 within their (and Hénon and Heiles') approximations. This question brings out an inherent difficulty in the method. It is possible, for example, that one of the initial conditions examined in Fig. 5.7, which appeared to be on a regular trajectory, actually resides in a narrow band of stochasticity bounded by KAM curves. Just as such a trajectory is

5.3. Determination of Liapunov Exponents and KS Entropy

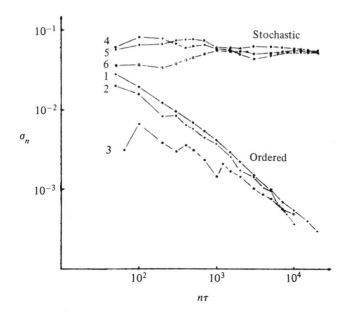

Figure 5.7. Behavior of σ_n at the intermediate energy $E = 0.125$ for initial points taken in the ordered (curves 1–3) or stochastic (curves 4–6) regions (after Benettin et al., 1976).

difficult to detect directly, it is difficult to detect by the present method, for which the leveling of σ_n to a very small constant value could lie beyond the number of iterations of the mapping actually computed. It is also possible that the region of energy near $E = 0.1$, where σ_∞ appears to be dropping to zero, may be a consequence of the appearance of KAM surfaces more finely dividing the phase space. We shall return to these question to discuss the rather more subtle case of more than two degrees of freedom in Chapter 6.

Calculation of All Liapunov Exponents. Following Benettin et al. (1980), we show how to calculate the entire set of Liapunov exponents for an M-dimensional flow. It is clear that any attempt to find σ_2, σ_3, etc., by choosing the initial tangent vector w along $\hat{e}_2, \hat{e}_3$, etc. (see Fig. 5.2b), must fail due to numerical error, which eventually brings $w(t)$ into alignment with $\hat{e}_1$. Instead, we choose an initial set of p orthonormal tangent vectors and numerically calculate the p-dimensional volume $V_p(t)$ defined by these vectors. The Liapunov exponent $\sigma_1^{(p)}$ of order p is then found using (5.2.14). Doing this for $p = 1, 2, \ldots, M$, we use (5.2.15) to find the Liapunov exponents $\sigma_1, \sigma_2, \ldots, \sigma_M$. A further difficulty now appears. As the tangent vectors evolve, the angle between any two vectors generally becomes too small for the numerical computations. Thus in addition to renormalizing the

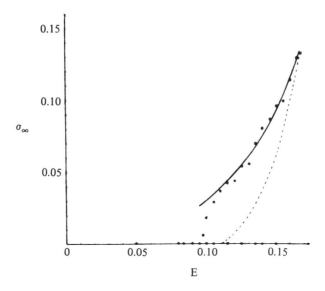

Figure 5.8. Nonvanishing values found for σ_∞ at different energies in the stochastic region (asterisks), vanishing values found in the ordered region (dots), exponential curve interpolating the nonvanishing values (continuous line), and tentative curve for KS entropy as a function of energy (dotted line) (after Benettin et al., 1976).

w's after each time interval τ, we must replace the evolved vectors by a new set of orthonormal vectors. These new vectors must be chosen to span the same subspace as the evolved set.

The above procedure can be systemized to calculate the evolution of all p-dimensional volumes at one time by computing the evolution of just M vectors and making the special choice of orthonormalizing by means of a Gram–Schmidt procedure. Recall $w_{k-1}(\tau)$ is the evolved tangent vector $w_{k-1}(0)$ along the trajectory from $x((k-1)\tau)$ to $x(k\tau)$. Using this, we first calculate for each time interval τ the quantities

$$d_k^{(1)} = \|w_{k-1}^{(1)}(\tau)\|, \tag{5.3.12a}$$

$$w_k^{(1)}(0) = \frac{w_{k-1}^{(1)}(\tau)}{d_k^{(1)}}. \tag{5.3.12b}$$

We then successively calculate, for $j = 2, \ldots, M$, the quantities

$$u_{k-1}^{(j)}(\tau) = w_{k-1}^{(j)}(\tau) - \sum_{i=1}^{j-1} [w_k^{(i)}(0) \cdot w_{k-1}^{(j)}(\tau)] w_k^{(i)}(0), \tag{5.3.12c}$$

$$d_k^{(j)} = \|u_{k-1}^{(j)}(\tau)\|, \tag{5.3.12d}$$

$$w_k^{(j)}(0) = \frac{u_{k-1}^{(j)}(\tau)}{d_k^{(j)}}. \tag{5.3.12e}$$

5.3. Determination of Liapunov Exponents and KS Entropy

During the $(k-1)$st time interval, the volume V_p increases by a factor of $d_k^{(1)} d_k^{(2)} \ldots d_k^{(p)}$. The definition (5.2.14) then implies

$$\sigma_1^{(p)} = \lim_{n \to \infty} \frac{1}{n\tau} \sum_{i=1}^{n} \ln(d_i^{(1)} d_i^{(2)} \ldots d_i^{(p)}). \tag{5.3.13}$$

Subtracting $\sigma_1^{(p-1)}$ from $\sigma_1^{(p)}$ and using (5.2.15), we obtain the pth Liapunov exponent

$$\sigma_p = \lim_{n \to \infty} \frac{1}{n\tau} \sum_{i=1}^{n} \ln d_i^{(p)}. \tag{5.3.14}$$

This is the relation that is actually used for the numerical calculations.

Benettin et al. (1980) have calculated the complete set of Liapunov exponents for several different Hamiltonian systems, including four and six-dimensional maps. We show here only their results for the three-degree-

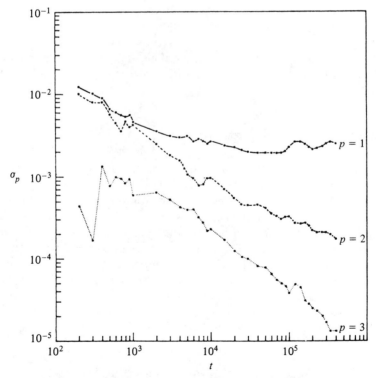

Figure 5.9. The three largest Liapunov exponents for the Contopoulos Hamiltonian (see text) with the initial point in a "small sea" (after Benettin et al., 1980).

of-freedom Hamiltonian of Contopolous et al. (1978)

$$H = \tfrac{1}{2}(q_1^2 + p_1^2) + \frac{\sqrt{2}}{2}(q_2^2 + p_2^2) + \frac{\sqrt{3}}{2}(q_3^2 + p_3^2) + q_1^2 q_2 + q_1^2 q_3.$$

The motion for this system is bounded when $H < 0.097$. For $H = 0.09$, the phase space is found to be divided roughly into three regions: a large stochastic region ("big sea") with $\sigma_1 \approx 0.03$, $\sigma_2 \approx 0.008$ and $\sigma_3 \approx 0$; a regular region ("islands") with $\sigma_1 = \sigma_2 = \sigma_3 = 0$; and an intermediate region ("small sea"). Figure 5.9 shows the calculation of the first three Liapunov exponents for an initial condition in the small sea. The last three exponents are found from (5.2.13). We see in the figure an apparent convergence of $\sigma_1 \approx 3 \times 10^{-3}$ and $\sigma_3 \approx 0$, the latter corresponding to the direction along the flow. The behavior of σ_2 is not completely clear.

As we describe in Section 6.2, these numerical computations for the small sea may be deceptive. For a system with three degrees of freedom, we expect the small and large seas to be connected, with the weak mechanism of Arnold diffusion transporting the phase point back and forth between them. After a sufficient time, an initial phase point in the small sea will be found in the large sea. Thus presumably $\sigma_1 \approx 0.03$, $\sigma_2 \approx 0.008$, for the small sea also, not in agreement with Fig. 5.9. This points out a major difficulty in numerically evaluating the Liapunov exponents: there is no *a priori* condition for determining the number of iterations n that must be used. Thus the numerical results must be supplemented by other techniques such as surface of section plots to clarify the basic mechanism of the chaotic motion in each case.

*5.4. Diffusion in Action Space

In regions of the phase space that are stochastic, or mostly stochastic with small isolated adiabatic islands, it may be possible to describe the evolution of the distribution function in action space (or velocity space) alone. This is, in fact, the problem of most practical interest. In the Fermi acceleration problem, for example, the motivation was to find a possible mechanism for acceleration of cosmic rays. The variations in the phases of the particles with respect to their accelerating fields are of little interest except as they are required for determining the heating rates and the final energy distribution. Similarly, for electron or ion cyclotron resonance heating, the heating rate and the energy distribution are the physically significant quantities.

In the following subsections, we consider the perturbed twist mapping

$$u_{n+1} = u_n + \epsilon f(u_{n+1}, \psi_n), \quad (5.4.1a)$$

$$\psi_{n+1} = \psi_n + 2\pi\alpha(u_{n+1}) + \epsilon g(u_{n+1}, \psi_n), \quad (5.4.1b)$$

with u and ψ the action-angle variables of the unperturbed system. In Section

5.4. Diffusion in Action Space

5.4a, we derive the Fokker–Planck equation for the distribution function $P(u, n)$ and discuss its validity. In Section 5.4b we make the random phase assumption to obtain the dynamical friction and diffusion coefficients, using the simplified and exact Fermi mappings of Section 3.4a to illustrate the calculations, and then in Section 5.4c we obtain the steady state and transient distributions. Higher-order corrections to the dynamical friction and diffusion coefficients, due to phase correlations are treated in Section 5.5.

*5.4a. The Fokker–Planck Equation

In this section we investigate in what sense the evolution of the distribution function $P(u, n)$ can be described by a stochastic process in the action u alone. Clearly we must confine our attention to a globally stochastic region of the phase space in which adiabatic islands do not exist or occupy negligible phase space volume. An example (see Section 3.4a, b and Figs. 1.14 and 3.16) is the region $u \lesssim u_s$ for the simplified Fermi problem

$$u_{n+1} = |u_n + f(\psi_n)|, \qquad (5.4.2a)$$

$$\psi_{n+1} = \psi_n + \frac{\Theta M}{u_{n+1}}, \quad \text{mod } \Theta, \qquad (5.4.2b)$$

where Θ is the phase recurrence interval (1 or 2π). In such a region, it may be possible to express the evolution of $P(u, n)$, the distribution in u alone, in terms of a Markov process in u (Wang and Uhlenbeck, 1945):

$$P(u, n + \Delta n) = \int P(u - \Delta u, n) W_t(u - \Delta u, n, \Delta u, \Delta n) \, d(\Delta u), \qquad (5.4.3)$$

where $W_t(u, n, \Delta u, \Delta n)$, the transition probability, is the probability that an ensemble of phase points having an action u at "time" n suffers an increment in action Δu after a "time" Δn. We make the additional assumption that $\Delta n \gg 1$ and that $\Delta u \ll (P^{-1} dP/du)^{-1}$; i.e., there exists a Δn such that in terms of (5.4.1)

$$1 \ll \Delta n \ll \left(\frac{\epsilon f_{\max}}{P} \frac{dp}{du}\right)^{-2}. \qquad (5.4.4)$$

Then we can expand the first argument of the integrand PW_t in (5.4.3) to second order in Δu to obtain (suppressing the unneeded arguments)

$$P(u - \Delta u) W_t(u - \Delta u) = P(u) W_t(u) - \frac{\partial (PW_t)}{\partial u} \Delta u + \frac{1}{2} \frac{\partial^2 (PW_t)}{\partial u^2} (\Delta u)^2.$$

Using this in (5.4.3) and noting that

$$\int W_t(u, n, \Delta u, \Delta n) \, d(\Delta u) = 1,$$

we obtain the Fokker–Planck equation

$$\frac{\partial P}{\partial n} = -\frac{\partial}{\partial u}(BP) + \frac{1}{2}\frac{\partial^2}{\partial u^2}(DP), \qquad (5.4.5)$$

where the frictional coefficient is

$$B(u) = \left(\frac{1}{\Delta n}\right)\int \Delta u\, W_t(u, n, \Delta u, \Delta n)\, d(\Delta u) \qquad (5.4.6)$$

and the diffusion coefficient is

$$D(u) = \left(\frac{1}{\Delta n}\right)\int (\Delta u)^2\, W_t(u, n, \Delta u, \Delta n)\, d(\Delta u). \qquad (5.4.7)$$

Relation Between Friction and Diffusion. Following Landau (1937), we show that for Hamiltonian systems

$$B = \frac{1}{2}\frac{dD}{du}, \qquad (5.4.8)$$

allowing the Fokker–Planck equation to be written as a diffusion equation

$$\frac{\partial P}{\partial n} = \frac{\partial}{\partial u}\left(\frac{D}{2}\frac{\partial P}{\partial u}\right). \qquad (5.4.9)$$

To prove (5.4.8), we write the change in action u to second order in the time

$$u(t + \Delta t) = u(t) + \dot{u}\Delta t + \tfrac{1}{2}\ddot{u}(\Delta t)^2,$$

where

$$\dot{u} = -\frac{\partial H}{\partial \psi},$$

$$\ddot{u} = -\frac{\partial^2 H}{\partial \psi^2}\dot{\psi} - \frac{\partial^2 H}{\partial \psi\, \partial u}\dot{u} - \frac{\partial^2 H}{\partial \psi\, \partial t}.$$

Using Hamilton's equations,

$$\ddot{u} = -\frac{\partial^2 H}{\partial \psi^2}\frac{\partial H}{\partial u} + \frac{\partial^2 H}{\partial \psi\, \partial u}\frac{\partial H}{\partial \psi} - \frac{\partial^2 H}{\partial \psi\, \partial t}$$

$$= -\frac{\partial}{\partial \psi}\left(\frac{\partial H}{\partial \psi}\frac{\partial H}{\partial u}\right) + \frac{\partial}{\partial u}\left(\frac{\partial H}{\partial \psi}\right)^2 - \frac{\partial}{\partial \psi}\left(\frac{\partial H}{\partial t}\right).$$

Thus $\Delta u = u(t + \Delta t) - u(t)$ is given to second order in Δt by

$$\Delta u = -\frac{\partial H}{\partial \psi}\Delta t + \tfrac{1}{2}(\Delta t)^2\left[\frac{\partial}{\partial u}\left(\frac{\partial H}{\partial \psi}\right)^2 - \frac{\partial}{\partial \psi}\left(\frac{\partial H}{\partial \psi}\frac{\partial H}{\partial u} + \frac{\partial H}{\partial t}\right)\right]. \qquad (5.4.10)$$

5.4. Diffusion in Action Space

Averaging over ψ, we obtain

$$\langle \Delta u \rangle_\psi = \tfrac{1}{2}(\Delta t)^2 \frac{\partial}{\partial u} \left\langle \left(\frac{\partial H}{\partial \psi}\right)^2 \right\rangle_\psi, \qquad (5.4.11)$$

where the other terms vanish since H is periodic in ψ. Similarly, to second order in Δt,

$$\Delta u \Delta u = \dot{u}^2 (\Delta t)^2 = \left(\frac{\partial H}{\partial \psi}\right)^2 (\Delta t)^2,$$

and averaging over ψ, we obtain

$$\langle \Delta u \Delta u \rangle_\psi = (\Delta t)^2 \left\langle \left(\frac{\partial H}{\partial \psi}\right)^2 \right\rangle_\psi. \qquad (5.4.12)$$

Comparing $\langle \Delta u \rangle_\psi$ and $\langle \Delta u \Delta u \rangle_\psi$, we immediately find the relation (5.4.8).

Validity of Fokker–Planck Equation. W_t is actually a function of the initial phase distribution as well as the initial action. However, we expect that a correlation "time" n_c (measured in number of collisions) exists, such that any reasonably smooth initial phase distribution relaxes to a uniform phase distribution after approximately n_c collisions. Provided Δn can be chosen considerably larger than n_c, W_t will be independent of the initial phase distribution. We illustrate that the general procedure is valid by showing that the direction of the exponentially growing eigenvector is primarily along the phase. This allows condition (5.4.4) to be fulfilled. For the simplified Fermi problem we use (3.3.48) together with the linearized matrix of (5.4.2)

$$\mathbf{A} = \begin{pmatrix} 1 & f' \\ -\rho & 1 - \rho f' \end{pmatrix},$$

with $\rho = \Theta M / u^2$, to obtain the eigenvector directions

$$(\Delta u, \Delta \psi) \propto (f', \lambda_{1,2} - 1).$$

For $\lambda_1 \gg 1$, the growing direction is primarily along the phase. From (3.3.52), $\lambda_1 \approx f'\rho$; $\lambda_1 \gg 1$ becomes increasingly well satisfied for u below the critical value of $u_s = \tfrac{1}{2}(\Theta M)^{1/2}$. To estimate n_c, we use $\mathbf{A}$ to obtain

$$\Delta u_{n+1} = \Delta u_n + f' \Delta \psi_n,$$
$$\Delta \psi_{n+1} = -\rho \Delta u_n + (1 - \rho f') \Delta \psi_n. \qquad (5.4.13)$$

Below the stochastic transition velocity u_s, ρ is greater than 2. Assume that the initial phases are spread over a small interval $\Delta \psi(0)$, and that $\Delta u(0) = 0$. Provided $\rho \gg 1$, the dominant terms in (5.4.13) then yield $\Delta \psi(n) \approx \rho^n \Delta \psi(0)$ and $\Delta u(n) \approx f' \times \rho^{n-1} \Delta \psi(0)$. Setting the phase spread $\Delta \psi(n)$ equal to the phase interval Θ (1 or 2π), we find

$$n_c = \frac{\ln[\Theta/\Delta\psi(0)]}{\ln \rho}, \qquad (5.4.14)$$

showing the weak logarithmic dependence of n_c on the initial phase interval and thus on the form of the initial phase distribution. In contrast, since $\Delta u(n_c) \ll u$, the action distribution remains localized while phase randomization occurs. Thus, provided $\Delta n \gg n_c$ we expect the Fokker–Planck description of the time evolution of P to be valid, and the Fokker–Planck coefficients B and D can then be obtained, approximately, from (5.4.6) and (5.4.7). Residual phase correlations will, however, exist over a single mapping period. A rigorous way to take these into account will be presented in Section 5.5a.

For $u > u_s$, invariants exist that relate action and phase independent of time. Such invariants exist within the adiabatic islands that cannot therefore be described by (5.4.5). In the sea surrounding the adiabatic islands, the process may be stochastic, but the random phase assumption is clearly not appropriate, as all phases are not available at a given action.

*5.4b. Transport Coefficients

We now calculate B and D for some simple mappings. Assuming that phase randomization occurs, we can average Δu over a uniform distribution of phases in (5.4.7) to obtain

$$D(u) = \frac{1}{2\pi} \int_0^{2\pi} d\psi [\Delta u(\psi)]^2. \qquad (5.4.15)$$

$B(u)$ is then obtained directly from (5.4.8). We illustrate this calculation for the simplified Fermi map (3.4.4) with sinusoidal velocity, for which $\Delta u = \sin \psi$, obtaining $D = \frac{1}{2}$ and $B = 0$. For the sawtooth velocity (3.4.7), using $\Delta u = \psi/2\pi - \frac{1}{2}$, we obtain similarly $D = \frac{1}{12}$ and $B = 0$.

In Fig. 5.10, we compare this latter result, obtained using the random-phase assumption, to the value of D obtained numerically as a function of n for 1000 particles placed at various initial values of velocity u. The sawtooth equations (3.4.7) are used for the computation. With $M = 10{,}000$, a stochastic transition is predicted at $u_s = 50$. For $u = 10, 20, 30$, and 40 and initial uniformly distributed phases, the phase correlation is found to be negligible so that $D(u, n) = D(u, 1) = \frac{1}{12}$ for $n > 1$. For $u = 50$, on the other hand, there is strong phase correlation so that D is not independent of n, even for $n \gtrsim 200$. For $u = 60$, another process also enters, as a number of the particles are initially trapped in adiabatic regions and do not take part in the diffusion process. Finally, if particles are not initially spread over all phases, there is

5.4. Diffusion in Action Space

Figure 5.10. Diffusion coefficient D as a function of n for the sawtooth wall velocity (3.4.7). Here $M = 10,000$, with 1,000 initial conditions placed at various initial velocities u; —, uniform initial phases; ---, $\Delta\psi = 0.001$ initially, $u = 30$ (after Lieberman and Lichtenberg, 1972).

a transient behavior for the first few n_c collisions during which phase randomization occurs, as seen from the dashed curve in Fig. 5.10 corresponding to $u = 30$ with an initial phase spread of $\Delta\psi = 0.001$. Calculating (5.4.14) for this case, we find $n_c \approx 6$, which corresponds roughly to the region over which the dashed curve differs significantly from $D = \frac{1}{12}$.

For the exact Fermi mapping (3.4.1) in which the physical displacement of the wall is included, we have seen in Section 3.4a that (u, ψ) are not the appropriate canonical variables. For this case we convert to the canonical pair (E, θ). This is most easily seen for the sinusoidal velocity impulse $f = \sin \psi$, $F = -\cos \psi$. Using these in (3.4.2), we obtain the area-preserving mapping for the moving wall

$$E_{n+1} = E_n + 2\sqrt{E_{n+1}} \sin \theta_n,$$

$$\theta_{n+1} = \theta_n + \frac{2\pi M}{\sqrt{E_{n+1}}} + \frac{\cos \theta_n}{\sqrt{E_{n+1}}}. \tag{5.4.16}$$

Then, letting $\bar{P}(E, n)$ be the energy distribution, we write the Fokker–Planck equation

$$\frac{\partial \bar{P}}{\partial n} = -\frac{\partial}{\partial E}(\bar{B}\bar{P}) + \frac{1}{2}\frac{\partial^2}{\partial E^2}(\bar{D}\bar{P}), \tag{5.4.17}$$

where

$$\bar{D} = \frac{1}{2\pi}\int_0^{2\pi} 4E \cos^2 \theta \, d\theta = 2E \tag{5.4.18}$$

and, from (5.4.8)

$$\bar{B} = \frac{1}{2}\frac{d\bar{D}}{dE} = 1. \tag{5.4.19}$$

The transport coefficients for the standard mapping (3.4.6) are $D_I = K^2/2$ and $B_I = 0$. Since this mapping locally approximates a generic mapping in the neighborhood of a period 1 fixed point, we can relate to D_I to D for that mapping. For example, the simplified Fermi map (3.4.4) with sinusoidal velocity can be linearized about $u_0 = M/m$ as in Section 4.1b to obtain the standard mapping (3.4.6) with $K = 2\pi M/u_0^2$ and $I_n = -K(u_n - u_0)$. Since $\Delta I = -K\Delta u$, the diffusion coefficients are related by $D_I = K^2 D$ or

$$D(u) = D_I(K(u))/K^2(u). \tag{5.4.20}$$

*5.4c. Steady-State and Transient Solutions

To obtain a steady-state solution to the Fokker–Planck equation, we assume perfectly reflecting barriers at $u = 0$ and $u = u_b$. Setting $\partial/\partial n = 0$ in (5.4.9) and taking the net flux to be zero,

$$\frac{D}{2}\frac{\partial P}{\partial u} = 0, \tag{5.4.21}$$

we obtain a uniform invariant distribution in velocity $P(u) = \text{const.}$ for the simplified map (3.4.4). For the exact map (5.4.16), we obtain similarly a uniform invariant distribution in energy $\bar{P}(E) = \text{const.}$ Introducing the velocity distribution $P(u)$ for the exact map through

$$P(u) \, du = \bar{P}(E) \, dE \tag{5.4.22}$$

and using $dE = u \, du$, we see that $P(u) = \text{const.} \times u$ for the exact map. In

5.4. Diffusion in Action Space

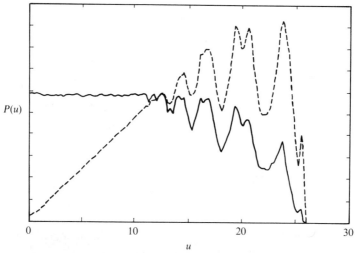

Figure 5.11. Comparison of velocity distribution $P(u)$ for the simplified Fermi map (3.4.4) [solid line] and the exact Fermi map (5.4.16) [dashed line].

Fig. 5.11, we compare the numerically calculated distributions for $M = 100$ and 5×10^6 iterations with these predictions. In the region below $u_s = (\pi M/2)^{1/2} \approx 12.5$, the predictions are verified. Above u_s, the distributions both fall off due to the presence of islands and higher-order correlations in the phase space, with the dips near the island centers. We return to these phenomena in Section 5.5.

We can also solve the transient Fokker–Planck equation. For the simplified Fermi mapping, substituting $D = \frac{1}{2}$ in (5.4.9), we have

$$\frac{\partial P}{\partial n} = \frac{1}{4} \frac{\partial^2 P}{\partial u^2}, \qquad (5.4.23)$$

which is a standard diffusion equation. With initial conditions of a δ-function at $u = 0^+$, we can solve (5.4.23) to obtain

$$P(u, n) = \frac{2}{(\pi n)^{1/2}} \exp\left(-\frac{u^2}{n}\right). \qquad (5.4.24)$$

This time development only holds, of course, until the particles begin to penetrate into the region with islands, $u > u_s$. The complete dynamics, including the transition region with adiabatic islands embedded in a stochastic sea, is very complicated and can only be solved numerically. We return to this in Section 5.5a.

5.5. The Effects of Correlations

As seen for generic twist mappings in Figs. 3.6 and 3.8 (Section 3.2d) and for a perturbed linear oscillator in Fig. 3.26 (Section 3.6b), the island structure embedded in a stochastic region near a KAM curve can be exceedingly complex. We might expect this structure to lead to long time correlation of stochastic orbits in the neighborhood of KAM curves, and this is, indeed, what happens. The transport coefficients in Section 5.4b were determined using the random phase assumption applied to the single step jump in the action $\Delta u_1 = u_1 - u_0$. However, as pointed out in Section 5.4a, the Fokker–Planck description of the motion is valid only in the limit $n \gg n_c$, where n_c is the number of steps for phase randomization to occur. We should therefore consider the jump $\Delta u_n = u_n - u_0$, where $n > n_c$. In Section 5.5a, we develop expressions for the transport coefficients in terms of Δu_n, thus obtaining, in the limit of large stochasticity parameter K, higher-order corrections to the single-step transport coefficients.

An alternative method for determining the diffusion near K_{crit} is based on calculation of the flux through just-broken KAM tori, or *cantori*. This approach, which provides a different perspective on the diffusion process, is described in Section 5.5b. Other corrections to D, arising from the long-time correlations near the edges of the islands, usually have minor effect on the diffusion, because the fastest diffusing particles dominate the overall diffusion. We consider these effects in Section 5.5c. However, we saw in Section 4.1b that there are ranges of K in the standard map which have stable fixed points that jump by $2\pi m$, m integer, in action on each mapping iteration. Correlations near these modes, called *accelerator modes*, cause streaming in phase space resulting in enhanced diffusion. We treat these phenomena in Section 5.5d.

5.5a. Higher-Order Transport Corrections

In this subsection, we obtain corrections to the single-step transport coefficients. The corrections were first obtained using Fourier techniques for the standard mapping in the limit of large K by Rechester and White (1980) and for any K by Rechester et al. (1981). However, these methods involve the introduction of external stochasticity into the canonical mapping equations. We defer consideration of their results to Section 5.6, where the effects of external stochasticity are considered. The Fourier technique can also be used without introducing external stochasticity, as has been done by Abarbanel and Crawford (1981), and we follow their treatment here. Abarbanel has also shown how the method can be applied to the wave-particle interaction problem (see Section 2.2b) in the limit of large wave fields. Since the standard mapping locally approximates wide classes of mappings, as we have discussed in Section 4.1b (see also Lichtenberg et al.,

5.5. The Effects of Correlations

1980), we discuss how the results obtained here can be applied to more general mappings.

Fourier Technique. We develop the Fourier technique using the standard mapping

$$I_{n+1} = I_n + K \sin \theta_n,$$
$$\theta_{n+1} = \theta_n + I_{n+1}.$$

For one step

$$\Delta I_1 = K \sin \theta_0, \tag{5.5.1}$$

and the transport coefficients are

$$F_{QL} = F = \frac{1}{2\pi} \int_0^{2\pi} \Delta I_1 \, d\theta_0 = 0, \tag{5.5.2a}$$

$$D_{QL} = \frac{D}{2} = \frac{1}{4\pi} \int_0^{2\pi} (\Delta I_1)^2 \, d\theta_0 = \frac{K^2}{4}. \tag{5.5.2b}$$

These are often referred to as the quasilinear results.[1]

To obtain the higher-order corrections, we calculate the diffusion coefficient

$$D_n = \frac{\langle (\Delta I_n)^2 \rangle_{I_n, \theta_n}}{2n}, \tag{5.5.3}$$

in terms of the conditional probability density W that an initial state (I_0, θ_0) at $n = 0$ evolves to a final state (I_n, θ_n) at step n,

$$D_n = \frac{1}{2n} \int W(I_n, \theta_n, n | I_0, \theta_0, 0)(I_n - I_0)^2 \, dI_n \, d\theta_n. \tag{5.5.4}$$

We need the recursion property

$$W(I, \theta, n | I_0, \theta_0, 0) = \int dI' \, d\theta' \, W(I, \theta, n | I', \theta', n - 1)$$
$$\times W(I', \theta', n - 1 | I_0, \theta_0, 0), \tag{5.5.5}$$

where, from the mapping equations,

$$W(I, \theta, n | I', \theta', n - 1) = \delta(I - I' - K \sin \theta') \delta(\theta - \theta' - I' - K \sin \theta'). \tag{5.5.6}$$

An attempt to calculate D_n directly from (5.5.4) to (5.5.6) by repeated iterations of the mapping becomes rapidly very tedious. The procedure can

---

[1] The factor of 1/2 which appears in the Fokker–Planck equation (5.4.5) is absorbed in the definition of the quasilinear diffusion coefficient.

be carried out for $n = 2$, giving interesting oscillations in D_2 with I_0, but it is not clear what the applicability of such a result is. For larger n we must turn to a technique that becomes simpler as n becomes large. We shall see that representing the diffusion in a Fourier transform space satisfies this requirement.

Expanding W in a Fourier series in θ and a Fourier integral in I,

$$W(I, \theta, n | I_0, \theta_0, 0) = \sum_m \int dq \, \exp(im\theta + iqI) a_n(m, q), \quad (5.5.7)$$

where the Fourier coefficient a_n is also a function of I_0 and θ_0.

We are interested in D_n for n large. Since I_n varies as $\sqrt{n}$, we see that only the quadratic term in I_n survives in (5.5.4); then for $n \gg 1$,

$$D_n = \frac{1}{2n} \int d\theta \, dI \, I^2 W(I, \theta, n | I_0, \theta_0, 0). \quad (5.5.8)$$

Using (5.5.7) in (5.5.8), the integral over θ picks out the $m = 0$ term of $a_n(m, q)$. After integrating twice by parts on I, to eliminate the I^2 term in (5.5.8), we have

$$D_n = -\frac{2\pi}{2n} \int dq \, dI \, \frac{\partial^2}{\partial q^2} [a_n(0, q)] e^{iqI}.$$

The integral over I yields $2\pi\delta(q)$. The q integration then just gives the integrand

$$D_n = -\frac{4\pi^2}{2n} \frac{\partial^2 a_n(0, q)}{\partial q^2} \bigg|_{q=0}. \quad (5.5.9)$$

We now obtain the recursion relation for the a_n's, which we need to evaluate (5.5.9). From (5.5.7),

$$a_n(m, q) = \frac{1}{(2\pi)^2} \int d\theta \, dI \, \exp(-im\theta - iqI) W(I, \theta, n | I_0, \theta_0, 0). \quad (5.5.10)$$

For $n = 0$, putting

$$W = \delta(I - I_0)\delta(\theta - \theta_0),$$

$$a_0 = \frac{1}{(2\pi)^2} \exp(-iqI_0 - im\theta_0). \quad (5.5.11a)$$

For $n > 0$, using (5.5.5) in (5.5.10)

$$a_n(m, q) = \frac{1}{(2\pi)^2} \int d\theta \, dI \, \exp(-im\theta - iqI) \int d\theta' \, dI' \delta(I - I' - K \sin \theta')$$

$$\times \delta(\theta - \theta' - I' - K \sin \theta')$$

$$\times \int dq' \sum_{m'} \exp(im'\theta' + iq'I') a_{n-1}(m', q'). \quad (5.5.11b)$$

5.5. The Effects of Correlations

We use the δ-functions to do the θ and I' integrations

$$a_n(m, q) = \frac{1}{(2\pi)^2} \sum_{m'} \int dq' \, d\theta' \, dI \, a_{n-1}(m', q')$$
$$\times \exp[-im(\theta' + I) - iqI + iq'(I - K \sin \theta') + im'\theta']. \quad (5.5.12)$$

Integrating over I yields a factor

$$2\pi\delta(-m - q + q')$$

so that the q' integration can be performed, yielding

$$a_n(m, q) = \frac{1}{2\pi} \sum_{m'} \int d\theta' a_{n-1}(m', q') \exp[i(m' - m)\theta' - iq'K \sin \theta'], \quad (5.5.13)$$

where

$$q' = q + m.$$

Using[2]

$$\exp(iq'K \sin \theta') = \sum_{l=-\infty}^{\infty} \mathcal{J}_l(|q'|K) \exp(il\theta' \, \text{sgn} \, q'),$$

where $\mathcal{J}_l$ is the ordinary Bessel function, the integration over θ' yields[3]

$$a_n(m, q) = \sum_{l, m'} \mathcal{J}_l(|q'|K)\delta_k(m' - m - l \, \text{sgn} \, q')a_{n-1}(m', q'). \quad (5.5.14a)$$

Doing the sum over m' then yields the recursion relation

$$a_n(m_n, q_n) = \sum_{l_n} \mathcal{J}_{l_n}(|q_{n-1}|K)a_{n-1}(m_{n-1}, q_{n-1}), \quad (5.5.14b)$$

where the δ_k-function in (5.5.14a) implies

$$m_n = m_{n-1} - l_n \, \text{sgn} \, q_{n-1}, \quad (5.5.15a)$$

and from (5.5.13)

$$q_n = q_{n-1} - m_n. \quad (5.5.15b)$$

Iterating (5.5.14b) n times yields a_n in terms of a_0:

$$a_n(m_n, q_n) = \sum_{l_n \ldots l_1} \mathcal{J}_{l_n} \mathcal{J}_{l_{n-1}} \cdots \mathcal{J}_{l_1} a_0(m_0, q_0). \quad (5.5.16)$$

Fourier Paths. The set of n integers $\{l_n \ldots l_1\}$ defines a path in the (m, q) Fourier space through the relations (5.5.15). For the calculation of D_n, (5.5.9) shows that the path must end at $m_n = 0$ and $q_n = 0$. A typical path is shown in Fig. 5.12a.

The arguments of the Bessel functions in (5.5.16) are $K|q_i|$. Thus for

---

[2] $\text{sgn}(x) = 1$ if $x \geq 0$ and $\text{sgn}(x) = -1$ if $x < 0$.
[3] $\delta_k(x) = 1$ if $x = 0$; otherwise $\delta_k(x) = 0$.

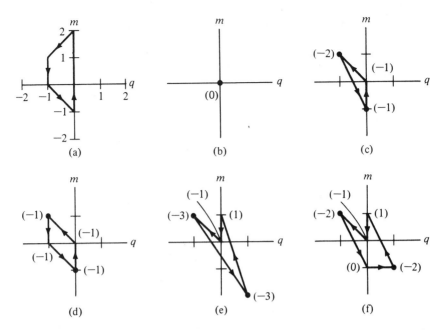

Figure 5.12. Paths in Fourier space (m, q). (a) A typical path that ends near the origin at $(m, q) = (0, 0+)$. (b) The quasilinear path that remains at $(0, 0+)$. (c) The order $K^{-1/2}$ excursion from $(0, 0+)$. (d)–(f): The order K^{-1} excursions from $(0, 0+)$. The numbers in parentheses give the values of l_n, the Bessel function modulus, corresponding to (m_{n+1}, q_{n+1}) at the vertices.

$K \to \infty$, the Bessel functions are small, falling off as $K^{-1/2}$, unless $q_i \to 0$. Thus, for large K, the dominant terms have all $q_i \to 0$. In this limit, the Bessel functions have small argument, and one term dominates with that term having all $l_i = 0$. From (5.5.15), this corresponds to a path of n steps that stays at the origin in Fourier space, as shown in Fig. 5.12b. For this path,

$$a_n(0, q) = \frac{1}{(2\pi)^2} [\mathcal{J}_0(Kq)]^n \exp(-iqI_0). \tag{5.5.17}$$

Using (5.5.9) and expanding $\mathcal{J}_0$ to quadratic terms,

$$D_n = -\frac{1}{2n} \frac{\partial^2}{\partial q^2} \left(1 - \frac{K^2 q^2}{4}\right)^n (1 - iqI_0 + \cdots)\Big|_{q=0}$$

or

$$D_n = -\frac{1}{2n} \frac{\partial^2}{\partial q^2} \left(1 - \frac{nK^2 q^2}{4} + \cdots\right)(1 - iqI_0 + \cdots)\Big|_{q=0}.$$

5.5. The Effects of Correlations

Table 5.2 Relation Between m and q

Step	m	q
r	0	q_r
$r+1$	m_1	$q_r - m_1$
$r+2$	m_2	$q_r - m_1 - m_2$
$r+3$	m_3	$q_r - m_1 - m_2 - m_3$
$r+4$	m_4	$q_r - m_1 - m_2 - m_3 - m_4$
$r+5$	m_5	$q_r - m_1 - m_2 - m_3 - m_4 - m_5$

Finally in the limit of n large,

$$D_n \to D_{QL} = \frac{K^2}{4},$$

which is the random phase or quasilinear result (5.5.2b).

Consider now the paths that leave the origin. To find the next correction to the quasilinear result, we want to stay near the origin except for one short excursion away. There are two paths that leave the origin at step r, make a three-step excursion, and return to the origin at stop $r+3$. To find these, we use the information in Table 5.2, generated by successive applications of (5.5.15). From this table, at step $r+3$, we require $m_3 = 0$ and $q = q_r$; hence,

$$m_1 + m_2 = 0$$

with the two solutions[4] $m_1 = \pm 1$, $m_2 = \mp 1$. The path corresponding to $m_1 = 1$, $m_2 = -1$ is shown in Fig. 5.12c. The l values for $q_r = 0+$ are obtained from (5.5.15a) and are indicated on the figure. The contribution to a_n follows immediately from (5.5.14) or its iterated form (5.5.16)

$$a_n = \frac{1}{(2\pi)^2} [\mathcal{J}_0(Kq)]^{n-3} \mathcal{J}_{-1}(K|q_r|) \mathcal{J}_{-2}(K|q_{r+1}|) \mathcal{J}_{-1}(K|q_{r+2}|) \exp(-iqI_0).$$

With $|q_r| = |q_{r+2}| = q$ and $|q_{r+1}| = 1$, and making the Bessel function orders positive, we have

$$a_n = \frac{1}{(2\pi)^2} [\mathcal{J}_0(Kq)]^{n-3} [\mathcal{J}_1(Kq)]^2 \mathcal{J}_2(K) \exp(-iqI_0). \qquad (5.5.18)$$

The second path with three steps is a 180° rotation of the path in Fig. 5.12c, and yields exactly the same result (5.5.18). For n large, there are $2n$ paths, corresponding to $r = 1, 2, \ldots, n$ and the rotations by 180°. Summing

---

[4] Solutions with $|m_1| > 1$ yield higher order corrections to D_n.

these and expanding (5.5.18) to retain q^2 terms only,

$$a_n(0, q) = \frac{2n}{(2\pi)^2} \frac{K^2 q^2}{4} \mathscr{J}_2(K).$$

Combining this result with the path that always stays at the origin, we find, using (5.5.9)

$$D_n = \frac{K^2}{2} [\tfrac{1}{2} - \mathscr{J}_2(K)]. \tag{5.5.19}$$

The second term, involving $\mathscr{J}_2$ is the correction to the quasilinear result in the limit of large K and n. Expanding $\mathscr{J}_2$,

$$D_n = \frac{K^2}{2} \left[\frac{1}{2} - \left(\frac{2}{\pi K}\right)^{1/2} \cos\left(K - \frac{5\pi}{4}\right) \right], \tag{5.5.20}$$

yielding corrections of order $K^{-1/2}$.

Continuing this procedure, we find by using Table 5.2 the paths that appear to contribute to the order K^{-1} corrections. These consist of the two four-step paths shown in Figs. 5.12d and e, the five-step path shown in Fig. 5.12f, and their 180° rotations. The contributions to a_n from these paths follow from (5.5.14) or (5.5.16) as before. To order K^{-1}, the final result is

$$D_n = \frac{K^2}{2} [\tfrac{1}{2} - \mathscr{J}_2(K) - \mathscr{J}_1^2(K) + \mathscr{J}_2^2(K) + \mathscr{J}_3^2(K)]. \tag{5.5.21}$$

This result was first obtained by Rechester and White (1980), but with the $\mathscr{J}_2^2$ term missing. Note also that $\mathscr{J}_3^2 - \mathscr{J}_1^2$ is of order K^{-2}; thus both terms can be omitted to order K^{-1}.

Rechester and White made a numerical study of the standard mapping, calculating the diffusion coefficient D_{50} for 3000 particles. Their results are shown in Fig. 5.13. The dots give the numerically calculated value of D_{50}, normalized to the quasilinear value $\tfrac{1}{4} K^2$, as the stochasticity parameter K is varied. The solid curve shows the analytical result with the $\mathscr{J}_2^2$ term omitted.[5] The oscillations in D were first noted by Chirikov (1979). For $K < 4$, we expect the effects of KAM surfaces to become increasingly important, requiring the calculation of more Fourier paths. An additional difficulty is the existence of nondiffusing trajectories that lie inside the stable islands. This can be resolved by introducing a small amount of extrinsic noise, such that the result is an average over diffusing and nondiffusing trajectories. A comparison of the Fourier path analysis (carried out numerically), and results from iterating the mapping, for $K < 4$, is shown in Fig. 5.14. For K

---

[5] For large K, the effect of including the $\mathscr{J}_2^2$ term is to shift the solid curve up a little. Also note that the D_{50} was numerically determined in the presence of a small extrinsic stochasticity. See Section 5.6 for further details.

5.5. The Effects of Correlations

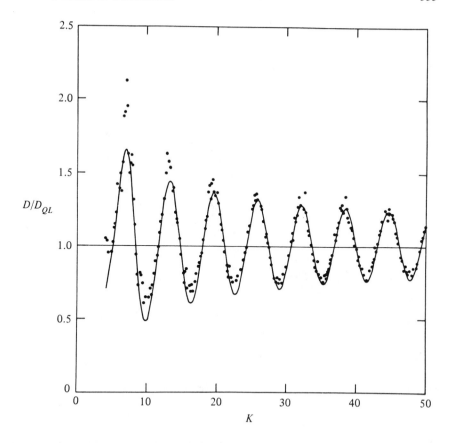

Figure 5.13. Plot of D/D_{QL} versus stochasticity parameter K. The dots are the numerically computed values and the solid line is the theoretical result in the large K limit (after Rechester and White, 1980).

near $K_c \approx 0.9716$, one finds numerically that

$$D_\infty \approx 0.1(K - K_c)^3. \qquad (5.5.22)$$

For $K < K_c$, a KAM barrier exists and there is no long-time diffusion in the absence of extrinsic noise. A discussion of the noise parameter, σ in Fig. 5.14, is left for Section 5.6.

Maps Not Periodic in Action. The Fourier path method depends on the peculiar periodicity in action of the standard map to evaluate the Fourier integrals in the limit of long times ($\Delta n \to \infty$, $D_n \to D_\infty$). Because of this, the method cannot be applied in the long time limit to maps without this periodicity. For generic maps in which the phase is randomized within a region of the action space for which the local approximation gives a

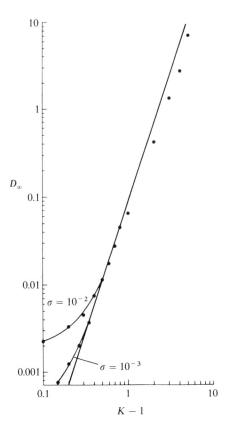

Figure 5.14. $D_\infty(K)$ versus K. The solid line has a slope of 3 (A. B. Rechester and R. B. White, private communication; see Murray et al., 1985).

near-constant stochasticity parameter K, it is possible to derive a local (in action) diffusion coefficient (Murray et al., 1985). An expansion procedure for the simplified Fermi map (3.4.4) gives the condition

$$(1/P)(\partial P/\partial u) \ll K(u)/2\pi. \qquad 5.5.23$$

With this satisfied, the diffusion coefficient becomes, using (5.4.20),

$$D_\infty(u) = \frac{1}{K^2(u)} D_\infty(K(u)). \qquad (5.5.24)$$

For a sharply peaked distribution, we expect (5.5.24) to hold only for times exceeding the time required for the distribution to broaden over many primary resonances, $n \gg 1/(K^2 D_\infty(u))$. If this n is sufficient to spread the distribution over the oscillations in $D_\infty(K)$ seen in Fig. 5.13, then the distribution function will be smoothed.

5.5. The Effects of Correlations

If we are interested in using the Fermi map to model a heating mechanism, then particles will generally start at low velocities, where the stable islands have negligibly small area. As the particles are heated they enter regions of phase space within which large islands exist. Without extrinsic stochasticity the particles will not penetrate the islands. Hence, although the equilibrium distribution is uniform in the ergodic phase space surrounding the embedded islands, the phase-averaged distribution $P(u)$ is not uniform (see Fig. 5.11). Thus to extract the diffusion of the untrapped species, alone, from the result (5.5.21) that includes occupation of all embedded islands, we divide the diffusion coefficient by the fraction of phase space occupied by stochastic orbits. If we denote by $P_s(K, I)$ (normalized by one) the stochastic equilibrium distribution for the standard map, then the relevant diffusion coefficient for the Fermi map is

$$D(u) = \frac{D_\infty(K(u))}{K^2(u)\langle P_s(K(u), I(\Delta u))\rangle_I}, \qquad (5.5.25)$$

where $K(u) = 2\pi M/u^2$ and $I(\Delta u) = -K\Delta u$. The average $\langle \ \rangle$ over a 2π interval in I ignores rapid variations in the diffusion coefficient, which is consistent with (5.5.23).

To compare the solution $P(u, n)$ of the Fokker–Planck equation to the numerically calculated distribution $P_s(u, n)$, we must also account for the effect of the exclusion of the embedded islands on P. Since the chaotic motion does not enter the islands, the phase-averaged invariant distribution $P_s(u)$ is not uniform, but has dips in the regions of u where islands exist (see Fig. 5.11). The observed distribution function $P_s(u, n)$ is then related to the solution of the Fokker–Planck equation $P(u, n)$ by

$$P_s(u, n) = \frac{P(u, n)P_s(u)}{\int_0^{u_b} P(u', n)P_s(u')\, du'}, \qquad (5.5.26)$$

where $\int_0^{u_b} P\, du = 1$. Clearly, as $n \to \infty$ and $P(u, n)$ becomes uniform, $P_s(u, n) \to P_s(u)$.

In addition to comparing $P_s(u, n)$ directly with distributions obtained by iterating the mapping equations, it is useful to calculate the second moment, or variance, of $P_s(u, n)$. This variance can be compared to the variance measured by iterating the map. We define the theoretical variance

$$\chi_t^2(u_0, n) = \frac{1}{n} \frac{\int (u - u_{\text{ave}})^2 P_s(u, n)\, du}{\int P_s(u, n)\, du}, \qquad (5.5.27)$$

where

$$u_{\text{ave}}(u_0, n) = \frac{\int u P_s(u, n)\, du}{\int P_s(u, n)\, du}.$$

We integrate the Fokker–Planck equation using the theoretical diffusion coefficient (5.5.25) and a delta function at action u_0 as an initial condition. This yields the distribution function $P(u, n)$ (the dependence on the initial

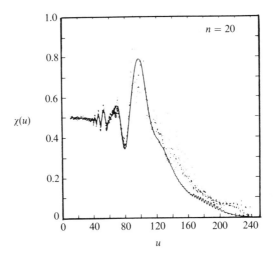

Figure 5.15. The variance plotted as a function of initial action. Each dot corresponds to a measurement of the diffusion obtained by iterating the Fermi map for 20 iterations. The solid line is the theoretical variance obtained by integrating the Fokker–Planck equation (after Murray et al., 1985).

action u_0 is suppressed). The final distribution function is obtained from (5.5.26). We compare the variance of this result to the numerical variance

$$\chi^2(u_0, n) = \frac{1}{nm} \sum_{i=1}^{m} (u_i(n) - u_{\text{ave}})^2. \tag{5.5.28}$$

The results after an iteration time $n = 20$ are shown as dots in Fig. 5.15. The solid line obtained by integrating the Fokker–Planck equation is a linear interpolation of several hundred calculations of the variance, each at a different initial u_0. The variances in Fig. 5.15 both show characteristic oscillations for $K > 4$ and fall to zero as K approaches one, as in the standard map. However, the oscillations for the Fermi map occur in action space rather than in parameter space. For $u < 40$, both the theoretical and measured variances no longer exhibit oscillations. This is because diffusing particles experience different local diffusion rates which average the rapid variations in $D(u)$ to the quasilinear value of $\frac{1}{2}$.

We can estimate the range of u for which the quasilinear diffusion coefficient can be used. We expect that quasilinear diffusion is adequate if large islands do not exist ($u < u_s$) and if particles diffuse over a range of action Δu comparable to or larger than the local period of the oscillations in $D(u)$. For large K the diffusion oscillates as $\cos K$ so we expect averaging when $(\partial K/\partial u)\Delta u \gtrsim \pi/2$. Using $\Delta u = \sqrt{nD_{QL}}$ and $\partial K/\partial u = 4\pi M/u^3$ for the Fermi map, we find quasilinear diffusion for

$$u < (32M^2 n)^{1/6} \leq u_s. \tag{5.5.29}$$

The validity of this criterion has been studied numerically for $10^2 < M < 10^8$

5.5. The Effects of Correlations

and $10 < n < 1000$. For actions satisfying it, the variance is within five percent of the quasilinear value (Murray et al., 1985).

5.5b. Diffusion near Critical K

For K near the critical value, $K_c \approx 0.9716$, it becomes increasingly difficult to use the Fourier path method for calculating diffusion. The reason is that correlations in the neighborhood of just-broken KAM curves need increasingly long Fourier paths to capture their effects. An alternative picture, developed by MacKay et al. (1984), can be used to calculate the diffusion in this parameter regime.

In Section 4.4 we explored the process by which island chains, or resonances corresponding to the continued fraction expansion of the golden mean, led to destruction of the last phase-spanning KAM torus. The perturbed KAM curve becomes increasingly convoluted, finally losing its integrity at $K = K_c$. The resulting structure with golden mean winding number has been called a *cantorus* since it originates from a torus and breaks up on a surface of section into a set of discrete points with complicated spacings, described by a Cantor set (Percival, 1979a). We discuss the properties of Cantor sets in Section 7.1d. We know that the cantori replace all destroyed tori, corresponding to all the irrationals. However, for a K value at which the last phase-spanning KAM surface is destroyed, most other cantori have become fairly transparent to diffusion.

If a single just-broken KAM surface or cantorus were responsible for limiting diffusion, we might be able to determine the diffusion rate by calculating the exchange of phase space (flux) through a cantorus. For example, consider the structure of a cantorus as shown in Fig. 5.16. As on a homoclinic orbit, neighboring hyperbolic fixed points get closer together when iterated forward or backward in time. Thus, if we take the largest gap labeled with a subscript zero, iterating the end points of that gap forward gives a decreasing set of gaps S_1, S_2, etc. Iterating backward also gives a

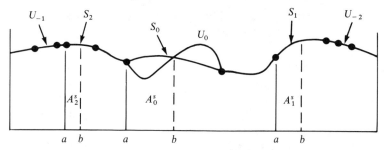

Figure 5.16. Structure of gaps in a phase-spanning cantorus. The dots are hyperbolic fixed points.

decreasing set of gaps U_{-1}, U_{-2}, etc. Now consider that this process is repeated until small gaps are found in each direction, S_n and U_{-n}. These gaps can be closed by straight-line segments which are then iterated back to the main gap, producing curves S_0 and U_0, as shown in the figure. The interchange of phase space between these two curves is called a *turnstile* and represents flux across the gap (MacKay *et al.*, 1984). The total area underneath the curves in the main gap (0) is, of course, the same. Although the flux can be calculated directly by this procedure, it is not convenient. A better procedure is to use the formula relating the change in phase space area in a mapping step to the difference in the generating functions at the edges of the curves, which we developed in Section 3.1c. From (3.1.40)

$$A_1 - A_0 = F(b) - F(a) = \Delta F_0. \tag{5.5.30}$$

Here $F(b) - F(a)$ is the difference in the generating function F_1 taken at one edge of the gap and at the center of the gap where the orbits S_0 and U_0 cross. (For simplicity of notation, we have dropped the subscript 1 on F, which designates the type of generating function.) The procedure is to sum all areas forward and backward until the areas become negligible. Iterating forward we have

$$\begin{aligned} A_0^s - A_1^s &= -\Delta F_0, \\ A_1^s - A_2^s &= -\Delta F_1, \\ &\vdots \\ A_{n-1}^s - A_n^s &= -\Delta F_{n-1}, \end{aligned} \tag{5.5.31}$$

and iterating backward

$$\begin{aligned} A_0^u - A_{-1}^u &= \Delta F_{-1}, \\ A_{-1}^u - A_{-2}^u &= \Delta F_{-2}, \\ &\vdots \\ A_{-(n-1)}^u - A_{-n}^u &= \Delta F_{-n}. \end{aligned} \tag{5.5.32}$$

Summing the two sets of equations, we see that all of the intermediate values of A cancel. Subtracting (5.5.31) from (5.5.32) we find the flux

$$\Delta A = A_0^u - A_0^s = \sum_{n=-\infty}^{\infty} \Delta F_n, \tag{5.5.33}$$

where we have dropped the negligible A_n's, $n \to \infty$. The reader may well ask if this sum is actually done in practice, as the end points for calculation of ΔF_1 are not easy to locate, and the number of terms in the sum required for a given accuracy is not known. In fact, the sum is actually done by successive calculations of the flux through the rational iterates of a just-broken KAM curve obtained from a partial fraction expansion (see Section 4.4), until sufficient accuracy is obtained. The end points of the central gap correspond to the x-points of the central island of each chain, and the center point of

5.5. The Effects of Correlations

the gap corresponds to the island center which has gone unstable. The island centers and x-points can be located by various techniques, e.g., as in Section 2.6b. Since the flux through the n-island chain is obtained as a sum over the n islands,

$$\Delta A = \sum_{j=1}^{n} \Delta F_j. \tag{5.5.34}$$

The process described above has been explored in a series of papers, in which the scaling of the transport near K_c has been calculated (MacKay et al., 1984) and an approximate diffusion rate has been calculated as a function of K (Meiss and Ott, 1986). The actual numerical calculation is rather tedious, but is aided by symmetries of the map. The theoretical treatment has also been used to obtain better understanding of the role that resonances play in the diffusion process (MacKay et al., 1987). The average number of iterations T for a phase point to cross the golden mean cantorus for $K > K_c$ is computed from the flux by

$$T = \frac{A_{\text{stoch}}}{\Delta A},$$

where A_{stoch} is the total accessible stochastic area. For the standard mapping the diffusion rate in action is then obtained from

$$D = \frac{(2\pi)^2}{T} = \frac{(2\pi)^2 \Delta A}{A_{\text{stoch}}}. \tag{5.5.35}$$

Since A_{stoch} changes only slowly with K, the proportionality of D near critical K can be found from the proportionality of ΔA. This was found by MacKay et al. (1984) to be

$$\Delta A = a(K - K_{\text{crit}})^{3.0},$$

which has the same scaling as (5.5.22), obtained from the Fourier path technique. There is also a theoretical prediction of this scaling with power law $3.01 \cdots$ from renormalization arguments (Greene et al., 1986), consistent with these numerical values.

The complete problem of calculating diffusion by this technique is much more complicated. There is a hierarchy of cantori surrounding island tori. Trajectories in the neighborhood of a KAM torus must penetrate through an increasingly fine grid of cantori gaps. On the way they may also be derailed to shadow a second-order island chain which itself has an increasingly fine grid of cantori gaps, and so on. This infinite regressive structure leads to a tree model for calculating the flux (Meiss and Ott, 1986), where computational difficulties lead to a consideration of only a few branches of the tree. Therefore, an accurate diffusion calculation cannot be made using the method, except locally, near a just-destroyed torus.

5.5c. Long-Time Correlations and Averaging

Correlations. For any dynamical quantity q, we define an autocorrelation

$$C(\tau) = \langle q(t-\tau)q(t)\rangle = \int_\mu q(t-\tau)q(t)\,d\mu, \qquad (5.5.36)$$

where the integration is over the phase space. For a fully stochastic phase space we expect exponential decay of C with τ. This was proved by Bunimovich and Sinai (1980) for the velocity autocorrelation of a particle reflecting off a periodic configuration of scattering centers (Lorentz gas), obtaining the bound

$$C(\tau) \leqslant \text{const.} \times \exp(-\alpha\tau^\gamma), \qquad 0 < \gamma \leqslant 1, \qquad (5.5.37)$$

and was confirmed by the numerical investigation of Casati et al. (1982).

However, in a divided phase space numerous numerical investigations (Meiss et al., 1983; Karney, 1983; Chirikov and Shepelyansky, 1984, 1986) have indicated a power law tail to C. The reason for this tail was discussed in the previous subsection. Numerical investigations of a related quantity, the probability that an initial phase space position in the neighborhood of an island remains in that neighborhood, gave a power law dependence

$$P(\tau) \propto \tau^{-\gamma} \qquad (5.5.38)$$

with $\gamma \approx 1.4$ for a small, isolated island (Karney, 1983) and $\gamma \approx 1.45$ for the main island of the standard mapping (Chirikov and Shepelyansky, 1984, 1986), indicating the universal nature of the phenomenon. These authors also show that if $C(\tau) \propto \tau^{-\eta}$, then $\eta = \gamma - 1$. Using the idea of flux penetration through cantori, and taking those cantori on golden surfaces which are most difficult to destroy, Meiss and Ott (1986) obtain $\gamma = 1.96$ which is considerably faster than the $\gamma = 1.45$ observed numerically. In their calculation they used the renormalization scaling of Greene et al. (1986) that the critical $K = K_c$ scales as $\Delta K_c \propto (\Delta\alpha_n)^{1/2}$, where $\Delta\alpha_n = |\alpha - \alpha_n|$, the distance in frequency from the locally most irrational frequency α to neighboring irrationals. Murray (1991) has shown both theoretically and numerically that the renormalization scaling is only accurate very close to the locally most irrational surface, while over a broader frequency range the scaling $\Delta K_c \propto \Delta\alpha_n$ holds. With this latter scaling, Murray finds that $\gamma \approx 1.5$, in close agreement with the numerical value. Chirikov and Shepelyansky (1984, 1986) also obtain this value with a more heuristic derivation. For further details the reader should consult Murray's paper.

Averaging over Distributions. From the arguments of the authors cited above, calculating diffusion including long-time correlations can be a complicated undertaking. A formalism, developed by Cary et al. (1981) and Cary and

5.5. The Effects of Correlations

Meiss (1981) for doubly periodic mappings, is to write the diffusion coefficient as a sum of momentum space correlations of the action impulse f, assumed to depend only on the phase variable θ,

$$C_j \equiv \langle f(\theta_i) f(\theta_{i+j}) \rangle_\mu. \tag{5.5.39}$$

Cary et al. have shown that the diffusion coefficient is then

$$D = \epsilon^2 \left[\tfrac{1}{2} C_0 + \lim_{k \to \infty} \sum_{j=1}^{k-1} \left(1 - \frac{j}{k}\right) C_j \right] \approx \epsilon^2 \left[\tfrac{1}{2} C_0 + \sum_{j=1}^{\infty} C_j \right], \tag{5.5.40}$$

where the last expression is valid when the C_j fall off rapidly enough with the number of mapping iterations j. The angular brackets denote an average over the doubly periodic phase space (mod $2\pi \times 2\pi$) region μ. If we expand f in a Fourier series,

$$f(\theta) = \sum_{l=-\infty}^{\infty} f_l e^{il\theta}, \tag{5.5.41}$$

and substitute (5.5.41) in (5.5.39), the correlations become

$$C_0 = \langle f^2(\theta_0) \rangle_\mu, \tag{5.5.42a}$$

$$C_j = \sum_{m,n} f_m f_n \chi_j(m, 0, \ldots, 0, n), \tag{5.5.42b}$$

where

$$\chi_j(m_0, \ldots, m_j) = \left\langle \exp\left(i \sum_{k=0}^{j} m_k \theta_{n+k}\right) \right\rangle_\mu$$

are the *characteristic functions*. The term C_0 yields the "quasilinear" value for D. The next term C_1 then yields the first-order correction to the diffusion. In principle, the higher-order terms can be summed to determine the effect of long-time correlations. This has been discussed by Karney (1983); as with the Fourier path method, the calculation becomes increasingly difficult as higher-order terms are desired.

An alternative argument can be used, invoking the asymptotic dependence of the correlations, together with a simple averaging argument, to show that (except in special cases) the long-time power law tails to the correlations do not significantly affect diffusion. Consider the standard mapping with an rms random walk step $(\Delta I)_{\text{rms}} = K/\sqrt{2}$, such that the diffusion coefficient $D = (\Delta I)^2_{\text{rms}} / 2\tau_{\text{step}} = K^2/4$, where $\tau_{\text{step}} = 1$ is just the mapping period, itself. If, now, τ_{step} has a probability distribution given by some $P(\tau)$ for the same $(\Delta I)_{\text{rms}}$, then we can write D as an integration over the probability

$$D = \int_1^\infty \frac{K^2}{4\tau} P(\tau) \, d\tau, \tag{5.5.43}$$

where $P(\tau) = \delta(\tau - 1)$ describes the uncorrelated mapping. If we take $P(\tau)$

as having the proportionality of (5.5.38) for some $\tau > \tau_{LT}$ and divide D into two parts, corresponding to short-time and long-time correlations, then the long-time portion of the diffusion can be integrated to obtain

$$D = D_{ST} + D_{LT} = \int_1^{\tau_{LT}} \frac{K^2}{4\tau} P(\tau)\, d\tau + \frac{K^2 a}{4\gamma} \tau_{LT}^{-\gamma}, \tag{5.5.44}$$

where a is a proportionality constant. From (5.5.44) it is clear that for $\gamma > 0$, the long-time correlation effects vanish with increasing τ_{LT}. Since $\gamma \cong 1.4$, numerically, this is, indeed, the case.

The situation is quite different in the presence of accelerator modes (see Section 4.1b). In that situation, the trajectories that closely shadow the mode step in action a distance proportional to their correlation times $(\Delta I)_{rms} \approx K\tau$. Thus the diffusion arising from the long-time tail of the distribution function has the proportionality

$$D_{LT}(\text{acc. mode}) \propto \int_{\tau_{LT}}^{\infty} \frac{(K\tau)^2}{\tau} P(\tau)\, d\tau. \tag{5.5.45}$$

Again, using $P(\tau)$ from (5.5.38) we find

$$D_{LT}(\text{acc. mode}) \propto \frac{1}{2-\gamma} \tau^{2-\gamma} \Big|_{\tau_{LT}}^{\infty}. \tag{5.5.46}$$

For $\gamma < 2$, which is found numerically, the part of the diffusion arising from the long-time tail of the correlations becomes infinite. This explains the numerical peaks found in the diffusion coefficient, in Fig. 5.13, for those values of K for which accelerator modes exist.

The physical explanation for the contrary results in (5.5.44) and (5.5.46) is quite clear. Diffusion is dominated by the fast diffusing trajectories. If the long-time correlations are inhibiting diffusion, then the overall diffusion coefficient is only slightly modified by the removal of some of the diffusing phase space. If, on the other hand, the long-time correlations produce streaming in the action space (accelerator modes), then these particles may dominate the diffusion. Whether they do or not depends on the rate of decay of correlations. The decay is slow enough in a divided phase space to cause the singularities in the standard map diffusion when accelerator modes are present. In more generic mappings, for which the standard map is only a local approximation, the accelerator modes become quasimodes which appear and disappear as the action varies. In this case, long-time correlations do not lead to singularities in the diffusion. We consider the diffusion under these circumstances in Section 5.5d.

Averaging over Action Space. We now consider the distinction between averaging over classes of diffusing particles and averaging a given class over the action space. In the former case, the D's sum directly; in the latter case, the inverse D's are summed. In the Fourier path method, these distinctions

5.5. The Effects of Correlations

do not arise because a global diffusion coefficient D was found. Application to the Fermi acceleration map gave local variations in D with respect to action, but these variations tend to be smoothed in solving a diffusion problem. We have seen in Section 5.5b, that for regions in which D is constricted, as at the turnstile of a just-broken phase-spanning KAM torus, the diffusion can be dominated by these constrictions. More generally, the continuity of steady-state flow in action leads to the consequence that the gradient of the density becomes steeper where the diffusion rate is slower and more gradual where the diffusion rate is high. Consider a region ΔI of steady-state flow Γ in action space (fixed source and sink), but variable diffusion coefficient. Within the region we have for the distribution function P

$$-\Gamma = D(I)\frac{dP}{dI} \equiv \langle D \rangle \left\langle \frac{dP}{dI} \right\rangle, \tag{5.5.47}$$

where the last equality defines an average D and an average dP/dI over the region. Thus

$$\frac{1}{\langle D \rangle}\frac{dP}{dI} = \frac{1}{D(I)}\left\langle \frac{dP}{dI} \right\rangle, \tag{5.5.48}$$

and, averaging this over the region ΔI,

$$\frac{1}{\langle D \rangle}\frac{1}{\Delta I}\int \frac{dP}{dI}dI = \left\langle \frac{dP}{dI} \right\rangle \frac{1}{\Delta I}\int \frac{dI}{D(I)},$$

or

$$\frac{1}{\langle D \rangle} = \frac{1}{\Delta I}\int_{\Delta I} \frac{dI}{D(I)}. \tag{5.5.49}$$

This inverse relation indicates that $\langle D \rangle$ is dominated by the regions in which $1/D(I)$ is largest, and therefore $D(I)$ is smallest. A somewhat different treatment of this phenomenon can be found in Cohen et al. (1981).

The above phenomenon has important consequences in determining the overall rate of diffusion in mappings for which quasi-accelerator modes are present over some range of values of the action (Lichtenberg et al., 1987), as described in the next subsection. In Section 5.6 we shall see that for situations in which extrinsic noise plays an important role, appropriate averaging also becomes a key to practical calculations.

5.5d. Quasiaccelerator Modes

Diffusion in generic systems is affected only slightly by long-time correlations. However, for systems containing accelerator modes, the long-time correlations can dominate the diffusion or can lead to nondiffusive transport. As discussed in Section 4.1b, for the standard map (4.1.3), accelerator modes are bounded by KAM curves in a doubly periodic (I, θ) phase space. The modes (islands) are found in a window of stability (in K) for

$$2\pi l < K < \sqrt{(2\pi l)^2 + 16} \qquad (5.5.50)$$

for integer $l \geq 1$. A mode is born at an inverse tangent bifurcation at $K = 2\pi l$ and period doubles at $K = \sqrt{(2\pi l)^2 + 16}$. The first mode ($l = 1$) has the largest stable range of K values and the largest maximum island size, and consequently is most important for modifying the local diffusion. The fixed points are located at $I_l = 2\pi m$, $K \sin \theta_l = 2\pi l$. For each I_l within the window, there are two θ_l's, corresponding to two islands side-by-side whose centers stream up and down respectively in the phase space, advancing by $\Delta I = \pm 2\pi l$ each iteration. Hence, phase points, trapped within the stable islands about these fixed points, stream rather than diffuse in action, leading to a singularity in the usually defined diffusion. The singularity can be removed by adding noise to the standard map. This has been done by Karney et al. (1982) to study the enhanced diffusion over ranges of K values for which accelerator modes exist.

For generic mappings that are not periodic in the action, such as the simplified Fermi map (4.1.1), the locally approximated standard mapping has a K value which is a function of the action u. This was shown in Section 4.1b by linearizing around a given action to obtain

$$K(u) = 2\pi M/u^2 \qquad (5.5.51)$$

and

$$I_n = -K\Delta u_n. \qquad (5.5.52)$$

From (5.5.51) we see that K depends on u_l and therefore the effective K of the mapping of a given trajectory changes as it diffuses or streams. Since K changes, the accelerator modes of the Fermi map are not truly stable modes of the standard map, and may be designated as quasiaccelerator modes. The larger the mapping parameter M, the longer the mapping remains near a given value of K over a range of u-values.

For quasiaccelerator modes there are two mechanisms by which diffusion is enhanced.

(1) Particles on locally unstable orbits become trapped on locally stable orbits as they are transported in the direction of increasing stable island size (phase space area). They then stream through values of u corresponding to increasing and then decreasing island size until they are detrapped at values of u having the same island size at which they were originally trapped.
(2) Trajectories near stable islands are only weakly unstable, and therefore shadow the stable trajectories over many mapping periods. This leads to long-time correlations in the phase which again result in streaming.

This latter effect has been considered by Karney (1983) to qualitatively explain the peaks in the diffusion for the standard map for the case in which

5.5. The Effects of Correlations

no orbits have initial phase space coordinates lying within stable accelerator-mode islands.

For generic maps, Lichtenberg et al. (1987) show that the first effect usually dominates over the second. We can estimate the effective diffusion arising from the quasiaccelerator mode due to trapping, in the original u-space, from the relation

$$D_T = \frac{(\Delta u)^2}{2\tau} g(K), \qquad (5.5.53)$$

where τ is the average streaming time, Δu is the average distance over which streaming takes place,

$$g(K) = \frac{A_I(K)}{(2\pi)^2} \qquad (5.5.54)$$

is the fraction of participating phase space, and A_I is the island area in $I - \theta$ space. Since, for a single iteration, I steps by $2\pi l$, $\Delta I = 2\pi l \tau$; we then have, using (5.5.52),

$$\Delta u \approx \tau. \qquad (5.5.55)$$

From (5.5.51) we obtain, by differentiation,

$$\Delta u = \frac{(2\pi M)^{1/2}}{2K^{3/2}} \Delta K. \qquad (5.5.56)$$

From (5.5.50) for large K, we calculate the value of $\Delta K = K - 2\pi l$ at the middle of the mode to be:

$$\Delta K_{\text{ave}} = 4/K. \qquad (5.5.57)$$

The island area can be estimated from the perturbation Hamiltonian in the neighborhood of the stable fixed point, or by numerical calculation. In the limit of large K, Lichtenberg et al. show that $A_I(K_{\text{ave}}) \approx 65/K^2$. Using this along with (5.5.54)–(5.5.57) in (5.5.53), we obtain

$$D_T \approx \frac{1.6(2\pi M)^{1/2}}{K^{9/2}}. \qquad (5.5.58)$$

For the lowest-order mode, $K \approx 2\pi$ and $D_T \approx 10^{-3} M^{1/2}$. Since the untrapped particles have $D_U \approx \frac{1}{2}$, and the D's add for these two classes of particles, as in Section 5.5c, we see that the trapped particles dominate the overall diffusion for $M \gtrsim 2.5 \times 10^5$.

We now calculate the diffusion within a Fokker–Planck formalism. In principle, if we are interested in times long compared to the streaming time and distances large compared to the mode width Δu, then we can sum the D's for the trapped and untrapped particles. To obtain the distribution more locally, we treat the streaming explicitly as sources and sinks in the Fokker–Planck equation. This is a natural description when only the trapping–detrapping mechanism is considered. The diffusion coefficient is then that of the nonstreaming particles, as previously obtained. The Fokker–Planck equation, including sources and sinks, can be written

$$\frac{\partial P}{\partial n} = \frac{\partial}{\partial u}\left(\tfrac{1}{2}D(u)\frac{\partial P}{\partial u}\right) + S(u) - L(u), \tag{5.5.59}$$

where $D(u)$ is given by (5.5.25). The sources and sinks S and L are straightforwardly determined, in principle, from the amount of phase space captured or lost from the islands. Since there are two symmetric islands at a given u in the mode near $\theta = \pm\pi/2$, transporting phase space in opposite directions, the loss and source terms are closely related, as $L = a'(u)P(u,n)$, $S = a'(u)P(\bar{u},\bar{n})$, where $a'(u) = |A'(u)/2\pi|$ and $A'(u)$ is the derivative with respect to u of the island area $A(u)$. The barred quantities are defined by $A(\bar{u}) = A(u)$, and for the steady state, P is independent of n. Taking A' antisymmetric about some maximum value A_m at u_m, and making the change of variable $u = u_m - x$, $\bar{u} = u_m + x$, we expand $P(u)$ and $P(\bar{u})$ in Taylor series about $u = u_m$ to obtain

$$S - L = a'(x)P'(0)2x. \tag{5.5.60}$$

Substituting this into the Fokker–Planck equation (5.5.59), we have, in the variable x

$$\frac{1}{2}\frac{d}{dx}\left(D\frac{\partial P}{\partial x}\right) + a'(x)P'(0)2x = 0. \tag{5.5.61}$$

To obtain a solution of (5.5.61) we need to have explicit expressions for $D(u)$ and $A(u)$. In general, $A(u)$ is a complicated function allowing only numerical calculations. However, because the solution of the Fokker–Planck equation depends only on $\int_0^{\Delta u} a\,dx$, the exact form of the variation of the island area is of little importance, and a simple parabolic fit can be used.

For comparison of the analytic solution (5.5.61) with numerical simulations, a source and sink were placed on opposite sides of the mode (Lichtenberg et al., 1987). To insure continuity of flux, the theoretical and numerical slopes are matched to $1/D(u)$, from (5.5.25), at the value of u for which the mode is born at an inverse tangent bifurcation ($K = 2\pi$). The comparison of theory and experiment is shown in Fig. 5.17. We see that agreement is quite reasonable for that portion of the mode in which the stable island is growing or nearly constant, but has significant deviations where the detailed variation of the island structure becomes significant.

As we have seen, accelerator modes significantly enhance diffusion only if the action space is nearly uniform (high M in the Fermi map). However, there has been considerable interest in the phase space structure. The bifurcation processes that give rise to the modes in the standard map and the resulting diffusion have been studied in a series of papers by Ichikawa and associates (see Ichikawa et al., 1987), including the multiple period

5.6. The Effect of Extrinsic Stochastic Force

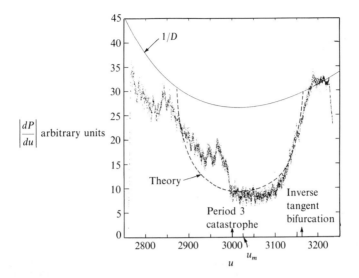

Figure 5.17. Comparison of theoretical and numerical relative values of $|dP/du|$. The theoretical value is shown as a dashed curve within the quasiaccelerator mode and a solid curve ($\propto 1/D$) outside of the mode ($M = 10^7$) (after Lichtenberg et al., 1987).

accelerator modes that exist for $1 < K < 2\pi$. The interesting period 3 bifurcation of the accelerator mode, which changes the local topology of the mode (and therefore the diffusion) was discussed briefly by Lichtenberg et al. (1987) and considered in detail by Ichikawa et al. (1989).

5.6. The Effect of Extrinsic Stochastic Forces

5.6a. Introduction

In Section 5.4 we found that strong resonance overlap led to *intrinsic diffusion* at a rate equal to that which would exist if the phase of the term driving the action is a random variable. This is equivalent to strong *extrinsic diffusion* driven by a random source external to the Hamiltonian phase flow. For wave-particle interactions, this result corresponds to a wave perturbation consisting of many large amplitude, uncorrelated waves, and is thus equivalent to the diffusion arising in quasilinear theory. These results indicate that in the limit of strong stochasticity, intrinsic and extrinsic diffusion lead to very similar behavior.

For extrinsic random variations that are small compared to the regular variations of the Hamiltonian flow, the combined effect of the Hamiltonian flow with the diffusive motion must be considered. The structure of the phase space of Hamiltonian systems has certain stability properties in the presence of either regular or random perturbations. We have already used one of these, that of KAM stability of nonresonant trajectories under small perturbation. If a weak stochastic perturbation is added to the motion on a KAM stable orbit, we cannot expect the trajectory to remain on a smooth orbit. Nevertheless, in a mostly regular region of the phase space, far from resonances, inherent stability properties exist that only allow diffusion across the trajectory on a time scale associated with the weak stochastic perturbation itself. This behavior has important consequences, both for physical problems that are always subject to noise, and for computer calculations that have inherent accuracy errors.

The basis for the above stability is the property that weak stochastic perturbations of trajectories with coordinates initially close together lead to subsequent separation of trajectories in time that are along, rather than across, nested phase space curves. Thus, while trajectories with initial conditions starting close together may spread rather rapidly away from each other on nearby KAM surfaces, they diffuse only slowly from one surface to another.

To show this for regular orbits, we use the twist map (3.1.8)

$$J_{n+1} = J_n,$$
$$\theta_{n+1} = \theta_n + 2\pi\alpha(J_{n+1}),$$

such that nearby initial conditions, as shown in Fig. 5.18, separate according to the relations, at J and at $J + \Delta J$,

$$\theta_1(J) = \theta_0 + 2\pi\alpha(J),$$
$$\theta_1(J + \Delta J) = \theta_0 + 2\pi\alpha(J + \Delta J),$$
$$= \theta_0 + 2\pi\alpha(J) + 2\pi\alpha'\Delta J.$$

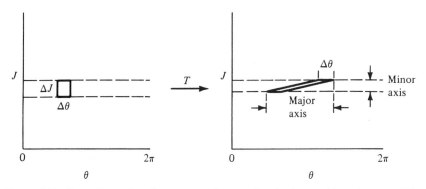

Figure 5.18. Spreading of a phase space element for the integrable twist map. The element spreads linearly along the dashed invariant curves.

5.6. The Effect of Extrinsic Stochastic Force

Thus, for the parallelepiped in the figure, the "major axis" is

$$\theta_1(J + \Delta J) - \theta_1(J) = 2\pi\alpha'\Delta J$$

and grows linearly with n, while the "minor axis" ΔJ stays fixed. Greene (1979a) has carried through this argument in the tangent space, showing that an initial small circle of error transforms to a large aspect ratio ellipse with the major axis oriented along the invariant curve, in agreement with the above reasoning.

The above feature allows the numerical computation of the motion for millions of iterations of the mapping, without excessive diffusion of the KAM curves. Because slow numerical diffusion does occur across KAM surfaces, however, this stochastic component sets a limit to the accuracy of determining the transition to global stochasticity.

For two-dimensional mappings, Lieberman and Lichtenberg (1972) numerically investigated the local slow diffusion due to a stochastic force by introducing a stochastic component into the simplified Ulam version of the Fermi acceleration mapping. They modified (3.4.7) to

$$u_{n+1} = |u_n + \psi_n - \tfrac{1}{2}|, \qquad (5.6.1a)$$

$$\psi_{n+1} = \psi_n + \frac{M}{u_{n+1}} + \Delta\psi, \quad \text{mod } 1, \qquad (5.6.1b)$$

where $\Delta\psi$ is a random phase shift. If $\Delta\psi$ is allowed to take on all phases between 0 and 1, we could expect the motion governed by these equations to reduce to the usual random walk problem, independent of the phase shift M/u_{n+1}, and this is indeed what is observed computationally. For a more restricted allowable region for $\Delta\psi$, which corresponds to a weak stochastic force, the adiabatic regions are filled in, but on a slower time scale than that required for generating the island structure itself. In Fig. 5.19, we illustrate this behavior for $-0.005 < \Delta\psi < 0.005$. For this small random component, the time scale for diffusion into the islands is longer than the time scale for the delineation of the adiabatic regions. As we see from Fig. 5.19 the smaller islands have been considerably filled in, while the larger islands have only been slightly modified. It is interesting to note that more densely occupied bands appear within the adiabatic region, corresponding to particles that have penetrated the adiabatic region due to the random phase fluctuations, but have subsequently primarily followed the adiabatic orbits. These denser bands also appear in the stochastic portion of the phase space near an island where the particle trajectories are almost regular. On a sufficiently long time scale, governed by the statistics of the occupation numbers, one expects these bands to disappear.

Figure 5.19. Phase space $u - \psi$ for the sawtooth Fermi mapping (5.6.1) with an additional weak stochastic force. Here $M = 10$ and the random phase jump $\Delta\psi$ is uniformly distributed in the range $|\Delta\psi| < 0.005$. (a) 10,240 and (b) 20,480 iterations of a single initial condition (after Lieberman and Lichtenberg, 1972).

5.6b. Diffusion in the Presence of Resonances

On a global scale, an important resonance structure in the phase space can have a very significant effect on the role of diffusion arising from a weak extrinsic random source. That is, if the phase space has large (but not overlapping) primary islands, a small jump in action can transport the phase point from a nonresonant KAM curve to a resonant or island curve. Considering the time between jumps in action to be long compared to the phase oscillation period, then a second jump in action can take place on the other side of the island and transport the trajectory to another nonresonant KAM curve on that side. The characteristic "resonant" diffusion coefficient, in the neighborhood of the island, is then strongly enhanced,

$$D \sim \frac{(\Delta I)^2}{2\tau}, \tag{5.6.2}$$

where ΔI is the characteristic half-island width and τ the time between random jumps in action. However, if significant portions of the space do not have large islands, then, as we shall see, the weak rate of nonresonant diffusion strongly inhibits the overall global diffusion rate.

Diffusion Coefficient. The diffusion rates for the standard mapping in the presence of a weak stochastic perturbation have been calculated analytically by Rechester and White (1980) and Rechester et al. (1981). The latter work introduced the Fourier path methods described in the previous section. We now show how to modify the method of Section 5.5a to include the presence of a stochastic perturbation. We modify the standard mapping to include the effect of a random jump ξ in the phase variable,

$$I_{n+1} = I_n + K \sin \theta_n, \tag{5.6.3a}$$

$$\theta_{n+1} = \theta_n + I_{n+1} + \xi_n, \tag{5.6.3b}$$

where ξ has a Gaussian probability distribution with mean-square deviation σ

$$p(\xi) = \frac{1}{\sqrt{2\pi\sigma}} \exp\left(-\frac{\xi^2}{2\sigma}\right). \tag{5.6.4}$$

The one-step conditional probability (5.5.6) is modified to

$$W(I, \theta, n | I', \theta', n-1) = \int d\xi \, p(\xi) \, \delta(I - I' - K \sin \theta')$$
$$\times \delta(\theta - \theta' - I' - K \sin \theta' - \xi). \tag{5.6.5}$$

Using this in (5.5.11b) and doing the θ and I' integrations over the δ-functions in (5.6.5), we obtain (5.5.12) multiplied by the factor

$$\int_{-\infty}^{\infty} \exp(-im\xi)p(\xi)\,d\xi = \exp(-\tfrac{1}{2}m^2\sigma). \tag{5.6.6}$$

This factor is carried through to the end of the calculation, leading to the recursion relation

$$a_n(m_n, q_n) = \sum_{l_n} \mathscr{J}_{l_n}(|q_{n-1}|K) \exp(-\tfrac{1}{2}m_n^2\sigma) a_{n-1}(m_{n-1}, q_{n-1}) \tag{5.6.7}$$

in place of (5.5.14b). Using the Fourier paths shown in Fig. 5.12(b)–(f), we obtain the diffusion coefficient

$$D_n = \frac{K^2}{2}\left[\frac{1}{2} - \mathscr{J}_2(K)e^{-\sigma} - \mathscr{J}_1^2(K)e^{-\sigma} + \mathscr{J}_2^2(K)e^{-2\sigma} + \mathscr{J}_3^2(K)e^{-3\sigma}\right], \tag{5.6.8}$$

which is the generalization of (5.5.21), including the effect of the extrinsic stochasticity.

The more interesting case is the limit $K \ll 1$. In this case, Rechester et al. (1981) show that only paths involving the Bessel functions $\mathscr{J}_0$ and $\mathscr{J}_1$ contribute the lowest order in K, and furthermore, for $n \gg (2/K)^2$, these paths must start and end at $m = 0$ and $q = 0$. Summing these paths, they obtain the result to lowest order in K

$$\frac{D_n}{D_{QL}} = \tanh\frac{\sigma}{2}, \tag{5.6.9}$$

where $D_{QL} = K^2/4$ is the quasilinear result. The paths have to be summed numerically to find the higher-order corrections in K. Doing this, their results using the Fourier method (solid lines) are compared with direct numerical computations of D_n (circles) in Fig. 5.20. For small K the results asymptote to (5.6.9), while for large K and either large or small σ the curves asymptote to D_{QL}. The interesting region for our present consideration is for $K \lesssim 1$ for which KAM curves prevent intrinsic global stochasticity, but for which island amplitudes are significant.

We can understand these results intuitively from a simple calculation, based on some of the important features of the phase space curves of Fig. 4.4, reproduced in Fig. 5.21. We note first that on an isolating KAM curve a jump in θ due to the jump in ξ, as given in (5.6.3), leads to a jump in action δI for the particular KAM curve, given, for small ξ, by

$$\delta I = |I_m \cos\theta|\xi \tag{5.6.10}$$

such that when averaged over the distribution (5.6.4)

$$(\delta I)^2_{\text{rms}} = \tfrac{1}{2}I_m^2\sigma. \tag{5.6.11}$$

5.6. The Effect of Extrinsic Stochastic Forces

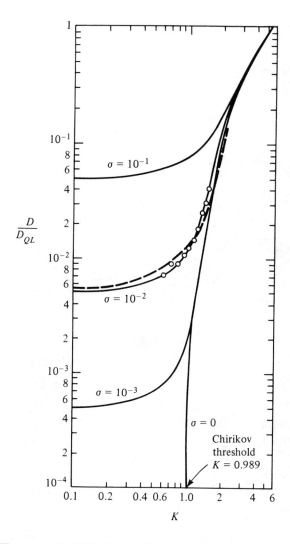

Figure 5.20. The normalized diffusion coefficient D/D_{QL} versus K for various values of external stochasticity σ; —, summation of Fourier paths; ---, heuristic theory; ○, numerical result for $\sigma = 10^{-2}$ (after Rechester et al., 1981).

From the standard Hamiltonian, as in (4.1.27), but taking the rotation rather than the libration orbits, a typical amplitude is $I_m \approx K/2$, giving a diffusion

$$D \approx \frac{(\delta I)^2_{\text{rms}}}{1} = \frac{K^2}{8} \sigma, \tag{5.6.12}$$

which is the result of the path integral method (5.6.9) for small σ and K.

However, for significant islands, as in Fig. 5.21, the results must be

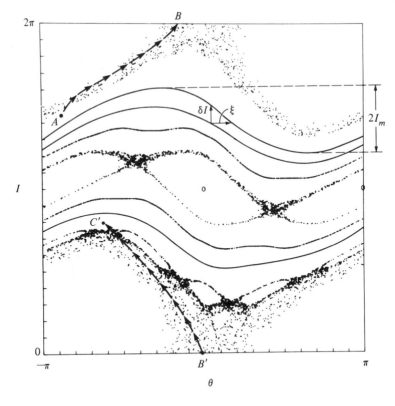

Figure 5.21. Phase plane for the standard mapping, illustrating a streaming trajectory that produces enhanced extrinsic diffusion for $K \lesssim 1$.

modified by removing the phase space for which island streaming can strongly enhance the diffusion. For example, considering a phase point that scatters into the stochastic region of the main island, shown as point A, then the subsequent trajectory shown by arrows $A \to B(B') \to C'$ takes the phase point around the island to C' where it may again be scattered onto an isolating KAM curve. If the time for oscillation T, obtained from (3.5.23), is small compared to $1/\sigma$ [which would be generally satisfied for small σ and $K = \mathcal{O}(1)$, since $T = \mathcal{O}(K^{-1/2})$], then the phase space in which diffusion occurs is shrunk by some factor proportional to the island width. Note that the diffusion from the trajectory $A \to C'$ can be either outward as described above, or inward onto the libration orbits surrounding the main island. These latter orbits do not contribute to a fast diffusion. Thus there are two classes of particles present in the system that contribute to the overall diffusion rate; a fast class that streams across the main island and slow class that diffuses into the island. We take as a simple estimate that the diffusion rate of particles diffusing fast to those diffusing slow is inversely proportional to the square

5.6. The Effect of Extrinsic Stochastic Forces

of the action distance traversed:

$$\frac{D_f}{D_s} = \frac{(2\pi)^2}{[2\pi - 2K^{1/2}]^2}, \qquad (5.6.13)$$

where, using (4.1.29), $2K^{1/2}$ is the width of the large island in Fig. 5.21, and we ignore the effects of other islands. We further consider that half the trajectories incident on A are trapped in the half-island oscillation from A to C', and that half of these are detrapped at C'. Thus one-fourth of the trajectories are in the fast class, streaming across from A to C', such that the effective diffusion is

$$D_{\text{eff}} = \tfrac{1}{4}D_f + \tfrac{3}{4}D_s. \qquad (5.6.14)$$

Applying (5.6.14) together with (5.6.12) and (5.6.13), we obtain the result shown as the dashed line in Fig. 5.20. We note that this heuristic calculation gives reasonable agreement even for K somewhat greater than 1, where the intrinsic stochasticity is global and would proceed for $\sigma = 0$. Near KAM surfaces that existed for $K < 1$, the trajectories still tend to follow these just-destroyed surfaces over long times. Thus the overall diffusion rate is still governed by the value of σ across these regions of the phase plane.

It is possible, even if significant portions of the space do not have large islands, to have strongly enhanced diffusion, with the island amplitude the characteristic step size. This can occur if a parameter, or a coordinate in another degree of freedom, diffuses, resulting in the diffusion of the resonant value of the action and thus the island center. This important diffusion mechanism inherently involves more than two degrees of freedom, and we defer its discussion to Chapter 6.

5.6c. Diffusion on Two Space and Time Scales

We have seen in the previous subsection that classes of particles, diffusing differently, are summed to obtain the overall diffusion rate. However, the diffusion is complicated by the requirement that the flux is conserved over regions of the action space for which diffusion is governed by slow extrinsic processes. This leads to an overall diffusion rate, as described in Section 5.6b, keyed to the slow diffusion rate but significantly enhanced by the island structure. The same situation was also found in Section 5.5b to exist in noise-free systems with K slightly greater than K_c. These are examples of diffusion on two space and time scales, in which the macroscopic distribution averaged over many 2π intervals in action space is slowly varying, while the local diffusion properties are quite nonuniform.

Another example of two space and time scale diffusion is the class of mappings treated in Section 3.6, which has both rotational and translational symmetry. As with the standard map, the calculation of diffusion applies both for purely intrinsic diffusion and with added extrinsic stochasticity. We consider the mapping of (3.6.5) which is the lowest-order four-step approximation to (3.6.3) with $q = 4$ (four kicks per gyroperiod). The global separatrix web has the macroscopically uniform structure shown in Fig. 3.23, while the structure within the web is quite nonuniform, as seen in Fig. 3.26. The macroscopic step size is $L_{\text{tile}} = \pi\sqrt{2}$, between adjacent cell centers when a separatrix is crossed. The macroscopic step time is complicated, and must be calculated as described below. The short time and space scales are described by the complicated dynamics of the separatrix mapping (3.6.9) and (3.6.10), but because the phase space is locally nearly uniform within the web, a phase space argument can be used to describe the average motion, obviating the need to calculate higher-order correlations near the $K = K_c$ border of the separatrix stochasticity.

Diffusion Without Extrinsic Stochasticity. Expanding the mapping (3.6.10) about a periodic point $\tau(\bar{w}) = l$, l integer, we obtain the standard mapping

$$\Delta I = K_s \sin \theta,$$
$$\Delta \theta = I, \tag{5.6.15}$$

where

$$K_s = \frac{\pi}{K_\alpha} \frac{w_0}{\bar{w}}. \tag{5.6.16}$$

At the stochastic border the Chirikov criterion, $K_s \approx 1$, gives

$$\bar{w} = w_1 = \pi w_0/K_\alpha. \tag{5.6.17}$$

Substituting (3.6.11) in (5.6.17), with $Q_0 = \pi/K_\alpha$, we have

$$w_1 = \frac{20\pi}{K_\alpha^3} \exp\left[-\frac{\pi^2}{2K_\alpha}\right]. \tag{5.6.18}$$

The border of stochasticity has been found numerically, agreeing well with (5.6.18) (Lichtenberg and Wood, 1989).

The important physics is in the determination of the number of mapping periods that are required, on average, for a single macroscopic random-walk step, L_{tile}. We make the following assumptions:

(1) The stochastic region is ergodic. This implies that the stochastically available phase space out to the KAM border is uniformly occupied in the asymptotic (long-time) limit.
(2) The effect of islands on the calculation can be ignored, resulting in a uniform phase space from the separatrix ($w = 0$) to the value $w = w_1$.
(3) In the small K_α limit, the lowest-order motion can be described by the continuous Hamiltonian (3.6.7), and the stochastic layer is sufficiently

5.6. The Effect of Extrinsic Stochastic Forces

narrow that the period in the stochastic separatrix is everywhere governed by (3.6.13).

With these assumptions we calculate the rms spreading from

$$L_{rms} = \frac{L_{tile} T^{1/2}}{(N\tau_{av})^{1/2}}, \qquad (5.6.19)$$

where T is the number of iterations, τ_{av} is the average rotation period, and N is the average number of rotations per separatrix crossing. The average period within the separatrix layer is

$$\tau_{av} = \frac{1}{w_1} \int_0^{w_1} P(w)\tau(w)\,dw,$$

where w_1 is the border of stochasticity (5.6.18). Since w is the canonical action, by assumption (1), $P(w) = 1$. Using this and $\tau(w)$ from (3.6.13), the integration yields

$$\tau_{av} = \frac{2}{K_\alpha}(3 - \ln w_1). \qquad (5.6.20a)$$

The logarithmic singularity at $w = 0$ is sufficiently weak that $\tau_{av} \approx \tau(w_1)$ as seen by comparing (5.6.20a) to (3.6.13). Substituting (5.6.18) in (5.6.20a) we obtain

$$\tau_{av} = \frac{\pi^2}{K_\alpha^2} + \frac{2}{K_\alpha}[3 - \ln(20\pi) + 3\ln K_\alpha]. \qquad (5.6.20b)$$

For small values of K_α the first term dominates and $\tau_{av} \propto 1/K_\alpha^2$. Equation (5.6.20) has been checked numerically by Lichtenberg and Wood (1989), giving excellent agreement.

Because of the correlations near KAM boundaries it is not possible to calculate separatrix crossings directly from a random walk. However, a simple phase space argument can be invoked to calculate the fraction of the phase space of the stochastic region that crosses the separatrix on each step of the separatrix map. The energy steps are given by $w_0 \sin\theta$, from (3.6.9), such that the phase space within $w_0 \sin\theta$ of the separatrix, with $\sin\theta$ negative, crosses the separatrix on each mapping step. Phase randomization then ensures that the phases are equally populated on each time step, such that the phase space crossing the separatrix on each step is $\int_\pi^{2\pi} w_0 \sin\theta = 2w_0$. Since the available phase space is $2\pi w_1$, a fraction $w_0/\pi w_1$ of the total phase space crosses on each separatrix mapping period. Since there are four such mapping periods in the original map (3.6.4) per period of the reduced map (3.6.5), the fraction of the phase space crossing on each iteration of the original map is just one-quarter of this value. The number of rotation periods

N for an initial phase point to take a random-walk step (separatrix crossing) is the inverse of this fraction, giving

$$N = \pi^2/4K_\alpha. \tag{5.6.21}$$

We substitute (5.6.20) and (5.6.21) in (5.6.19) to obtain the value of $L_{\rm rms}$ for any T. The global diffusion rate definition $D_{\rm web} \equiv L_{\rm rms}^2/T$ then gives

$$D_{\rm web} = \frac{8K_\alpha^3}{\pi^2 + 2K_\alpha[3 - \ln(20\pi) + 3\ln K_\alpha]}. \tag{5.6.22}$$

The result shows $D_{\rm web}$ scaling with K_α^3 to lowest order. This is slower than the usual quasilinear scaling that we found in Section 5.4. However, it is fast compared to the diffusion that occurs *along* stochastic layers in higher dimensions, as we shall see in Chapter 6.

The diffusive nature of the spreading of initial conditions is seen numerically from the basic $L_{\rm rms} \propto T^{1/2}$ dependence. The calculation of the number N of local oscillations (revolutions around a given mesh) for each separatrix crossing, from (5.6.21), was compared with numerical results, obtaining good agreement which improved as the number of iterations was increased from 10^5 to 5×10^5 (Lichtenberg and Wood, 1989). The difference between theory and experiment is qualitatively due to the correlations near the border of stochasticity and the deviations of the available phase space from the estimate based on (5.6.17). Lower iteration numbers give shorter separatrix crossing times, resulting from the difficulty in penetrating into the phase space close to the KAM barrier.

Using (5.6.22), $L_{\rm rms}$ (solid line) is compared to numerical calculations for 1000 initial conditions and for $T = 65{,}000$ iterations per initial condition (dots), for each K_α, in Fig. 5.22. We also calculate $L_{\rm rms}$, using the numerical separatrix crossing times (triangles). The difference between the dots and triangles for the two numerical procedures is indicative of the breakdown of ergodicity on short time scales. That is, phase space trajectories, from which the numerical diffusion rate is calculated employing 65,000 iterations, do not penetrate the phase space near the border of stochasticity as well as the trajectories from which the separatrix crossings are calculated, employing 5×10^5 iterations. We see that the long-time correlations near the stochastic borders, characteristic of a divided phase space, do not impede the diffusion. Rather, the inability of trajectories, on shorter time scales, to fill the regions of phase space in which strong correlations exist, tends to increase the rate of intermediate-time-scale diffusion over the asymptotic value.

Diffusion Including Extrinsic Stochasticity. We now introduce extrinsic stochasticity into the mapping equations through random jumps in both v_r and u_r in (3.6.5), each uniformly distributed between $\pm l$, imparted each mapping period. The extrinsic diffusion alone ($K_\alpha \to 0$), acting on the mapping, is itself a two space and time scale process, when considered over many tiles. Local

5.6. The Effect of Extrinsic Stochastic Forces

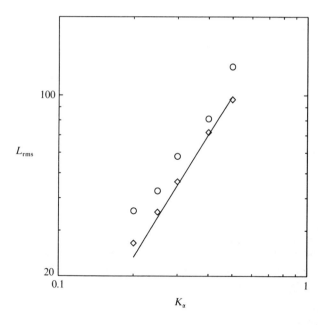

Figure 5.22. Comparison of theoretical (circles) and numerical values (diamonds and fitted line) of $L_{rms} = \Delta I_{rms} \propto D_{web}^{1/2}$ (for a fixed number of iterations).

diffusion within each tile results in separatrix crossings. Because of the rotation of particles within each tile on a fast time scale the result is to step the crossing particles the distance between tile centers. The global extrinsic diffusion D_{ex}, is then governed by the step size of a tile and a characteristic time for the exchange of tile phase space across the bounding separatrices,

$$D_{ex} = \frac{L_{tile}^2}{\tau_{ex}}, \quad (5.6.23)$$

where $L_{tile} = \sqrt{2\pi}$ for the particular mapping being considered. We can characterize τ_{ex} by

$$\tau_{ex} = (f_{ex} P_{ex})^{-1}, \quad (5.6.24)$$

where f_{ex} is the fraction of phase space that is accessible to crossing the separatrix on each mapping step due to the extrinsic stochasticity, and P_{ex} is the probability for that phase space to actually cross. Lichtenberg and Wood (1989a) use geometric arguments to calculate these quantities. The accessible phase space consists of four edge strips of length L_{tile} and width $l\sqrt{2}$, while the entire tile phase space is L_{tile}^2. The fraction of accessible phase space is then $f_{ex} = 4\sqrt{2}l/L_{tile}$. For uniformly distributed u_r and v_r, the probability of a step a distance δ across curves of constant Hamiltonian can be directly calculated to be

$$g(\delta) = (\tfrac{1}{2}l^2)(\sqrt{2l} - |\delta|).$$

The phase space crossing probability in the accessible layer is then

$$P_{\text{ex}} = \frac{1}{\sqrt{2l}} \int_0^{\sqrt{2l}} dx \int_x^{\sqrt{2l}} g(\delta) \, d\delta = \tfrac{1}{12}.$$

Substituting this into (5.6.24) results in

$$\tau_{\text{ex}} = \frac{3 L_{\text{tile}}}{\sqrt{2l}}. \tag{5.6.25}$$

The addition of extrinsic noise to the intrinsically diffusing particles can either increase or decrease the diffusion. For small noise amplitudes the primary effect of the noise is to make the entire phase space accessible to the diffusing particles by filling the interior of each tile. The diffusion rate is then reduced by a factor

$$f = \frac{4 w_1 L_{\text{tile}}}{L_{\text{tile}}^2}, \tag{5.6.26}$$

which is the ratio of the area of intrinsically stochastic phase space to the entire phase space. The diffusion due to extrinsic stochasticity becomes equal to this reduced intrinsic diffusion for $\tau_{\text{ex}} = \tau_{\text{tile}}$, where $\tau_{\text{tile}} = N\tau_{\text{tile}}/f$ and where $N\tau_{\text{av}}$ is the intrinsic separatrix crossing time. For larger l at a given K_α (or smaller K_α at a given l) the extrinsic diffusion becomes increasingly dominant.

The competition is illustrated numerically in Fig. 5.23 in which for a fixed N, L_{rms} is plotted against K_α, with noise as a parameter. The case with no noise (circles), with $L_{\text{rms}} \propto K_\alpha^{3/2}$, is repeated from Fig. 5.22, including two more values at large K_α. The diffusion with noise lies above the diffusion without noise for values of K_α at which the extrinsic diffusion is larger than the web diffusion. For $K_\alpha = 0.1$, we expect the diffusion to be governed by the extrinsic noise alone. For noise $l = \tfrac{1}{20}$, we calculate using (5.6.23) after 65,000 iterations that $L_{\text{rms}} = 82$ compared to the value $L_{\text{rms}} = 60$, found numerically. An interesting feature of Fig. 5.23 is the break in the slope of L_{rms} with noise. This should occur when $\tau_{\text{tile}} \approx \tau_{\text{ex}}$. Using the leading term in τ_{av} from (5.6.20) and N from (5.6.21) to determine τ_{tile}, we obtain

$$\frac{\pi^4}{4 K_\alpha^3 f(K_\alpha)} = \frac{3\pi}{l}, \tag{5.6.27}$$

which agrees with the positions of the slope changes in Fig. 5.23. Beyond this value of K_α the values from (5.6.23) of D_{ex} with extrinsic noise should be roughly a factor f below the values from (5.6.22) of D_{web} without noise. For $l = \tfrac{1}{20}$ this break occurs at $K_\alpha \approx 0.7$ for which $D_{\text{ex}}/D_{\text{web}} \approx 0.3$ in

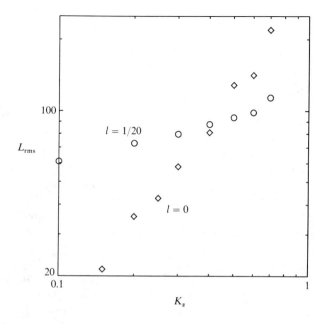

Figure 5.23. Diffusive spread of the distribution versus K_α with l as a parameter, after 2^{16} iterations for 1,000 initial conditions in the web (after Lichtenberg and Wood, 1989a); circles, $l = 1/20$; diamonds, $l = 0$.

reasonable agreement with the phase space area ratio $f \approx 0.3$. Since w_1 and therefore f, from (5.6.26) increases exponentially with K_α, D_{ex}/D_{web} increases rapidly with increasing K_α beyond the break.

5.7. Slow Separatrix Crossings

We consider the change in action and the resulting phase space diffusion for a slowly varying nonlinear oscillator described by the Hamiltonian

$$H = \tfrac{1}{2}I^2 + A(\epsilon t) \cos(q + \psi(\epsilon t)), \qquad (5.7.1)$$

where $\epsilon \ll 1$ indicates that the amplitude A and frequency modulation ψ vary slowly with respect to the fundamental frequency $A^{1/2}$. With $\epsilon = 0$, the phase space trajectories are just those of the pendulum, as given in Fig. 1.4. For ϵ finite, the separatrix position changes in time by changing the fixed point position through changing ψ, by changing the amplitude through changing A, or by changing both. For this "adiabatic" problem, away from any separatrix the action is an adiabatic invariant, as described in Section 2.3. However, because the frequency of the oscillator goes to zero on the separatrix, the adiabatic assumption must break down there, such that the action for orbits crossing the separatrix is not conserved.

This problem has been considered in various contexts, particularly for particle motion in a plasma acted upon by a wave with a modulated amplitude. It has been considered both numerically and in various analytical approximations by many authors (e.g., Aamodt and Jaeger, 1974; Timofeev, 1978; Tennyson, 1979; Chirikov et al., 1980; Cary et al. 1986; and Bruhwiler and Cary, 1989). There are two points of view for looking at the effect of slow variation. One, following Chirikov, is to obtain the region of overlap of nonlinear resonances. The second, following Timofeev (1978) and Cary et al. (1986), is to directly calculate the action jump and the resulting diffusion in the phase space region which has been traversed by a separatrix. Although reasonable estimates of diffusion have been made over some parameter ranges from the resonance overlap viewpoint (e.g., Ram et al., 1986), the latter point of view must be employed for arbitrary slow passage through resonance.

Preliminary to our calculations, we first show that a slow variation of ψ (frequency modulation) is equivalent to sweeping a separatrix, in action, through phase space. To see this we transform (5.7.1), using the generating function

$$F_2(q, \bar{I}, t) = [q - \psi(t)]\bar{I} + \int^t \dot\psi^2 \, dt' - \int^t A(t') \, dt', \qquad (5.7.2)$$

to obtain a new Hamiltonian

$$\bar{H} = \tfrac{1}{2}(\bar{I} - \dot\psi)^2 - A(1 - \cos \bar{q}), \qquad (5.7.3)$$

where $\bar{I} = I$ and $\bar{q} = q - \psi(t)$. The fixed point in action is just $\bar{I} = \dot\psi$, and thus at constant A the separatrix traverses the phase space as $\dot\psi$ varies.

Resonance Overlap. Following the general method of Chirikov et al. (1980), we illustrate the formation of a region of resonance overlap and stochasticity for the Hamiltonian

$$H = \tfrac{1}{2}I^2 - k \cos(\theta + \lambda \sin \Omega t). \qquad (5.7.4)$$

Here the oscillator is phase modulated with peak amplitude $\lambda = \Delta\omega/\Omega$ at the slow modulation frequency Ω, such that the frequency modulation amplitude $\Delta\omega$ is held fixed as Ω is varied. Expanding the second term in (5.7.4) in a Fourier series yields

$$\cos(\theta + \lambda \sin \Omega t) = \sum_{n=-\infty}^{\infty} \mathscr{J}_n(\lambda) \cos(\theta + n\Omega t). \qquad (5.7.5)$$

The Bessel functions $\mathscr{J}_n$ have significant amplitude provided $|n| \lesssim \lambda$. The result is the formation of a multiplet layer of driving resonances of width approximately $2\Omega\lambda$ centered about the frequency $\omega = I = 0$.

5.7. Slow Separatrix Crossings

The motion within the multiplet is described by the Hamiltonian

$$H = \tfrac{1}{2}I^2 - k \sum_n \mathcal{J}_n(\lambda) \cos(\theta + n\Omega t). \tag{5.7.6}$$

Proceeding as in Section 2.4a, we have $G = 1$ and $F = k\mathcal{J}_n(\lambda)$. The full separatrix width in action for each resonance in the multiplet is, from (2.4.31),

$$2\Delta I_{max} = 4(F/G)^{1/2} = 4|k\mathcal{J}_n(\lambda)|^{1/2}. \tag{5.7.7}$$

The spacing between resonances in frequency is

$$\delta\omega = \Omega.$$

Using the two-thirds rule (see Section 4.7), the overlap condition in frequency space is

$$\frac{2\Delta\omega_{max}}{\Omega} > \frac{2}{3}, \tag{5.7.8}$$

where $\Delta\omega_{max}$ is the resonance width in frequency space. Noting that $\Delta\omega = G\Delta I$ (with $G = 1$), substituting (5.7.7) in (5.7.8), and approximating the Bessel function by its rms value $(\pi\lambda)^{-1/2}$, and substituting $\lambda = \Delta\omega/\Omega$, the overlap condition becomes

$$k \gtrsim \frac{\Omega^{3/2}(\Delta\omega)^{1/2}}{20}. \tag{5.7.9}$$

We have the surprising result that the overlap occurs for weaker perturbations if the modulation is slower, i.e., the minimum k for overlap is related to the modulation frequency by $k \propto \Omega^{3/2}$. We note, however, that the passage through exact resonance, which leads to stochastic motion, occurs only twice every period $2\pi/\Omega$ of the slow modulation. As $\Omega \to 0$, the time scale for stochastic diffusion tends to infinity. Since the width of the multiplet layer is proportional to $\lambda\Omega$ and thus for $\lambda = \Delta\omega/\Omega$, $\Delta\omega$ = constant, the width of the layer also remains constant. In the limit $\Omega \to 0$, the number of resonances increases faster than the resonance amplitude decreases, thus always leading to a stochastic layer. This layer was investigated by Tennyson (1979) in connection with storage ring dynamics. The result is illustrated in Fig. 5.24 for the mapping equivalent to (5.7.3):

$$\begin{aligned} I_{n+1} &= I_n + K \sin \theta_n, \\ \theta_{n+1} &= \theta_n + I_{n+1} + \Delta\omega \cos(\Omega n \Delta t), \end{aligned} \tag{5.7.10}$$

where $K = 0.007$ and $\Omega\Delta t \ll 1$. For $\Delta\omega = 0$, the single resonance is seen in Fig. 5.24a with $2\Delta I_{max} = 4K^{1/2}$. For $\Delta\omega = 0.63$, three cases are shown for successively smaller values of Ω. In Fig. 5.24b the multiplet layer, without overlap, is seen, having a total width somewhat larger than $2\Delta\omega$. In Fig. 5.24c there is some, but not total, resonance overlap, and in Fig. 5.24d the

366 5. Stochastic Motion and Diffusion

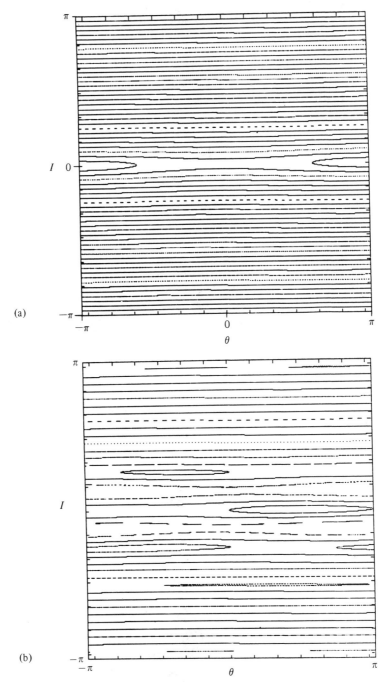

Figure 5.24. The effect of a slow modulation of the frequency (phase shift per mapping period) for the standard mapping: (a) no modulation; (b), (c), and (d) decreasing modulation frequency while holding modulation amplitude constant.

5.7. Slow Separatrix Crossings

(c)

(d)

Figure 5.24—continued

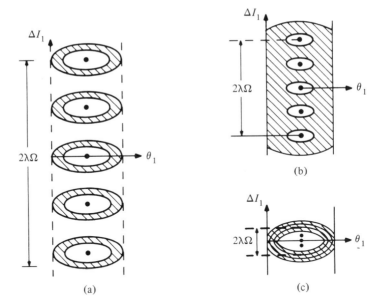

Figure 5.25. Three regimes in the formation of an overlapping layer of modulational resonances as the modulation frequency Ω is varied. (a) High Ω leads to nonoverlapping resonances within the multiplet. There is Arnold diffusion but no modulational diffusion. (b) Intermediate Ω leads to overlapping of modulational resonances and strong modulational diffusion. (c) Low Ω leads to formation of a trapping regime and weak modulational diffusion (after Lieberman and Tennyson, 1982).

resonance overlap is complete. The width ΔI of the region of overlap is the region crossed by the slowly varying separatrix. From (5.7.3) and using $\dot{\psi} = \lambda\Omega \cos \Omega t$, we obtain $\Delta I \approx 2\lambda\Omega$, in agreement with the width determined from the Bessel function amplitudes in (5.7.5).

If the phase modulation λ is held constant as Ω is varied, rather than the frequency modulation $\Delta\omega$ held constant, then the layer becomes narrower as Ω becomes smaller. Although overlap still occurs, for sufficiently small Ω, we may have $\Delta\omega_{\max} > \lambda\Omega$, occurring when

$$k \gtrsim \frac{\Omega^2 \lambda^{5/2}}{13}. \qquad (5.7.11)$$

The multiplet then merges into a single-resonance trapping layer that corresponds to a weaker diffusion. The three regimes defined by the two inequalities (5.7.9) and (5.7.11) (thin isolated layers, thick layer, and trapping) are shown schematically in Fig. 5.25.[6]

---

[6] In the third regime the width of the merged layer may be better approximated by $\Delta I_{\max} = 4K^{1/2}$. In this case, (5.7.11) becomes $k \gtrsim \Omega^2 \lambda^2 / 16$.

5.7. Slow Separatrix Crossings

The multiplet layer formed by modulation of a single degree of freedom can lead to enhanced diffusion *along* resonances if that degree of freedom is coupled to another degree of freedom. Chirikov et al. (1980) have treated this problem for phase modulation with constant λ, and we shall return to it in Section 6.2d.

Change in Action for One Resonance Crossing. For slow separatrix crossing, the diffusion is most straightforwardly calculated from a direct determination of the change in action for a single crossing. The general case of both time varying A and ψ has been considered (Cary et al., 1986). We summarize their results for the simpler case of variation of A alone. For this case, the crossings are between rotation and libration and there are no crossings from rotation above to below the separatrix. We define the action for libration for (5.7.1) with A and ψ constant as

$$J = \tfrac{1}{2} \oint I \, dq, \tag{5.7.12}$$

and for rotation as

$$J = \int_0^{2\pi} I \, dq, \tag{5.7.13}$$

so that J remains constant when crossing from rotation to libration for $\dot{A} \to 0$. Then Timofeev (1978) and Cary et al. (1986) have calculated the change in action for a separatrix crossing to be

$$\delta J = -(4\dot{A}/A) \ln \left| 2 \sin \frac{\pi h A^{1/2}}{4\dot{A}} \right|, \tag{5.7.14}$$

where the order-ϵ, phase-dependent crossing parameter h is the difference in energy between the separatrix energy and the energy at the crossing of the unstable fixed point line $q = 0$ (outside the separatrix) that is closest to the separatrix crossing. Note that for the choice of sign in (5.7.1), the unstable fixed point is at $q = 0$. Cary et al. (1986) have shown that $0 < h < |4\dot{A}/A^{1/2}|$ and thus the logarithm is well behaved as $\dot{A} \to 0$. The quantities A and $\dot{A}$ are evaluated at the time that the initial action coincides with the separatrix action, but this need not be determined exactly.

We take A explicitly to be

$$A = A_0 - A_1 \cos \epsilon t \tag{5.7.15}$$

with $A_1 < A_0$. The separatrix action J_{sx} is found from (5.7.13) to be

$$J_{sx} = 8A^{1/2}. \tag{5.7.16}$$

For those trajectories that cross the separatrix, the crossing period is just

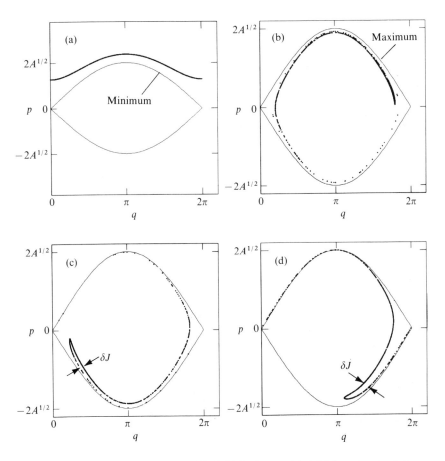

Figure 5.26. The spread in action with initial phases: (a) initial state at minimum separatrix width; final state at maximum separatrix width for (b) $\epsilon = 0.1$, (c) $\epsilon = 0.4$, (d) $\epsilon = 0.8$ (after Bruhwiler and Cary, 1989).

half the period of oscillation in (5.7.15)

$$\tau = \pi/\epsilon. \quad (5.7.17)$$

Because h is phase dependent, the values of δJ depend on the phases of the initial conditions on any particular curve of constant adiabatic invariant J, away from the separatrix. This is illustrated in Fig. 5.26 for a set of initial conditions at the time of minimum separatrix width shown in (a) that cross the separatrix, as it expands to its maximum width for $\epsilon = 0.1$, 0.4, and 0.8 in (b), (c), and (d), respectively. As ϵ becomes larger, the maximum δJ increases, with $\delta J \propto \epsilon$ as shown.

Diffusion Coefficient. The local quasilinear diffusion coefficient is defined in

5.7. Slow Separatrix Crossings

the limit of random steps in J as

$$D_{QL}(J) = \frac{\langle(\delta J)^2\rangle}{2\tau}, \tag{5.7.18}$$

where the $\langle \ \rangle$ denotes a phase-average over the values of h in (5.7.14). Cary et al. (1986) show that, to order ϵ, a uniform distribution of initial conditions over an adiabatic invariant annulus $(J_i, J_i + \delta J_i)$ corresponds to a uniform distribution of crossing parameters h. We can therefore average over a uniform distribution of h to evaluate (5.7.18). From (5.7.14)

$$\langle(\delta J)^2\rangle \cong \left(\frac{4\dot{A}}{A}\right)^2 \frac{1}{h_{\max}} \int_0^{h_{\max}} \left(\ln\left|2\sin\frac{\pi h A^{1/2}}{4\dot{A}}\right|\right)^2 dh, \tag{5.7.19}$$

where $h_{\max} = |4\dot{A}/A^{1/2}|$. This can be integrated to obtain

$$\langle \delta J^2 \rangle = \frac{4\pi^2}{3}\left(\frac{\dot{A}}{A}\right)^2. \tag{5.7.20}$$

Since from (5.7.15)

$$\dot{A}^2 = \epsilon^2[A_1^2 - (A - A_0)^2], \tag{5.7.21}$$

we see that $\langle(\delta J)^2\rangle \propto \epsilon^2$. The period between kicks is inversely proportional to ϵ, such that the quasilinear diffusion coefficient, using (5.7.16) and (5.7.17) is

$$D_{QL} = \frac{2\pi}{3}\epsilon^3 \frac{A_1^2 - ((J/8)^2 - A_0)^2}{(J/8)^4}. \tag{5.7.22}$$

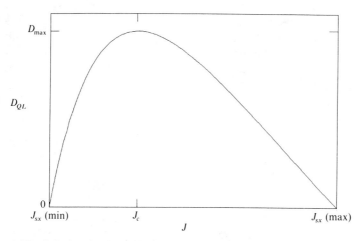

Figure 5.27. Calculated value of D_{QL} as a function of J for a given ϵ (after Bruhwiler and Cary, 1989).

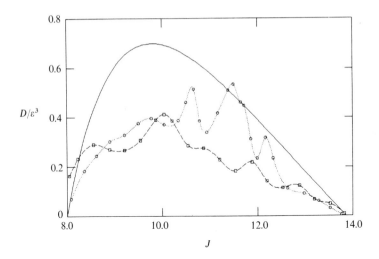

Figure 5.28. Comparison of theory for D_{QL} (solid line) and numerical values, verifying the ϵ^3 scalings; □ $\epsilon = 0.1$; ○ $\epsilon = 0.02$ (after Bruhwiler and Cary, 1989).

D_{QL} is plotted versus J in Fig. 5.27. Bruhwiler and Cary (1989) have checked the ϵ^3 scaling and found it to hold for $\epsilon < 0.4$. This is shown in Fig. 5.28, where D_{QL}/ϵ^3 from the theory is compared to a numerical calculation at $\epsilon = 0.1$ and 0.02. The two numerical results have roughly the same magnitude, thus verifying the ϵ^3 scaling. However, numerical results are as much as a factor of two lower than the theory predicts, indicating that phase correlations significantly affect the size and shape of D_{QL}.

Cary and Skodje (1988) have obtained corrections to quasilinear diffusion due to phase correlations on successive crossings. They show that there is roughly a twenty percent variation in D that is highly oscillatory in ϵ with a period ΔJ of order $1/\epsilon$. It is reasonable to infer that the effect of correlations should also lead to oscillations of D_{QL} with J, as seen in Fig. 5.28. Analogous oscillations for fast resonance crossing were obtained by Murray et al. (1985); (see also Fig. 5.15).

CHAPTER 6

Three or More Degrees of Freedom

*6.1. Resonance in Multidimensional Oscillations

We have seen that resonances between oscillators in two degrees of freedom lead to the formation of a dense set of resonance layers in the action space. Within each layer, stochastic motion exists. However, energy conservation prevents large excursions of the motion along the layer. Only motion across the layer is important. For a near-integrable system with a weak perturbation, the stochastic layers are isolated by KAM surfaces.

For three or more degrees of freedom, two new effects appear: (1) Resonance layers are no longer isolated by KAM surfaces. Generically, the layers intersect, forming a connected web dense in the action space. (2) Conservation of energy no longer prevents large stochastic motions of the actions along the layers over long times. As a result, excursions of the actions along resonance layers are *generic* in systems with three or more degrees of freedom. Furthermore, the interconnection of the dense set of layers ensures that the stochastic motion, stepping from layer to layer, can carry the system arbitrarily close to any region of the phase space consistent with energy conservation, although the time required to explore the phase space may be very long.

The basic mechanism of intrinsic diffusion along resonance layers for autonomous systems of three or more degrees of freedom (or nonautonomous systems with two degrees of freedom) is called *Arnold diffusion* after V. I. Arnold who first described its existence (Arnold, 1964). Arnold diffusion is universal in that there is no critical perturbation strength for its presence (although the diffusion rate vanishes with vanishing perturbation strength). There have been a number of numerical investigations of this phenomenon (e.g., Froeschlé, 1971, 1972; Chirikov, 1971; Froeschlé and Scheidecker, 1973;

at least for simple cases of the interaction of only three resonances, there have been comparisons with theoretical models (Chirikov, 1971, 1979; Gadiyak *et al.*, 1975).

Although Arnold diffusion is usually associated with nonoverlapping resonances (see Chirikov, 1979), it can equally well be associated with groups of overlapping resonances, which strongly enhance the rate of diffusion. A model problem illustrating both these regimes is that of a ball bouncing between a flat wall and a wall periodically rippled in two dimensions. This configuration, analogous to the Fermi mapping with an additional degree of freedom, was used by Tennyson *et al.* (1979) to calculate the diffusion rate and will be used as our basic example to illustrate the calculation method and the results.

A particular example of diffusion along layers of overlapping resonances is that of *modulational diffusion*, which we treat in Section 6.2d. In this situation, a slow variation of a fundamental frequency with time creates sidebands that can overlap in certain parameter ranges. Modulational diffusion is not universal; i.e., there is a critical perturbation strength below which the sideband resonances do not overlap. It is of particular interest in that the overlap occurs if the modulation frequency is slow compared to the frequency being modulated. This result is somewhat counterintuitive, i.e., one might expect adiabatic behavior in this limit. Modulational diffusion effects seem to be important for determining the beam characteristics in high-energy colliding beam storage rings (Tennyson, 1979; Israilev *et al.*, 1977).

In addition to intrinsic stochasticity, the presence of extrinsic noise also produces chaotic motion along resonance layers in multidimensional systems. In Section 6.3 we examine the effect of weak, external stochasticity. If the motion in the absence of external noise is an oscillation within a resonance layer, then the classical transport due to external noise or dissipation can be strongly enhanced along the layer. This diffusion was treated by Chirikov (1979a) and by Tennyson (1982). An important example of diffusion in a system with three degrees of freedom in the presence of extrinsic stochasticity is particle confinement in a toroidal magnetic field. We treat this problem in Section 6.4, comparing the theoretical predictions with numerical simulations.

In Section 6.5, we review briefly the limit where the number of degrees of freedom becomes large. We find examples for which the motion is essentially regular, and other examples for which the motion is essentially stochastic.

*6.1a. Geometric Relations

We consider an integrable Hamiltonian system with N degrees of freedom. In action-angle form,

$$H_0 = H_0(\boldsymbol{I}),$$

6.1. Resonance in Multidimensional Oscillations

where I is the N-tuple of actions. The motion in the $2N$-dimensional phase space (I, θ) is on an N-dimensional torus defined by the N-tuple of angles θ conjugate to I:

$$I(t) = I_0, \qquad \theta(t) = \omega(I)t + \theta_0, \qquad (6.1.1)$$

where

$$\omega_j(I) = \frac{\partial H_0}{\partial I_j} \qquad (6.1.2)$$

is a component of the N-vector of unperturbed frequencies.

Action Space. Figure 6.1 shows the N-dimensional action space. For the unperturbed system, the actions are conserved. One may define an $N - 1$ dimensional energy surface by the condition

$$H_0(I) = \alpha.$$

For example, for free particle motion in N-dimensions,

$$H_0 = \sum_{j=1}^{N} I_j^2;$$

the energy surfaces are spheres, as shown in the figure.

One may also define an $N - 1$ dimensional resonance surface by the condition

$$\mathbf{m} \cdot \boldsymbol{\omega}(I) = 0, \qquad (6.1.3)$$

where $\mathbf{m}$ is called the resonance vector and has integer components. Since there is a resonance surface for each resonance vector, these surfaces are dense in the action space. For the free particle, several resonance surfaces are shown as the flat planes in Fig. 6.1.

We consider now the effect of a small perturbation, periodic in θ:

$$H = H_0(I) + \epsilon H_1(I, \theta),$$

$$H_1 = \sum_k V_k(I) \exp(i\mathbf{m}_k \cdot \boldsymbol{\theta}), \qquad (6.1.4)$$

where k represents the sum over all resonance vectors $\mathbf{m}_k$. The motion in action space is

$$\dot{I} = \frac{-\partial H}{\partial \boldsymbol{\theta}} = -i\epsilon \sum_k \mathbf{m}_k V_k \exp(i\mathbf{m}_k \cdot \boldsymbol{\theta}). \qquad (6.1.5)$$

We see that each component k drives an oscillation in I in the direction $\mathbf{m}_k$. For most k's the oscillation will be nonresonant, $\mathbf{m}_k \cdot \boldsymbol{\theta}(t) \neq \text{const.}$, and the amplitude of the oscillation in I will be of order ϵ. However, for some value

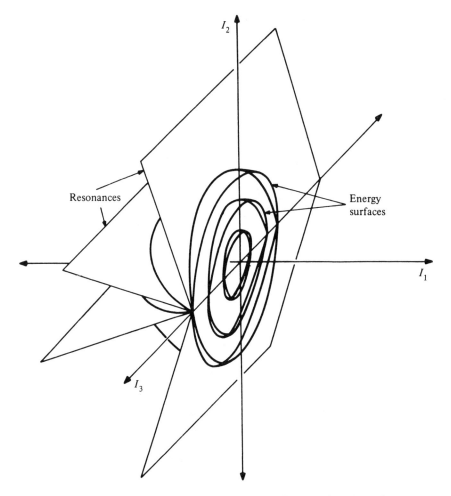

Figure 6.1. The action space showing energy surfaces (spheres) and resonance surfaces (planes) for the unperturbed, free particle Hamiltonian (after Lieberman and Tennyson, 1982).

$k = R$, we may find a resonant motion

$$\boldsymbol{m}_R \cdot \boldsymbol{\theta}(t) = \theta_R = \text{const.}, \tag{6.1.6}$$

where θ_R is the resonance phase. In the direction of $\boldsymbol{m}_R$, which we define to be the direction of the resonance action I_R, we have seen by use of secular perturbation theory in Section 2.4, that the amplitude of the oscillation is of order $\epsilon^{1/2}$.

As an example, Fig. 6.2 shows some resonance surfaces and energy surfaces

6.1. Resonance in Multidimensional Oscillations

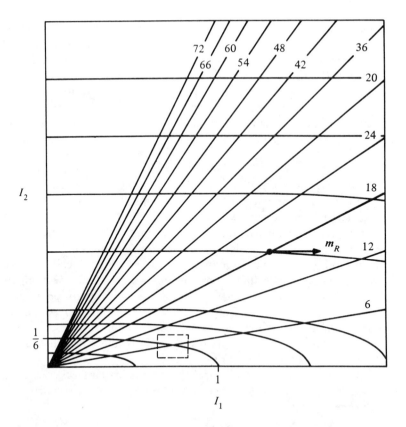

Figure 6.2. Resonance curves (lines) and energy contours (ellipses) in two-dimensional action space. The Hamiltonian is $H_0(I) = I_1^2 + (6I_2)^2$. The resonance labels are the values of m_1, where $\omega_1 m_1 + \omega_2 = 0$. The region in the dashed box is enlarged in Fig. 6.15 (after Tennyson, 1982).

for the two-degree-of-freedom Hamiltonian

$$H_0 = I_1^2 + (6I_2)^2. \tag{6.1.7}$$

The resonance surfaces, from (6.1.3), are lines in the action space given by

$$m_1 I_1 + 36 m_2 I_2 = 0. \tag{6.1.8}$$

Some of these (for $m_2 = 1$) are plotted in the figure. Note that since

$$m_R \cdot \frac{\partial H_0}{\partial I} = 0 \tag{6.1.9}$$

at resonance, the resonance vector m_R lies in the unperturbed energy surface, as shown in the figure. In general, m_R, the direction of the resonance action excursion, is not perpendicular to the resonance surface. For a resonance vector that intersects a resonance surface at an oblique angle, as shown in the dashed box, strongly enhanced diffusion can occur, as we shall describe in Section 6.3. It can be seen from Fig. 6.2 that even for arbitrary m_1 and m_2, resonance surfaces do not intersect on a constant (nonzero) energy surface. This property is generic for systems with two degrees of freedom.

For three or more degrees of freedom, resonance surfaces generically intersect. We illustrate this in Fig. 6.3a for a free particle in three dimensions

$$H_0 = \tfrac{1}{2}(I_1^2 + I_2^2 + I_3^2).$$

The resonance surfaces are planes that pass through the origin in action space. The two planes intersect at nonzero actions along a line, as shown in the figure. The resonance surfaces also intersect a spherical energy surface $H_0(I) = \alpha$ in great circle meridians (for this example, the resonance vectors m_R happen to lie perpendicular to the resonance planes). An energy-conserving motion from one resonance to another is possible. The motion may proceed along a meridian of one resonance to an intersection, turn sharply, and then move along a new meridian. This type of motion is generic to systems with three or more freedoms. The intersection of resonances in the constant energy surface generates a dense interconnected network, the so-called Arnold web. The web for this example is illustrated in Fig. 6.3b, with all resonances shown for which $|m_j| \leq 2$.

In the $2N$-dimensional phase space, the resonance surfaces, defined by (6.1.3) have dimension $2N - 1$. The KAM surfaces, being perturbed tori defined by the condition

$$I = \text{const.},$$

are N-dimensional. The interconnection of resonance layers into the Arnold web can then be understood geometrically. For $N \geq 3$, the $(2N - 1)$-dimensional resonance surfaces cannot be isolated from each other by N-dimensional KAM surfaces (see Fig. 1.16). We have also seen a resonance layer in a three-dimensional projection in Fig. 1.17, taking as the resonant action $J_1 = I_R$ and an additional action variable $J_2 = I_S$. The layer extends along the unprojected action I_S (although its properties, such as its thickness, may vary with I_S). The KAM surfaces near exact resonance appear as elliptical tubes within the layer. In this figure, I_S represents one of the $N - 1$ action variables (excluding the resonance action I_R), which define motion *along* a layer.

The essential feature of the motion along resonance layers can be illustrated from the following argument. Assume an energy-conserving change in the actions such that the Hamiltonian stays fixed,

$$\Delta H = \Delta H_0 + \epsilon \Delta H_1 = 0,$$

with the resonance action I_R confined to the resonance layer

$$\Delta I_R = \mathcal{O}(\epsilon^{1/2}); \qquad (6.1.10)$$

6.1. Resonance in Multidimensional Oscillations

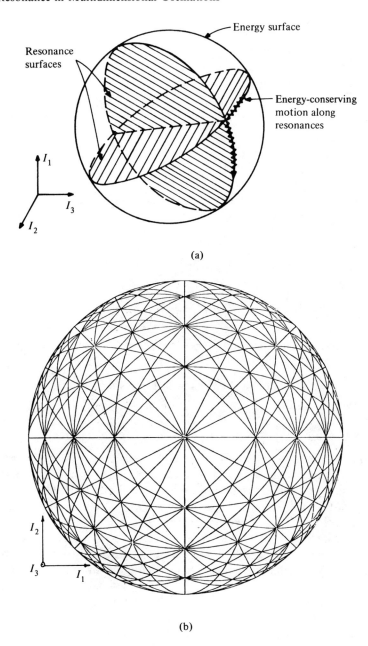

Figure 6.3. Illustrating Arnold diffusion. (a) Intersection of two resonance surfaces in an action space having three degrees of freedom. An energy-conserving motion (wiggly line) from one resonance surface to another is possible (after Lieberman and Tennyson, 1982). (b) The Arnold web for the free particle Hamiltonian. Only some of the intersecting resonances are shown (after Tennyson et al., 1979).

then for three degrees of freedom,

$$\frac{\partial H_0}{\partial I_R}\Delta I_R + \frac{\partial H_0}{\partial I_S}\Delta I_S + \frac{\partial H_0}{\partial I_T}\Delta I_T = \mathcal{O}(\epsilon). \tag{6.1.11}$$

Even though (6.1.10) holds, large excursions in the two actions I_S and I_T along the resonance are possible; i.e., ΔI_S and ΔI_T can both be large with (6.1.10) only constraining

$$\frac{\partial H_0}{\partial I_S}\Delta I_S + \frac{\partial H_0}{\partial I_T}\Delta I_T = \mathcal{O}(\epsilon^{1/2}).$$

If, on the other hand, there are only two degrees of freedom, then $\Delta I_T \equiv 0$, and (6.1.11) together with (6.1.10) implies

$$\Delta I_S = \mathcal{O}(\epsilon^{1/2}),$$

and large excursions along resonance layers are forbidden.

*6.1b. Examples of Arnold Diffusion

The existence of diffusion along a resonance was demonstrated rigorously by Arnold (1964) for the specific Hamiltonian

$$H = \tfrac{1}{2}(I_R^2 + I_S^2) + \epsilon(\cos\theta_R - 1)(1 + \mu\sin\theta_S + \mu\cos t).$$

For $\mu = 0$ and $\epsilon \neq 0$, there are two integrals of the motion, $I_S = $ const. and

$$H_R = \tfrac{1}{2}I_R^2 + \epsilon\cos\theta_R = \text{const.},$$

with H_R describing the motion of the resonance action I_R. The perturbation terms ($\mu \neq 0$) lead to the formation of a stochastic layer around the separatrix of the resonance. Since the perturbation drives both I_R and I_S, a chaotic motion along the stochastic layer (along I_S) occurs. The third degree of freedom here is the time t and its action $-H$.

Arnold conjectured that diffusion along resonances was generic to many-dimensional nonlinear oscillations, but no rigorous proof was known. Holmes and Marsden (1981), applying a method due to Melnikov (1963; see Section 7.7), have established Arnold diffusion for a large class of near-integrable Hamiltonian systems with at least three degrees of freedom.

Numerical studies of chaotic motion in systems with at least three degrees of freedom were performed by Froeschlé and his associates. Froeschlé (1971; 1972) studied numerically the number of isolating integrals in three-dimensional systems, concluding that, as the initial conditions are varied, either two or zero integrals (beside the usual energy integral) exist. This result is in agreement with Arnold's theoretical prediction that the motion is either on a KAM torus or in the stochastic web. The numerical techniques involved the use of projections of thin slices of the four-dimensional surface

6.1. Resonance in Multidimensional Oscillations

of section onto a flat plane, calculation of approximate measures for finite orbit segments, and studies of the maximal Liapunov exponent (see Section 5.2). In later studies (Froeschlé and Scheidecker, 1973), the variation of the entire set of Liapunov exponents was determined numerically, and a simple "Gambler's ruin" model was used to estimate the (Arnold) diffusion rate. Applying this technique to both three- and four-degree-of-freedom systems (respectively four- and six-dimensional mappings), the results were found to be in agreement with Froeschlé's conjecture that a dynamical system with N degrees of freedom has in general $N - 1$ or zero integrals (beside the energy integral). These two cases correspond, respectively, to either regular motion on a KAM torus or to chaotic motion within the Arnold web in action space.

The Billiards Problem. Froeschlé's four- and six-dimensional mappings were coupled systems of two-dimensional Chirikov–Taylor standard mappings. To illustrate the calculation of Arnold diffusion, we consider a simpler but related mapping that approximately describes the three-dimensional motion of a ball bouncing between a smooth wall at $z = h$ and a fixed wall at $z = 0$, which is rippled in two dimensions x and y. The surface of section is given in terms of the ball positions in the x_n and y_n directions and the trajectory angles $\alpha_n = \tan^{-1} v_x/v_z$ and $\beta_n = \tan^{-1} v_y/v_z$, evaluated just before the nth collision with the rippled wall. The ball motion is shown schematically in Fig. 6.4a, and the definitions of the variables in the x, z plane are shown in Fig. 6.4b. Assuming that the ripple is small, the rippled wall may be replaced by a flat wall at $z = 0$ whose normal vector is a function of x and y, analogous to the idea of a Fresnel mirror. The simplified difference equations exhibit the general features of the exact equations, and may be written in explicit form

$$\alpha_{n+1} = \alpha_n - 2a_x k_x \sin k_x x_n + \mu k_x \gamma_c,$$
$$x_{n+1} = x_n + 2h \tan \alpha_{n+1},$$
$$\beta_{n+1} = \beta_n - 2a_y k_y \sin k_y y_n + \mu k_y \gamma_c,$$
$$y_{n+1} = y_n + 2h \tan \beta_{n+1},$$
(6.1.12)

where $\gamma_c = \sin(k_x x_n + k_y y_n)$, a_x and a_y are the amplitudes of the ripple in the x and y directions, respectively, and μ is the amplitude of the diagonal ripple and represents the coupling between the x and y motions.

If $\mu = 0$, the system breaks into two uncoupled parts describing motion in $x - z$ and $y - z$ separately. Figure 6.5 shows the motion in the $\alpha - x$ surface of section for the uncoupled case. A number of different orbits are shown, each with different initial conditions. Each particle was run for 1000 iterations. The plot displays the usual features of a system with two degrees of freedom: (1) regular (KAM) orbits, (2) resonance island orbits, and (3)

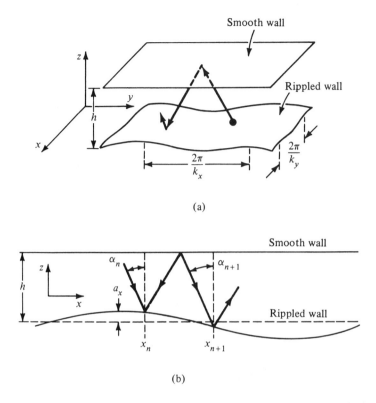

Figure 6.4. The three-dimensional billiards problem. (a) A point particle bounces back and fourth between a smooth and a periodically rippled wall. (b) Motion in two degrees of freedom, illustrating the definition of the trajectory angle α_n and the bounce position x_n just before the nth collision with the rippled wall (after Tennyson et al., 1979).

stochastic orbits. The island orbits are examples of "higher-order" KAM curves. The central resonance at $\alpha = 0$, $x = 0$ corresponds to a stable motion for which the ball bounces up and down along z in the valley of the rippled wall. The island orbits encircling this resonance correspond to "adiabatic" motion in the valley with a small oscillation back and forth in x occurring over many bounce times in z. There are two major stochastic orbits visible in Fig. 6.5. The thick stochastic layers for α near $\pm \pi/2$ are regions of stochasticity produced by all overlapping resonances having one bounce period in z equal to one or more periods along x. Physically, these motions correspond to grazing angle trajectories, as shown in Fig. 6.5. Isolated from the thick layer by KAM curves spanning the space in x is the thin stochastic layer that has formed near the separatrix associated with the central resonance. Physically, the separatrix orbit corresponds to a motion in x for which the ball is either just reflected or is just transmitted over a hill.

6.1. Resonance in Multidimensional Oscillations

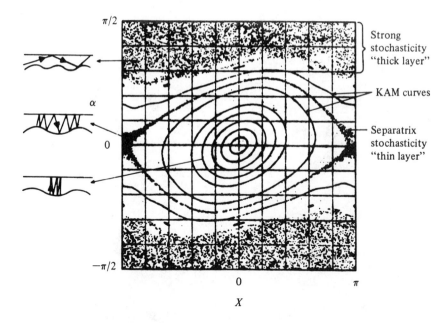

Figure 6.5. Motion in the $\alpha - x$ surface of section for the uncoupled billiards problem. The parameters are $\mu = 0$; $\lambda_x : h : a_x$ as $100:10:2$; $\lambda_x = 2\pi/k_x$. Fifteen initial conditions at $x = 0$ are each iterated for 1000 collisions with the rippled wall (after Tennyson et al., 1979).

A typical numerical calculation showing Arnold diffusion in the coupled system is given in Fig. 6.6. The surface of section for the system is four-dimensional (α, x, β, y), which we represent in the form of two, two-dimensional plots (α, x) and (β, y). Thus, two points, one in (α, x) and one in (β, y), are required to specify a point in the four-dimensional section. In Fig. 6.6, the two plots are superimposed for convenience, and x and y have been normalized to their respective wavelengths $2\pi/k_x$ and $2\pi/k_y$. The initial condition, as shown in Fig. 6.6, has been chosen on an island encircling the central resonance in x and within the thin separatrix layer for y. This corresponds to an initial adiabatic motion in x, well-confined in the valley, while in y the motion just reaches or passes over a hill. We observe numerically that the y motion is confined to its separatrix layer until the x motion reaches its own separatrix layer. The successive stages of the diffusion of the $\alpha - x$ motion are shown in Fig. 6.6b–d. In the absence of coupling ($\mu = 0$), the motion in the $\alpha - x$ plane should be confined to a smooth closed curve encircling the central resonance. For a finite coupling, α and x diffuse slowly due to the small randomizing influence of the stochastic $\beta - y$ motion. The $\alpha - x$ diffusion is the motion along the $\beta - y$ stochastic layer; i.e., it is the Arnold diffusion. The diffusion is shown for 1.5×10^5, 3.5×10^6, and 10^7 iterations of the mapping. At this time the $\alpha - x$ motion has diffused

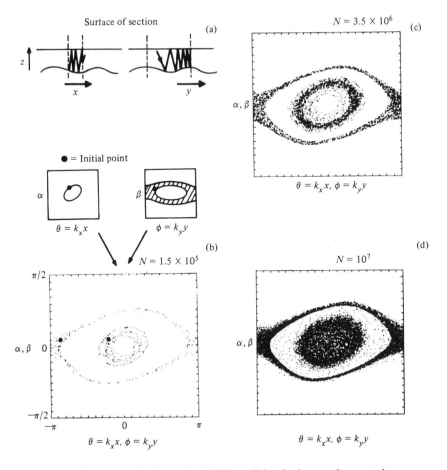

Figure 6.6. Thin-layer diffusion. The initial condition is close to the central resonance in the $\alpha - x$ space and within the separatrix stochastic layer in the $\beta - y$ space. The parameters are $\mu/h = 0.004$; $\lambda_x:h:a_x$ and $\lambda_y:h:a_y$ as $100:10:2$ (after Tennyson et al., 1979).

out to its own thin separatrix layer. Continued iteration of the mapping shows that the trajectory point diffuses over most of $\alpha - x$ plane. In particular, the change of direction from diffusion along the $\beta - y$ separatrix layer to diffusion along the $\alpha - x$ separatrix layer has been observed numerically. Similarly, the change of direction from diffusion along a separatrix layer to diffusion along a thick layer has been observed. Figure 6.7 shows these effects in the (α, β) action space (near $x = y = 0$) for the initial condition of Fig. 6.6, after 5×10^7 iterations of the mapping. The trajectory has wandered randomly along thin and thick layers in the action space, as shown, spending much of its time in the region of thick layers for both $\alpha - x$ and $\beta - y$ motion.

6.1. Resonance in Multidimensional Oscillations

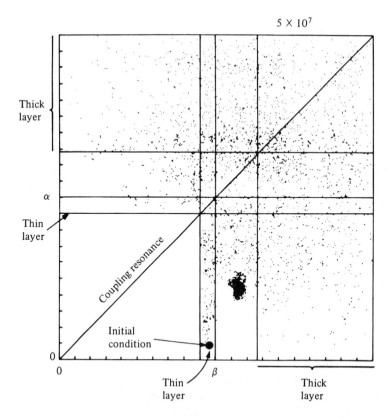

Figure 6.7. Projection of the motion in the $\alpha - \beta$ action space for $x \approx 0$, $y \approx 0$. The parameters and initial conditions are the same as for Fig. 6.6. After 5×10^7 iterations, the orbit has wandered in and out of the thick and thin layers of both the $\alpha - x$ and $\beta - y$ motions (after Lieberman and Tennyson, 1982).

All of this, however, is just part of the story. Recall that there exists a dense set of resonance surfaces in the action space. In particular, consider a coupling resonance, where physically the motion is "adiabatically" confined to a valley in both the x and y directions as the ball rapidly bounces along z. The ball executes small-amplitude oscillations in both the x and y directions. If the oscillation frequencies ω_x and ω_y satisfy

$$m_1 \omega_x + m_2 \omega_y = 0,$$

then we have a resonance with its stochastic separatrix layer also a part of the Arnold web. Including only a single coupling resonance $m_1 = -m_2 = 1$ adds the diagonal line seen in Fig. 6.7. For initial conditions such that the system is placed in the separatrix of the coupling resonance, and thus within the Arnold web, the billiard motion initially appears to be stable, consisting of a fast bounce motion in z and slower, small amplitude oscillations in x and y; the motion appears to be adiabatically confined to a small neighborhood near $x = y = 0$. However, this is not the case. After a sufficient time, the billiard can be found executing grazing angle motion in both the x and y directions. The diffusion typically proceeds first along the coupling resonance, then along the thin layer in x or y, and finally along the corresponding thick layer. The billiard motion will rarely return to the neighborhood of $x = y = 0$, because the overwhelming fraction of the Arnold web consists of the thick stochastic layers, with a negligible (but dense!) fraction of the web in regions such as the coupling resonances, where the motion appears adiabatic. This remarkable behavior is characteristic of the phenomenon of Arnold diffusion.

6.2. Diffusion Rates Along Resonances

A theoretical calculation of the Arnold diffusion rate was first performed by Chirikov (1971, 1979), and his collaborators. For the billiards problem, described in Section 6.1b, the diffusion has been calculated by Tennyson *et al.* (1979) and by Lieberman (1980). The basic theoretical procedure is to break the original three-degree-of-freedom system into two two-degree-of-freedom systems that are successively solved. We refer to this decomposition as the *stochastic pump model*, with the simplest decomposition that of considering only three resonances. In this model, the guiding resonance, along which the Arnold diffusion proceeds, is associated with (say) degree-of-freedom 2. The coupling between degrees-of-freedom 1 and 2, described by the Hamiltonian

$$H_{\text{across}}(I_1, I_2, \theta_1, \theta_2) = \text{const.}, \tag{6.2.1}$$

generates the strongly chaotic motion across the separatrix layer. The weaker Arnold diffusion is then obtained from

$$H_{\text{along}}(I_2, I_3, \theta_2, \theta_3) = \text{const.}, \tag{6.2.2}$$

which describes the coupling between degrees-of-freedom 2 and 3. The motions described by (6.2.1) and (6.2.2) are solved successively, with (6.2.1) first yielding the stochastic variations of $\theta_2(t)$ and $I_2(t)$ in the separatrix layer. These are inserted into (6.2.2), which is then solved to obtain the stochastic variation $I_3(t)$ describing the Arnold diffusion.

The major difficulty in applying this procedure is determining which resonances are important in driving the chaotic motion across and along the layers. For a three-resonance calculation, the first (guiding) resonance can be chosen arbitrarily and defines the local region within the Arnold web where the diffusion is to be determined. The strongest remaining resonance in the

6.2. Diffusion Rates Along Resonances

system is taken to drive the chaotic motion across the layer. The remaining strongest resonance is assumed to drive the Arnold diffusion.

*6.2a. Stochastic Pump Diffusion Calculation

We illustrate these ideas by calculating the rate of Arnold diffusion for the billiards problem given by (6.1.12). We consider three different regimes of successively slower diffusion rates. The first describes the diffusion of α along the thick stochastic layer of the $\beta - y$ motion. The quantity α experiences diffusive fluctuations that result from the small coupling to the random y motion. The second process is similar to the first, except that α now diffuses along the thin separatrix layer of the $\beta - y$ motion. The third process occurs near a coupling resonance between the x and y motions, chosen so that the periods of oscillation around the central resonance for the x and y motions are the same. Thick-layer diffusion tends to be much faster than thin-layer diffusion due to the greater randomness of the y motion in the former case. The coupling resonance diffusion is very slow.

The first step in the calculation is to find a Hamiltonian that will generate the surface of section mappings, (6.1.12). As in Section 3.1c, the difference equations are transformed to differential equations by introduction of a δ-function force, which then are easily integrated to obtain the "kicked" Hamiltonian

$$H(\alpha, x, \beta, y, n) = 2h \ln \sec \alpha + 2h \ln \sec \beta$$
$$- 2\delta_1(n)C(x, y), \quad (6.2.3)$$

where

$$C(x, y) = a_x \cos k_x x + a_y \cos k_y y$$
$$- \tfrac{1}{2}\mu \cos(k_x x + k_y y), \quad (6.2.4)$$

and

$$\delta_1(n) \equiv \sum_{m=-\infty}^{+\infty} \delta(n - m) = 1 + 2 \sum_{q=1}^{\infty} \cos(2\pi n q). \quad (6.2.5)$$

Note that H in (6.2.3) is a nonautonomous Hamiltonian in two degrees of freedom. It is related to the net energy in the x and y motion and is not conserved.

Thick-Layer Diffusion. The initial conditions appropriate to thick-layer diffusion have β and y within the thick stochastic layer, with α and x chosen to yield small amplitude libration near the central resonance. In the absence of coupling ($\mu = 0$), the motion in the $\alpha - x$ plane is confined to a smooth closed curve like those seen close to the center of Fig. 6.5. For a finite coupling, α and x diffuse slowly due to the small randomizing influence of the stochastic $\beta - y$ motion.

For the initial conditions appropriate to thick-layer diffusion, we decompose $H = H_x + H_y$, with new canonical variables $\theta = k_x x$, $\phi = k_y y$, $\bar{\alpha} = \alpha/k_x$, $\bar{\beta} = \beta/k_y$, to obtain

$$H_y = 2h \ln \sec \beta - 2\delta_1(n) a_y \cos \phi \qquad (6.2.6a)$$

and

$$H_x = h\alpha^2 - 2a_x \cos \theta + \mu \cos[\theta + \phi(n)], \qquad (6.2.6b)$$

where for convenience in later calculations we retain the old momenta in the new Hamiltonian. We also set $\ln \sec \alpha \simeq \alpha^2/2$ since $\alpha^2 \ll 1$, and ϕ in (6.2.6b) is now considered to be an explicit function of n. This decomposition is a big assumption, as it neglects the coupling term in (6.2.6a); we also set $\delta_1 = 1$ in (6.2.6b), consistent with a slow libration in x. By this means, we obtain two nonautonomous Hamiltonians, each in one degree of freedom. We solve first for the $\beta - \phi$ motion across the layer, the "stochastic pump," and, substituting this motion into (6.2.6b), find the $\alpha - \theta$ motion along the layer, whose diffusive component is the Arnold diffusion.

In the thick layer, in which there are many overlapping resonances, ϕ randomizes on the time scale of a single step. Therefore, to a good approximation, we assume that ϕ makes a sudden random jump to a new phase whenever n is an integer. The evolution of H_x from (6.2.6b) is, using Hamilton's equations,

$$\frac{dH_x}{dn} = \frac{\partial H_x}{\partial n}.$$

Taking the partial derivative of (6.2.6b) with ϕ an explicit function of n and integrating once by parts,

$$\frac{\partial H_x}{\partial n} = \frac{d}{dn}[\mu \cos(\theta + \phi)] + \mu \frac{d\theta}{dn} \sin[\theta + \phi(n)]. \qquad (6.2.7)$$

The first term contributes only a small oscillation with no net change over long periods of time. For small amplitude libration in the $\alpha - x$ plane, we have

$$\theta = \theta_0 \cos(\omega_x n + \chi_0), \qquad (6.2.8)$$

where

$$\omega_x = \frac{2\pi}{T_x} = 2k_x(a_x h)^{1/2}.$$

Using this, we integrate the second term in (6.2.7) over the "time" interval from m to $m + 1$:

$$\Delta H_x = \int_m^{m+1} dn \, \mu \theta_0 \omega_x \sin[\omega_x n + \chi_0] \sin[\theta + \phi(n)].$$

For $\omega_x \ll 1$, the integrand is constant, yielding

$$\Delta H_x = \mu \theta_0 \omega_x \sin[\omega_x m + \chi_0] \sin[\theta + \phi(m)]. \qquad (6.2.9)$$

6.2. Diffusion Rates Along Resonances

We square this and average over both χ_0 and ϕ to get

$$\langle \Delta H_x^2 \rangle = \tfrac{1}{4}\mu^2\theta_0^2\omega_x^2,$$

where we have used the assumption that ϕ is randomized at $m = $ integer. The ensemble average over χ_0 is equivalent to a time average over many uncorrelated jumps. The thick-layer diffusion rate is then

$$D_1 = \tfrac{1}{2}\langle \Delta H_x^2 \rangle = \tfrac{1}{8}\mu^2\theta_0^2\omega_x^2. \tag{6.2.10}$$

The parameters μ and ω_x will remain constant as H_x diffuses. The quantity θ_0, however, increases with H_x, resulting in an increase in the diffusion rate as the x oscillations grow.

In Fig. 6.8a–c, the theoretical values of D_1 (solid lines) are compared with measurements (triangles) obtained from the direct iteration of the difference equations. For each experiment, 100 particles were started with identical initial conditions on a libration curve of the $\alpha - x$ plane and with random initial conditions in the thick stochastic layer of the $\beta - y$ plane. The motion was followed for 500 wall collisions, and the rms value of the normalized energy $(\alpha^2)_{\text{rms}} = [(h^{-2})\langle\Delta H_x^2\rangle n]^{1/2}$ is compared with the theory. Figure 6.8a shows the variation with coupling strength μ, Fig. 6.8b the variation with period T_x, and Fig. 6.8c the variation with the number of iterations n. Each triangle represents the average of four separate runs. The theoretical predictions, although consistently a little high, are quite good. The discrepancy probably reflects a small deviation from true random phase.

Thin-Layer Diffusion. Here the initial conditions remain close to the central resonance of the $\alpha - x$ space, but are now chosen inside the thin stochastic layer surrounding the separatrix of the $\beta - y$ space. The diffusion of the $\alpha - x$ motion is again caused by the small coupling to the stochastic y motion, but since thin-layer trajectories are considerably less "random" than thick-layer trajectories, the diffusion is significantly weaker.

To calculate the Arnold diffusion rate, we again decompose the Hamiltonian (6.2.3) into (6.2.6a) and (6.2.6b), where now, in (6.2.6a), only the $q = 0$ and $q = 1$ terms in $\delta_1(n)$ need be kept. We also put $\ln \sec \beta \simeq \beta^2/2$ for the separatrix motion, yielding, in place of (6.2.6a)

$$H_y = h\beta^2 - 2a_y \cos \phi - 4a_y \cos 2\pi n \cos \phi. \tag{6.2.11}$$

In (6.2.11), the first two terms exhibit the separatrix associated with the central $\beta - y$ resonance, and the third term generates the thin stochastic layer surrounding the separatrix. The procedure, as before, is to first solve (6.2.11), substitute the solution for $\phi(n)$ into (6.2.6b), and then find the energy change ΔH_x as ϕ swings between $-\pi$ and π. Starting with (6.2.7) and again neglecting the first term we have

$$\frac{dH_x}{dn} = \mu\frac{d\theta}{dn}\sin[\theta + \phi(n)]. \tag{6.2.12}$$

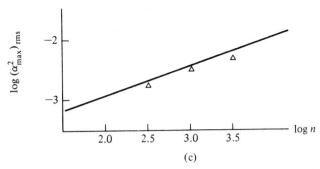

Figure 6.8. Thick-layer diffusion. Comparison of the theoretical diffusion with the results of simulation experiments. The rms value of α_{max}^2 is plotted versus (a) the coupling strength μ; (b) the libration period T_x; and (c) the number of iterations n. The parameters (except for those varied) are $\mu/h = 0.0002$; $n = 500$; $\lambda_x : h : a_x$ as $100:10:1$; and $\lambda_y : h : a_y$ as $100:10:1.7$. The statistical spread of the 100 initial conditions is within the height of the triangles (after Tennyson et al., 1979).

6.2. Diffusion Rates Along Resonances

As before, $\theta(n)$ corresponds approximately to small librations, given by (6.2.8). But instead of randomizing $\phi(n)$ with each bounce, we now assume that it evolves like the phase on a pendulum separatrix (see Section 1.3a)

$$\phi(n) = 4\tan^{-1}[\exp(\omega_y n)] - \pi. \tag{6.2.13}$$

Here, as before, ω_x and ω_y are the frequencies for small oscillations about the central fixed points of the $\alpha - x$ and $\beta - y$ motions, respectively,

$$\omega_x = 2k_x\sqrt{a_x h}, \qquad \omega_y = 2k_y\sqrt{a_y h}.$$

Defining $s \equiv \omega_y n$, $Q_0 \equiv \omega_x/\omega_y$, writing $\chi_0 = Q_0 s_0 - \pi/2$ in (6.2.8), substituting into (6.2.12), and integrating, we have

$$\Delta H_x = \mu \theta_0 Q_0 \int_{-\infty}^{+\infty} ds\, \mathcal{I}(s),$$

where

$$\mathcal{I} = \cos[Q_0(s + s_0)] \sin\{\theta_0 \sin[Q_0(s + s_0)] + \phi\}.$$

Using $\theta_0 \ll 1$,

$$\mathcal{I} = \cos[Q_0(s + s_0)] \sin\phi. \tag{6.2.14}$$

We have already considered integrals of this type in Section 3.5a, where we found that they were of the form of Melnikov–Arnold integrals

$$\mathcal{A}'_m = \lim_{s_1 \to \infty} 2\int_0^{s_1} \cos\left[\frac{m}{2}\phi(s) \pm Q_0 s\right], \tag{6.2.15}$$

where $\phi(s)$ is the separatrix phase motion. These integrals are improper; i.e., no limit formally exists. However, as discussed in Section 3.5a, they are the sum of a rapidly oscillating part and a "jump." The oscillating part may be large compared with the jump, but produces only a bounded oscillation in H_x, which is not randomized on the time scale of the separatrix motion and averages to zero. The jump can be evaluated giving, for our problem with $m = 2$,

$$\Delta H_x = \tfrac{1}{2}\mu\theta_0 Q_0 \sin(Q_0 s_0)[\mathcal{A}_2(-Q_0) - \mathcal{A}_2(Q_0)], \tag{6.2.16}$$

where (see Section 3.5)

$$\mathcal{A}_2(\pm Q_0) = \frac{4\pi Q_0 \exp(\pm \pi Q_0/2)}{\sinh(\pi Q_0)}. \tag{6.2.17}$$

We then have

$$\Delta H_x = -\frac{4\pi\mu\theta_0 Q_0^2 \sin(Q_0 s_0) \sinh(\pi Q_0/2)}{\sinh(\pi Q_0)}. \tag{6.2.18}$$

We know from the separatrix mapping (see Section 3.5) that the phase within the separatrix layer $Q_0 s_0$ is randomized after every half-period of $\phi(n)$. We

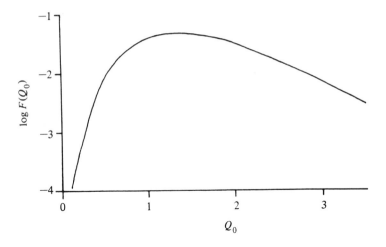

Figure 6.9. Plot of $F(Q_0)$ for the dependence of thin-layer diffusion on $Q_0 = \omega_x/\omega_y$ (after Tennyson et al., 1979).

can therefore phase average ΔH_x^2 over this time period to get

$$\langle \Delta H_x^2 \rangle_{so} = 8\pi^2 \mu^2 \theta_0^2 F(Q_0), \tag{6.2.19}$$

where

$$F(Q_0) = \frac{Q_0^4 \sinh^2(\pi Q_0/2)}{\sinh^2(\pi Q_0)}. \tag{6.2.20}$$

A plot of $F(Q_0)$ is shown in Fig. 6.9. It is sharply peaked close to $Q_0 = 1$; if the characteristic frequencies of the separatrix and libration motion differ by as much as a factor of four, the diffusion will be reduced by two orders of magnitude.[1]

To obtain the diffusion coefficient, we need to know the mean half-period $\bar{T}_y$ of the motion in the thin stochastic layer. The half period of a true pendulum that follows a trajectory very close to the separatrix is, from (3.5.23),

$$T_y = \frac{1}{\omega_y} \ln \left| \frac{32}{w} \right|,$$

where $w \equiv (H_y - H_s)/H_s \ll 1$ and $H_s = 2a_y$ is the separatrix energy. Chirikov (1979) has shown that the average half-period inside the stochastic layer may be computed by simply integrating the half-period over the energy interval

---

[1] The Melnikov–Arnold integral is calculated in the limit $\omega_y T_y \gtrsim 2\pi/Q_0$. Hence, for $Q_0 \ll 1$ in (6.2.20) the equation is only valid if the layer is very thin.

of the layer, obtaining

$$\bar{T}_y = \frac{1}{\omega_y} \ln \left| \frac{32e}{w_1} \right|, \tag{6.2.21}$$

where w_1 is the relative energy at the barrier transition defining the edge of the layer, which we have calculated in (4.2.22) with $K_2 = 1$, and e is the natural base. (The separatrix width is not appreciably affected by the small coupling, $\mu \ll a_y$.)

Combining (6.2.19) and (6.2.21), we get the thin-layer diffusion coefficient:

$$D_2 = \frac{\langle \Delta H_x^2 \rangle_{so}}{2\bar{T}_y} \tag{6.2.22}$$

or

$$D_2 = \frac{4\pi^2 \mu^2 \theta_0^2 \omega_y F(Q_0)}{\ln(32e/|w_1|)}. \tag{6.2.23}$$

In Fig. 10a–c, the theoretical thin-layer diffusion is compared with computational measurements. Each triangle represents the final spread of 100 particles that have been started with identical initial conditions in the $\alpha - x$ space and slightly different initial conditions in the thin stochastic layer of the $\beta - y$ space. The theoretical curves were calculated from (6.2.23) with $w_1 = 0.191$ determined numerically. Again, the theoretical values fall slightly above the experimental values, probably due to the fact that the y motion phase $\phi(n)$ is not completely randomized with each successive half-period of the separatrix motion.

6.2b. Coupling Resonance Diffusion

It is also possible to calculate the diffusion rate for a coupling resonance of the billiards problem, such as $\omega_x = \omega_y$ ($m_x = -m_y$). The kicked Hamiltonian must first be transformed to the neighborhood of this resonance, and then expanded in that neighborhood. This rather complicated calculation was carried out by Lieberman (1980). Rather than do this here, we introduce a simpler system, after Chirikov (1979), which has been used to illustrate both coupling resonance and many-resonance Arnold diffusion (Chirikov et al., 1980):

$$H = \tfrac{1}{2}p_x^2 + \tfrac{1}{2}p_y^2 + \tfrac{1}{4}x^4 + \tfrac{1}{4}y^4 - \mu xy - \epsilon x f(t). \tag{6.2.24}$$

Here t is the time, and the quantities μ and ϵ are taken to be small parameters. This Hamiltonian describes the motion of two nonlinear oscillators that are coupled quadratically with strength μ. Oscillator x is driven by a periodic function of time f with strength ϵ. We take the system to be in the neighborhood of the coupling resonance

$$\omega_x - \omega_y \approx 0, \tag{6.2.25}$$

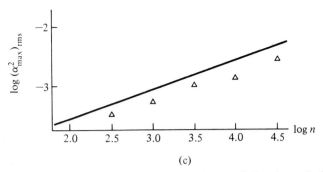

Figure 6.10(a, b, c). Thin-layer diffusion. Comparison of the theoretical diffusion with the results of simulation experiments. The three graphs show the variation of the rms value of α_{max}^2 with μ, T_x, and n. The parameters (except for those varied) are $\mu/h = 0.0002$; $n = 2000$; $\lambda_x:h:a_x$ as $100:10:1$; and $\lambda_y:h:a_y$ as $100:10:1.8$. The statistical spread of the 100 initial conductions is within the height of the triangles (after Tennyson et al., 1979).

6.2. Diffusion Rates Along Resonances

where ω_x and ω_y are the unperturbed frequencies of the two nonlinear oscillators (not to be confused with the linear frequencies ω_x and ω_y for the billiards problem).

Action-Angle Transformation. In order to calculate the Arnold diffusion, we must first transform to action-angle variables of the unperturbed system $\mu = \epsilon = 0$. The unperturbed Hamiltonian consists of the two nonlinear uncoupled oscillators

$$H_0 = \tfrac{1}{2}p_x^2 + \tfrac{1}{2}p_y^2 + \tfrac{1}{4}x^4 + \tfrac{1}{4}y^4. \tag{6.6.26}$$

For this system each oscillator has its own constant energy E_x and E_y, respectively, and the action variable is formed in the usual way

$$I_x = \frac{1}{2\pi}\oint p_x\, dx = \frac{2}{\pi}\int_0^{x_M} (2E_x)^{1/2}\left(1 - \frac{x^4}{4E_x}\right)^{1/2} dx, \tag{6.2.27}$$

where $x_M = (4E_x)^{1/4}$ is the amplitude of the x oscillation. By making the change of variable $\xi = x/(4E_x)^{1/4}$, (6.2.27) becomes

$$I_x = \frac{4}{\pi} E_x^{3/4} \int_0^1 (1 - \xi^4)^{1/2}\, d\xi. \tag{6.2.28}$$

From (6.2.28) we obtain the relation between the action I and the energy E for each oscillator ($E \propto I^{4/3}$). The integral is of the elliptic type with a value $(\sqrt{2}/3)\mathcal{K}(1/\sqrt{2})$, where $\mathcal{K}(1/\sqrt{2}) \approx 1.85$ is the complete elliptic integral of the first kind. Solving for E (or the Hamiltonian), we have

$$\bar{H}_0 = A(I_x^{4/3} + I_y^{4/3}), \tag{6.2.29}$$

where

$$A = \left(\frac{3\pi}{4\sqrt{2}\mathcal{K}(1/\sqrt{2})}\right)^{4/3} \approx 0.87,$$

and the frequencies are

$$\omega_{x,y} = \frac{\partial \bar{H}_0}{\partial I_{x,y}} = \tfrac{4}{3}AI_{x,y}^{1/3}. \tag{6.2.30}$$

It is also possible to find the solution as a function of time for each unperturbed oscillator in terms of elliptic functions. This has been done by Chirikov (1979), obtaining

$$\frac{x(t)}{x_M} = \mathrm{cn}(\omega t) = \frac{\pi\sqrt{2}}{\mathcal{K}(1/\sqrt{2})}\sum_{n=1}^{\infty}\frac{\cos[(2n-1)\omega t]}{\cosh[\pi(n - \tfrac{1}{2})]}$$

$$\approx 0.95 \cos\omega t + \frac{\cos 3\omega t}{23} + \frac{\cos 5\omega t}{(23)^2} + \cdots. \tag{6.2.31}$$

From (6.2.31) we see that there is very little harmonic content in the oscillation. It is therefore reasonable to keep only the first term in (6.2.31). Defining an angle variable $\theta \equiv \omega t$, the Hamiltonian for the complete system can then be written

$$\bar{H} = A(I_x^{4/3} + I_y^{4/3}) - \mu x_M(I_x) y_M(I_y) \cos \theta_x \cos \theta_y$$
$$- \epsilon x_M(I_x) \cos \theta_x f(t), \tag{6.2.32}$$

where x_M and y_M are the x and y amplitudes. Comparing the unperturbed Hamiltonians (6.2.26) and (6.2.29), we see that

$$x_M = (4A)^{1/4} I_x^{1/3}, \qquad y_M = (4A)^{1/4} I_y^{1/3}. \tag{6.2.33}$$

Near the coupling resonance, $\theta_x - \theta_y$ is a slowly varying function of time. To exhibit the slow oscillation about the resonance and its associated chaotic separatrix layer, we transform to new coordinates

$$\psi_1 = \theta_x - \theta_y, \qquad \psi_2 = \theta_x + \theta_y, \tag{6.2.34a}$$

using the generating function

$$F_2 = (\theta_x - \theta_y) I_1 + (\theta_x + \theta_y) I_2,$$

such that

$$I_x = \frac{\partial F_2}{\partial \theta_x} = I_1 + I_2, \qquad I_y = \frac{\partial F_2}{\partial \theta_y} = -I_1 + I_2. \tag{6.2.34b}$$

This transformation is similar to that used in Section 2.4, but brings out more explicitly the symmetry of the problem.

Near the coupling resonance, $I_x \approx I_y$ so $I_1 \ll I_2$. Inserting (6.2.34) into Hamiltonian (6.2.32) and expanding the first term to quadratic order in I_1, we obtain, approximately, the new Hamiltonian

$$K = 2AI_2^{4/3} + \tfrac{1}{2} G I_1^2 - F \cos \psi_1 - F \cos \psi_2$$
$$- \epsilon x_M(I_2) \cos\left(\frac{\psi_1 + \psi_2}{2}\right) f(t), \tag{6.2.35}$$

where

$$G(I_2) = \tfrac{8}{9} A I_2^{-3/2} \tag{6.2.36}$$

and

$$F(I_2) = \tfrac{1}{2} \mu x_M(I_2) y_M(I_2). \tag{6.2.37}$$

Averaging over the fast variable ψ_2 yields

$$\langle K \rangle = 2AI_2^{4/3} + \tfrac{1}{2} G I_1^2 - F \cos \psi_1, \tag{6.2.38}$$

such that $I_2 \approx \text{const.}$, $\omega_2 = 2\omega_x$, and I_1, ψ_1 execute slow oscillations about the coupling resonance $I_1 = 0$, $\psi_1 = 0$ with frequency near resonance

$$\omega_1 = \sqrt{FG} \propto \sqrt{\mu}. \tag{6.2.39}$$

6.2. Diffusion Rates Along Resonances

Three-Resonance Interaction. We now choose a driving term

$$f(t) = \cos \Omega_1 t + \cos \Omega_2 t, \tag{6.2.40}$$

such that the last term in (6.2.35) is near resonance for both frequencies, and consider the ordering

$$\delta\omega = (\omega_x - \Omega_2) > (\Omega_1 - \omega_x) \gtrsim 0. \tag{6.2.41}$$

We assume that these two resonances dominate the behavior, with the stronger (Ω_1) resonance driving the chaotic motion across the layer, and the weaker (Ω_2) resonance driving the Arnold diffusion along the layer. The diffusion rate can then be calculated in a relatively straightforward way (see Chirikov, 1979, Section 7.5). The nonlinearity also introduces many other higher-order resonances, but these contribute very weakly to the diffusion. We therefore drop the fourth term in (6.2.35) and decompose K as

$$K = K_{\text{across}} + K_{\text{along}}$$

with

$$K_{\text{across}} = \tfrac{1}{2} G I_1^2 - F \cos \psi_1 - \epsilon x_M \cos\left(\frac{\psi_1 + \omega_2 t}{2}\right) \cos \Omega_1 t \tag{6.2.42}$$

and

$$K_{\text{along}} = 2AI_2^{4/3} - \epsilon x_M \cos\left(\frac{\psi_1(t) + \psi_2}{2}\right) \cos \Omega_2 t. \tag{6.2.43}$$

These equations are analogous to (6.2.11) and (6.2.6b), respectively, and one proceeds as previously to calculate the Arnold diffusion rate from dK_{along}/dt. The result (after Chirikov, 1979) is

$$D = \frac{\pi^2 x_M^2 \omega_x^2}{\omega_1} \frac{\exp(-\pi Q_0)}{\ln|32e/w_1|}, \tag{6.2.44}$$

where $Q_0 = \delta\omega/\omega_1$ is the frequency ratio.

The Arnold diffusion rate is quite similar for the billiards problem (Lieberman, 1980), although the calculation is more complicated. The rate of diffusion becomes exponentially small as the driving resonances become further apart and the coupling becomes weaker ($\delta\omega/\omega_1$ becomes large). We emphasize that the above calculation is local, and varies continuously as the system point moves along the coupling resonance, nearer or further from the resonances that drive it.

6.2c. Many Resonance Diffusion

The previous calculations of the Arnold diffusion rate are analytically derived using only three resonances. These calculations seem to agree with numerical simulations provided the perturbation is not too weak. However, for sufficiently weak perturbations, many resonances are important, and the three-resonance theory predicts diffusion rates that are much lower than those calculated from numerical simulations. The many resonance regime is called the Nekhoroshev region, after the Soviet mathematician who first derived a rigorous upper bound on the diffusion rate there (Nekhoroshev, 1977). However, Nekhoroshev's upper bound is generally many orders of magnitude larger than the actual diffusion rate.

The many-resonance regime has been examined numerically and some analytical estimates made (Chirikov, 1979; Chirikov et al., 1979) for the Hamiltonian of (6.2.24) with $f(t)$ given by

$$f(t) = \frac{\cos vt}{1 - C \cos vt} = \sum_m \frac{2e^{-\sigma m}}{\sigma} \cos mvt, \quad (6.2.45)$$

where $\sigma \simeq (1 - C^2)^{1/2}$. Since the most important dependence is in the exponential factor, the numerical results were normalized to emphasize this factor. That is, defining D^* by

$$D = \frac{\pi^2 x_M^2 \omega_x^2}{\omega_1 \ln|32e/w_1|} D^*,$$

for the three-resonance result in (6.2.44), log D^* is plotted versus Q_0 for a wide range of parameters, obtaining the results shown in Fig. 6.11. For all numerical results, $\omega_x = \omega_y = 4.5v$ and $\sigma = 0.1$. Comparison of an analytic calculation (dashed line) using (6.2.44), for the three resonances $\omega_x = 4v$, $\omega_x = 5v$, and $\omega_x = \omega_y$, gives very good agreement with the numerical results for Q_0 small; but the theory severely underestimates the diffusion for Q_0 large.

The discrepancy for Q_0 large can be resolved if one considers the harmonic resonances $mv - k\omega_x$ between the driving term and the x oscillator. The $k > 1$ resonances appear when the full expansion (6.2.31) is used to give the x motion. Although these resonances are excited with very small amplitudes, they can strongly contribute to the overall Arnold diffusion rate because they may lie close to the initial condition $\omega_x = \omega_y = 4.5v$; i.e., $\delta\omega = mv - k\omega_x$ can be small such that $Q_0 \sim 1$ for these resonances.

Chirikov et al. (1979) assume D^* to be of the form

$$D^* = A \exp(-BQ_0^\gamma) \quad (6.2.46)$$

and fit the parameters to the numerical data finding the best fit for $\gamma = 1/2$ ($-\log D \propto \mu^{-1/4}$). The upper-bound estimate by Nekhoroshev (1977) gives a considerably smaller γ (Chirikov, 1979). For the general Hamiltonian

$$H(\boldsymbol{I}, \boldsymbol{\theta}) = H_0(\boldsymbol{I}) + \mu H_1(\boldsymbol{I}, \boldsymbol{\theta}),$$

where μ is a small parameter that couples the degrees of freedom, and for which H_0 has lowest-order terms given by a quadratic form $H_0 = \tfrac{1}{2} I_0^2 + \cdots$, then Nekhoroshev's upper bound is given for N degrees of freedom as

$$|\dot{\boldsymbol{I}}| \leqslant |\boldsymbol{\omega}| |\boldsymbol{I}| \mu^{1+q} \exp(-1/\mu^q), \quad (6.2.47a)$$

6.2. Diffusion Rates Along Resonances

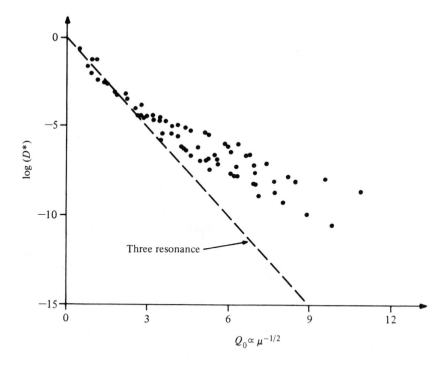

Figure 6.11. Normalized Arnold diffusion coefficient D^* versus Q_0. The dots are numerical calculations; the dashed line is the prediction of the three-resonance theory (after Chirikov et al., 1979).

with

$$q(N) = \frac{2}{3N^2 - N + 8}, \qquad (6.2.47b)$$

where the diffusion rate is proportional to $|\dot{I}|^2$. Since $Q_0 \propto 1/\omega_1$ and from (6.2.39), $\omega_1 \propto \sqrt{\mu}$, we see by comparing (6.2.46) and (6.2.47a) that $\gamma = 2q$. Using $N = 3$ in (6.2.47b), we find $\gamma = \frac{1}{8}$, and the upper-bound diffusion rate decays very slowly with μ, in contrast to the numerical results in Fig. 6.11. Chirikov (1979) argues that a better estimate for q in the many resonance regime is $q = 1/N$; i.e. $\gamma = \frac{2}{3}$ for $N = 3$. This value is in reasonable agreement with the numerical results.

6.2d. Modulational Diffusion

We turn now to modulational diffusion for which the chaotic motion is driven along a band of overlapping resonances, caused by a slow modulation of the driving perturbation. Following Chirikov et al. (1980, 1982, 1985), we

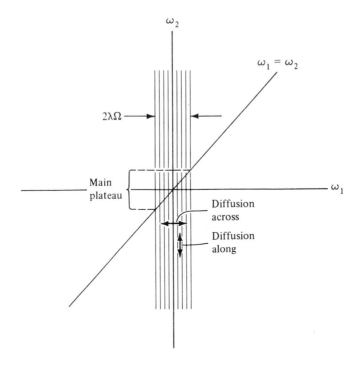

Figure 6.12. Frequency space, illustrating the various aspects of modulational diffusion.

illustrate this mechanism for the Hamiltonian

$$H = \tfrac{1}{2}I_1^2 - k\cos(\theta_1 + \lambda \sin \Omega t) + \tfrac{1}{2}I_2^2 - \epsilon \cos(\theta_1 - \theta_2). \quad (6.2.48)$$

Here oscillator 1 is phase modulated with amplitude λ at the slow modulation frequency Ω.

Expanding the second term in (6.2.48) in a Fourier series yields the Hamiltonian for oscillator 1, alone,

$$H_{\text{across}} = \tfrac{1}{2}I_1^2 - k \sum_n \mathcal{J}_n(\lambda) \cos(\theta_1 + n\Omega t). \quad (6.2.49)$$

The Bessel functions $\mathcal{J}_n$ have significant amplitude provided $|n| \lesssim \lambda$. The result is the formation of a multiplet layer of driving resonances of width approximately $2\Omega\lambda$ centered about the frequency $\omega_1 = I_1 = 0$. The multiplet is shown as a vertical set of lines in the ω_2, ω_1 frequency space in Fig. 6.12. In order to have a wide band of stochasticity, the resonances within the multiplet layer should overlap. That is, the width of a single resonance should at least equal two-thirds of the distance between resonances (satisfying the $K = 1$ overlap criterion for the standard mapping). The overlap criterion for the multiplet was obtained in Section 5.7.

6.2. Diffusion Rates Along Resonances

Within the multiplet when the motion is coupled to a third degree of freedom, then the onset of modulational diffusion is given by (5.7.9). Below this level of perturbation strength k the weak single resonance (or thin-layer) Arnold diffusion will occur. The diffusion along the merged layer of resonances, in the I_2 direction, can be calculated by the stochastic pump method. For motion along the resonance layer, the Hamiltonian is

$$H_{\text{along}} = \tfrac{1}{2}I_2^2 - \epsilon \cos(\theta_1(t) - \theta_2). \qquad (6.2.50)$$

Taking the time derivative and integrating by parts once as in (6.2.7),

$$\frac{dH_{\text{along}}}{dt} = -\epsilon \frac{d}{dt}(\cos(\theta_1 - \theta_2)) + \epsilon \sin(\theta_1 - \theta_2)\frac{d\theta_2}{dt}.$$

The first term leads to only a small oscillation in H_{along} and is ignored. Thus

$$\Delta H_{\text{along}} \approx \epsilon \int \sin\phi \, \frac{d\theta_2}{dt} \, dt, \qquad (6.2.51)$$

where the phase $\phi(t) = \theta_1 - \theta_2$.

To evaluate the phase, we need the unperturbed orbits for θ_1 and θ_2. The θ_2 motion is found from (6.2.50) with $\epsilon = 0$:

$$\theta_2 = \omega_2 t - \chi_0 - \frac{\pi}{2}. \qquad (6.2.52)$$

The θ_1 motion described by (6.2.49) is difficult to obtain analytically, and an approximate description is necessary. We need $\theta_1(t)$ to first order in k, so we write

$$H_0 = \tfrac{1}{2}I_1^2 \qquad (6.2.53a)$$

and

$$H_1 = -k \sum_n \mathscr{J}_n(\lambda) \cos(\theta_1 + n\Omega t). \qquad (6.2.53b)$$

Applying canonical perturbation theory as in Section 2.2b, we note $\langle H_1 \rangle = 0$. Then from (2.2.44), $\bar{H} = H_0$, such that $\bar{I}_1 = I_0 = \text{const.}$ and $\bar{\theta}_1 = I_0 t$. From (2.2.45), the first-order generating function is found to be

$$S_1 = k \sum_n \frac{\mathscr{J}_n(\lambda)}{n\Omega + \bar{I}_1} \sin(\theta_1 + n\Omega t). \qquad (6.2.54)$$

To first order, putting $\theta_1 = \bar{\theta}_1 = I_0 t$ in (6.2.53b) and using (6.2.54),

$$\theta_1(t) = I_0 t + k \sum_n \frac{\mathscr{J}_n(\lambda)}{(n\Omega + I_0)^2} \sin[(I_0 + n\Omega)t]. \qquad (6.2.55)$$

This expression for θ_1 is a function of I_0. We know that over long times the system samples all values of $|I_0| < \lambda\Omega$ within the modulational layer, so we shall average over these I_0's in what follows. For large λ, the Bessel function in (6.2.62) is small unless $|n| \lesssim \lambda$. We therefore take $n \approx \lambda$ in (6.2.55) to find

$$\phi(t) \approx (I_0 - \omega_2)t + \chi_0 + \frac{\pi}{2} + kR \frac{\mathcal{J}_\lambda(\lambda)}{(I_0 + \lambda\Omega)^2} \sin(I_0 + \lambda\Omega)t, \quad (6.2.56)$$

where the fitting parameter R has been introduced to represent an "effective" number of terms for $|n| \lesssim \lambda$ in the sum (6.2.55). Equation (6.2.56) is our basic approximation for the phase motion.

The modulational diffusion coefficient is given by

$$D(\omega_2) = \lim_{T \to \infty} \frac{1}{T} \langle [\Delta H_{\text{along}}(T)]^2 \rangle_{I_0, \chi_0}. \quad (6.2.57)$$

Substituting (6.2.51) in (6.2.57), with the average over I_0 shown explicitly,

$$D(\omega_2) = \lim_{T \to \infty} \frac{1}{2T} \frac{\epsilon^2}{2\lambda\Omega} \left\langle \int_{-\lambda\Omega}^{\lambda\Omega} dI_0 \int_{-T}^{T} dt'' \omega_2 \sin \phi(t'') \int_{-T}^{T} dt' \omega_2 \sin \phi(t') \right\rangle_{\chi_0}. \quad (6.2.58)$$

Using (6.2.56) and again expanding in Bessel functions, we have

$$\sin \phi(t) = \sum_j A_j(I_0) \cos\{[(j+1)I_0 + j\lambda\Omega - \omega_2]t + \chi_0\}, \quad (6.2.59\text{a})$$

where

$$A_j(I_0) = \mathcal{J}_j\left[\frac{kR\mathcal{J}_\lambda(\lambda)}{(I_0 + \lambda\Omega)^2}\right]. \quad (6.2.59\text{b})$$

Substituting (6.2.59a) in both time integrals in (6.2.58), we first perform the t'' integration to obtain

$$\int_{-\infty}^{\infty} dt'' \sin \phi(t'') = \sum_j A_j(I_0) \cos \chi_0 \frac{2\pi}{j+1} \delta\left(I_0 + \frac{j\lambda\Omega}{j+1} - \frac{\omega_2}{j+1}\right). \quad (6.2.60)$$

We substitute (6.2.59a) and (6.2.60) into (6.2.58) and do the I_0 integration. Because of the δ-function in (6.2.60), the argument of the cosine in (6.2.59a) is independent of t. We then have

$$D(\omega_2) = \frac{\epsilon^2 \omega_2^2}{2\lambda\Omega} \sum_j \frac{2\pi}{j+1} A_j^2\left(\frac{\omega_2 - j\lambda\Omega}{j+1}\right) \langle \cos^2 \chi_0 \rangle \frac{1}{2T} \int_{-T}^{T} dt',$$

where the sum over j is now restricted to the values of j given by the δ-function in (6.2.60),

$$j = \frac{\omega_2 - I_0}{I_0 + \lambda\Omega}.$$

6.2. Diffusion Rates Along Resonances

Since j is an integer, as I_0 varies from $\lambda\Omega$ to $-\lambda\Omega$, j varies from

$$j = l(\omega_2) = \text{integer part} \left\{ \frac{1}{2}\left(1 + \frac{\omega_2}{\lambda\Omega}\right) \right\} \tag{6.2.61}$$

to $j = \infty$. For the usual case of small Bessel function argument in (6.2.59b), the dominant term in the sum is $j = l$. Dropping all other terms, averaging over χ_0 and t', and using (6.2.59b), we obtain the final result

$$D(\omega_2) = \frac{\pi}{2} \frac{\epsilon^2 \omega_2^2}{(l+1)\lambda\Omega} \mathcal{J}_l^2 \left[\frac{kR(l+1)^2 \mathcal{J}_\lambda(\lambda)}{(\omega_2 + \lambda\Omega)^2} \right]. \tag{6.2.62}$$

As ω_2 increases from zero, l changes discontinuously as seen from (6.2.61). The diffusion coefficient then varies in a series of descending steps as ω_2 increases from zero, with the main plateau ($l = 0$) in the range $0 < \omega_2 < \lambda\Omega$, and the lth plateau in the range

$$(2l - 1)\lambda\Omega < \omega_2 < (2l + 1)\lambda\Omega. \tag{6.2.63}$$

In the usual case, the Bessel function argument in (6.2.62) is small [Note $\mathcal{J}_\lambda(\lambda) \approx \lambda^{-1/3}/2$ and, numerically, $R \lesssim 10$]. For $l = 0$, we then have the main plateau value

$$D_{pl} = \frac{\pi}{2} \frac{\epsilon^2 \omega_2^2}{\lambda\Omega}. \tag{6.2.64}$$

Physically, we expect strong diffusion in the main plateau regime for $\omega_2 < \lambda\Omega$, since the exact resonance condition between the two oscillators

$$\omega_1 = I_0 = \omega_2$$

does occur within the modulational layer (see Fig. 6.12).

In Fig. 6.13, after Chirikov et al. (1982), the numerically determined normalized diffusion coefficient

$$D_n = \frac{D}{(\epsilon^2 \omega_2^2 / \Delta\omega)} \tag{6.2.65}$$

is plotted (dots) as a function of $\omega_2/\Delta\omega$. Here, the actual numerical half-width of the modulational layer $\Delta\omega$ has been used in place of $\lambda\Omega$; for the data in the figure, $\Delta\omega \approx 1.3\lambda\Omega$, with $\lambda = 10$, $\Omega = 10^{-2}$, and $k = 5 \times 10^{-4}$. The main plateau regime is clearly seen, with the average numerical value $D_n = 1.60$ agreeing well with the value 1.57 calculated from (6.2.64).

Numerically, for $\omega_2 > \Delta\omega$, the diffusion coefficient drops abruptly, and then falls off in a steplike fashion with increasing ω_2. This is precisely the behavior predicted by the theoretical result (6.2.62). To compare theory and numerical calculations, the effective number of terms R must be determined. This was done by fitting (6.2.62) to the data at the edges of the last two

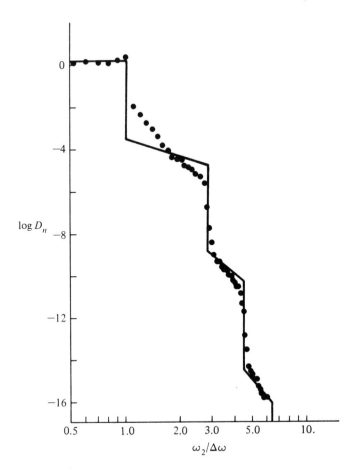

Figure 6.13. The normalized modulational diffusion coefficient D_n versus $\omega_2/\Delta\omega$. The dots are the numerical computations with $\lambda = 10$, $\Omega = 10^{-2}$, $k = 5 \times 10^{-4}$, and $\Delta\omega = 1.3\lambda\Omega$. The solid line is the theory using (6.2.69). Note the constant main plateau for $\omega_2/\Delta\omega < 1$ and the geometric variation of the higher-order plateaus in this doubly logarithmic plot (after Chirikov et al., 1982).

plateaus ($l = 2, 3$), yielding $R \approx 5.3$. With this value, the solid curve then gives the theoretical diffusion coefficient (6.2.62). The agreement is quite reasonable, considering that the single term expression for the phase (6.2.56) was used instead of the exact expression (6.2.55). Note that the theory predicts a sharp drop after each plateau, but that the subsequent ($l \geq 1$) plateaus are not level. From (6.2.62), for small Bessel function argument,

$$D_n(\omega_2) \propto \left[1 + \frac{\omega_2}{\Delta\omega}\right]^{-4l},$$

6.2. Diffusion Rates Along Resonances

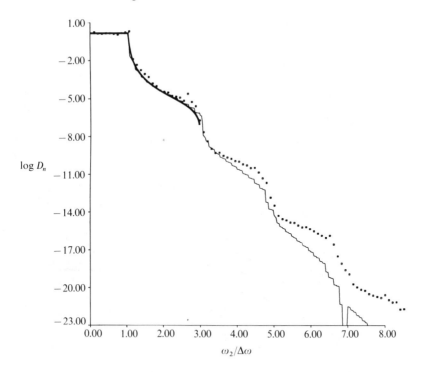

Figure 6.14. Normalized modulational diffusion rate; $\lambda = 10.00$, $\Omega = 0.01$, $k = 5 \times 10^{-4}$ (after Chirikov et al., 1985).

where l is given by (6.2.61). For $l \gg 1$, D_n falls geometrically on the plateau as

$$D_n \propto \left(\frac{\omega_2}{\Delta\omega}\right)^{-4l}.$$

These predictions are in reasonable agreement with the numerical results.

The preceding analysis can be extended by keeping the full sum over n in (6.2.55), rather than introducing the fitting parameter R as in (6.2.56) [see Chirikov et al., 1985)]. In this case the theory has no adjustable parameters. The analysis is complicated and we just give the final comparison of theory and numerical results in Fig. 6.14, which shows the normalized diffusion rate (6.2.65) on a logarithmic scale versus $\omega_2/\Delta\omega$. The dots give the numerical results, and the heavy solid line gives an analytical result for the main plateau and first secondary plateau, obtained by neglecting some of the terms in the sum (6.2.55). The thin solid line shows the theoretical result obtained by numerically performing the sum over all first-order terms that were neglected in obtaining the strictly analytical result. We see that the first-order theory still underestimates the diffusion for the secondary plateaus, although the general behavior of D is successfully predicted.

6.3. Extrinsic Diffusion

We now turn to near-integrable systems that are subject to an externally generated transport process, such as a classical diffusion due to "noise." The interaction of the noise with the resonance structure can lead to enhanced diffusion. We have already examined, in Section 5.6b (see Fig. 5.21), the enhancement of classical diffusion on the action space due to transport across a separatrix separating rotation from libration. This process occurs in systems with one or more degrees of freedom, when an extrinsic diffusion is present.

In this section we consider enhanced classical transport along resonance layers. Enhanced diffusion along layers is possible when a non-energy-conserving extrinsic diffusion is imposed on an otherwise autonomous system having two (or more) degrees of freedom. The extrinsic diffusion essentially adds a third degree of freedom to the system, thus allowing motion along resonances, as described in the previous sections. For extrinsic diffusion on the action space, this process has been analyzed by Tennyson (1982), and we adopt his general point of view in Section 6.3a. A related process for which there is extrinsic diffusion of a parameter, leading to diffusion of the resonance center, has been described by Chirikov (1979a) and Cohen and Rowlands (1981) and is treated in Section 6.3b.

6.3a. Resonance Streaming

Geometric Construction. We return to the example of the unperturbed Hamiltonian (6.1.7)

$$H_0 = I_1^2 + (6I_2)^2,$$

whose resonance surfaces and energy surfaces were shown in Fig. 6.2. We replot in Fig. 6.15 the section within the small dashed box in Fig. 6.2 and assume that a perturbation

$$H_1 = V_R(I_1, I_2) \cos(6\theta_1 - \theta_2) \tag{6.3.1}$$

is present. The resonance vector for the single resonance

$$\mathbf{m}_R = (6, -1)$$

is tangent to the energy surface, as shown in Fig. 6.15. Thus the island oscillation lies along the energy surface as shown. The dashed lines indicate the full resonance width obtained from secular perturbation theory (Section 2.4a). Depending on the initial conditions, the actual libration amplitude can lie anywhere within these dashed lines. The action averaged over the libration oscillation is at the center of the resonance at A.

Now, consider a non-energy-conserving, extrinsic diffusion, which we characterize by an instantaneous jump in action from point "a" to point "b"

6.3. Extrinsic Diffusion

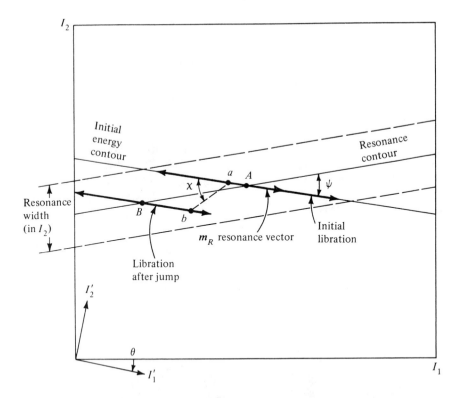

Figure 6.15. Oscillation center displacement inside a resonance layer. The phase point, oscillating along the energy contour, makes a jump a–b due to external noise or dissipation. The oscillation center makes a corresponding jump A–B (after Tennyson, 1982).

as shown in Fig. 6.15. The oscillation center or average action has jumped from A to B, and the magnitude of this displacement in action space is

$$|A - B| = |a - b| \sin \chi \csc \psi. \tag{6.3.2}$$

We see that the oscillation center jump $|A - B|$ lies along the resonance surface and is equal to the component of the action jump perpendicular to the energy surface, $|a - b| \sin \chi$, multiplied by the factor $\csc \psi$. Over many random jumps in various directions we average $\sin^2 \chi$, which is of order unity. However, $\csc \psi$ is locally fixed, and if ψ is small, the diffusion of the oscillation center is greatly enhanced. We call the oscillation center diffusion along the resonance contour *resonance streaming*.

If there are R jumps per unit time, then we define the oscillation center diffusion coefficient along the resonance as

$$D^{\text{osc}} = \tfrac{1}{2} R \langle |A - B|^2 \rangle.$$

The classical extrinsic diffusion coefficient perpendicular to the energy surface is

$$D_{\perp E} = \tfrac{1}{2} R \langle |a - b|^2 \sin^2 \chi \rangle. \tag{6.3.3}$$

From the above argument,

$$D^{\text{osc}} = D_{\perp E} \csc^2 \psi. \tag{6.3.4}$$

Diffusion Calculation. If we consider an isotropic extrinsic diffusion D_0, such that I_1 and I_2 diffuse equally, then obviously $D_{\perp E} = D_0$ and $D^{\text{osc}} = D_0 \csc^2 \psi$. If the diffusion on the action space is anisotropic, then $D_{\perp E}$ is found from the elements of the diffusion tensor. Assuming I_1 and I_2 are the principal axes for the extrinsic diffusion such that the diffusion tensor is

$$D = \begin{bmatrix} D_1 & 0 \\ 0 & D_2 \end{bmatrix},$$

then an initial δ-function distribution in action diffuses to the distribution

$$F = \frac{1}{4\pi t (D_1 D_2)^{1/2}} \exp\left(-\frac{I_1^2}{4 D_1 t} - \frac{I_2^2}{4 D_2 t}\right) \tag{6.3.5}$$

after a time t (Chandrasekhar, 1943). We introduce a new system I_1', I_2' rotated from the old through an angle θ such that I_2' is in the direction perpendicular to the energy contour

$$\begin{pmatrix} I_1 \\ I_2 \end{pmatrix} = \begin{pmatrix} \cos \theta & -\sin \theta \\ \sin \theta & \cos \theta \end{pmatrix} \begin{pmatrix} I_1' \\ I_2' \end{pmatrix}. \tag{6.3.6}$$

Substituting (6.3.6) in (6.3.5), we obtain the distribution F' in the new coordinates. If we integrate F' over I_1', we obtain

$$\int F'(I_1', I_2', t) \, dI_1' \propto \exp\left(-\frac{I_2'^2}{4 D_{\perp E} t}\right),$$

where

$$D_{\perp E}^{-1} = \frac{\tfrac{1}{2}(D_1^{-2} + D_2^{-2}) \sin^2 2\theta + D_1^{-1} D_2^{-1} \cos^2 2\theta}{D_1^{-1} \sin^2 \theta + D_2^{-1} \cos^2 \theta} \tag{6.3.7}$$

is the diffusion coefficient along I_2'. Given $D_{\perp E}$ from (6.3.7), we can then calculate D^{osc} from (6.3.4).

An alternate procedure to treat an anisotropic diffusion tensor is to transform to a new action space $\bar{I}$ such that the diffusion tensor is isotropic. That is, the components of the metric of the new space g^{ij} are chosen such that

$$D^{ij} = \bar{D}_0 g^{ij}, \tag{6.3.8}$$

where the D^{ij} are the old diffusion tensor components and $\bar{D}_0$ is the isotropic

6.3. Extrinsic Diffusion

extrinsic diffusion coefficient in the new space. If, for example, we consider the diffusion tensor to be given by

$$\mathbf{D} = \bar{D}_0 \begin{bmatrix} 1 & 0 \\ 0 & 36 \end{bmatrix}, \tag{6.3.9}$$

then the components of length in the I_1, I_2 space are taken to have the metric

$$ds_1 = dI_1, \qquad ds_2 = \tfrac{1}{6} dI_2, \tag{6.3.10}$$

such that equal increments in the space $ds_1 = ds_2$ cover six units of I_2 for each unit of I_1 covered. Thus the space is compressed in I_2 relative to I_1. Additionally a change in action coordinates can be made

$$\bar{I}_1 = I_1, \qquad \bar{I}_2 = \tfrac{1}{6} I_2. \tag{6.3.11}$$

Then, in the new coordinates the diffusion tensor is isotropic. This last step is not strictly necessary to see the $\csc^2 \psi$ enhancement, as the coordinates can be stretched without relabeling. These relations are illustrated in Fig. 6.16 for a Hamiltonian with unperturbed part

$$H_0 = I_1^2 + I_2^2 \tag{6.3.12}$$

and a coupling resonance $\omega_1 = \omega_2$, with the diffusion tensor given by (6.3.9). In (a) the diffusion in the original space is shown. In (b) the space is stretched as in (6.3.10) to make the diffusion isotropic. In (c) the coordinates are relabeled by the transformation (6.3.11). The relabeled Hamiltonian is obtained by substituting (6.3.11) in (6.3.12) to obtain

$$\bar{H}_0 = \bar{I}_1^2 + (6\bar{I}_2)^2. \tag{6.3.13}$$

The new variables are related to the old through the generating function

$$F_2 = \theta_1 \bar{I}_1 + 6\theta_2 \bar{I}_2, \tag{6.3.14}$$

from which the new angles are related to the old by

$$\bar{\theta}_1 = \frac{\partial F_2}{\partial \bar{I}_1} = \theta_1, \qquad \bar{\theta}_2 = \frac{\partial F_2}{\partial \bar{I}_2} = 6\theta_2.$$

The old resonance condition $\omega_1 - \omega_2 = 0$ is replaced by the new condition

$$6\bar{\omega}_1 - \bar{\omega}_2 = 0 \tag{6.3.15}$$

or in action space, the line

$$\bar{I}_2 = \tfrac{1}{6} \bar{I}_1. \tag{6.3.16}$$

We have, in fact, reproduced the model unperturbed Hamiltonian (6.1.7) of our example. We can explicitly solve for $\csc^2 \bar{\psi}$ from (6.3.13) and (6.3.16) for any $\bar{H}_0$. For example, with $\bar{H}_0 = 1$, we obtain an enhancement factor

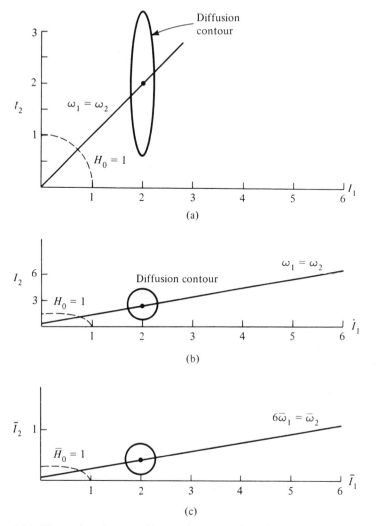

Figure 6.16. Illustrating the stretching and compressing of the phase space to make the extrinsic diffusion isotropic.

$\csc^2 \bar{\psi} = 9.2$ such that

$$\bar{D}^{\text{osc}} = 9.2 \bar{D}_0.$$

To obtain the oscillation center diffusion in the old space, we must transform back to the original action space.

To illustrate resonance streaming numerically (following Tennyson, 1982), we take the Hamiltonian

$$H = I_1^2 + (6I_2)^2 + V_R \cos(6\theta_1 - \theta_2), \tag{6.3.17}$$

6.3. Extrinsic Diffusion

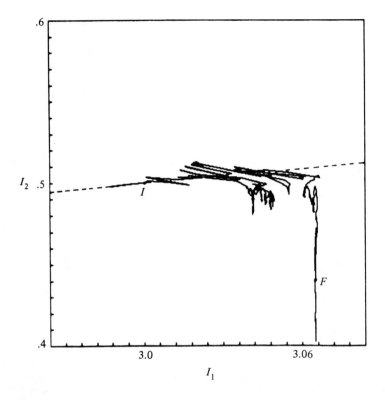

Figure 6.17. Resonance diffusion for the Hamiltonian of (6.3.17) with a random step in I_2 of $\Delta I_2 = 10^{-5} \sin(r_n)$, where r_n is a random number between 0 and 2π chosen at time steps $\Delta t = 1$. The trajectory is the connection of phase points obtained by averaging over 500 time steps, so that the oscillation center position is found (after Tennyson, 1982).

with $V_R = 10^{-5}$, and choose an external diffusion $\Delta I_2 = 10^{-5} \sin(r_n)$ [ΔI_1 is chosen to be zero for simplicity]. Here r_n is a random number between 0 and 2π taken at time steps $\Delta t = 1$. The trajectory shown in Fig. 6.17 is that of the oscillation center with successive time-averaged ($\Delta t = 500$) positions connected by line segments. The initial position I is at the center of the resonance, with the particle eventually diffusing away from the resonance, the run ending at F. The dashed line gives the resonance contour. Three types of diffusive motion can be seen in the figure. The motion along the resonance contour is that of oscillation center diffusion. The motion at a sharp angle to the resonance contour represents enhanced diffusion in the libration direction across the resonance, of the type illustrated in Fig. 6.15. This motion is nearly along a constant energy contour for positions near but outside the resonance separatrix. Finally, the vertical motion near F is the long-time classical diffusion of I_2 alone, well outside of the resonance region.

For a fixed extrinsic diffusion, oscillation center diffusion can only exist for sufficiently large resonance widths such that the time it takes for classical diffusion across the resonance

$$\tau_D = \frac{(\Delta I_R)^2}{2D_{\perp R}}$$

is large compared to the oscillation period within the resonance:

$$\hat{\omega}_R \tau_D > 2\pi. \qquad (6.3.18)$$

Here ΔI_R and $D_{\perp R}$ are the full resonance width and the extrinsic diffusion coefficient perpendicular to the resonance surface, and $\hat{\omega}_R$ is the libration frequency within the resonance. For the example of (6.3.17), $\Delta I_R \approx 3 \times 10^{-3}$, $D_{\perp R} \approx 10^{-8}$, $\tau_D \approx 450$, $\hat{\omega}_R \approx 0.03$, and the condition $\hat{\omega}_R \tau_D \approx 12 > 2\pi$ is satisfied.

The weakly collisional regime of oscillation center transport, satisfying (6.3.18), has a counterpart in particle diffusion in toroidal magnetic fields known as *banana* diffusion. The name follows from the banana shape of the resonance islands, when the action-angle coordinates are presented in polar coordinates. At higher collisionality (smaller resonance widths), for which (6.3.18) is not satisfied, there is a transition to a regime independent of the randomization time, which is therefore known as the *plateau* regime. Chirikov (1979a) and Cohen and Rowlands (1981) treat this collisional regime using a mapping representation, which we consider in Section 6.3b. A description of plateau diffusion is deferred to Section 6.4, where we consider the example of particle diffusion in toroidal magnetic fields in the presence of resonances. Finally, at still higher collisionality, the resonance structure is unimportant, and we obtain a third regime of strictly classical diffusion, also described in Section 6.4. The average rate of diffusion in a system with many separated resonances depends on the fraction of phase space covered by the resonances and the collisionality regime for each resonance.

The resonance streaming process can be generalized to systems with many degrees of freedom (Tennyson, 1982). The extrinsic diffusion in the action variable $I_{\perp E}$, perpendicular to the energy surface, drives diffusion of the oscillation center along the resonance surface, in the action direction $I_{\| R}$ defined by the projection of the resonance vector m_R onto the resonance surface.

Oscillation center diffusion of the type discussed above has been postulated by Tennyson (1982) as the mechanism for beam blow-up in storage rings. It is also implicit in the enhancement of diffusion in various magnetic confinement and heating devices. Tennyson has also pointed out that for two-dimensional dissipative systems in which both I_1 and I_2 would tend to decrease with time, resonance trapping and consequent streaming can actually rapidly increase one of the actions. He likened this phenomenon to the motion of a boat on ice runners skating against the wind. In Fig. 6.15 the boat, constrained to run along the resonance contour, is pushed by a

6.3. Extrinsic Diffusion

wind, the dissipative force, along ab on the "sail" fixed on the energy contour. If the resonance contour in Fig. 6.15 has a negative slope, the resultant force along AB will drive the boat "upwind" in I_2, although the total energy of the system continually decreases.

6.3b. Diffusion of a Parameter

Mapping Representation. With the proper transformation, it is possible to represent a class of resonance streaming problems in a simpler form. We consider the dynamical system to be described by a mapping of the general form

$$J_{n+1} = J_n + f(\mu) \sin \theta_n, \tag{6.3.19a}$$

$$\theta_{n+1} = \theta_n + 2\pi\alpha(J_{n+1}, \mu), \tag{6.3.19b}$$

where J and θ are canonical action and angle variables, as described in Section 3.1, and μ is an additional parameter unaffected by the dynamical process of the mapping. Equations (6.3.19) satisfy condition (3.1.16) for area preservation, and thus are derivable from a Hamiltonian by introduction of the periodic δ-function as in (3.1.34). Since the Hamiltonian is nonautonomous, it is equivalent to a two-degree-of-freedom system. If the parameter μ is described by an independent dynamical equation of the form

$$\mu_{n+1} = \mu_n + \zeta_n, \tag{6.3.20}$$

where $\mu = \mu_{n+1}$ in (6.3.19), then, if ζ_n is a random variable, μ generates an extrinsic diffusion analogous to that described in Section 6.3a. This formulation has been used by Chirikov (1979a) and by Cohen and Rowlands (1981) to calculate transport in magnetic confinement systems. The use of $\sin \theta$ in (6.3.19a) already implies that a single Fourier component has been extracted as the primary resonance under consideration.

In order to simplify the analysis, the mapping can be linearized in both J and μ, as done in Section 4.1b, to obtain an augmented standard mapping. To do this, we choose a $\mu = \mu_0$ in (6.3.19) and expand about a fixed point (J_0, θ_0), with $\theta_0 = 0$ and J_0 given by

$$\alpha(J_0, \mu_0) = k, \quad k \text{ integer.}$$

We obtain

$$I_{n+1} = I_n + K \sin \theta_n, \tag{6.3.21a}$$

$$\theta_{n+1} = \theta_n + I_{n+1} + P_{n+1}, \tag{6.3.21b}$$

$$P_{n+1} = P_n + \xi_n/\tau^{1/2}, \tag{6.3.21c}$$

where

$$I_n = 2\pi \frac{\partial \alpha}{\partial J_0} (J_n - J_0), \qquad (6.3.22a)$$

$$P_n = 2\pi \frac{\partial \alpha}{\partial \mu_0} (\mu_n - \mu_0), \qquad (6.3.22b)$$

$$K = 2\pi f(\mu_0) \frac{\partial \alpha}{\partial J_0}, \qquad (6.3.23a)$$

and

$$\xi_n = 2\pi \frac{\partial \alpha}{\partial \mu_0} \tau^{1/2} \zeta_n. \qquad (6.3.23b)$$

K is the stochasticity parameter and ξ_n is the normalized random variable. We assume that $\langle \zeta_n \rangle = 0$ and $\langle \zeta_n^2 \rangle = \sigma^2$ for the random process over one mapping iteration. We choose τ so that $\langle \xi_n^2 \rangle = 1$; thus τ is a dimensionless parameter

$$\tau = \left(2\pi\sigma \frac{\partial \alpha}{\partial \mu_0}\right)^{-2}, \qquad (6.3.24)$$

which gives the number of mapping periods required to diffuse P through a unit amplitude:

$$\langle P^2 \rangle = \frac{n}{\tau}. \qquad (6.3.25)$$

When $K \gg 1$, the resonances overlap and the diffusion in the action I is just the quasilinear value computed in Section 5.4

$$D_{QL} = \frac{\langle \Delta I^2 \rangle}{2n} = \frac{K^2}{4}. \qquad (6.3.26)$$

In this regime the angle variable is fully stochastic, and the diffusion in action is unaffected by the extrinsic diffusion of μ. Similarly, when $\tau \ll 1$ the phase is randomized by the extrinsic diffusion for each step, and the limiting diffusive rate (6.3.26) holds for any K. The case of interest here, corresponding to resonant streaming studied in the previous section, is for

$$K \ll 1, \qquad \tau \gg 1. \qquad (6.3.27)$$

As a consequence of (6.3.27), the changes in I and P over one iteration are small, and we can approximate the difference equations (6.3.21) with the differential equations

$$\frac{dI}{dn} = K \sin \theta, \qquad (6.3.28a)$$

$$\frac{d\theta}{dn} = I + P. \qquad (6.3.28b)$$

6.3. Extrinsic Diffusion

and

$$\frac{dP}{dn} = \frac{\xi}{\tau^{1/2}}. \tag{6.3.28c}$$

Diffusion of the Oscillation Center. The Hamiltonian, formed in the usual way from (6.3.28a) and (6.3.28b),

$$H = \frac{I^2}{2} + P(n)I + K\cos\theta \tag{6.3.29}$$

has an oscillation center at $I = -P$. Thus, using (6.3.25),

$$\langle I^2 \rangle = \langle P^2 \rangle = \frac{n}{\tau}. \tag{6.3.30}$$

The local diffusion coefficient within the resonance is then

$$D_r = \frac{\langle I^2 \rangle}{2n} = \frac{1}{2\tau}. \tag{6.3.31}$$

In order for the diffusion within the resonance to be of the oscillation center type, given by (6.3.31), the phase point must remain in resonance over many oscillation periods. That is, the parameter

$$S \equiv \omega_0 \tau_d \gg 1, \tag{6.3.32a}$$

where ω_0 is the oscillation frequency and τ_d is the time it takes the oscillation center to diffuse a distance equal to the resonance width. Secular perturbation theory shows that the width of the oscillation governed by Hamiltonian (6.3.29) is proportional to $K^{1/2}$ and thus from (6.3.30), the oscillation center diffuses a distance equal to its amplitude in a time

$$\tau_d \approx K\tau.$$

The frequency of the oscillation is given approximately by

$$\omega_0 \approx K^{1/2}.$$

Substituting these values in (6.3.32a), the condition for oscillation center diffusion becomes

$$S \approx K^{3/2}\tau \gg 1. \tag{6.3.32b}$$

In fact, a change in variables

$$P' = K^{-1/2}P, \qquad I' = K^{-1/2}I, \qquad n' = K^{1/2}n,$$

shows that Eqs. (6.3.28) are characterized by the single parameter S.

Although the local diffusion coefficient may be of some interest, the diffusion averaged over both resonant and nonresonant portions of the phase space is usually more physically significant. We now estimate this average. As described in Section 5.6b, assuming the diffusion rate outside of the resonance layer is negligible, the average diffusion rate is proportional to the fraction of time spent in resonance, which can be taken as the fraction of the resonant phase space to the total phase space. Using the Hamiltonian (6.3.29), evaluated on the separatrix, to calculate the resonance area, we have the resonance fraction

$$f_r = \frac{4(2K)^{1/2} \int_0^\pi (1 - \cos\theta)^{1/2}\, d\theta}{(2\pi)^2} = \frac{4K^{1/2}}{\pi^2}$$

and the average diffusion is then

$$\langle D \rangle = D_r f_r = \frac{2}{\pi^2} \frac{K^{1/2}}{\tau}. \tag{6.3.33}$$

The diffusion of the oscillation center is not the only mechanism of particle diffusion within the resonance. There is a second, phase-dependent term that spreads particles within the resonance. This phase-dependent effect has been calculated by Cohen and Rowlands (1981). For $K \ll 1$, they obtain a correction to (6.3.31)

$$D_r = \frac{1}{2\tau}\left[1 + \frac{1}{2 - K/2}\right]. \tag{6.3.34}$$

However, it is not clear whether this correction applies to long-time diffusion over many resonance widths. In any case, the correction factor is of the order of 50%, and we are making other errors of this order by not properly using the motion near the separatrix in the calculation. We therefore take (6.3.33) as a reasonable estimate of the average diffusion in the action space.

Relation to Resonance Streaming. We now consider the relation between the calculation of D^{osc} for resonance streaming in Section 6.3a and the calculation of D_r in this subsection. What we shall show is that, with some simplifying assumptions, the two problems can be brought into correspondence. In Fig. 6.18a the calculation of D^{osc} is illustrated in the action space (I_1, I_2). The resonance line is crossed by the energy surface. The resonance vector m, along which the libration occurs, is tangent to the energy surface. The extrinsic diffusion perpendicular to the energy surface $D_{\perp E}$ drives the oscillation center diffusion along the resonance surface, with

$$D^{\text{osc}} = D_{\perp E} \csc^2 \psi. \tag{6.3.4}$$

In Fig. 6.18b we have rotated the action space to new variables I'_1, I'_2 such that $D_{\perp E}$ lies along I'_2 and m lies along I'_1. The magnitude of $D_{\perp E}$ and the angle ψ remain unchanged.

In Fig. 6.18c we transform again to new variables (P, I), stretching I'_2 by a factor of $\cot \psi$, while leaving I'_1 unchanged:

$$P = I'_2 \cot \psi, \qquad I = I'_1.$$

6.3. Extrinsic Diffusion

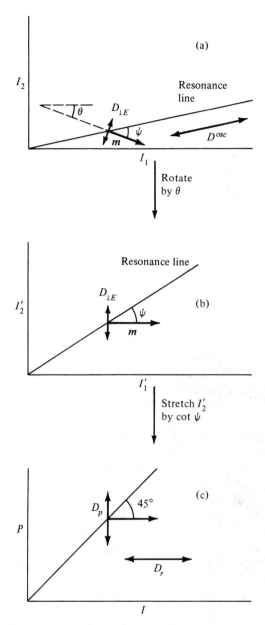

Figure 6.18. Action space transformations to relate resonance streaming (described in Section 6.3a) to the diffusion of a parameter (described in Section 6.3b).

The resonance line now cuts the $P - I$ plane at 45°, and the extrinsic diffusion coefficient increases to

$$D_p = D_{\perp E} \cot^2 \psi. \tag{6.3.35}$$

Considering P as a parameter, and using (6.3.30), we find the component of the oscillation center diffusion along I to be $D_r = D_p$. Comparing (6.3.35) and (6.3.4)

$$D_r = D^{osc} \cos^2 \psi. \tag{6.3.36}$$

We note from Fig. 6.18a that D_r is the diffusion coefficient for the oscillation center, projected along the libration direction. This shows the corresondence between the point of view in Section 6.3a and in this subsection.

For many degrees of freedom, the corresponding transformations can also be made, with the general transformation procedure having been described under the rubric *orthogonal metrics* by Chirikov (1979).

6.4. Diffusion in Toroidal Magnetic Fields

An example of considerable importance for plasma confinement, which exhibits both intrinsic and extrinsic diffusion, is that of particle motion in toroidal magnetic fields. The diffusion can be either of the magnetic field lines themselves, or of particles moving across the field lines. Of particular interest is the case of particles diffusing across magnetic field lines in the presence of both a wave field and extrinsic stochasticity (particle collisions or noisy fields). For this case, depending on the relation between the wave field, and the particle motion, there can be either diffusion in a single action, governed by the restrictions discussed in Section 5.6b, or diffusion in two actions, leading to the resonance streaming of Section 6.3. In Section 6.4a we consider the basic processes of resonance island formation in toroidal magnetic fields. In Section 6.4b we discuss qualitatively the effect of particle drifts and the resultant extrinsic diffusion. We contrast the situation in which the resonant island centers are fixed with the case in which the island centers diffuse. In Section 6.4c we treat an example of the latter case, illustrating the theory developed in Section 6.3b. In Section 6.4d we introduce the subject of self-consistent motion, where the field driving the particle motion depends in part on the particle motion itself.

6.4a. Magnetic Islands

Magnetic Field Configurations. The simplest toroidal magnetic field configuration is that produced by a long current-carrying conductor. (Here the torus is in configuration space, rather than phase space.) This magnetic field

6.4. Diffusion in Toroidal Magnetic Fields

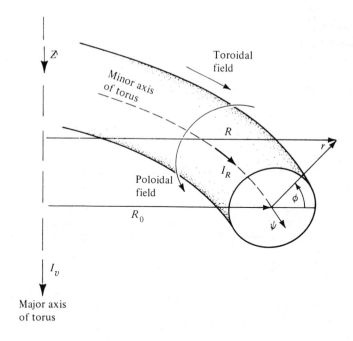

Figure 6.19. Toroidal magnetic field configuration.

does not confine particles, due to drifts across field lines. However, the addition of a toroidal current gives rise to a second field component, known as a poloidal field, at right angles to the toroidal field. This results in a helical field pattern around the torus as shown in Fig. 6.19. The field lines are similar in structure to the trajectories for the two-dimensional oscillator, as given in Fig. 3.1a. The complete conditions for the confinement and stability of a plasma, which has collective modes of behavior, will not be considered here. These conditions have led to a variety of current configurations, including a rigid toroidal current-carrying conductor at the minor axis $r = 0$, together with a vertical field (the levitron), an external helical conductor at $r = a$ around a physical torus (the stellarator), and a toroidal plasma current centered on the minor axis (the tokamak).

For these configurations the equations of the magnetic field lines can be put in Hamiltonian form (see, e.g., Morozov and Solov'ev, 1966; Rosenbluth et al., 1966; and Freis et al., 1973). For configurations that have azimuthal symmetry $\partial/\partial \psi \equiv 0$ (the tokamak and levitron), the equations have the form of a one-degree-of-freedom nonlinear oscillator, with the invariant being the flux enclosed within a magnetic surface (see below for definitions of these quantities). The breaking of this azimuthal symmetry effectively introduces an explicit dependence of the "timelike" variable ψ into the Hamiltonian, and thus all the consequent complexity of a two-degree-of-freedom system.

Magnetic Surfaces. The equations of the field line are defined in (r, ϕ, ψ) coordinates by

$$\frac{dr}{B_r} = \frac{r\, d\phi}{B_\phi} = \frac{R\, d\psi}{B_\psi}, \qquad (6.4.1)$$

where r and $R = R_0 + r\cos\phi$ are the local minor and major radii of the field line (see Fig. 6.19), and B_r, B_ϕ, and B_ψ are the three components of the magnetic field. In these coordinates, B_r and B_ϕ are poloidal field components and B_ψ is the toroidal field component. For a tokamak with $B_\psi = B_0$ at $R = R_0$, the field can be approximated by (Solov'ev and Shafranov, 1970)

$$\boldsymbol{B} = \left(0, \frac{\bar{B}_\phi(r)}{h_s}, \frac{B_0}{h_s}\right), \qquad (6.4.2)$$

where $h_s = 1 + (r/R_0)\cos\phi$ gives the field strength variation from the outside to the inside of the torus, and contains all the ϕ variation. In the levitron a vertical field is required for the plasma equilibrium. This introduces an additional ϕ dependence that is much stronger than the toroidally produced variation. Neglecting the latter, $\boldsymbol{B}$ can be written as

$$\boldsymbol{B} = \left(B_v \sin\phi, \frac{B_0}{\beta r} + B_v \cos\phi, B_0\right), \qquad (6.4.3)$$

where B_v is the vertical field (assumed upward in Fig. 6.19), and β relates the ring current to the vertical current, $\beta = I_v/I_R R_0$. The vertical field weakens the poloidal field on the inside of the torus, giving a poloidal field null ($B_\phi = 0$) at $\phi = \pi$ for $r = B_0/\beta B_v$. If the field given by (6.4.2) or (6.4.3) is substituted into (6.4.1), the field lines trace out a set of concentric surfaces having toroidal symmetry and nested about the minor axis of the torus. These are called magnetic surfaces. For the levitron field, a set of magnetic surfaces in a $\psi = $ constant plane is shown in Fig. 6.20. Note the presence of a magnetic surface with a separatrix, the x-point corresponding to the poloidal field null.

As we shall see, the only quantity of importance in characterizing an unperturbed magnetic surface is the value of the rotational transform (angular frequency of ϕ per passage around the major axis)

$$\iota = \int_0^{2\pi} \frac{d\phi}{d\psi}\, d\psi = 2\pi\alpha, \qquad (6.4.4)$$

where α is the rotation number, as previously defined. For the tokamak, h_s tends to be near unity, with a small oscillatory component. We can obtain an unperturbed Hamiltonian by introducing the canonical action variable

6.4. Diffusion in Toroidal Magnetic Fields

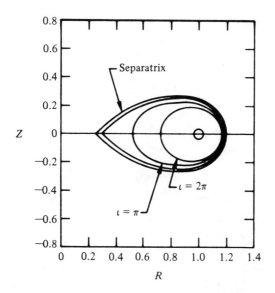

Figure 6.20. Magnetic flux surfaces for the levitron projected onto a plane $\psi = $ const.; $R_0 = 1$ (after Freis et al., 1973).

$\zeta = r^2/2$, and setting $h_s = 1$. The equations of motion (6.4.1) are then

$$\frac{d\zeta}{d\psi} = 0, \qquad \frac{d\phi}{d\psi} = \frac{\iota(\zeta)}{2\pi},$$

which can be derived from the Hamiltonian

$$H_0(\zeta) = \frac{1}{2\pi} \int_0^\zeta \iota(\zeta)\, d\zeta = \text{const.}, \qquad (6.4.5)$$

which is clearly in action-angle form. Similarly, for the levitron field, from (6.4.3), setting $B_0 = 1$ for convenience,

$$\begin{aligned}
\frac{d\zeta}{dz} &= B_v \sqrt{2\zeta} \sin \phi, \\
\frac{d\phi}{dz} &= \frac{1}{2\beta\zeta} + \frac{B_v}{\sqrt{2\zeta}} \cos \phi,
\end{aligned} \qquad (6.4.6)$$

where $z = \psi/2\pi$. Equations (6.4.6) can be derived from a Hamiltonian

$$H = -\frac{1}{\beta}\left[\ln\left(\frac{8}{\sqrt{2\zeta}}\right) - 2\right] + B_v \sqrt{2\zeta} \cos \phi.$$

For this case, before we consider the perturbation, we must transform to action-angle variables, as in Section 1.2,

$$J = \frac{1}{2\pi}\int_0^{2\pi} \zeta\, d\phi, \tag{6.4.7a}$$

$$\bar{\phi} = \frac{\partial S(J, \phi)}{\partial J}, \tag{6.4.7b}$$

where S is the generating function. Using perturbation theory as in Section 2.2 or 2.5 to second-order,

$$H_0 = \frac{1}{\beta}\ln\left(\frac{8}{\sqrt{2J}} - 2\right) - 2B_v^2\beta J - \tfrac{3}{2}B_v^4\beta^3 J^2 \tag{6.4.7c}$$

with a rotational transform $\iota = dH_0/dJ$. A given value of ι specifies a particular magnetic surface for the unperturbed system. We emphasize that with azimuthal (ψ) symmetry, the Hamiltonian for the magnetic field lines describes a system with a single degree of freedom, and therefore the field line trajectory is completely integrable.

From the above examples it is apparent that two-dimensional magnetic fields have a Hamiltonian form. This follows from the divergence-free property of the field as can be seen explicitly in x, y, z coordinates with no z-variation

$$\frac{\partial B_x}{\partial x} + \frac{\partial B_y}{\partial y} = 0. \tag{6.4.8}$$

The field line equations can then be written in Hamiltonian form

$$\frac{dx}{dz} = \frac{B_x}{B_z} = -\frac{\partial H}{\partial y}, \qquad \frac{dy}{dz} = \frac{B_y}{B_z} = \frac{\partial H}{\partial x},$$

which satisfies (6.4.8). For field variation in three dimensions there is still a Hamiltonian form for the field line equations, but the representation is not as direct (Boozer, 1983, 1984).

Magnetic Islands. As a model for the magnetic field perturbation, we consider an additional term in the Hamiltonian

$$H = H_0(J) + \epsilon H_1(J, \bar{\phi}, \psi) \tag{6.4.9}$$

and expanding about the unperturbed orbits, $J = J_0 + \Delta J$ and $\bar{\phi} = (\iota/2\pi)\psi + \Delta\bar{\phi}$ (where $J_0 = \zeta_0$ for the tokamak approximation), we obtain a pair of perturbation equations of the form

$$\frac{d\Delta J}{d\psi} = \epsilon \sum_{m,n} A_{mn}\cos(m\bar{\phi} - n\psi + \chi_{mn}),$$

$$\frac{d\Delta\bar{\phi}}{d\psi} = \frac{1}{2\pi}\frac{\partial \iota}{\partial J}\Delta J, \tag{6.4.10}$$

6.4. Diffusion in Toroidal Magnetic Fields

where the A_{mn} are the Fourier components of $\partial H_1/\partial \phi$. Choosing unperturbed magnetic surfaces (or J_0) such that the rotational transform gives a resonance between the poloidal and toroidal field line motion

$$m\bar{\phi} - n\psi = 2\pi k, \quad k \text{ integer},$$

than a transformation to a rotating frame, as in Section 2.4,

$$\hat{\phi} = m\bar{\phi} - n\psi$$

gives the Hamiltonian for the perturbed motion

$$\hat{H} = \frac{m^2}{2\pi} \frac{d\iota}{dJ} \frac{\Delta \hat{J}^2}{2} + \epsilon A_{mn} \cos \hat{\phi}, \qquad (6.4.11)$$

where $\Delta \hat{J} = \Delta J/m$. The half-width of the island separatrix is then, from (2.4.31),

$$\Delta \hat{J}_{\max} = \frac{2}{m} \left| \frac{\epsilon A_{mn}}{(1/2\pi)(d\iota/dJ)} \right|^{1/2}. \qquad (6.4.12)$$

As an illustration of island formation, a ψ-dependent perturbation was introduced into the levitron magnetic field by tilting the ring current (Freis et al., 1973). Using (6.4.12) with the known value of A_{mn}, they compared the perturbation result with exact numerical integration of the field line equations (6.4.1). The agreement for the $m = n = 1$ island ($\iota = 2\pi$), was excellent, provided the perturbation did not exceed the threshold for global stochasticity. For larger perturbation, the break-up into chains of secondary islands was observed, as expected from the theory of Sections 2.4 and 4.3. In Fig. 6.21a the theoretical (solid lines) and numerical (dots) island widths are given showing the island forming on the inside of the $\iota = 2\pi$ magnetic surface. The banana shape of the island in physical space is evident. The local rotation number around the island is $\alpha = 1/5.6$, and second-order island structure is not in evidence. In Fig. 6.21b the perturbation amplitude (ring tilt) has been increased until the local rotation number $\alpha = \frac{1}{4}$ (the island number nearest the elliptic singular point is 4). The theoretically predicted island has broken up into four unstable second-order island clumps, and the long-time fieldline motion is, in fact, stochastic. A second-order island calculation, as in Section 2.4b, reveals that the $\alpha = \frac{1}{4}$ and $\alpha = \frac{1}{5}$ islands overlap, leading to the observed stochasticity.

The same form of the island perturbation was obtained for a helical external current by Rosenbluth et al. (1966) and Filonenko et al. (1967); for general current perturbations in tokamaks by Rechester and Stix (1976) and Finn (1977); and for the $m = n = 1$ mode helical current by Lichtenberg (1984). For all of these cases, the perturbation acts directly on the magnetic surfaces and breaks the resonant surfaces into finite width islands. For the $m = n = 1$ mode, the assumed single helicity perturbation emphasizes the interaction that couples the helical mode to the $m = 1$, $n = 0$ toroidal

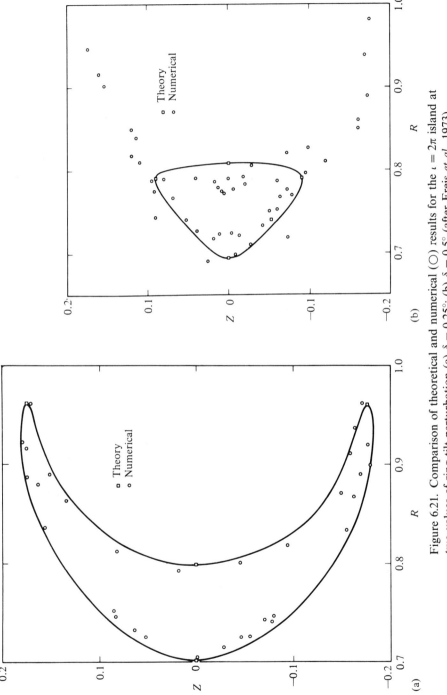

Figure 6.21. Comparison of theoretical and numerical (○) results for the $\iota = 2\pi$ island at two values of ring tilt perturbation (a) $\delta = 0.25°$; (b) $\delta = 0.5°$ (after Freis et al., 1973).

symmetry which generates second-order islands. The width of the stochastic region surrounding the main island, due to overlap of second-order islands, was compared to the numerically calculated stochastic width giving excellent agreement. In these problems it is assumed that the charged particles in the system follow the field lines exactly, and thus the finite Larmor radius of the particles plays no role. Since we are considering a time-independent Hamiltonian in two degrees of freedom, the intrinsic diffusion resulting from island overlap is that derived in Chapter 5, and diffusion along resonances is absent.

For the problem of externally applied helical fields, using the stellarator concept to achieve a plasma equilibrium, the applied fields generate magnetic islands that are particularly severe near the outside of the plasma volume, as analyzed originally by Rosenbluth et al. (1966) and Filonenko et al. (1967). Since these fields are somewhat under the control of the machine designer, the question can be asked whether the fields can be designed to increase the plasma volume having good flux surfaces; i.e., small islands and little stochasticity. The obvious answer is to minimize those harmonic components in the field that supply the principal contributions to the most harmful edge islands. Because of the complexity of stellarator fields, and the lack of clear identification of the most destructive islands, the procedures can be quite complicated. Two related approaches have been developed to do this: Cary (1982) varied components until the adjacent edge island pairs did not overlap (good KAM magnetic surfaces exist); and Hanson and Cary (1986) used a simpler procedure of examining the island residues. These results showed that the plasma volume contained by good magnetic surfaces could be increased by almost a factor of two.

6.4b. Drift Surfaces and Diffusion in Static Fields

Drift Surfaces. Finite temperature charged particles gyrating around magnetic field lines in nonuniform magnetic or electric fields are not completely tied to the field lines, but drift slowly across them. Fourier components of the nonuniform field variation having the major radius periodicity can resonate with Fourier components of the minor radius periodicity, leading to drift surface islands. The drift equations for particle motion in a current-free field are given by (see, e.g., Schmidt, 1979, Section 2.2)

$$v_D = \frac{dr}{dt} = \frac{F \times B}{eB^2} + v_\parallel \frac{B}{B}, \qquad (6.4.13a)$$

$$\frac{dv_\parallel}{dt} = -\frac{e}{M}\frac{\partial \Phi}{\partial s} - \frac{\mu}{M}\frac{\partial B}{\partial s}, \qquad (6.4.13b)$$

where e and M are the particle charge and mass,

$$F = -\left(\mu + \frac{Mv_\parallel^2}{B}\right)\nabla B - e\nabla\Phi$$

is the force averaged over a gyro-orbit, $\mu = \frac{1}{2}Mv_\perp^2/B$ the magnetic moment, Φ the electric potential, $v_\parallel$ the particle velocity along the field line, s the coordinate along the field line, and r the guiding center position of the gyrating particle. In the drift approximation, μ is an adiabatic invariant and is assumed to be well conserved (see Section 2.3b). If B and Φ are not explicit functions of time, then the motion is described by a reduced Hamiltonian that is autonomous and has two degrees of freedom. In this case, the time can be eliminated in favor of s as the independent variable by using

$$\frac{d}{dt}(s) = v_\parallel,$$

where $v_\parallel$ is related to the constants of the motion μ and E through the relation

$$v_\parallel = \left(\frac{2}{M}\right)^{1/2}(E - \mu B)^{1/2}, \tag{6.4.14}$$

and E is the total energy. The resulting drift motion then occurs in a system with two degrees of freedom, which is completely analogous to the magnetic field line motion described in the previous section.

Nonresonant Motion. Consider first a magnetic field gradient alone. For magnetic surfaces that have no ϕ-variation (see Fig. 6.19), F is perpendicular to the magnetic surface, and v_D, as seen from (6.4.13), lies in the surface. However, for toroidal fields of the levitron or tokamak type, the lack of ϕ symmetry gives rise to radial drifts. The time scale for this drift motion is generally long compared to the time scale for motion along the field. Thus neglecting high-order resonances, the radial motion can be described by a one-dimensional autonomous Hamiltonian and is thus integrable.

There are two kinds of orbits depending on whether or not $v_\parallel \ll v_\perp$. For $v_\parallel \ll v_\perp$, particles are trapped on a field line near the outside of the torus ($\phi \approx 0$) and bounce back and forth between high field regions located on the inside of the torus. The drift motion then appears as a "banana" when projected onto a $\psi = \text{const.}$ plane. The width of the banana Δr at $\phi = 0$ is proportional to the particle's Larmor radius ρ_L but enhanced by the ratio of the major-to-minor radii of the torus

$$\Delta r \sim \left(\frac{R}{a}\right)^{1/2}\frac{\rho_L}{\iota}. \tag{6.4.15}$$

The banana shape is similar to that seen in Fig. 6.21a except that it is thinner

6.4. Diffusion in Toroidal Magnetic Fields

(because of the ρ_L proportionality) and is on the outside of the torus. These banana orbits exist on any magnetic surface for trapped particles (see Solov'ev and Shafranov, 1970 for a complete discussion), and therefore do not depend on resonances between the ϕ and ψ motion.

For the case of passing particles, the particles drift on a surface that is similar in shape to and within a distance $\pm \rho_L/\iota$ of the magnetic surface. The orbits are just slightly perturbed versions of the magnetic surfaces.

Drift Islands. If we now allow B and Φ to vary in both ϕ and ψ, passing particles experience resonances between the ϕ and ψ motion that lead to drift islands. For magnetic gradients, because of the ρ_L dependence, these drift islands are small compared to the magnetic islands arising from the same perturbation. If a static electric field exists, such as a drift wave, with Fourier expansion

$$\Phi = \sum \Phi_{mn} \exp[i(m\phi - n\psi)],$$

then drift islands are formed due to resonance between the ϕ and ψ motion. The island widths are proportional to the potential and thus the drift islands are formed independently of the amplitude of any magnetic islands that may exist due to field line motion. A set of equations of the form of (6.4.10) can then be obtained for the perturbed motion, with the Fourier amplitudes ϵA_{mn} proportional both to the Fourier coefficients of the potential Φ_{mn} and to the Larmor radius ρ_L. This calculation was performed by Brambilla and Lichtenberg (1973), to obtain an island half-width, analogous to (6.4.12) but measured in physical space,

$$\Delta r = 2\left(\frac{e\Phi_{mn}}{kT}\frac{R}{a}\frac{\rho_L}{d\iota/dr}\right)^{1/2}. \tag{6.4.16}$$

Here kT is the characteristic particle energy (k is Boltzmann's constant and T is the temperature).

Diffusion in Static Fields. The width of the resonant islands given by (6.4.16) may be large compared to the width of the nonresonant banana motion given by (6.4.15). Nevertheless, for static fields, the nonresonant bananas are usually more important in giving rise to extrinsically driven diffusion. The reason for this apparent contradiction is easy to see. For static potentials, the resonant drift surface islands are centered on fixed magnetic surfaces, $m\omega_\phi - n\omega_\psi = 0$. Since $\omega_\phi/\omega_\psi = d\phi/d\psi$, a function of r alone, the locations of the resonance centers are independent of $v_\parallel$ and μ. The diffusion arising from weak interparticle collisions, which causes $v_\parallel$ and μ to vary in a random manner, is therefore of the type discussed in Section 5.6b. (If the drift islands overlap, then the resulting intrinsic global stochasticity generally determines the diffusion rate; however, because of the ρ_L dependence in the island amplitude, the islands rarely overlap, and we ignore intrinsic diffusion in the following argument.) For large but nonoverlapping islands, there is enhanced diffusion across each island width. However, the overall diffusion rate is limited to much smaller values by the nonresonant drift surfaces between islands.

In contrast, the nonresonant banana orbits of the trapped particles exist on any magnetic surface and are therefore not inhibited in this manner. Collisional changes of $v_\parallel$ and μ scatter the particles from trapped to passing orbits. Each such scatter either transfers the average radial particle position from its instantaneous position on a passing orbit to the center of the banana on a trapped orbit, or conversely. Depending on the collisionality, three distinct types of behavior are observed when banana orbits are present.

1. For low collisionality, the particles step on the average a distance of a banana width in the time it takes for $v_\parallel$ to diffuse across a resonance width. Since this time is proportional to the collision time, the diffusion rate is proportional to the collision frequency.
2. For intermediate collisionality, a resonant particle does not execute a full banana oscillation before it is detrapped. The step size is thus reduced to a fraction of a banana width. However nonresonant particles near the separatrix now also have oscillations of order this step size, and thus contribute to the overall diffusion. Furthermore the effective resonance width is increased, and thus the time to diffuse across the resonance region is also increased. Including the three effects of increased diffusion time, increased fraction of particles, and decreased step size, the result is a diffusion coefficient independent of collision frequency. This case is therefore usually called the plateau regime.
3. For high collisionality, a fluid model must be used to calculate the diffusion. The resulting value of Δr, the step size in the diffusion coefficient, is enhanced by a factor of $1/\alpha$ [see (6.4.4)] over the gyroradius, and the consequent diffusion, known as "Pfirsch-Schlüter" diffusion, is again proportional to collision frequency.

The characteristic step sizes in all cases are proportional to the Larmor radius $\rho_L = v_T/\Omega$; thus the diffusion coefficient D has the classical scaling $D \propto \rho_L^2 \propto 1/B^2$ and is therefore known as *neoclassical diffusion*. The reader interested in detailed calculations of this process is referred to a review by Galeev and Sagdeev (1979). As mentioned in Section 6.3a, these three regimes also exist in resonance streaming problems.

6.4c. Time-Varying Fields

In this section we consider a problem related to neoclassical diffusion in that the perturbed orbits are banana shaped and can exist on any magnetic surface. However, the perturbation to be treated is resonant, and thus the "bananas" are really islands whose widths can be strongly enhanced over

6.4. Diffusion in Toroidal Magnetic Fields

those for neoclassical diffusion. The reason that these islands can exist on any magnetic surface is the inclusion of a time variation in the potential. This introduces another degree of freedom into the perturbed Hamiltonian such that the energy E is no longer conserved. Thus one cannot eliminate $v_{\parallel}$ from the motion using (6.4.14), and the resonance condition becomes

$$m\omega_\phi(r, v_\parallel) + n\omega_\psi(r, v_\parallel) + l\omega = 0,$$

where ω is the frequency of the potential variation. This condition yields for the location of a resonance center (with fixed m, n, and l)

$$r = r(v_\parallel, \omega).$$

If there is continuous extrinsic diffusion of $v_\parallel$, then the resonance center will continuously diffuse in r. The diffusion is thus an example of the resonance streaming developed in Section 6.3 and can, in particular, be transformed directly into the mapping form described in Section 6.3b.

If the potential arises from a field which is time varying, then the phase of the potential for a particle position r is given by

$$\theta = \boldsymbol{k} \cdot \boldsymbol{r} - \omega t, \tag{6.4.17}$$

where $\boldsymbol{k}$ is the wave vector and ω the wave frequency. Within the drift approximation the particle velocity is parallel to the field lines, i.e., drift velocities are assumed slow compared to particle velocities. The rate of change of phase is therefore given by

$$\frac{d\theta}{dt} = k_\parallel(r) v_\parallel - \omega, \tag{6.4.18}$$

where $k_\parallel$ is the wavenumber parallel to the magnetic field. As described above, the parallel particle velocity is no longer ignorable, and the resonance center diffuses as $v_\parallel$ diffuses. This problem has been treated by Gell et al. (1975) and in more detail by Nevins et al. (1979).

Construction of the Mapping Equations. We simplify the magnetic field geometry by using rectangular coordinates with x corresponding to r. The coordinate system used is shown in Fig. 6.22. Here $\boldsymbol{k} = k_0 \hat{y}$ is a fixed wavevector, and $\boldsymbol{B}(x)$ lies in the $y - z$ plane with $B_y = 0$ at $x = 0$. We define the shear length L_S by

$$B_y(x) = B \frac{x}{L_S},$$

where for a torus $L_S^{-1} = (R/a)(d\iota/dr)$. The component of $\boldsymbol{k}$ parallel to $\boldsymbol{B}$ is then given by

$$k_\parallel(x) = k_\perp \frac{x}{L_S}, \tag{6.4.19}$$

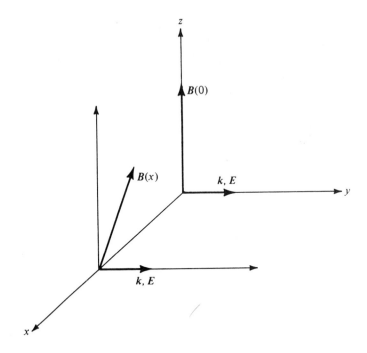

Figure 6.22. The configuration of the model for calculation of resonance drift island diffusion.

where to first order we set $k_0 = k_\perp$. Taking the potential of the wave to be of the form

$$\Phi = \Phi_0 \cos \theta, \tag{6.4.20}$$

then (6.4.13a) and (6.4.13b) can be written in the form

$$\frac{dx}{dt} = \frac{k_\perp}{B} \Phi_0 \sin \theta, \tag{6.4.21}$$

$$\frac{dv_\parallel}{dt} = \frac{e}{M} k_\parallel \Phi_0 \sin \theta. \tag{6.4.22}$$

Equations (6.4.21) and (6.4.22), together with (6.4.18), describe the perturbed motion in the absence of collisions. We assume that at a given $x = x_0$ resonance occurs for a velocity $v_{\parallel 0}$ such that (6.4.18) becomes

$$k_{\parallel 0} v_{\parallel 0} - \omega = 0, \tag{6.4.23}$$

where we have used the notation $k_\parallel(x_0) = k_{\parallel 0}$. At resonance, with $\theta = 0$, the particle motion is unperturbed by the potential. We follow the usual procedure of considering the perturbed motion about resonance by linearizing

6.4. Diffusion in Toroidal Magnetic Fields

$x = x_0 + \Delta x$ and $v_{\|} = v_{\|0} + \Delta v_{\|}$ to obtain

$$\Delta \dot{x} = \frac{k_\perp}{B} \Phi_0 \sin \theta, \tag{6.4.24}$$

$$\Delta \dot{v}_{\|} = \frac{e}{M} k_{\|0} \Phi_0 \sin \theta + \zeta, \tag{6.4.25}$$

and

$$\dot{\theta} = \frac{k_\perp}{L_s} v_{\|0} \Delta x + k_{\|0} \Delta v_{\|}, \tag{6.4.26}$$

where to obtain (6.4.26) we have expanded (6.4.19) about $x = x_0$ and inserted this into (6.4.18). The dots are the total time derivatives. We have further added to (6.4.25) the random velocity component ζ due to interparticle collisions, such that, for collisions alone,

$$\langle \Delta v_{\|}^2 \rangle = \frac{v_T^2}{\tau_c} t, \tag{6.4.27}$$

where v_T is the thermal velocity and τ_c is the collision time. A change in ζ changes $v_{\|}$ without changing x, and thus changes the resonance center of the perturbed motion according to (6.4.23). We must therefore transform our equations to the appropriate form to decouple the diffusive motion from the resonance motion, as illustrated in Fig. 6.18, i.e., transform Eqs. (6.4.24)–(6.4.26) to the form of (6.3.28). To do this we introduce a new variable y for which the dynamical part of (6.4.25) is eliminated. Multiplying (6.4.24) by $(e/M) k_{\|0}(B/k_\perp)$ and subtracting this from (6.4.25), we have

$$\Delta \dot{v}_{\|} - \frac{\Omega k_{\|0}}{k_\perp} \Delta \dot{x} = \zeta. \tag{6.4.28}$$

Defining y as

$$y = \Delta v_{\|} - \frac{\Omega k_{\|0}}{k_\perp} \Delta x, \tag{6.4.29}$$

we have a new form of (6.4.25) for the random motion alone

$$\dot{y} = \zeta. \tag{6.4.30}$$

Solving (6.4.28) for $\Delta v_{\|}$ in terms of y and Δx and substituting into (6.4.26), we have

$$\dot{\theta} = k_{\|0} y + \frac{k_{\|0}^2}{k_\perp} \Omega(1 + S) \Delta x, \tag{6.4.31}$$

where S, here, is a parameter measuring the importance of the shear

$$S = \frac{k_\perp^2}{k_{\|0}^2} \frac{v_{\|0}}{L_S \Omega}. \tag{6.4.32}$$

When $S \ll 1$, shear is unimportant, and the island width is limited by the ratio of $k_\|$ to $k_\perp$ at a fixed position. When $S \gg 1$, the island width is limited by the shear. A simple scale change in the variables now transforms (6.4.24), (6.4.31), and (6.4.30) into the standard form of (6.3.28). Setting

$$I = \frac{k_{\|0}^2}{k_\perp} \Omega(1 + S)\Delta x \tag{6.4.33a}$$

and

$$P = k_{\|0} y, \tag{6.4.33b}$$

then

$$\dot{I} = K \sin \theta, \tag{6.4.34a}$$

$$\dot{\theta} = I + P, \tag{6.4.34b}$$

$$\dot{P} = \xi, \tag{6.4.34c}$$

where

$$K = k_{\|0}^2 v_T^2 (1 + S) \frac{e\Phi_0}{kT}, \tag{6.4.35a}$$

$$\xi = k_{\|0} \zeta, \tag{6.4.35b}$$

and we have defined v_T by $Mv_T^2 = kT$. The island width scales in the usual manner as $K^{1/2}$.

Diffusion Calculation. We are now in a position to calculate the resonant diffusion in I, as in Section 6.3b. Using the definition of ζ from (6.2.25) and (6.2.27) in (6.4.34c) with ξ given by (6.4.35b) we obtain

$$\langle P^2 \rangle = \frac{k_{\|0}^2 v_T^2}{\tau_c} t.$$

The diffusion coefficient is then

$$D_r = \frac{\langle I^2 \rangle}{2t} = \frac{\langle P^2 \rangle}{2t} = \tfrac{1}{2} k_{\|0}^2 v_T^2 / \tau_c. \tag{6.4.36}$$

Assuming that only resonant diffusion is important, then the diffusion, in I space, is given as in (6.3.33) by

$$\langle D \rangle = D_r f_p, \tag{6.4.37}$$

where f_p is the fraction of resonant particles. Taking the characteristic range

6.4. Diffusion in Toroidal Magnetic Fields

of $v_\parallel$ to be v_T, then the range of P, from (6.4.33), is $k_{\parallel 0} v_T$ and

$$f_p \approx \frac{K^{1/2}}{k_{\parallel 0} v_T} \tag{6.4.38}$$

giving

$$\langle D \rangle = \tfrac{1}{2} k_{\parallel 0}^2 v_T^2 (1+S)^{1/2} \left(\frac{e\Phi_0}{kT}\right)^{1/2} \left(\frac{1}{\tau_c}\right). \tag{6.4.39}$$

In terms of the physical variable Δx, the diffusion is obtained by substituting the scaling from (6.4.32):

$$D_x = \frac{1}{2} \frac{k_\perp^2}{k_{\parallel 0}^2 (1+S)^{3/2}} \left(\frac{e\Phi_0}{kT}\right)^{1/2} \frac{\rho_L^2}{\tau_c}. \tag{6.4.40}$$

The result in (6.4.40) has been obtained by Nevins et al. (1979) in a different manner in which the step length for diffusion is taken to be an island width, as in the two degrees of freedom problem, and an effective collision time is used

$$\tau_{\text{eff}} = \left(\frac{v_{\text{trap}}}{v_T}\right)^2 \tau_c, \tag{6.4.41}$$

where v_{trap} is estimated as

$$v_{\text{trap}}^2 = (1+S) \frac{e\Phi_0}{M}.$$

However, the physical mechanism of resonance streaming is hidden in this presentation. The exact numerical value of the diffusion is not obtained from (6.4.40), because of the approximation of (6.4.38). The diffusion described by (6.4.40) has sometimes been called *pseudo-classical diffusion*, as it has the classical scaling $D_x \propto \rho_L^2/\tau_c$, but is enhanced by a coefficient that depends on the amplitude Φ_0 of a resonant perturbation.

Comparison of Analytical and Numerical Results. Nevins et al. (1979) compared all of the scalings of (6.4.40) with computer simulations using the exact equations of single particle motion and a Monte Carlo collision operator. The code follows typically 500 to 1000 particles to obtain good statistics. All of the scalings checked reasonably well, verifying the theory. Here we reproduce only two of these results. In Fig. 6.23 the effect of the shear parameter is studied by comparing the numerical results, labeled D^*, with the analytic formula for D_x in (6.4.40), with a parameter inserted for best fit. They find $D^* = 0.8 D_x$, which is good agreement, although there are some additional subtleties. In the limit $S \ll 1$, using a more exact transport theory derived in the absence of shear, Nevins and co-workers compare the numerical and analytic calculations of diffusion as a function of collision frequency. In Fig. 6.24, we see a plateau regime (so-called because it is

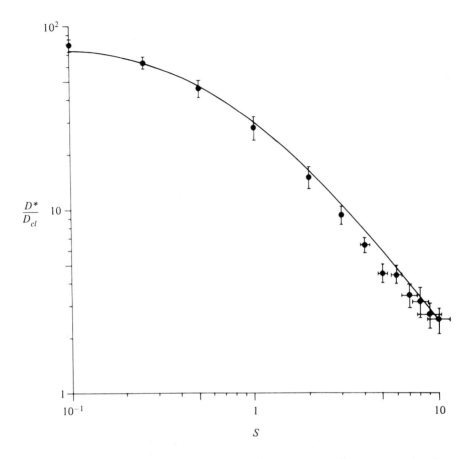

Figure 6.23. The diffusion coefficient D^* normalized to $D_{cl} = \rho_L^2/\tau_c$ versus S, showing the transition between the regimes of small and large shear. The quantities $e\Phi_0/kT$, $k_\parallel/k_\perp$, $\omega/k_\parallel v_T$, and $k_\perp \rho_L$ are held constant at 0.01, 0.03, 0.5, and 4.3×10^{-3}, respectively, while $k_\perp L_S$ is varied. The solid line is $0.8 D_x$, with D_x given by (6.4.40). The horizontal error bars reflect the finite spatial resolution of the measured diffusion coefficient from the numerical simulation (after Nevins et al., 1979).

independent of v_c) and an interesting transitional regime near $v_{\text{eff}}/\omega_0 = 1$, where ω_0 is the frequency of the resonant oscillation of x. The solid line is the prediction of the more exact calculation for the banana regime without shear. The plateau line can be determined by making the simple assumption that for $v_{\text{eff}} > \omega_0$ the distance a particle drifts in x between collisions, which is the random step in the diffusion process, shortens with increasing collisionality at just the rate to keep

$$\omega_0 \tau_{\text{eff}} \simeq 1. \tag{6.4.42}$$

6.4. Diffusion in Toroidal Magnetic Fields

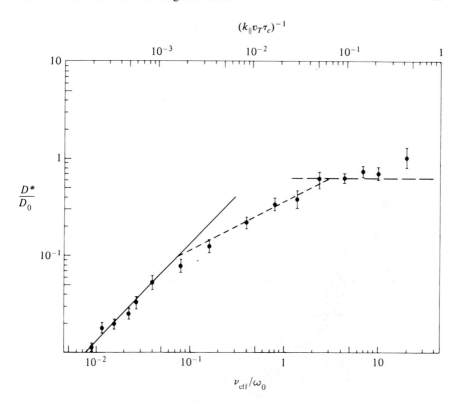

Figure 6.24. Measured values of the enhancement in the diffusion coefficient normalized to $D_0 = (k_\parallel v_T)^{-1}(k_\perp \Phi_0/B)^2$ versus the collision frequency. The solid line is given by $D^* = 1.3 D_x$ with D_x given by (6.4.40). The dashed line shows the self-consistent diffusion rate given by (6.4.43). The quantities $e\Phi_0/kT$, $k_\parallel/k_\perp$, $\omega/k_\parallel v_T$ and $k_\perp \rho_L$ were held constant at 0.08, 0.2, 0, and 4.3×10^{-3}, respectively. The vertical error bars here represent fluctuations in D^* from the numerical simulations (after Nevins et al., 1979).

Using (6.4.42) with $\omega_0 = K^{1/2}$ obtained from (6.4.35a) and τ_{eff} from (6.4.41) to substitute for τ_c in (6.4.40), we obtain the result for the plateau region, independent of τ_c,

$$D_{x(\text{pl})} = \frac{k_\perp^2}{k_\parallel^2} k_\parallel v_T \left(\frac{e\Phi_0}{kT}\right)^2 \rho_L^2. \qquad (6.4.43)$$

The result, obtained in the limit of no shear, agrees within a small numerical factor with a kinetic theory treatment (Sagdeev and Galeev, 1969). We point out, however, that the use of (6.4.42), while reasonable, is not within the scope of the theory of oscillation center diffusion.

6.4d. The Self-Consistent Problem

Diffusion in toroidal magnetic fields illustrates a very important factor in many real problems, which has not explicitly been considered in this monograph, that of self-consistent fields; that is, the fields within which particles move may be generated in part by the collective motion of the particles themselves. In this situation the Hamiltonian that determines the dynamics is not *a priori* known. These problems have generally been attacked by numerical simulation of the complete particle and field equations, as illustrated below.

Tearing Modes and Disruptions in Tokamaks. For a tokamak device, as described in Section 6.4a, the poloidal component of the magnetic field arises from a toroidal current created by the plasma particles that are assumed to be moving on regular flux surfaces. However, the plasma can be subject to resistive tearing instabilities (e.g., White *et al.*, 1977; Carreras *et al.*, 1981), which lead to helical perturbations of the current, i.e., a current having a basic symmetry $l\phi - n\psi = $ const.; l and n integers. These perturbations break the azimuthal symmetry, and the harmonics of the perturbation appear as islands on the rational surfaces. With cylindrical symmetry a single helical mode would result in a single resonant island; it would therefore not lead to island overlap and stochasticity. However, the inclusion of toroidal effects adds new major resonances. For example, a primary helically excited mode with $l = 2$, $n = 1$ excites a major two-island structure on the $\iota = \pi$ magnetic surface. The toroidal symmetry adds a significant component of a three-island structure on the $\iota = 2\pi/3$ rational surface. These rational surfaces commonly occur within the plasma of tokamaks, with consequent large magnetic islands that affect the plasma motion. In Fig. 6.25 the magnetic surface structure is shown arising from the numerical simulation of a saturated $l = 2$, $n = 1$ mode, giving rise to the $\iota = \pi$ resonant islands. The coupling through the toroidal terms also gives rise to resonant islands on the $\iota = 2\pi/3$ rational surface. Despite the large island size, little stochastic wandering of field lines was found in this case. However, if an $l = 2$, $n = 2$ helical mode is also present, then large regions of stochasticity are observed. The evolution of two such modes has been followed by numerical simulation, integrating the self-consistent field and particle equations, to obtain the result shown in Fig. 6.26. In the first frame the $\iota = \pi$ and $\iota = 2\pi/3$ islands are clearly seen. In the second frame the interaction between the current components of the $l = 2$, $n = 1$ and $l = 2$, $n = 2$ symmetries has led to stochastic wandering of most of the field lines near the $\iota = \pi$ island chain. In the third frame the interaction has spread to engulf the $\iota = 2\pi/3$ island. The resulting drastic modification of the original helical current pattern has led to the destruction of the central flux surfaces in the fourth frame. The abrupt change in the current distribution, as the islands overlap, is thought to be the cause of current disruptions in tokamaks.

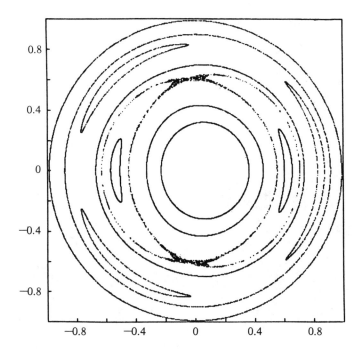

Figure 6.25. Magnetic field line plot showing a saturated 2/1 magnetic island and its 3/1 toroidal satellite (after Carreras et al., 1981).

6.5. Many Degrees of Freedom

There has been considerable interest in understanding systems with a large number of degrees of freedom. The motivation, on one hand, has been concerned with the behavior of nonlinear partial differential equations, and, on the other, with the relation to the statistical mechanics of the many-body problem. In Section 6.1a we have formally considered a many-dimensional system within which a particular set of resonance surfaces can be singled out for study. A single, many-dimensional resonance has a simple resonance width, defined in the proper coordinates, and this width may be projected onto any single action of the multidimensional system. This geometric interpretation was developed further in Section 6.3, and mathematical details of the coordinate transformations to exhibit the resonance locally can be found in Chirikov (1979). From this perspective the question of a large number of degrees of freedom appears to resolve itself into the question of whether the density of important resonances, as projected onto a single action, increases faster than the widths of the resonances decrease, as the system energy is spread over more degrees of freedom. If this happens, then we would expect resonance overlap and strongly chaotic motion to occur for N degrees of freedom, as $N \to \infty$.

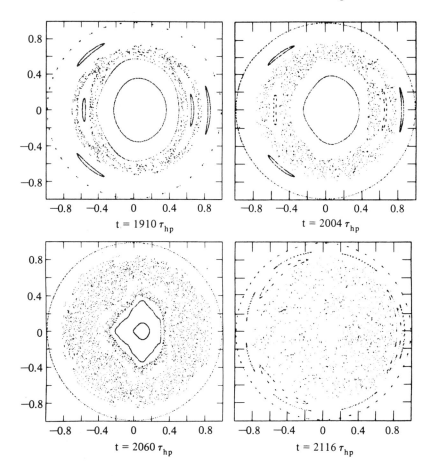

Figure 6.26. Magnetic field line configurations at four different times. In regions where the magnetic field is stochastic, the intersections of a field line with a poloidal plane appear as a dotted region. Where flux surfaces are well defined, they appear as closed curves. Here τ_{hp} is the characteristic MHD time (after Carreras *et al.*, 1981).

Another question, which we considered in Section 6.2c, is the dependence of Arnold diffusion on the number of degrees of freedom N. There, we presented a rigorous upper bound for the diffusion rate for small perturbations, obtained by Nekhoroshev (1977), and a more heuristic closer estimate obtained by Chirikov (1979). Comparison with a numerical example having many resonances but with just three degrees of freedom appeared to be consistent with the Chirikov estimate. The scaling with N has not been directly explored. Related to the diffusion in phase space, Benettin *et al.*, (1987a, b) argue that low-frequency modes will preferentially be excited, thus effectively "freezing out" the high frequencies over exponentially long times.

6.5. Many Degrees of Freedom

For infinite-degree-of-freedom nonlinear systems described by nonlinear partial differential equations, nonlinear normal mode (soliton) solutions are known to exist for some equations (see, e.g., Zabusky, 1962; Kruskal and Zabusky, 1964; Lamb, 1980). However, these solutions can be unstable to perturbations (see Driscoll and O'Neil, 1976, 1978, and Section 6.5b). From the existence of an infinite number of constants of the motion for these equations, it would be expected that these instabilities lead to recurrences as is seen for low-energy-per-degree-of-freedom (weak nonlinearity) in a chain of coupled oscillators (Fermi et al., 1955; Bivins et al., 1973).

There appear to be two competing tendencies in systems with many degrees of freedom at a fixed total energy as $N \to \infty$. On one hand, the separation of resonances becomes increasingly small. On the other hand, the widths of the resonances also tend to become small. Depending on the average ratio of the resonance width to distance between resonances, it is possible for a system to approach either essentially complete stochasticity or essentially regular motion as the number of degrees of freedom increases. An example of the former case is the Lennard–Jones gas. An example of the latter case is the limit of a continuous system, such as a vibrating string at low energy. Although there is no known rigorous demarcation between these two types of behavior, a comparison of the resonance width to resonance spacing for a few major resonances, as N increases at fixed total system energy, may give a good indication for the large N behavior. A crucial question is the distribution of the fixed energy among the modes, which must be known in order to calculate resonance overlap. In contrast to two-degree-of-freedom Hamiltonian systems and related mappings, systems with many degrees of freedom are still poorly understood. Major questions such as the relation between Arnold diffusion and energy equipartition remain unanswered.

To thread our way through the variety of problems in this field, we will first, in Section 6.5a, consider some results for strongly nonlinear systems in which the energy might be expected to be spread rather uniformly among the degrees of freedom and which appear to be quite stochastic for large N. We will then discuss generic oscillator chains in which the behaviour is more complex, exhibiting transitions between mostly regular and mostly stochastic motion, even at large N. In Section 6.5b we discuss linear oscillators with nonlinear nearest-neighbor coupling first studied by Fermi et al. (1955), and known as the FPU model. In the continuum limit, the equations can be put into correspondence with the modified Korteveg–de Vries (mKdV) equation (Zabusky, 1967) which is integrable. In Section 6.5c we consider similar issues for a chain of nonlinear pendula coupled by linear springs, which in the continuum limit corresponds to another integrable system, described by the sine–Gordon equation. This latter chain has the additional feature that time discretization puts it into correspondence with a set of coupled standard maps, which is useful both conceptually and for numerical calculations.

6.5a. Strongly Nonlinear Coupled Systems

Some insight into this problem may be gained from generalizing the Fermi acceleration mapping to that in which the oscillating wall has N modes of oscillation, with the mean energy E of the wall oscillation held fixed. This was done for two modes by Howard et al. (1982), using the mapping equations

$$u_{n+1} = u_n + \frac{A_s \sin s\phi_n + A_r \sin r\phi_n}{(A_s^2 + A_r^2)^{1/2}}$$

$$\phi_{n+1} = \phi_n + \frac{4\pi M}{(r+s)u_{n+1}},$$

such that the energy (amplitude squared) is interchanged between oscillators as the A's are varied. Without repeating the details of their results, it is clear by analogy with the single frequency mapping results (Section 3.4e) that the resonance width in action (and frequency) space for N modes is proportional to $A^{1/2}$. Since $\sum_i A_i^2 \approx NA^2 \approx$ const. for equal-amplitude oscillators, we find $\Delta I_{max} \propto N^{-1/4}$. If the oscillators are randomly spaced as projected on a fixed action surface, we might expect a distribution of island separations between any two resonances with an rms value $\delta I_{rms} \propto N^{-1/2}$; thus the overlap criterion, as given by the ratio of island width to separation, is

$$\frac{\Delta I_{max}}{\delta I_{rms}} \propto N^{1/4}. \tag{6.5.1}$$

Comparing one and two modes, Howard et al. (1982) found an average effect at least this large, but the dependence on N was not explored. The implication of (6.5.1) is that as the number of modes increases without bound, with constant total energy shared among the modes, the stochastic motion envelops the entire phase space.

The Model of Attracting Sheets. The model of many bodies constrained to move along a line has also been used by Froeschlé and Scheidecker (1975) to investigate the increasing stochasticity with increasing degrees of freedom. They considered gravitating attracting sheets of mass m (gravitational constant G), giving the Hamiltonian

$$H(\mathbf{u}, \mathbf{x}) = \tfrac{1}{2}m \sum_{i=1}^{N} u_i^2 + 4\pi Gm \sum_{i=1}^{N} \sum_{j>i}^{N} |x_j - x_i|.$$

They explored the change, in time, of the distance between initially neighboring orbits in phase space. The initial positions of the sheets were chosen in a random manner, over a number of runs, keeping the total system energy the same for each run. The system equations were analytically integrated until two sheets crossed, when a discontinuity in the force required a new

6.5. Many Degrees of Freedom

Table 6.1. Variation of the Number of Integrable and Ergodic Orbits for 100 Experiments when the Number of Sheets Increases

N	Number of integrable orbits	Number of ergodic orbits
2	100	0
3	96	4
4	14	86
5	1	99
6	0	100
7	0	100

integration. The log of the distance between two orbits was plotted against time, with a linear relation indicating exponential growth or instability (see Section 5.3). Their results are shown in Table 6.1, indicating a rapid development of stochasticity over the phase space as the number of sheets N increases. Since the number of degrees of freedom is $N - 1$, the case of two sheets is integrable. Wright and Miller (1984) have looked at the statistical distribution of velocities and concluded that the distribution is essentially ergodic for $N > 4$.

The interpretation of these results for small N is somewhat confusing, as the discontinuity in the force as sheets pass through one another may violate the KAM theorem requirements of smoothness of the forcing function (see Section 3.2a). This type of violation leads to global diffusion for all nonisland orbits in the Fermi acceleration problem with a sawtooth forcing function (Section 3.4). In that problem KAM orbits can exist locally, although they are not isolating in the usual sense. If the fraction of island orbits decreases rapidly as the number of degrees of freedom increases, one would expect a rapidly increasing number of globally stochastic orbits, in agreement with the observed results.

The Lennard–Jones Potential. Another system that has been numerically studied is that of the Lennard–Jones potential

$$V(r_{ij}) = 4\epsilon \left[\left(\frac{\sigma}{r_{ij}}\right)^{12} - \left(\frac{\sigma}{r_{ij}}\right)^{6} \right],$$

where r_{ij} is the interaction distance between particles i and j and σ is the scale length of the interaction, such that a weak attracting force for $r_{ij} \gtrsim \sigma$ is bucked by a strong repelling force for $r_{ij} \lesssim \sigma$. Here ϵ is a small parameter. The Hamiltonian for the system takes the form

$$H = \frac{1}{2} \sum_{i=1}^{N} p_i^2 + \sum_{i=1}^{N} \sum_{j<i}^{N} V(r_{ij}).$$

Galgani and associates (see Galgani and LoVecchio, 1979) studied this problem, with the masses distributed along a single line, separating the motion into near-normal modes, as was done for the FPU nonlinear spring. They generally found increasing energy dispersion among modes, with increasing N and fixed total system energy, where initially the energy was placed primarily in a single mode.

Stoddard and Ford (1973) have examined the behavior of a system of particles interacting according to the Lennard–Jones potential but in a plane, rather than along a line. For this case the potential is the smoothed analogue to the hard sphere (really disk) gas that has been proved to be ergodic and mixing by Sinai (see Sections 1.4a and 5.2). They considered 100 particles with various initial states to represent different densities and temperatures in real fluids. The rate of increase of phase space distance for neighboring points was studied, with exponential rates indicating the existence of stochasticity. For this large number of degrees of freedom ($2N$ with $N = 100$), they found exponential growth for all initial conditions. This would clearly be expected for the strong-coupling, strongly nonlinear Lennard–Jones interaction potential.

Symplectic Map Lattice. Kaneko and Konishi (1989), Konishi (1989), Konishi and Kaneko (1990) treat the coupled mappings

$$p_i(t+1) = p_i(t) + \frac{K}{2\pi}\{\sin[2\pi(x_{i+1}(t) - x_i(t))] - \sin[2\pi(x_i(t) - x_{i-1}(t))]\},$$

$$x_i(t+1) = x_i(t) + p_i(t+1), \quad \mod 1,$$

where $i = 1, 2, \ldots, N$ with periodic boundary conditions which they call the symplectic map lattice. They find that after an initial transient, for $0.2 \lesssim K \lesssim 1$, the diffusion in p can be fitted to the form of a "stretched exponential" in the perturbation strength

$$D \propto K \exp(-\xi(1/K)^q), \tag{6.5.2}$$

where $q \approx 0.45$ for $N \geq 4$ and ξ is a constant of order unity. As discussed in Section 6.2c this is close to a simple exponential dependence $D \propto \exp(-cQ_0)$ where $Q_0 \propto K^{-1/2}$ is the frequency ratio as used in the three-resonance model of Section 6.2. This result has been shown to hold for two weakly coupled standard maps (Wood et al., 1990). The result is significantly different from that for many-resonance diffusion in three degrees of freedom as found by Chirikov (see Section 6.2c), where the estimate was $q \approx 0.33$. In both cases the diffusion rate is much slower than with fully overlapped resonances in all degrees of freedom. We shall see in Section 6.5b that relaxation times to energy equipartition also fit a "stretched exponential," in the perturbation strength with an exponent power of $q \approx 0.5$.

A simple way of looking at this diffusion is to note that particles diffuse in stochastic regions but not in regular ones. If, in a single degree of freedom,

6.5. Many Degrees of Freedom

the fraction of stochastic to total phase space area is a_k, then considering N orthogonal degrees of freedom, the regular normalized phase space volume would be approximated by

$$V_{\text{reg}} \approx \prod_{k=1}^{N} (1 - a_k).$$

Ordering a_k by, for example, mode frequency, we see that unless the a_k fall off sufficiently rapidly with k, the regular phase space volume $V_{\text{reg}} \to 0$ as $N \to \infty$, i.e, the phase space is almost entirely stochastic. Nevertheless, because the numerically observed diffusion follows an exponential scaling law, it appears to be primarily along resonances, rather than across them.

6.5b. The Fermi–Pasta–Ulam Model and the Korteveg–de Vries Equation

A model of a high-dimensional system, which has been closely examined, is that of a number of masses constrained to move along a line, with a specified force law governing their interaction. For example, to investigate the partial differential equation for a nonlinear string

$$\frac{\partial^2 y}{\partial t^2} - \frac{\partial^2 y}{\partial x^2}\left[1 + 3\beta\left(\frac{\partial y}{\partial x}\right)^2\right] = 0, \tag{6.5.3}$$

Fermi, Pasta, and Ulam (FPU) (Fermi et al., 1955) approximated the partial differential equation with a set of nonlinear, ordinary, coupled differential equations (equimass particles connected by nonlinear springs). This is done with the discretization of $y(x, t)$ as $y_j(t)$,

$$\frac{\partial y}{\partial x} = \frac{y_{j+1} - y_j}{\Delta x} \quad \text{or} \quad \frac{y_j - y_{j-1}}{\Delta x},$$

$$\frac{\partial^2 y}{\partial x^2} = \frac{y_{j+1} - 2y_j + y_{j-1}}{(\Delta x)^2},$$

where $\Delta x = L/N$, with L the length of the string and $N - 1$ the number of oscillators. Then (6.5.3) becomes

$$\ddot{y}_j = \frac{(y_{j+1} - 2y_j + y_{j-1})}{(\Delta x)^2}$$
$$\times \left\{1 + \beta\left[\frac{(y_{j+1} - y_j)^2}{(\Delta x)^2} + \frac{(y_j - y_{j-1})^2}{(\Delta x)^2} + \frac{(y_{j+1} - y_j)(y_j - y_{j-1})}{(\Delta x)^2}\right]\right\}.$$
$$\tag{6.5.4}$$

The original numerical work and also analytic investigation by Izrailev and Chirikov (1966) were with fixed end points at $j = 0$ corresponding to $x = 0$, and $j = N$ corresponding to $x = L$, such that $j = 1, 2, \ldots, N - 1$ for the oscillators. The numerical results with 64 particles at low energy indicated that energy did not equipartition among the oscillators but rather a beat phenomenon existed with regular recurrences of initial conditions. This interesting result, contrary to the original expectation of the authors, stimulated a number of investigations of the system (6.5.4) (see Bivins et al., 1973).

We note that choosing a different string length L' at fixed N yields an equation of the form (6.5.4) with $\Delta x \to \Delta x' = L'/N$ and $y_j \to y'_j$. Introducing the change of variables $y'_j \to y_j L'/L$ and $t' \to tL'/L$ leads to (6.5.4) again. Since $dy'_j/dt' = dy_j/dt$, the energy per mode is unchanged. Thus increasing N by adding oscillators to the end of the chain at fixed Δx is equivalent to adding oscillators by subdividing the chain at fixed L, provided that the time and displacement are rescaled. Hence, without loss of generality, we can take $\Delta x = L/N = 1$ in (6.5.4).

For an analytical investigation, a transformation to the normal coordinates of the linear system ($\beta = 0$) is useful (e.g., Ford and Waters, 1963; Izrailev and Chirikov, 1966). Izrailev and Chirikov use the transformation

$$y_j = \left(\frac{2}{N}\right)^{1/2} \sum_{k=1}^{N-1} a_k \sin \frac{\pi k j}{N}, \tag{6.5.5}$$

where only the sine series is used due to the fixed ends. We emphasize that these are not true modes of the nonlinear system and thus interchange of energy among these quasimodes will occur even in the absence of stochasticity. In this representation the equations of motion (6.5.4) can be written in the form

$$\ddot{a}_k + \omega_k^2 a_k \left[1 + \frac{3\beta}{2N} \omega_k^2 a_k^2\right] = \text{interaction term}, \tag{6.5.6}$$

where

$$\omega_k = 2 \sin\left(\frac{\pi k}{2N}\right) \tag{6.5.7}$$

is the unperturbed quasimode frequency, the self-interaction term has been moved to the left-hand side of (6.5.6) and represents a nonlinear frequency shift, and the interaction term on the right-hand side is a sum over terms, cubic in the a_m's with at least one of the a_m's having $m \neq k$. If several quasimodes are excited, then it is possible to apply the standard method of overlap of neighboring resonances to compute the border of stochasticity. Izrailev and Chirikov estimate the stochasticity condition for low- and high-frequency modes, respectively, to be

$$\beta \left(\frac{\partial y}{\partial x}\right)^2_{\max} \gtrsim \frac{1}{k}, \qquad k \ll N, \tag{6.5.8a}$$

$$\beta \left(\frac{\partial y}{\partial x}\right)^2_{\max} \gtrsim \frac{\pi^3}{N^2} \left(\frac{k}{N}\right)^2, \qquad N - k \ll N. \tag{6.5.8b}$$

6.5. Many Degrees of Freedom

These results can be obtained within a small numerical factor by equating the nonlinear frequency shift $\Delta\omega_k$ of a single quasimode to the spacing $\delta\omega_k$ between adjacent quasimodes. From (6.5.6), the nonlinear frequency for the mode is $\omega_{k(NL)}^2 = \omega_k^2[1 + (3\beta/2N)\omega_k^2 a_k^2]$, which gives, for small frequency shifts,

$$\Delta\omega_k \approx (3\beta/4N)\omega_k^3 a_k^2. \tag{6.5.9}$$

Expanding (6.5.7), such that $\omega_k \approx \pi k/N$, $k \ll N$, and $\omega_k \approx 2$, $N - k \ll N$, we obtain the separation between quasimodes

$$\delta\omega_k = -\begin{cases} \pi/N, & k \ll N, \\ (\pi^2/2N^2)(N - k - \tfrac{1}{2}), & N - k \ll N. \end{cases} \tag{6.5.10}$$

By requiring that $\Delta\omega_k/\delta\omega_k \gtrsim 1$, we obtain the overlap condition

$$\frac{3\beta}{4N}\omega_k^2 a_k^2 \gtrsim \begin{cases} 1/k, & k \ll N, & (6.5.11\text{a}) \\ (\pi^2/2N^2)(N - k - \tfrac{1}{2}), & N - k \ll N. & (6.5.11\text{b}) \end{cases}$$

With the identification

$$\left(\frac{\partial y}{\partial x}\right)_{\max}^2 \approx \frac{2\pi^2 k^2 a_k^2}{N^3},$$

(6.5.11) is, outside of small numerical factors, equal to (6.5.8), provided we make the approximation, used in obtaining (6.5.8), that $N - k \approx 1$ in the (b) parts. Alternatively, we identify the mode energy as

$$E_k \simeq \omega_k^2 a_k^2$$

and again assuming that essentially only one driving mode is excited

$$\frac{E_k}{N} = \begin{cases} 4/3\beta k, & k \ll N, & (6.5.12\text{a}) \\ (2\pi^2/3\beta N^2)(N - k - \tfrac{1}{2}), & N - k \ll N, & (6.5.12\text{b}) \end{cases}$$

with the left-hand side considered as the system energy density. If a set of Δk modes are simultaneously excited, then the right-hand side of (6.5.6) has some Δk dependence, leading to a Δk dependence in (6.5.12). This is, in fact, often done in numerical experiments, as discussed below.

Relations (6.5.12) indicate that for low-frequency modes the criterion for the energy density at overlap is independent of N but decreases with increasing mode frequency, while at high frequency, overlap is essentially independent of mode frequency and decreases with N. On the other hand, Galgani and associates (1979; Benettin et al., 1987a, b) conclude, on the basis of numerical experiments and some general arguments, that the dependence of overlap energy density should be relatively constant with mode frequency at low frequency and increase at the higher frequencies; i.e., the oscillators at higher frequency should have larger critical energies than the lower-frequency oscillators. They explain this result by showing that the rate of transfer of energy into the higher modes can be exponentially slow.

Numerical investigations of stochasticity and equipartition for a periodic FPU lattice with $\Delta x = 1$ were performed by Livi et al. (1985) and Pettini and Landolfi (1990). For the periodic case, (6.5.7) is modified to give the mode frequencies $\omega_k = 2\sin(\pi k/N)$, such that the frequencies are symmetric about $k = N/2$. They considered the entropy-like quantity

$$h(t) = -\sum_{k=1}^{N/2} P_k(t) \ln P_k(t) \geq 0, \tag{6.5.13}$$

where $P_k(t) = E_k/E$, E_k is the energy of a normal mode with wave number $2\pi k/N$, and $E = \sum_i E_i$ is the total energy. The quantity $h(t) = 0$ if all the energy is in one mode. Its maximum value is

$$h_{\max} = \ln N/2 \tag{6.5.14}$$

for energy equilibrium. Since $h_{\max}$ depends on N, they introduce the normalized quantity

$$\eta = \frac{h_{\max} - h(\infty)}{h_{\max} - h(0)}, \tag{6.5.15}$$

where $h(\infty)$ is the "asymptotic" value of $h(t)$, assuming that $h(t)$, suitably averaged over a finite interval in t, is monotonically increasing. The normalized quantity η is then bounded between zero and one, with zero corresponding to equipartition.

Choosing a range of N from 64 to 512 and picking a cluster of Δk modes centered at some $\bar{k}$ with $\Delta k/N$ fixed at $\frac{1}{16}$, Pettini and Landolfi (1990) fitted $\eta(t)$ to the functional relation

$$\eta(t) = \begin{cases} \exp[-(t/\tau)^\nu], & t < \tau_R, \tag{6.5.16a} \\ \exp[-(\tau_R/\tau)^\nu], & t > \tau_R, \tag{6.5.16b} \end{cases}$$

where τ is a time constant. Varying the energy per mode, they found that (6.5.16) fits the data quite well, with $\ln \tau_R$ versus $\ln(E/N)$ falling on a universal curve, for a fixed Δk and $\bar{k}/N$, with $0.3 \lesssim \nu \lesssim 0.5$. However, τ_R decreased somewhat as Δk was increased. One way of understanding this result is to consider that for times greater than τ_R, $\eta(t)$ is approximately equal to $\eta(\infty)$, which is dependent on N through fluctuations in mode energies that keep $h(\infty)$ from equaling $h_{\max}$. From (6.5.16b), we can write τ_R as

$$\tau_R = \tau[\ln 1/\eta(\infty)]^{1/\nu}.$$

With $\nu = $ const., and taking the distribution of oscillator energies to be of the Boltzmann form $f_i \propto \exp[-\beta\omega_i I_i]$, where $\omega_i I_i$ is the unperturbed mode energy, i.e., maximum entropy at fixed total energy, we find $h(\infty) = \ln(N/2) - 0.423$ and the corresponding $\eta(\infty)$ to be

$$\eta(\infty) = 0.423/\ln \frac{N}{2\Delta k}.$$

6.5. Many Degrees of Freedom

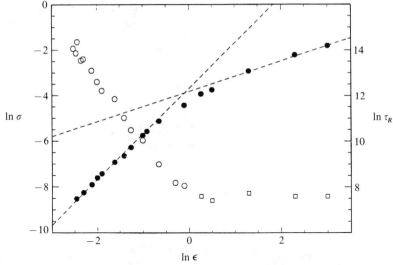

Figure 6.27. The relaxation time τ_R and the maximum Liapunov exponent σ versus the energy per mode $\epsilon = E/N$, for $N = 128$ and $\Delta k = 4$. The dashed lines are power laws ϵ^2 and $\epsilon^{2/3}$. Open circles and squares are relaxation times, and closed circles are the Liapunov exponents (after Pettini and Landolfi, 1990).

For most of the numerical work, $\Delta k/N$, and thus $\eta(\infty)$, were held fixed, and therefore $\tau_R \propto \tau$, which then gives the rate of diffusion of the initial state, through the Arnold web, toward a more probable state where the fraction of phase space volume occupied maximizes the entropy. Pettini and Landolfi also obtained the fastest-growing Liapunov exponent σ, which settles to a reasonably asymptotic value after a relatively long time (compared to τ_R). For a particular choice of N, $\bar{k}$ and Δk, the results shown in Fig. 6.27 are obtained. They interpret the transition in the slope from $\sigma \propto \epsilon^2$ to $\sigma \propto \epsilon^{2/3}$ with increasing E/N, as a transition to strong stochasticity. They present a model of a random process which predicts the $\sigma \propto \epsilon^{2/3}$ scaling, lending credence to this point of view. If the above interpretation of $\tau_R \propto \tau$ is correct, then the transition to a constant τ_R, with increasing ϵ shown in Fig. 6.27, might be interpreted as a transition to a many-mode-overlap condition, such that the diffusion is across resonances, rather than along resonances. Above the transition value of ϵ, there would be only a weak ϵ dependence. Below this value, the measured variation with ν indicates a diffusion process that scales intermediate between the three resonance diffusion, $\nu \sim \frac{1}{2}$, and a numerically determined many resonance diffusion model (see Section 6.2c).

Because of the different N-scaling between the numerically observed equipartition and analytic expressions such as (6.5.12) for mode overlap, a simple mode overlap criterion appears to be neither necessary nor sufficient for equipartition. This conclusion also holds for the discretized sine–Gordon equation considered in Section 6.5c.

Instability in the mKdV Equation.

Another aspect of the dynamics of continuous (infinite-degree-of-freedom) systems is that the (integrable) soliton solutions can be unstable to a perturbation in the initial conditions.

Putting the FPU chain in the form

$$\frac{\partial^2 y_j}{\partial t^2} = (y_{j+1} - y_j) - (y_j - y_{j-1}) + \tfrac{1}{3}[(y_{j+1} - y_j)^3 - (y_j - y_{j-1})^3], \quad (6.5.17)$$

where the displacements of the lattice sites have been rescaled to obtain the nonlinear coefficient $\tfrac{1}{3}$, then Zabusky and Kruskal (1965) have shown that making y into a continuous variable $y(t, x)$, a change of variable

$$u \equiv -\frac{y_t}{2h} + \frac{1}{2}\int_0^{y_x}(1 + h^2\eta^2)^{1/2}\, d\eta,$$

transforms (6.5.17) in lowest order in $\Delta x \equiv h = L/N$ into the mKdV equation

$$u_\tau + 12u^2 u_\xi + u_{\xi\xi\xi} = 0, \quad (6.5.18)$$

where L is the chain length, N is the number of oscillators, the time and length variables have been rescaled by $\tau = h^3 t/24$ and $\xi = x - ht$, and the subscripts t, x, τ, ξ denote differentiation with respect to that variable.

The periodic solutions of (6.5.18), stationary in the frame $\xi - C\tau$, may be obtained by integrating (6.5.18) twice. This gives

$$\tfrac{1}{2}u_\xi^2 + u^4 - \tfrac{1}{2}Cu^2 - Bu + A \equiv \tfrac{1}{2}u_\xi^2 + P(u) = 0, \quad (6.5.19)$$

where A and B are constants of integration. For a periodic lattice the mean of u must be zero which implies $B = 0$. Equation (6.5.19) is in the form of a one-degree-of-freedom Hamiltonian which is therefore integrable. Solutions have been obtained in terms of the Jacobi elliptic functions (or cnoidal waves) $\operatorname{cn}(\xi, q)$, with q^2 (the modulus) taken as a parameter with $0 \leqslant q \leqslant 1$. Driscoll and O'Neil (1983, 1985) examined the solutions for stability, both analytically for long wavelength perturbations, and numerically. They determined the unstable modes of any solution u_0 by numerically solving the linearized eigenvalue problem

$$iv v - Cv_\xi + 12(u_0^2 v)_\xi + v_{\xi\xi\xi} = 0, \quad (6.5.20)$$

where $u(\xi, \tau) = u_0(\xi) + v(\xi)\exp(iv\tau)$. The waves are found to be stable if all four roots of the associated polynomial $P(u)$ are real, and unstable if two roots are real and two are complex. Although the normalized time τ is scaled by $h \propto 1/N$, the complete rescaling of the normalized equation back to unnormalized variables give growth rates independent of N. Driscoll and O'Neill then compare the growth rates with those found from numerical integration of the equations of motion for the FPU chain with various values of N. For wavenumber $\pi k/L$, they obtain the growth rate γ as a function of the modulus q^2 of the elliptic function as shown in Fig. 6.28. In this situation the growth of unstable modes in the continuous limit is found to be an upper bound on the mode growth of the finite chain.

6.5. Many Degrees of Freedom

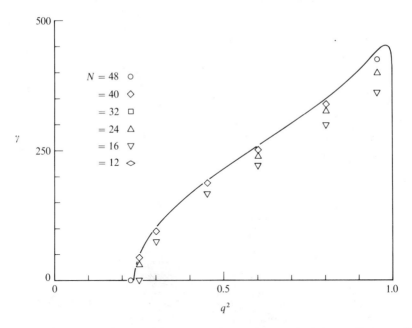

Figure 6.28. Scaled lattice instability rate γ (data points) and mkdV prediction (solid curves) for cnoidal waves versus modulus q^2 (after Driscoll and O'Neill, 1978).

The instability found in the continuous system may, therefore, explain the rapid distribution of energy among modes that is found in the discrete chain over short periods of time. For example, the instability may explain the results of Bivins *et al.* (1973) who focus on the coupling of a few Fourier modes, with one large amplitude mode driving the growth of neighboring modes, which is concerned with relatively short periods of time. Bivins *et al.* also observe transitions from the characteristic oscillatory energy exchange at small nonlinearity to seemingly chaotic behavior at larger nonlinearity, over short time scales. It is not clear in the short time scale studies whether the distribution of energy among many modes indicates instability on integrable phase space trajectories, with implied long-time recurrences, as in the continuous case, or true stochastic behavior. In the next subsection, in which we treat the discretized sine–Gordon equation, we consider some of these questions further.

Approximation of the Nonlinear Schrödinger Equation. Berman and Kolovskii (1984) have examined the FPU chain from another perspective. They show that in the high frequency limit of $\Delta k/\bar{k} \ll 1$ there is a well-defined condition for which the FPU equations can be approximated by the nonlinear Schrödinger (NLS) equation, which is integrable. The validity of the approximation changes with time due to energy interchange among modes. For low energy the mode dispersion remains bounded, that is, $\Delta k(t) < \Delta k_{\max}$, for all t, with Δk_{max} a function of the energy density E/N. Without going into details, they find that the NLS equation exhibits an instability (similar to that described above for the mkdV equation) if

$$\frac{E}{N} \geq \frac{\pi^2}{3\beta}\left(\frac{1}{N}\right)^2. \tag{6.5.21}$$

This causes mode spreading which destroys the NLS approximation and therefore its integrability. Berman and Kolovskii numerically find that the nonlinear wave solution does break up when the energy is sufficiently above the threshold given by (6.5.21).

6.5c. Discretized Sine–Gordon Equation

The sine–Gordon equation

$$y_{tt} - y_{xx} + \sin y = 0, \tag{6.5.22}$$

can be discretized in space in the same manner as the nonlinear spring (6.5.3) to obtain the system Hamiltonian

$$H = \sum_{i=1}^{N} \tfrac{1}{2}p_i^2 + \sum_{i=1}^{N} (1 - \cos y_i) + \sum_{i,j=1}^{N} A_{ij} y_i y_j, \tag{6.5.23}$$

where the coupling matrix A_{ij} is given by

$$A_{ij} = \frac{(2\delta_{ij} - \delta_{i,j-1} - \delta_{i,j+1})}{(\Delta x)^2}, \tag{6.5.24}$$

$p_i = \dot{y}_i$, $\Delta x = L/N$, and δ_{ij} is the Kronecker δ.

This Hamiltonian has the following simple physical interpretation: the first two terms correspond to N identical pendula with linearized frequencies normalized to unity, and the last term represents harmonic coupling between nearest neighbors. Since the linear part of this Hamiltonian is exactly solvable, we formulate the problem in terms of the normal modes of the harmonic springs, which, for fixed boundary conditions at $y = 0$ and $y = L$, are

$$e_{jr} = g^{-1} \sin\left(\frac{\pi jr}{N+1}\right), \tag{6.5.25}$$

where $g = [(N + 1)/2]^{1/2}$ is a normalization factor. A transformation similar to that of (6.5.5)

$$y_j = g \sum_{r=1}^{N} u_r e_{jr}, \qquad p_j = g \sum_{r=1}^{N} v_r e_{jr}, \tag{6.5.26}$$

6.5. Many Degrees of Freedom

yields the Hamiltonian in terms of the linearized normal modes

$$H = \sum_{r=1}^{N}\left[\frac{v_r^2}{2} + \frac{\omega_r^2}{2}u_r^2\right] + g^{-2}\sum_{j=1}^{N}\left[1 - \cos\left(g\sum_{r=1}^{N}u_r e_{jr}\right)\right] = H_0 + H_1. \quad (6.5.27)$$

The ω's, which are the eigenvalues of the linear problem, are given by

$$\omega_r = \frac{2(N+1)}{L}\sin\left(\frac{\pi r}{2(N+1)}\right). \quad (6.5.28)$$

As with the mKdV equation, the sine–Gordon equation can be solved in terms of cosine-amplitude Jacobi elliptic functions

$$y = 4\tan^{-1}(A\,\text{cn}(\mu x, k_x)\,\text{cn}(vt, k_t)), \quad (6.5.29)$$

with $\mu = (4m+2)\mathcal{K}/L$, m integer

$$K = \int_0^{\pi/2} \frac{d\theta}{\sqrt{1 - k_x^2 \sin^2\theta}}$$

$$v^2 - \mu^2 = \frac{1 - A^2}{1 + A^2},$$

$$k_x^2 = \frac{A^2}{\mu^2(1+A^2)^2}(1 + \mu^2(1+A^2)),$$

$$k_t^2 = \frac{A^2}{v^2(1+A^2)^2}(-1 + v^2(1+A^2)),$$

such that for a given mode number m, the solution is set by the amplitude A alone. The effect of discreteness on these normal modes can be investigated by comparing the solution (6.5.29) to the evolution of y from applying Hamilton's equations to (6.5.23).

To greatly increase computational speed, and therefore the ability to explore long times and large parameter ranges, Goedde et al. (1990) also discretize in time by multiplying the pendulum potential term in (6.5.16) by a periodic δ-function $\sum_{i=-\infty}^{\infty} \delta(t/T - i)$, as we have previously done, to describe the Hamiltonian of a mapping. The result is a set of coupled maps

$$\bar{u}_s = u_s \cos \omega_s T + \frac{1}{\omega_s}\left[v_s - g^{-2}T\sum_{i=1}^{N} e_{is}\sin\left(g\sum_{r=1}^{N} u_r e_{ir}\right)\right]\sin \omega_s T, \quad (6.5.30a)$$

$$\bar{v}_s = -\omega_s u_s \sin \omega_s T + \left[v_s - g^{-2}T\sum_{i=1}^{N} e_{is}\sin\left(g\sum_{r=1}^{N} u_r e_{ir}\right)\right]\cos \omega_s T, \quad (6.5.30b)$$

which forms a $2N$-dimensional symplectic map. If the mapping frequency is fast compared to all natural frequencies of the N-oscillator chain, then the dynamics is indistinguishable from the continuous time system over short times. For lower mapping frequencies, resonances between the mapping frequency and the ω_r's can lead to parametric instabilities (Goedde *et al.*, 1990).

Some Numerical Results. We first look at a few numerical results to get a qualitative picture of the time evolution. Consider an oscillator chain of $N + 1 = 64$ and $L = 32$ such that from (6.5.28) the highest frequency is $\omega_{max} = 4 \sin \pi/2 = 4$, and the lowest frequency is $\omega_{min} = \pi/32$. Taking $T = 0.1$ such that there is no low-order resonance with the mapping frequency, we can examine the interaction over a range of energies for various distributions of energy among the modes. At low energy ($E \approx 4$), putting all the energy in modes 1 and 3 results in a fairly rapid distribution of energy among a few low-frequency modes, but no further spreading over $t = 50,000$ (500,000 time steps of the mapping) as shown in Fig. 6.29a. In contrast, with higher energy $E \approx 128$, a mode distribution approximating equipartition is fairly rapidly obtained, as shown in Fig. 6.29b, again at $t = 50,000$. If, on the other hand, this same high energy is put in modes 37 and 39, the distribution is far from equipartition at time $t = 50,000$, as shown in Fig. 6.29c. The corresponding initial and final amplitude distributions $y(x)$ for (b) and (c) are shown in Fig. 6.30. (The lower-energy case corresponding to Fig. 6.29a remains spatially near-sinusoidal with period 2.) This result demonstrates the rapid mixing of modes at higher energies when the energy is placed in low modes, and the resistance to this mixing when the energy is placed in high modes.

A further understanding of this behavior can be obtained from a two-mode approximation, which holds over a short period of time when the initial energy is primary in two modes. If the two-coupled-mode system exhibits stochastic behavior, we can then be fairly confident that stochasticity is present in the N-mode dynamics. For the energy primarily in two modes we use the approximation

$$1 - \cos\left(g \sum_r e_{ir} u_r\right) \approx 1 - \cos(g(e_{1a} u_a + e_{1b} u_b)),$$

which, for $a = 1$ and its strongest resonance $b = 3$, gives the approximate two-mode Hamiltonian

$$H = \frac{v_1^2 + v_3^2}{2} + \frac{\omega_1^2 u_1^2 + \omega_3^2 u_3^2}{2}$$

$$+ 2[1 - \mathcal{J}_0(u_1)\mathcal{J}_0(u_3) + 2\mathcal{J}_3(u_1)\mathcal{J}_1(u_3) - 2\mathcal{J}_6(u_1)\mathcal{J}_2(u_3) + \cdots]. \quad (6.5.31)$$

The Bessel functions appear from the lowest-order expansions of the sinusoids. In this approximation the N dependence appears only in the frequency. The phase space for mode 1 in a surface of section of mode 3 is shown in Fig. 6.31a, b, c with (b) and (c) corresponding to the respective

6.5. Many Degrees of Freedom

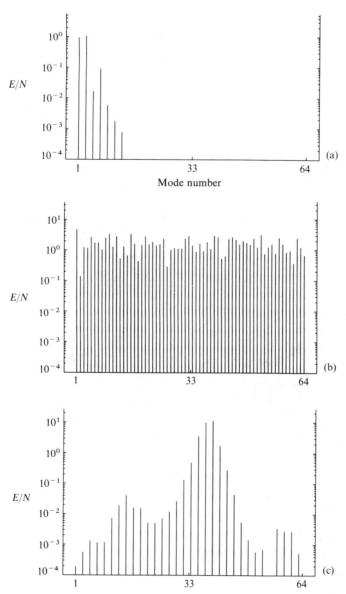

Figure 6.29. Energy spectrum at time $t = 50,000$, $N = 64$, $L = 32$. (a) $E/N = \frac{1}{16}$, initial energy in modes 1 and 3; (b) $E/N = 2$, initial energy in modes 1 and 3; (c) $E/N = 2$, initial energy in modes 37 and 39.

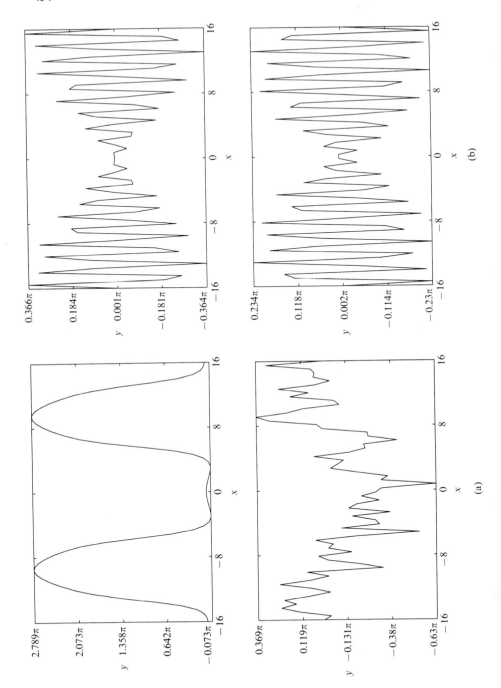

Figure 6.30. Initial and final (at $t = 50{,}000$) spatial configurations corresponding to the cases shown in Figs. 6.29b, c: $N = 64$, $L = 32$, $E/N = 2$. (a) Initial energy in modes 1 and 3; (b) initial energy in modes 37 and 39.

6.5. Many Degrees of Freedom

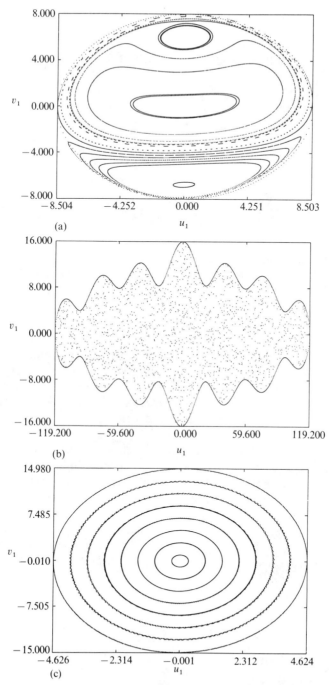

Figure 6.31. Surface of sections showing the (u_1, v_1) phase space of the two-dimensional Hamiltonian (6.5.31); $N = 64$, $L = 32$. (a) $E/N = \frac{1}{2}$, modes (1, 3); (b) $E/N = 2$, modes (1, 3); (c) $E/N = 2$, modes (37, 39).

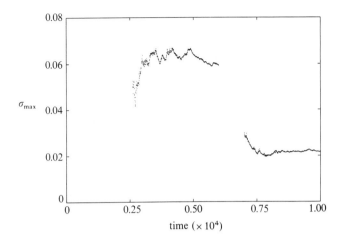

Figure 6.32. Maximum Liapunov exponent early in time (1) and later in time (2) for $E/N = 2$. The break in the curve is an artifact from restarting the averaging process.

cases of the long-time spectrum (b) and (c) in Fig. 6.29, and Fig. 6.31a corresponding to an intermediate energy of $E \approx 32$. Clearly Fig. 6.31b shows strong overlap of the two modes which we might expect would lead to rapid energy spread among many modes, as observed in Fig. 6.29b. The relatively smooth curves in Fig. 6.31a and c do not imply integrability, as there are still narrow channels through which particles can diffuse by Arnold diffusion. Also, many-mode interactions may broaden the channels and, possibly, cause overlap.

As in the FPU chain, the largest Liapunov exponent in the N-mode mapping can be calculated. The result for the case of large energy initially in two low-frequency modes (corresponding to Fig. 6.29b) is shown in Fig. 6.32. For short times with strong two-mode overlap, as characterized by Fig. 6.31b, $\sigma_{\max}$ is large. However, after a relatively short time, enough energy is pumped into other modes that the two-mode overlap criterion is not satisfied. There is a fairly abrupt transition to a lower value of $\sigma_{\max}$ in which the spreading in phase space is due to a more complicated mode interaction rather than a simple two-mode overlap.

CHAPTER 7

Bifurcation Phenomena and Transition to Chaos in Dissipative Systems

7.1. Simple and Strange Attractors

For motion in dissipative systems, the phase space volume contracts onto an attractor of lower dimensionality than the original space. Furthermore, as a system parameter changes, the motion on the attractor may change from regular to chaotic. Although our knowledge of chaotic behavior in dissipative systems is by no means complete, the broad features are now clear, and certain techniques to calculate the properties of the motion have emerged.

In the first section of this chapter, we review the basic properties of dissipative systems, such as the contraction of phase space volume and the concept of simple attractors and their bifurcations, on which the motion is regular. We then introduce the phenomenon of chaotic motion in dissipative systems, which appears in the form of a *strange attractor*. An example of a strange attractor was presented in Section 1.5. Two other examples of dissipative systems exhibiting strange attractors are given here: the Rössler system and the Hénon mapping. In each case we point out the features of the chaotic motion that suggest the importance of embedded nearly-one-dimensional maps and the geometric nature of the strange attractor. This geometry is described in terms of Cantor sets and their fractal dimension. We discuss the calculation of fractal dimension and its relation to the Liapunov exponents.

In Section 7.2 we consider the dynamics of noninvertible one-dimensional maps, looking first at the periodic behavior, including fixed points, their linear stability, and their bifurcation structure. We introduce the concept of period-doubling renormalization from both an algebraic and functional point of view. The chaotic behavior is considered next and is described in terms of Liapunov exponents, asymptotic distributions, power spectra, and their invariant properties. Section 7.3 describes bifurcation phenomena in two-dimensional maps and related three-dimensional flows. We review the mechanism of period doubling to chaos in both dissipative and area-preserving maps, introduce the dissipative standard map, and give a unified picture of the transition between Hamiltonian and dissipative maps due to period doubling. In Section 7.4, we consider one-dimensional maps on a circle, and describe mode-locking and the transition to chaos in mode-locked systems, using as an example the sine-circle map. In Section 7.5, we describe another route to chaos, than of intermittency, and show its relation to tangent bifurcations. In Section 7.6, we introduce the phenomena of crises, at which strange attractors and their basins are created or modified, and we show how such crises occur in one- and two-dimensional maps. In Section 7.7, we describe the method of Melnikov to obtain the transition from regular to chaotic motion near separatrices, and illustrate the method for the examples of a driven damped nonlinear oscillator and the dissipative standard map.

Excellent reviews of work on chaotic motion in dissipative systems are available. They include reviews emphasizing one-dimensional maps (May, 1976; Feigenbaum, 1980; Collet and Eckmann, 1980), and general reviews of chaotic effects in dissipative systems (Holmes, 1977; Rabinovich, 1978; Treve, 1979; Ruelle, 1980; Helleman, 1980; Shaw, 1981; Ott, 1981). Books complementary to the results presented here include of those of Guckenheimer and Holmes (1983), Schuster (1988), and Parker and Chua (1989).

7.1a. Basic Properties

We consider the motion described by N first-order differential equations of the form

$$\frac{dx}{dt} = V(x), \tag{7.1.1}$$

where x and V are N-vectors and V is explicitly independent of time. The *phase space* for this system is N-dimensional, with coordinates x_i, $i = 1, N$. If V is smooth, then a *solution flow* $x(x_0, t)$ exists for all t. A point in phase space specifies a unique state of the system (7.1.1). In Section 1.2b we described the construction of a *Poincaré surface of section* or *return map* for Hamiltonian systems. Return maps for dissipative systems have some, but not all, of the features of the surface of section already discussed for Hamiltonian systems. Referring to Fig. 1.3a, which shows a surface Σ_R in the phase space, the Poincaré map is found by choosing a point x_n on Σ_R and integrating (7.1.1) to find the next intersection x_{n+1} of the orbit with Σ_R. In this way we construct the map

$$x_{n+1} = f(x_n), \tag{7.1.2}$$

7.1. Simple and Strange Attractors

where x and f are vectors of dimension $N - 1$. If V is smooth and Σ_R is everywhere transverse to V, then it can be shown that the Poincaré map f is also smooth. Furthermore, since the solution $x(x_0, t)$ exists for all times, f is *invertible*; i.e. one can solve (7.1.2) to find

$$x_n = f^{-1}(x_{n+1}). \tag{7.1.3}$$

This corresponds, in Fig. 1.3a, to generating f^{-1} by reversing the direction of time and integrating (7.1.1) from x_{n+1} to x_n.

Contraction of Phase Space Volume. By Liouville's theorem (see Section 1.2b), the motion for a Hamiltonian system conserves the volume in phase space. In contrast, dissipative systems give rise to motions that, on the average, contract phase space volumes. (For bounded motions, phase space volume cannot, on the average expand.)

We calculate the rate of change of a small rectangular volume $\Delta\tau$ about x_0, writing $\Delta\tau$ as a product

$$\Delta\tau(x_0, t) = \prod_i \Delta x_i, \tag{7.1.4}$$

where

$$\Delta x_i(x_0, t) = \frac{\partial x_i(x_0, t)}{\partial x_{i0}} \Delta x_{i0}. \tag{7.1.5}$$

The rate of change of $\Delta\tau$ is

$$\Lambda(x) \equiv \frac{1}{\Delta\tau} \frac{\partial(\Delta\tau)}{\partial t} = \sum_i \frac{1}{\Delta x_i} \frac{\partial(\Delta x_i)}{\partial t}. \tag{7.1.6}$$

But from (7.1.5),

$$\frac{\partial(\Delta x_i)}{\partial t} = \frac{\partial}{\partial x_{0i}} \frac{\partial x_i(x_0, t)}{\partial t} \Delta x_{i0}. \tag{7.1.7}$$

Using (7.1.1) with $dx_i/dt \equiv \partial x_i(x_0, t)/\partial t$ in (7.1.7) and evaluating near $t = 0$, we obtain from (7.1.6) the instantaneous rate of change of the volume as

$$\Lambda = \sum_i \frac{\partial V_i}{\partial x_i} = \text{div } V. \tag{7.1.8}$$

The rate of change of volume is a local quantity, depending on $x(t)$, and may be positive (expanding) or negative (contracting). However, what we mean here by a dissipative system is that, averaged over an orbit, the volume must contract. Writing the average contraction rate as

$$\Lambda_0(x_0) = \lim_{t \to \infty} \frac{1}{t} \ln \left| \frac{\Delta\tau(x_0, t)}{\Delta\tau(x_0, 0)} \right|, \tag{7.1.9}$$

we must have $\Lambda_0 < 0$ for all x_0.

For an N-dimensional map, the local volume $\Delta\tau$ contracts by a factor $|\det \mathbf{M}(x)|$ for each iteration, where $\mathbf{M}$ is the Jacobian matrix of the map. This factor represents a rate of volume contraction of

$$\Lambda(x) = \frac{1}{\Delta\tau}\frac{\partial(\Delta\tau)}{\partial n} = \ln|\det \mathbf{M}(x)|, \qquad (7.1.10)$$

where n is the iteration number. Averaging this contraction rate along an orbit as in (7.1.9), we obtain $\Lambda_0(x_0)$.

Not much is known about general dissipative systems, where the dependence of Λ on x is arbitrary. For many of the problems that have been studied,

$$\Lambda(x) = -c,$$

where c is a positive constant, expressing the fact that in these systems the volume uniformly contracts everywhere in the phase space. Obviously $\Lambda_0 = -c$ for such systems.

Liapunov Exponents. The average volume contraction rate can be written in terms of the Liapunov exponents, which were defined for Hamiltonian systems in Section 5.2b. Exactly the same definition applies for dissipative systems. For an N-dimensional phase space, there are N real exponents that can be ordered as

$$\sigma_1 \geqslant \sigma_2 \geqslant \cdots \geqslant \sigma_N,$$

with $\sigma_1 = \sigma_{\max}$ being the largest exponent, and one of the exponents representing the direction along the flow, being zero. From (5.2.14) and (5.2.16), the average volume contraction rate is

$$\Lambda_0 = \sum_{i=1}^{N} \sigma_i. \qquad (7.1.11)$$

The Liapunov exponents and the definitions of Λ_0 in (7.1.9) and (7.1.11) apply to maps as well as flows, including the Poincaré map for a given flow. The $N - 1$ exponents σ_i of the Poincaré map are proportional to the N exponents of the flow with the zero exponent deleted, as given in (5.2.20).

Chaotic motion, as in Hamiltonian systems, is associated with exponential divergence of nearby trajectories; i.e., $\sigma_1 > 0$ for chaotic motion. However, the volume must contract. These two facts show that one- and two-dimensional flows cannot give rise to chaotic motion.[1] In two dimensions, $N = 2$, the Poincaré map is one-dimensional (and invertible) and (7.1.11) yields $\Lambda_0 = \sigma_1$. Such a map cannot be both dissipative ($\Lambda_0 < 0$) and chaotic

---

[1] Actually, the Poincaré-Bendixson theorem states that chaotic motion is impossible for two-dimensional flows, whether dissipative or nondissipative. We have already seen this to be the case for a Hamiltonian system with one degree of freedom (two phase space dimensions).

7.1. Simple and Strange Attractors

($\sigma_1 > 0$). The simplest dissipative systems displaying chaotic behavior are thus three-dimensional flows and their related two-dimensional maps. The maps must have $\sigma_1 > 0$ and $\sigma_1 + \sigma_2 < 0$ for the motion to be chaotic.

One-dimensional noninvertible maps arise in certain limiting cases from $N \geq 2$ invertible maps. Their behavior is examined in Section 7.2. For these maps, the link between volume expansion and exponential divergence of nearby trajectories is broken. For $\sigma_1 > 0$, one observes bounded chaotic motion. As will be seen, these maps are, in an approximate sense, embedded in higher-dimensional dissipative systems.

7.1b. Simple Attractors and Their Bifurcations

Since the phase space volume shrinks to zero for a dissipative system, the stable, steady state motion for an N-dimensional system must lie on a "surface" of dimension less than N. Loosely speaking, this surface is called an attractor. There is no universally accepted definition; following Lanford (1981), we call a subset X of the phase space an attractor if:

(1) X is invariant under the flow;
(2) there is an (open) neighborhood around X that shrinks down to X under the flow;
(3) no part of X is transient;
(4) X cannot be decomposed into two nonoverlapping invariant pieces.

The *basin of attraction* of X is the set of states in phase space that approach X as $t \to \infty$. Often there are a finite number of attractors $X_1, \ldots, X_M$ for an N-dimensional flow, although cases are known that have infinitely many attractors. Except for a set of measure zero, all initial states lie in the basin of one of the M attractors. We illustrate these relationships in Fig. 7.1.

For an N-dimensional flow with a single parameter μ,

$$\dot{x} = V(x, \mu), \qquad (7.1.12)$$

the equilibrium points $x(\mu)$ are given by

$$V(x, \mu) = 0. \qquad (7.1.13)$$

For a given equilibrium, we can shift the origin of the coordinates so that the equilibrium occurs at $x = 0$. Linearizing about this equilibrium, we obtain

$$\dot{x}_i = \sum_j J_{ij} x_j, \qquad (7.1.14)$$

where

$$J_{ij} = \left(\frac{\partial V_i}{\partial x_j}\right)_{x=0} \qquad (7.1.15)$$

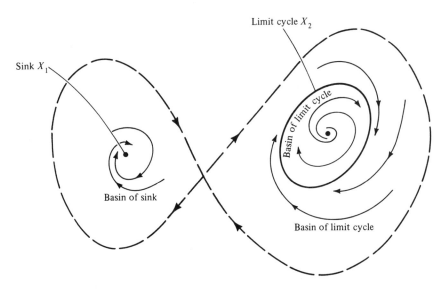

Figure 7.1. Attractors in phase space and their basins of attraction. A sink and a limit cycle are shown.

is the Jacobian matrix of V. Introducing a coordinate change $x \to x'$ to diagonalize J_{ij}, we obtain

$$x'_i(t) = x'_{i0} e^{v_i t}, \tag{7.1.16}$$

where the v_i are the eigenvalues of J_{ij}. The v's are real or occur as complex conjugate pairs. A condition for the equilibrium to be an attractor is asymptotic stability, i.e., Re $v_i < 0$ for all i.

The real part of one or more of the eigenvalues may pass though zero at a critical parameter value, leading to *bifurcation phenomena*. We can shift the origin of μ so that the bifurcation occurs at $\mu = 0$. If a single real eigenvalue passes through zero, then the bifurcation is essentially one dimensional; i.e., the bifurcation can be found in one-dimensional flows or on one-dimensional manifolds embedded in higher-dimensional flows.

Letting x be the single coordinate and normalizing x and μ (e.g., see Crawford, 1989), a one-dimensional bifurcation can be put in a *normal form*. The generic case is called a *tangent* or *saddle-node* bifurcation and occurs when $(\partial V/\partial \mu)_0 \neq 0$ and $(\partial^2 V/\partial x^2) \neq 0$. The normal form is

$$V_N(x, \mu) = \Theta_1 \mu + \Theta_2 x^2, \tag{7.1.17}$$

where the normalized variables are found as

$$\tfrac{1}{2}\mu \left| \frac{\partial V}{\partial \mu} \frac{\partial^2 V}{\partial x^2} \right| \to \mu, \tag{7.1.18}$$

$$\tfrac{1}{2}x \left| \frac{\partial^2 V}{\partial x^2} \right| \to x, \tag{7.1.19}$$

where $\Theta_1 = \text{sgn}(\partial V/\partial\mu)_0$, and $\Theta_2 = \text{sgn}(\partial^2 V/\partial x^2)_0$. There are four subcases depending on whether Θ_1 and Θ_2 are ± 1. For $\Theta_1 = +1$ and $\Theta_2 = -1$, $V = \mu - x^2$ has fixed points at $dx/dt = 0$ or $V = -\partial U/\partial x = 0$, which occur at $\pm\sqrt{\mu}$. Graphing the potential $U(x, \mu)$ versus x for fixed x and $\mu > 0$, we see that there is a stable root $x = +\sqrt{\mu}$ and an unstable fixed point at $x = -\sqrt{\mu}$. For $\mu < 0$, there are no fixed points for real x. The creation of these two fixed points as μ passes through zero is illustrated in Fig. 7.2a. A solid (dashed) line denotes a stable (unstable) equilibrium. The other three subcases are found by reflection about $x = 0$, $\mu = 0$, or both.

If $(\partial V/\partial\mu)_0 = 0$ then we have a *transcritical* bifurcation or *exchange of stability*. Introducing normalized variables, we have

$$V_N(x, \mu) = \Theta_1 x\mu + \Theta_2 x^2, \tag{7.1.20}$$

with $\Theta_1 = \text{sgn}(\partial^2 V/\partial\mu\partial x)_0$ and Θ_2 is as before. There are again four subcases. For $\Theta_1 = 1$ and $\Theta_2 = -1$, we have $V = \mu x - x^2$. As shown in Fig. 7.2b, the stable equilibrium or sink (solid line) passes from the fixed point at $x = 0$ to the fixed point at $x = \mu$ as μ passes through zero from left to right. The other three cases are again found by reflection about $x = 0$, $\mu = 0$, or both. Finally, if $(\partial V/\partial\mu)_0 = 0$ and $(\partial^2 V/\partial x^2)_0 = 0$, we have a *pitchfork bifurcation* and, in normalized variables,

$$V_N(x, \mu) = \Theta_1 \mu x + \Theta_2 x^3, \tag{7.1.21}$$

with $\Theta_1 = \text{sgn}(\partial^2 V/\partial\mu\partial x)_0$ and $\Theta_2 = \text{sgn}(\partial^3 V/\partial x^3)_0$. For $\Theta_1 = 1$ and $\Theta_2 = -1$, $V = \mu x - x^3$. As shown in Fig. 7.2c, the sink at $x = 0$ is destroyed and two new sinks at $x = \pm\sqrt{\mu}$ are created. In addition, there can be an *inverted pitchfork* bifurcation, illustrated in Fig. 7.2d for $\Theta_1 = -1$ and $\Theta_2 = 1$, and there are two other subcases corresponding to reflection about $\mu = 0$.

The nongeneric nature of the latter two bifurcations is seen by adding a perturbation δV, yielding Fig. 7.2e or f for the exchange of stability (Fig. 7.2b), Fig. 7.2g or h for the pitchfork (Fig. 7.2c), and Fig. 7.2i or j for the inverted pitchfork (Fig. 7.2d). In all cases, the transcritical or pitchfork bifurcations either disappear or are converted into one or more tangent bifurcations. We note in Fig. 7.2h the possibility of hysteresis phenomena as μ is varied. This will be illustrated further in Section 7.6. Figure 7.2a–j gives all possible bifurcations present in one-dimensional flows.

When the real part of a pair of complex conjugate eigenvalues passes through zero, then a two-dimensional *Hopf bifurcation* can occur. For this bifurcation, an equilibrium $V(x_1, x_2, \mu) = 0$ is converted into a periodic orbit or *limit cycle* which is a simple closed curve $x_2 = x_2(x_1)$ (see the curve X_2 in Fig. 7.1); i.e., there is an onset of time-dependent, nonlinear behavior. The normal form, in polar coordinates, is

$$\dot{r} = \Theta_1 \mu r + \Theta_2 r^2, \tag{7.1.22a}$$

$$\dot{\theta} = \omega_0, \tag{7.1.22b}$$

464 7. Bifurcation Phenomena and Transition to Chaos in Dissipative Systems

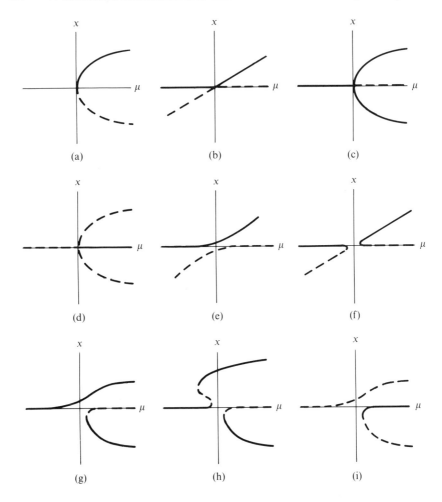

Figure 7.2. Bifurcations in one and two-dimensional flows. (a) Tangent; (b) exchange of stability; (c) pitchfork; (d) inverted pitchfork; (e) and (f) symmetry breaking for exchange of stability; (g) and (h) symmetry breaking for pitchfork; (i) and (j) symmetry breaking for inverted pitchfork; (k) Hopf; (l) inverted Hopf; (m) and (n) period-doubling.

where $v = \mu \pm j\omega_0$ is the pair of eigenvalues with $\omega_0 \neq 0$. For $\Theta_1 = 1$, $\Theta_2 = -1$, and $\mu < 0$, the right-hand side of (7.1.22a) is always negative, and the motion spirals into the sink at $r = 0$. For $\mu > 0$, the right-hand side is positive near $r = 0$ and the fixed point ceases to be attracting. In this case we note that the right-hand side is positive for $r < \mu$, leading to an increase in r with time, and negative for $r > \mu$, leading to a decrease in r with time.

7.1. Simple and Strange Attractors

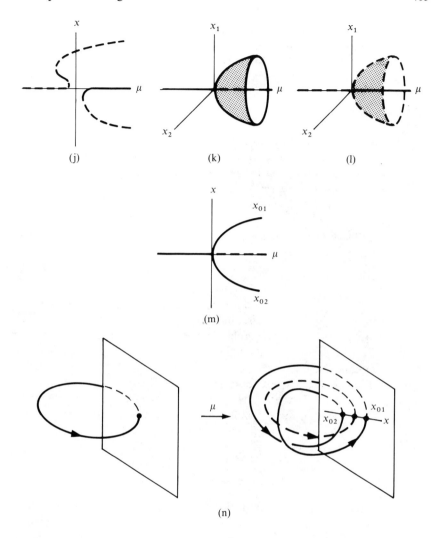

Figure 7.2 (cont.)

Thus the motion is attracted to the limit cycle $r(t) = \mu$, $\theta(t) = \omega_0 t + \theta_0$, yielding stable motion on a circle of radius μ. The change from sink to limit cycle as μ passes through zero is illustrated in Fig. 7.2k. The *inverted Hopf* bifurcation, $(\Theta_1 = -1, \Theta_2 = 1)$. Fig. 7.2l is also possible. Two other subcases are obtained by reflection about $\mu = 0$. Figure 7.2a–l shows the only bifurcations that are present in two-dimensional flows.

Attractors and their bifurcations in mappings are quite similar, but also have some significant differences to the corresponding phenomena in flows.

466   7. Bifurcation Phenomena and Transition to Chaos in Dissipative Systems

In order to bring out these similarities and differences we present the correspondence in tabular form in Table 7.1. Considering a map to represent the surface of section of a flow, then a (period-1) fixed point of a map, $x_e = f(\mu, x_e)$, corresponds to a periodic orbit (limit cycle) in the flow. As shown in the table, the normal forms V_N for one-dimensional fixed point bifurcations of maps are the same as those for bifurcations of equilibria in flows.

We note two important differences between the bifurcations in maps and those in flows, which give rise to two additional phenomena. A mapping can have an eigenvalue $\lambda = -1$ for which there is a new type of bifurcation, that of *period-doubling*. This can only occur when f is an odd function of x. In this case, the (period-1) fixed point at $x = 0$ bifurcates to become a periodic orbit (period-2 fixed point), with the motion alternating between x_{01} and x_{02} on successive iterations, as shown in Fig. 7.2m. If we look at the twice iterated map

$$f^{(2)}(x, \mu) - f(f(x, \mu), \mu), \qquad (7.1.23)$$

then the fixed point of $f^{(2)}$ at $x = 0$ undergoes the regular pitchfork bifurcation ($\lambda = 1$), having the bifurcation diagram of Fig. 7.1c. This bifurcation produces a new stable limit cycle in the corresponding flow having twice the period of the original limit cycle (see Fig. 7.1n); hence the term *period-doubling*.

In two dimensions, when a complex conjugate pair of eigenvalues crosses the unit circle $|\lambda| = 1$, then a Hopf bifurcation can occur, analogous to (7.1.22). Provided the nonresonance conditions $\lambda^3 \neq 1$ and $\lambda^4 \neq 1$ are met (Crawford, 1989), the normal form for the bifurcation is

$$r_{n+1} = \Theta_1 \mu r_n + r_n + \Theta_2 r_n^3, \qquad (7.1.24a)$$

$$\theta_{n+1} = \theta_n + \omega. \qquad (7.1.24b)$$

The bifurcation is similar to that illustrated in Fig. 7.1k or l, but has added complexity because the phase equation (7.1.24b) is unstable to θ-dependent perturbations, leading to such phenomena as mode-locking, which we describe in Section 7.4. Considering that the fixed point of $x = 0$ represents a periodic orbit (limit) cycle of a three-dimensional flow, then for the Hopf bifurcation, the limit cycle sheds a torus as it goes unstable. In the generic case, the attracting tori are not structurally stable as the parameter μ is varied because of the mode-locking phenomenon (see Lanford, 1981, for details).

The preceding picture of the bifurcations of nonchaotic attractors is a simplified view of bifurcation theory. The problems of general interest often occur on surfaces embedded in a higher-dimensional space. The linear dynamics near the equilibrium of an N-dimensional flow is described by the Jacobian matrix whose eigenvalues determine the stability, with the corresponding eigenvectors giving the principal directions associated with each eigenvalue. The modes are stable, unstable, or neutral, depending on whether

7.1. Simple and Strange Attractors

Table 7.1. Relations Between Maps and Flows

	Flows	Maps
Nonlinear equation	$\dot{x} = V(\mu, x)$	$x_{n+1} = f(\mu, x)$
Linear equation	$\dot{x}_i = \sum_j \dfrac{\partial V_i}{\partial x_j} x_j$	$x_{i,n+1} = \sum_j \dfrac{\partial f_i}{\partial x_j} x_{j,n}$
Diagonalization $x \to x'$ to obtain eigenvalues	$\dfrac{\partial V_i}{\partial x_j} \to \begin{pmatrix} v_1 & & & \\ & v_2 & & \\ & & \ddots & \\ & & & v_N \end{pmatrix}$	$\dfrac{\partial f_i}{\partial x_j} \to \begin{pmatrix} \lambda_1 & & & \\ & \lambda_2 & & \\ & & \ddots & \\ & & & \lambda_N \end{pmatrix}$
Linear solution	$x'_i = x'_i(0) e^{v_i t}$	$x_{in} = x_{i0} \lambda_i^n$
Asymptotic stability	$\operatorname{Re} v_i < 0$ all i	$|\lambda_i| < 1$ all i
Normal forms $x \to x$	$\dot{x} = V_N(\mu, x)$	$x_{n+1} = x_n + V_N(\mu, x_n)$
One-dimensional bifurcations	$v = 0$	$\lambda = 1$ or $\lambda = -1$
Hopf bifurcations (complex conjugate pair of eigenvalues)	$\operatorname{Re} v_i = 0$	$\lambda = e^{\pm i 2\pi\theta}$

Re $v_i < 0$, Re $v_i > 0$, or Re $v_i = 0$, respectively. If we divide the eigenvectors into these three classes, then each class forms a subspace of the N-dimensional space that are tangent spaces of the corresponding *invariant manifolds*, which we can designate as W_s, W_u, and W_c, respectively. We have already introduced two of these in Section 3.2b, as H^+ and H^-, to designate the stable and unstable manifolds at a hyperbolic fixed point. W_s and W_u are unique and have the simple dynamical properties that $x(t) \to 0$ as $t \to \infty$ if $x(t) \in W_s$ and $x(t) \to 0$ as $t \to -\infty$ if $x(t) \in W_u$. The remaining manifold, W_c, is known as the *center manifold* and plays an important role in bifurcation theory. In fact, any trajectory near a center manifold, that does not diverge away from it, will be locally attracted onto it, and thus the long-time behavior is described by the motion within the center manifold. The difficult problem is how to obtain the dynamics in the reduced space. The basic technique is to find the tangent space of the center manifold in the neighborhood of μ for which the bifurcation is to be examined (usually transformed to the origin $x = 0, \mu = 0$), and then to deform the tangent space, nonlinearly, onto the center manifold. The procedure is described, for example, in Guckenheimer and Holmes (1983, Chap. 3). A review has been given by Crawford (1989).

Our interest here is to examine relatively low-dimensional systems, where the reduction to the center manifold to examine simple bifurcations is not an issue. Rather, we emphasize that even in a relatively low dimension, the resulting motion can be very complex, resulting in a chaotic or strange attractor as described in Section 1.5 and in more detail in the following subsection.

7.1c. Examples of Strange Attractors

The lowest dimensionality flow that exhibits a strange attractor is $N = 3$, as in the Lorenz system described in Section 1.5. A simpler system that exhibits similar features was considered by Rössler (1976a), and we describe this example below. The counterpart of a two-dimensional mapping ($N = 3$ for the corresponding flow) exhibits other features of interest, which we shall illustrate using a mapping devised by Hénon (1976).

The Rössler Attractor. We consider the set of nonlinear equations (Rössler, 1976a)

$$\dot{X} = -(Y + Z),$$
$$\dot{Y} = X + \tfrac{1}{5}Y, \qquad (7.1.25)$$
$$\dot{Z} = \tfrac{1}{5} + Z(X - \mu).$$

An analog computer simulation of (7.1.25) for $\mu = 5.7$ is shown in Fig. 7.3a (Shaw, 1981). The strange attractor, projected onto the X–Y plane, bears a marked resemblance to the structure shown in Fig. 1.18b as viewed from

7.1. Simple and Strange Attractors

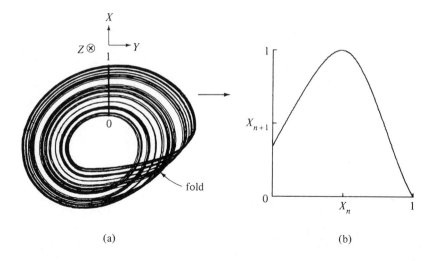

Figure 7.3. The Rössler attractor. (a) Projection of the motion onto the X–Y plane; (b) one-dimensional map, generated by successive intersections of the projected flow with the surface of section $Y = 0$ (after Shaw, 1981).

above. If we take a cut across the attractor, as shown in Fig. 7.3a, and plot X at the $(n + 1)$st crossing as a function of X at the nth crossing, then we obtain approximately the one-dimensional noninvertible map shown in Fig. 7.3b. The chaotic motion of the Rössler attractor is approximately described by the dynamics of the map for X.

The reduction from a three-dimensional flow to a one-dimensional map is approximate, and the line in Fig. 7.3b should be thickened, corresponding to the structure in the layer sectioned at $Y = 0$, which actually contains infinitely many leaves. However, the volume contraction rate (7.1.8) for this and many other strange attractors is so large that all the leaves appear merged into one in any simulation. Thus generally the one-dimensional map preserves most of the behavior of the original flow.

The transition from simple to strange attractor sometimes proceeds via a sequence of period-doubling bifurcations that converge at some limiting parameter value. Beyond this value, one generally finds chaotic behavior superimposed on a reverse bifurcation sequence with period halving on successive bifurcations. Figure 7.4, obtained by Crutchfield et al. (1980) on an analog computer, illustrates these bifurcations for the Rössler attractor starting at (a) $\mu = 2.6$ where the attractor is a simple limit cycle. The next two period-doubling bifurcations (b) and (c) are shown, along with their power spectra that are sharply peaked, a characteristic of regular motion. The successive frequency halvings (period doublings) are clearly visible. The transition to chaotic motion takes place near $\mu_\infty = 4.20$. The strange attractors for $\mu > \mu_\infty$ appear in (d), (e), and (f) as bands of chaotic behavior

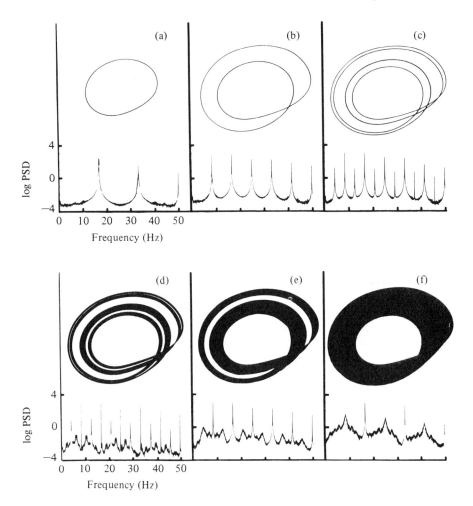

Figure 7.4. Period-doubling bifurcations of the Rössler attractor projected onto the $X-Y$ plane, along with the corresponding power spectral density of $Z(t)$. (a) $\mu = 2.6$; (b) $\mu = 3.5$; (c) $\mu = 4.1$; (d) $\mu = 4.23$; (e) $\mu = 4.30$; (f) $\mu = 4.60$ (after Crutchfield et al., 1980).

that lie between the regular orbits shown previously. The corresponding power spectra are a superposition of broadband noise due to the chaotic component and sharp peaks due to the periodic component of the motion. The reverse bifurcations and their frequency doublings are clearly seen.

We shall see that the transition to chaotic motion via period-doubling bifurcations is often present for a wide class of dissipative systems, including one-dimensional noninvertible maps, and higher-dimensional invertible maps and flows. The scaling of the bifurcation sequence as a parameter is varied,

7.1. Simple and Strange Attractors

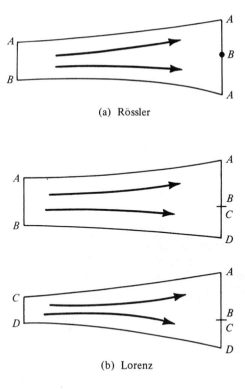

Figure 7.5. Topological structure of the Rössler and Lorenz attractors. All line segments with the same label (AB or CD) are joined together.

and the shape of the power spectrum, are universal near the transition. These topics are treated in Sections 7.2 and 7.3.

Topologically, as shown in Fig. 7.5a, the Rössler attractor has the simple structure of a sheet that has been stretched, folded over once, and joined from right to left edge. The Lorenz attractor is more complicated, consisting of two such sheets, with the stretched edge of each sheet, on the right, divided into two parts, as illustrated in Fig. 7.5b. The right edge of each sheet joins both sheets on their left. Other topological structures for strange attractors can also be imagined.

Although the Rössler attractor is topologically simpler than the Lorenz attractor, the imbedded one-dimensional map, Fig. 7.3b, has regions where the slope has a magnitude less than unity. As described in Section 7.2, it is not so easy to prove that this map is chaotic. We have already noted that the approximate one-dimensional map for the Lorenz attractor has a slope whose magnitude is everywhere greater than unity. Such maps have been proved to be chaotic.

The Hénon Attractor. The leaved structure across the layer can be clearly seen in an example due to Hénon (1976), who considered the simple case of a two-dimensional quadratic mapping

$$x_{n+1} = y_n + 1 - ax_n^2,$$
$$y_{n+1} = bx_n. \tag{7.1.26}$$

The mapping is invertible and may be thought of as the Poincaré section for some three-dimensional flow. For one iteration the area contracts by the factor $|\det \mathbf{M}|$, where

$$\det \mathbf{M} = \frac{\partial(x_{n+1}, y_{n+1})}{\partial(x_n, y_n)} = -b \tag{7.1.27}$$

is the Jacobian of the mapping. One can show (Hénon, 1976) that (7.1.26) is the most general quadratic mapping with constant Jacobian. For x_0 very large, the quadratic term makes $|x_n| \to \infty$. However, for (x_0, y_0) within some finite area near the origin, the solution converges toward an attractor. The mapping has two fixed points

$$x_{0,1} = (2a)^{-1}\{-(1-b) \pm [(1-b)^2 + 4a]^{1/2}\},$$
$$y_{0,1} = bx_{0,1}, \tag{7.1.28}$$

provided

$$a > a_0 = -\frac{(1-b)^2}{4}. \tag{7.1.29a}$$

The point x_1 is always linearly unstable, while x_0 is unstable for

$$a > a_1 = \frac{3(1-b)^2}{4}. \tag{7.1.29b}$$

(This is seen by writing $x = x_{0,1} + \Delta x$, $y = y_{0,1} + \Delta y$, and linearizing (7.1.26) in the usual way). When a is increased beyond a_1, the point attractor x_0 is numerically observed to undergo a cascade of period-doubling bifurcations. The attractor is still simple, consisting of a periodic set of $p = 2^n$ points, with $p \to \infty$ as $a \to a_2$. Beyond this value, for $a_2 < a < a_3$, the numerical evidence strongly suggests the presence of a strange attractor for most values of a in this range. For $a > a_3$, most points escape to infinity.

To see leaves of the strange attractor, Hénon chose $b = 0.3$, small enough to give a good folding, but large enough to see the fine structure. For $b = 0.3$, $a_0 \approx -0.1225$, $a_1 \approx 0.3675$, $a_2 \approx 1.06$, and $a_3 \approx 1.55$. Choosing $a = 1.4$, the strange attractor is seen in Fig. 7.6a, obtained from 10^4 iterations of the mapping starting at the unstable fixed-point x_0, y_0. When the area within the small square in Fig. 7.6a is magnified, then 10^5 iterations yield the more detailed view of the attractor shown in Fig. 7.6b. When the square in Fig. 7.6b is magnified, 10^6 iterations yield still more detail of the structure across

7.1. Simple and Strange Attractors

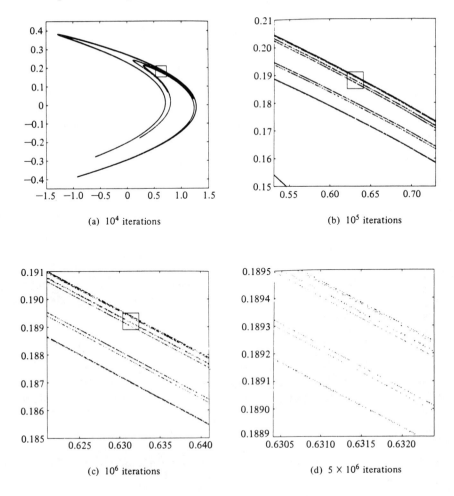

Figure 7.6. The leaves of the Hénon attractor. (a) Initial condition at the unstable fixed point x_0, y_0; (b) enlargement of the small square in (a); (c) enlargement of the small square in (b); (d) enlargement of the small square in (c). Note the scale invariance of the leaved structure (after Hénon, 1976).

the leaves, as shown in Fig. 7.6c. A final magnification in Fig. 7.6d, for 5×10^6 iterations, shows again the scale invariance of this attracting structure: Figures 7.6b–d, obtained by repeated magnification, all look identical. This similarity, across the layer, corresponds to the structure of a *Cantor set*, as will be described in Section 7.1d. The exponential divergence of initial, close trajectories has been verified computationally (Feit, 1978; Curry, 1979; Simó, 1980), strongly suggesting that the motion is chaotic on the attractor.

That a strange attractor always exists for $a_2 < a < a_3$ is an oversimplification. In fact, there exist many small subintervals within which the motion is periodic, with period 3, 4, 5, ..., and these motions generally undergo period-doubling bifurcations (Simó, 1980). These phenomena are studied in Section 7.2.

It has not been possible to prove mathematically the chaotic nature of the Hénon attractor. The introduction of a discontinuity into the mapping makes this proof possible. Lozi (1978) has described such an attractor, replacing x_n^2 with $|x_n|$ in the map (7.1.26), which have been proved to be chaotic by Misiurewicz (1980).

7.1d. Geometric Properties of Strange Attractors

We have seen from the numerical results on the Hénon map that the many-leaved structure of the attractor repeats itself on finer and finer scales. This leaved structure can be put in correspondence with a Cantor set, whose general properties then give considerable insight into the nature of the attractor. In this section we introduce the dimensionality of a Cantor set for which we need a general definition of fractal dimension. We then illustrate how a Cantor set may be constructed and determine its dimension. Finally, we discuss briefly several techniques to calculate or measure the fractal dimension of strange attractors, with particular attention given to the relationship between the dimension and the Liapunov exponents. In our treatment here we follow, in part, the reviews of Treve (1978) and Ott (1981). We defer to Section 8.3 a detailed discussion of the various definitions of fractal dimension, the spectrum of fractal dimensions, and procedures for this calculation.

Cantor Sets and Fractal Dimension. For our purposes, the definition of fractal dimension is taken to be the *capacity*

$$D_0(S) = \lim_{\epsilon \to \infty} \frac{\ln M(\epsilon)}{\ln(1/\epsilon)}, \qquad (7.1.30)$$

where S is a subset of an N-dimensional space, and $M(\epsilon)$ is the minimum number of N-dimensional cubes of side ϵ needed to cover the subset. For small ϵ this definition implies

$$M(\epsilon) \sim K\epsilon^{-D_0}. \qquad (7.1.31)$$

The capacity of a point, a line, and an area embedded in a two-dimensional space are the usual values 0, 1, and 2, respectively, i.e., if we take squares of side ϵ the number required to cover the point is proportional to $1/\epsilon^0$, to cover the line $M \propto 1/\epsilon^1$, and to cover the surface $M \propto 1/\epsilon^2$.

A Cantor set is a compact metric space that is totally disconnected, uncountable, and may have zero measure. It has typically a fractional dimension $D_0 \neq$ integer and displays rescaling invariance; i.e., a subset of

7.1. Simple and Strange Attractors

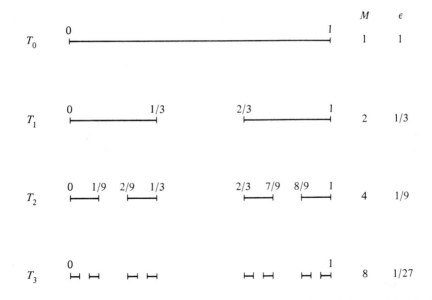

Figure 7.7. The construction of a Cantor set, with some values of the number M of line segments of length ϵ required to cover the set T_n.

the set, when properly magnified, looks like the original set. An example known as the "middle-thirds Cantor set" that illustrates the construction of a Cantor set is shown in Fig. 7.7. We mark the points $\frac{1}{3}$ and $\frac{2}{3}$ on the closed interval [0, 1] and delete the open (excluding end points) interval $(\frac{1}{3}, \frac{2}{3})$ called the "middle third," thus forming the set T_1. We then delete the middle thirds of the two remaining segments in T_1, thus forming T_2. We continue indefinitely in this way constructing $T_3, T_4, \cdots$. The Cantor set T is the intersection of all the T_n's. (Loosely speaking, the intersection is the "limiting T_n.")

From the construction, we see that T_n consists of $M = 2^n$ disjoint intervals, each of length $\epsilon = (1/3)^n$. Applying (7.1.30), the capacity of T is

$$D_0 = \lim_{n \to \infty} \frac{\ln 2^n}{\ln 3^n} = 0.630.$$

It is easy to see that T has measure zero

$$\lim_{n \to \infty} \epsilon M = (2/3)^n \to 0.$$

To see that T is uncountable, we note that any real number x between 0 and 1 can be written in base 3 as

$$x = a_1(1/3) + a_2(1/9) + \cdots + a_n(1/3)^n + \cdots,$$

476   7. Bifurcation Phenomena and Transition to Chaos in Dissipative Systems

where a_n is one of the three values 0, 1, or 2. When is x a member of T? If any $a_n = 1$, then by construction x is excluded from T_n,[2] hence x is not in T. Thus all x's contained in T can be written as a sequence of 0's and 2's; hence, they can be put in a one-to-one correspondence with the binary representation of the real numbers between 0 and 1, which is uncountable.

Relation Between Fractal Dimension and Liapunov Exponents. Kaplan and Yorke (1979) have introduced a fractal dimension that is related to the spectrum of Liapunov exponents:

$$D_{KY} = j + \frac{\sum_{i=1}^{j} \sigma_i}{-\sigma_{j+1}}, \qquad (7.1.32)$$

where we assume the Liapunov exponents are ordered in the usual way

$$\sigma_1 > \sigma_2 > \cdots > \sigma_N,$$

and where j is the largest integer such that

$$\sigma_1 + \sigma_2 + \cdots + \sigma_j > 0.$$

Russell et al. (1980) have compared (7.1.30) with (7.1.32) for several two-dimensional maps and three-dimensional flows. For two-dimensional maps with $\sigma_1 > 0 > \sigma_1 + \sigma_2$, representing one expanding direction with area contraction, (7.1.32) reduces to

$$D_{KY} = 1 - \frac{\sigma_1}{\sigma_2}. \qquad (7.1.33)$$

For three-dimensional flows with $\sigma_1 > 0 > \sigma_3$, $\sigma_2 = 0$,[3] and $\sigma_1 + \sigma_3 < 0$,

$$D_{KY} = 2 - \frac{\sigma_1}{\sigma_3}. \qquad (7.1.34)$$

For the direct numerical calculation of D_0 based on (7.1.30), the space is divided into boxes of size ϵ, and the map is iterated until the initial transients have died away, and the motion is on the attractor. For subsequent iterations, a list is kept of those boxes containing at least one point on the attractor. Each newly generated point is checked to see if its box is on the list; if not, it is added to the list. After many iterations, the number of boxes approaches $M(\epsilon)$. From (7.1.30), a graph of $\ln M$ versus $\ln(1/\epsilon)$ will be linear. Thus D_0 is determined by fitting a straight line to this graph for several small values of ϵ, and extrapolating the result to $\epsilon \to 0$.

---
[2] We exclude sequences such as $0.12222 \cdots = 0.20000 \cdots$.
[3] Recall that one of the exponents must be zero for a flow.

7.1. Simple and Strange Attractors

Table 7.2. Calculation of Fractal Dimensions[a]

System	D_{KY}	D_0
Hénon map $a = 1.2, b = 0.3$	1.200 ± 0.001	1.202 ± 0.003
Hénon map $a = 1.4, b = 0.3$	1.264 ± 0.002	1.261 ± 0.003
Kaplan and Yorke map $\alpha = 0.2$	1.4306766	1.4316 ± 0.0016
Zaslavskii map $\Gamma = 3.0, \epsilon = 0.3,$ $v = 10^2 \times 4/3$	1.387 ± 0.001	1.380 ± 0.007
Ordinary differential equations describing wave–wave interaction	2.317 ± 0.001	2.318 ± 0.002

[a] After Russell et al. (1980).

Table 7.2, after Russell et al. (1980), shows a comparison[4] of this direct calculation of D_0 with the value D_{KY} determined using (7.1.32) for three different two-dimensional maps and one three-dimensional flow. The maps are the Hénon map (7.1.26), the (noninvertible) Kaplan and Yorke map

$$x_{j+1} = 2x_j, \qquad \text{mod } 1,$$
$$y_{j+1} = \alpha y_j + \cos 4\pi x_j, \qquad (7.1.35)$$

and the Zaslavskii map

$$x_{j+1} = [x_j + v(1 + \mu y_j) + \epsilon v\mu \cos 2\pi x_j], \qquad \text{mod } 1,$$
$$y_{j+1} = \exp(-\Gamma)(y_j + \epsilon \cos 2\pi x_j). \qquad (7.1.36)$$

To find the Liapunov exponents for the Hénon and Zaslavskii maps, σ_1 is numerically calculated and σ_2 is determined using the known area contraction for these maps (7.1.10)

$$\sigma_1 + \sigma_2 = \ln|\det \mathbf{M}|, \qquad (7.1.37)$$

where $\det \mathbf{M} = -b$ for the Hénon map and $e^{-\Gamma}$ for the Zaslavskii map. For the Kaplan and Yorke map, σ_1 and σ_2 are found analytically to be $\sigma_1 = \ln 2$

---

[4] The fractal dimension of Kaplan and Yorke is more related to the *information dimension*, which has been shown to be bounded from above by the *capacity* given in (7.1.30). Russell et al. (1980) numerically determine the capacity. Apparently the capacity is very close to the information dimension for the results given in Table 7.1. See Section 8.3 for further details.

and $\sigma_2 = \ln \alpha$. For the flow system, $\sigma_1 + \sigma_3$ is known analytically and σ_1 is calculated numerically. The agreement between (7.1.30) and (7.1.32) is good in all cases. Numerically, the calculation of D_{KY} from (7.1.32) is much less costly than the calculation of D_0 from (7.1.30).

The various forms of fractal dimensions and ways of calculating them analytically and numerically are a well-explored subject. Some of the key developments can be found in Packard *et al.* (1980), Froehling *et al.* (1981), Farmer (1982), Grassberger and Procaccia (1983), Grassberger (1985), and Halsey *et al.* (1986). We consider these ideas, together with the more general concept of a *multifractal* (Mandelbrot, 1974) in Section 8.3, where we also give other relevant references.

7.2. One-Dimensional Noninvertible Maps

7.2a. Basic Properties

We consider here the properties of one-dimensional maps, of the form

$$x_{n+1} = f(x_n, C), \qquad (7.2.1)$$

where C is a parameter. As we have seen, such maps are often[5] embedded in N-dimensional dissipative dynamical systems. They also arise naturally as simple models of system behavior. For example, x_n may represent the population of an insect species in a forest at year n, with f describing different environments that are characterized by the parameter C. Suppose the initial population x_0 is specified. How does the subsequent population evolve, and how does this evolution vary as we change the environment through C?

We are going to restrict our attention to f's of a particular form. If f is linear with x, then the solution of (7.2.1) is trivial. The case where f is invertible (monotonic with x) is also simple; the steady state solutions are periodic and do not have a sequence of period-doubling bifurcations or chaotic behavior. As we shall see, such behavior is associated with a local region of x near $f'(x) = 0$. Accordingly, we shall consider the simplest interesting case where f is a map having a single maximum (or minimum). Typically f may look something like that shown in Fig. 7.8. For x small, x will grow; while for x large, x will decrease. This form of f might model an environment in which limited food supply and the effects of pollution limit the population.

The evolution of x starting initially at x_0 is found by iterating (7.2.1). We can graphically determine the evolution as follows:

(a) Starting at x_0 move vertically to $x_1 = f(x_0)$.

---

[5] This embedding holds if the N-dimensional map has a single positive Liapunov exponent.

7.2. One-Dimensional Noninvertible Maps

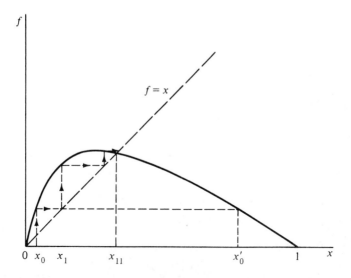

Figure 7.8. A typical one-dimensional noninvertible map.

(b) Transfer x_1 to the x-axis by moving horizontally to the 45° line $f = x$.

(c) Repeat (a) and (b) to determine x_2, x_3, etc. The procedure is shown for initial conditions x_0 or x'_0 in Fig. 7.8. We see that f is noninvertible. That is, two values x_0 and x'_0 generate the same value f_1 and therefore a single value x_1 for the next iteration. It is clear, therefore, that the map cannot be uniquely inverted to give x_0 (or x'_0) once x_1 has been determined.

For the case illustrated in Fig. 7.8, no matter what x_0 is chosen, the solution converges to the point of zero population growth, labeled x_{11} in the figure. It is easy to see that x_{11} is a period 1 fixed point of the map:

$$x_{11} = f(x_{11}). \tag{7.2.2}$$

Clearly x_{11} is a stable (attracting) fixed point. Another period 1 fixed point lies at $x = 0$, but is unstable (repelling).

Quadratic Maps. As mentioned, the transition to chaos for a one-hump map is determined by the local behavior of the map near its extremum. In the generic case for which $f' = 0$ and $f'' \neq 0$ at the extremum, the map is locally quadratic. A Taylor series expansion about the extremum leads to consideration of the general quadratic map

$$z_{n+1} = a + bz_n + cz_n^2. \tag{7.2.3}$$

With a linear transformation, we obtain the normalized form, which we call

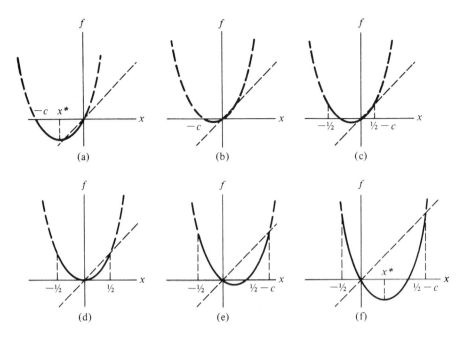

Figure 7.9. Parameter ranges for the quadratic map $x_{n+1} = 2Cx_n + 2x_n^2$. The solid line shows the region that is mapped into itself. Case I: $-C < x < 0$; (a) $C > \frac{1}{2}$; (b) $0 < C < \frac{1}{2}$. Case II: $-\frac{1}{2} < x < \frac{1}{2} - C$; (c) $0 < C < \frac{1}{2}$; (d) $C = 0$; (e) $-\frac{1}{2} < C < 0$; (f) $-1 < C < -\frac{1}{2}$.

the *quadratic map*:

$$x_{n+1} = f(x_n), \qquad (7.2.4a)$$

with

$$f(x_n) = 2Cx_n + 2x_n^2. \qquad (7.2.4b)$$

There are two parameter ranges of interest. For $0 < C < 2$ the interval $-C < x < 0$ is mapped into itself, as shown in Figs. 7.9a, b. However, negative values of C are also of interest. For $-1 < C < \frac{1}{2}$, the interval $-\frac{1}{2} < x < \frac{1}{2} - C$ is mapped into itself. This range of C is illustrated in Figs. 7.9c–f. Note that the parameter ranges overlap, as illustrated by Figs. 7.9b, c which are similar except for the interval in x which maps into itself. Of particular importance is the extremum in f, which we denote by x^*:

$$x^* = -C/2.$$

Another useful form for the quadratic map is found by putting $x = -(\mu/2)y$ and $C = \mu/2$ in (7.2.4), leading to the logistic map

$$y_{n+1} = \mu y_n(1 - y_n). \qquad (7.2.5)$$

7.2. One-Dimensional Noninvertible Maps

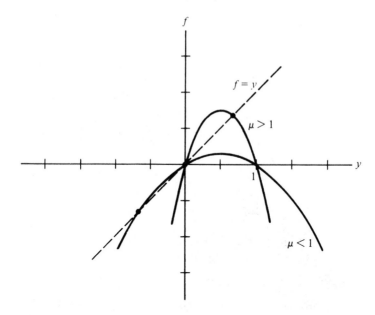

Figure 7.10. The logistic map $f = \mu y(1 - y)$ for $0 < \mu < 4$; the interval $0 < y < 1$ is mapped into itself.

For $0 < \mu < 4$, the interval $0 < y < 1$ is mapped into itself. This corresponds to the positive range of C values $0 < C < 2$ depicted in Figs. 7.9a, b. The logistic map has been used for studies in population biology as well as for examining the general behavior of one-dimensional maps. The logistic map is shown in Fig. 7.10 for two different values of μ. Both (7.2.4) and (7.2.5) are equivalent forms for the general quadratic map (7.2.3).

Mirror Symmetry. Quadratic maps have a symmetry that yields identical behavior for two different parameter values. Putting the linear transformation

$$x = \bar{x} + \tfrac{1}{2} - C \tag{7.2.6}$$

into the quadratic map (7.2.4) yields

$$\bar{x}_{n+1} = 2\bar{C}\bar{x}_n + 2\bar{x}_n^2, \tag{7.2.7}$$

where the new parameter is

$$\bar{C} = 1 - C. \tag{7.2.8}$$

Since (7.2.7) and (7.2.4) have the same form, it follows that the properties of the quadratic map with parameters C and $1 - C$ are identical. Thus all phenomena appear twice at C values symmetric about $\tfrac{1}{2}$. Similarly, putting

$$y = (1 - \mu^{-1}) - (1 - 2\mu^{-1})\bar{y} \qquad (7.2.9)$$

into the logistic map (7.2.5) yields

$$\bar{y}_{n+1} = \bar{\mu}\bar{y}_n(1 - \bar{y}_n), \qquad (7.2.10)$$

where

$$\bar{\mu} = 2 - \mu, \qquad (7.2.11)$$

i.e., symmetry around $\mu = 1$.

7.2b. Periodic Behavior

We consider the fixed points of the quadratic map (7.2.4) and their linear stability. The period 1 fixed points are found from

$$x_1 = 2Cx_1 + 2x_1^2, \qquad (7.2.12)$$

corresponding to the intersection of the curve $f(x)$ given in (7.2.4b) with the 45° line. There are two roots, one at $x_{10} = 0$ and the other at

$$x_{11} = \tfrac{1}{2} - C. \qquad (7.2.13)$$

The stability of the motion near these fixed points is obtained by linearizing the mapping as in Section 3.3. Putting

$$x_n = x_1 + \Delta x_n \qquad (7.2.14)$$

into (7.2.4) and keeping only linear terms yields

$$\Delta x_n = \lambda_1^n \Delta x_0, \qquad (7.2.15)$$

where the eigenvalue $\lambda_1 = f'(x_1)$. Thus x_1 is a stable (attracting) fixed point provided $|\lambda_1| < 1$ and is unstable (repelling) if $|\lambda_1| > 1$. We see that $x_{11} = \tfrac{1}{2} - C$ is stable for

$$|f'(x_{11})| = |2 - 2C| < 1$$

or when C lies in the range $\tfrac{1}{2} < C < \tfrac{3}{2}$. Similarly, the fixed point x_{10} is stable provided

$$|f'(x_{10})| = |2C| < 1$$

or when $-\tfrac{1}{2} < C < \tfrac{1}{2}$.

Bifurcation Phenomena. Let us consider the stability of the map as the parameter C decreases, starting from a value of $C > \tfrac{1}{2}$ for which x_{11} is stable and x_{10} is unstable. At $C_0 = \tfrac{1}{2}$, x_{11} becomes unstable, and x_{10} becomes stable. The latter root remains stable until C is decreased further to $C_1 = -\tfrac{1}{2}$.

To understand what happens when C is decreased below $-\tfrac{1}{2}$, we must

7.2. One-Dimensional Noninvertible Maps

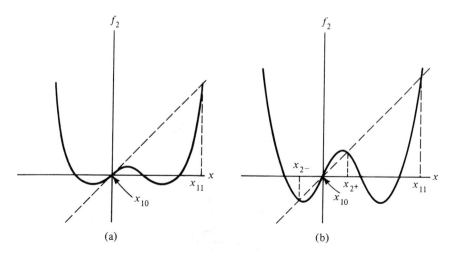

Figure 7.11. The creation of a pair of period 2 fixed points $x_{2\pm}$ for the quadratic map. The graphs show $f_2(x) = f(f(x))$ versus x. (a) $C > -\frac{1}{2}$; (b) $C \lesssim -\frac{1}{2}$.

consider the period 2 fixed points, given by

$$x_2 = f(f(x_2)). \qquad (7.2.16)$$

These correspond to the intersection of the function

$$f_2(x) = f(f(x)) \qquad (7.2.17)$$

with the 45° line, as shown in Fig. 7.11a, b for $C \gtrsim -\frac{1}{2}$ and $C \lesssim -\frac{1}{2}$. For $C \gtrsim -\frac{1}{2}$, the two intersections correspond to the period 1 fixed points $x_{11} = \frac{1}{2} - C$ and $x_{10} = 0$, which obviously satisfy (7.2.16). The slope of f_2 at x_{10} is less than one, since the chain rule applied to (7.2.17) yields

$$f'_2(x_{10}) = \lambda_1^2,$$

and $|\lambda_1| < 1$ for $C > -\frac{1}{2}$. As C is decreased below $-\frac{1}{2}$, the slope increases beyond 45° and a pair of period 2 fixed points are born, as shown in Fig. 7.11b. These points satisfy

$$x_{2+} = f(x_{2-}) = f_2(x_{2+}), \qquad (7.2.18a)$$

$$x_{2-} = f(x_{2+}) = f_2(x_{2-}). \qquad (7.2.18b)$$

Later we shall obtain explicit expressions for x_{2+} and x_{2-} as functions of C.

The stability of the motion near $x_{2\pm}$ can be obtained by writing

$$x_{2,n} = x_{2\pm} + \Delta x_{2,n}$$

and linearizing the mapping. We find near x_{2-} that

$$\Delta x_{2,n} = \lambda_{2-}^n \Delta x_0,$$

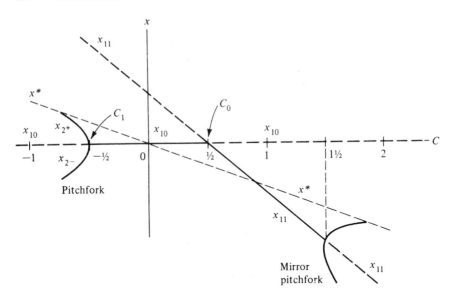

Figure 7.12. The fixed points x_{10}, x_{11}, x_{2+}, and x_{2-} versus C. Stable fixed points are shown as solid lines, unstable fixed points, as dashed lines. The light dashed line shows the extremum of f at $x^* = -C/2$.

where by the chain rule applied to (7.2.17)

$$\lambda_{2-} = f'_2(x_{2-}) = f'(x_{2-})f'(x_{2+}). \tag{7.2.19}$$

Similarly, for the motion near x_{2+},

$$\lambda_{2+} = f'(x_{2+})f'(x_{2-}) = \lambda_{2-}. \tag{7.2.20}$$

Thus the slopes (eigenvalues) are the same at each of the pair of period 2 fixed points. As noted in Section 3.3a, this was also a general property of all families of Hamiltonian fixed points of arbitrary period.

For C just below $C_1 = -\frac{1}{2}$, it is clear from Fig. 7.11b that the slope λ_2 at x_{2-} (and x_{2+}) satisfies

$$|\lambda_2| < 1.$$

Thus just after the period 1 fixed point $x_{10} = 0$ goes unstable, a pair of stable period 2 fixed points (a period 2 cycle) appears. The situation is illustrated in Fig. 7.12, in which the locations of the fixed points are plotted versus the parameter C, with a solid line indicating stability and a dashed line indicating instability. The birth of the pair of stable period 2 fixed points as C decreases below $-\frac{1}{2}$ is an example of a *period-doubling bifurcation* (as described previously in Section 7.1b), with x_{10} the handle, and x_{2-}, x_{2+} the tines of

7.2. One-Dimensional Noninvertible Maps

the pitchfork (see Fig. 7.2m). The mirror pitchfork given by (7.2.7) is also shown.

As C decreases further, the pair of period 2 fixed points goes unstable. Since the maps near x_{2-} and x_{2+} appear locally quadratic, we should expect that each of these solutions bifurcates again, giving birth to a stable period 4 motion when the period 2 motion goes unstable, and that these again bifurcate as C is decreased. Is the entire parameter range filled with a stable solution or one of its (stable) bifurcations? The answer is no. The sequence of period-doubling bifurcations terminates at infinite period but at a finite value of the parameter $C = C_\infty$. Beyond this value lie regions of chaotic behavior.

Algebraic Renormalization. To understand the period-doubling sequence, quantitatively, we must consider the renormalization of a one-hump map as we pass from one bifurcation to the next. An exact renormalization theory for period doubling in all one-humped maps exists in the asymptotic limit of a large number of period doublings (Feigenbaum, 1978; Collet and Eckmann, 1980). The exact theory can be developed in terms of a functional equation whose solution yields the exact scaling of the period-doubling bifurcations, at its fixed points in the function space. The resulting functional equations must be solved numerically. Another approach is to take the mapping itself to be the form of the mapping at the fixed point and find the mapping parameter required to transform the rescaled mapping into itself by the period doubling operation. If, in addition, we locally approximate the one-hump map by the quadratic map we can then develop an approximate renormalization theory using algebraic rather than functional equations. The procedure, developed by Helleman (1980, 1981), is similar in concept to that used in Section 4.2b to obtain corrections to the island overlap criterion. We first employ this procedure to calculate the rescaling constants. This is followed by a brief summary of the analysis using the functional equations.

Writing the one-dimensional quadratic map in normalized form, we have

$$x_{n+1} = 2Cx_n + 2x_n^2. \tag{7.2.4}$$

We have already found that the stable period 1 solution $x_{10} > 0$ is first created with decreasing C at $C_0 = \frac{1}{2}$, and that the stable period 2 solution is created at $C_1 = -\frac{1}{2}$. The period 2 orbit is found by solving (7.2.18) to obtain

$$x_{2\pm} = a \pm b, \tag{7.2.21a}$$

with

$$2a = -\tfrac{1}{2} - C, \tag{7.2.21b}$$

$$4b^2 = (C + \tfrac{1}{2})(C - \tfrac{3}{2}), \tag{7.2.21c}$$

where $b > 0$ in the range where the period 2 orbits exist. Substituting

$$x = x_{2\pm} + \Delta x$$

486   7. Bifurcation Phenomena and Transition to Chaos in Dissipative Systems

into (7.2.4) we obtain, exactly,
$$\Delta x_{n+1} = d\Delta x_n + 2\Delta x_n^2,$$
$$\Delta x_{n+2} = e\Delta x_{n+1} + 2\Delta x_{n+1}^2,$$
where
$$d = 2C + 4x_{2+}, \tag{7.2.22}$$
$$e = 2C + 4x_{2-}, \tag{7.2.23}$$

and the initial conditions are chosen such that the even subscripts in Δx give the motion near x_{2+}, and the odd subscripts give the motion near x_{2-}. Eliminating Δx_{n+1} and keeping up to quadratic terms in Δx, for n even,
$$\Delta x_{n+2} = de\Delta x_n + 2(e + d^2)\Delta x_n^2. \tag{7.2.24}$$

Rescaling by putting
$$x' = \alpha \Delta x \tag{7.2.25}$$

in (7.2.24), we obtain
$$x'_{n+2} = 2C'x'_n + 2x_n'^2 \tag{7.2.26}$$
with
$$C' = de/2 = -2C^2 + 2C + 2, \tag{7.2.27}$$
$$\alpha = e + d^2 = 16b^2 - 12b, \tag{7.2.28}$$

with b given by (7.2.21c). Since (7.2.26) has the same form as the original mapping (7.2.4), its fixed points undergo a period-doubling bifurcation at $C' = -\frac{1}{2}$. Using (7.2.27), this corresponds to
$$C_2 = \frac{1 - \sqrt{6}}{2} \approx -0.72474,$$

the value of C for creation of a stable period 4 solution. At this value of C, using (7.2.21), $x_{2+} \approx 0.466$ and $x_{2-} \approx -0.241$. We note that one root, x_{2+}, lies near the extremum in f at $x^* = -C/2$, a fact which we use below in calculating the rescaling of the bifurcations of the plus and minus tines.

The cascade of bifurcations will accumulate at the value $C' = C = C_\infty$. Using (7.2.27)
$$-2C_\infty^2 + 2C_\infty + 2 = C_\infty$$
or
$$C_\infty = \frac{1 - \sqrt{17}}{4} \approx -0.781. \tag{7.2.29}$$

By iterating the quadratic map, we numerically determine the accumulation

7.2. One-Dimensional Noninvertible Maps

point

$$C_\infty \approx -0.78497$$

in good agreement with the renormalized value.

We can also find an approximate convergence of C to C_∞ by assuming an asymptotic expansion in the form of a geometrical series

$$C_k \sim C_\infty + A\delta^{-k}. \tag{7.2.30}$$

Substituting this into (7.2.27) and noting that when k bifurcations have taken place off the new orbit, then $k + 1$ bifurcations have taken place off the old orbit,

$$C_k = -2C_{k+1}^2 + 2C_{k+1} + 2,$$

we find

$$\delta = -4C_\infty + 2 = 1 + \sqrt{17} \approx 5.12. \tag{7.2.31}$$

The exact renormalization equation that applies to all one-hump maps can be numerically solved to obtain the value

$$\delta \approx 4.6692,$$

which was first obtained by Feigenbaum (1978). Note that the geometric scaling law (7.2.30) can be expressed as

$$\frac{C_{k+1} - C_k}{C_k - C_{k-1}} = \frac{1}{\delta}. \tag{7.2.32}$$

We can see that δ is independent of the parametrization by introducing an arbitrary parameter P given by

$$P = g(C).$$

Assuming g is invertible near C, we can expand P in this neighborhood

$$P_k - P_\infty = g'(C_\infty)(C_k - C_\infty). \tag{7.2.33}$$

Solving for C_k and inserting into (7.2.32) yields the universal relation

$$\frac{P_{k+1} - P_k}{P_k - P_{k-1}} = \frac{1}{\delta}.$$

Actually, the exact renormalization theory shows that δ is universal for the parameter variation of all one-hump maps. Assuming that the asymptotic expansion (7.2.30) is valid even for $k = 0$ and $k = 1$, we find a useful estimate for the accumulation point in terms of the zero and first bifurcation

$$P_\infty \approx P_0 + \frac{\delta}{\delta - 1}(P_1 - P_0)$$

$$\approx P_0 + (1.13)(P_1 - P_0). \tag{7.2.34}$$

Finally, let us consider the rescaling parameter α in (7.2.25). From (7.2.28) and (7.2.21c), we obtain α at the accumulation point C_∞ given by (7.2.29):

$$\alpha = 16b^2 - 12b \approx -2.24, \qquad (7.2.35)$$

while the exact renormalization theory yields

$$\alpha \approx -2.5029,$$

in agreement with numerical iteration of the map. Physically, α asymptotically rescales the variable x on successive bifurcations; i.e., if we magnify by α the x-interval near the extremum $x^* = -C/2$ and look at the next bifurcation, it will appear the same as the previous one. Thus the separation of the tines for successive bifurcation solutions near x^* scales as

$$\frac{x_{k+} - x_{k-}}{x_{k+1,+} - x_{k+1,-}} = \alpha. \qquad (7.2.36)$$

These are the solutions that grow out of x_{2+}.

Since f is locally quadratic, the first iterates of a cluster of points near x^* must rescale as α^2; i.e., if the x_k's are the bifurcation solutions near x^*, then their first iterates satisfy

$$\frac{f(x_{k+}) - f(x_{k-})}{f(x_{k+1,+}) - f(x_{k+1,-})} = \alpha^2. \qquad (7.2.37)$$

These are the solutions that grow out of x_{2-}. Thus half the elements of a period 2^k solution rescale as α and the other half as α^2.

We illustrate this behavior in Fig. 7.13a, b. In Fig. 7.13a, we note that a separation Δx_1 near the minimum of the map transforms to a separation $|\Delta x_2| \propto \Delta x_1^2$. However, Δx_2 then transforms back to an interval Δx_3 near the maximum with $|\Delta x_3| \propto \Delta x_2$. Thus the solution jumps back and forth between the tines of the pitchfork from linear to quadratic scaling in α. We note also the reversal of the sense of Δx on each traversal of the map. The resulting bifurcation structure for the first three bifurcations is shown in Fig. 7.13b, indicating the widths of the various tines after each bifurcation. The order of traversing the tines for the period 4 and the period 8 cycles, starting on the upper tine, is shown in parentheses in the figure. We note the difference in the separation of the tines, with the separation of the x_{2+} tines (near x^*) being a factor of α greater than the separation of the x_{2-} tines. When another bifurcation takes place, the separations are as shown in the figure.

Any invertible transformation from x to some new variable y preserves (7.2.36) and (7.2.37) with x replaced by y. Thus the constant α is also universal, in the same sense as δ.

In addition to the bifurcation sequence as C decreases, the mirror bifurcation sequence defined by (7.2.6)–(7.2.8) also occurs. Here C increases from $\frac{1}{2}$ until the accumulation point is reached. The accumulation point for the

7.2. One-Dimensional Noninvertible Maps

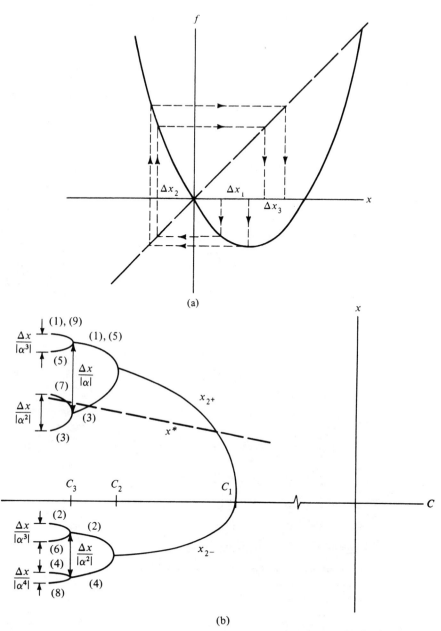

Figure 7.13. Illustrating the scaling of an interval in x for the quadratic map. (a) Variation of an initial interval Δx_1 near the extremum of the quadratic map with successive iterations. Note that $\Delta x_2 \sim (\Delta x_1)^2$ but that $\Delta x_3 \sim \Delta x_2$. (b) The order of traversal and separation of the pitchfork tines for the first three bifurcations of the quadratic map. The order of traversal is shown in parentheses for a period 4 and a period 8 orbit. The alternation in the separation of the pitchfork tines from linear to quadratic scaling with α should be noted.

490 7. Bifurcation Phenomena and Transition to Chaos in Dissipative Systems

mirror sequence is found from (7.2.8) and (7.2.79):

$$\bar{C}_\infty = (3 + \sqrt{17})/4 \approx 1.7808.$$

For the logistic map (7.2.5), the usual range of parameters considered is $0 < \mu < 4$. This corresponds to the mirror sequence for the quadratic map. Using (7.2.8) and noting $\mu = 2\bar{C}$, the first bifurcation occurs at $\bar{\mu}_1 = 3$ and the accumulation point is

$$\bar{\mu}_\infty = (3 + \sqrt{17})/2 \approx 3.5616,$$

which compares well with the value found by direct iteration of the logistic map $\bar{\mu}_\infty = 3.5700$.

The existence of period doubling through very large periods is clearly shown in Fig. 7.14 (van Zeyts, 1981), obtained by numerical iteration of the quadratic map. The bifurcation sequence x_k versus C is plotted on a doubly logarithmic scale. The constant rate of convergence δ along the C-axis, and the constant rate of convergence α of the bifurcation solutions along the x_k-axis are clearly seen.

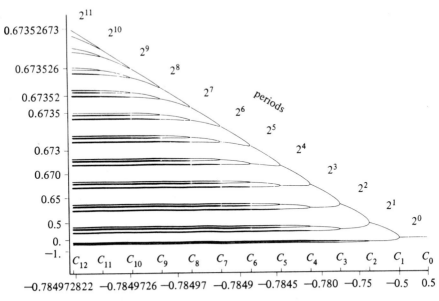

Figure 7.14. Period-doubling bifurcation sequence for the quadratic map. Vertically plotted are the x_k of the attractor of period 2^k, splitting off at C_k from an attractor of period 2^{k-1}, which continues as a repellor, not plotted here. Note the constant rates δ (at which the C_k converge) and α (at which the orbits converge) in this doubly logarithmic plot. To the left of C_∞ an infinite number of repellors remain (after van Zeyts, 1981; see Helleman, 1981).

7.2. One-Dimensional Noninvertible Maps

Functional Renormalization Theory. Following Feigenbaum (1978, 1979) and Schuster (1988) we let $f(\mu, x)$ be a general map having a quadratic maximum at $x = 0$, with a general parameter μ. We have seen that at some μ_∞, f is self-similar near $x = 0$ when rescaled by α. This implies that the universal function

$$g^*(x) = \lim_{n \to \infty} \alpha^n f^{(2^n)}\left(\mu_\infty, \frac{x}{\alpha^n}\right) \tag{7.2.38}$$

exists; hence, g^* must satisfy the rescaling equation

$$g^*(x) = \alpha g^*(g^*(x/\alpha)) \triangleq Tg^*, \tag{7.2.39}$$

where T is the period-doubling operator. We normalize by choosing $g^*(0) = 1$ and note that since f has a quadratic maximum, g^* is an even function of x; hence $g^{*\prime}(0) = 0$. With these specifications, the functional equation (7.2.39) has a unique solution for g^* and α, which, however, can only be found numerically. To estimate α analytically, we approximate $g^* \approx 1 + ax^2$ near $x = 0$. Inserting this into (7.2.39) and neglecting the x^4 terms, we obtain, by equating the constant and x^2 terms separately, $a = \alpha/2$ and $\alpha^2 + 2\alpha - 2 = 0$, which yields the solution $\alpha = -(1 + \sqrt{3}) \approx -2.732$. This is close to the true value $\alpha \approx -2.5029$ found by solving (7.2.39) numerically.

We note the interpretation that g^* is a fixed point in function space of the period-doubling operator T. We can investigate the stability of this fixed point by introducing a perturbation

$$g_\lambda = g^* + \epsilon h_\lambda. \tag{7.2.40}$$

Applying T to (7.2.40), and linearizing, we obtain

$$Tg_\lambda = g^*(x) + \epsilon\alpha[g^{*\prime}(g^*(x/\alpha))h_\lambda(x/\alpha) + h_\lambda(g^*(x/\alpha))] \triangleq g^*(x) + \epsilon L_{g^*}h_\lambda, \tag{7.2.41}$$

where L_{g^*} is the linearized period-doubling operator and a prime denotes a derivative with respect to the argument. The stability of g^* is determined by the spectrum of eigenvalues of L_{g^*}:

$$L_{g^*}h_\lambda = \lambda h_\lambda, \tag{7.2.42}$$

which yields the equation

$$\alpha[g^{*\prime}(g^*(x/\alpha))h_\lambda(x/\alpha) + h_\lambda(g^*(x/\alpha))] = \lambda h_\lambda(x) \tag{7.2.43}$$

for the eigenvalues λ and their eigenvectors h_λ in function space. We note by applying T to g_λ n times and linearizing that

$$T^n g_\lambda = g^* + \epsilon \lambda^n h_\lambda. \tag{7.2.44}$$

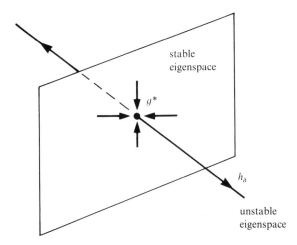

Figure 7.15. The fixed point g^* of the period-doubling operator T in function space, along with the stable and unstable eigenspaces.

Hence, g^* is unstable in the eigenvector direction h_λ if $|\lambda| > 1$. It can be shown (Feigenbaum, 1978, 1979) that (7.2.43) has a single unstable eigenvalue $\lambda \triangleq \delta \approx 4.669$. The situation is shown schematically in function space in Fig. 7.15.

We can estimate δ by choosing the approximate form $h_\lambda \approx h_0$ (a constant) and using $g^* \approx 1 + \alpha x^2/2$, found previously, in (7.2.43) to obtain, from the constant terms in (7.2.43) alone, $\lambda = \alpha^2 + \alpha \triangleq \delta$. Using $\alpha = -(1 + \sqrt{3})$ found previously, we obtain $\delta = 3 + \sqrt{3} \approx 4.732$, which is close to the true value 4.6692 found by numerical solution of (7.2.39) and (7.2.43).

In order to understand how the parameter μ changes at successive period-doublings and how this change is related to the unstable eigenvalue δ of the linearized period-doubling operator L_{g^*}, we let μ_n be the value of μ for which $x = 0$ is a member of a stable 2^n-cycle:

$$f^{(2^n)}(\mu_n, 0) = 0. \tag{7.2.45}$$

Since the eigenvalue λ_n of the map for the period 2^n-cycle is the product of the slopes $\partial f^{(2^n)}/\partial x$ of each member of the 2^n-cycle and $(\partial f^{(2^n)}/\partial x)_0 = 0$ because f has a quadratic maximum at $x = 0$, we see that $\lambda_n = 0$. At this value μ_n, the cycle is called *superstable*. Since a stable 2^n-cycle is born at $\lambda_n = +1$ and goes unstable at $\lambda_n = -1$, a superstable cycle lies "midway" within the parameter range μ where a stable 2^n-cycle exists.

To see how μ_n scales with successive period-doublings, we linearize f about μ_∞:

$$f(\mu_n, x) = f(\mu_\infty, x) + (\mu_n - \mu_\infty)h(x), \tag{7.2.46}$$

where $h(x) = (\partial f(\mu, x)/\partial \mu)_{\mu = \mu_n}$. Applying the period-doubling operator n

7.2. One-Dimensional Noninvertible Maps

times, linearizing, and taking the limit $n \to \infty$, we obtain

$$\lim_{n \to \infty} T^n f(\mu_n, x) = g^*(x) + (\mu_n - \mu_\infty) L_{g^*}^n h(x). \tag{7.2.47}$$

Operating n times on h using L_{g^*} as in (7.2.42), only the unstable eigenvalue $\lambda = \delta$ survives in the limit $n \to \infty$:

$$\lim_{n \to \infty} T^n f(\mu_n, x) = g^*(x) + (\mu_n - \mu_\infty) \delta^n h(x). \tag{7.2.48}$$

We now let $x = 0$ and note that

$$\lim_{n \to \infty} T^n f(\mu_n, 0) = \alpha^n f^{(2^n)}(\mu_n, 0) = 0, \tag{7.2.49}$$

because of (7.2.45). Hence, (7.2.48) becomes

$$0 = g^*(0) + (\mu_n - \mu_\infty) \delta^n h(0)$$

from which we obtain the scaling

$$\mu_n - \mu_\infty \sim \delta^{-n}, \tag{7.2.50}$$

as found in (7.2.30) for the specific quadratic map with $\mu \equiv C$.

Fourier Spectra. The complicated dynamic behavior of a bifurcating system is often studied experimentally through its Fourier spectrum. The universal character of the Fourier spectrum of a one-dimensional map, as the accumulation point C_∞ is approached, has been calculated by Feigenbaum (1979a), and we follow his analysis here.

For periodic solutions, the Fourier spectra are discrete. As period-doubling bifurcations occur, subharmonics of the basic mapping frequency appear in the spectrum of x. For our analysis, we replace the iteration number by the continuous time variable t and denote by $x^{(n)}(t)$ the solution for the nth bifurcation, with T_n the period of the solution. To calculate the spectrum at each bifurcation, we need a general relation for the positions of the attractors. From the arguments following (7.2.36) and (7.2.37), and the accompanying Fig. 7.13b, the tines of the pitchfork for the $(n + 1)$st bifurcation are divided into two groups whose separation distances can be calculated iteratively from

$$x^{(n+1)}(t) - x^{(n+1)}(t + T_n)$$

$$= [x^{(n)}(t) - x^{(n)}(t + T_{n-1})] \begin{cases} \dfrac{1}{\alpha} & 0 \leqslant t < T_{n-1}, \\ \dfrac{1}{\alpha^2} & T_{n-1} \leqslant t \leqslant T_n, \end{cases} \tag{7.2.51}$$

where $t = 0$ is taken on the uppermost (plus) branch of the tines.

The lth Fourier component of $x^{(n+1)}(t)$ is

$$X_l^{(n+1)} = \frac{1}{T_{n+1}} \int_0^{T_{n+1}} dt\, x^{(n+1)}(t) \exp\left(\frac{-2\pi i l t}{T_{n+1}}\right).$$

The integration may be performed over T_n by shifting the variable in the second half-period

$$X_l^{(n+1)} = \frac{1}{2T_n} \int_0^{T_n} dt [x^{(n+1)}(t) + (-)^l x^{(n+1)}(t + T_n)] \exp\left(\frac{-\pi i l t}{T_n}\right). \quad (7.2.52)$$

For $l = 2k$ even,

$$x^{(n+1)}(t) = x^{(n+1)}(t + T_n) = x^{(n)}(t)$$

and (7.2.52) yields

$$X_{2k}^{(n+1)} = X_k^{(n)}.$$

Thus a component at a given frequency, once it has come into existence for some n, remains unchanged during subsequent bifurcations. For $l = 2k + 1$ odd, substituting (7.2.51) into (7.2.52) yields

$$X_l^{(n+1)} = \left(\frac{1}{\alpha}\int_0^{T_{n-1}} + \frac{1}{\alpha^2}\int_{T_{n-1}}^{T_n}\right)\frac{dt}{2T_n}[x^{(n)}(t) - x^{(n)}(t + T_{n-1})]\exp\left(\frac{-\pi i l t}{T_n}\right).$$

Shifting the limits of the second integral,

$$X_l^{(n+1)} = \frac{1}{2\alpha}\left[1 + i\frac{(-1)^k}{\alpha}\right]$$

$$\times \int_0^{T_{n-1}} \frac{dt}{2T_{n-1}}[x^{(n)}(t) - x^{(n)}(t + T_{n-1})]\exp\left(-\pi i \frac{l}{2}\frac{t}{T_n}\right). \quad (7.2.53)$$

Using the Fourier expansion

$$x^{(n)}(t) = \sum_{k'} X_{k'}^{(n)} \exp\left(\frac{2\pi i k' t}{T_n}\right)$$

in (7.2.53), we obtain the Fourier coefficients for the $(n + 1)$st bifurcation in terms of the coefficients for the nth bifurcation:

$$X_{2k+1}^{(n+1)} = -\frac{1}{2\alpha}(1 - i(-1)^k)\left(1 + \frac{i}{\alpha}(-1)^k\right)S, \quad (7.2.54)$$

where

$$S = \frac{1}{\pi i}\sum_{k'}\frac{1}{(2k'+1) - \frac{1}{2}(2k+1)} X_{2k'+1}^{(n)}. \quad (7.2.55)$$

In the limit of large n, setting $2k + 1 = \xi$ and $2k' + 1 = \xi'$, Feigenbaum (1979a) converts the sum in (7.2.55) to an integral over ξ', obtaining the

7.2. One-Dimensional Noninvertible Maps

estimate

$$S \approx \tfrac{1}{2} X^{(n)}\left(\frac{\xi}{2}\right).$$

Using this result in (7.2.54) yields the scaling

$$|X^{(n+1)}(\xi)| = \gamma^{-1}\left|X^{(n)}\left(\frac{\xi}{2}\right)\right|, \qquad (7.2.56)$$

where

$$\gamma = 2\sqrt{2}\alpha^2/(1+\alpha^2)^{1/2}.$$

Using the exact value $\alpha \approx -2.509$, we find $\gamma \approx 6.57$. According to (7.2.56), to obtain the amplitude of the fully developed subharmonics that first appear at the $(n+1)$st bifurcation, we rescale the amplitude of the subharmonics that first appear at the nth bifurcation down by a factor of $6.57 = 8.18$ dB.

Near the transition to chaos, we can obtain a better estimate for γ by assuming that the $X^{(n)}_{2k-1}$ in (7.2.55) have roughly equal amplitudes and random (uncorrelated) phases. Then squaring (7.2.55) and averaging over the phases yields

$$|S|^2 \approx \frac{4}{\pi^2}\left|X^{(n)}\left(\frac{\xi}{2}\right)\right|^2 \sum_{k'} \frac{1}{(4k'-2k+1)^2} \approx \tfrac{1}{2}\left|X^{(n)}\left(\frac{\xi}{2}\right)\right|^2. \qquad (7.2.57)$$

Using this result in (7.2.54), we obtain

$$\gamma = 2\alpha^2/(1+\alpha^2)^{1/2}, \qquad (7.2.58)$$

which yields $\gamma \approx 4.65$ for $\alpha \approx -2.509$. A more careful examination of (7.2.55) using renormalization theory yields $\gamma \approx 4.578$ for the scaling between the mean values (averaged over k' and k) of $|X^{(n)}_{2k'+1}|$ and $|X^{(n+1)}_{2k+1}|$ (Nauenberg and Rudnick, 1981). Thus, the mean amplitude separation of successive, fully developed subharmonics should be $4.578 = 6.6$ dB. This prediction has been verified computationally (Nauenberg and Rudnick, 1981) and is also seen experimentally. We defer a description of the experimental results to Section 8.6.

Other Periodic Cycles. We have seen how the fixed point x_{10} gives rise to a cascade of period-doubling bifurcations as C decreases from C_0 toward C_∞. For C less than C_∞, we encounter further parameter regimes having periodic behavior. These new cycles are born in pairs, one unstable, the other stable, via tangent bifurcations. We illustrate the birth of a period 3 cycle in Fig. 7.16, where the plot for the quadratic map (7.2.4)

$$f^{(3)}(x) = f(f(f(x)))$$

versus x, along with the 45° line, for two parameter values. The period 3 roots satisfy

$$x_3 = f^{(3)}(x_3).$$

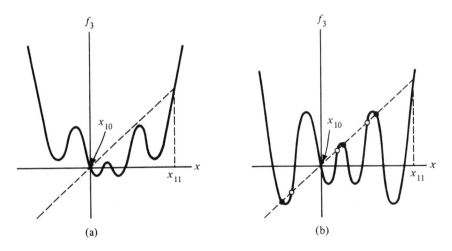

Figure 7.16. Birth of a pair of period 3 cycles by a tangent bifurcation. The solid circles show the stable period 3 orbit; the open circles show the unstable period 3 orbit: (a) $C \gtrsim C_0^{(3)}$, (b) $C \lesssim C_0^{(3)}$.

For $C \gtrsim C_0^{(3)}$, there are only two roots, corresponding to the period 1 roots x_{10} and x_{11}. For $C \lesssim C_0^{(3)}$, the hills and valleys in $f^{(3)}$ steepen until simultaneously the first valley and the last two hills touch the 45° line, and then intercept it in six new points, corresponding to two distinct cycles of three points each. By considering the slopes of $f^{(3)}$ at the intersection points, it is easy to see that one cycle, consisting of the three solid dots in Fig. 7.16, is born stable, while the other cycle, shown as the open circles, is always unstable. Solving for x_3, the two period 3 cycles are born at

$$C_0^{(3)} = (1 - \sqrt{8})/2 \approx -0.9142$$

with $x_{30} \approx 0.4296$, $x_{31} \approx -0.4164$, and $x_{32} \approx 1.1081$.

Since the map $f^{(3)}$ is locally quadratic near the roots of the stable period 3 cycle, we expect that this cycle undergoes a cascade of period-doubling bifurcations through periods 3, 6, 12, 24, ..., as C decreases below $C_0^{(3)}$. The accumulation point $C_\infty^{(3)}$ can be found from the renormalization procedure given in this subsection or by direct iteration of the mapping. The latter yields $C_\infty^{(3)} \approx -0.92475$. The mirror sequence period 3 cycle at $\bar{C} = 1 - C$ also exists.

In the same way, stable cycles with basic period $n = 4, 5, 6, \ldots$, can be found by considering $f^{(4)}$, $f^{(5)}$, etc. Each cycle has its own bifurcation sequence from $C_0^{(n)}$ to its accumulation point $C_\infty^{(n)}$. The way in which these basic cycles are arranged along the interval of parameter values is universal for all one-hump quadratic maps. Through period 6, for example, the basic

7.2. One-Dimensional Noninvertible Maps

cycles are ordered with C decreasing from C_0 as having periods

$$1, 6, 5, 3, 5, 6, 4, 6, 5, 6.$$

The total number of stable cycles and their order of appearance can be found (see May, 1976). For example, there are 202 basic cycles with period less than or equal to 11; their ordering in parameter values has been given by Metropolis *et al.* (1973). Geisel and Nierwetberg (1981) have shown that there is a universal renormalization structure to the widths and ordering of these cycles.

7.2c. Chaotic Motion

The stable cycles, which represent regular motion, are dense and occupy a finite fraction of the parameter interval. The motion for the remaining parameter values is unstable and densely fills a finite interval in x for an arbitrary initial x_0. Nearby trajectories diverge exponentially. We refer to this type of motion as chaotic (see Li and Yorke, 1975; May and Oster, 1976). The possibility of divergent but bounded trajectories arises because one-hump noninvertible maps act by first stretching the mapped interval and then folding a portion of this interval back on itself. Since folding cannot take place for one-dimensional invertible maps, such maps do not display chaotic behavior. Chaotic motion in one-dimensional maps can be characterized using the techniques for Hamiltonian maps described in Chapter 5. Of particular usefulness is the Liapunov exponent σ and the *equilibrium invariant distribution* $P(x)$.

Liapunov Exponent. From (5.2.8), for one dimension, there is a single Liapunov exponent

$$\sigma(x_0) = \lim_{N \to \infty} \frac{1}{N} \sum_{i=1}^{N} \ln \left| \frac{df}{dx_i} \right|. \tag{7.2.59}$$

As illustrated in Fig. 7.17, σ will be positive if the slope f' has a magnitude greater than unity when averaged over an orbit. Except for a set of measure zero, σ is independent of the initial value x_0. For $\sigma > 0$, we have a chaotic orbit; for $\sigma < 0$, a stable (attracting) cycle exists, and the orbit, after an initial transient, is periodic. Since the stable cycles are dense in the parameter space of interest, the curve of σ versus the parameter C must be highly convoluted. Figure 7.18 (after Shaw, 1981) shows the results of a numerical experiment on the quadratic map, in which σ has been computed using (7.2.59). The curve is composed of 300 points equally spaced in C, each point representing 100,000 iterations. The low-period attracting orbits, which have relatively wide parameter windows, are visible as regions where σ is negative. Other higher-period attracting orbits, which densely fill the C interval, are invisible,

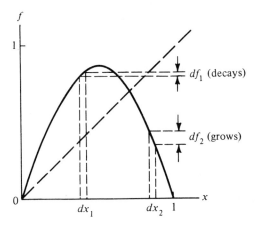

Figure 7.17. Relation of the slope df/dx of a one-dimensional map to the Liapunov exponent. The exponent is the average of $\ln|df/dx|$ over the orbit.

because their parameter windows are narrower than the spacing of the points along C. The chaotic orbits are found in the regions where σ is positive. Near the critical parameter C_∞ where the Liapunov exponent turns positive, Huberman and Rudnick (1980) have shown that σ scales as $|C - C_\infty|^\eta$, where $\eta = \ln 2/\ln \delta \approx 0.4498$.

We can show (Oseledec, 1968) that σ is independent of the specific form in which the map is written. Suppose we make an invertible coordinate transformation

$$\bar{x} = g(x) \tag{7.2.60}$$

with $g' \neq 0$. Then the original map

$$x_{n+1} = f(x_n) \tag{7.2.61}$$

is transformed to the new map

$$\bar{x}_{n+1} = \bar{f}(\bar{x}_n). \tag{7.2.62}$$

The Liapunov exponent for the new map is

$$\bar{\sigma} = \lim_{N \to \infty} \frac{1}{N} \sum_{i=1}^{N} \ln \left| \frac{d\bar{f}}{d\bar{x}_i} \right|. \tag{7.2.63}$$

But $d\bar{f} = d\bar{x}_{n+1}$, and the chain rule applied to (7.2.60) then yields

$$\frac{d\bar{x}_{i+1}}{d\bar{x}_i} = \frac{(dg/dx_{i+1}) \, dx_{i+1}}{(dg/dx_i) \, dx_i}. \tag{7.2.64}$$

When (7.2.64) is used in (7.2.63), we obtain (7.2.59), i.e., $\bar{\sigma} = \sigma$.

Invariant Distributions. We say that $P(x)$ is an invariant distribution of the

7.2. One-Dimensional Noninvertible Maps

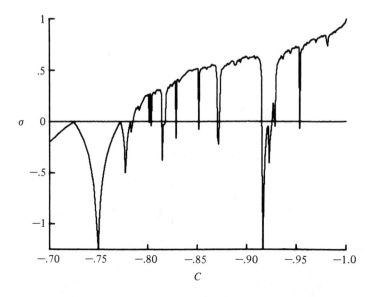

Figure 7.18. Graph of the Liapunov exponent σ versus C for the quadratic map $x_{n+1} = 2Cx_n + 2x_n^2$. For $\sigma > 0$, the motion is chaotic; for $\sigma < 0$, the motion is periodic. This numerically generated graph has been smoothed, being composed of 300 points equally spread in C. Actually, the periodic orbits ($\sigma < 0$) densely fill the C-interval (after Shaw, 1981).

mapping T if

$$P(x) = TP(x). \tag{7.2.65}$$

Other names are *invariant measure* or *probability distribution*. We assume that P integrates to unity:

$$\int P(x)\, dx = 1. \tag{7.2.66}$$

In general there are many distributions invariant to a given mapping. However, a unique equilibrium distribution is generally singled out by repeated iteration of the map such that the time average is equal to the space average over the distribution for almost all initial conditions x_0. For a parameter value corresponding to a stable, period n cycle, $P(x)$ is discrete, consisting of n δ-functions at the n stable fixed points of the map, each δ-function having integrated value $1/n$. Except for a set of measure zero, every x_0 leads under repeated iteration of the map to this unique steady-state distribution. Similarly, for parameters yielding chaotic motion, all initial conditions, except for a set of measure zero, yield a unique equilibrium distribution. This invariant distribution may be discontinuous in x, but it is typically nonzero over a finite range of x values.

7. Bifurcation Phenomena and Transition to Chaos in Dissipative Systems

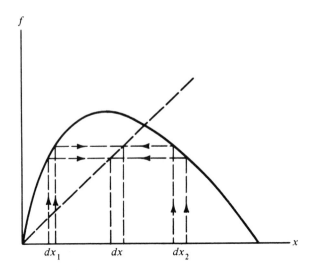

Figure 7.19. The construction of the invariant distribution $P(x)$. The number of orbits within dx is equal to the number within the two pre-images dx_1 and dx_2.

Numerically, $P(x)$ can be constructed from (7.2.65). As illustrated in Fig. 7.19, the number of trajectories $P(x)\,dx$ within some small interval dx in x is equal to the number within the corresponding intervals at the inverse points of the mapping. Since there are two inverse points x_1 and x_2 for a one-hump map,

$$P(x)\,dx = P(x_1)\,dx_1 + P(x_2)\,dx_2. \tag{7.2.67}$$

Writing

$$\frac{dx}{dx_1} = \left|\frac{df}{dx}\right|_{x_1}, \quad \text{etc.,}$$

we have

$$P(x) = \frac{P(x_1)}{|df/dx|_{x_1}} + \frac{P(x_2)}{|df/dx|_{x_2}}. \tag{7.2.68}$$

This is a functional equation for P that can rarely be solved analytically. However, a numerical solution is easily obtained by iterating (7.2.68) as follows:

(a) Choose an initial guess $P_i = P_1(x)$.
(b) Evaluate the right-hand side using P_i to obtain the next iterate P_{i+1}.
(c) Repeat (b) until convergence is achieved.

The procedure is illustrated in Fig. 7.20 (after Shaw, 1981) for the logistic map with $\mu = 4$

$$f(x) = 4x(1-x).$$

7.2. One-Dimensional Noninvertible Maps

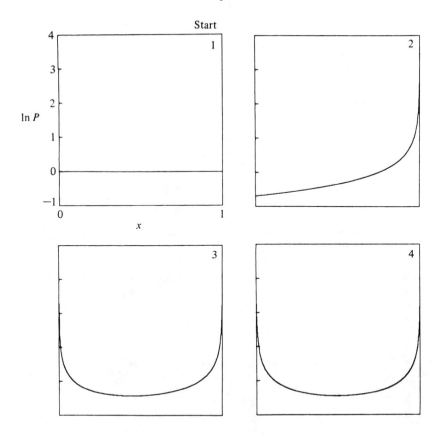

Figure 7.20. Numerical calculation of the invariant distribution $P(x)$ for the logistic map with $\mu = 4$, showing the initial guess and the first three iterations (after Shaw, 1981).

The first four iterations in (7.2.68) are shown with the initial guess $P_1(x) = 1$. The procedure converges rapidly to the invariant distribution

$$P(x) = \frac{1}{\pi}[x(1-x)]^{-1/2},$$

which will be found analytically below. This method gives the invariant distributions for the logistic map with $\mu = 3.8$ and $\mu = 3.825$, as shown in Fig. 7.21. In these cases, the motion appears chaotic and the distribution is nonzero over a finite range of x.

Use of the invariant distribution allows one to replace time averages by spatial averages over the invariant distribution. For example, the Liapunov

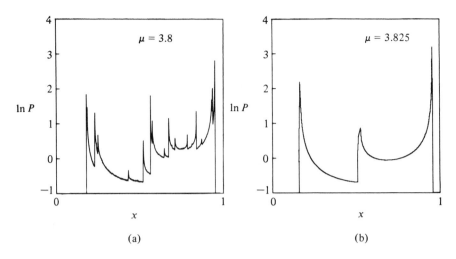

Figure 7.21. Two direct numerical calculations of the invariant distribution $P(x)$ for the logistic map. (a) $\mu = 3.8$; (b) $\mu = 3.825$. Note the discontinuous structure of $P(x)$ and the evidence of a reverse bifurcation sequence as μ increases (after Shaw, 1981).

exponent may be found using

$$\sigma = \int dx P(x) \ln \left| \frac{df}{dx} \right|. \tag{7.2.69}$$

If an invertible transformation $\bar{x} = g(x)$ is introduced, then the invariant distribution transforms so as to conserve the number of trajectories within any small interval in x

$$\bar{P}(\bar{x}) \, d\bar{x} = P(x) \, dx. \tag{7.2.70}$$

Equation (7.2.70) yields the new invariant distribution in terms of the old.

The Tent Map. We illustrate some of the preceding ideas using as a simple example the symmetric tent map shown in Fig. 7.22. The map has a single maximum $f(\frac{1}{2}) = a$ but is not a quadratic map. The slope f' is $+2a$ for the left half of the map and $-2a$ for the right half. It is clear that the motion is always chaotic for $a > \frac{1}{2}$, since all trajectories diverge exponentially (see Fig. 7.17). The invariant distribution is found from (7.2.68):

$$P(x) = \frac{1}{2a} \left[P\left(\frac{x}{2a}\right) + P\left(1 - \frac{x}{2a}\right) \right].$$

For $a = 1$, this has the obvious solution $P(x) = 1$. The Liapunov exponent

7.2. One-Dimensional Noninvertible Maps

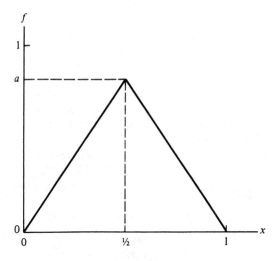

Figure 7.22. The symmetric tent map.

for this value of a is found from (7.2.69):

$$\sigma = \int_0^1 dx \ln 2 = \ln 2.$$

Since $\sigma > 0$, the motion is chaotic.

Now let us consider the logistic map with $\mu = 4$:

$$f(x) = 4x(1 - x). \tag{7.2.71}$$

If we introduce the transformation

$$\bar{x} = \left(\frac{2}{\pi}\right) \sin^{-1} \sqrt{x}, \tag{7.2.72}$$

then (7.2.71) is transformed to the symmetric tent map with $a = 1$:

$$\bar{f}(\bar{x}) = \begin{cases} 2\bar{x}, & 0 < \bar{x} < \tfrac{1}{2}, \\ 2 - 2\bar{x}, & \tfrac{1}{2} < \bar{x} < 1. \end{cases} \tag{7.2.73}$$

Using (7.2.70) with $\bar{P}(\bar{x}) = 1$, we obtain the invariant distribution for the logistic map (7.2.71):

$$P(x) = \frac{d\bar{x}}{dx} = \frac{1}{\pi} [x(1-x)]^{-1/2}. \tag{7.2.74}$$

This distribution was found numerically in Fig. 7.20. If we evaluate the

Liapunov exponent from (7.2.69) for the logistic map of (7.2.71), we obtain

$$\sigma = \frac{1}{\pi} \int_0^1 \frac{\ln|4(1-2x)|}{[x(1-x)]^{1/2}} dx = \ln 2.$$

We should not be surprised that (7.2.71) and (7.2.73) have the same Liapunov exponent, since σ is invariant under coordinate transformation. Similarly, any invertible transformation of the quadratic map (7.2.4) preserves the spectrum $\sigma(C)$ shown in Fig. 7.18.

The exact solutions for the motion for the logistic map (7.2.71) with $\mu = 4$ and the "mirror" solution for the quadratic map (7.2.4) with $C = -1$ are known, and given, respectively, by the minus and plus solutions:

$$x_n = \tfrac{1}{2} \mp \tfrac{1}{2} \cos(2^{n+1} \phi_0), \tag{7.2.75}$$

where ϕ_0 is determined by the initial condition on x_0. These solutions have been shown to be ergodic and mixing, with exponential separation of nearby trajectories (Ulam and Von Neumann, 1947). This follows from the phase advance equation $\phi_{n+1} = 2\phi_n$ (mod 2π), as described in Section 5.2d.

Reverse Bifurcation of Chaotic Bands. We have seen in Section 7.2b that the stable period 1 motion for the quadratic map (7.2.4) bifurcates repeatedly as C decreases from $C_0 = \tfrac{1}{2}$ to $C_\infty = -0.78497\ldots$, leading to the bifurcation tree shown in Fig. 7.14. What is the nature of the motion as C becomes increasingly negative beyond $C = C_\infty$?

The qualitative features are seen in Fig. 7.23 after Lorenz (1980), Collet and Eckmann (1980), and Helleman (1981). The steady state x_n's ($1000 < n < 4000$) are shown as a vertical set of dots for each value of C on the horizontal axis. Linear scales are used for both axes. We see clearly the presence of chaotic bands of motion for $C < C_\infty$. As C becomes increasingly negative below C_∞ the bands merge, giving a reverse set of bifurcations at various values C_k^* noted on the figure. We also see the presence of higher-period bifurcation trees, embedded in the chaotic region, for $C < C_\infty$. The stable 6, 5, and 3 cycles and their bifurcation trees (with decreasing C) are visible. The same scalings δ and α are found for the reverse bifurcations of the stochastic bands as for the bifurcations of the periodic attractors. These results, obtained by Grossmann and Thomae (1977), can also be derived from the renormalization theory of the previous subsection (Helleman, 1981).

Power Spectra. The "noisy" motion in the chaotic regime can be characterized by its power spectrum $G(\omega, C)$, where ω is the frequency and C is the parameter. As with the periodic spectra, the continuous power spectra are easily measured experimentally, in contrast to other properties of chaotic systems such as the Liapunov exponent. Of particular interest is the spectrum at some reverse bifurcation parameter C_k^* near the onset of chaotic motion

7.2. One-Dimensional Noninvertible Maps

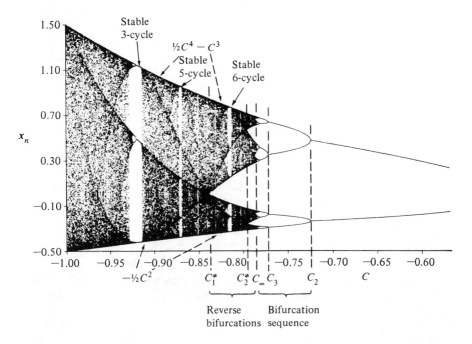

Figure 7.23. Numerical calculation of the reverse bifurcation sequence for the quadratic map. For each value of C, 3000 x_n's with $1000 < n < 4000$ have been plotted. The chaotic bands of motion for $C < C_\infty$ are clearly seen, with the band mergings occurring at C_k^*. The stable 6, 5, and 3 cycles (with decreasing C) embedded within the predominantly chaotic region of C are clearly seen (after Collet and Eckmann, 1980).

($C \lesssim C_\infty$), where the band mergings take place. We decompose the chaotic motion in the bands into periodic and noisy components:

$$x_n = \sum_J A_j \exp(i\omega_j n) + r(n), \qquad (7.2.76)$$

where $\omega_j = 2\pi j/2^k$ and the Fourier coefficients are

$$X_N(\omega, C) = \lim_{N \to \infty} \frac{1}{N} \sum_{n=1}^{N} x_n \exp(-i\omega n). \qquad (7.2.77)$$

The correlation function $\mathscr{C}(n)$ is defined as

$$\mathscr{C}(n) = \lim_{N \to \infty} \frac{1}{N} \sum_{i=1}^{N} (x_i - \langle x \rangle)(x_{i+n} - \langle x \rangle),$$

$$\langle x \rangle = \lim_{N \to \infty} \frac{1}{N} \sum_{i=1}^{N} x_i.$$

The power spectrum G is the Fourier transform of $\mathscr{C}$:

$$G(\omega, C) = \frac{1}{2\pi} \sum_{n=-\infty}^{\infty} \mathscr{C}(n) \cos \omega n.$$

Using (7.2.76), we obtain

$$G(\omega, C) = \sum_j |A_j|^2 \delta(\omega - \omega_j) + |r(\omega, C)|^2. \tag{7.2.78}$$

The spectrum consists of a set of sharp peaks (the periodic motion from band to band) superimposed on top of broadband noise (the chaotic motion within the bands). The part of the spectrum associated with the noise has a frequency dependence that when integrated yields a total noise power that scales with the parameter C.

First, summarizing Huberman and Zisook (1981), we exhibit the universal scaling for the integrated noise power spectrum $\mathscr{N}(C)$ near the accumulation point C_∞. The motion for $C = -1$ has been shown to be mixing. Further, the motion has been shown to be δ-correlated (see Grossman and Thomae, 1977); i.e., $\mathscr{C}(n) = 0$ for $n \neq 0$. This represents a very strong chaotic property, i.e., a complete decorrelation of the motion in one mapping period.[6]

Similarly, the motion at each reverse bifurcation parameter value C_k^* is mixing and appears to be approximately δ-correlated; namely, the action of $f^{(2^k)}(x)$ on any fixed band at $C = C_k^*$, has an invariant distribution which is a scaled down version of that of the map $f(x)$ at $C = -1$. Therefore, the frequency spectrum for the noise $r(n)$ corresponds approximately to white noise, and has a correlation function (Grossmann and Thomae, 1977)

$$\mathscr{C}(n) = \begin{cases} W^2, & n = 0, \\ 0, & n > 0, \end{cases} \tag{7.2.79}$$

where W^2 is a constant proportional to the average of the square of the width of one of the 2^k bands. As described previously the chaotic bands are images of the periodic attractors, and thus half the bands have a width that scales as $1/\alpha$, and the remaining half have a width that scales as $1/\alpha^2$ [see also (7.2.51)] as C_{k+1}^* decreases to C_k^*. The power spectrum then scales as

$$W_{k+1} = \left(\frac{1}{2} \frac{1}{\alpha^2} + \frac{1}{2} \frac{1}{\alpha^4} \right)^{1/2} W_k, \tag{7.2.80a}$$

which is geometric scaling of the form

$$W_k = W_0 \beta^{-k}. \tag{7.2.80b}$$

---

[6] We note that for many systems exhibiting stochastic or chaotic motion, the decay of the correlations with increasing n is not generally rapid; in fact, in many systems $\mathscr{C}(n)$ decays as a power of n (see Casati et al., 1981).

7.2. One-Dimensional Noninvertible Maps

Substituting (7.2.80a) in (7.2.80b), we find

$$\beta = \frac{\sqrt{2}\alpha^2}{\sqrt{\alpha^2 + 1}} \approx 3.27.$$

We note that $\sqrt{2}\beta = \gamma$ given by (7.2.58). The integrated noise power below the fundamental driving frequency 2π is

$$\mathcal{N}(C) = \int_0^{2\pi} d\omega\, |r(\omega, C)|^2. \tag{7.2.81}$$

Using Parseval's identity, $\mathcal{N} \propto \mathcal{C}(0)$, given in (7.2.79). Substituting (7.2.80b) into (7.2.79), we see that the integrated noise power scales geometrically as

$$\mathcal{N}(C_k^*) = \mathcal{N}_0 \beta^{-2k}. \tag{7.2.82}$$

But the reverse bifurcation sequence itself scales geometrically as

$$C_k^* - C_\infty \approx (C_0^* - C_\infty)\delta^{-k}. \tag{7.2.83}$$

Eliminating k from the last two equations, we find that the integrated noise power as a function of the parameter C scales as

$$\mathcal{N}(C) = \text{const.}(C_\infty - C)^\sigma, \tag{7.2.84}$$

where σ is a universal constant

$$\sigma = \frac{2 \ln \beta}{\ln \delta} = 1.525.$$

Huberman and Zisook (1981) have checked this result against numerical calculations of the noise power for the logistic map, finding excellent agreement with the theoretical value.

The frequency dependence of the noise power has also been calculated (Wolf and Swift, 1981). The calculation is similar to that of Feigenbaum (1979) for the discrete spectrum of the regular motion (see Section 7.2b). Ignoring the periodic component A_j in (7.2.78), they obtain for the noise spectrum

$$r(\omega, C_k^*) = g_k(\omega)\tilde{r}(2^k\omega), \tag{7.2.85}$$

where $\tilde{r}(\omega)$ is the (approximately) white noise spectrum associated with the map f at $C = -1$, $2^k\omega$ is taken modulo 2π, and

$$g_k(\omega) = \sum_{j=1}^{2^k} W_{jk} e^{-i\omega t} \tag{7.2.86}$$

is the Fourier spectrum associated with the sequence of the 2^k widths W_{jk} of the chaotic bands. The g_k are computed similarly to the periodic spectrum,

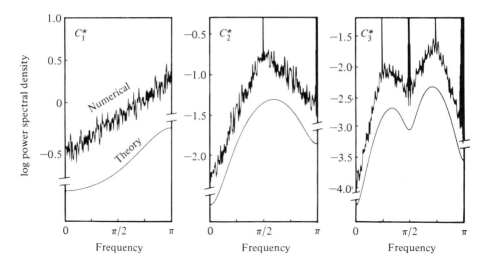

Figure 7.24. Noisy power spectrum at the reverse bifurcation values C_1^*, C_2^*, and C_3^* for the quadratic map. The sharp peaks and the white noise are subtracted from the theory. The theory is displaced downward for clarity (after Wolf and Swift, 1981).

obtaining the recursion relation (Wolf and Swift, 1981)

$$g_{k+1}(\omega) = \frac{1}{\sqrt{2\alpha}}\left(1 - \frac{e^{i\omega}}{\alpha}\right)g_k(2\omega). \qquad (7.2.87)$$

Using $\tilde{r}(2^k\omega)$ approximately flat and the normalization $g_0 = 1$ allows (7.2.85) to be evaluated. Figure 7.24 shows the numerically computed noisy power spectra at C_1^*, C_2^*, and C_3^*. The corresponding universal power spectra $|g_k|^2$ obtained from (7.2.87) are shown as the solid curves, shifted down for clarity. The white noise factor $|\tilde{r}|^2$ and the sharp peaks $|A_j|^2$ are also not shown. The agreement between theory and numerical calculation is quite good.

7.3. Bifurcation Phenomena in Two-Dimensional Maps

We have seen that chaotic motion may appear in dissipative flows having at least three dimensions, or in corresponding (invertible) Poincaré maps having two or more dimensions. In this section, we consider the simplest such systems of two-dimensional dissipative maps. As we shall see, the two-dimensional dissipative map has as its two limiting cases the one-dimensional noninvertible map, for the Jacobian of the map tending to

7.3. Bifurcation Phenomena in Two-Dimensional Maps

zero, and the two-dimensional area-preserving map, for the Jacobian tending to unity, leading to a unified picture of the bifurcations. In Section 7.3a, the algebraic renormalization technique is applied to two-dimensional invertible maps, and successive period-doubling bifurcations are shown to converge geometrically to a limiting parameter value, beyond which one obtains locally chaotic motion. In Section 7.3b, we extend this analysis to the area-preserving (Hamiltonian) case. In Section 7.3c, we show the unity of the bifurcation phenomena, including both period-doubling to chaos and the strange attractors that are formed in the chaotic regime.

The phenomena that we treat here, although universal, are localized to the immediate phase space neighborhood of a fixed point of the map. We defer global descriptions of the behavior of two-dimensional dissipative maps to subsequent sections.

7.3a. Dissipative Maps

We now show that a cascade of period-doubling bifurcations is the mechanism by which a transition from regular to chaotic motion occurs in a large class of two-dimensional invertible maps. Furthermore, near the transition, the dynamics can be described locally by a one-dimensional noninvertible map. These results have been proved using the exact renormalization theory (Collet et al., 1980). However, following the treatment of Helleman (1980, 1981), we continue our description using an approximate theory and consider the successive bifurcations of a fixed point (period 1 orbit)[7] of a two-dimensional map T. After the first bifurcation, the fixed point is unstable. We expand the map to quadratic terms, near the unstable fixed point, obtaining expressions for the deviations u, v from the fixed point of the form

$$\begin{pmatrix} u_{n+1} \\ v_{n+1} \end{pmatrix} = \begin{pmatrix} \Lambda_{11} & \Lambda_{12} \\ \Lambda_{21} & \Lambda_{22} \end{pmatrix} \begin{pmatrix} u_n \\ v_n \end{pmatrix} + \begin{pmatrix} \Gamma_{11} & \Gamma_{12} & \Gamma_{13} \\ \Gamma_{21} & \Gamma_{22} & \Gamma_{23} \end{pmatrix} \begin{pmatrix} u_n^2 \\ u_n v_n \\ v_n^2 \end{pmatrix}. \tag{7.3.1}$$

We assume that the Jacobian for (7.3.1) is a constant B whose magnitude is less than unity. If the Jacobian is not constant, it generally becomes a constant locally near the renormalization transition; hence the transition is much the same as when the Jacobian is constant.

Equation (7.3.1) can be put in the standard form

$$x_{n+1} + Bx_{n-1} = 2Cx_n + 2x_n^2 \tag{7.3.2}$$

by the following means (Helleman, 1981):

---

[7] For a period k orbit, we consider the map T^k.

(1) Transform to variables u', v' which diagonalize Λ in (7.31). Then Λ' has eigenvalues λ_1, λ_2 (both real since the fixed point is unstable) and Γ is transformed to Γ'.
(2) Impose the condition of constant Jacobian B; this determines all elements of Γ' in terms of $\Gamma'_{11}, \Gamma'_{13}$, and the eigenvalues.
(3) Introduce sum and difference variables

$$s = u'\sqrt{\Gamma'_{11}} + v'\sqrt{\Gamma'_{13}},$$
$$d = u'\sqrt{\Gamma'_{11}} - v'\sqrt{\Gamma'_{13}}.$$

(4) Rescale s by introducing

$$x = \frac{\sqrt{\Gamma'_{11}}(\lambda_1 - \lambda_2)}{2\lambda_1} s,$$

This procedure yields (7.3.2) with the parameter

$$C = \frac{\lambda_1 + \lambda_2}{2}. \tag{7.3.3}$$

The standard form (7.3.2) can be obtained directly for several simple maps. For example, the dissipative generalized standard map

$$u_{n+1} = (1 - \delta)u_n + f(v_n),$$
$$v_{n+1} = v_n + u_{n+1}, \tag{7.3.4}$$

with

$$f(v) = Av + Dv^2$$

yields (7.3.2) with the substitutions $x = Dv/2$, $C = (2 - \delta + A)/2$, and $B = 1 - \delta$. Hénon's dissipative map (7.1.14) can also be immediately cast in standard form.

Quadratic Renormalization. Proceeding as in Section 7.2b, we examine the behavior near the fixed point $x_{10} = 0$ as C decreases. The fixed point is stable for

$$|C| < \frac{1 + B}{2}$$

and is unstable for

$$C < -\frac{1 + B}{2}, \tag{7.3.5}$$

giving birth to a period-2 stable orbit $x_{2\pm}$ as illustrated in Fig. 7.12. The two roots are found by writing

$$x_{2\pm} = a \pm b \tag{7.3.6a}$$

7.3. Bifurcation Phenomena in Two-Dimensional Maps

and iterating (7.3.2) twice, yielding, analogous to (7.2.21b, c),

$$2a = -\frac{1+B}{2} - C, \tag{7.3.6b}$$

$$4b^2 = \left[C + \frac{1+B}{2}\right]\left[C - \frac{3(1+B)}{2}\right]. \tag{7.3.6c}$$

Substituting

$$x = x_{2\pm} + \Delta x$$

into (7.3.2), we obtain, exactly,

$$\Delta x_n + B\Delta x_{n-2} = e\Delta x_{n-1} + 2\Delta x_{n-1}^2, \tag{7.3.7a}$$

$$\Delta x_{n+1} + B\Delta x_{n-1} = d\Delta x_n + 2\Delta x_n^2, \tag{7.3.7b}$$

$$\Delta x_{n+2} + B\Delta x_n = e\Delta x_{n+1} + 2\Delta x_{n+1}^2, \tag{7.3.7c}$$

where d and e are given by (7.2.22) and (7.2.23), repeated here:

$$d = 2C + 4x_{2+}, \tag{7.3.8a}$$

$$e = 2C + 4x_{2-}. \tag{7.3.8b}$$

The even subscripts in Δx give the motion near x_{2+}, and the odd subscripts give the motion near x_{2-}. Multiplying (7.3.7a) by B, (7.3.7b) by e, and adding the resulting two equations to (7.3.7c), we obtain

$$\Delta x_{n+2} + B'\Delta x_{n-2} = 2C'\Delta x_n + 2e\Delta x_n^2 + 2[\Delta x_{n+1}^2 + B\Delta x_{n-1}^2], \tag{7.3.9}$$

where

$$B' = B^2, \tag{7.3.10a}$$

$$C' = \tfrac{1}{2}de - B = -2C^2 + 2(1+B)C + 2B^2 + 3B + 2. \tag{7.3.10b}$$

The square brackets in (7.3.9) also contain a term proportional to Δx_n^2, which can be seen as follows. Writing $r = \Delta x_{n+1}/\Delta x_{n-1}$, the term in brackets becomes

$$\Delta x_{n+1}^2 + B\Delta x_{n-1}^2 = (r^2 + B)\Delta x_{n-1}^2. \tag{7.3.11}$$

However, neglecting quadratic terms, (7.3.7b) yields

$$(r + B)\Delta x_{n-1} \approx d\Delta x_n. \tag{7.3.12}$$

Substituting (7.3.12) in the right-hand side of (7.3.11),

$$\Delta x_{n+1}^2 + B\Delta x_{n-1}^2 \approx \frac{d^2(r^2 + B)}{(r + B)^2} \Delta x_n^2. \tag{7.3.13}$$

We expect $r \approx 1$, i.e., $|\Delta x_{n+1}|$ to be near $|\Delta x_{n-1}|$ because of the quadratic nature of the pitchfork bifurcation. The right-hand side of (7.3.13) has an extremum with respect to r at $r = 1$. Since the value of (7.3.13) is insensitive to small variations around $r = 1$, we set $r = 1$ to obtain

$$\Delta x_{n+1}^2 + B\Delta x_{n-1}^2 \approx \frac{d^2}{1+B}\Delta x_n^2. \tag{7.3.14}$$

Substituting (7.3.14) in (7.3.9) and introducing the rescaling

$$x' = \alpha \Delta x, \tag{7.3.15}$$

we have

$$x_{n+2}' + B'x_{n-2}' = 2C'x_n' + 2x_n'^2, \tag{7.3.16}$$

where

$$\alpha = e + \frac{d^2}{1+B}. \tag{7.3.17}$$

Equation (7.3.16) is identical in form to the original map (7.3.2), indicating that the fixed points of the rescaled mapping undergo bifurcation at parameter values given by (7.3.5) with B' and C' replacing B and C.

The cascade of bifurcations described by the renormalizations (7.3.10) will accumulate at the values $B' = B = B_\infty$ and $C' = C = C_\infty$. For a dissipative map, $|B| < 1$ and, from (7.3.10a), $B_\infty = 0$. Thus all dissipative maps look locally one-dimensional near the accumulation point; i.e., compare (7.3.16) with $B' = 0$ to (7.2.26). It should also be no surprise, putting $B = B_\infty = 0$ in (7.3.10b), that the condition $C' = C = C_\infty$ yields the same value

$$C_\infty = \frac{1 - \sqrt{17}}{4} \approx -0.781 \tag{7.2.29}$$

as was found in the one-dimensional case. The parameter value at the kth bifurcation converges geometrically to C_∞ as

$$C_k - C_\infty \sim A\delta^{-k}$$

with the same convergence factor as in the one-dimensional case $\delta = 1 + \sqrt{17} \simeq 5.12$. The rescaling parameter $\alpha \approx -2.24$ given by (7.3.17) is identical to (7.2.35). These results indicate the universal character of all dissipative systems near the transition to chaotic behavior and have been verified computationally for many one-, two-, or higher-dimensional maps. However, it must be emphasized that the transition is local, i.e., specific to a particular fixed point and its bifurcation sequence. In general, a dissipative system has many fixed points, each of which must undergo period-doubling bifurcations before a global transition to the chaotic behavior of a strange attractor takes place.

7.3b. Area-Preserving Maps

In contrast to these results for dissipative systems, a two-dimensional Hamiltonian map has a more complicated bifurcation cascade. Since $B = 1$

7.3. Bifurcation Phenomena in Two-Dimensional Maps

for area preservation in (7.3.2), the renormalization limit has $B' = B = B_\infty = 1$ (for $B = -1$, we consider the square of the map; see Helleman, 1981, for details). Thus while a cascade of period-doubling bifurcations does occur, a Hamiltonian map retains its two-dimensional character near the accumulation point (see Bountis, 1981, for numerical evidence). As a result, the scaling parameters δ and α, as well as the value C_∞, while universal for all two-dimensional Hamiltonian maps, are *different* from those of dissipative maps. Furthermore, another universal scaling parameter β exists for Hamiltonian maps, which gives the rescaling of the additional dimension that persists in the renormalization limit.

For the Hamiltonian map [$B = 1$ in (7.3.2)]

$$x_{n+1} + x_{n-1} = 2Cx_n + 2x_n^2, \tag{7.3.18}$$

the results (7.3.5)–(7.3.17) are still valid. However, from (7.3.10a),

$$B' = B^2 = B_\infty = 1,$$

and putting $C' = C = C_\infty$ in (7.3.10b) with $B = 1$ then yields the accumulation point

$$-2C_\infty^2 + 4C_\infty + 7 = C_\infty$$

or

$$C_\infty = \frac{3 - \sqrt{65}}{4} \approx -1.2656. \tag{7.3.19}$$

This is close to the calculated numerical value[8] -1.2663 and different from the value for dissipative maps, $C_\infty \approx -0.78$.

Assuming that C_k converges geometrically to C_∞ as

$$C_k - C_\infty \sim A\delta_H^{-k}$$

and substituting this into (7.3.10b) with $B = 1$,

$$C_k = -2C_{k+1}^2 + 4C_{k+1} + 7,$$

we find

$$\delta_H = 4C_\infty + 4 = 1 + \sqrt{65} \approx 9.06. \tag{7.3.20}$$

Numerically, $\delta_H \approx 8.72$ for Hamiltonian maps, different from the value $\delta \approx 4.66$ for dissipative maps.

Finally, the rescaling parameter α in (7.3.15) is given by (7.3.17) with $B = 1$:

$$\alpha = e + \tfrac{1}{2}d^2.$$

---

[8] An exact renormalization theory has been developed for two-dimensional Hamiltonian maps (Eckmann et al., 1981) which gives values of the rescaling parameters in very close agreement with those obtained numerically from the mapping.

Using e and d from (7.3.8) with a and b given by (7.3.6),

$$\alpha \approx -4.096, \qquad (7.3.21)$$

whereas $\alpha \approx -4.018$ by numerical calculation. This value of α is markedly different from the value $\alpha \approx -2.5$ for dissipative systems.

In Fig. 7.25, we illustrate these bifurcations for the map (7.3.18). A sequence of orbits in the x_{n+1} versus x_n phase plane is shown near the fixed point $(0, 0)$ as the parameter C is varied. The parameter A near each plot indicates the magnification of the region around $(0, 0)$ for that plot. Reading from top to bottom and from left to right, the $k = 1, 2, 3,$ and 4 bifurcations are clearly seen in the figure.

A more complete study of two-dimensional Hamiltonian maps reveals the existence of an additional rescaling parameter β (Collet et al., 1981; Greene et al., 1981). Following Greene et al. (1981), we illustrate this by reexpressing (7.3.18) in the form of a quadratic DeVogelaere map

$$\begin{aligned} x_{n+1} &= -y_n + g(x_n), \\ y_{n+1} &= x_n - g(x_{n+1}), \end{aligned} \qquad (7.3.22)$$

where

$$g = Cx + x^2.$$

By writing the map in DeVogelaere form, we bring out the symmetries in the bifurcation phenomena. The period 1 fixed point at $(0, 0)$ in (7.3.22) goes unstable at $C = -1$, leading to the usual bifurcation tree, as shown in Fig. 7.26. Numerically, successive bifurcations are found to converge geometrically with C at a rate $\delta_H \approx 8.72$, roughly in agreement with (7.3.20). The accumulation point $C_\infty \approx -1.2663$ is in good agreement with (7.3.19).

However, we now have an additional dimension in y, which also must have a rescaling parameter. This is seen in Fig. 7.27. The circles are the period 2 orbits that arise when the period 1 fixed point (the square) goes unstable, the triangles are the period 4 orbits, and the dots are the period 8 orbits. One can see evidence of self-similar behavior in this figure. The pattern of orbit positions, centered on the square, repeats itself when centered on the left circle (after flipping the pattern over in the x-direction about the left circle), but on a reduced scale. Numerically, the patterns can be brought into correspondence by magnifying the x-axis by a factor of $\alpha \approx -4.018$, in good agreement with (7.3.21), and the y-axis by a factor of $\beta \approx 16.36$. Actually, these factors are exact only in the renormalization limit. In fact, not only does the pattern of periodic points repeat itself in this limit, but it can be shown from renormalization theory that the map at $C = C_\infty$ is universal and repeats itself on squaring and then rescaling by factors of α (along x) and β (along y) (see Greene et al., 1980), Thus, if T_∞ is the universal map, then

$$T_\infty = S T_\infty^2 S^{-1} \triangleq R T_\infty, \qquad (7.3.23)$$

7.3. Bifurcation Phenomena in Two-Dimensional Maps

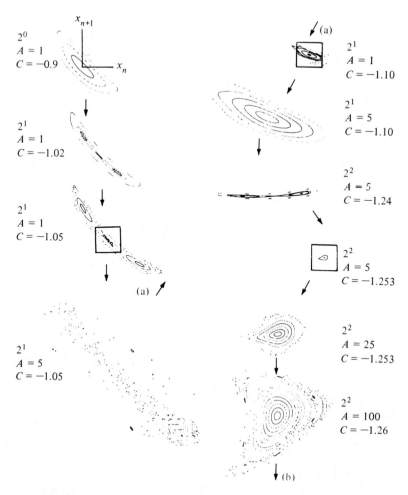

Figure 7.25 (*above and next page*). Surface of section plots of the period-doubling sequence for the Hamiltonian map (7.3.18). Here x_{n+1} is plotted versus x_n for various values of the parameter C. The parameter A indicates the magnification of the plot (after van Zeyts, 1981; see Helleman, 1981).

where R is the period doubling renormalization operator on area-preserving maps and

$$S = \begin{bmatrix} \alpha & 0 \\ 0 & \beta \end{bmatrix}. \quad (7.3.24)$$

We note the interpretation that T_∞ is the fixed point of the operator R in function space.

We have already found expressions for C_∞, δ, and α by an approximate

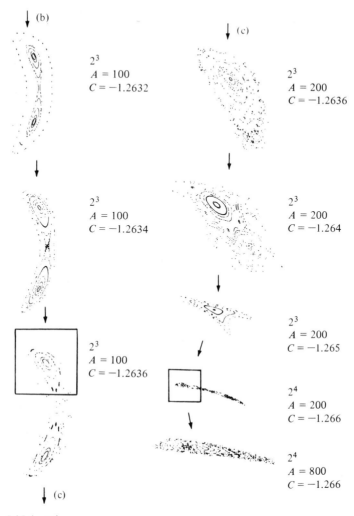

Figure 7.25 (cont.)

quadratic renormalization. An approximate calculation of the second rescaling parameter β can also be obtained. This has been done by MacKay (in Helleman, 1981, Appendix C), and we adapt his point of view here. Using (7.3.22), we write

$$x_{n+1} = -y_n + Cx_n + x_n^2, \qquad (7.3.25a)$$

$$y_{n+1} = x_n - Cx_{n+1} - x_{n+1}^2. \qquad (7.3.25b)$$

The period 1 orbit is at $x_{10} = y_{10} = 0$. The period 2 orbit, which bifurcates

7.3. Bifurcation Phenomena in Two-Dimensional Maps

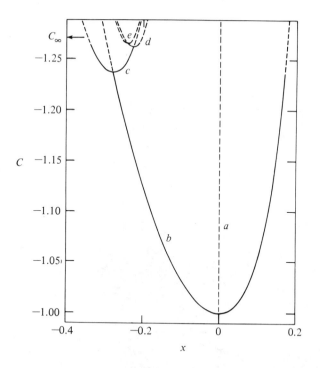

Figure 7.26. Bifurcation tree for the quadratic DeVogelaere map (7.3.22). The solid (dashed) lines show the stable (unstable) fixed points (after Greene et al., 1981).

from it, has $y_{2\pm} = 0$ and $x_{2\pm}$ given by (7.3.6) as before. Introducing

$$x = x_{2\pm} + \Delta x, \quad y = y_{2\pm} + \Delta y,$$

into (7.3.25), we obtain exactly

$$\Delta x_{n+1} = -\Delta y_n + \frac{d}{2} \Delta x_n + \Delta x_n^2, \tag{7.3.26a}$$

$$\Delta y_{n+1} = \Delta x_n - \frac{e}{2} \Delta x_{n+1} - \Delta x_{n+1}^2, \tag{7.3.26b}$$

with d and e given by (7.3.8) as before. Iterating (7.3.26a) once,

$$\Delta x_{n+2} = -\Delta y_{n+1} + \frac{e}{2} \Delta x_{n+1} + \Delta x_{n+1}^2. \tag{7.3.27}$$

Subtracting (7.3.27) from (7.3.26b) yields

$$\Delta x_{n+2} = -\Delta x_n + e\Delta x_{n+1} + 2\Delta x_{n+1}^2. \tag{7.3.28}$$

Using (7.3.26a) to eliminate the terms in Δx_{n+1} in (7.3.28) and keeping only

518   7. Bifurcation Phenomena and Transition to Chaos in Dissipative Systems

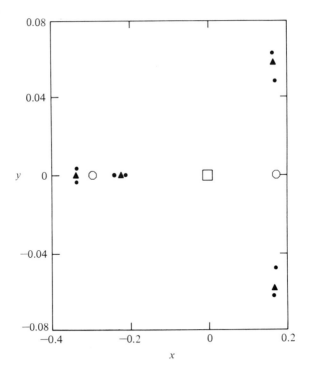

Figure 7.27. Positions of the (unstable) fixed points for $C = C_\infty$ (after Greene et al., 1981).

linear terms in Δx and Δy and the quadratic term in Δx, we obtain

$$\Delta x_{n+2} = -e\Delta y_n + C'\Delta x_n + \alpha \Delta x_n^2, \tag{7.3.29}$$

where

$$C' = \tfrac{1}{2}de - 1 \tag{7.3.30}$$

as in (7.3.10b) with $B = 1$, and

$$\alpha = e + \tfrac{1}{2}d^2 \tag{7.3.31}$$

as in (7.3.17) with $B = 1$. Rescaling x and y by the factors

$$x' = \alpha \Delta x, \tag{7.3.32}$$

$$y' = \alpha e \Delta y, \tag{7.3.33}$$

we obtain

$$x'_{n+2} = -y'_n + C'x'_n + x'^2_n, \tag{7.3.34}$$

which is the same as (7.3.25a) with a renormalized coefficient C'. The same result is easily found for the rescaling of (7.3.25b). From (7.3.33), the second rescaling parameter is then

$$\beta = \alpha e = e^2 + \tfrac{1}{2}ed^2 \approx 16.91, \qquad (7.3.35)$$

which is close to the numerically determined value of $\beta \approx 16.36$.

The essential correctness of this picture has been shown, numerically, for many two-dimensional area-preserving maps. All such maps look identical near the accumulation point of the bifurcation sequence when viewed in the appropriate coordinate system (see Greene et al., 1981).

7.3c. Transition Between Hamiltonian and Dissipative Chaos

The preceding descriptions of period-doubling give rise to a unified picture of bifurcation phenomena in two-dimensional dissipative maps and their two limiting cases of one-dimensional noninvertible maps and two-dimensional area-preserving maps. It also proves possible to use renormalization methods to determine the nature of the chaotic attractors that exist in the regime beyond the period-doubling accumulation point.

Bifurcations in the Dissipative Standard Map. To illustrate these phenomena we introduce the dissipative standard map (Zaslavskii, 1978):

$$I_{n+1} = JI_n + K \sin \theta_n, \qquad (7.3.36a)$$

$$\theta_{n+1} = \theta_n + I_{n+1}, \qquad (7.3.36b)$$

where $J = 1 - \delta$ is the Jacobian of the map, For $J = 1$ we recover the standard map. For $J = 0$ we obtain the one-dimensional map,

$$\theta_{n+1} = \theta_n + K \sin \theta_n, \qquad (7.3.37)$$

a special case of the sine–circle map, treated in Section 7.4, which is noninvertible for $K > 1$.[9] Whereas the standard map (4.1.3) is doubly periodic (in both I and θ), the dissipative standard map ($J \neq 1$) is periodic in θ alone. Hence, the range of I is $-\infty < I < \infty$.

Following Schmidt and Wang (1985), we consider the stability of periodic orbits in the K–J plane, concentrating on the fixed point at $\theta_1 = \pi$, $I_1 = 0$, and its bifurcation sequence. For $J = 1$ and $K < 4$, this corresponds to the stable fixed point for the main island shown in Fig. 4.3. Stability is determined from the characteristic equation for the Jacobian matrix of (7.3.36),

$$\lambda^2 - (1 - K + J)\lambda + J = 0. \qquad (7.3.38)$$

---

[9] For a physical interpretation of the dissipative standard map, we note that I can be eliminated from (7.3.36) to yield $\theta_{n+1} - 2\theta_n + \theta_{n-1} + \delta(\theta_n - \theta_{n-1}) = K \sin \theta_n$, which is a finite-difference form of the dynamics for a driven pendulum with damping coefficient δ.

520 7. Bifurcation Phenomena and Transition to Chaos in Dissipative Systems

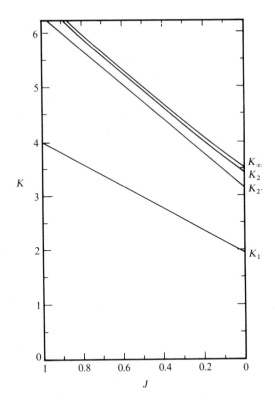

Figure 7.28. Parameter-space diagram of stable periodic orbits. Period 2 bifurcates into 2' (after Schmidt and Wang, 1985).

The orbit is stable for $|\lambda| < 1$, yielding

$$0 < K < 2(1 + J), \tag{7.3.39}$$

with the upper limit shown as K_1 on the K versus J plot in Fig. 7.28. At this limit both eigenvalues are equal to -1, and hence the orbit period-doubles (see Section 4.1b for details). The period 2 orbit and its stability range can be found, as in Section 4.1b for the $J = 1$ case, giving

$$2(1 + J) < K < \pi(1 + J), \tag{7.3.40}$$

with the upper limit shown as K_2' in Fig. 7.28. At this limit in K, both eigenvalues are equal to 1; hence the next bifurcation is a pitchfork (see Section 7.1b) that gives rise to two period 2 orbits rather than a period 4 orbit.[10]

---

[10] This bifurcation, interrupting the period-doubling sequence, is due to a special symmetry in the standard map, and is not generic.

7.3. Bifurcation Phenomena in Two-Dimensional Maps

The range of stability for each of these orbits is found to be

$$\pi(1 + J) < K < [2(1 + J^2) + \pi^2(1 + J)^2]^{1/2}.$$

At the upper limit K_2, $\lambda = -1$ and both period 2 orbits double into period 4 orbits. The subsequent period-doubling bifurcation sequence for each of the two orbits is the same and can be obtained numerically, as described in Section 7.3a. At the accumulation point $K_\infty(J)$, a transition to chaos occurs in the local neighborhood of the (θ_1, I_1) fixed point. $K_\infty(J)$ is plotted in Fig. 7.28, with $K_\infty(1) \simeq 6.6$ and $K_\infty(0) \approx 3.5$. The same type of stability diagram (K versus J) can be constructed for all periodic orbits, not just the orbit $(\pi, 0)$. It appears that all stable periodic orbits of the Hamiltonian map connect continuously all the way to the one-dimensional map.

Returning to the primary fixed point at $(\theta, I) = (\pi, 0)$, we ask what happens for $K > K_\infty$ for the general case $0 < J < 1$. We find that the general structure of the chaos for the one-dimensional case $J = 0$ described in Section 7.2c survives for K near but slightly exceeding K_∞. That is, a reverse bifurcation sequence of chaotic bands is present similar to that shown in Fig. 7.23, where $K_k^*(J)$ (analogous to C_k^* in the figure) gives the transition from 2^k to 2^{k-1} bands of chaos as K increases. For $J \neq 0$, however, these 2^k chaotic bands become 2^k-piece strange attractors, each piece having the typical structure of a strange attractor similar to that shown in Fig. 7.6 for the Hénon map.

However, the curves $K_k^*(J)$, giving the boundaries between 2^k and 2^{k-1}-piece attractors do not span the range $0 \leq J \leq 1$, as do the boundary curves for the stable periodic orbits. The typical structure is shown in Fig. 7.29, where, for ease of viewing, $K_k^* - K_\infty$ is plotted logarithmically versus J. The 1-piece boundary K_1^* spans the entire range of J, but the 2, 4, 8, ... piece boundaries run from the $J = 0$ line and terminate on the K_1^* boundary at a finite value $J_k < 1$.[11] Hence, the sequence of strange attractors terminates at some finite 2^k and becomes a 1-piece attractor. For example, Fig. 7.29 shows that for $J \approx 0.85$, the sequence will be

$$\cdots \to \text{32-piece} \to \text{16-piece} \to \text{8-piece} \to \text{1-piece}$$

as K increases from K_∞. The 4- and 2-piece attractors are not present.

To complete the picture, we can ask what happens if K increases much beyond the 1-piece boundary K_1^*. We emphasize that the preceding picture is local, near a given fixed point of the mapping. For the dissipative standard map (7.3.36), choosing a fixed point at a different value of I would result in a bifurcation sequence with different values of K_k. Of course, universality requires that the same bifurcation ratios hold at the accumulation points, for both the forward and inverse bifurcation sequences. However, the endpoints of the inverse bifurcation sequences, resulting in a 1-piece attractor,

---

[11] There are two different boundaries K_2^* and $K_2'^{2*}$ shown for $k = 2$, arising from a symmetry in the map, just as for the stable period 2 orbits that bifurcate to two period 2 orbits in Fig. 7.28.

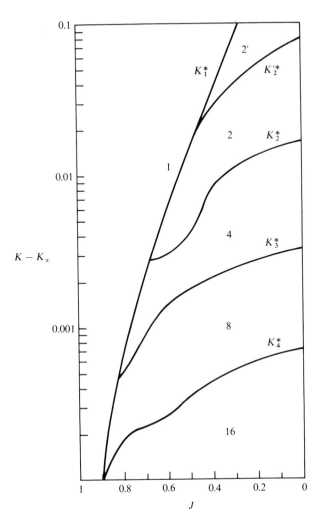

Figure 7.29. $(K - K_\infty)$ is plotted versus J for multiple-piece strange attractors (after Schmidt and Wang, 1985).

also will occur at their individual values of K_1^*. For a general map, there is another boundary (call it K_0^*) lying above K_1^*. Beyond this boundary, the local 1-piece attractor becomes connected to another attractor (either regular or chaotic) that exists outside of the local region; i.e., the local strange attractor ceases to exist as an independent structure.

The mechanism for the transition from a 2^k to a 2^{k-1}-piece strange attractor can be understood in terms of the stable and unstable manifolds of the unstable periodic orbits. Consider elements of the 2^k and 2^{k-1} orbits with their two manifolds, as shown in Fig. 7.30. It can be shown that the

7.3. Bifurcation Phenomena in Two-Dimensional Maps

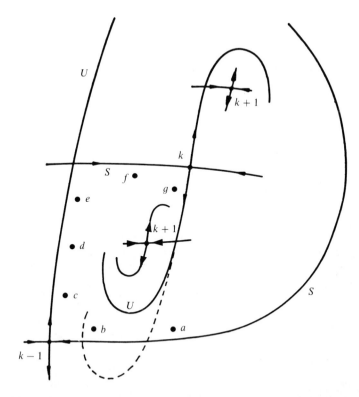

Figure 7.30. Schematic representation of stable (S) and unstable (U) manifolds of period 2^{k-1}, 2^k, and 2^{k+1}. By increasing K the unstable manifold of period 2^k intersects the stable manifold of period 2^{k-1} as indicated by the dashed line (after Chen et al., 1987).

stable manifold of the 2^k orbit always intersects the unstable manifold of the 2^{k-1} orbit, but the converse is not necessarily true. A point near the stable manifold of the 2^{k-1} orbit will map successively closer to the orbit, depart along the unstable manifold, and then can be carried to the stable and then unstable manifold of the 2^k orbit (sequence of points a–g). The trajectory has no way of returning to the manifold of the 2^{k-1} orbit. Hence we have a 2^k-piece attractor. However, as K is increased, a heteroclinic intersection will take place (dashed curve), opening up a path for the trajectory to pass from the unstable manifold of the 2^k orbit to the stable manifold of the 2^{k-1} orbit. This leads to a 2^{k-1}-piece attractor. This sudden transition from a 2^k to a 2^{k-1}-piece attractor as a parameter is varied is an example of an *interior crisis*. We discuss the general phenomenon of crises in Section 7.6.

The structure of a 2^k-piece strange attractor and the transitions to a 2^{k-1}-piece or a 1-piece attractor are shown symbolically in Fig. 7.31. The arrows indicate flows between stable and unstable manifolds of period 2^k orbits, $k = 0, 1, 2, \ldots$ Horizontal arrows show the flow from stable (S) to

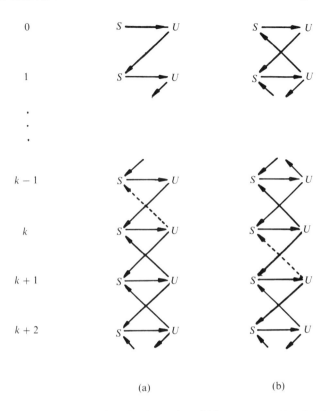

Figure 7.31. Diagram explaining the dynamics within a strange attractor. The arrows indicate flows between stable (S) and unstable (U) manifolds of periodic points of period 2^k, $k = 0, 1, 2, \ldots$. Horizontal arrows show an omnipresent flow from the stable manifold to the unstable one of a periodic point. Oblique arrows are consequences of heteroclinic intersections. (a) For small Jacobians J upward links are switched on one after the other when K is increased. (b) Slightly below the critical J_k only one upward link is missing. The switch closes if J_k is increased slightly (after Chen et al., 1987).

unstable (U) manifold for each periodic orbit. Diagonal arrows give the heteroclinic intersections. In (a), the 2^k-piece attractor consists of all unstable periodic orbits having periods 2^k or greater. The basin for this attractor is the entire set of periodic orbits shown. When the heteroclinic intersection (dashed arrow) occurs, then a 2^{k-1}-piece attractor is formed. A further increase in K leads to the next heteroclinic intersection between U_{k-1} and S_{k-2}, reducing the attractor to one with 2^{k-2} pieces. For J close to unity, the heteroclinic intersections do not proceed monotonically from the greatest number to the smallest number of pieces. This is shown schematically in Fig. 7.31b for which a 2^{k+1}-piece attractor is transformed directly into a 1-piece attractor when the heteroclinic intersection between U_{k+1} and S_k is formed.

7.4. Circle Maps

The quantitative values of $K(J)$ for which these transitions occur were given in Fig. 7.29.

Renormalization About T_∞. One might expect that the structure of many-piece strange attractors and their reverse bifurcation sequence is universal near $K = K_\infty$ and $J = 1$, which is the fixed point of the universal Hamiltonian map T_∞ given in (7.3.23). Renormalization theory shows that this is indeed the case (Chen *et al.*, 1987). This can be seen by linearizing about the fixed point T_∞ in function space defined in (7.3.23). Writing $T = T_\infty + \epsilon \tilde{T}$, then application of the period doubling operator followed by linearization yields

$$R(T_\infty + \epsilon \tilde{T}) = T_\infty + \lambda \epsilon \tilde{T}, \tag{7.3.41}$$

where the $\tilde{T}$'s are the eigenvectors in function space and the λ's are the eigenvalues. There are an infinity of eigenvalues and eigenvectors that can be ordered as $\lambda \simeq 8.722$, $\lambda = 2$, $\lambda = 1$, $\lambda = 0.117$, etc. Only two eigenvalues are unstable. The largest eigenvalue $\lambda = \delta_H \approx 8.72$ gives the accumulation rate for Hamiltonian period-doubling bifurcations near T_∞, such that the corresponding eigenvector $\tilde{T}$ is Hamiltonian. The second largest eigenvalue $\lambda = 2$ gives the accumulation rate for dissipative perturbations $\tilde{T}$, corresponding to perturbations of the Jacobian of the map from unity. Hence we have the renormalization scalings,

$$K_k^* - K_\infty \propto \delta_H^{-k} \tag{7.3.42}$$

and

$$1 - J_k \propto 2^{-k}. \tag{7.3.43}$$

Chen *et al.* (1987) have extended this local renormalization analysis into the nonlinear region to show that the values of the Jacobian J_k at which $K_k^*(J)$ terminates on the K_1^* boundary are universal for all maps in the large k limit. The scaling of J_k was determined to be

$$J_k = J_{k+1}^2 \tag{7.3.44}$$

with, for example, $J_3 \approx 0.82$ found numerically, a universal constant, for four different dissipative maps. According to renormalization theory, the strange attractors themselves are universal; i.e., if one chooses a K and J such that the map produces a 2^k-piece attractor, then iteration of the map and rescaling as in (7.3.23) produces a 2^{k-1}-piece strange attractor that is found at $J' = J^2$ and $K' - K_\infty = \delta_H(K - K_\infty)$.

7.4. Circle Maps

Two-parameter maps of a periodic variable onto itself, called *circle maps*, are very simple yet can display a variety of important phenomena exhibited by more complicated systems. In the parameter ranges for which these maps are invertible, we observe quasiperiodic and periodic behavior, including mode locking. For noninvertible maps, period doubling and chaotic behavior is found, similar to that described in the previous sections.

7.4a. Basic Formalism

Circle mappings are usually defined on the unit interval, $0 \leq x \leq 1$,

$$x_{n+1} = h(x_n) = x_n + \Omega_0 + g(x_n), \qquad (7.4.1)$$

where g is periodic modulo 1; i.e., $g(x + 1) = g(x)$ and Ω_0 is the unperturbed ($g \equiv 0$) winding number, which can be limited to the interval $0 \leq \Omega_0 \leq 1$. A useful paradigm is the sine-circle map,

$$x_{n+1} = x_n + \Omega_0 - (K/2\pi) \sin 2\pi x_n, \qquad (7.4.2)$$

which can be derived from a dissipative standard map,

$$\begin{aligned} I_{n+1} &= (1 - \delta)I_n + K \sin \theta_n, \\ \theta_{n+1} &= \theta_n + 2\pi\Omega_0 + I_{n+1} \quad (\text{mod } 2\pi). \end{aligned} \qquad (7.4.3)$$

The Jacobian of (7.4.3) is $1 - \delta$, which measures the rate of area contraction. In the limit of $\delta = 1$, the phase space collapses onto a single dimension, yielding (7.4.2) with $\theta = 2\pi x + \pi$.

The sine-circle map is invertible for $K < 1$. At $K = 1$ the map develops a cubic singularity at $x = 0$, and for $K > 1$ the map has a local maximum and is noninvertible. The $K > 1$ regime exhibits chaos as described in Section 7.2c.

For $K \leq 1$, an important quantity describing the dynamics is the winding number

$$\Omega(\Omega_0, K) = \lim_{n \to \infty} \frac{x_n - x_0}{n}. \qquad (7.4.4)$$

If the map is smooth and invertible, then Ω is well defined and independent of x_0. For $K = 0$, the motion is a rotation with winding number Ω_0. For $\Omega = P/Q$ rational with P, Q relatively prime, the orbit is periodic; for Ω irrational an ergodic orbit covers the circle. Since the rationals have measure zero the typical orbit is ergodic. For $0 < K \leq 1$ the map has a unique attractor for each Ω_0 that is either a periodic orbit, $\Omega = P/Q$ rational, or ergodic, $\Omega =$ irrational. For $0 < K \leq 1$, Ω is monotonic with Ω_0 (mod 1) and forms what is known as a *devil's staircase*. That is, around each $\Omega_0 = P/Q$, there is a range of Ω_0 for which $\Omega = P/Q$. These fixed values of Ω are connected by irrational values for the intervening values of Ω_0. The flat regions of Ω_0 for which Ω is rational are called *mode locking* regions. Since there are an infinite number of rationals interspersed between the rationals, this gives interspersed flat regions on all scales, leading to the colorful

7.4. Circle Maps

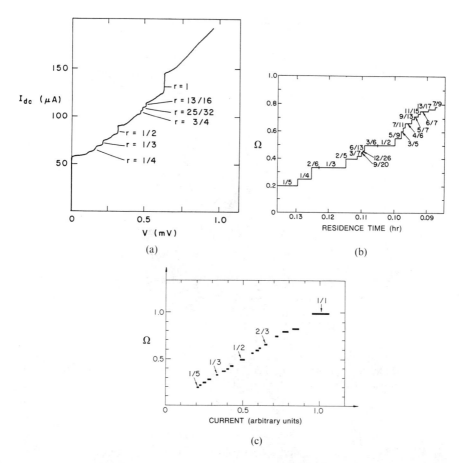

Figure 7.32. Dynamical-system staircases. The fractions are the ratios of competing frequencies that have locked to form the plateaus. (a) Current-voltage characteristics of a niobium Josephson junction driven by a 295-GHz microwave current. The voltage is a measure of the frequency. (b) Frequency of oscillations in a complex chemical reaction as a function of the reaction rate. (c) Frequency of voltage oscillations in a superionic conductor, barium sodium niobate, driven by a direct current (after Bak, 1986).

designation as a devil's staircase. Some experimental staircases are shown in Fig. 7.32. We note the direct analogy with the island structures that we have already described for Hamiltonian maps. The flat regions are analogous to the island with average fixed rotation number $\Omega = P/Q$, while the intervening regions correspond to the KAM surfaces. Here, however, the phenomenon has been collapsed onto a single dimension. This allows a two-dimensional representation of the regions of rational winding number as a function of K and Ω_0, as shown in Fig. 7.33. The regions of $\Omega = P/Q$ coalesce to the values

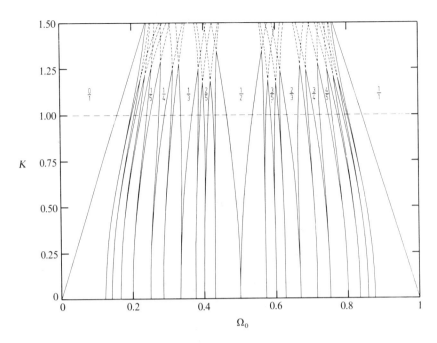

Figure 7.33. Winding numbers as a function of coupling K and unperturbed winding number Ω for the sine–circle map. The Arnold tongues are labeled with their rational values (after Bak et al., 1988).

of $\Omega_0 = P/K$ at $K = 0$ and widen with increasing K. The structure shown in Fig. 7.33 has been studied by Arnold (1983) and the regions of $\Omega = P/Q$ are known as Arnold tongues.

For $K \to 0$ at a fixed Q, the width of the tongues scales as $\Delta\Omega_0 \propto K^Q$, while for $Q \to \infty$ at fixed K, $\Delta\Omega_0 \propto Q^{-3}$ (Ecke et al., 1988). Although there is a dense set of rationals, for $0 < K < 1$, both the rational segments and the irrationals between them have finite measure on the Ω_0 interval and have dimension $D_0 = 1$. Figure 7.34 shows the measure μ_0 of the quasiperiodic (ergodic) set as a function of K, which follows the empirical relation $\mu_0(K) = (1 - K)^{0.314}$ (Jensen et al., 1984; Ecke et al., 1988). The quasiperiodic set is a so-called *fat fractal* (Umberger and Farmer, 1985) due to the rational segments embedded within the set; i.e., although the fractal dimension $D_0 = 1$, the coarse-grained measure $\mu(\epsilon)$, where ϵ is the size of the coarse graining, scales as

$$\mu(\epsilon) = \mu_0 + A\epsilon^\beta,$$

where $\beta \approx 0.68 \pm 0.05$ (Ecke et al., 1988). This is analogous to the stochastic phase space regions of a Hamiltonian map, which are also fat fractals, that arise from the fractal structure of the embedded islands.

7.4. Circle Maps

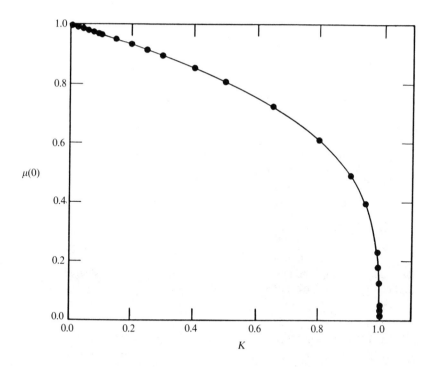

Figure 7.34. The measure of ergodic orbits $\mu(0)$ versus the nonlinearity parameter K. The solid curve is a plot of $(1 - K)^{0.314}$ (after Ecke et al., 1988).

7.4b. Critical Behavior and Universality

The sine–circle map becomes noninvertible at the critical value $K = 1$. At this point, the measure of the quasiperiodic set vanishes, and a Cantor set is formed. Jensen et al. (1984) determined the fractal dimension (capacity) of this set by finding numerically the total width $S(\epsilon)$ of the mode locked intervals $\Delta\Omega_0(P/Q)$ having sizes greater than ϵ. Thus, the average number of remaining holes of smaller size is

$$M(\epsilon) = (1 - S(\epsilon))/\epsilon. \tag{7.4.5}$$

Then plotting $\log M(\epsilon)$ against $\log 1/\epsilon$ and using (7.1.30), they obtained $D_0 = 0.8700$. Thus, at $K = 1$, the mode-locked regions have full measure on the line, and the devil's staircase is complete, leaving a fractal remnant of quasiperiodic points. Actually, this Cantor set is a multifractal and can be further characterized by an infinite number of generalized dimensions and a corresponding scaling index spectrum (see Section 8.3).

At criticality, the structure of the devil's staircase is *self-similar* and *universal*. By self-similar we mean that a magnification of any part of the devil's staircase looks just like the entire staircase. This is illustrated in

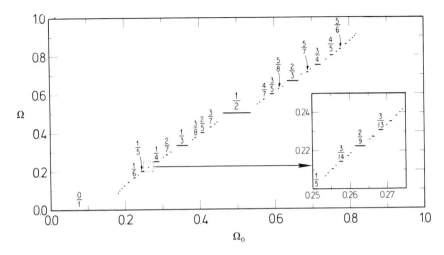

Figure 7.35. Winding number at the transition to chaos. The self-similarity under magnification is illustrated (after Bak et al., 1988).

Fig. 7.35. We have already met this self-similar property in connection with higher-order islands of the standard mapping and with the leaves of the Hénon attractor. By universal we mean that at criticality D_0 is the same for any periodic function on the interval in (7.4.1).

The transition from quasiperiodic motion to chaos was examined numerically by Shenker (1982), who investigated the scaling properties of the rational iterates of the golden mean winding number $\Omega_{GM} = (\sqrt{5} - 1)/2$. As we recall from Section 4.4a, this corresponds to the continued fraction expansion

$$\Omega_{GM} = \cfrac{1}{1 + \cfrac{1}{1 + \cdots}},$$

such that the iterates are ratios of the Fibonacci numbers

$$\Omega_i = F_i/F_{i+1}$$

defined by $F_{i+1} = F_i + F_{i-1}$ with $F_0 = 0$ and $F_1 = 1$. The convergence rate of the sequence is

$$\lim_{i \to \infty} \frac{\Omega_{i+1} - \Omega_i}{\Omega_i - \Omega_{i-1}} = -\Omega_{GM}^2. \tag{7.4.6}$$

For a given K, each value of Ω_i corresponds to a range of values of the unperturbed winding number Ω_{0i} within an Arnold tongue. To determine a

7.4. Circle Maps

particular $\Omega_{0i}(K)$, we choose the orbit that passes through $x_{0i} = 0$, defined by

$$h^{F_i}(x_{0i} = 0, \Omega_{0i}, K) = F_{i-1},$$

where from (7.4.1), h^{F_i} is the F_ith iterate of the map. This is called the *superstable orbit* because it passes through an extremum of $h(x)$ in the limit $K \to 1$; i.e., $dh/dx = 0$ at $x = 0$ and $K = 1$. Recall that an orbit is stable provided

$$|\lambda| = \left| \prod_{i=1}^{F_i} \left(\frac{\partial h}{\partial x} \right)_{x_{0i}} \right| < 1,$$

with the orbit for $\lambda = 0$ having maximum stability. If we define an equivalent set of ratios of the Ω_{0i}

$$\delta_{GM}(K) = \frac{\Omega_{0,i+1}(K) - \Omega_{0i}(K)}{\Omega_{0i}(K) - \Omega_{0,i-1}(K)} \quad (7.4.7)$$

then, as expected for $0 < K < 1$,

$$\lim_{i \to \infty} \delta_{GM}(K) = -\Omega_{GM}^{-2}. \quad (7.4.8)$$

However, in the limit $K = 1$, Shenker found numerically that

$$\lim_{\substack{i \to \infty \\ K \to 1}} \delta_{GM}(K) = -\Omega_{GM}^{-2.164\cdots} \equiv \delta_{GM} \approx 2.833. \quad (7.4.9)$$

This more rapid decay corresponds to the underlying fractal structure of the quasiperiodic orbits at criticality. Shenker also found that the point on the attractor nearest to $x = 0$ has a scaling exponent α_{GM}; i.e.,

$$\lim_{\substack{i \to \infty \\ K \to 1}} h^{F_i - 1}(0, \Omega_{0,i+1}, K)/h^{F_i}(0, \Omega_{0i}, K) = \alpha_{GM} \approx -1.28. \quad (7.4.10)$$

Furthermore, Shenker found the values of δ_{GM} and α_{GM} to be universal, in that they applied to any circle mapping with a cubic inflection point at criticality that had winding number Ω_{GM}. Figure 7.36 shows the critical attractor for the sine circle map at the golden mean winding number. At criticality, the points eventually cover the circle, and hence we find $D_0 = 1$. Nevertheless, this attractor is actually a multifractal. We describe its properties in detail in Section 8.3.

The scaling coefficients δ_{GM} and α_{GM} have been determined from a renormalization group analysis by Feigenbaum et al. (1982) and Rand et al. (1982). Similar universal scalings are found for the other irrational winding numbers with periodic continued fraction expansions, with limiting values of δ and α at criticality differing slightly from the golden mean scaling numbers (Shraiman, 1984). For the more typical irrationals with nonperiodic continued fraction expansions, there is no simple scaling behavior, but an average behavior can be defined (Jensen et al., 1984; Farmer and Satija, 1985).

532   7. Bifurcation Phenomena and Transition to Chaos in Dissipative Systems

Figure 7.36. The orbit with golden mean winding number of the sine–circle map (after Bak et al., 1988).

7.4c. Supercritical Behavior

A complete picture of the dynamics in the noninvertible regime ($K > 1$ for the sine–circle map) has yet to be developed. Beyond criticality, the tongues overlap, with high-Q (thin) tongues overlapping near $K = 1$ and low-Q (thick) tongues overlapping at larger values of K. In this regime, the motion can be chaotic. The motion can also remain mode locked, although hysteresis effects appear. In this case, the unique rotation numbers that exist for $K < 1$ are replaced by rotation intervals, such that any rotation number within the interval can occur, depending on the initial value of x.

Figure 7.37 shows the structure of a typical tongue beyond criticality (Bak et al., 1988). A unique mode number persists in the central part of the tongue beyond $K = 1$. A rotation interval with hysteresis is found near the sides. A cascade of bifurcations develops within the tongue for K sufficiently large. These include not only the period doubling bifurcations shown, but also $\Omega_0 = 3P/3Q$, $5P/5Q$, etc., and their period doubling bifurcations (Glass and Perez, 1982). These sequences terminate at a "mushroom cap," beyond which one obtains a region of mainly chaotic motion. For Q small, the mushroom sticks a considerable distance above the critical surface $K = 1$; for Q large, the mushroom is small and sticks up only a little bit (MacKay and Tresser, 1986). For the limiting tongue at the golden mean winding number, we obtain chaos for $K > 1$. The Liapunov exponent for this case is found to scale as

7.4. Circle Maps

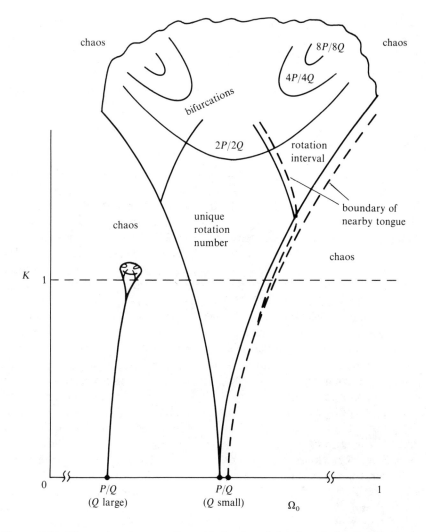

Figure 7.37. Illustrating the complicated behavior of the sine–circle map above the invertibility boundary at $K = 1$.

(Kaneko, 1984; Jensen and Procaccia, 1985)

$$\lambda_{GM} = (K - 1)^{\gamma_{GM}}, \tag{7.4.11}$$

where

$$\gamma_{GM} = \frac{\ln \Omega_{GM}}{\ln \alpha_{GM}^2} \simeq 0.948. \tag{7.4.12}$$

The preceding scenario for the transition to chaos is analogous to that in Hamiltonian systems. The resonance overlap condition $K = 1$ for the sine–

circle map corresponds to the transition to global stochasticity at $K \approx 0.9716$ for the standard map, where the last KAM curve at the golden mean winding number is broken. The periodic motion that persists for $K > 1$ within low-order tongues for the sine–circle map corresponds to the persistence of low-order islands for $K > 0.9716$ in the standard map.

7.4d. Physical Examples

Invertible phenomena such as mode locking are common in physical systems. Strongly damped periodically driven systems, such as the damped driven pendulum, have this behavior. If the damping is strong, the derivative of the phase becomes locked to the phase, and the dynamics acts as if it were one dimensional. A stroboscopic view of the behavior at fixed time intervals then generates a circle mapping. Examples of this behavior can be found in a Josephson junction (V. N. Belykh et al., 1977), the Belusov–Zabotinsky chemical reaction (Maselko and Swinney, 1985), and the oscillations in an ionic conductor (Martin and Martienssen, 1986). In these examples the underlying equations are only known approximately. Thus complicated physical systems have dynamics with similar qualitative properties. Many other examples of this type of behavior can be found (see Bak, 1986).

Although invertible phenomena such as Arnold tongue structures are common in dynamical systems, the noninvertible dynamics of the circle map in the supercritical region are generally found only as an approximation. This is because most physical systems have reversible dynamics. As an exception, we discuss a practical example (Bernstein and Lieberman, 1990) that has the form of a circle map, namely, a first-order digital phase locked loop. Since the primary purpose of a phase locked loop is synchronization of signals, the phase (mode) locking is itself the desired result. Phase locked loops, either first or higher order, play an important part in communication systems, and knowledge of the dynamical behavior of the underlying circle map allows better design procedures. Second-order digital loops correspond to two-dimensional dissipative mappings and have interesting chaotic transients (see Section 8.1).

A block diagram for a first-order nonuniformly sampling digital phase locked loop is shown in Fig. 7.38. From the input signal $s(t)$, the sample and hold (SH) block produces $s_n = s(t_n)$, where t_n is the sampling time. A variable frequency oscillator (VFO) sets the period of the oscillator according to

$$T_{n+1} = t_{n+1} - t_n = g(s_n), \qquad (7.4.13)$$

which is used to clock the sample and hold. We consider an ideal sample and hold, a VFO with a characteristic frequency given by $f = f_0 + bs_n$, where f_0 is the frequency without applied signal and a sinusoidal signal $s(t) = A_m \sin(\omega_1 t + \theta_0)$ is applied. The equation for the period T_{n+1} in terms

7.4. Circle Maps

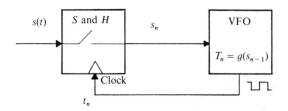

Figure 7.38. Block diagram of a nonuniformly sampling first-order digital phase locked loop (DPLL) (after Bernstein and Lieberman, 1990).

of the nth sample of the input s_n is

$$T_{n+1} = \frac{1}{f_0 + bs_n} \qquad (7.4.14)$$

and the corresponding circle map equation is

$$\phi_{n+1} = \phi_n + \frac{\omega_1}{f_0 + bA_m \sin \phi_n}. \qquad (7.4.15)$$

Letting $k = \omega_1/f_0$ and $A = -bA_m/f_0$ and substituting in (7.4.15) we obtain the circle map

$$\phi_{n+1} = \phi_n + \frac{k}{1 - A \sin \phi_n}. \qquad (7.4.16)$$

For A small, the denominator can be expanded to give the sine-circle map (mod 2π) (7.4.2) with k and Ak corresponding to $2\pi\Omega_0$ and K, respectively.

For an experimental realization of the loop, shown in Fig. 7.38, the greatest departure from the ideal comes from the finite time needed to sample a signal to a given accuracy. For the experimental results, the sampling errors were less than 1%. In Fig. 7.39 the experimental and numerical bifurcation diagrams are compared. In Fig. 7.39a, successive iterates of $s_n = A_m \sin \phi_n$ from the experiment are plotted versus k, holding $A = 0.28$. In Fig. 7.39b, we show the corresponding bifurcation diagram obtained from (7.4.16) over the same parameter range. There is good agreement between theory and experiment. The critical k is $k \approx 1/0.28$; the figure shows some of the interesting dynamics that the loop can generate, such as the period doubling and intermittency routes to chaos.

Figure 7.40 shows the regions of chaos for the map (7.4.16) in the $A - k$ parameter space, determined by calculating the Liapunov exponent on a grid within this parameter space. The regions of significantly positive ($\lambda \geq 0.001$) exponent are shown as black squares. The large unoccupied region in the lower left is bounded by the exact invertibility edge (critical curve), given approximately by $A = 1/k$ for $A \ll 1$. We observe the lack of

536   7. Bifurcation Phenomena and Transition to Chaos in Dissipative Systems

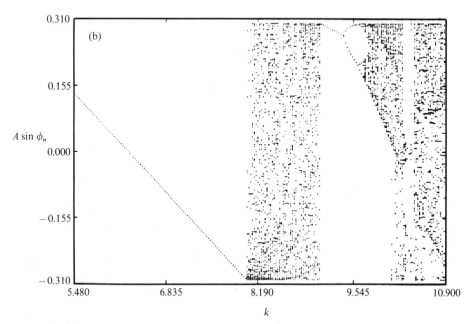

Figure 7.39. Bifurcation diagram of the DPLL with $A = 0.28$; (a) experimental; (b) calculated (after Bernstein and Lieberman, 1990).

7.5. Intermittency

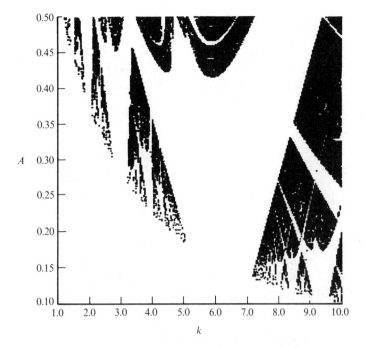

Figure 7.40. Plot of significantly positive Liapunov exponents σ for the map describing the experimental DPLL. Black squares indicate regions of parameter space where $\sigma \geq 0.001$; 20,000 iterations were used to calculate σ (after Bernstein and Lieberman, 1990).

chaos below criticality and the nonchaotic structure within the low-order tongues that penetrate deeply above the critical curve. We also note small regions of regular motion embedded within the chaos far above the critical curve. For the experiment, it is much more time consuming to do reconstruction from the data series. Consequently, only a few parameter values were analyzed for the experiment and checked against the numerical results, with reasonable agreement between theory and experiment. Many unresolved questions remain regarding supercritical behavior in such maps.

7.5. Intermittency

We have seen, in Section 7.2, that by changing a parameter of a noninvertible map, there is a route to chaos through a period doubling sequence. Another route to chaotic dynamics is through intermittency, a phenomenon in which regular motion of varying lengths is interspersed with bursts of chaotic motion (Pomeau and Manneville, 1980). The process has been observed in a number of experiments, as a parameter of the experiment is varied. It is also of particular interest as a mechanism for generating $1/f$ noise, which is sometimes observed in driven systems, e.g., current fluctuations in a resistor or semiconductor with voltage applied. It is found that self-similarity of the regular portions of the motion, as criticality is approached, can have a Fourier spectrum exhibiting a $1/f$ singularity (e.g., see Procaccia and Schuster, 1983; Ben Misrachi et al., 1985). Intermittency has also played an important role in the study of nonlinear systems because the renormalization-group analysis near criticality has an analytic solution that can be compared with other approximation solutions and with numerical results (Hu and Rudnick, 1982).

In the following subsections we first consider the basic mechanism that gives rise to intermittency, exhibit an approximate solution for the behavior, and classify the types of bifurcations that give rise to intermittency. We then develop the renormalization solution and compare the result to numerical observations. Finally, we briefly consider the relationship to $1/f$ noise and comparisons with experimental results.

7.5a. Intermittency Mechanisms

Consider the mapping

$$y_{n+1} = y_n + ay_n^2 + \epsilon. \tag{7.5.1}$$

The trajectory (dots) through the (y_{n+1}, y_n) space is then shown, for $\epsilon > 0$, in Fig. 7.41a. There is no fixed point, but for ϵ small the trajectory remains in the neighborhood of the value of $y_n = 0$ for a number of iterations, producing nearly regular motion. As ϵ is reduced, a tangency occurs at $\epsilon = 0$, and for $\epsilon < 0$ a pair of stable and unstable fixed points appears, as shown in Fig. 7.41b. This behavior is that of a tangent bifurcation, as shown in Fig. 7.2a, and the intermittency resulting from it was classified by Pomeau and Manneville (1980) as Type I. It is found in many mapping embodiments of physical systems. For example, the one-dimensional quadratic map

$$x_{n+1} = f(x_n) = 2Cx_n + 2x_n^2 \tag{7.2.4}$$

has a critical value $C_0^{(3)} = (1 - \sqrt{8})/2$ where a period 3 cycle $x_{3i} = 0.4296$, $-0.4164, 1.1081, i = 0, 1, 2$, is born at a tangent bifurcation, as described in Section 7.2b (see Fig. 7.16). Here $f^{(3)}(x_{3i}) = x_{3i}$ and $f'(x_{3i}) = 1$. Expanding $f^{(3)}(x)$ about $x - x_{3i}$ to quadratic order and about $C - C_0^{(3)}$ to linear order, we obtain (7.5.1) with $y = x - x_{3i}$, $a = 6$ and $\epsilon = 8x_{30}x_{31}x_{32}(C - C_0^{(3)}) \approx -1.5858(C - C_0^{(3)})$. The same type of behavior was found, numerically, by Pomeau and Manneville (1980) to be embedded in the Lorenz model (1.5.2), for the appropriate parameters, as shown in Fig. 7.42.

7.5. Intermittency

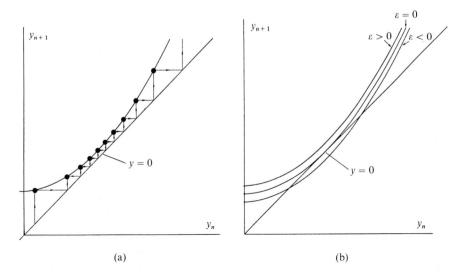

Figure 7.41. (a) Slow passage through the channel region; (b) tangent bifurcation near $\epsilon = 0$ (after Pomeau and Manneville, 1980).

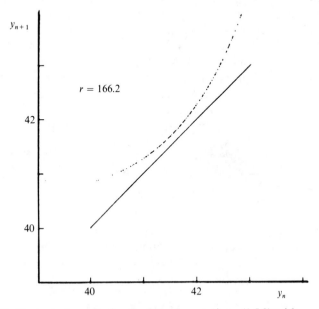

Figure 7.42. Numerical results for the Lorenz equations (1.5.2) with $\sigma = 10$, $b = \frac{8}{3}$, as usually chosen and $r = 166.2$ ($r_c = 166.06$) (after Pomeau and Manneville, 1980).

The scaling with ϵ of the average length of time τ spent in the neighborhood of the near tangency can be readily found. In this neighborhood the change in y per iteration is small, so that the difference equation can be replaced by a differential equation

$$\frac{dy}{dt} = ay^2 + \epsilon, \tag{7.5.2}$$

where t is measured in units of the mapping period. Integrating (7.5.2), we obtain

$$t = \frac{1}{(a\epsilon)^{1/2}} \left[\tan^{-1}\left(\frac{y_{\text{out}}}{(\epsilon/a)^{1/2}}\right) - \tan^{-1}\left(\frac{y_{\text{in}}}{(\epsilon/a)^{1/2}}\right) \right], \tag{7.5.3}$$

where y_{in} and y_{out} are the (nominal) positions at which the trajectory enters and exits the regular region. If $y_{\text{out}}/(\epsilon/a)^{1/2}$ is large and y_{in} is equally probable for all $|y_{\text{in}}| < y_{\text{out}}$, then the first term yields $\pi/2$ and the second term averages to zero to obtain

$$\tau = \frac{\pi}{2} \frac{1}{(a\epsilon)^{1/2}}. \tag{7.5.4}$$

This result was checked numerically by Hirsch et al. (1981) and found to hold over the range $10^{-6} < \epsilon < 10^{-3}$.

Intermittency has been classified by Pomeau and Manneville (1980) into three types, depending on the type of bifurcation and its eigenvalue transition. The return map dynamics, bifurcations and eigenvalues are shown in Fig. 7.43. All have tangencies at the critical value but differ in the dynamical properties of the flows. For one-dimensional noninvertible maps, only Types I and III intermittency exist. Type I intermittency is born at a tangent bifurcation and is characterized, for $\epsilon > 0$, by a monotonically increasing signal $y_{n+1} > y_n$ in the regular region.

Type II intermittency is born at an inverse pitchfork bifurcation, as illustrated in Fig. 7.2d. The most general mapping model is

$$y_{n+1} = -(1 + \epsilon)y_n + ay_n^2 - by_n^3, \tag{7.5.5}$$

where a and b are constants such that $a^2 - b < 0$ and ϵ is the parameter. This bifurcation is nongeneric because the constant term ($\propto \epsilon$), usually obtained by expansion of a general map about the fixed point $y = 0$, is zero. For $\epsilon > 0$, the laminar phase of the intermittent signal alternates in sign, with $|y_{n+1}| > |y_n|$. Solving for $|y|$ in the differential equation limit as in (7.5.2), we obtain

$$1 + \frac{\epsilon}{by_{\text{out}}^2} = \left(1 + \frac{\epsilon}{by_{\text{in}}^2}\right) e^{-2\epsilon t},$$

7.5. Intermittency

Type	Typical map dynamics		Bifurcation	Eigenvalues at tangency
	$\varepsilon < 0$ (stable)	$\varepsilon > 0$ (unstable)		
I				+1
II				+1, +1
III				−1

Figure 7.43. Three types of intermittency: Type I, Eq. (7.5.1); Type II, Eq. (7.5.6); Type III, Eq. (7.5.5) (after Schuster, 1988).

from which, averaging over y_{in} as in (7.5.4), we find

$$\tau = \frac{1}{y_{\text{max}}(b\epsilon)^{1/2}} \cot^{-1}\left(\frac{1}{y_{\text{max}}}\left(\frac{\epsilon}{b}\right)^{1/2}\right),$$

where y_{max} signals the end of the laminar phase of the motion. For $\epsilon/b \ll y_{\text{max}}^2$ we obtain the scaling $\tau \propto \epsilon^{-1/2}$, as in (7.5.4), but for $\epsilon/b \gg y_{\text{max}}^2$ we find $\tau \propto 1/\epsilon$, as verified numerically by Lahiri and Nag (1989).

In addition to Types I and III intermittency, Type II intermittency is found in invertible maps of two or more dimensions, and their corresponding flows having three or more dimensions. The inverted Hopf bifurcation illustrated in Fig. 7.21 is responsible for this behavior. A simple model is

$$\begin{aligned} r_{n+1} &= (1 + \epsilon)r_n + ar_n^3, \\ \theta_{n+1} &= \theta_n + \Omega, \end{aligned} \quad (7.5.6)$$

where r and θ are polar coordinates for a two-dimensional rectangular map. For $\epsilon > 0$, the laminar motion for r continuously increases, and the motion has a spiral character.

7.5b. Renormalization-Group Analysis

We have seen in the renormalization-group treatment for a quadratic map f at a period-doubling accumulation point (see Section 7.2b) that successive iterates $f^{(2^n)}$ when properly rescaled converge to a universal function $g^*(x)$, given by the fixed point of the period-doubling operator T:

$$g^*(x) = Tg^*(x) = \alpha g^*(g^*(x/\alpha)). \tag{7.5.7}$$

For the boundary conditions $g^*(0) = 1$ and $g^{*\prime}(0) = 0$, appropriate to the calculation in Section 7.2b, a solution is found with $\alpha \approx -2.5029$, the universal scaling parameter. Both α and $g^*(x)$ are obtained by solving (7.5.7) numerically. Introducing a perturbation from g^*,

$$f_\lambda = g^* + \epsilon h_\lambda, \tag{7.5.8}$$

then leads to an exponential divergence from the fixed point along the unstable direction given by

$$Tf_\lambda(x) = g^*(x) + \lambda \epsilon h_\lambda(x), \tag{7.5.9}$$

where h_λ is the eigenfunction and $\lambda = \delta \approx 4.6692$ is the eigenvalue for the period-doubling rate of instability of T along h_λ. Although g^*, h, α, and δ are uniquely determined from the functional equations, their solution, for period doubling, could only be found numerically.

For mapping dynamics at a tangency, the equation (7.5.1) has the form (we revert to the customary x, rather than y, as the variable)

$$x_{n+1} = f(x_n) = x_n + ax_n^2, \tag{7.5.10}$$

which is shown in Fig. 7.44a. If we examine the second iterate $f^{(2)}$ near $x = 0$

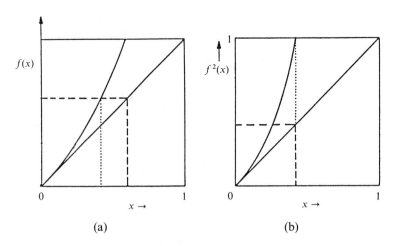

Figure 7.44. $f(x)$ and $f^2(x)$ versus x for $x \geq 0$. The second iterate $f^2(x)$ looks similar to the original $f(x)$ in the dotted square (after Schuster, 1988).

7.5. Intermittency

we see that it appears like a rescaled version of f (Fig. 7.44b). Similarly, $f^{(4)}$ looks like a rescaled version of $f^{(2)}$. We are then led to ask whether the successive iterates $f^{(2^n)}$ when properly rescaled converge to a universal function g^* as in (7.5.7), except that the boundary conditions at the tangency are now $g^*(0) = 0$ and $g^{*\prime}(0) = 1$, rather than the boundary conditions $g(0) = 1$ and $g^{*\prime}(0) = 0$ for period doubling. Surprisingly, this change of boundary conditions allows an analytical solution of the fixed point equation (7.5.7) and the linearized eigenvalue equation.

First, following Hirsch et al. (1982), the class of tangent maps can be enlarged by introducing a more general tangency exponent z, which determines the "universality class,"

$$f(x) = x + a|x|^z. \tag{7.5.11}$$

For $z = 2$, (7.5.7) has a fixed point solution

$$g^*(x) = 1/(1 - ax), \tag{7.5.12}$$
$$\alpha = 2,$$

which can be verified by substituting (7.5.12) into (7.5.7). For arbitrary z, Hirsch et al. find the solution to the fixed point equation (7.5.7) by expanding it to third order, and then obtain the solution to the eigenvalue problem, also by expansion.

Hu and Rudnick (1982) were able to find explicit forms for both g^* and for the perturbation h_λ, and we outline their method below. For our presentation we follow the more detailed treatment in Schuster (1988). Hu and Rudnick found that the recursion relation

$$x' = g(x) \tag{7.5.13}$$

could also be expressed in the implicit form

$$G(x') = G(x) - a. \tag{7.5.14}$$

The fixed point equation (7.5.7), after rescaling by α can be written as

$$\alpha g^*[g^*(x)] = g^*(\alpha x)$$

or

$$\alpha x''(x) = x'(\alpha x). \tag{7.5.15}$$

Operating on this with G we have

$$G(\alpha x'') = G[x'(\alpha x)] = G(\alpha x) - a. \tag{7.5.16}$$

From (7.5.14) we obtain

$$G(x'') = G(x') - a = G(x) - 2a$$

or

$$\tfrac{1}{2}G(x'') = \tfrac{1}{2}G(x) - a. \tag{7.5.17}$$

544 7. Bifurcation Phenomena and Transition to Chaos in Dissipative Systems

Comparing (7.5.17) with (7.5.16) we see that, for consistency, G^* must satisfy

$$\tfrac{1}{2}G^*(x) = G^*(\alpha x). \tag{7.5.18}$$

This simpler form of the fixed point equation allowed Hu and Rudnick to find the solution

$$G^*(x) = |x|^{-(z-1)},$$
$$\alpha = 2^{1/(z-1)}, \tag{7.5.19}$$

which for $z = 2$ gives the scaling factor in (7.5.12). Inverting (7.5.14) and using (7.5.19), we have the fixed point solution

$$g^*(x) = G^{*-1}[G^*(x) - a] = \left(\frac{1}{|x|^{z-1}} - (z-1)a\right)^{-1/(z-1)}. \tag{7.5.20}$$

The simpler implicit form of the fixed point equation also allowed Hu and Rudnick to obtain an explicit solution to the linear eigenvalue problem. To determine the eigenfunctions h_λ and their eigenvalues λ, we apply T to (7.5.8), which yields, for $\epsilon \ll 1$,

$$Tf_\lambda = g^*(x) + \epsilon\alpha[g^{*\prime}(g^*(x/\alpha))h_\lambda(x/\alpha) + h_\lambda(g^*(x/\alpha))]. \tag{7.5.21}$$

Equating (7.5.21) to (7.5.9) and letting $x/\alpha \to x$, we obtain the equation

$$\alpha[g^{*\prime}(g^*(x))h_\lambda(x) + h_\lambda(g^*(x))] = \lambda h_\lambda(\alpha x). \tag{7.5.22}$$

Hu and Rudnick introduce implicit functions

$$G_\lambda(x) = G^*(x) + \lambda H_\lambda(x), \tag{7.5.23}$$

and employing a calculation similar to that used for the fixed point functions G^* to relate $h_\lambda(x)$ to $H_\lambda(x)$, they obtain the analytic solution

$$h_\lambda(x) = \frac{1}{ap}\left[\frac{1}{|x|^{z-1}} - a(z-1)\right]^{-z/(z-1)}\left[\frac{1}{x^p} - \left(\frac{1}{|x|^{z-1}} - a(z-1)\right)^{p/(z-1)}\right], \tag{7.5.24}$$

$$\lambda = 2^{p-z-1/(z-1)}, \tag{7.5.25}$$

where the integer p specifies the eigenfunction of interest. For the generic case $z = 2$ and choosing $h_\lambda(0) = $ const., as in (7.5.1), then evaluating (7.5.24) near $x = 0$ yields $p = 3$, such that $\lambda = 4$ and

$$h_\lambda = \frac{1}{3a}\left(\frac{1}{x} - a\right)^{-2}\left[\frac{1}{x^3} - \left(\frac{1}{x} - a\right)^3\right]. \tag{7.5.26}$$

In general $p = 2z - 1$ for $h_\lambda(0) = $ const., and hence

$$\lambda = 2^{z/(z-1)}. \tag{7.5.27}$$

7.5. Intermittency

We note that by applying T to f_λ n times, and using (7.5.22) that

$$T^n f_\lambda = g^* + \epsilon \lambda^n h_\lambda, \qquad (7.5.28)$$

such that the perturbed solution diverges away from g^* at the rate λ.

The renormalization solutions can be used, for example, to determine the scaling of the average length of time τ spent in the neighborhood of the tangency. Since using the map f_λ, an initial point spends twice as many iterations in the laminar region as using the map f_λ^2, we have the relation $\tau(f_\lambda) = 2\tau(Tf_\lambda)$ and, similarly, for 2^n iterations

$$\tau(f_\lambda) = 2^n \tau(T^n f_\lambda).$$

Choosing

$$\epsilon \lambda^n \approx 1 \qquad (7.5.29)$$

in (7.5.28) to define the laminar region, such that $\tau(T^n f_\lambda) = \text{const.}$, then $\tau \sim 2^n$. Substituting (7.5.27) in (7.5.29), we obtain $\epsilon (2^n)^{z/(z-1)} \approx 1$ such that $\tau \approx \epsilon^{-(z-1)/z}$. This agrees with (7.5.4) for $z = 2$. The scaling of τ can also be modified to include the effect of noise (Hirsch et al., 1982). A somewhat different approach to the scaling problem has been taken by a group at Kyushu University (e.g., see Kobayashi et al. (1989), and references therein).

7.5c. Power Spectra and Experimental Results

Power spectra can be determined from the probability of time spent in the regular regions near a tangency. Defining the power spectrum S_f in the usual way in terms of the autocorrelation function $C(m)$,

$$S_f \propto \lim_{N \to \infty} \frac{1}{N} \sum_{m=0}^{N} \cos(2\pi m f) C(m), \qquad (7.5.30)$$

where

$$C(m) = \lim_{N \to \infty} \frac{1}{N} \sum_{n=0}^{N} x_{n+m} x_n, \qquad (7.5.31)$$

then the $C(m)$ can be evaluated for an ideal model that assumes short bursts of chaotic signal of unit amplitude interspersed by quiet (zero amplitude) periods given by the laminar motion. $C(m)$ is then proportional to the probability of finding a signal at time m, given a signal at time zero. Evaluating these probabilities using renormalization theory (e.g., see Schuster, 1988), one obtains S_f for the various universality classes characterized by the power law z of the tangency. Spectral powers have been obtained for $z < 3$ by Procaccia and Schuster (1983) and for $z \geqslant 3$ by Ben-Mizrachi et al. (1985); a few values are listed in (7.5.32):

$$\lim_{f \to 0} S_f \propto \begin{cases} f^{-1/(z-1)}, & 2 < z < 3, \\ \dfrac{1}{f |\log f|^2}, & z = 2, \\ f^{-(2z-3)/(z-1)}, & \tfrac{3}{2} < z < 2. \end{cases} \qquad (7.5.32)$$

For $z \approx 2$ a $1/f$ spectrum is predicted, although the results display an additional logarithmic singularity exactly for $z = 2$, the generic case. The predictions are in reasonable agreement with numerically determined spectra from the mappings, in the range of interest, as shown in Fig. 7.45.

The low-frequency divergence of the power spectrum arises because arbitrarily long regions of regular behavior occur with finite probability for the exactly tangent map. If the map is not exactly tangent, then we saw that the average regular region scales as $\tau \propto \epsilon^{-(z-1)/z}$. We would therefore expect

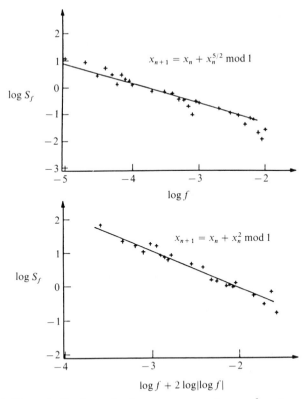

Figure 7.45. Numerically determined power spectra for $z = \tfrac{5}{2}$ and $z = 2$ compared to (7.5.32) (after Procaccia and Schuster, 1983).

7.5. Intermittency

to find a cutoff in the $1/f$ spectrum at

$$f_c \propto \tau^{-1} \propto \epsilon^{(z-1)/z}.$$

This scaling was, in fact, found numerically, for $z = \tfrac{5}{2}$ by Procaccia and Schuster (1983). This result leads to two important observations:

(1) The characteristic $1/f$ spectrum should only be observable near criticality and thus should be very sensitive to parameter changes.
(2) Extrinsic noise is equivalent to introducing a random set of ϵ's into the mapping and therefore would also lead to a cutoff in the $1/f$ spectrum.

Both of these phenomena would lead us to conclude that the $1/f$ spectrum generated from intermittency is not robust, and therefore is probably not the origin of observed current fluctuations in resistors, for example. On the other hand, intermittency has been observed in many experiments, with an accompanying $1/f^\mu$ ($\mu \sim 1$) spectrum, at critical parameter values, for which there is a transition to chaos.

The probability distribution $p(t)$ for the laminar time intervals is of interest for comparison with experimental measurements. We can write

$$p(t) = p(x) \left| \frac{dx}{dt} \right|, \tag{7.5.33}$$

where $p(x)$ is the (uniform) probability density for injection into the laminar region at x and, using (7.5.3) for Type I intermittency,

$$t = \frac{1}{(a\epsilon)^{1/2}} \left[\tan^{-1} \frac{x_{\max}}{(\epsilon/a)^{1/2}} - \tan^{-1} \frac{x}{(\epsilon/a)^{1/2}} \right], \tag{7.5.34}$$

where $x_{\max}$ signals the end of the laminar region. Using (7.5.34) to evaluate the derivative in (7.5.33), we obtain

$$p(t) = \frac{\epsilon}{2x_{\max}} \left\{ 1 + \tan^2 \left[\tan^{-1} \frac{x_{\max}}{(\epsilon/a)^{1/2}} \right] - (\epsilon a)^{1/2} t \right\}. \tag{7.5.35}$$

One can also obtain (Schuster, 1988)

$$p(t) \sim \frac{\epsilon^2 e^{4\epsilon t}}{(e^{4\epsilon t} - 1)^2} \tag{7.5.36}$$

for Type II; and

$$p(t) \sim \frac{\epsilon^{3/2} e^{4\epsilon t}}{(e^{4\epsilon t} - 1)^{3/2}} \tag{7.5.37}$$

for Type III intermittency.

Type I intermittency, being generic, is both the most observed and the most studied using mapping approximations. Observations of intermittency of the fluid velocity in a Rayleigh–Bénard cell (Bergé et al., 1980) led to the

548   7. Bifurcation Phenomena and Transition to Chaos in Dissipative Systems

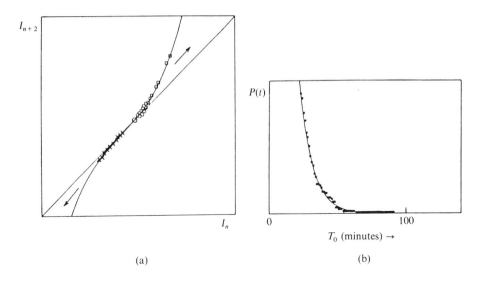

Figure 7.46. Type III intermittency for Rayleigh–Bénard convection. (a) Poincaré map I_{n+2} versus I_n for $\epsilon = 0.098$; (b) Number of laminar lengths with $t > T_0$ (after Dubois et al., 1983).

pioneering investigation of the Lorenz model by Pomeau and Manneville. Other experimental studies have been made using the Belousov–Zhabotinsky reaction (Pomeau et al., 1981) and the Josephson junction (Yeh and Kao, 1983). It is also possible to construct systems of nonlinear electronic circuit elements for which the analysis indicates that an embedded mapping exists, exhibiting the characteristic tangency resulting in intermittency. This has been done by Jeffries and Perez (1982), and their results show the intermittency scaling.

Type III intermittency has been observed by Dubois et al. (1983) in a Rayleigh–Bénard convection cell. They observed the growth of a subharmonic and a transition to the oscillatory burst behavior characteristic of Type III intermittency. Plotting maxima I_n of both the subharmonic mode (even n, crosses) and the fundamental mode (odd n, squares), the Poincaré map shown in Fig. 7.46a is obtained, corresponding to the Type III intermittency return map in Fig. 7.43. In Fig. 7.46b, the quantity $\int_{T_0}^{\infty} p(t) \, dt$ is plotted for the experimental data, showing good agreement with (7.5.37) (solid line).

7.6. Crises and Basin Boundaries

In previous sections, we examined the period doubling, quasiperiodicity, and intermittency routes to chaos. There is another route, that involves the

7.6. Crises and Basin Boundaries

simultaneous creation of a basin of attraction and its chaotic attractor as a parameter a is varied. Suppose that a chaotic attractor and its basin exist for $a < a_c$; then as $a \to a_c$ from below, the attractor collides with an unstable periodic orbit that lies on the basin boundary, and disappears. Such a collision has been called a *crisis* by Grebogi et al. (1983). For $a > a_c$, the motion can be either regular or chaotic on a new attractor with a new basin of attraction. If regular, then the motion is transiently chaotic (see Section 8.1). In this section, we describe the crises and the changes in structure of the basin boundaries. There are two types of crises. At a *boundary crisis*, the chaotic attractor and its basin are destroyed, after which the motion is transiently chaotic and ultimately tends toward another attractor which may either be nonchaotic (e.g., a fixed point or limit cycle) or may be chaotic. At an *interior crisis*, the chaotic attractor collides with an unstable periodic orbit which is in the interior of its basin of attraction, and the attractor undergoes a sudden explosion in its size. In an interior crisis the basin of attraction does not change.

7.6a. Crises

One-Dimensional Maps. Simple examples of boundary and interior crises are seen for the one-dimensional quadratic map

$$x_{n+1} = a - x_n^2, \qquad (7.6.1)$$

which is related to (7.2.4b) by a simple change of variables in (7.6.1): $x \to -(2x + C)$ and $a \to C^2 - C$. The basin boundary for (7.6.1) is trivial (two points x_{1b} and x_{2b}), and we can concentrate on the crisis mechanism itself. Figure 7.47 shows the bifurcation diagram from the map. For $a < -\frac{1}{4}$, all initial conditions $|x| < \infty$ tend to an attractor at $x = -\infty$. At $a = -\frac{1}{4}$, a second attractor and its basin are created by a tangent bifurcation at $x = -\frac{1}{2}$. For $a > -\frac{1}{4}$, the basin boundary of this new attractor is defined by the unstable fixed point

$$x = -x^* = -\tfrac{1}{2} - (\tfrac{1}{4} + a)^{1/2},$$

shown as the dashed line in Fig. 7.47, with the basin given by $|x| < x^*$. As a increases, the attractor period doubles to chaos at a_∞. At $a = a_c = 2$, the attractor and its basin are destroyed by a boundary crisis. It can be seen from the figure that the mechanism is a collision of the attractor with the unstable fixed point $-x^*$. The unstable fixed point defines one boundary of the basin of attraction, as it represents a "potential hill" for the dynamics. For $x < -x^*$, the trajectory is attracted to $x = -\infty$. Since the size of the attractor is expanding faster than the basin with increasing a, for a slightly greater than 2 a trajectory anywhere in the chaotic region will eventually find itself to the left of the hill ($x < -x^*$) and thus be attracted to $x = -\infty$. The motion is therefore transiently chaotic. For a only slightly greater than

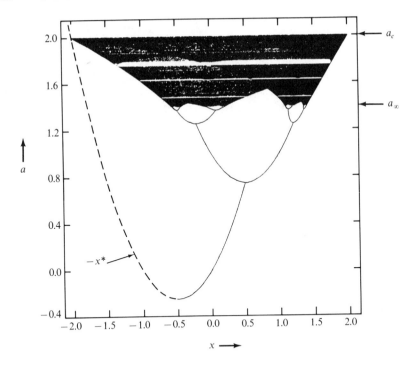

Figure 7.47. Bifurcation diagram for the map (7.6.1). The dashed curve is the unstable fixed point. For a given value of a, an initial condition x is chosen and its orbit is plotted after the first several thousand iterates. The figure is generated by doing this for many different values of a (after Grebogi et al., 1983).

2, the mean time for the chaotic transient can be estimated to be $\langle \tau \rangle \sim (a - 2)^{-1/2}$ (see Section 8.1).

An example of an interior crisis is seen in a blow-up of a section of Fig. 7.47 given in Fig. 7.48. Here a period 3 orbit is born at $a = a_3 \approx 1.75$ and period doubles to chaos. The unstable periodic orbit born in the tangent bifurcation is shown as the dashed line. Between $a \approx 1.78$ and 1.79 the chaotic bands are narrow. They widen abruptly at the interior crisis $a = a_3^* \approx 1.79$ where the period 3 chaotic attractor collides with the unstable periodic orbit. For $a > a_3^*$, the period 3 chaotic attractor disappears and the motion reverts to the period 1 chaotic attractor within its basin $[-x^*, x^*]$.

The physical mechanism is similar to that described above for the exterior crisis, except that in this case, before the crisis occurs, the empty internal regions are part of the attractor basin. Concentrating on the left-hand blank region in Fig. 7.48 with $a < a_3^*$, for an initial value at, say, $x = -0.6$, the trajectory is rapidly attracted into the three-piece attractor. With $a > a_3^*$ the attracted initial condition can explore a region of x which crosses over the unstable orbit (dashed line), so returning the trajectory, occasionally, to

7.6. Crises and Basin Boundaries

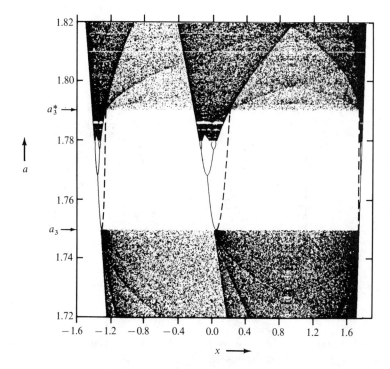

Figure 7.48. Blow-up of the bifurcation diagram of Fig. 7.47 in the region of the period 3 tangent bifurcation. The dashed curves denote the unstable period 3 orbit created at the tangent bifurcation (after Grebogi et al., 1983).

the neighborhood of the initial condition. The three-piece attractor has thus merged into a single piece, ranging over the former three-piece basin. Note from Fig. 7.48, however, that for a slightly greater than a_3^*, the fraction f of the time spent in the formerly empty regions of x is small. Grebogi et al. (1983) obtain an approximate scaling

$$f \approx (a - a_3^*)^{1/2} g(\ln(a - a_3^*)), \tag{7.6.2}$$

where g is a periodic function.

Higher-Dimensional Systems. Basin boundaries can be either smooth or have a fractal structure. If d is the dimension of a map or the phase space for a flow, then the Hausdorff dimension D_{b0} of the boundary must satisfy $D_{b0} \geq d - 1$ in order to divide the surface of section (or phase space). Following Grebogi et al., we describe crises in these systems using as an example the two-dimensional, invertible Hénon map (7.1.26), for which

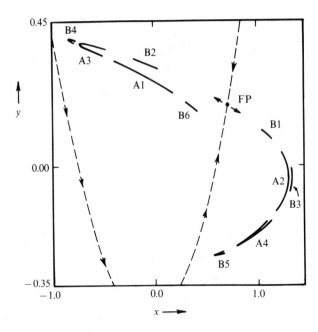

Figure 7.49. Orbits on the 4-piece attractor cycle from segment A1 to A2 to A3 to A4 and back to A1. Orbits on the 6-piece attractor cycle (B1 → B2 → ··· → B6 → B1). The stable manifold of the unstable saddle fixed point FP is shown as a dashed line and the unstable directions are indicated (after Grebogi et al., 1983).

$D_{b0} \geq 1$:

$$x_{n+1} = y_n + 1 - ax_n^2,$$
$$y_{n+1} = bx_n.$$
(7.1.26)

Figure 7.49 shows the y–x section for $b = 0.3$ and $a = 1.0807$. For this value of a there is a 4-piece and a 6-piece chaotic attractor, each having a fractal character and its own basin. An unstable fixed point (saddle) is located at $x \approx 0.6911$, $y \approx 0.2073$. Both attractors appear to lie on the unstable manifold of the fixed point. The dashed line shows a part of the stable manifold. Figure 7.50 shows the basins for the 6-piece and 4-piece attractors in black and white, respectively, along with the 4-piece attractor itself. This figure was constructed by calculating the Liapunov exponent for a grid of 1.09×10^4 initial conditions and determining whether $\sigma \to 0.1391$ for the 4-piece attractor, or $\sigma \to 0.1097$ for the 6-piece attractor. Comparing Figs. 7.49 and 7.50, it appears that the basin boundary is the stable manifold of the fixed point and that the basin boundaries have a fractal structure, i.e., a blow-up of the boxed region in Fig. 7.50 has a rescaled version near the center of the box.

7.6. Crises and Basin Boundaries

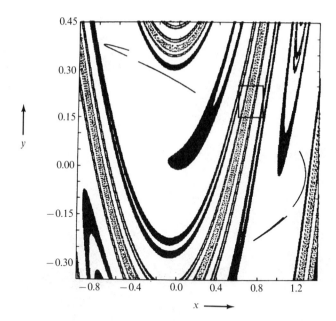

Figure 7.50. Basins of attraction for the 6-piece chaotic attractor (black) and the 4-piece chaotic attractor (blank) for $a = 1.0807$ and $b = 0.3$. Also shown is the 4-piece chaotic attractor (after Grebogi et al., 1983).

The 6-piece attractor undergoes a boundary crisis as a increases. The attractor is created at $a \approx 1.062$ by a tangent (saddle-node) bifurcation. The stable orbit period doubles to chaos at $a \approx 1.076$ and the chaos disappears at $a \approx 1.081$ due to a collision between the attractor and the unstable orbit, which lies on the boundary separating the 6-piece and 4-piece basins, in the same manner as shown in Fig. 7.47 for the one-dimensional case. Figure 7.51a shows how the saddle on the basic boundary collides with a segment (B5 in Fig. 7.49) of the 6-piece attractor at the critical value $a_c \approx 1.081$. The collision of the attractor and the saddle is clear. We also see (Fig. 7.51b) that a "finger" from the 4-piece basin reaches down and "touches" the 6-piece attractor. There are, in fact, an infinite number of such fingers, too narrow to be shown in the figure. This is because the stable manifold (basin boundary) of the saddle just touches, but does not cross, the unstable manifold on which the attractor lies, at the crisis, yielding a homoclinic tangency (Newhouse, 1980). For $a > a_c$ there is transient chaos, with $\langle \tau \rangle \sim (a - a_c)^{-\gamma}$, where the scaling exponent γ has been given in terms of the Liapunov exponents of the map by Grebogi et al. (1986). We describe this in detail in Section 8.1.

An example of an interior crisis can be seen for small $|b|$. Since $b = 0$

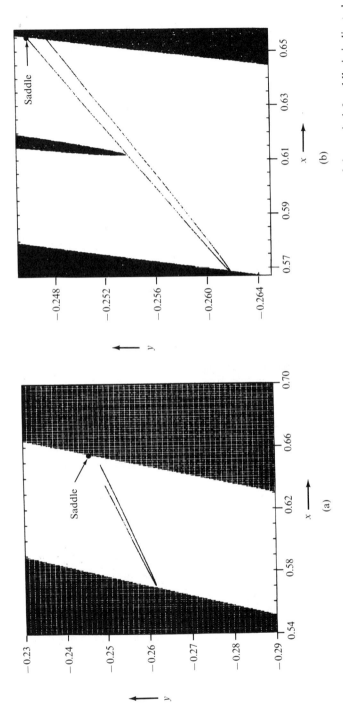

Figure 7.51. A segment of the 6-piece attractor in its basin with $b = 0.3$. The location of one component of the period 6 saddle is indicated. The dark region denotes the basin of the 4-piece attractor. (For this value of a each segment of the "6-piece" attractor consists of two pieces so it is really a 12-piece attractor.) (a) $a = 1.0800000$; (b) $a = 1.0807430$ (after Grebogi et al., 1983).

7.6. Crises and Basin Boundaries

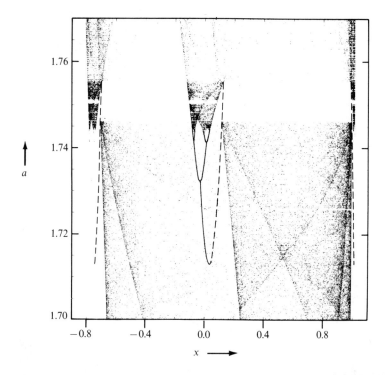

Figure 7.52. Bifurcation diagram for $b = -0.015$ (after Grebogi et al., 1983).

corresponds to the quadratic map, we can view the small $|b|$ limit as a two-dimensional perturbation of the one-dimensional map. Figure 7.52 shows the a–x bifurcation diagram for $b = -0.015$ in the range of a between 1.70 and 1.76 where a 3-piece attractor is born at $a \approx 1.713$, period doubles to chaos, and disappears at $a_c \approx 1.756$ due to an interior crisis. At the crisis, the period 3 unstable saddle, which lies on the basin boundary, collides with the 3-piece attractor itself. The 1-piece attractor also has a boundary crisis at $a_c^* \approx 1.746$ when it collides with the period 3 unstable saddle.

We note a crucial difference from the quadratic map ($b = 0$). Comparing Figs. 7.48 and 7.52, we see that there are *two* attractors present in the range $1.713 \lesssim a \lesssim 1.746$ for $b = -0.015$, while only *one* attractor can be present for the one-dimensional case. The coexistence of multiple attractors is characteristic of multidimensional systems and is an example of *hysteresis*. That is, an initial condition on the 1-piece attractor will jump to the 3-piece attractor as a increases beyond $a_c^* \approx 1.746$ where the 3-piece attractor disappears. However, if a is then decreased through a_c^* the system remains on the 3-piece attractor until $a \lesssim 1.713$, whereupon it jumps back to the original 1-piece attractor.

Actually, there are different period 3 bifurcation sequences as a increases,

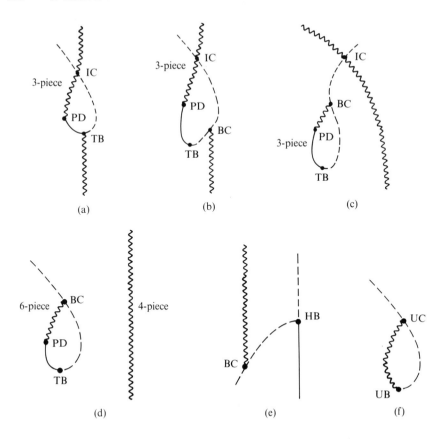

Figure 7.53. Diagrams for various types of crises as described in the text (after Grebogi et al., 1983).

depending on the value of b. We illustrate these in Fig. 7.53 using the "crisis diagrams" introduced by Grebogi et al. (1983). In these diagrams, a increases vertically, a solid line represents a stable periodic orbit, a dashed line represents an unstable periodic orbit, and a wavy line represents an attracting chaotic orbit. For $b > b_* \approx -10^{-3}$, shown in Fig. 7.53a, the original bifurcation sequence for the one-dimensional map is observed; namely, the chaotic attractor disappears at a tangent (saddle-node) bifurcation (TB), the 3-piece attractor period doubles (PD) to chaos and is destroyed by an interior crisis (IC), and the original attractor reemerges. For $b_* > b \gtrsim b'_* \approx -0.02$, the sequence given in Fig. 7.52, the crisis diagram is shown in Fig. 7.53b. It consists of a tangent (saddle-node) bifurcation to create a period 3 attractor coexisting with the original attractor. The latter is then destroyed

7.6. Crises and Basin Boundaries

by a boundary crisis (BC), the period 3 attractor period doubles to chaos and disappears at an interior crisis, and the original attractor re-emerges. For $b < b'_*$, the bifurcation sequence shown in Fig. 7.53c is similar to (b) except that the 3-piece attractor is destroyed by a boundary crisis and the original attractor suddenly widens at an interior crisis, rather than *vice versa*. Figure 7.53d shows the boundary crisis for the 6-piece attractor at $b = 0.3$ shown in Fig. 7.49.

Other examples of crises have been found analytically or strongly implied from numerical results, including the Lorenz attractor (1.5.2) (Kaplan and Yorke, 1979) and the driven damped pendulum (Huberman and Crutchfield, 1979). Figure 7.53e shows the crisis diagram for the Lorenz system, in which the unstable periodic orbit that destroys the attractor is an unstable limit cycle that arises from a Hopf bifurcation (HB) of a stable fixed point.

Another type of crisis is found in two-dimensional noninvertible maps, or invertible maps having three or higher dimensions or their corresponding four- and higher-dimensional flows (Grebogi et al., 1985). For such maps, the basin boundary can be a fractal curve; i.e., a continuous curve everywhere nondifferentiable. (The basin boundaries in two-dimensional invertible maps, which are the stable manifold of an unstable saddle, are fractals but are not fractal curves; i.e., they are not homeomorphic to a circle.) A fractal curve can be generated by an unstable pair bifurcation, in which two unstable orbits A and B are created as a parameter a decreases through the critical value a_c. For a two-dimensional noninvertible map, orbit B (a "repeller") has two unstable directions, and lies on the fractal curve that is the basin boundary. Orbit A is a saddle, and the chaotic attractor that lies within the fractal curve boundary is the unstable manifold of A. Conversely, as a increases through a_c, the orbit A in the attractor and B on the basin coalesce, and there is a crisis. Beyond the crisis, for $a > a_c$, there is superpersistent transient chaos, with the mean lifetime of the chaotic transient scaling as (Grebogi et al., 1985).

$$\langle \tau \rangle \sim k_1 \exp[k_2(a - a_c)^{1/2}].$$

This will be discussed further in Section 8.1. The diagram for this unstable pair bifurcation (UB) and subsequent unstable pair crisis (UC) is shown in Fig. 7.53f.

7.6b. Fractal Basin Boundaries

An example of a fractal boundary is shown in Fig. 7.50 for the Hénon map with $b = 0.3$ and $a = 1.0807$. The metamorphosis of the boundary from smooth to fractal is shown in Fig. 7.54 for two different values of $a = 1.150$ and 1.395 (Grebogi et al., 1987). A similar metamorphosis is found for the driven damped pendulum. For these a's, the Hénon attractor is either a period 1 or period 2 fixed point. One of the basins is shown as the white

558   7. Bifurcation Phenomena and Transition to Chaos in Dissipative Systems

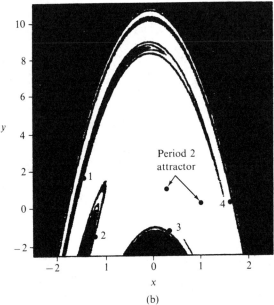

Figure 7.54. Basin of attraction of infinity in black for the Hénon map with $b = -0.3$. (a) $a = 1.150$; (b) $a = 1.395$ (after Grebogi et al., 1986).

7.6. Crises and Basin Boundaries

region in the figure. At $a = 1.150$, the basin boundary is the smooth stable manifold of the period 1 saddle. As a increases through a critical value $a_{sf} \approx 1.315$, the basin boundary suddenly becomes fractal, with $D_{bo} \approx 1.530 \pm 0.006$ determined numerically. At $a = 1.395$, the boundary is the closure of the stable manifold of the period 4 saddle shown in Fig. 7.54b. There is another change in the basin boundary at $a_{ff} \simeq 1.396$ in which the boundary remains fractal but has become the closure of the stable manifold of a period 3 saddle. Grebogi et al. (1986) find that these sudden changes in the basin boundary are associated with the *accessibility* of the various period n unstable saddles from a point in the basin of the attractor. A boundary point is called accessible from a point in the basin if one can draw a finite length curve from the boundary point to the point in the basin without crossing the boundary. At the transition a_{sf} between Figs. 7.54a and b, the stable and unstable manifolds of the period 1 saddle are tangent. For $a > a_{sf}$, these manifolds generate a homoclinic tangle that "buries" the period 1 saddle, rendering it inaccessible to the attractor basin. Similarly, at a_{ff} the stable and unstable manifolds of the period 4 orbit are tangent, and, for $a > a_{ff}$, this orbit becomes inaccessible from the attractor basin. The basin boundary suddenly shifts to the stable manifold of the period 3 orbit, which becomes the accessible orbit on the boundary. Since a chaotic attractor at crisis collides with the saddle on the basin boundary, this must be an accessible saddle. To determine the basin boundary by some analytical means, we must then find the accessible saddle.

Final State Sensitivity. The transition from a smooth to a fractal boundary implies an increase in the sensitivity of the final state to uncertainties in the initial conditions (Grebogi et al., 1983b). Suppose a smooth boundary line divides a finite phase space area into two basins with attractors A and B (see Fig. 7.55). Suppose all initial conditions are uncertain to within a circle of radius ϵ; two such initial conditions are shown in Fig. 7.55. Let f_u be the fraction of phase space that is uncertain; i.e., an initial condition within f_u may go to either attractor A or B. Clearly, for a smooth boundary line, $f_u \sim \epsilon$. However, if the boundary is fractal, then we can have

$$f_u \sim \epsilon^\alpha \tag{7.6.3}$$

with $\alpha < 1$. In this case, reducing ϵ by a factor of two yields less than a factor of two reduction in f_u.

To relate α for a d-dimensional phase space or surface-of-section to the capacity dimension D_{bo} of the basin boundary,

$$D_{bo} = \lim_{\epsilon \to 0} \frac{\ln M(\epsilon)}{\ln(1/\epsilon)}, \tag{7.6.4}$$

we note that the volume of the uncertain region of phase space is the total volume of all the $M(\epsilon)$ d-dimensional cubes needed to cover the boundary.

560 7. Bifurcation Phenomena and Transition to Chaos in Dissipative Systems

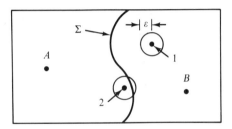

Figure 7.55. A region of phase space divided by the basin boundary Σ into basins of attraction for the two attractors A and B. 1 and 2 represent two initial conditions with uncertainty ϵ (after Grebogi et al., 1983b).

Since the volume of one cube is ϵ^d, the total volume is $\epsilon^d M(\epsilon)$. Since (7.6.4) implies $M(\epsilon) \sim \epsilon^{-D_{b0}}$, we obtain (7.6.3) with

$$\alpha = d - D_{b0}. \tag{7.6.5}$$

For the basin boundary to divide the phase space, we must require $D_{b0} \geq d - 1$, which implies using (7.6.5) that $\alpha \leq 1$. Grebogi et al. (1983b) have verified numerically the scaling (7.6.3) and the relation (7.6.5) for a two-dimensional map. For the basic boundary of the Hénon attractor shown in Fig. 7.54b, with $D_{b0} \approx 1.530$, we obtain $f \sim \epsilon^{0.47}$. As another example, the Lorenz system (1.5.2) for $13.93 \lesssim r \lesssim 24.06 = r_1$ has two nonchaotic attractors, representing steady clockwise or counterclockwise flow. Grebogi et al. (1983b) give the estimate $\alpha \sim 0.1$ at $r_1 - r \approx 4$ and $\alpha \sim 10^{-3}$ at $r_1 - r \sim 0.8$, with $\alpha \to 0$ as $r \to r_1$. Hence there is extreme sensitivity to initial conditions as $r \to r_1$. For some nonlinear systems, the basin boundaries appear to be characterized by not just one but a finite number of values of D_{b0} in different regions of the boundary (Grebogi et al., 1987a, b), which can be intermingled.

7.7. Determination of Homoclinic Intersection

We have seen that if the stable and unstable manifolds of an unstable periodic orbit intersect transversely, then a homoclinic tangle is generated. The motion near the separatrix becomes exceedingly complicated, having embedded orbits of arbitrarily large period (Smale, 1967). At the very least, this leads to transient chaotic behavior near the separatrix, although a trajectory may be ultimately drained into a nonchaotic attractor such as a stable periodic orbit. Transient chaos is the topic of Section 8.1. We have also seen in Section 7.3c and Section 7.6 that strange attractors can undergo sudden transitions called crises at the onset of homoclinic (or heteroclinic) intersections. For these reasons, it is of interest to determine analytically the onset of a transverse homoclinic intersection.

7.7. Determination of Homoclinic Intersection

We describe a method due to Melnikov (1963) for analyzing this motion near separatrices of near-integrable systems. The method yields a criterion for the onset of a homoclinic intersection near the separatrix of an integrable system which undergoes a dissipative perturbation. We have seen (see Section 3.2b and Fig. 3.4) that a generic Hamiltonian perturbation always yields chaotic motion in a layer surrounding a separatrix. For a dissipative perturbation, the motion near the separatrix is not necessarily chaotic either transiently or in the steady state. It is therefore important to predict under what conditions a homoclinic intersection first appears.

Melnikov's method has been applied to the study of perturbed dynamical systems by Morosov (1973, 1976), McLaughlin (1979), and Holmes (1979, 1980); in particular, Morosov and Holmes have studied Duffing's equation using the method. We adapt Holmes' general point of view (see Greenspan and Holmes, 1981) in our description here. For convenience, and to illustrate the basic approach, we consider a two-dimensional autonomous system that has a single hyperbolic fixed point and is perturbed by a periodic function of time:

$$\dot{x} = f_0(x) + \epsilon f_1(x, t), \qquad (7.7.1)$$

where $x = (x_1, x_2)$ and f_1 is periodic in t with period T. The unperturbed system is taken to be integrable and is assumed to possess a hyperbolic fixed point X_0 and an integrable separatrix orbit $x_0(t)$ such that

$$\lim_{t \to \pm \infty} x_0(t) = X_0.$$

The system is illustrated in phase space (x_1, x_2) in Fig. 7.56a. The stable and unstable orbits $x^s(t)$ and $x^u(t)$ are labeled in the figure and smoothly joined to each other. There is generally an elliptic fixed point within the separatrix orbit.

When the system is perturbed, then the phase space must be extended to three dimensions (x_1, x_2, t), and we most conveniently view the motion in a surface of section $t = \text{const.} \pmod{T}$. As described in Section 3.2b, the stable and unstable orbits in the surface of section do not smoothly join together when the system is perturbed. For a Hamiltonian perturbation the orbits generically intersect, leading to an infinite number of homoclinic points and chaotic motion. For a more general (dissipative) perturbation, there are three possibilities (Chillingworth, 1976), shown in Fig. 7.56. In Fig. 7.56b, the unstable orbit always lies outside the stable orbit. In Fig. 7.56c, the unstable orbit always lies inside the stable orbit. In Fig. 7.56d, the stable and unstable orbits intersect transversely, leading to an infinite number of intersections. It is only in the last case that chaotic motion appears.

Melnikov's Method. To find the condition for intersection, we calculate, using perturbation theory, the distance D from the unstable to the stable orbit at time t_0. For $D < 0$ for all t_0, we have Fig. 7.56b, for $D > 0$, we have

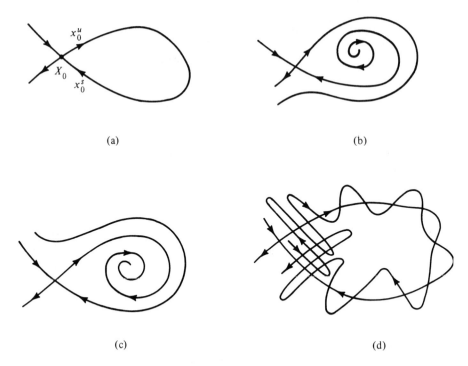

Figure 7.56. Stable and unstable orbits for the hyperbolic fixed point X_0. (a) Integrable system; the orbits smoothly join. (b) Perturbed system; the unstable orbit lies outside the stable orbit. (c) Perturbed system; the stable orbit lies outside the unstable orbit. (d) Perturbed system; the stable and unstable orbits intersect.

Fig. 7.56c, and if D changes sign for some t_0, then we have the chaotic motion of Fig. 7.24d.

To calculate D, we need the stable and unstable orbits x^s and x^u to first order in ϵ. Writing

$$x^{s,u}(t, t_0) = x_0(t - t_0) + \epsilon x_1^{s,u}(t, t_0), \tag{7.7.2}$$

where t_0 is an arbitrary initial time, and inserting (7.7.2) into (7.7.1), we obtain to first order

$$\frac{dx_1^{s,u}}{dt} = \mathbf{M}(x_0) \cdot x_1^{s,u} + \epsilon f_1(x_0(t - t_0), t), \tag{7.7.3}$$

where

$$\mathbf{M}(x_0) = \begin{pmatrix} \dfrac{\partial f_{01}}{\partial x_{01}} & \dfrac{\partial f_{01}}{\partial x_{02}} \\ \dfrac{\partial f_{02}}{\partial x_{01}} & \dfrac{\partial f_{02}}{\partial x_{02}} \end{pmatrix}$$

7.7. Determination of Homoclinic Intersection

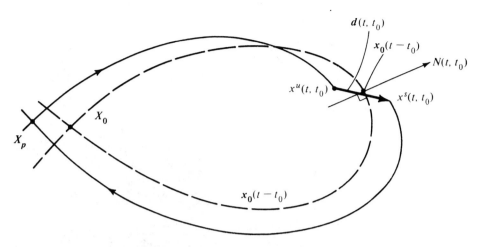

Figure 7.57. Illustrating the definition of the Melnikov distance $D = N \cdot d$. The dashed curve is the unperturbed separatrix, emanating from the hyperbolic point X_0. Under perturbation, the hyperbolic point moves to X_p.

is the Jacobian matrix of f_0 evaluated at $x_0(t - t_0)$, and where the second subscripts denote the components of f_0 and x_0. We must solve (7.7.3) for x^s for $t > t_0$, and for x^u for $t < t_0$, with the condition that

$$x^s(t \to \infty) = x^u(t \to -\infty) = X_p,$$

where X_p is the perturbed position of the hyperbolic fixed point. The two solutions differ by

$$\begin{aligned} d(t, t_0) &= x^s(t, t_0) - x^u(t, t_0) \\ &= x_1^s(t, t_0) - x_1^u(t, t_0). \end{aligned} \qquad (7.7.4)$$

The Melnikov distance $D(t, t_0)$ is defined as

$$D(t, t_0) = N \cdot d, \qquad (7.7.5)$$

which is the projection of d along a normal N to the unperturbed orbit x_0 at t (see Fig. 7.57). From (7.7.1) (with $\epsilon = 0$), a normal to $x_0(t - t_0)$ is[12]

$$N(t, t_0) = \begin{pmatrix} -f_{02}(x_0) \\ f_{01}(x_0) \end{pmatrix}. \qquad (7.7.6)$$

Introducing the wedge operator

$$x \wedge y = x_1 y_2 - x_2 y_1$$

---

[12] Although N is not normalized to unity, this is not essential to determine the condition for intersection of the separatrices.

and substituting (7.7.6) in (7.7.5), we can write

$$D(t, t_0) = f_0 \wedge d. \tag{7.7.7}$$

To find an explicit expression for D, we use (7.7.4) to write

$$D = D^s - D^u \tag{7.7.8a}$$

with

$$D^{s,u}(t, t_0) = f_0 \wedge x_1^{s,u}. \tag{7.7.8b}$$

Taking the time derivative of (7.7.8b),

$$\dot{D}^s = \dot{f}_0 \wedge x_1^s + f_0 \wedge \dot{x}_1^s$$
$$= \mathbf{M}(x_0) \cdot \dot{x}_0 \wedge x_1^s + f_0 \wedge \dot{x}_1^s. \tag{7.7.9}$$

Using $\dot{x}_0 = f_0$ and also (7.7.3) in (7.7.9),

$$\dot{D}^s = \mathbf{M}(x_0) \cdot f_0 \wedge x_1^s + f_0 \wedge \mathbf{M}(x_0) \cdot x_1^s + f_0 \wedge f_1. \tag{7.7.10}$$

The first two terms in (7.7.10) combine to give

$$\dot{D}^s = \operatorname{Tr} \mathbf{M}(x_0) f_0 \wedge x_1^s + f_0 \wedge f_1$$
$$= \operatorname{Tr} \mathbf{M}(x_0) D^s + f_0 \wedge f_1, \tag{7.7.11}$$

where $\operatorname{Tr} \mathbf{M}$ is the trace of the Jacobian matrix of f_0. Since D^s follows the stable orbit, we must integrate (7.7.11) from t_0 to ∞. Rather than do this in general, we specialize to the case of an unperturbed Hamiltonian system, for which $\operatorname{Tr} \mathbf{M} = 0$ on the separatrix. Integrating (7.7.11) then yields

$$D^s(\infty, t_0) - D^s(t_0, t_0) = \int_{t_0}^{\infty} f_0 \wedge f_1 \, dt.$$

But

$$D^s(\infty, t_0') = f_0(x_0(\infty - t_0)) \wedge x_1^s = 0$$

because $f_0(X_0) = 0$. Thus,

$$D^s(t_0, t_0) = -\int_{t_0}^{\infty} f_0 \wedge f_1 \, dt. \tag{7.7.12}$$

Proceeding similarly to calculate D^u, we obtain

$$\dot{D}^u = \operatorname{Tr} \mathbf{M}(x_0) D^u + f_0 \wedge f_1.$$

Since D^u follows the unstable orbit, we integrate from $-\infty$ to t_0 to obtain, for an unperturbed Hamiltonian system,

$$D^u(t_0, t_0) = -\int_{-\infty}^{t_0} f_0 \wedge f_1 \, dt. \tag{7.7.13}$$

7.7. Determination of Homoclinic Intersection

Using (7.7.12) and (7.7.13) in (7.7.8a), we obtain finally

$$D(t_0, t_0) = -\int_{-\infty}^{\infty} f_0 \wedge f_1 \, dt. \tag{7.7.14}$$

Evaluating D, we examine its behavior as a function of t_0. If D changes sign at some t_0, then the case in Fig. 7.56d obtains and chaotic motion is present near the separatrix.

Application to Duffing's Equation. A dynamical system that has played an important role in the development of chaos theory is Duffing's equation

$$\ddot{x} - x + x^3 = -\epsilon\delta\dot{x} + \epsilon\gamma\cos\omega t, \tag{7.7.15}$$

which describes the motion of a nonlinear oscillator in the presence of small damping δ and a periodic driving term with amplitude γ. Some of the earliest observations of chaotic attractors were made on this equation, which exhibits a rich variety of chaotic phenomena in various parameter regimes (see Ueda, 1980). Here, we use this example, following Holmes (1979), to determine the transition to chaotic motion. Rewriting (7.7.15) in the form (7.7.1)

$$\dot{x} = v, \tag{7.7.16a}$$

$$\dot{v} = x - x^3 + \epsilon[\gamma\cos\omega t - \delta v], \tag{7.7.16b}$$

we have the unperturbed Hamiltonian

$$H_0 = \tfrac{1}{2}v^2 - \tfrac{1}{2}x^2 + \tfrac{1}{4}x^4 \tag{7.7.17}$$

with constant energy curves $H_0 = \text{const.}$ as shown in Fig. 7.58. There is a

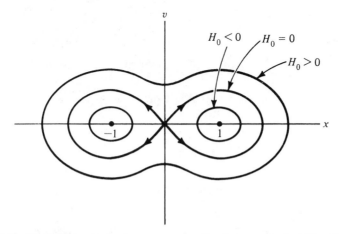

Figure 7.58. Illustrating constant energy curves for the unperturbed Hamiltonian in Duffing's equation (weak spring case).

single hyperbolic fixed point at $x = v = 0$, with the integrable separatrix orbit for $H_0 = 0$. Setting $H_0 = 0$ in (7.7.17), solving for $v(x)$, and inserting into (7.6.16a), we obtain

$$\frac{dx}{dt} = -x\left(1 - \frac{x^2}{2}\right)^{1/2}$$

with the solution

$$x_0(t) = \sqrt{2}\,\text{sech}\,t, \qquad (7.7.18a)$$

$$v_0(t) = -\sqrt{2}\,\text{sech}\,t\,\tanh t. \qquad (7.7.18b)$$

Comparing (7.7.16) and (7.7.1), we have

$$f_{01} = v, \qquad f_{11} = 0,$$
$$f_{02} = x - x^3, \qquad f_{12} = \gamma \cos \omega t - \delta v.$$

Then

$$\mathbf{f}_0 \wedge \mathbf{f}_1 = v_0[\gamma \cos \omega t - \delta v_0]$$

and from (7.7.14),

$$D = -\int_{-\infty}^{\infty} dt[\gamma v_0(t - t_0) \cos \omega t - \delta v_0^2(t - t_0)]. \qquad (7.7.19)$$

Using (7.7.18b) in (7.7.19) and changing variables to $\tau = t - t_0$,

$$D = \sqrt{2}\gamma \sin t_0 \int_{-\infty}^{\infty} d\tau\,\text{sech}\,\tau\,\tanh \tau \sin \omega \tau + 2\delta \int_{-\infty}^{\infty} \text{sech}^2 \tau\,\tanh^2 \tau\,d\tau.$$

The second integral is elementary:

$$\int_{-\infty}^{\infty} \text{sech}^2 \tau\,\tanh^2 \tau\,d\tau = \left.\frac{\tanh^3 \tau}{3}\right|_{-\infty}^{\infty} = \frac{2}{3}.$$

The first integral can be evaluated by the method of residues to yield

$$\int_{-\infty}^{\infty} d\tau\,\text{sech}\,\tau\,\tanh \tau \sin \omega \tau = \pi\omega\,\text{sech}\left(\frac{\pi\omega}{2}\right).$$

Thus,

$$D(t_0, t_0) = \sqrt{2}\pi\gamma\omega\,\text{sech}\left(\frac{\pi\omega}{2}\right)\sin \omega t_0 + \frac{4\delta}{3}. \qquad (7.7.20)$$

The condition for transverse intersection and chaotic separatrix motion is that D change sign at some t_0, which occurs, from (7.7.20), for

$$\delta < \delta_c = \frac{3\sqrt{2}\pi}{4}\gamma\omega\,\text{sech}\left(\frac{\pi\omega}{2}\right). \qquad (7.7.21)$$

7.7. Determination of Homoclinic Intersection

Persistence of Chaotic Motion. It must be emphasized that the condition for transverse intersection (7.7.21) is a local criterion for stochasticity, valid near the unperturbed separatrix. In general, this criterion does not signal the appearance of a strange attractor, which represents persistent chaotic motion over a global domain of phase space. For the Duffing problem with zero dissipation, $\delta = 0$, the perturbation is Hamiltonian, and steady state chaos always occurs. We know that this motion is confined within a separatrix layer bounded by KAM curves. For $\delta > 0$, the KAM curves are all destroyed and the homoclinic orbit, which may be chaotic near the unperturbed separatrix, may wander far from the separatrix, becoming trapped in an attracting periodic orbit or "sink." Such behavior has been observed by Holmes (1979) in analog computer simulations of Duffing's equation. Thus all that one can say is that meeting the condition for transverse intersection (7.7.21) causes solutions to "wander" in an apparently irregular manner for some time before approaching an attracting set, which may be either a periodic attractor or a strange attractor. We treat these phenomena in Section 8.1.

In fact, the appearance of a strange attractor in Duffing's equation seems to be associated, numerically, with the accumulation point of the cascade of bifurcations of the two sinks near $x = \pm 1$, $v = 0$ (see Fig. 7.58). Holding δ and ω fixed, Holmes studied the behavior of the solutions as γ was increased from zero using an analog computer. His results are summarized in Fig. 7.59 for a particular value of ω and δ. For $\gamma < 0.76$, only regular motion of the type in Fig. 7.56c was observed. For $0.76 < \gamma < 0.95$, there was transient chaotic behavior, but the orbits all approached one of the two sinks as $t \to \infty$. For $0.95 < \gamma < 1.08$, a cascade of successive bifurcations occurred for the sinks. For $1.08 < \gamma < 2.45$, computational results suggest the appearance of a strange attractor, except for a small interval $1.15 < \gamma < 1.2$ where a period five orbit was seen. Thus although both the intersection of separatrices and the completion of the bifurcation cascade for the two sinks are necessary for the appearance of a strange attractor in this system, these conditions are not sufficient.

Figure 7.59. Typical behavior of Duffing's equation as the amplitude γ of the periodic driving term is increased, with the damping δ and the frequency ω held constant.

568   7. Bifurcation Phenomena and Transition to Chaos in Dissipative Systems

Application to Mappings. Melnikov's method can be applied to two-dimensional dissipative maps. For example, the dissipative standard map (7.3.4) can be written (putting I for u and θ for v)

$$\frac{dI}{dn} = [-\delta I + f(\theta)]\delta_1(n),$$

$$\frac{d\theta}{dn} = I,$$

(7.7.22)

where $\delta_1(n)$ is the periodic δ-function (3.1.33)

$$\delta_1(n) = 1 + 2 \sum_{q=1}^{\infty} \cos 2\pi q n.$$

Putting, for example, $f = K \sin \theta$ and assuming that both δ and the $q = 1$ harmonic are of order ϵ, with all other harmonics being negligible,

$$\frac{dI}{dn} = K \sin \theta + \epsilon[-\delta I + 2K \sin \theta \cos 2\pi n],$$

$$\frac{d\theta}{dn} = I,$$

(7.7.23)

which is in the form (7.7.1). The zero-order separatrix solution for the pendulum motion is, from (1.3.21),

$$\theta_0(t) = 4 \tan^{-1} \exp(\sqrt{K} t) - \pi,$$

$$I_0(t) = 2\sqrt{K} \sin\left(\frac{\theta_0(t)}{2}\right).$$

(7.7.24)

Using (7.7.24) in (7.7.14), we evaluate the Melnikov distance as

$$D(n_0) = \int_{-\infty}^{\infty} I_0(n - n_0)\{-\delta I_0(n - n_0) + 2K \sin[\theta_0(n - n_0)] \cos 2\pi n\} \, dn.$$

(7.7.25)

The first integral in (7.7.25) is elementary, and the second can be written in terms of the MA integrals defined in (3.5.15). The result is

$$D(n_0) = \frac{8\pi}{Q_0}\left[Q_0^3 \frac{\sinh(\pi Q_0/2)}{\sinh \pi Q_0} \sin 2\pi n_0 - 2\delta\right],$$

(7.7.26)

where $Q_0 = 2\pi/K^{1/2}$. Thus the condition for chaotic separatrix motion is

$$\delta < \frac{Q_0^3}{2} \frac{\sinh(\pi Q_0/2)}{\sinh \pi Q_0},$$

(7.7.27)

7.7. Determination of Homoclinic Intersection

or, for $Q_0 \gg 1$,

$$\delta < \frac{Q_0^3}{2} \exp\left(-\frac{\pi Q_0}{2}\right).$$

Melnikov's method can be extended to many-dimensional systems, and also to the case of a homoclinic cycle between two or more hyperbolic fixed points (Holmes, 1980). Using the latter, one can study the separatrix motion for secondary island chains around a primary fixed point.

CHAPTER 8
Chaotic Motion in Dissipative Systems

8.1. Transient Chaos

In Section 7.6 we showed how a crisis can cause a strange attractor to disappear, leading to a motion that can be transiently chaotic. A necessary condition for transient chaos, that the motion near a perturbed separatrix be chaotic, was described in Section 7.7. In this section, we consider the phenomenon of transient chaos, including a calculation of the transient distribution using a Fokker–Planck equation, and a calculation of the absorption rate into stable attracting fixed points. We also describe the transition from transient chaos to a steady state chaotic attractor due to a crisis. We defer consideration of steady state distributions for chaotic attractors to Section 8.2.

Transient chaos has been studied for a variety of dissipative maps, both on a 2-torus and on a cylinder (unbounded in action). For example, Chirikov and Izrailev (1976, 1981) studied the quadratic map on a 2-torus

$$p_{n+1} = \{p_n + Kg(x_n) - \delta(p_n - \tfrac{1}{2})\}, \qquad (8.1.1a)$$

$$x_{n+1} = \{x_n + p_{n+1} - \tfrac{1}{2}\}, \qquad (8.1.1b)$$

where $g(x) = x^2 - x + \tfrac{1}{6}$ and the braces $\{\ \}$ denote modulo 1. This map has a Jacobian $1 - \delta$, where δ is the dissipation parameter. For the Hamiltonian limit $\delta = 0$ with $K \gg 1$, the phase space ($0 \leqslant p < 1, 0 \leqslant x < 1$) is found to be mostly chaotic, with small embedded islands. For finite but small δ, the chaotic motion becomes transient, and all initial conditions ultimately are attracted into embedded islands. There are two types of behavior in the regime of transient chaos. (1) For weak dissipation, $0 < \delta < \delta_s, \delta_s \ll 1$, the mean rate of decay $\bar{\nu}$ of the chaotic transient, averaged over an ensemble of initial conditions, is an increasing function of δ. The

8.1. Transient Chaos

transient behavior in this case is determined by the underlying Hamiltonian dynamics and the rate of phase space contraction $1 - \delta$. (2) For $\delta_s < \delta < \delta_{cr}$, $\bar{v}$ decreases with δ and tends to zero as $\delta \to \delta_{cr}$. This behavior is dominated by a boundary crisis at $\delta = \delta_{cr}$ (Grebogi et al., 1986a; see Section 7.6a). For $\delta \geqslant \delta_{cr}$ all stable periodic orbits of the map are "disconnected" from an underlying attractor, which then exists as an independent attracting set. In Section 8.1a, we consider the parameter range $\delta < \delta_s$, and in Section 8.1b we consider transient chaos near crises.

8.1a. Weak Dissipation

We consider chaotic motion in weakly dissipative twist maps of the form (5.4.1)

$$u_{n+1} = u_n + \epsilon f(\psi_n, u_{n+1}, \delta),$$
$$\psi_{n+1} = \psi_n + 2\pi\alpha(u_{n+1}) + \epsilon g(\psi_n, u_{n+1}, \delta), \qquad (8.1.2)$$

where f and g are periodic in ψ with period 2π and $\delta \ll 1$ is the dissipation strength. The map (8.1.2) is Hamiltonian when $\delta = 0$, with u_n and ψ_n the unperturbed ($\epsilon = 0$) action-angle variables. The usual picture of Hamiltonian chaos then ensues, with intermingled regions of chaotic and regular motion, as shown in Fig. 1.14 for the example of the simplified Fermi map (3.4.4). For $0 < \delta \ll 1$, the fixed points at the centers of the islands in Fig. 1.14 become attracting centers (sinks), and the KAM curves no longer exist; numerical iteration of the map suggests that all persistent chaotic motion is destroyed. The complete destruction of persistent chaos when a very small dissipation is added to near-integrable Hamiltonian systems is typical behavior in such systems.

For mappings on a 2-torus with weak dissipation, the phase space distribution is nearly uniform over the transiently stochastic region, and, if δ is also a constant, then the rate of decay can be estimated to be

$$\bar{v} \approx \frac{A_I \delta}{A_s}, \qquad (8.1.3)$$

where A_I/A_s is the ratio of the total phase space area occupied by the stable islands to the area occupied by the stochastic trajectories in the underlying Hamiltonian ($\delta = 0$) system. This estimate can be understood as follows: In one iteration of the map, the area of each stable island A_k is reduced by the amount $A_k \delta$, capturing this much phase space from the surrounding stochastic sea. For all islands a phase space $A_I \delta = \sum_k A_k \delta$ is captured. For δ small the stochastic phase space continually redistributes itself to be uniform such that the density decay per mapping period is given by

$$\frac{\Delta P}{P} = -\frac{A_I \delta}{A_s},$$

where A_s is the stochastic phase space area. The resulting exponential contraction has an exponent given by (8.1.3).

For mappings on the cylinder, the calculation is more complicated in that the phase space distribution in action must be calculated to determine weighting factors for the density at each island. Taking the action variable to be u, as in the Fermi acceleration problem that has isolated primary stable islands in the region $u_s < u < u_b$ (see Section 3.4), then the decay rate for the transiently chaotic region is estimated as the density-weighted sum

$$\bar{v} = \sum_k v_k, \qquad (8.1.4a)$$

where

$$v_k = P_Q(u_k) A_k \delta / A_s \qquad (8.1.4b)$$

and the sum is over all stable primary fixed points u_k in the region $u_s < u < u_b$. Here P_Q is the time-independent part of an assumed quasistatic phase space distribution

$$P(u, n) = P_Q(u) \exp(-\bar{v} n), \qquad (8.1.5)$$

and n is the iteration number. P_Q is determined from the steady state solution of the Fokker–Planck equation

$$-\frac{\partial}{\partial u}(BP) + \frac{1}{2}\frac{\partial}{\partial u}\left(D\frac{\partial P}{\partial u}\right) = 0, \qquad (8.1.6)$$

where $D(u)$ is the diffusion coefficient for the underlying Hamiltonian part of the perturbation, and $B(u)$ is the friction coefficient due to the dissipative part (see Section 5.4a). Setting $\delta = 0$ in (8.1.2), the diffusion coefficient D is determined as in (5.4.7). Considering δ as perturbative, the friction coefficient can be determined using the random phase assumption

$$B(u) = \frac{\epsilon}{2\pi} \int_0^{2\pi} d\psi \, [f(\psi, u, \delta) - f(\psi, u, 0)]. \qquad (8.1.7)$$

The above results have been verified in a number of examples. In addition to the example given in (8.1.1), Izrailev et al. (1987) studied the generalized Hénon map, on a 2-torus

$$x_{n+1} = [y_n + d + ax_n^2], \qquad (8.1.8a)$$

$$y_{n+1} = [bx_n], \qquad (8.1.8b)$$

where [ ] denotes modulo 4; i.e., $|x| \leq 2$ and $|y| \leq 2$. The Jacobian of this map is $-b$. The mean decay rate $\bar{v}$ of the transient chaos was verified numerically for $a = 1$, $\delta = 1 - |b| = 0.01, 0.02$, and 0.05, and $0.1 \leq d \leq 1.0$. Furthermore, $\bar{v}$ was found to decrease rapidly as δ approached some critical value δ_{cr} and for $\delta > \delta_{cr}$, a strange attractor was observed.

A practical example of similar behavior arises in a nonuniformly sampling, digital phase-locked loop (Bernstein et al., 1989). The mapping equations

8.1. Transient Chaos

for this system on the 2-torus are

$$I_{n+1} = \{I_n - r\delta \sin \phi_n + \delta \sin(\phi_n - I_n)\}, \quad (8.1.9a)$$

$$\phi_{n+1} = \{\phi_n + I_{n+1}\}, \quad (8.1.9b)$$

where $\{\ \}$ denotes modulo 2π. This map has the additional feature that the Jacobian

$$J = 1 - \delta \cos(\phi - I)$$

can be less than or greater than one depending on the phase space coordinates. However, since the island of interest is centered at $\phi = 0, I = 0$, the cosine has a value near unity so that the Jacobian can be considered constant for the calculation, $J \approx 1 - \delta$, as in the previous examples. The map was studied numerically for $0.58 \leq \delta \leq 0.85$ and $3.5 \leq r \leq 4.5$. Over this parameter range the underlying Hamiltonian system contains a single relatively large stable island. The number of phase points undergoing transient chaotic motion decays exponentially with iteration number n, having mean decay rate $\bar{\nu}$. Expanding the term $\delta \sin(\phi_n - I_n)$ in (8.1.9), the lowest-order part, independent of I_n, combines with the remaining mapping, to give the underlying Hamiltonian map

$$I_{n+1} = I_n - (r-1)\delta \sin \phi_n, \quad (8.1.10a)$$

$$\phi_{n+1} = \phi_n + I_{n+1}, \quad (8.1.10b)$$

which is the standard map with stochasticity parameter $K = (r-1)\delta$. A grid of 25×25 initial conditions yields the typical surface of section shown in Fig. 8.1, consisting of a large stable island having area A surrounded by a stochastic sea.

Because the island area is rather large, a somewhat better approximation to the decay rate than (8.1.3) is used. If A contracts by an amount δA, then the attracting region captures a fraction $\delta A/(A_0 - A)$ of the transiently chaotic phase space, where $A_0 = 4\pi^2$. Equating $e^{-\bar{\nu}}$ to the fraction not captured, we obtain

$$\bar{\nu} = -\ln\left[1 - \frac{\delta A}{A_0 - A}\right]. \quad (8.1.11)$$

We recover (8.1.3) by expanding the logarithm to lowest order. A typical case studied was $r = 4.0$, $\delta = 0.688$, such that $K = 2.06$. The measured decay rate was 0.15/iteration, and the theoretical decay rate from (8.1.11) was 0.18/iteration, with this error typical over the cases studied (Bernstein et al., 1989).

The action variable is not always defined on a torus (e.g., the Fermi acceleration map), in which case the distribution P is a function of action. As described above, the fraction of initial conditions captured by each island depends on the value of the distribution function near that island, which must be determined by solving the Fokker–Planck equation.

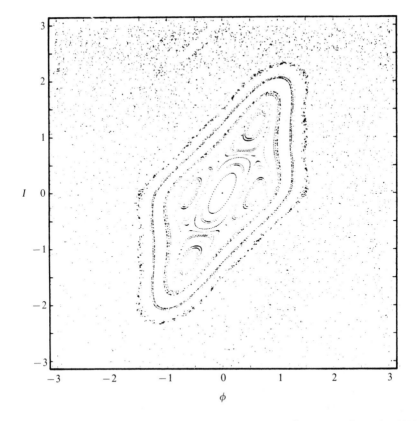

Figure 8.1. I versus ϕ for the standard map at $K = 2.13$ (after Bernstein *et al.*, 1989).

We illustrate the calculation procedure using the example of the dissipative Fermi map (Tsang and Lieberman, 1986). The dissipation is introduced by assuming that the ball suffers a fractional loss δ in velocity upon collision with the fixed wall. The map is then

$$\bar{u} = (1 - \delta)u_n - \sin \psi_n, \qquad (8.1.12a)$$

$$\bar{\psi} = \psi_n + 2\pi M/\bar{u}, \qquad (8.1.12b)$$

$$(\psi_{n+1}, u_{n+1}) = (\bar{\psi}, \bar{u}) \operatorname{sgn} \bar{u}. \qquad (8.1.12c)$$

The function $\operatorname{sgn} \bar{u} = \pm 1$ for $\bar{u}$ positive or negative, respectively, is introduced to maintain $u_{n+1} \geqslant 0$ for low velocities $u_n < (1 - \delta)^{-1}$, while preserving the continuity of the map near $u = 0$. The Jacobian of the map is $1 - \delta$. The primary fixed points of the map are found as in Section 3.4 by setting $u_{n+1} = u_n$ and $\psi_{n+1} = \psi_n \pmod{2\pi}$ in (8.1.12) to obtain

$$(u_k, \psi_k) = (M/k, \sin^{-1}(-u_k\delta)), \qquad (8.1.13)$$

8.1. Transient Chaos

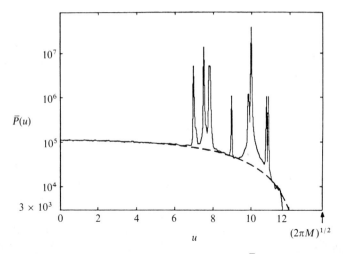

Figure 8.2. Cumulative, phase-averaged distribution $\bar{P}$ versus u, for $M = 30$, $\delta = 0.003$, and $N = 5 \times 10^4$ iterations. The solid curve shows the numerical result; the dashed curve shows the quasistatic theory (after Tsang and Lieberman, 1986).

where k is an integer. For $u_k \delta \ll 1$, as in the Hamiltonian map, there are two fixed points for each k at $\psi_k \approx 0$ and $\psi_k \approx \pi$, with $\psi_k \approx \pi$ stable for $u_k > u_s = (\pi M/2)^{1/2}$ and $\psi_k \approx 0$ always unstable. For weak dissipation, $0 < \delta < \delta_{cr}$, the fixed points of the Hamiltonian map become attractors (sinks), and all persistent chaotic motion is destroyed, with transient chaotic motion surrounding the sinks below $u_b = (2\pi M)^{1/2}$. In Fig. 8.2 we plot the cumulative phase-integrated distribution

$$\bar{P}(u) = R \int_0^N dn \int_0^{2\pi} d\psi \, P(u, \psi, n),$$

for $M = 30$, $\delta = 0.003$, after $N = 5 \times 10^4$ iterations, for $R = 100$ initial conditions at low velocities chosen randomly. We see evidence of attracting sinks between u_s and u_b near the primary resonances at $k = 3$ (a period 1 and a period 5 sink coexist) and at $k = 4$ (a period 1 and a period 3 sink coexist). The density leaving the stochastic region flows into these sinks, forming spikes in the figure. The location and structure of these sinks correspond to the Hamiltonian ($\delta = 0$) structure of the stable fixed points (8.1.13) of the Fermi map. The period 3 and period 5 sinks correspond to secondary fixed points encircling the period 1 primary fixed points.

To compare the numerical results with theory we solve the Fokker–Planck equation (8.1.6) for the invariant distribution, using $B = -u\delta$ obtained from (8.1.7), and a value of D which can be approximated over the entire range $0 < u < u_b$ as

$$D = \tfrac{1}{2}(1 - u^2/u_b^2)^2. \tag{8.1.14}$$

With the quasi steady state assumption that the net flux is zero,

$$\tfrac{1}{4}\left(1 - \frac{u^2}{u_b^2}\right)^2 \frac{dP}{du} + u\delta P = 0,$$

the solution is

$$P_Q(u) \propto \exp\left[-\frac{2\delta u_b^2 u^2}{u_b^2 - u^2}\right]. \tag{8.1.15}$$

This distribution, scaled to the value of $\bar{P}$ at $u = 0$, is plotted as the dashed line in Fig. 8.2 for $M = 30$ and $\delta = 0.003$. Equally good agreement is found for all other cases studied. Using (8.1.4), Tsang and Lieberman compared the numerical and analytical values of the decay rate $\bar{v}$ for the range of parameters $0.0003 \leqslant \delta \leqslant 0.01$ and $30 \leqslant M \leqslant 300$, with the theory and experiment typically agreeing to within a factor of two. As examples, for $\delta = 0.01$ and for $M = 300$, $\bar{v} = 1.1 \times 10^{-5}$ numerically and 0.9×10^{-5} analytically. For $M = 30$, the corresponding values are $\bar{v} = 2.7 \times 10^{-5}$ and 6.6×10^{-5}, respectively. The large deviation in the second case is mainly due to local significant variation from the approximation used for the stable island areas.

Local Description. A local, but more complete description of transient chaos has been put forth by various authors (Grebogi et al., 1986a; Widom et al., 1983; Kantz and Grassberger, 1985; Szepfalusy and Tel, 1986). Because the number of phase points in a region of transient chaos decays exponentially with iteration number n, there exist initial conditions having arbitrarily long, transiently chaotic trajectories. Each such limiting trajectory is a fractal object having zero measure, which can be called a *chaotic repeller*, and which is part of the boundary of the basin of the (nonchaotic) attractor (see Section 7.6). On each of these repellers embedded in phase space, one generally finds a unique invariant distribution and one can define topological and metric entropies, Liapunov exponents and various fractal dimensions, just as these are defined for the persistent chaos on a chaotic attractor. In some cases, initial conditions that are "close" to a chaotic repeller are immediately repelled. In other cases, nearby initial conditions are first attracted to the repeller, and repulsion occurs at much later times, in directions transverse to the attracting ones. Such chaotic repellers may be called *chaotic semi-attractors*.

For a repeller in a one-dimensional map, such as the quadratic map (7.6.1), $x_{n+1} = a - x_n^2$, Kantz and Grassberger (1985) propose that the decay rate $\bar{v}$ of the chaotic transient near a repeller can be written as

$$\bar{v} = (1 - D_1)\sigma, \tag{8.1.16}$$

where D_1 is the information dimension (see Section 8.3) and σ is the (positive)

8.1. Transient Chaos

Liapunov exponent on the repeller. Expression (8.1.16) for the decay rate $\bar{v}$ can be understood in terms of the fractal dimension (capacity) D_0, which closely bounds D_1 from above. Partition an interval l covering the repeller into equal subintervals of length $\epsilon \ll l$. The number M of subintervals required to cover the repeller within l is $M(\epsilon) = M_0 \epsilon^{-D_0}$, where D_0 is the fractal dimension. The total length of these subintervals is $R = M_0 \epsilon^{1-D_0}$. After one iteration, the interval l is mapped to a new interval l' that covers the repeller, and the interval and subintervals are increased in length by the factor e^σ, where σ is the Liapunov exponent. Subintervals having length $\epsilon e^{-\sigma}$ within l are mapped to subintervals having length ϵ within l'. The total length of the subintervals within l that lie on the repeller within the interval l' is

$$l'_R = M(\epsilon e^{-\sigma})\epsilon = M_0 \epsilon^{1-D_0} e^{\sigma D_0}.$$

Thus after one iteration, the fraction $e^{-\bar{v}}$ of phase points remaining on the repeller is

$$e^{-\bar{v}} = \frac{l'_R/l'}{l_R/l} = e^{-\sigma(1-D_0)},$$

which is the result in (8.1.16) with $D_0 \to D_1$.

Generalizing to higher-dimensional systems, one may write

$$\bar{v} = \sum_{i=1}^{d}{}' \sigma_i(1 - D_{1i}), \qquad (8.1.17)$$

where the D_{1i} are the information dimensions along the directions defined by the d Liapunov exponents σ_i, and the sum is over the unstable directions ($\sigma_i > 0$) only. The D_{1i}'s are related to the total information dimension and the Liapunov exponents as described in Section 8.3. The decay rate (8.1.17) has been verified for three different two-dimensional maps (Krantz and Grassberger, 1985).

Returning to the map (8.1.1), we illustrate this behavior in Fig. 8.3, showing the attractors and repellers for $K = 3.5$ and $\delta = 0.1$. The heavy dots represent a stable period 6 cycle, whose attractor basin is extremely small. The main attractor is the fixed point marked by a cross. The fourfold structure around this fixed point is a semi-attractor, consisting of a chaotic period 4 cycle. Phase space in its neighborhood is attracted rapidly onto the structure and is then repelled, leading to a slow decay of the density at a rate v_1. Figure 8.3 also shows a repeller characterized by the dots scattered all over the figure. Trajectories escape from this repeller by either going directly towards the fixed point, or by being attracted first towards the semi-attractor, with the latter found to be the more frequent. The decay rate v_2 of the large repeller was found to be fast. If we assume that almost all the phase space from the large repeller is first attracted to the semi-attractor then the two structures

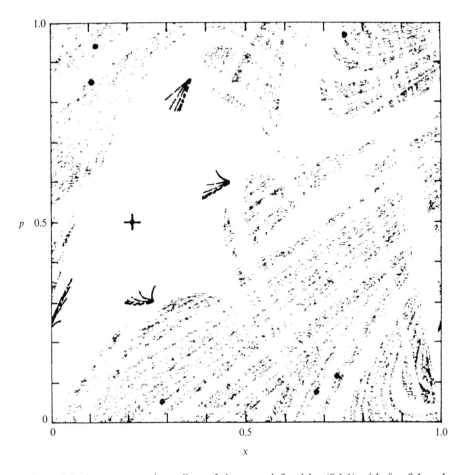

Figure 8.3. Attractors and repellers of the map defined by (8.1.1) with $\delta = 0.1$ and $K = 3.5$. The cross at $(x, p) = (0.2133, 0.5)$ is the main attractor. A second attractor with a very small basin is made up by the six heavy dots. The four fans around the main attractor represent the semi-attractor which governs the long-time transients. The remaining points form a repeller with shorter lifetime (after Kantz and Grassberger, 1985).

are in series. The average lifetime of the chaotic transient is then the sum of the individual lifetimes, $T = v_1^{-1} + v_2^{-1}$.

In general, the repeller structure may be very complicated, with repellers and semi-attractors connected in complicated patterns. This is particularly true in weakly dissipative systems, where many interwoven basins of attraction exist, each having its own connections to local repeller structures. It is presumably the combination of these complicated local structures that yields the overall global decay rate, which must be equivalent to the absorption rate into the sinks, as calculated from (8.1.4).

8.1b. Transient Chaos Near a Crisis

In the strongly dissipative regime $\delta_s < \delta < \delta_{cr}$, the transient behavior is dominated by a boundary crisis as described in Section 7.6. The decay rate $\bar{v}$ tends toward zero as δ approaches δ_{cr}, and a strange attractor exists for $\delta > \delta_{cr}$. For values of δ slightly below δ_{cr}, it is often found (Grebogi et al., 1986a; see Section 7.6a) that $\bar{v} \sim (\delta_{cr} - \delta)^{\gamma}$, where γ is the critical exponent of the chaotic transient.

The one-dimensional quadratic map (7.6.1) exhibits this behavior near the critical value $a_{cr} = 2$. For $a < a_{cr}$, a chaotic attractor exists over the range $-x_1 < x < x_2$, which itself lies within a basin of attraction $-x^* \leq x \leq x^*$, where

$$x^* = \tfrac{1}{2} + (\tfrac{1}{4} + a)^{1/2} \tag{8.1.18}$$

and where $-x^*$ is the location of the unstable fixed point of the map. As $a \to 2$, $-x_1 \to -x^*$, the strange attractor collides with its basin boundary (the "crisis") and is destroyed. For $a > 2$, the region $-x_1 < x < x_2$ formerly occupied by the chaotic attractor is a region of transient chaos, with the orbit ultimately tending to an attractor at $x = -\infty$. The decay rate of the chaotic transient is found to scale as

$$\bar{v} \sim (a - a_{cr})^{1/2}. \tag{8.1.19}$$

This scaling results because, for $a > a_{cr}$, there is an escape region $|x| < (a - x^*)^{1/2}$ within the region of transient chaos, which first maps into $x > x^*$, then maps into $x < -x^*$, and thereafter on repeated iteration to the attractor at $x = -\infty$. Writing $a = a_{cr} + \Delta a$ and using (8.1.18), we obtain

$$|x| < [\tfrac{2}{3}(a - a_{cr})]^{1/2}$$

for the size of the escape region, and the scaling (8.1.19) immediately follows.

Similar phenomena are observed in higher dimensions (Grebogi et al., 1986a). There appear to be two possible crises in strictly dissipative ($J < 1$ everywhere) two-dimensional maps. These are illustrated in Fig. 8.4. In (a), the stable manifold of an unstable periodic orbit (B) on the basin boundary collides with the unstable manifold of an unstable periodic orbit (A) on the strange attractor. In (b), the stable and unstable manifolds of an unstable periodic orbit (B) on the basin boundary collide. In both cases, the chaotic attractor is the closure of the branch of the unstable manifold of (B) that points into the basin of attraction. The critical exponents for these crises have been derived by Grebogi et al. (1986a) and verified for the Hénon map [(8.1.8), but not mod 4]. For type (a)

$$\gamma_a = \tfrac{1}{2} + \frac{\ln|\lambda_1|}{|\ln|\lambda_2||}, \tag{8.1.20a}$$

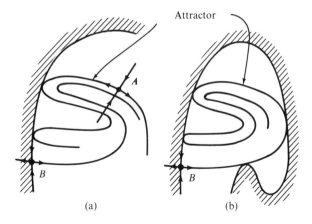

Figure 8.4. (a) Schematic illustration of heteroclinic tangencies of the stable manifold of the unstable periodic orbit B and the unstable manifold of the unstable periodic orbit A. (For simplicity, we take the periods of A and B to be 1.) Crosshatching denotes the basin of another attractor. (b) Schematic illustration of homoclinic tangencies of the stable and unstable manifolds of the unstable periodic orbit B (from Grebogi et al., 1986a).

and for type (b),

$$\gamma_b = \frac{\ln|\lambda_2|}{\ln|\lambda_1 \lambda_2|^2}. \tag{8.1.20b}$$

Here, λ_1 and λ_2 are the contracting and expanding eigenvalues, respectively, of the periodic orbits A and B.

8.2. Invariant Distributions on Strange Attractors

For area-preserving maps, the invariant distribution on any chaotic subset in the surface of section is trivially a constant. For chaotic motion on a strange attractor in a dissipative system, the invariant distribution $P(x)$ is not known *a priori* and must be found for each attractor of interest. A common procedure for one-dimensional maps, based on iterating an initial distribution using (7.2.68), was described in Section 7.2c. For two-dimensional maps that are strongly attracting, the strange attractors can be reduced to an approximate one-dimensional form, and the invariant distribution in one dimension determined as in Section 7.2c. We describe this procedure in Section 8.2a. For weakly dissipative maps, such as (8.1.2), the invariant distribution can be calculated reasonably accurately from a few iterations of

8.2. Invariant Distributions on Strange Attractors

an initial phase space distribution, provided the initial distribution is chosen efficiently. We shall consider this case in Section 8.2b and show that the phase-averaged distribution, obtained from solving the Fokker–Planck equation, is such an efficient choice.

As mentioned in Section 7.2c, $P(x)$ must have the property that it is invariant under the flow:

$$P(x) = TP(x), \qquad (8.2.1)$$

where T is the mapping or the transformation between Poincaré sections. Although there are many invariant distributions for an attractor, a particular one (the equilibrium distribution) is singled out by the fact that the time average over almost any orbit is equal to the ensemble average for this distribution. For convenience we refer to the equilibrium invariant distribution as the invariant distribution, ignoring all others.

By use of the invariant distribution, we may replace time averages by space averages in calculating the steady state value of a physical observable. If $G(x)$ is an observable function on phase space, then the time average of G over the orbit starting at x_0 is

$$\bar{G}(x_0) = \lim_{n \to \infty} \frac{1}{n} \sum_{i=0}^{n-1} G(T^i x). \qquad (8.2.2)$$

For almost all x_0 in the basin of a given attractor, $\bar{G}$ is the same, independent of x_0, i.e.,

$$\bar{G} = \int dx\, P(x) G(x), \qquad (8.2.3)$$

where $P(x)$ is the equilibrium distribution for the attractor. The calculation of $\bar{G}$ using (8.2.3) is often more convenient than using (8.2.2). The basic procedure for finding P consists of iterating (8.2.1)

$$P^{(i+1)}(x) = TP^{(i)}(x), \qquad (8.2.4)$$

starting with an initial guess $P^{(0)}(x)$ [assuming $P^{(0)}(x)$ is zero for x not contained in the attracting basin], to obtain[1]

$$P(x) = \lim_{i \to \infty} P^{(i)}(x).$$

8.2a. Reduction to a One-Dimensional Map

Some progress in reducing the computation required to obtain the invariant distribution was made by Bridges and Rowlands (1977), who considered

---

[1] In calculating a practical average of an observable, it is usually sufficient to use a coarse-grained invariant distribution, in which one averages over the fine-scaled features.

two-dimensional maps that can be approximately described in a one-dimensional limit. Considering maps of the form

$$\bar{x} = F(x, y), \tag{8.2.5a}$$

$$\bar{y} = bG(x, y), \tag{8.2.5b}$$

with b a small parameter, we assume in zero order that $y = 0$ in (8.2.5a) to obtain the one-dimensional map

$$\bar{x} = F(x, 0) = F_0(x). \tag{8.2.6}$$

The y-variation of the attractor is obtained by setting $y = 0$ in (8.2.5b) and inverting to find

$$x = G_0^{-1}\left(\frac{\bar{y}}{b}\right).$$

Inserting this into (8.2.6) yields the zero-order equation for the attractor

$$\bar{x} = F_0\left(G_0^{-1}\left(\frac{\bar{y}}{b}\right)\right). \tag{8.2.7}$$

To proceed to higher order in the expansion for the attractor, we write (x, y) for $(\bar{x}, \bar{y})$ in (8.2.7) and solve for y to obtain

$$y = bG_0(F_0^{-1}(x)). \tag{8.2.8}$$

Using this in (8.2.5a) as before yields the first-order equation for the attractor. The Hénon map (7.1.26) is of the form (8.2.5) with $G(x, y) = x$. The structure of the attractor calculated by the above procedure agrees surprisingly well with numerical calculations.

Although Bridges and Rowlands did not determine the invariant distribution, we can write in zero order

$$P^{(0)}(x, y) = P_1(x)\delta(y - y(x)), \tag{8.2.9}$$

where $P_1(x)$ is the invariant distribution associated with the one-dimensional map (8.2.6) and $y(x)$ is given by (8.2.8). For $b \ll 1$ this is already a good approximation for P.

It is common to view a three-dimensional flow as an approximate one-dimensional noninvertible map and then numerically calculate the invariant distribution for the map (Shaw, 1981; Ott, 1981). Two examples have already been given: the Lorenz attractor in Section 1.5 and the Rössler attractor in Section 7.1c. However, few direct comparisons between the actual distribution and the one-dimensional approximation have been made. Izrailev et al. (1981) have compared a numerical calculation of $P_1(x)$ with a

8.2. Invariant Distributions on Strange Attractors

numerical calculation of

$$\int P(x, y)\, dy$$

for the three-dimensional flow of a parametrically excited nonlinear oscillator, finding good agreement. They also point out that the small parameter in (8.2.5b) is related to the fractional part of the fractal dimension $D_{KY} = \sigma_1/|\sigma_2|$:

$$b \sim \exp[|\sigma_2|(D_{KY} - 1)].$$

8.2b. Direct Iteration of an Initial Distribution

The basic assumption is that the invariant distribution can be found iteratively from an initial distribution that is phase averaged, and thus determined from a Fokker–Planck equation in the action alone. Successively better approximations to the invariant distribution are then found by repeated iteration of the distribution over the map, with typically only a few iterations required to give a reasonable representation of the distribution. This method is most easily applied in the limit when the volume contraction rate is near zero, $|\sigma_1 + \sigma_2| \approx 0$, where the method of Bridges and Rowlands is inapplicable.

Phase-Averaged Distribution. The phase-averaged distribution can be found from the steady state Fokker–Planck equation. If there is no net flux, (8.1.6) becomes

$$-BP + \tfrac{1}{2}D\frac{dP}{du} = 0. \qquad (8.2.10)$$

Taking as an example the dissipative Fermi map (8.1.12), with $\delta \geqslant 0.01$, the distribution is mainly confined to $u \ll u_b$ such that the quasilinear value $D \approx \tfrac{1}{2}$ is a good approximation. Using (8.1.7), $B = -\delta u$ is obtained from (8.1.12). Substituting these values in (8.2.10) and integrating,

$$P(u) = \left(\frac{8\delta}{\pi}\right)^{1/2} \exp(-2\delta u^2). \qquad (8.2.11)$$

In Fig. 8.5, this analytical expression (solid line) is compared with the numerical results for $\delta = 0.1$ and various values of M in the range 20–500. The numerical points all lie along the same straight line, independent of M, as predicted by (8.2.11).

Phase-averaged invariant distributions can also be found when D is a function of the action, and in the presence of sources and sinks. For area-preserving maps, a solution to the Fokker–Planck equation with a

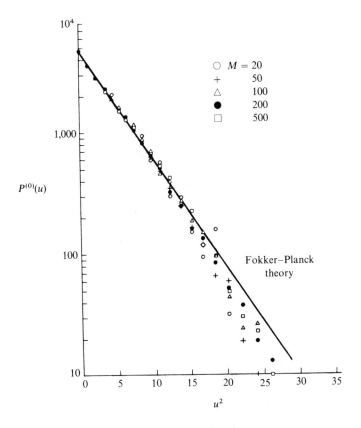

Figure 8.5. Comparison of the numerically calculated invariant distribution integrated over phase ψ with the Fokker–Planck solution $P^{(0)}(u)$, for $\delta = 0.1$ and various values of M (after Tsang and Lieberman, 1984a).

source at u_1 and a sink at u_2 is

$$P(u) = \int_u^{u_2} \frac{A}{D(u)}\,du, \tag{8.2.12}$$

with A proportional to the source strength. For dissipative systems there is a competition between the dissipation and the flow which determines the invariant distribution. If we consider a source at low velocity, then for the dissipative Fermi mapping (8.1.12), the time-independent Fokker–Planck equation, using D from (8.1.14) is

$$\frac{1}{2}\left(1 - \frac{u^2}{u_B^2}\right)^2 \frac{dP}{du} - u\delta P = \Gamma. \tag{8.2.13}$$

8.2. Invariant Distributions on Strange Attractors

Dividing by $\frac{1}{4}(1 - u^2/u_B^2)^2$, the use of the integrating factor $\exp[\int 4u\delta/(1 - u^2/u_B^2)^2]$ yields the solution

$$P = \exp\left[-\int^u \frac{4\delta u\, du}{(1 - u^2/u_B^2)^2}\right]\int_u^{u_2} \frac{du\, 4\Gamma}{(1 - u^2/u_B^2)^2} \exp\left[\int^u \frac{4\delta u\, du}{(1 - u^2/u_B^2)^2}\right]. \quad (8.2.14)$$

For flow-dominated behavior (weak dissipation) we can approximate the solution as obtained from (8.2.12), while for strong dissipation the solution approximates that obtained from (8.2.11). The transition region can be important in some pnysical problems, such as the stochastic acceleration of electrons in the r.f. sheaths of a capacitive discharge, in which the dissipation arises from interactions with the background gas (Goedde et al., 1988).

Iterating the Phase-Averaged Distribution. If we start with $P^{(0)}(\psi, n) = P(u)$, to find successively higher-order approximations we iterate $P^{(0)}$ successively by the map. Letting $z' = (\psi', u')$ be the pre-image of $z = (\psi, u)$, and $z_{n+1} = Tz_n$, where T is the map (8.1.2), then the i and $i + 1$ approximations are related by

$$P^{(i+1)}(\psi, u)\, d\psi\, du = P^{(i)}(\psi', u')\, d\psi'\, du', \quad (8.2.15)$$

which yields

$$P^{(i+1)}(z) = JP^{(i)}(z'(z)), \quad (8.2.16)$$

where $J(T^{-1}z) = |\partial(\psi', u')/\partial(\psi, u)|$ is the Jacobian of the inverse map $T^{-1}z = z'(z)$. The first-order invariant distribution is then

$$P^{(1)}(z) = J(T^{-1}z)P^{(0)}(T^{-1}z). \quad (8.2.17)$$

By repeatedly applying (8.2.16) we obtain the nth-order approximation

$$P^{(n)}(z) = P^{(0)}(T^{-n}z)\prod_{i=1}^{n} J(T^{-i}z). \quad (8.2.18)$$

Equation (8.2.18) has been verified for strange attractors in both the dissipative Fermi map (8.1.12) and the dissipative standard map (7.3.36) (Tsang and Lieberman, 1984). As an example, for the Fermi map with $M = 100$, numerical computations suggest[2] that a strange attractor exists

---

[2] An estimate for the regime of δ and M where a strange attractor exists is found by requiring that all primary fixed points $u_k = M/k$ be unstable in the phase space region $0 \leqslant u \leqslant \delta^{-1}$ where the attractor can exist. A rough estimate for this is found by requiring that the kth-order fixed point disappears (inverse tangent bifurcation), $M\delta/k = 1$, at the same M and δ for which the $(k + 1)$st-order point loses stability, $\pi(k + 1)[(k + 1)^2 - M^2\delta^2]^{1/2} \approx 2M$. Eliminating k and solving for δ in the limit of large k yields the condition $\delta \gtrsim (2/\pi^2 M)^{1/3}$ for a strange attractor [$\delta \gtrsim 0.13$ at $M = 100$]. Actually, δ is not a smooth function of M, and thin regions of stability exist for larger values of δ. The observation of chaotic motion for smaller values of δ, e.g., as shown in Fig. 8.6, does not demonstrate the existence of a strange attractor. The observed quasistationary chaotic motion may be transiently chaotic with a lifetime much exceeding the iteration time, because, from (8.2.11), the phase-averaged distribution is (exponentially) small in the accessible phase space region where stable islands can exist.

for $\delta > 0.03$. A typical case is $\delta = 0.1$. Figure 8.6a shows a portion of the (u, ψ) surface of section in the range $4.4 \leq u \leq 4.8$. The leaved structure of the attractor is quite evident. There are finer structures within the leaves. Here the region has been divided into 100 intervals along u and 100 intervals along ψ, forming 10,000 cells. The map is iterated 3×10^6 times for a single initial condition, and the number inside each cell (not readily seen) is a logarithmic measure of the number of occupations. If the number of occupations per cell is summed over the phases ψ at a fixed u, the phase-averaged invariant distribution $P(u)$ is obtained numerically, as plotted in Fig. 8.5.

To determine P analytically, to some low order, the steady state solution (8.2.11) to the phase-averaged Fokker–Planck equation is used as the zero-order distribution. The first-order distribution is then calculated from (8.2.17) with $J = (1 - \delta)^{-1}$, and T^{-1} is found by inverting the Fermi map (8.1.12) [for $u > 0$],

$$\psi = \bar{\psi} - \frac{2\pi M}{\bar{u}},$$

$$u = \frac{\bar{u} + \sin[\bar{\psi} - (2\pi M/\bar{u})]}{1 - \delta},$$

to obtain

$$P^{(1)}(\psi, u) = \frac{1}{1-\delta}\left(\frac{8\delta}{\pi}\right)^{1/2} \exp\left\{-\frac{2\delta}{(1-\delta)^2}\left[u + \sin\left(\psi - \frac{2\pi M}{u}\right)\right]^2\right\}.$$

(8.2.19)

To compare this with the numerically calculated invariant distribution $P(\psi, u)$, the expected occupation number is plotted in Fig. 8.6b using $P^{(1)}$, in the same expanded region of the surface of section as for the numerical calculation of P in Fig. 8.6a. The same number of total occupations (3×10^6) and the same logarithmic measure for the number of occupations in each of the 100×100 cells are used. The band structure seen in the magnified image of $P^{(1)}$ corresponds closely to the numerically determined bands seen in Fig. 8.6a. The higher-order expressons $P^{(2)}$ and $P^{(3)}$ were determined (Tsang and Lieberman, 1984) and are plotted in Figs. 8.6c and 8.6d, showing successively better agreement with the numerical calculation.

Any initial distribution $P^{(0)}(\psi, u)$ can be iterated numerically to obtain successively better approximations to the invariant distribution on the attractor, as was done in Section 7.2c for a one-dimensional map. However, the choice of $P^{(0)}(\psi, u)$ as the solution of the Fokker–Planck equation greatly facilitates the convergence. If an analytic approximation is desired such as $P^{(1)}(\psi, u)$ in (8.2.19), then the choice of a good initial distribution is an essential first step to make the method tractable.

8.2. Invariant Distributions on Strange Attractors

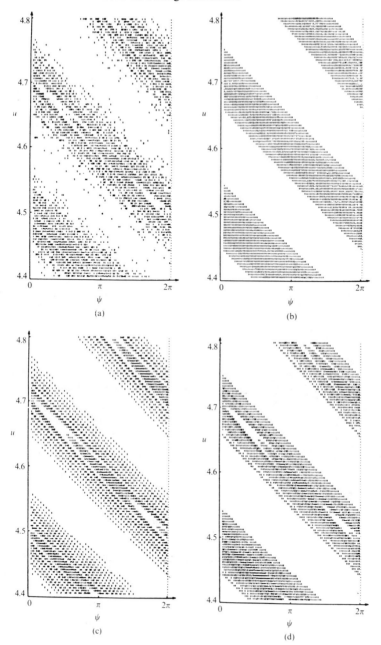

Figure 8.6. A portion of the surface of section $u - \psi$ for the dissipative Fermi map (8.1.12) with $M = 100$ and $\delta = 0.1$. (a) One particle with 3,000,000 collisions of which 9382 occupations appear in the range $4.4 \leq u < 4.8$; (b) first-order analytical result $P^{(1)}(\psi, u)$; (c) second-order result $P^{(2)}(\psi, u)$; (d) third-order result $P^{(3)}(\psi, u)$; the functions are plotted by computer with the scales and effective number of iterations the same as for the numerical result (a) (after Tsang and Lieberman, 1984a).

The general procedure described above can also be used as the dissipation becomes stronger, but the Fokker–Planck solution becomes an increasingly poor first approximation. In fact, the inaccuracy arises not from the phase averaging, but from the approximations by which the Fokker–Planck differential equation is derived from the integral equations that describe the Markov process. Tsang and Lieberman (1984a) used the more basic relations to obtain a good approximation to the phase-averaged distribution for the dissipative standard map (7.3.36) for large dissipation strengths. They were then able to obtain analytically the successively finer structure of the attractor distribution by successive iterations of the map, as we have shown in Fig. 8.6 for the case of weak dissipation.

8.3. Multifractals and Their Properties

In Section 7.1d we described a simple *homogeneous fractal* or Cantor set and, using as an example the "middle-thirds Cantor set," calculated a fractal dimension, called its *capacity*. However, the scaling structures in dynamical systems are generally more complicated than those of homogeneous fractals. A more general concept is that of a *multifractal* (Mandelbrot, 1974), which is an interwoven structure of infinitely many fractals. Multifractals are characterized by an infinite number of dimensions and a related scaling index spectrum.

An example of a multifractal is the attractor for the circle map (7.4.2) with the golden mean winding number at the onset of chaos $K = 1$. Figure 7.36 shows that there is considerable variation in the density of points along the attractor.

8.3a. Generalized Dimensions

To quantify these observations we first introduce a probability density on a fractal set. Consider N points on a fractal embedded in a d-dimensional box of size l_0. On the coarsest scale there are N points in volume l_0^d. At the first level we partition the box into M boxes of size $(l_0/s_i)^d$, $i = 1, \ldots, M$, with box i containing N_i points. The probability density of this level is $p_i = N_i/N$. (Note that boxes can have the same value of s.) At the second level we partition each of the M boxes in a manner similar to the first level partition, obtaining M^2 boxes, and so on. A simple example (following Halsey et al., 1986) is shown in Fig. 8.7. Here, we erase an open segment from each "mother" segment at one level to produce two "daughter" segments at the next level, with each left-side (right-side) daughter having a length $\frac{1}{4}$ ($\frac{2}{5}$), and probability $\frac{3}{5}$ ($\frac{2}{5}$) times that of the mother. This probability is proportional

8.3. Multifractals and Their Properties

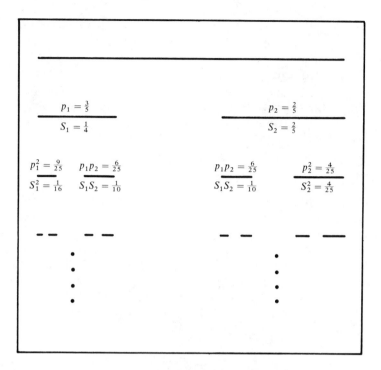

Figure 8.7. The construction of a two-scale Cantor set (after Halsey et al., 1986).

to the number of times that an orbit visits a particular segment in a dynamical system.

Definitions. Following Hentschel and Procaccia (1983), we generalize the concept of fractal dimension by introducing the quantity

$$D_{Cq} = \frac{1}{q-1} \lim_{\epsilon \to 0} \frac{\ln \sum_{i=1}^{M(\epsilon)} p_i^q}{\ln \epsilon}. \tag{8.3.1}$$

The attractor is covered by $M(\epsilon)$ boxes of size ϵ, and p_i is the probability density in the ith box. For $q = 0$, (8.3.1) yields the *capacity* D_{C0}. Taking the limit $q \to 1$, we obtain

$$D_{C1} = \lim_{\epsilon \to 0} \frac{\sum_i p_i \ln p_i}{\ln \epsilon}, \tag{8.3.2}$$

which is the *information dimension* (Renyi, 1959). For $q = 2$, we obtain the *correlation dimension* D_{C2} (Grassberger and Procaccia, 1983). The generalized dimension D_{Cq} can be evaluated for $-\infty < q < \infty$. It can be shown that $D_{Cq} \geq D_{Cq'}$ for any $q' > q$, where the equality generally obtains only for a homogeneous fractal. Hence, $D_{C0} \geq D_{C1}$, which is a reasonable result since all points are counted equally in determining D_{C0}, and there are generally more lower probability points than higher probability points on an attractor.

Related to the capacity is a concept, previously introduced, called the *Hausdorff dimension* (Hausdorff, 1918). To define it we cover the attractor with boxes of variable size $\epsilon_i < \epsilon$ and introduce the partition sum

$$\Gamma_{H0} = \sum_i \epsilon_i^D. \tag{8.3.3}$$

We vary the ϵ_i (for a fixed ϵ) to minimize Γ_{H0} and then take the limit $\epsilon \to 0$. It can be shown that this limit is either zero or infinity, except for one particular value of $D = D_{H0}$, which is the Hausdorff dimension. The capacity D_{C0} is an "unoptimized" version of the Hausdorff dimension, and hence examples can be constructed having $D_{C0} > D_{H0}$. However, all chaotic attractors that have been tested have $D_{C0} = D_{H0} \equiv D_0$ (Grebogi et al., 1988), which we use as our subsequent notation for both quantities.

The Hausdorff dimension has been generalized by Grassberger (1985) and Halsey et al. (1986), by introducing a partition sum

$$\Gamma_{Hq} = \sum_i \frac{p_i^q}{\epsilon_i^{\tau(q)}} \tag{8.3.4}$$

with

$$\tau(q) = (q-1)D_q. \tag{8.3.5}$$

For a fixed q, optimizing over the ε_i, and taking the limit as $\epsilon \to 0$, we obtain a unique $\tau(q)$, hence D_{Hq}, for which Γ_{Hq} is neither zero nor infinity. Again, it can be shown that $D_{Hq} \geq D_{Hq'}$ for $q' > q$. We expect $D_{Cq} = D_{Hq} \equiv D_q$ for all chaotic attractors.

An Analytic Calculation. In Section 7.1d we found the capacity of the middle-thirds Cantor set. Extending this calculation to the generalized capacity dimensions (8.3.1) for a multifractal seems straightforward, but this is not the case, because for any fixed ϵ, $M(\epsilon)$ is the number of boxes needed to cover the Cantor set itself, and not the number needed to cover the set at a certain level. $M(\epsilon)$ is very difficult to find analytically; e.g. see Tsang (1986) for further discussion and an example.

On the other hand, the generalized Hausdorff dimension (8.3.4) can often be found analytically. We illustrate this for the multifractal of Fig. 8.7. The partition sum for the nth level is $\Gamma_{Hq}^{(n)} = g_q^n$, where, from (8.3.4),

$$g_q = \frac{(\frac{3}{5})^q}{(\frac{1}{4})^\tau} + \frac{(\frac{2}{5})^q}{(\frac{2}{5})^\tau} \tag{8.3.6}$$

8.3. Multifractals and Their Properties

is the *fractal generator* for constructing the Cantor set. For $\Gamma_{Hq}^{(n)}$ to remain finite and $n \to \infty$ we must have $g_q = 1$, which determines $\tau(q)$. Setting $q = 0$ and $g_q = 1$ in (8.3.6) and solving numerically, we obtain $\tau(0) \approx -0.6110$ and hence, from (8.3.5), $D_0 \approx 0.6110$. Fixing any value of q in (8.3.6), we can obtain $\tau(q)$ and hence D_q in a similar manner. The case $q = 1$ is special because $q - 1 = 0$; hence $\tau(1) = 0$ and (8.3.5) yields an indeterminate solution for D_1. Writing $D_q = \tau(q)/(q-1)$ and differentiating numerator and denominator, we obtain

$$D_1 = \lim_{q \to 1} \frac{d\tau}{dq}. \tag{8.3.7}$$

Evaluating $d\tau/dq$ from (8.3.6) with $g_q = 1$, we obtain $D_1 \approx 0.5616$.

In Section 7.1d we have already noted the Kaplan–Yorke conjecture (7.1.32) that relates the Liapunov exponents to a "dimension" D_{KY}. There is considerable analytical and numerical evidence that $D_{KY} = D_1$, the information dimension, although this has not been proved in general (Grebogi et al., 1988).

Time Series Analysis. In principle, the generalized capacity dimensions can be determined from a time series of N d-dimensional points generated by a known d-dimensional map or system of differential equations. To do this we partition the space in which the attractor is embedded into M boxes of size ϵ. Within each box we have $p_i = N_i/N$, where N_i is the number of points in box i. We then evaluate the sum in (8.3.1) [or (8.3.2)]. Plotting $\ln(\text{sum})$ versus $\ln \epsilon$ and fitting a straight line, we determine $(q-1)D_q$. In practice, however, this procedure may be computationally or experimentally too costly. Furthermore, effects of extrinsic noise may obscure the result, and the embedding dimension may not be known; e.g., if the time series arises from an experiment.

Grassberger and Procaccia (1983) give an alternate procedure to more efficiently determine the correlation dimension D_2 from a time series. They introduce a quantity

$$C(\epsilon) = \frac{1}{N^2} \times \{\text{number of pairs } (i,j) \text{ whose separation is less than } \epsilon\} \tag{8.3.8}$$

and show that

$$C(\epsilon) \sim \sum_{i=1}^{M(\epsilon)} p_i^2, \tag{8.3.9}$$

which determines D_2 in (8.3.1). Typically, 20,000 points suffice to evaluate (8.3.8), whereas 200,000 points would be required for the equivalent box-counting of (8.3.9). For the logistic map (7.2.5) at the onset of chaos

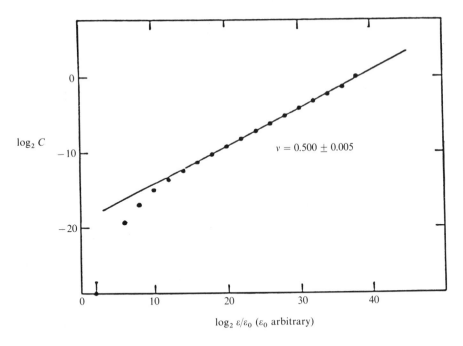

Figure 8.8. Correlation integral for the logistic map (7.2.5) at the onset of chaos $\mu = 0.356\ldots$. The starting point was $y_0 = \frac{1}{2}$, the number of points was $N = 30{,}000$ (after Grassberger and Procaccia, 1983).

($\mu \approx 0.356994$) using 30,000 points, Grassberger and Procaccia (1983) plot $\log_2 C(\epsilon)$ versus $\log_2 \epsilon/\epsilon_0$ (reproduced in Fig. 8.8) to obtain $D_2 \approx 0.500 \pm 0.005$. They also obtain an analytic bound, using a renormalization approach, of $0.4926 < D_2 < 0.5024$. For this example, the renormalization also yields $0.5376 < D_0 < 0.5386$ and $D_1 \approx 0.517097$. Grassberger and Procaccia have also obtained good results for attractors embedded in higher-dimensional maps and flows.

8.3b. Scaling Index Spectrum

A characterization of multifractals that is related to their generalized dimensions is the spectrum of scaling indices $f(\alpha)$, described by Halsey et al. (1986). Following their argument, this is defined by introducing the maximum length scale at a given level $\epsilon = \max\{\epsilon_i\}$ and assuming that the probability of the ith piece of the attractor scales as

$$p_i \sim \epsilon^{\alpha_j}, \qquad (8.3.10)$$

which defines the scaling index α_j. Halsey et al. suggest that the number of

8.3. Multifractals and Their Properties

times α takes on a value between α' and $\alpha' + d\alpha'$ will be of the form

$$d\alpha' \rho(\alpha') \epsilon^{-f(\alpha')}, \tag{8.3.11}$$

where $f(\alpha')$ is a continuous function and ρ is a nonsingular weighting function. Then the sum of any function over the pieces can be converted to an integral over α'. In particular, for the qth power of (8.3.10), we obtain

$$\sum_i p_i^q = \int d\alpha' \, \rho(\alpha') \epsilon^{-f(\alpha')} \epsilon^{\alpha' q}. \tag{8.3.12}$$

We can then relate $f(\alpha)$ and D_q by converting the partition sum (8.3.4) over i into an integral over α. Substituting (8.3.12) in (8.3.4), we obtain

$$\Gamma_{Hq} = \epsilon^{-\tau} \int d\alpha' \, \rho(\alpha') \epsilon^{q\alpha' - f(\alpha')}. \tag{8.3.13}$$

In the limit $\epsilon \to 0$, the minimum value of the exponent of ϵ within the integral dominates; this occurs when

$$\frac{d}{d\alpha'} [q\alpha' - f(\alpha')]_{\alpha' = \alpha(q)} = 0, \tag{8.3.14}$$

which yields

$$\frac{df}{d\alpha} = q. \tag{8.3.15}$$

Furthermore, Γ_{Hq} in (8.3.13) is neither zero nor infinity only for

$$\tau(q) = q\alpha - f. \tag{8.3.16}$$

Equations (8.3.15) and (8.3.16) define a Legendre transformation between $f(\alpha)$ and $\tau(q)$. Thus given $f(\alpha)$, (8.3.15) determines q and then (8.3.16) determines τ. D_q is then found from (8.3.5). Inverting the transformation using (8.3.15) and (8.3.16), we obtain

$$\frac{d\tau}{dq} = \alpha. \tag{8.3.17}$$

Hence given D_q (for all q), (8.3.5) yields $\tau(q)$, (8.3.17) yields α, and (8.3.16) yields f. We note (Halsey et al., 1986) that the spectrum is convex, $d^2f/d\alpha^2 < 0$, so that f has an absolute maximum which from (8.3.15) corresponds to $q = 0$. Setting $q = 0$ in (8.3.5) and (8.3.16) and equating yields $D_0 = f$ at this maximum. From these equations with $q \to +\infty(-\infty)$ the left-most (right-most) values of α are $\alpha_{\min} = D_\infty$ ($\alpha_{\max} = D_{-\infty}$), at which point from (8.3.15) the $f(\alpha)$ curve has infinite slope. From (8.3.4), the values $D_\infty(D_{-\infty})$ arise from the largest (smallest) p_i, corresponding to the most concentrated (rarified) part of the multifractal. For the multifractal example shown in Fig. 8.7, Fig. 8.9 shows $f(\alpha)$, which is obtained by solving (8.3.6),

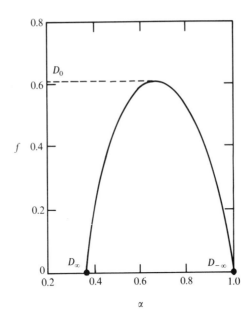

Figure 8.9. The $f(\alpha)$ spectrum for the two-scale Cantor set in Fig. 8.7 (after Halsey et al., 1986).

numerically, for τ and then using the prescription for finding α and f. For $q = \pm\infty$ we can solve (8.3.6) analytically to obtain

$$\alpha_{\min} = \frac{\ln(\frac{3}{5})}{\ln(\frac{1}{4})} \approx 0.3684 \quad \text{and} \quad \alpha_{\max} = \frac{\ln(\frac{2}{5})}{\ln(\frac{2}{5})} = 1,$$

which agree with the values found numerically in Fig. 8.9.

Some Results. The spectrum for the period-doubling attractor at the transition to chaos has been calculated by Bensimon et al. (1986), and is shown in Fig. 8.10 along with experimental measurements for forced Rayleigh–Bénard convection in mercury (Glazier et al., 1986). Some typical error bars are shown for the data, which are largest near $\alpha_{\max}$ where the attractor is the most rarified and the statistics are the poorest. Similar results for the transition to chaos due to quasiperiodicity at the golden mean winding number are shown in Fig. 8.11 (Jensen et al., 1985). Here $D_0 = 1$ (apparent from physical principles, as the attractor covers the circle), $\alpha_{\min} \approx 0.6326$, and $\alpha_{\max} \approx 1.8980$. In both cases the agreement between theory and experiment is within numerical error. The quasiperiodicity $f(\alpha)$ spectrum has also been observed in a driven semiconductor experiment by Gwinn and Westervelt (1987). Using a large data set they obtained remarkably good agreement with the theory.

8.3. Multifractals and Their Properties

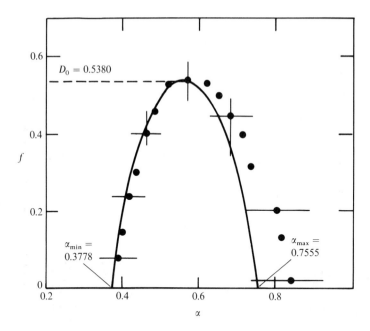

Figure 8.10. The $f(\alpha)$ spectrum for period-doubling. The dots represent the experimental measurements (after Glazier et al., 1986).

Although the $f(\alpha)$ spectra shown in Figs. 8.10 and 8.11 have generally the same shape, they have sufficiently different values of D_0, α_{min}, and α_{max} that the underlying mechanism for the transition to chaos in the two experiments can clearly be distinguished. This is not necessarily the case with other presentations of the experimental data. For example, Fig. 8.12 shows the experimental attractor for the quasiperiodic onset of turbulence. The temperature T_{n+1} at the bottom of the convection cell is plotted versus T_n. Although Figs. 8.12 and 7.36 look superficially different, they are actually the same multifractal seen in different coordinates; i.e., they have the same scaling spectrum.

Experimental data are often corrupted with noise, such that the underlying attractor is partially obscured. (We shall discuss attractor reconstruction more generally in Section 8.4.) Recognition of the underlying $f(\alpha)$ spectrum depends on having both a sufficient number of data points on the attractor and sufficiently low noise. The experimental spectra in Figs. 8.10 and 8.11 were found using approximately 2500 noisy data points (a time series) on the surface of section of the attractor. One way to find $f(\alpha)$ for these data is to first determine the recurrence time $m_i(\epsilon)$ for each point to return to within a distance ϵ of its initial value. We then estimate p_i as $p_i \sim 1/m_i(\epsilon)$ and use (8.3.10) to determine α_j. We then determine how many points have α's within a given range $d\alpha'$ and use (8.3.12) to determine f. Actually, this procedure is not very good for sparse noisy data. A different procedure was

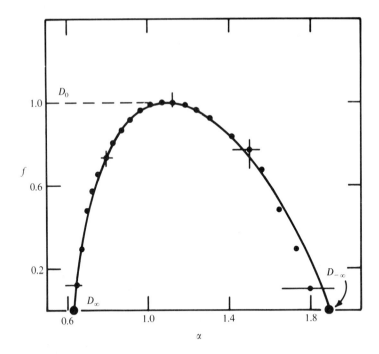

Figure 8.11. The $f(\alpha)$ spectrum for the golden mean attractor shown in Fig. 8.12. The dots correspond to the experimental measurements (after Jensen *et al.*, 1985).

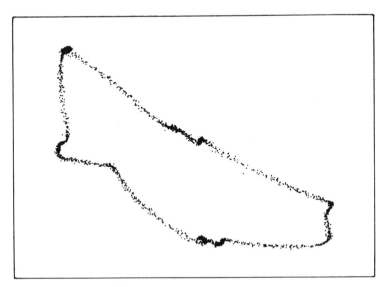

Figure 8.12. The experimental golden mean attractor from the forced Rayleigh–Bénard experiment. Plotted is a measure of successive iterates of the temperature T_{n+1} versus T_n (after Jensen *et al.*, 1985).

used (Jensen et al., 1985), based on the integral in (8.3.13) that arises from the sum

$$G = \sum_{i=1}^{M} p_i^q,$$

where $M(\varepsilon)$ is the number of boxes of size ε to cover the attractor. Hence, they evaluate

$$\begin{aligned} G(\varepsilon,q) &= \frac{1}{N} \sum_{i=1}^{M} N_i p_i^{q-1} \\ &= \frac{1}{N} \sum_{j=1}^{N} p_j^{q-1} \\ &= \langle m_j^{1-q} \rangle, \end{aligned} \qquad (8.3.18)$$

where $\langle \ \rangle$ denotes an average over the N data points. For Γ_{Hq} neither zero nor infinity, (8.3.13) requires $G \sim \varepsilon^{\tau(q)}$. Plotting $\ln G$ versus ϵ for various ϵ's and fitting a straight line, they determine $\tau(q)$. Equations (8.3.16) and (8.3.17) then yield $f(\alpha)$. Still another method for directly evaluating $f(\alpha)$ without resorting to the intermediate Legendre transform has been proposed by Chhabra and Jensen (1989).

The calculation of $f(\alpha)$ for strange attractors well within the chaotic regime, and not just at the onset of chaos, can be performed by first calculating a time series of data points from the map or differential equations and then applying the preceding methods. However, extensive computational resources are required. Tsang (1986) has given an approximate analytical procedure to determine $f(\alpha)$ directly from the mapping equations, and has applied the procedure to the dissipative standard and Hénon maps. Another approach relates the spectrum to the properties of the unstable periodic orbits on the attractor (Grebogi et al., 1988), which allows an efficient calculation of $f(\alpha)$ within the chaotic regime.

8.4. Reconstructing Dynamics

In the previous section we considered the types of information that could be obtained from a known dynamical system of equations whose solutions exhibit chaotic behavior. This information was primarily geometrical and statistical, such as the generalized dimensions and scaling spectrum of an underlying attractor. Often, however, the underlying equations are either unknown or else sufficiently complicated that they are of little help in understanding the character of the resulting dynamics. In these situations the approach most widely considered is analysis of a time series of some variable, in order to construct both statistical information about the dynamics and, possibly, a representation of the equations of motion, themselves. Most of the work to date has centered on development of the methods, rather than an analysis of dynamical systems whose equations are unknown. For example, the Hénon map (7.1.26) has been studied in some detail, while systems with truly unknown dynamics, such as stock market returns, have been examined only statistically and yielded little information (see Brock, 1988).

For an important class of problems, the equations of motion are infinite dimensional, with the problem being to find a description of the dynamics on a low-dimensional attractor for a given parameter regime. The onset of fluid turbulence has motivated many of these studies. However, a useful illustrative example is the Mackey–Glass differential delay equation (Mackey and Glass, 1977),

$$\frac{dx(t)}{dt} = \frac{ax(t-\tau_d)}{1 + [x(t-\tau_d)]^c} - bx(t), \tag{8.4.1}$$

which models the regeneration of blood cells in patients with leukemia. Here $x(t)$ represents the blood cell concentration at time t and $x(t-\tau_d)$ is the concentration at an earlier time $t-\tau_d$. For these patients, τ_d may be excessively large and (8.4.1) yields a chaotically fluctuating concentration $x(t)$. Note that the initial conditions for (8.4.1) must be specified over the entire range $-\tau_d < t < 0$; hence, the system is infinite dimensional.

Reconstruction for systems such as the Hénon and Mackey–Glass attractors has involved both determining statistical quantities such as the Liapunov exponents, the fractal dimension of the attractor and the embedding dimension in which it lies, and in developing local and global methods to predict the near-term future of the dynamics. We summarize the methods in the following subsections.

8.4a. Delay Coordinates and Embedding Dimension

Suppose that we know the time variation of only one component $x(t)$ of a multidimensional dynamical system $\mathbf{x}(t)$. To reconstruct the dynamics, we must first determine the embedding dimension d for the motion. Following Packard et al. (1980) and Takens (1981), we introduce the d-dimensional vectors

$$\boldsymbol{\zeta}_i = (x(t_i), x(t_i + \tau), \ldots, x(t_i + (d-1)\tau)), \tag{8.4.2}$$

where t_i is the time for the ith data point $\boldsymbol{\zeta}_i$ and τ is some fixed delay time. It has been shown by Eckmann and Ruelle (1985) that this formulation of the underlying vector space can be constructed from the more obvious phase space coordinates:

$$x(t), \frac{dx(t)}{dt}, \ldots, \frac{d^{d-1}}{dt^{d-1}}x(t).$$

8.4. Reconstructing Dynamics

The formation of the "embedding" vector ζ_i in (8.4.2) is known as *delay reconstruction*.

In principle, for an infinite sequence of infinitely precise points uncorrupted by noise, the choice of τ is not important and d must be chosen large enough so that the reconstruction transformation is invertible; i.e., points in the reconstructed phase space uniquely identify points in the original phase space. However, for a finite series of noisy data points, the reconstruction is noninvertible due to the noise, and its statistical measures, such as Liapunov exponents and generalized dimensions, are found to depend on the choice of d and τ. Thus the central task of delay reconstruction is to choose properly these values to minimize the noninvertibility and the effects of noise. If τ is chosen too small (with respect to the shortest period T of the attractor) then $x(t_i) \approx x(t_i + \tau) \approx \cdots \approx x(t_i + (d-1)\tau)$ and the attractor is stretched out along a diagonal in ζ-space, such that the attractor is obscured by noise in the data. If τ is chosen too large, then distant values in the time series are uncorrelated due to noise and the exponential divergence of trajectories.

A common but *ad hoc* choice for τ and d is to first compute the autocorrelation function

$$C(t) = \frac{1}{T} \int_0^T x(t' + t) x(t') \, dt', \qquad (8.4.3)$$

and choose τ to be a characteristic time such as the first inflection point or zero of C, or some fraction thereof (see Broomhead and King, 1986; and Fraser, 1989). Typically, $\tau \sim 0.2T$, where T is the fundamental period of the motion on the attractor. The choice of the dimension d of the embedding space is then made. It is clear that d must be sufficiently large to give a unique (invertible) presentation of the dynamics. This would certainly require d to be as large as the dimension D_0 of the attractor on which the data lie, and for a strange attractor of noninteger dimension, an embedding dimension of the next larger integer. Takens (1981) has shown that an embedding can always be accomplished in a space of dimension $d \geq 2D_0 + 1$, but in many numerical examples the weaker condition $d \geq \text{int}(D_0 + 1)$ is found to be sufficient. [Here $\text{int}(\cdot)$ is the integer part.] One usually numerically calculates D_0 from the data set for various embedding dimensions d, as described in Section 8.3a. One increases d until D_0 remains constant and identifies the minimum d for which D_0 remains constant as the embedding dimension. In practice, D_2 is often determined because it is easier to calculate. We give examples of this procedure in the next subsection.

Procedures for optimally choosing τ and d have been put on a firmer theoretical basis. Fraser and Swinney (1986), and Fraser (1989) introduce the *mutual information I* and its multidimensional generalization, the *redundancy* R_d^τ

$$R_d^\tau = \left\langle \log\left[\frac{p(x_0, x_1, \ldots, x_{d-1})}{p(x_0)p(x_1)\cdots p(x_{d-1})}\right]\right\rangle, \tag{8.4.4}$$

where $I(x, \zeta) = R_2^\tau$, $p(x_0, x_1, \ldots, x_{d-1})$ is the d-dimensional joint probability density for the ζ's, $p(x_i)$ is the one-dimensional probability density for the x_i's, and the brackets denote the ensemble average. If the coordinates are statistically independent, then

$$p(x_0, x_1, \ldots, x_{d-1}) = p(x_0)p(x_1)\cdots p(x_{d-1}) \tag{8.4.5}$$

and $R_d^\tau = 0$. If there is a strict functional dependence of any one of the x_i's on another x_j, then $R_d^\tau = \infty$. Thus R_d^τ is a measure of the statistical dependence of the reconstruction coordinates on each other. This is in contrast to the general notion of linear independence

$$\langle x_0 x_1, \ldots, x_{d-1}\rangle = \langle x_0\rangle\langle x_1\rangle \cdots \langle x_{d-1}\rangle \tag{8.4.6}$$

which corresponds to a zero of the autocorrelation function (8.4.3). Fraser and Swinney assert that it is much more important to minimize the statistical dependence, i.e., minimize R_d^τ, than it is to minimize the linear dependence. They give an $N \log N$ algorithm, where N is the number of points, to choose d and τ. First they introduce the *marginal redundancy*

$$R_d'^\tau = R_{d+1}^\tau - R_d^\tau, \tag{8.4.7}$$

which can be written, using (8.4.4) as

$$R_d'^\tau = \left\langle \log\left[\frac{p(x_d|x_0, x_1, \ldots, x_{d-1})}{p(x_d)}\right]\right\rangle. \tag{8.4.8}$$

The procedure (Fraser, 1989) is to increase d over a range of τ's until the conditional probability distribution $p(x_d|x_0, x_1, \ldots, x_{d-1})$ is sharply peaked; i.e., until $R_d'^\tau$ is large and gets no larger, and then to select τ so as to maximize a quantity defined as the "useful information"

$$Q(d, \tau) = (d-1)(A - h_k\tau) - R_d^\tau. \tag{8.4.9}$$

The first term in (8.4.9) insures that τ is small compared to A/h_k and the second term enforces the statistical independence of the coordinates ζ. Here the accuracy A is the average information that an isolated scalar measurement provides; i.e., $A = I(x, \zeta)$, and h_k given by (5.2.24) is the KS entropy.

Figure 8.13 shows the difference in reconstructions for the Roux attractor (Roux et al., 1983) for $d = 2$. At $T = 24.6$ s, the first zero of the autocorrelation function, much of the attractor lies in the dark line along the upper left edge, where the dynamics cannot be distinguished. At $T = 108.2$, the first minimum of R_2^τ, the structure of the attractor is more clearly seen. [The first

8.4. Reconstructing Dynamics

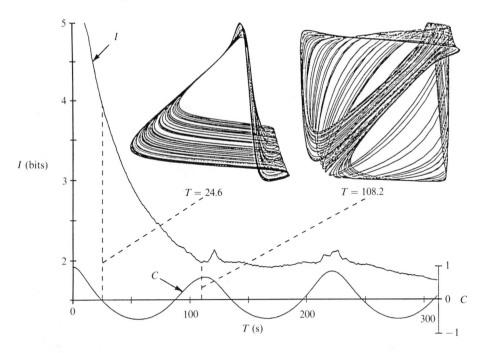

Figure 8.13. Phase portraits of the Roux attractor in the Belousov–Zhabotinskii reaction. The dependence of the mutual information I and the autocorrelation function C, on T, are shown for calculations over 32,768 points. The coordinates used in constructing the portrait on the left are linearly independent (zero autocorrelation), while the coordinates used in the portrait on the right are more generally independent (local minimum of mutual information) (after Fraser and Swinney, 1986).

term in (8.4.9) was not included for this early example of the reconstruction procedure, such that the first minimum in R_2^τ with respect to τ was selected.]

Broomhead and King (1986) give an alternative procedure for finding τ and d based on a linear, singular systems analysis. The basic idea is that the mean-squared distance between points on the reconstructed attractor should be maximized. Their procedure has the advantage that the optimal linear combinations of the x_i's, $i = 0, \ldots, d - 1$, are used to form the reconstruction coordinates, rather than choosing the individual x_i's as coordinates. Furthermore, there is a built-in filter to reduce the effects of noise on the reconstruction. However, the analysis is based on the use of a linear independence condition (8.4.6) rather than the more general notion of statistical independence (8.4.5). For further discussion and reconstruction examples, the papers of Broomhead and King (1986) and Fraser (1989) should be consulted.

8.4b. Statistical Quantities

In analyzing a time series one would certainly wish to determine if the largest Liapunov exponent is positive, i.e., if the system motion is chaotic. Furthermore, it has been shown numerically (Farmer, 1982) that the information dimension D_1 can be obtained from the Liapunov exponents by using the Kaplan and Yorke conjecture (see Fredrikson et al., 1983) that

$$D_1 = D_{KY} = j + \frac{\sum_{i=1}^{j} \sigma_i}{|\sigma_{j+1}|}, \qquad (7.1.32)$$

where the exponents are ordered in decreasing size with j exponents $\geqslant 0$. We have already discussed the numerical methods to find these exponents in Section 6.3b. Here we have the additional requirements to choose τ and find the proper embedding space. These requirements, together with the effects of noise, error analysis, etc., have been considered in the previous subsection and are discussed in detail by Wolf et al. (1985), Broomhead and King (1986), and Fraser (1989). For low-dimensional attractors, d can easily be determined by plotting the attractor on the plane ($d = 1$ or 2). For higher dimensions, d can be varied to successively higher values and the σ's recomputed until convergence is reached. Wolf et al. also give programs for computation, and their results for the Hénon, Lorenz, and Mackey–Glass systems are summarized in Table 8.1. Since the σ's yield D_1 through (7.1.32) and the values of D_0 and D_2 are usually close to D_1, the above procedure may give sufficient statistical information. Alternatively, D_0 and D_2 may be directly calculated, as described below.

To determine the generalized dimensions D_q of an attractor directly from a time series of a single variable, as with the Liapunov exponents, both the time delay τ and the embedding dimension d must be chosen. Again, d must be greater than D_0, but if d is chosen too large then the effects of noise and poor statistics yield uncertain values of the dimension. In the simplest procedure, often used in numerical studies, one chooses a τ that seems reasonable on physical grounds and tries successive values of d, rechoosing τ if the reconstructed attractor does not meet the basic requirements of the last subsection. Consider the procedure for finding D_2, given in Section 8.3a. For d chosen smaller than the proper embedding dimension, one finds $C(\epsilon) \sim \epsilon^d$. For d larger than this value, one obtains the expected scaling $C(\epsilon) \sim \epsilon^{D_2}$. Figure 8.14 shows the results of this procedure for a single-variable time series for the Hénon map (7.1.26), using the delay coordinates (x_i, x_{i+2}) for $d = 2$ and (x_i, x_{i+2}, x_{i+4}) for $d = 3$. For $d = 3$ and 20,000 points, we obtain $D_2 = 1.25 \pm 0.2$. For $d = 2$ and 30,000 points, we find $D_2 = 1.21 \pm 0.01$, showing good agreement with the reconstruction. For real systems with noise, the results are not so clear. The limit as $\epsilon \to 0$ cannot be approached, as the noise blurs the chaotic dynamics, yielding $D_2 \approx d$. If the noise is not too large, plots such as Fig. 8.14 can still be used to find D_2 within an appropriate range of ϵ-values.

8.4. Reconstructing Dynamics

Table 8.1. The Model Systems and Their Liapunov Spectra and Dimensions as Computed from the Equations of Motion (after Wolf et al., 1985)

System	Parameter values	Liapunov spectrum (bits/s)	Liapunov dimension[a]
Hénon $X_{n+1} = 1 - aX_n^2 + Y_n$ $Y_{n+1} = bX_n$	$a = 1.4$ $b = 0.3$	$\lambda_1 = 0.603$ $\lambda_2 = -2.34$ (bits/iter.)	1.26
Lorenz $\dot{X} = \sigma(Y - X)$ $\dot{Y} = X(R - Z) - Y$ $\dot{Z} = XY - bZ$	$\sigma = 16.0$ $R = 46.92$ $b = 4.0$	$\lambda_1 = 2.16$ $\lambda_2 = 0.00$ $\lambda_3 = -32.4$	2.07
Mackey–Glass $\dot{X} = \dfrac{aX(t - \tau_d)}{1 + [X(t - \tau_d)]^c} - bX(t)$	$a = 0.2$ $b = 0.1$ $c = 10.0$ $\tau_d = 31.8$	$\lambda_1 = 6.30E - 3$ $\lambda_2 = 2.62E - 3$ $|\lambda_3| < 8.0E - 6$ $\lambda_4 = -1.39E - 2$	3.64

[a] As defined in (7.1.32).

A more interesting example is the Mackey–Glass equation (8.4.1) with $a = 0.2$ and $b = 0.1$. Farmer (1981) has used this to investigate the change in dimension of an underlying chaotic attractor as the delay time τ_d is increased. He used the numerical solution of (8.4.1) to make phase plots of $x(t)$ versus $x(t - \tau_d)$. The choice of $\tau = \tau_d$ for the phase plots was arbitrary. His results, given in Fig. 8.15, show increasing complexity with increasing τ_d. In (a), with $\tau_d = 14$, the attractor is periodic and therefore not chaotic. In (b) the attractor appears chaotic at $\tau_d = 17$, with increasingly complex motion in (c) at $\tau_d = 30$. Analyzing D_0 by the box-counting algorithm using (7.1.30) and D_{KY} by computing the Liapunov exponents and using the Kaplan–Yorke conjecture (7.1.32), he obtained the results shown in Table 8.2. For $50 \lesssim \tau_d \lesssim 200$, Farmer finds the scaling $D_{KY} \approx 0.1\tau_d$. The box counting algorithm becomes increasingly difficult to use with increasing dimension, as described in Section 8.3a, and so these values become imprecise. The value of D_{KY} at $\tau_d = 30$ agrees with a similar determination by Wolf et al. (1985). Since the attractor can have a large fractal dimension for large τ_d, the embedding dimension d must be large. Table 8.2 gives the results of Grassberger and Procaccia (1983) for the embedding dimension d, based on the convergence of the correlation dimension D_2 as d is increased. The minimum d's for which $C \sim \epsilon^{D_2}$ are 3, 3, 4 and 10 for $\tau_d = 17, 23, 30,$ and 100, respectively. However, we see that a more accurate value of D_2 is

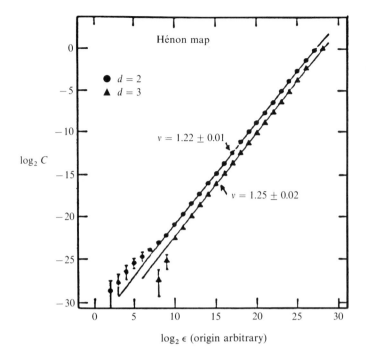

Figure 8.14. Modified correlation integrals for the Hénon map (7.1.26). The time series consisted of coordinates $x_n, x_{n+2}, x_{n+4}, \ldots$, and $\zeta = (x_n, x_{n+2}, x_{n+2(d-1)})$ for each d. For $d = 2$, $N = 30{,}000$; for $d = 3$, $N = 20{,}000$ (after Grassberger and Procaccia, 1983).

obtained for somewhat larger values of d. For example, at $\tau_d = 100$ we obtain reasonable convergence to $D_2 \approx 7.2$ at $d = 14$, while $D_2 \approx 5.8$ at $d = 10$. Although it was not done in this study, it is here that a more precise method of choosing d and τ, as considered by Fraser (1989), might be useful. For this higher-dimensional example, we see that the d at which D_2 has converged lies in the expected range $\text{int}(D_0 + 1) \leq d \leq \text{int}(2D_0 + 1)$ and is, in fact, close to the upper bound which is Taken's sufficient minimum embedding dimension.

8.4c. Short-Time Prediction

Local Reconstruction. In addition to characterizing the statistical properties of the attractor, we can also use known data to predict future values of a time series. Of course, exponential divergence of nearby trajectories, coupled with noise, will prevent long-time predictions. This behavior is normally apparent in the data itself; i.e., data points close together in the embedding

8.4. Reconstructing Dynamics

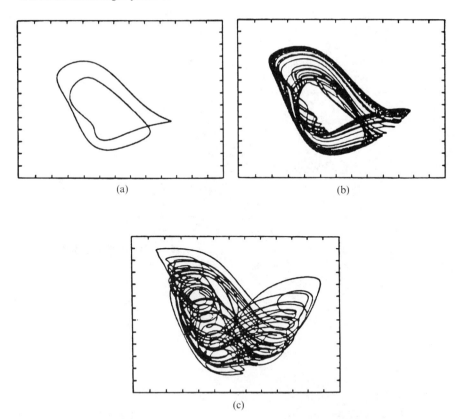

Figure 8.15. Phase plots obtained by plotting $x(t)$ versus $x(t - \tau_d)$, after letting transients relax; (a) $\tau_d = 14$, (b) $\tau_d = 17$, (c) $\tau_d = 30$ (after Farmer, 1982).

space will exponentially separate along the unstable direction when followed for a number of equal intervals in time. This behavior suggests that a *local* reconstruction might be an efficient mechanism for prediction. That is, follow a cluster of locally neighboring trajectories in time, centered about the unknown desired trajectory, such that the desired trajectory lies within the cluster, with a high probability, for the same time interval.

Using this basic principle, various similar but distinct techniques have been developed for analysis (Crutchfield and McNamara, 1987; Farmer and Sidorowich, 1987, 1988; Abarbanel *et al.*, 1989). The latter authors compare alternative, local reconstruction procedures, estimating the errors for each and comparing them with numerical results. We summarize their approach here. There are three basic steps:

(1) Pick a form for the local map; e.g., a polynomial, rational function etc., F, that maps $x(t)$ to $x(t + T)$.
(2) Assign a neighborhood to a point x.

Table 8.2. The Correlation Exponent D_2 for the Mackey–Glass Equation with $a = 0.2$, $b = 0.1$, as a Function of the Embedding Dimension d, Compared to D_0 and D_{KY}, for Increasing Values of τ_d (after Grassberger and Procaccia, 1983)

τ_d	D_2	D_0	D_{KY}
17.0	$1.95 \leftarrow 0.03$ $(d = 3)$ 1.95 ± 0.03 $(d = 4)$ 1.95 ± 0.03 $(d = 5)$	2.13 ± 0.03	2.10 ± 0.02
23.0	2.38 ± 0.15 $(d = 3)$ 2.43 ± 0.05 $(d = 4)$ 2.44 ± 0.05 $(d = 5)$ 2.42 ± 0.1 $(d = 6)$	2.76 ± 0.06	2.82 ± 0.03
30.0	2.87 ± 0.3 $(d = 4)$ 3.0 ± 0.2 $(d = 5)$ 3.0 ± 0.2 $(d = 6)$ 2.8 ± 0.2 $(d = 7)$	> 2.94	3.58 ± 0.04
100.0	5.8 ± 0.3 $(d = 10)$ 6.6 ± 0.2 $(d = 12)$ 7.2 ± 0.2 $(d = 14)$ 7.5 ± 0.15 $(d = 16)$		

(3) Vary the coefficients, e.g., of the polynomial, to achieve the least mean square error in mapping the neighborhood points to their known values.

Then use this best fit polynomial to predict the future of x at a time T later. Polynomials are a good form because their coefficients can be linearly fit to minimize least square deviations. Neighborhoods can be partitioned into disjoint sets (cubes), or overlapping balls. Choosing neighborhoods well is important to achieve good predictions.

Farmer and Sidorowich discuss two methods of forecasting over a time interval T, which they call direct and iterative. For *iterative forecasting* the data in the neighborhood of $x(t)$ are followed for a single short time step $\tau = T/n$, where τ may be a mapping period or an arbitrarily chosen interval for a differential equation. The least squares criterion is then used to predict $x(t + \tau)$. The process is repeated, using neighbors of $x(t + \tau)$, to predict $x(t + 2\tau)$, and so on for a time $T = n\tau$. For *direct forecasting*, the neighbors of $x(t)$ are followed continuously for the entire time interval T and $x(t + T)$ is obtained directly. Estimating the growth of the error for each method, they find, somewhat counterintuitively, that iterative forecasting is generally superior. They estimate the rms prediction errors e for each method to have the proportionalities

$$e \propto \begin{cases} N^{-q/D_1} e^{q\sigma_{max} T}, & \text{direct,} \\ N^{-q/D_1} e^{\sigma_{max} T}, & \text{iterative,} \end{cases} \quad (8.4.10)$$

8.4. Reconstructing Dynamics

where $q = K + 1$ and K is the degree of the polynomial, N is the number of neighboring points, and σ_{max} is the largest Liapunov exponent. We see that the predicted error grows faster by an exponential factor e^q for direct iteration. Farmer and Sidorowich check the scalings of N, q, and T, for both methods, using the Mackey–Glass equation (8.4.1) with $\tau_d = 17$ and $\tau = 6$. For this value of τ_d, $D_1 = 2.10$ as given in Table 8.2. The results for linear and quadratic polynomials, given in Fig. 8.16, agree with the postulated scalings. They also checked the scaling of e with N, by repeating the results in Fig. 8.16 for $N = 10{,}000$. Using data from a real experiment, on the other hand, and using a τ short compared to the natural period, Farmer and Sidorowich obtained lower errors for direct forecasting. Although not explored in their paper, the reason was probably due to a combination of the short iteration time τ coupled with the noise corruption, which tended to mask the underlying dynamics.

Abarbanel et al. (1989) also consider local reconstruction and use various methods of near-neighbor weighting to improve the predictive power. They

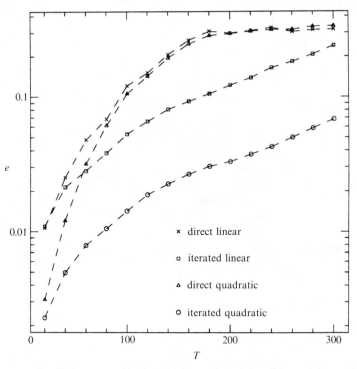

Figure 8.16. Prediction error for the Mackey–Glass delay differential equation as a function of extrapolation time, with $N = 500$ sampled at $\Delta t = 1$, $\tau_d = 17$, $\tau = 6$, and $d = 4$, using nearest neighborhoods with twice the minimum number of points needed to fit the coefficients by least squares. The fractal dimension of the underlying attractor is $D_1 = 2.1$ (after Farmer and Sidorowich, 1988).

also demonstrate for chaotic dynamics in the Hénon map and the Lorenz equations that important statistical quantities, such as the largest Liapunov exponent, are not predicted properly. Liapunov exponents and other constants, such as moments of the invariant distribution, can be imposed as additional constraints. Abarbanel *et al.* do this by optimizing with the additional parameters as constraints. By adding new constraints, the best fit to the data becomes slightly worse, as expected. The reconstructed dynamics are in some sense closer to the original dynamics, while simultaneously weakening their predictive power.

Global Reconstruction. Another approach to predicting the future is to find a *global* approximation to the function **F**. That is, we expand the "small" neighborhoods of the local reconstruction method to comprise the entire attractor, and even, if possible, points that are not on the attractor itself, but that lie in the basin of attraction. This approach has been described by Crutchfield and McNamara (1987), using nonorthogonal basis functions for **F**, and by Cremers and Hübler (1987) using orthogonal basis functions.

Consider, following Crutchfield and McNamara (1987), polynomials as *nonorthogonal basis vectors*. If, for example, we choose $d = 2$ and a polynomial of order $K = 2$, we construct the vector **F** from the complete set of components of x, each having six terms

$$\bar{x}_1 = a_{10} + a_{11}x_1 + a_{12}x_2 + a_{13}x_1^2 + a_{14}x_2^2 + a_{15}x_1x_2, \\ \bar{x}_2 = a_{20} + a_{21}x_1 + a_{22}x_2 + a_{23}x_1^2 + a_{24}x_2^2 + a_{25}x_1x_2, \tag{8.4.11}$$

where $\bar{x}$ is x_{n+1} when $x = x_n$. Due to lack of orthogonality, the terms increase rapidly with both d and K, with the number of terms in each of the d vectors given by

$$T = (d + K)!/d!\,K! \tag{8.4.12}$$

such that $T = 6$, for the example above. Although the equations are nonlinear in the variables x_1, x_2, they are linear in the coefficients a_{ij}, and thus we can use multivariate analysis to optimize the a_{ij} using a standard "goodness" fit. For example, we define the deviation from the dynamics as

$$\sigma_{\text{obs}} = (\sigma_0, \sigma_1, \ldots, \sigma_{d-1}) \tag{8.4.13}$$

where the ith component is given by

$$\sigma_i^2 = \frac{1}{N} \sum_{n=0}^{N-1} (y_{n,i} - F_i(\boldsymbol{x}_n))^2. \tag{8.4.14}$$

Here **F** is the reconstructed mapping, $\boldsymbol{y}_n$ is the observation following $\boldsymbol{x}_n$, such that $\boldsymbol{y}_n = \boldsymbol{x}_{n+1}$, and N is the number of observations. Minimization of $|\sigma_{\text{obs}}|^2$ then gives the best least squares fit to the data, given by

$$\hat{A} = YX^T(XX^T)^{-1}, \tag{8.4.15}$$

8.4. Reconstructing Dynamics

where $\hat{A}$ is the $T \times d$ matrix of the optimized coefficients a_{ij}, Y is the $d \times N$ matrix of the $y_{n,i}$, and X is the $T \times N$ matrix of the quantities $1, x_1, x_2, x_1^2, x_2^2, x_1 x_2, \ldots$. The variance–covariance matrix describing the goodness of the fit can also be determined and is given by

$$\hat{\Sigma} = (YY^T - \hat{A}XY^T)/(N - d). \tag{8.4.16}$$

The quantity $\hat{\Sigma}$ gives a measure of the observational noise which is the extrinsic noise amplified by the exponential separation of trajectories between each observation.

The systematic procedure, described above, is performed for a given d and K. Too small values of d or K will result in a poor fit to the data. Too large values result in extra complexity with little improvement in result. If the data lie on an attractor, d must be chosen large enough to embed the attractor, as described in the previous subsection. To incorporate these ideas Crutchfield and McNamara define a *model entropy* I for each best fit at a given d and K, given by

$$I \approx \sum_{i=1}^{d} \log_2 \sigma_i + dT(K) \tag{8.4.17}$$

which is to be minimized over d and K. The first term is just the error from (8.4.13) which we expect to decrease with increasing d and K. The second term would be expected to increase with the number of terms dT, as shown in (8.4.17). Further justification can be found in the original paper. The procedure is to scan both d and K to find a minimum of I.

As a rather simple, but nontrivial example, the Hénon map [(7.1.26) with $b = 0.3$ and $a = 1.4$]

$$x_{n+1} = y_n + 1 - 1.4x_n^2, \tag{8.4.18a}$$

$$y_{n+1} = 0.3x_n, \tag{8.4.18b}$$

is used to generate a time series. We have illustrated the attractor for these parameters in Fig. 7.6. To reconstruct the map, 1000 data points in the x variable were used, and τ was chosen to be a single iteration of the map. A minimum value of I, from (8.4.17), was found for $d = 2$, as expected. Keeping only a_{ik} with amplitudes larger than 10^{-3} the equations of motion were found to be

$$\bar{x}_{n+1} = 1.0 + 0.3\bar{y}_n - 1.4\bar{x}_n^2, \tag{8.4.19a}$$

$$\bar{y}_{n+1} = 1.0\bar{x}_n. \tag{8.4.19b}$$

Since we have only used x values in constructing (8.4.19), to compare (8.4.19) with (8.4.18), we use (8.4.18b) to eliminate y_n in (8.4.18a), obtaining

$$x_{n+1} = 0.3x_{n-1} + 1 - 1.4x_n^2. \tag{8.4.20}$$

Similarly, eliminating $\bar{y}_n$ in (8.4.19a) and identifying $\bar{x}$ with x, we see that (8.4.19) is identical to (8.4.18). We should not be surprised to obtain an exact reconstruction, since the Hénon map has only a quadratic nonlinearity, and all quadratic terms were included in the basis functions selected for the reconstruction.

Alternatively to the reconstruction described above, a set of *orthogonal basis functions* can be used to obtain the equations of motion. This eliminates all of the cross terms in describing the dynamics. Cremers and Hübler (1987) use this method, which they also found to be convenient for application of feedback to control the dynamics (Hübler *et al.*, 1988). As previously, the basic technique is to minimize the error as in (8.4.15), except that F is chosen from a set of orthogonal functions, such as Fourier modes or Legendre polynomials. The method has been applied to various dynamical systems with considerable success. For example, the Lorenz system (1.5.2) was explored with the values studied by Lorenz ($\sigma = 10$, $b = \frac{8}{3}$, $r = 28$) to obtain the chaotic attractor shown in Fig. 1.19. The coefficients of the terms of the flow vector field were determined to better than one percent with all other terms negligible.

Noise and Noise Reduction. An important consideration in reconstructing dynamics from experimental data is the corruption of the signal by noise. We have already mentioned this problem, briefly, in Section 8.4b, in connection with determination of dimension. Generally, the graphical presentation of data, as in Fig. 8.13, will be somewhat blurred by noise, arising either from measuring error or from noise within the dynamical process. Although the two types of noise are different, they can be considered within the same framework. Methods of processing the signal to reduce the noise have been explored by Farmer and Sidorowich (1988) and by Hammel (1990). The underlying process is closely connected to the shadowing theorem for hyperbolic systems (Anosov, 1967; Bowen, 1975), which states that for an orbit corrupted by a small amount of noise such that

$$\|x_{k+1} - f(x_k)\| < \delta, \qquad 0 \leqslant k \leqslant N - 1,$$

for all points on the orbit up to $N - 1$, where $\hat{x}_{k+1} = f(x_k)$ is the noiseless dynamics, then there exists an ϵ such that a noiseless orbit can be found,

$$\hat{y}_{k+1} = f(\hat{y}_k),$$

such that

$$\|\hat{y}_k - x_k\| \leqslant \epsilon, \qquad 0 \leqslant k \leqslant N - 1, \quad N \to \infty;$$

i.e., the shadowing orbit always remains within $\epsilon \sim \delta$ of the noisy orbit. This theorem is quite remarkable in that a noisy orbit exponentially diverges from the noiseless one starting with the same initial conditions.

The reason why such an orbit can be found, and the underlying scheme for noise reduction, can be visualized from the simple construction in Fig. 8.17. Consider the circles representing noisy measurements resulting from

8.4. Reconstructing Dynamics

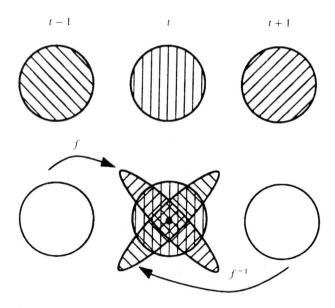

Figure 8.17. The circles represent noisy measurements of a deterministic trajectory at three different times. As successive measurements are transported to the same point in time, the associated noise probability distributions distort according to the local derivatives of the dynamical system (after Farmer and Sidorowich, 1988).

repeated iteration of an initial condition. If the underlying mapping f is known, the data within the circles are chosen such that at successive time steps the centers of the circles of noisy data (assumed well separated) transform into each other from left to right. Now, transforming $t - 1$ forward by f and $t + 1$ backward by f^{-1}, the circles are distorted to contract along the stable manifold and expand along the unstable manifold by f, and oppositely by f^{-1}. The result is an intersection whose uncertainty has been reduced. The proof of the shadowing theorem is based on this construction. The construction can be carried out for successive pieces of the phase space along the orbit, and the results iterated to obtain very large reductions in the noise after a few iterations. The numerics can be carried out in a variety of ways, as described in the referenced papers. The general procedure, applied to the Hénon map (7.1.26) is shown in Fig. 8.18. The noisy orbit is shown in Fig. 8.18a for a relative noise amplitude, $\Delta = 0.045$, injected on each step, sufficient to blur the attractor leaves. After applying a noise reduction of the general form described above, for 20 iterations of all neighborhoods, the results in Fig. 8.18b were obtained. In fact, only a few iterations are necessary to achieve this dramatic visual improvement. The method, however, does not work equally well on all points. From Fig. 8.17 it is clear that the improvement depends on the stretching and contracting being in distinct directions. This occurs at most locations along the attractor

612 8. Chaotic Motion in Dissipative Systems

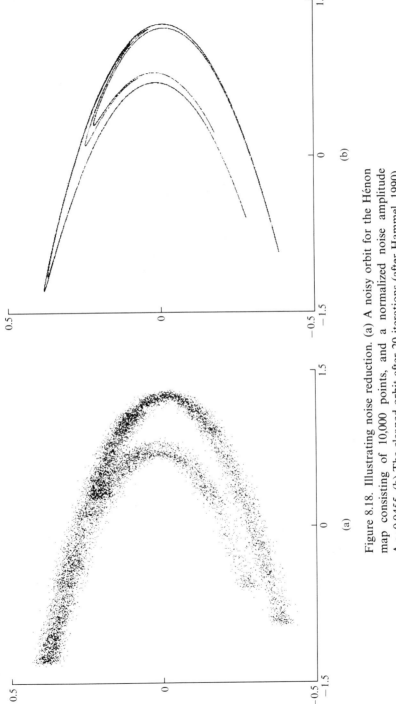

Figure 8.18. Illustrating noise reduction. (a) A noisy orbit for the Hénon map consisting of 10,000 points, and a normalized noise amplitude $\Delta \simeq 0.0455$. (b) The cleaned orbit after 20 iterations (after Hammel, 1990).

8.5. Spatio-Temporal Dynamics

but breaks down at inflection points, where the attractor is tangent to the stable manifold (homoclinic tangencies), leaving these regions essentially unimproved. These nonhyperbolic regions also destroy the $\epsilon - \delta$ estimate of the shadowing theorem. The size of such regions, however, is sufficiently small not to be noticeable in Fig. 8.18b.

The noise reduction techniques, although theoretically interesting and dramatic for finding the phase portraits of known maps that have been blurred by noise, have not yet been applied to data sets arising from unknown dynamics. The various means of analyzing the data sets are discussed in the cited papers. It should also be noted that the above procedure would not be expected to significantly improve short-term prediction of future values.

8.5. Spatio-Temporal Dynamics

8.5a. Maps on a Lattice

Coupled systems of maps on a lattice can be used to describe many space and time varying processes of physical interest. We illustrate the approach for a large class of systems that can be modeled by a homogeneous, periodically driven reaction–diffusion equation

$$\frac{\partial u}{\partial t} = D \frac{\partial^2 u}{\partial x^2} + G(t)R(u), \tag{8.5.1}$$

where $u(x, t)$ is some field quantity, x is position, t is time, D is a diffusion constant, R is a nonlinear reaction rate, and G is a periodic function of time. Following Keeler (1986), if we model the reaction as an infinite series of δ-functions

$$G(t) = \sum_{k=0}^{\infty} \delta(t - k\tau), \tag{8.5.2}$$

then we can first integrate (8.5.1) over the infinitesimal time interval $[n\tau - \epsilon, n\tau + \epsilon]$ to obtain

$$u(x, n\tau + \epsilon) = u(x, n\tau - \epsilon) + R(u(x, n\tau - \epsilon)). \tag{8.5.3}$$

Over the subsequent time interval $[n\tau + \epsilon, (n + 1)\tau - \epsilon]$, only the diffusion acts. Introducing the differencing $t = n\tau$ and $x = iw$ into (8.5.1), with w the space interval between the i and $i + 1$ sites, we obtain

$$\frac{u(iw, (n+1)\tau - \epsilon) - u(iw, n\tau + \epsilon)}{\tau}$$

$$= \frac{D}{w^2} [u((i-1)w, n\tau + \epsilon) - 2u(iw, n\tau + \epsilon) + u((i+1)w, n\tau + \epsilon)]. \tag{8.5.4}$$

614   8. Chaotic Motion in Dissipative Systems

Letting $u_n^i \triangleq u(iw, n\tau - \epsilon)$, we obtain the coupled lattice map system

$$u_{n+1}^i = f(u_n^i) + a_0 f(u_n^i) + a_L f(u_n^{i-1}) + a_R f(u_n^{i+1}), \qquad (8.5.5)$$

where the local map is

$$f(u) = u + R(u), \qquad (8.5.6)$$

and

$$\begin{aligned} a_0 &= -2\alpha^2, \\ a_L &= a_R = \alpha^2, \end{aligned} \qquad (8.5.7)$$

with $\alpha^2 = D\tau/w^2$ the coupling parameter. Keeler (1986) gives a more general derivation and has extended these results to nonnearest-neighbor coupling.

Lattice maps of the form (8.5.5) display a rich variety of phenomena, including domain formation (in x), coexistence of domains having regular and chaotic motion, random motion of domain boundaries, transient chaos, intermittency, spatial and temporal turbulence, etc. They have been widely studied; see conference proceedings in *Physica* 23D (1986) and reviews by Kaneko (1989) and Crutchfield and Kaneko (1987).

We illustrate some of the phenomena for the logistic map (7.2.5)

$$f(u) = \mu u(1 - u). \qquad (8.5.8)$$

As described in Section 7.2, as μ increases, the single logistic map undergoes a period doubling cascade, from a first period doubling at $\mu_1 = 3$ to chaos at $\mu_\infty \approx 3.59$. For $\mu > \mu_\infty$, chaotic bands are formed that have an inverse cascade, as seen in Fig. 7.23, with the final merging of the period 2 band into a single band of chaos at $\mu_1^* \approx 3.685$.

For N coupled maps on a circle (periodic boundary conditions) with $\mu < 3$, any set of initial conditions on the lattice asymptotically approaches a unique solution flat in x, which is the stable period 1 fixed point of the uncoupled map. For $\mu > 3$, however, the system can organize into domains in which u makes sharp upward (*kink*) or downward (*antikink*) transitions in x at domain walls (see Fig. 8.19a). The domain structure is fixed by the choice of initial conditions and, for μ not exceeding a critical value (roughly μ_∞), the domain walls are fixed in space. As μ increases from μ_1, period doubling within domains occurs (see Fig. 8.19b). To understand domain formation, we note that in the uncoupled system in the range of μ for which the uncoupled maps have period 2 fixed points ($\mu \gtrsim 3$), each map asymptotically approaches a stable period 2 solution, but depending on initial conditions, adjacent sites can have solutions with different phases; e.g., $u_1, u_2, u_1, \ldots$ for site i and $u_2, u_1, u_2, \ldots$ for site $i + 1$. For the coupled system, this structure persists, although the kinks are smoothed and extend over several lattice sites by the finite diffusion. The kinks and antikinks trade places on successive iterations, forming a stationary node that is centered at an unstable period 2^p cycle, $f^{(2^p)}(u^*) = u^*$, which has been stabilized by the finite coupling. Since

8.5. Spatio-Temporal Dynamics

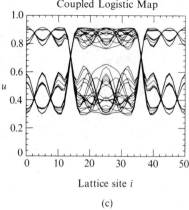

Figure 8.19. Period doubling to chaos for $\alpha^2 = \frac{1}{3}$. (a) $\mu = 3.328$, period 2 behavior is observed; (b) $\mu = 3.518$, the solution has bifurcated to period 4; (c) $\mu = 3.654$, chaos has appeared in the central region. In all cases a kink–antikink pair forms a domain wall (after Keeler and Farmer, 1986).

the domain structure is fixed and depends on the initial conditions, there are many different attractors coexisting in the system.

For μ close to μ_∞ (depending on the coupling strength), the motion within individual domains can be chaotic (see Fig. 8.19c). As μ increases, the large domains go chaotic before the smaller domains, so that the structure can consist of large chaotic and small regular domains. With further increase in μ, domain walls become unpinned and move in x. When two domain walls collide they annihilate, leaving behind a single larger domain having uniform phase (see Fig. 8.20a). Thus the coupled system can ultimately tend toward a single uniform phase attractor. As μ increases toward μ_1^*, the structure of

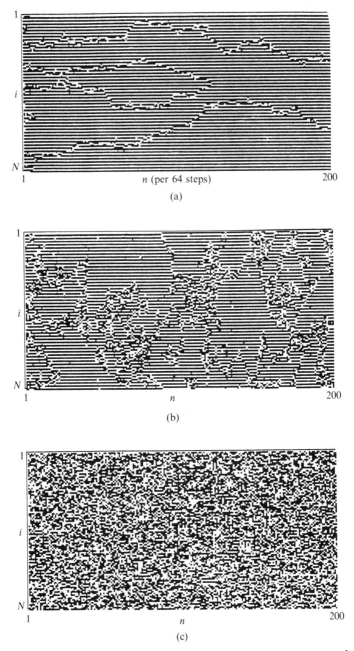

Figure 8.20. Space–time diagram for the coupled logistic lattice with $\alpha^2 = 0.05$, $N = 100$ and starting with a random initial condition. Every 8th or 64th time step is plotted from 0 to 200. If $u_n(i)$ is larger than u^* (unstable fixed point of the logistic map), the corresponding space–time pixel is painted as black, while it is left blank otherwise; (a) $\mu = 3.864$; (b) $\mu = 3.926$; (c) $\mu = 3.960$ (after Kaneko, 1989).

8.5. Spatio-Temporal Dynamics

domains separated by kinks breaks down because there are no longer well-defined phases for the solutions in different regions of the lattice; e.g., it is not possible to distinguish $u_1, u_2, u_1, \ldots$ from $u_2, u_1, u_2, \ldots$. Hence, as μ increases past μ_1^*, new kinks form continuously and annihilate spontaneously (see Fig. 8.20b), leading to global turbulence on the lattice. The regular regions eventually disappear with further increase in μ (see Fig. 8.20c).

Natural Spatial Wavelength. The smoothing of the kinks by diffusion leads to the notion of a natural spatial wavelength λ, which arises due to the competition between reaction and diffusion. In certain parameter regimes, this process can lead to the breakup of large chaotic domains into a series of smaller regular domains. We can roughly estimate λ by taking $u \sim \exp(i2\pi x/\lambda)$ in the diffusion term in (8.5.1) and setting $\partial u/\partial t \approx 0$ in (8.5.1) (reaction balances diffusion) to obtain

$$-\frac{4\pi^2 D u}{\lambda^2} + GR \approx 0. \tag{8.5.9}$$

Using (8.5.6) with the logistic map (8.5.8) yields $R(u) = (\mu - 1)u - \mu u^2$, which has a maximum $R_{\max} = (\mu - 1)^2/4\mu$ at $u_{\max} = (\mu - 1)/2\mu$. Using $R_{\max}$ and $u_{\max}$ in (8.5.9) and averaging the reaction rate in (8.5.9) over a period ($\langle G \rangle \approx 1$), we estimate

$$\lambda \approx 2\pi \left(\frac{2D}{\mu - 1}\right)^{1/2}. \tag{8.5.10}$$

The presence of spatial oscillations at the natural period within a stable domain is illustrated in Fig. 8.19c. The appearance of spatial oscillations can induce a transition to chaos within a domain that preempts the period doubling sequence; e.g., period 16 stable cycle → period 16 chaos directly. On the other hand, the breaking up of a large domain into a spatially oscillating solution can give rise to regular motion within the domain. Figure 8.21 shows an example of this behavior for $D = \alpha^2 = 0.05$, $\mu \approx 3.7$, such that, from (8.5.10), λ (in units of the lattice site i) is approximately $\lambda \approx 2\pi(0.1/2.7)^{1/2} \approx 1.2$. For the smaller μ (Fig. 8.21a), we can see the characteristic mixture of large chaotic and small regular domains. At the larger value of μ (Fig. 8.21b), large flat domains are no longer found, and large domains having a quasiperiodic period 2 spatial pattern have appeared. Again, a further increase in μ will unpin the domain boundaries, which will diffuse and annihilate.

Spatio-Temporal Intermittency. Resonance effects are found when the domain size is commensurate with the natural wavelength. In the regime $3.59 < \mu < 3.7$, if the domain size L is near a half-integral number of natural wavelengths, $L \approx m\lambda/2$, the motion is on a regular cycle with spatial structure at the natural wavelength. For nonresonance, $L \neq m\lambda/2$, the motion is chaotic. If

Figure 8.21. Space–amplitude plot for the coupled logistic lattice. Amplitudes $u_n(i)$'s are overlayed for 50 time steps, after the first 1000 iterations of the transient, with $\alpha^2 = 0.05$, $N = 100$, and random initial configurations. (a) $\mu = 3.600$; (b) $\mu = 3.750$ (after Kaneko, 1989).

the kinks are unpinned, then their slowly changing position produces a slowly varying change in L, and the motion on a given lattice site can be intermittently regular and chaotic. This form of intermittency, seen numerically by Kaneko (1984, 1989), has some similarity to that found near a tangent bifurcation (see Section 7.5a); i.e., the regular events are very weakly unstable (Keeler and Farmer, 1986). However, the intermittency persists over a wide range of μ, in contrast to that near a tangent bifurcation, which is found for only a narrow window of parameters. Taking a spatial average over lattice sites, Keeler and Farmer show numerically that this robust intermittency gives rise to a $1/f$ power spectrum.

8.5. Spatio-Temporal Dynamics

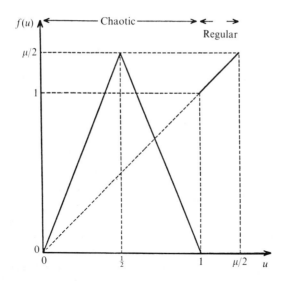

Figure 8.22. Local map $f(u)$ (after Chaté and Manneville, 1988).

Chaté and Manneville (1988) have described a simple mapping model that fulfills the "minimal" requirements for spatio-temporal intermittency, shown in Fig. 8.22. The map f for $1 < \mu < 2$ has only a chaotic dynamics on the unit interval in u (see discussion of the tent map in Section 7.2c for details). However, for $\mu > 2$, the iteration falls into a "hole" of size $1 - 2/\mu$ through which it passes onto a continuum of stable fixed points $1 < u < \mu/2$ (upper right-hand corner in Fig. 8.22). For $\mu > 2$, the single map is transiently chaotic only, but the coupled system (8.5.5) is found to display spatio-temporal intermittency if α^2 is sufficiently large; hence the local transient chaos is converted into a sustained globally turbulent state by the coupling alone. Chaté and Manneville find this behavior numerically and give an interpretation of these results in terms of a percolation theory model.

Other one-dimensional maps and other forms of coupling have also been studied. These include the circle map (see Section 7.4), and couplings such as unidirectional (flow) systems having $a_0 = -a_L$ and $a_R = 0$. For further details, the reader should consult the review by Crutchfield and Kaneko (1987). Coupled lattice systems can naturally arise for which the uncoupled map is M-dimensional and invertible. An example is the dissipative sine–Gordon equation which is discretized in space and time. This represents a system of N identical pendula driven by an impulse gravitational force. The pendula are coupled to nearest neighbors by linear springs and the pivots are frictionally damped. The behavior of this system, in the absence of damping, was described in Section 6.5c as an example of a symplectic lattice map.

8.5b. Self-Organized Criticality

Spatially extended dynamical systems can evolve toward a critical state, having no intrinsic time or length scales, by a self-organizing process (Bak et al., 1988a). A paradigm system is a sandpile that is built by randomly dropping sand onto the pile until an unstable state is formed and the sand slides off. Eventually a critical slope for the pile is reached such that avalanches on all space and time scales occur. The self-organized critical state is thus an attractor for the dynamics. The long-range correlations associated with the critical state can lead to the appearance of universal phenomena such as $1/f$ noise and self-similar fractal structure.

Following Bak et al. (1988a), we introduce these ideas by considering a sandpile model. For the simplest dynamics in one dimension, the system evolves to the least-stable metastable state, which is a trivial state having uninteresting properties. However, by extending the system to two dimensions, one finds numerically that a self-organized critical state arises exhibiting $1/f$ noise and finite size scaling. Kadenoff et al. (1989) have introduced more complicated one-dimensional dynamics and obtained self-organized critical states exhibiting multifractal structure, which we also describe. We also describe the related dynamics of a mechanical model of earthquakes consisting of coupled differential equations. The model also exhibits self-scaling behavior (Carlson and Langer, 1989a, b). We note experimental studies by Jaeger et al. (1989) showing that real sandpiles exhibit some evidence of self-organized critical behavior.

Coupled Mappings: Sandpiles. Figure 8.23 shows a one-dimensional sandpile of integer length L. We let

$$z_n = h(n) - h(n+1) \tag{8.5.11}$$

be the differences in heights between successive positions along the sandpile. If we add one "grain" (unit) of sand to position n, then

$$\begin{aligned} z_n &\to z_n + 1, \\ z_{n-1} &\to z_{n-1} - 1. \end{aligned} \tag{8.5.12}$$

When the height difference is larger than some fixed critical value z_c, then we let one grain fall to the lower level:

$$\left. \begin{aligned} z_n &\to z_n - 2, \\ z_{n\pm 1} &\to z_{n\pm 1} + 1 \end{aligned} \right\} \quad z_n > z_c. \tag{8.5.13}$$

We choose the boundary conditions such that sand can flow off of the right edge only:

$$\left. \begin{aligned} z_0 &= 0, \\ z_L &\to z_L - 1, \\ z_{L-1} &\to z_{L-1} + 1, \end{aligned} \right\} \quad z_n > z_c. \tag{8.5.14}$$

8.5. Spatio-Temporal Dynamics

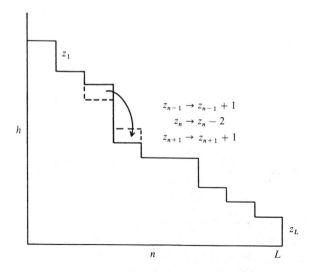

Figure 8.23. One-dimensional "sand-pile automaton." The state of the system is specified by an array of integers representing the height difference between neighboring plateaus; there is a wall at the left (after Bak et al., 1988a).

Equation (8.5.13) is a nonlinear discretized diffusion equation. Dropping one grain on this system at a random location will lead generally to an avalanche of events each described by (8.5.13), which continue until all z_n are below threshold. We can keep a record of the number of grains D that fall off the edge and the duration T [number of steps for which (8.5.13) is applied] for each avalanche. When an avalanche ends we can drop another grain onto a new site n, setting in motion a new avalanche, etc.

Note that a state is stable if $z_n < z_c$ for all n. Hence the number of stable states is z_c^L, which is very large for $z_c \geq 2$. For the simple dynamics (8.5.13), if sand is added to an initially empty system, then the pile will build up until $z_n = z_c$ for all n. This is the least stable of all the stable states. A grain added to any site n then falls down the pile and off the end at site L. This least stable state is an attractor of the dynamics, independent of the initial conditions; e.g., we could start with any unstable state and the system would relax to the least stable state, or we can start with any initial stable state, and the system will eventually build up to the least stable state.

For the least stable state, the probability distribution for D is trivial: $\rho(D) = 1$ for $D = 1$ and $\rho(D) = 0$ otherwise. The probability distribution for the duration T of the avalanche is also trivial: since the grains are dropped at random sites, $\rho(T) = 1/L$ for all T between 1 and L. Since $\rho(T)$ is uniform, we obtain a $1/f^2$ power spectrum for the flow down the slope. In contrast, the flow off the edge of the sand-pile has a white noise spectrum (Jensen et al., 1989).

In two dimensions, one generalization of (8.5.11)–(8.5.13) is the algorithm $z(x, y) = 2h(x, y) - h(x + 1, y) - h(x, y + 1)$, such that

$$z(x - 1, y) \rightarrow z(x - 1, y) - 1,$$
$$z(x, y - 1) \rightarrow z(x, y - 1) - 1, \quad (8.5.15)$$
$$z(x, y) \rightarrow z(x, y) + 2,$$

and

$$\left.\begin{array}{l} z(x, y) \rightarrow z(x, y) - 4, \\ z(x, y \pm 1) \rightarrow z(x, y \pm 1) + 1, \\ z(x \pm 1, y) \rightarrow z(x \pm 1, y) + 1, \end{array}\right\} \quad z(x, y) > z_c, \quad (8.5.16)$$

respectively, for a square pile with sites $1 \leq x, y \leq L$. A similar generalization can be made in three dimensions. One might expect, in analogy with the one-dimensional case, that the sandpile will build up to the least stable state for which $z(x, y) = z_c$ for all x, y. However, when two grains fall diagonally down according to (8.5.16), then the surrounding sites will become unstable, and the disturbance will spread to their neighbors and so on, forming an instability (large avalanche). Hence the least stable state is unstable with respect to small perturbations and cannot represent an attractor for the dynamics. Instead, the system will become marginally stable at the point where the perturbation can lead to any motion from a single grain lost to a large avalanche; i.e., the perturbation can grow with some probability over all length (and time) scales. This is the stationary, self-organized critical state. The z's are reduced compared to the z's of the least-stable metastable state.

Figure 8.24 shows the typical domain structures obtained by Bak et al., dropping a single grain onto a 100 × 100 array of sites that has relaxed to the critical state, for the case of closed boundary conditions

$$z(0, y) = z(x, 0) = z(L + 1, y) = z(x, L + 1) = 0. \quad (8.5.17)$$

During one "unit" of time, (8.5.16) is applied at each site for which $z(x, y) > z_c$, resulting, generally, in many grain slides, all updated in parallel during that unit of time. We see that domains exist from a single site up to an extent comparable to the system itself. If we let S be the "size" of the avalanche, i.e., the total number of grain slides [number of times (8.5.16) is applied during a single avalanche], induced by dropping a single grain, then one finds that the probability distribution of avalanche sizes is

$$\rho(S) \sim S^{-\tau}, \quad (8.5.18)$$

where, numerically, $\tau \approx 1.0$ for a two-dimensional sandpile, and $\tau \approx 1.37$ for a three-dimensional sandpile. Similarly, one finds that the distribution of avalanche times T (number of units of time for the avalanche) scales as

$$\rho(T) \sim T^{-\eta}, \quad (8.5.19)$$

8.5. Spatio-Temporal Dynamics

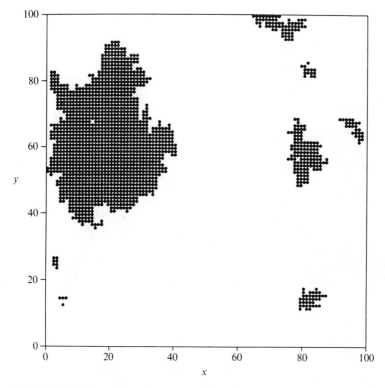

Figure 8.24. Typical domain structures resulting from several local perturbations for a 100×100 array. Each cluster is triggered by a single perturbation (after Bak et al., 1988a).

where $\eta \approx 0.43$ for two dimensions and $\eta \approx 0.92$ for three dimensions. Such a distribution of avalanche times, operating simultaneously and independently, leads to a power law frequency spectrum $\sim f^{-\beta}$, with $\beta \approx 1.57$ for two dimensions and $\beta \approx 1.08$ for three dimensions, determined numerically.

For each avalanche, the total time T is less than the size S, because the average growth rate γ of the avalanche is greater than unity; i.e., for a single increment in time, there are more than one simultaneous slides within the avalanche:

$$S \approx T^{1+\gamma}. \tag{8.5.20}$$

Bak et al. (1988) numerically obtain $\gamma \approx 0.57$ for two dimensions, and $\gamma \approx 0.71$ for three dimensions. The scaling exponents $\eta, \tau, \beta, \gamma$ are numerically found to be different for different boundary conditions (e.g., (8.5.17)) and diffusion rules (e.g., (8.5.16)).

For example, the open boundary conditions

$$\left.\begin{array}{l} z(L, y) \to z(L, y) - 3, \\ z(x, L) \to z(x, L) - 3, \\ z(L, L) \to z(L, L) - 2, \end{array}\right\}, \quad z > z_c \quad (8.5.21)$$

lead, numerically, to $\eta \approx 1.05$ for a two-dimensional system.

The scaling of the probability distributions with system size L has also been examined. Bak *et al.* find that their two-dimensional systems obey the finite-size scaling relation

$$\rho(T, L) = T^{-\eta} F(L^\sigma/T), \quad (8.5.22)$$

where F is a scaling function of the single scaled variable L^σ/T and σ is the critical exponent. Figure 8.25 shows the finite-size scaling plot for the open system, giving numerically $\sigma \approx 0.75$.

Kadanoff *et al.* (1989) have explored many different diffusion rules for one- and two-dimensional systems and have examined in detail the scaling of the probability distributions ρ with the system size L. For some rules, the

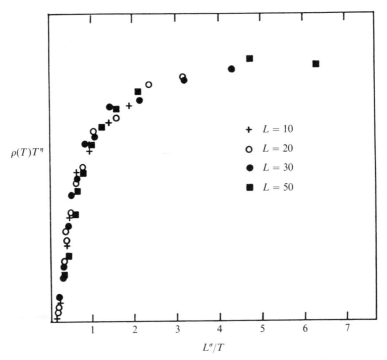

Figure 8.25. Finite-size scaling plot (after Bak *et al.*, 1988a).

8.5. Spatio-Temporal Dynamics

finite-size scaling relation (8.5.22) does not fit well, and the multifractal scaling

$$\frac{\log_{10} \rho(T, L)}{\log_{10}(L/L_0)} = f(\alpha), \tag{8.5.23}$$

where

$$\alpha = \frac{\log_{10}(T/T_0)}{\log_{10}(L/L_0)}, \tag{8.5.24}$$

provides a better fit. We have discussed $f(\alpha)$ in Section 8.3. The two forms (8.5.22) and (8.5.23) only agree when F is a simple power law and f is a linear function. In that case the probability is described by only two scaling indices (η and σ). Otherwise, there is a spectrum of scaling indices; i.e., all the values taken on by $df/d\alpha$. In one dimension, more complicated diffusion rules than that given in (8.5.13) yield nontrivial probability distributions, well described by multifractal scalings. In two dimensions, both finite-size and multifractal scaling agree well with the numerical results.

Jaeger et al. (1989) have experimentally studied the particle flow along a free surface in real sandpiles consisting of spherical glass beads or rough aluminum oxide particles. The average slope of the surface was varied either by turning a circular or semicircular drum partially filled with "sand" or by randomly adding particles to the top surface of a pile contained in a box with one open side. For continuous slow variation of the slope, the avalanches were found to start at an angle $\theta_m = \theta_r + \delta$ (where δ is typically a few degrees) and θ_r is the angle the sandpile relaxes to after the event. Both the time interval Δt between two avalanche pulses and the duration $\Delta \tau$ of the avalanches were measured. Because of the hysteresis, events occur at regular intervals and critical scaling was not observed. By vibrating the sandpile mechanically using a speaker coupled to the system, δ could be reduced to near zero. At the same time, θ_r is reduced to some θ_{ss}. For this case, the measured power spectra for Δt and $\Delta \tau$ are in qualitative agreement with the predictions of the criticality theory (Jensen et al., 1989).

Other physical systems may be found that exhibit self-organized critical behavior. For example, Bak et al (1988a) show that the basic one-dimensional sandpile dynamics also describe a set of pendula coupled by torsion springs. The heights $h(n)$ are analogous to the winding numbers of the springs, and the z_n are the spring forces acting on the pendula. When z_n exceeds z_c, so that the spring force exceeds the gravitational force, then the pendulum rotates one revolution, leading to the dynamics (8.5.13).

Coupled Differential Equations. An example of self-organized criticality is provided by a mechanical model of an earthquake fault (Carlson and Langer, 1989a, b). The system differs from that of a "sandpile" because it is described by a coupled set of differential equations rather than difference equations.

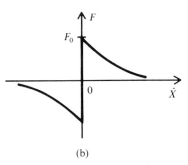

Figure 8.26. (a) Block and spring system and (b) friction force $F(\dot{X})$. For the numerical results given in Figs. 8.27 and 8.28 the parameters have values $m = 1$, $k_c = 2500$, $k_p = 4\cdot$, $a = 1$, and $v_0 = 0.01$; the friction force $F(\dot{X}) = F_0/(1 + \dot{X})$ for $\dot{X} > 0$ with $F_0 = 50$ (after Carlson and Langer, 1989a).

Furthermore, there is no explicit randomness in the model; instead, the system evolves deterministically from the specified initial conditions.

The model is illustrated in Fig. 8.26a. Each of N blocks of mass m located at position $X_j(t)$ is coupled to its two neighbors by linear springs of strength k_c. The blocks are separated by a distance a when the spring forces are zero. Each block is pulled forward by linear "leaf" springs of strength k_p that move at constant velocity v_0. Finally, each block is subject to a nonlinear friction force $F(\dot{X})$ shown in Fig. 8.26b. The equations of motion for the system are

$$m\ddot{X}_j = k_c(X_{j+1} - 2X_j + X_{j-1}) - k_p(X_j - v_0 t) - F(\dot{X}_j). \quad (8.5.25)$$

Simple solutions to (8.5.25) are uniform motion

$$X_j = v_0 t - \frac{1}{k_p} F(v_0),$$

and periodic motion where all blocks alternatively stick and slide. Both are unstable to perturbations for all $v_0 \neq 0$.

The system (8.5.25) was solved numerically for a wide range of parameters and initial conditions and for system sizes up to 200 blocks. The system appears to evolve to a statistical steady state for which one observes motions

8.5. Spatio-Temporal Dynamics

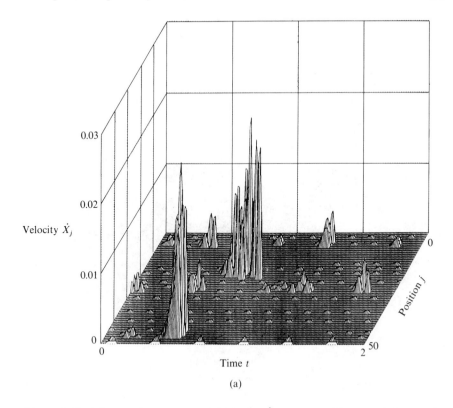

Figure 8.27. (*above and on next page*) Velocity $\dot{X}_j$ versus position j and time t. The smallest features in (a) are periodic creep events which involve only one block. The great event illustrated in (b) involves all of the blocks and essentially resets the system (after Carlson and Langer, 1989a).

on all space and time scales. In Fig. 8.27, $\dot{X}_j(t)$ is plotted to illustrate the space and time variation of the observed "earthquake" events. Events could be divided into three types: slipping of a single block, large but localized events, and global events that involve all of the blocks and essentially "reset" the system. The large localized events are found to be self-similar over a range of length scales which widens as the loading speed v_0 decreases. The "earthquake" magnitude M is defined to be

$$M = \ln\left[C \sum_j \int_{\text{event}} \dot{X}_j \, dt \right]. \quad (8.5.26)$$

Plotting the distribution of the number of earthquakes $N_E(M)$ from the numerical results yields the power law scaling shown in Fig. 8.28; i.e.,

$$\ln N_E \simeq A - BM, \quad (8.5.27)$$

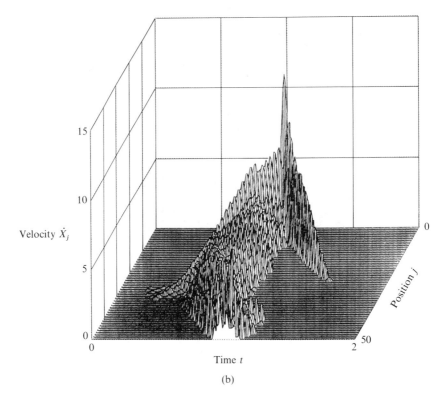

Figure 8.27 (cont.)

where A is a constant and $B \approx 1$. This scaling agrees with the Gutenberg–Richter law for real earthquakes.

Bak and Tang (1989) have pointed out that the coupled map (8.5.16) is a plausible model of earthquake fault dynamics in two dimensions. The blocks are coupled by springs along the fault horizontally and vertically. When the total force z on a block exceeds a maximum local pinning force z_c, then the block slips to a nearby position, as given by (8.5.16). Starting with no force, $z(x, y) = 0$ for all x, y, one increases the force by letting $z(x, y) \to z(x, y) + 1$ at a site chosen randomly. Eventually, the force exceeds the pinning force somewhere and an "earthquake" results. Bak and Tang show that this two-dimensional mapping dynamics also leads to the Gutenberg–Richter law.

8.6. The Fluid Limit

Just as problems in celestial mechanics have motivated much of the development of nonlinear Hamiltonian systems, the fluid medium has

8.6. The Fluid Limit

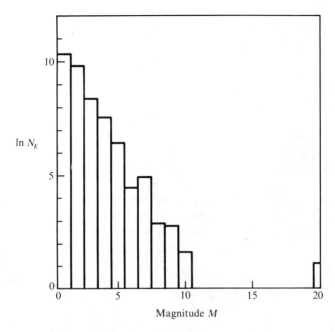

Figure 8.28. Distribution of earthquake magnitudes $N_E(M)$. For this result the number of blocks is $N = 200$ and the constant C in Eq. (8.5.26) is $10^6 e$ (after Carlson and Langer, 1989a).

motivated much of the development of dissipative systems. For incompressible flow, the fundamental equations are the Navier–Stokes equations:

$$\frac{\partial \boldsymbol{v}}{\partial t} + \boldsymbol{v} \cdot \nabla \boldsymbol{v} = -\frac{1}{\rho_0} \nabla p + \nu \nabla^2 \boldsymbol{v}, \qquad (8.6.1a)$$

$$\nabla \cdot \boldsymbol{v} = 0, \qquad (8.6.1b)$$

where $\boldsymbol{v}(\boldsymbol{x}, t)$ and $p(\boldsymbol{x}, t)$ are the fluid velocity and pressure, respectively, at point $\boldsymbol{x}$ and time t; ρ_0 is the (constant) density and ν is the viscosity. Flows are classified in terms of the Reynolds number $R = uL/\nu$, which is a nondimensional measure of the nonlinearity in (8.6.1), where u is a typical velocity and L is a typical length scale. If the Reynolds number is not too large, the flow is *laminar*, in the sense that it displays regular and predictable variations in both space and time. As the Reynolds number increases, flows typically undergo a sequence of instabilities until, at some large enough value of the Reynolds number, they become fully turbulent. Much work has focused on a high Reynolds number which corresponds to fully developed turbulence. In this regime, scaling arguments can be used to show that the spatial Fourier modes have an energy spectrum proportional to $k^{-5/3}$, where k is the mode wave number (Kolmogorov, 1941). Furthermore, this spectrum can be used with other dimensional arguments to show that the number of participating modes (degrees of freedom) should be proportional to $R^{9/4}$ (e.g., see Frisch and Orzag, 1990). In this situation we would not expect results from low-dimensional dynamics to be very relevant to fluids. At the transition to turbulence, on the other hand, the dimensional arguments are not adequate and the Reynolds number may be relatively small, leading to greater relevance of low-dimensional dynamics.

Although it has been known for hundreds of years that fluid systems undergo a transition from regular to turbulent behavior, detailed experimental measurements of the velocity flow, temperature distribution, etc., near the transition, have only been made during the last two decades. The onset of turbulence typically occurs rather abruptly, preceded by the observation of one or a few independent frequencies, their harmonics, and (sometimes) their subharmonics. The main experiments in which these observations have been made have been various versions of Rayleigh–Bénard convection and of Taylor vortices generated between concentric differentially rotating cylinders (circular Couette flow).

In Section 8.6a, we consider how finite-dimensional systems can be used to approximate infinite-dimensional fluids. Then in Section 8.6b, we consider experimental results from fluids to see if the low-dimensional mechanism for the transitions to turbulence can describe real fluids. We use the results of Rayleigh–Bénard convection and Couette flow experiments in this comparison. Finally, in Section 8.6c, we show that in some situations Hamiltonian dynamics can be used to describe fluid behavior.

8.6a. Fourier Mode Expansions

We consider motion in fluid systems or other infinite-dimensional systems described by partial differential equations of the form

$$\frac{\partial \boldsymbol{Q}(\boldsymbol{x}, t)}{\partial t} = \mathscr{L}(\boldsymbol{x})\boldsymbol{Q}(\boldsymbol{x}, t), \tag{8.6.2}$$

where $\boldsymbol{Q}$ is an M-dimensional vector that specifies the state of the fluid system [e.g., the components of $\boldsymbol{Q}$ are the pressure $p(\boldsymbol{x}, t)$, fluid velocity $\boldsymbol{v}(\boldsymbol{x}, t)$, mass density $\rho(\boldsymbol{x}, t)$, etc.], $\boldsymbol{x}$ is the configuration space (with components x, y, and z in three dimensions), and $\mathscr{L}(\boldsymbol{x})$ is a time-independent, nonlinear differential operator. An example is the set of equations (8.6.1). A common technique in studying the behavior of (8.6.2) is to Fourier analyze $\boldsymbol{Q}$ into mode amplitudes

$$\boldsymbol{Q}(\boldsymbol{x}, t) = \sum_{\boldsymbol{k}} \boldsymbol{q}_{\boldsymbol{k}}(t) e^{i\boldsymbol{k}\cdot\boldsymbol{x}}, \tag{8.6.3}$$

where

$$\boldsymbol{q}_{\boldsymbol{k}}(t) = \frac{1}{(2\pi)^3} \int d^3x \, \boldsymbol{Q}(\boldsymbol{x}, t) e^{-i\boldsymbol{k}\cdot\boldsymbol{x}}. \tag{8.6.4}$$

8.6. The Fluid Limit

Inserting (8.6.3) in (8.6.2) and using the orthogonality of the exponential functions $e^{i\mathbf{k}\cdot\mathbf{x}}$, we obtain for each mode $\mathbf{q}_k$ an equation of the form

$$\dot{\mathbf{q}}_k = V_k(\mathbf{q}_1,\ldots,\mathbf{q}_k), \tag{8.6.5}$$

where the overdot as usual means d/dt. If only the N "most important modes" in the sum (8.6.3) are kept, then (8.6.5) represents a set of MN, first-order ordinary differential equations describing the evolution in time of the mode amplitude components. This procedure of Fourier analysis followed by truncation is often called the Galerkin approximation.

We illustrate this procedure for the problem of Rayleigh–Bénard convection and derive from an extreme truncation, keeping only the three "most important" components, the equations (1.5.2) for the Lorenz attractor. The problem is illustrated in Fig. 8.29a. A fluid slab of thickness h is heated from below, maintaining a temperature difference ΔT between the bottom hot surface and the top cold surface. Gravity acts downward. The fluid motion is described by the Navier–Stokes equations. Assuming only two-dimensional motion ($\partial/\partial z \equiv 0$), we can define a stream function $\psi(x, y, t)$ for the motion, such that the fluid velocity $\mathbf{u}(x, y, t)$ is given by

$$\mathbf{u} = \nabla \times (\hat{z}\psi). \tag{8.6.6}$$

We also define $\Theta(x, y, t)$ to be the departure of the temperature profile $T(x, y, t)$ from a vertical profile that linearly decreases with the height y;

$$\Theta = T - \frac{\Delta T y}{h}. \tag{8.6.7}$$

Thus $\Theta = 0$ if the fluid has no convection. The fluid motion can be described in terms of these variables by means of two partial differential equations (Lorenz, 1963):

$$\frac{\partial}{\partial t}(\nabla^2 \psi) = \frac{\partial(\psi, \nabla^2 \psi)}{\partial(x, y)} + \nu \nabla^4 \psi + g\alpha \frac{\partial \Theta}{\partial y}, \tag{8.6.8a}$$

$$\frac{\partial}{\partial t}(\Theta) = \frac{\partial(\psi, \Theta)}{\partial(x, y)} + \frac{\Delta T}{h}\frac{\partial \psi}{\partial y} + \kappa \nabla^2 \Theta. \tag{8.6.8b}$$

Here ν is the kinematic viscosity, g the gravitational acceleration, α the thermal expansion coefficient, and κ the thermal conductivity. We assume free boundaries such that ψ and $\nabla^2\psi$ vanish at both surfaces.

For ΔT small, the stable equilibrium is $\psi = 0$, $\Theta = 0$. The fluid is at rest, with heat transported from the bottom to the top surface by thermal conduction. Lord Rayleigh studied the linear stability of this equilibrium and showed that above a critical value ΔT_c, the equilibrium is unstable. Just above ΔT_c, a new equilibrium is established characterized by a stable,

632   8. Chaotic Motion in Dissipative Systems

(b)

Figure 8.29. The Rayleigh–Bénard convection problem from which the Lorenz system is derived. (a) Fluid system with $\Delta T = T_1 - T_0 > 0$. (b) Steady convection for $\Delta T \gtrsim \Delta T_c$.

circulating fluid flow (see Fig. 8.29b) having

$$\psi = \psi_0 \sin\left(\frac{\pi a x}{h}\right) \sin\left(\frac{\pi y}{h}\right), \qquad (8.6.9a)$$

$$\Theta = \Theta_0 \cos\left(\frac{\pi a x}{h}\right) \sin\left(\frac{\pi y}{h}\right), \qquad (8.6.9b)$$

where a is a parameter characterizing the wavelength along x. The minimum value of ΔT_c occurs for $a = 1/\sqrt{2}$. We introduce the Rayleigh number as a dimensionless measure of the temperature difference

$$R_a = \frac{g\alpha h^3 \Delta T}{\nu\kappa}. \qquad (8.6.10)$$

The critical Rayleigh number for onset of steady convection is

$$R_c = \frac{\pi^4(1 + a^2)^3}{a^2}, \qquad (8.6.11)$$

having a minimum value $27\pi^4/4$ when $a = 1/\sqrt{2}$.

As R_a is increased still further above R_c, the steady convection equilibrium becomes linearly unstable. Experimentally, for this larger value of R_a, the

8.6. The Fluid Limit

convection takes place in an unsteady, irregular manner. To determine the motion in this case, we note that Eqs. (8.6.8) are in the form (8.6.2). Thus, following Saltzman (1962), we may expand ψ and Θ in a double Fourier series in x and y, with the Fourier coefficients functions of t alone. By truncating the series to a finite number of terms, one obtains the motion in the finite-dimensional phase space of Fourier coefficients. Saltzman numerically found solutions with chaotic behavior. Lorenz (1963) studied a simplified system in which only three Fourier coefficients were kept, writing

$$a(1 + a^2)^{-1}\kappa^{-1}\psi = \sqrt{2}X(t) \sin\left(\frac{\pi a x}{h}\right) \sin\left(\frac{\pi y}{h}\right),$$

$$\pi R_c^{-1} R_a(\Delta T)^{-1}\Theta = \sqrt{2}Y(t) \cos\left(\frac{\pi a x}{h}\right) \sin\left(\frac{\pi y}{h}\right) - Z(t) \sin\left(\frac{2\pi y}{h}\right). \quad (8.6.12)$$

Here X represents the amplitude of the convection motion, Y the temperature difference between ascending and descending currents, and Z the distortion of the vertical temperature profile from linearity. Substituting these forms into the partial differential equations (8.6.8) yields the Lorenz system:

$$\dot{X} = -\sigma X + \sigma Y, \quad (8.6.13a)$$

$$\dot{Y} = -XZ + rX - Y, \quad (8.6.13b)$$

$$\dot{Z} = XY - bZ, \quad (8.6.13c)$$

with dimensionless parameters σ, r, b, where $\sigma = \nu/\kappa$ is the Prandtl number, $r = R_a/R_c$ is the normalized Rayleigh number, $b = 4(1 + a^2)^{-1}$ is a geometrical factor, and the dot represents differentiation with respect to a normalized time

$$\tau = \pi^2 h^{-2}(1 + a^2)\kappa t.$$

The Lorenz system and the nature of its strange attractor were described in Section 1.5 and in previous sections of this and the preceding chapters. Here we enquire in what sense the exact solution to the Rayleigh–Bénard convection problem behaves like the Lorenz system; i.e., for the same initial flow, do both systems tend to the same type of final state as $t \to \infty$? At first sight this seems improbable, since the Galerkin expansion has been sharply truncated, leaving only three modes. Systematic extensions of the Lorenz equations to five, seven, and fourteen mode components exhibit behavior with some features similar to that of the Lorenz system, including the presence of strange attractors. However, the transition to chaotic behavior may proceed by a variety of bifurcation sequences (Curry, 1978; see also Helleman, 1980, for additional references). Indeed, numerical solutions of the Rayleigh–Bénard equations in two dimensions (8.6.8), from which the Lorenz equations are derived, do not exhibit turbulent behavior. This is in distinction to three-dimensional Rayleigh–Bénard convection, discussed below, for which turbulence is experimentally observed.

A property of the Navier–Stokes equations sheds some light on this question. It has been shown (Foias and Prodi, 1967) that the two-dimensional solutions to the Navier–Stokes equations can be described in the limit $t \to \infty$ by a finite number N_2 of Fourier modes. This result has been extended to the three-dimensional solutions to give an upper bound N_3 of modes (Foias and Treve, 1981). One is thus assured that for a fixed set of parameters and initial flow, and for some minimum N_2 (or N_3), the Galerkin expansion will yield all physically relevant properties of the actual flow. Treve (1981) has suggested a numerical test to determine *post facto* whether the number of modes used exceeds the required value of N_2 (or N_3). In a related problem, a rigorous upper bound on the dimension of the underlying attractor in Rayleigh–Bénard convection has been found by Foias *et al.* (1987), which can also be related to the number of modes required to describe the system. The authors also state that this upper bound is a good estimate of the attractor dimension.

Numerical calculations of two-dimensional Rayleigh–Bénard convection determined the number of modes required for the solutions of the equations to become independent of mode number (Maschke and Saramito, 1982). Their results indicate successive transitions with increasing R_a from steady convection through a Hopf bifurcation, a period doubing bifurcation, and a bifurcation onto a torus, but no transition to turbulence up to $R_a \approx 800 R_c$. In order to obtain the converged solution at the highest Rayleigh number investigated, 350 modes were required (Maschke, private communication).

8.6b. The Transition to Turbulence

In Chapter 7 we investigated a number of routes to chaos for low-dimensional systems. These included overlapping of phase-locked regions, period doubling, intermittency, and crises. Although we pointed out in the previous subsection that many-dimensional systems might not act precisely like the low-dimensional ones, nevertheless, we might expect to see some of these phenomena in higher dimensions, particularly if the attracting set is low dimensional.

The original picture of the transition to turbulence, suggested by Landau, argued that turbulence may be viewed as a hierarchy of instabilities. As a parameter, such as the Reynolds or Rayleigh number, increases from zero, a succession of unstable modes appear and saturate in nonlinear periodic states having frequencies $\omega_1, \omega_2, \omega_3, \ldots$, which are not rationally related. We should then observe the successive appearances of singly periodic, doubly periodic, triply periodic, etc., motion as the parameter increases. Thus we obtain a sequence of Hopf bifurcations with successive flows on a 1-torus, 2-torus, 3-torus, etc. The motion will appear more and more complicated as the number of frequencies increases. However, only after an infinite number

8.6. The Fluid Limit

of frequencies have appeared will the motion have a continuous spectrum and be truly chaotic. Although a few independent modes have been observed in experiments, with as many as four independent modes observed in Couette flow (see Gorman *et al.*, 1980), the generally abrupt transition to an aperiodic spectrum is not consistent with the Landau picture. Furthermore, it has been shown theoretically (see Eckmann, 1981) that a succession of Hopf bifurcations of the Landau type, and indeed multiply periodic motion itself, is nongeneric. From our previous studies of chaotic attractors, we might expect this mechanism to be the basis of more modern theories of turbulence, and this is, in fact, the case.

Ruelle and Takens (1971) proposed an alternate route to turbulence in which they adapted the first two Hopf bifurcations of the Landau picture, but conjectured that nonlinearities would destroy triply periodic motion, leading to the appearance of a chaotic attractor. The original conjecture required a four-dimensional flow. The conjecture that triply periodic motion is generically unstable has been proved, and the number of dimensions required has been reduced to three (Newhouse *et al.*, 1978). The correctness of this picture for some fluid problems is bolstered by the fact that the 14-mode Galerkin approximation to the Rayleigh–Bénard convection problem shows, contrary to the three-mode Lorenz system, that chaotic time dependence is preceded by doubly periodic motion on a two-dimensional surface that is embedded in the 14-dimensional phase space (Curry, 1978).

Some experimental results appear to confirm the Ruelle–Takens model. The power spectra of these systems exhibit first one, then two, and possibly three independent frequencies. Broadband noise suddenly appears just when the third frequency seems ready to appear. The broadband noise is evidence of the transition to chaotic behavior. Experiments have been performed on Taylor vortices between rotating cylinders (Fenstermacher *et al.*, 1978) and on Rayleigh–Bénard convection (Ahlers and Behringer, 1978). Figure 8.30, taken from Swinney and Gollub (1978), shows the power spectrum of the velocity between rotating cylinders driven at three different speeds (left) and the power spectrum of heat transport at three different heat inputs in Rayleigh–Bénard convection. In both cases, we sometimes see the appearance of first one, then two independent frequencies f_1 and f_2, followed by a transition to a broadband spectrum. However, the results are sometimes dependent on the initial state, and phase locking between f_1 and f_2 sometimes occurs. In another Couette experiment (Gorman *et al.*, 1980), at least four independent frequencies have been observed, indicating the Ruelle–Takens transition does not always take place after two Hopf bifurcations.

The Ruelle–Takens model has some similarities to the onset of chaos in circle maps. In the fluid, the two-frequency dynamics constitute a self-generated phase-locked state. An analogous driven phase-locked state can be obtained by applying a second frequency to the fluid in the parameter range for which the fluid itself exhibits a single frequency. This has been

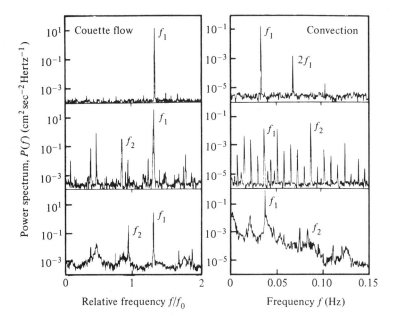

Figure 8.30. Power spectrum of velocity in rotating cylinders driven at three different speeds (left), and of heat transport at three different heating powers in Rayleigh–Bénard convection (right) (after Swinney and Gollub, 1978).

done in a Rayleigh–Bénard convection experiment by Stavans *et al.* (1985). In this configuration the complete mode-locked diagram was observed as described in Section 7.4. Forcing the two frequencies to be quasiperiodic led to earlier onset of chaotic fluid behavior than in the low-frequency ratio tongue, as is also observed in circle maps.

A third model of the transition to turbulence involves period-doubling bifurcation sequences. The transition begins from a stationary state that undergoes a Hopf bifurcation to a single periodic orbit with frequency f_1 (including its higher harmonics). For a further increase in the parameter, there is a period-doubling bifurcation, and we obtain the frequency $\frac{1}{2}f_1$ and higher harmonics. Repeated bifurcations yield repeated period doublings that give rise to the frequencies $\frac{1}{4}f_1$, $\frac{1}{8}f_1$, etc., accumulating at a critical value of the parameter, beyond which we have the onset of broadband noise and chaotic motion associated with a strange attractor. The mechanism of period doubling was described in Section 7.2 and was shown to have a universal character. Furthermore, near the accumulation point, the motion was shown to be essentially one-dimensional (see Section 7.3a). The model is computationally well established for many low-dimensional dynamical systems, such as the Rössler attractor, Hénon map, Duffing equation, etc., as described in Sections 7.1–7.3.

Several experiments on Rayleigh–Bénard convection show period-doubling

8.6. The Fluid Limit

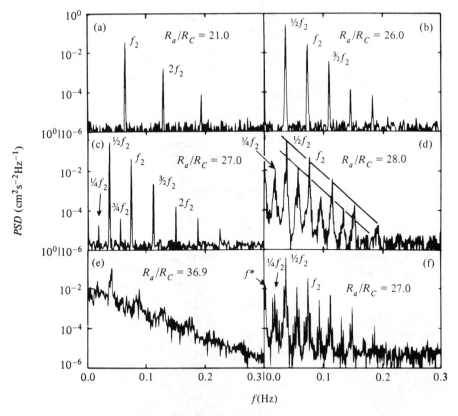

Figure 8.31. Period-doubling sequences in the velocity power spectral density of water in a Rayleigh–Bénard experiment. Note the creation of subharmonics at increasing Rayleigh numbers R_a, the creation of chaotic "bands" in (d) and a "continuous nonperiodic" spectrum in (e). Upon return to the R_a of (c) a second frequency f^* appears in (f) (after Gollub et al., 1980).

behavior and other evidence of universal bifurcation sequences. High-resolution power spectra in Rayleigh–Bénard experiments in water (Gollub et al., 1980; Giglio et al., 1981) exhibit a number of period-doubling bifurcations, as shown in Fig. 8.31. The Fourier amplitudes of the fully developed successive subharmonics have also been determined experimentally and compared with Feigenbaum's model. In Section 7.2b, the ratio of the fully developed successive subharmonics was estimated in (7.2.56) to be $\gamma \approx 6.6$.[3] In Fig. 8.31d, the upper straight line has been drawn through the $\frac{1}{2}f_2$ harmonic peaks, and the lower line has been drawn a factor of 6.6 below

---

[3] The original estimate (Feigenbaum, 1979a) was $\gamma \approx 6.6$; in later work (Nauenberg and Rudnik, 1981), $\gamma \approx 4.6$ was found. For further details, see the discussion regarding (7.2.56). The experimental results did not clearly distinguish between these two values.

the upper line. Note that most of the $\frac{1}{4}f_2$ harmonic peaks fall near this line, in agreement with the model. In another Rayleigh–Bénard experiment in liquid helium (Libchaber and Maurer, 1980), the successive subharmonics f_1, $\frac{1}{2}f_1$, $\frac{1}{4}f_1$, $\frac{1}{8}f_1$, and $\frac{1}{16}f_1$ were observed in the power spectrum of the temperature. The ratio of successive subharmonics gives roughly the constant factor 6.6 in agreement with theory. However, we have seen that similar experiments did not show these features. We might expect that the presence of external noise would wash out the finer features of the subharmonic structure, and thus, depending on the experiment, subharmonics may or may not be seen. This behavior has, in fact, been observed in model calculations by Crutchfield and Huberman (1980).

A fourth possible mechanism for the onset of turbulence is the model of Pomeau and Manneville (1980) that a transition to intermittent chaotic behavior occurs. We have described this process in Section 7.5 and have shown computational evidence of its existence. The Lorenz system also displays this behavior in certain parameter ranges (Manneville and Pomeau, 1980). Intermittent transitions to turbulence have been seen in many experimental systems including the Rayleigh–Bénard system (Libchaber and Maurer, 1980) and an experimental study on chemical turbulence (Roux et al., 1980; see Appendix A). In these cases there is an intermittent transition between two states of a bistable system, and it is not clear that this behavior is described by the Pomeau–Manneville model. These models are summarized in Table 8.3.

Since different transitions to turbulence are observed in real fluids, it is clear that a single mechanism cannot explain the transition. Nevertheless, if, for example, a system does show period doubling, then one can predict the rate of doubling as the parameter changes. But at present there is no way to tell in advance which systems will show period doubling. It must also be noted that these models describe only the onset of turbulence and say nothing about the fully developed turbulent state (e.g., see Matthaeus and Montgomery, 1980). Furthermore, real fluids have both spatial and temporal behavior. In Section 8.5, we considered spatially coupled dissipative systems which can display both spatial and temporal turbulence. In fact, some of the characteristics of fluids, such as transfer of energy into higher modes, is observed in these coupled systems. However, the theory of these more complicated systems is still in its early stages, and the connections to real fluids have not been made.

8.6c. Hamiltonian Chaos in Fluids

We saw in Section 6.4 that magnetic field lines had a Hamiltonian representation, and that the Hamiltonian form is a consequence of the divergence-free nature of the magnetic flux $\nabla \cdot \boldsymbol{B} = 0$. One of the three spatial dimensions served as a timelike variable, conjugate to the Hamiltonian. If

8.6. The Fluid Limit

Model	Mechanism					
(a) Landau	Stationary point $\xrightarrow{\text{(Hopf)}}$	Singly periodic orbit	$\xrightarrow{\text{(Hopf)}}$	Doubly periodic orbit	$\xrightarrow{\text{(Hopf)}}$	Triply periodic orbit $\xrightarrow{\text{(Hopf)}} \cdots \rightarrow$ Turbulent motion
(b) Ruelle–Takens–Newhouse	Stationary point $\xrightarrow{\text{(Hopf)}}$	Singly periodic orbit	$\xrightarrow{\text{(Hopf)}}$	Doubly periodic orbit	$\longrightarrow$	Strange attractor
(c) Feigenbaum	Stationary point $\xrightarrow{\text{(Hopf)}}$	Singly periodic orbit (period T)	$\xrightarrow{\text{(pitchfork)}}$	Singly periodic orbit (period $2T$)	$\xrightarrow{\text{(pitchfork)}}$	Singly periodic orbit (period $4T$) $\xrightarrow{\text{(pitchfork)}} \cdots \rightarrow$ Strange attractor
(d) Pomeau–Manneville	Stationary point $\xrightarrow{\text{(Hopf)}}$	Singly periodic orbit	$\xrightarrow{\text{Reverse tangent bifurcation}}$	Intermittent chaotic motion		

Table 8.3. Models for the transition to fluid turbulence and their mechanisms.

the Hamiltonian is independent of this variable, then the two other spatial dimensions correspond to the canonical variables for a single-degree-of-freedom system which is integrable.

The same formalism can be applied to an incompressible fluid for which the incompressibility implies that the Eulerian velocity field has no divergence,

$$\nabla \cdot \boldsymbol{v}(\boldsymbol{x}, t) = 0, \tag{8.6.14}$$

where $\boldsymbol{v}$ and $\boldsymbol{x}$ are the velocity and position of the fluid. The fluid motion is, however, explicitly a function of time, and therefore we must restrict our treatment to two dimensions, say $\boldsymbol{v} = (u, w)$, $\boldsymbol{x} = (x, z)$, to complete the analogy. Then, from the chain rule for derivatives, it is apparent from (8.6.14) that $\boldsymbol{v}$ is determined from a *stream function* $\psi(\boldsymbol{x}, t)$ by the equations [see (8.6.6)]

$$u(\boldsymbol{x}, t) = -\frac{\partial \psi}{\partial z}, \qquad w(\boldsymbol{x}, t) = \frac{\partial \psi}{\partial x},$$

such that ψ plays the role of a time-dependent Hamiltonian. Solution of the equations of motion

$$\dot{x} = -\frac{\partial \psi}{\partial z}, \qquad \dot{z} = \frac{\partial \psi}{\partial x}, \tag{8.6.15}$$

for given initial conditions x_0, z_0, then gives the Lagrangian trajectory

$$\boldsymbol{x}(t) = \boldsymbol{X}(\boldsymbol{x}_0, t).$$

For a given fluid problem, as a forcing parameter is changed, one might expect transitions from quiescent, to steady, to a time-periodic Eulerian fluid velocity. The fluid particles thus see both space- and time-dependent forces, which leads to stochastic trajectories near separatrices in the flow. This stochasticity is called *Lagrangian turbulence*.

Many systems have been shown to display Lagrangian turbulence. Aref (1984) used analytic and numerical techniques to study chaotic advection in a two-dimensional incompressible flow stirred by an agitating point vortex. Chaiken *et al.* (1986) demonstrated the existence of Lagrangian turbulence in an experimental study of a system containing two nonconcentric rotating cylinders with the gap containing a viscous fluid. Knobloch and associates (Knobloch, 1986; Dangelmayr and Knobloch, 1987; Knobloch and Weiss, 1987) studied two species fluid mixtures heated from below, which show transitions from regular to stochastic behavior when the fluid motion bifurcates from simple wave to modulated wave equilibria. Experimentally, both simple and modulated waves have been observed (Heinrichs *et al.*, 1987) and the transport associated with such waves has been investigated (Moses and Steinberg, 1986). In the following, we present a brief account of the work following Knobloch and Weiss (1987, 1989).

Knobloch and Weiss consider a binary fluid mixture heated from below, and confined between two horizontal plates at vertical positions $z = 0, 1$.

8.6. The Fluid Limit

The idealized boundary conditions of these surfaces are stress-free velocity and fixed temperature and concentrations. Periodic boundary conditions are taken in x, which results in a single wave number becoming unstable as the Rayleigh number is increased. The remaining horizontal direction is taken to have translational invariance to preserve the two-spatial-dimension character of the flow. Without presenting details, the solution of the equilibrium fluid motion is found to bifurcate from a nonconvecting state to steady overturning convection, followed by a Hopf bifurcation to oscillatory convection (traveling waves), as the Rayleigh number is increased. The two bifurcations may also coincide, in which case it becomes possible to analytically study a secondary Hopf bifurcation to modulated waves. This is the case studied by Knobloch and Weiss. They find a solution for the stream function near the secondary bifurcation of the form

$$\psi(\xi, \zeta, \tau) = \psi_0(\xi, \zeta) + \delta\psi_1(\xi, \zeta, \tau), \qquad \delta \ll 1, \tag{8.6.16}$$

with

$$\psi_0(\xi, \zeta) = -\zeta + R\cos\xi \sin\zeta,$$

$$\psi_1(\xi, \zeta, \tau) = \frac{1}{2}\left[\left(1 - \frac{2}{\alpha}\right)\cos(\xi + \alpha\tau + \theta) + \left(1 + \frac{2}{\alpha}\right)\cos(\xi - \alpha\tau - \theta)\right]\sin\zeta.$$

Here ζ is z normalized to lie from 0 to π, $\xi = \omega_0 t + kx$ is the x variable translated into the frame of the wave with frequency ω_0 and wavenumber k, normalized to 2π periodic, and $\tau = \omega_0 t$ is the time normalized to the wave frequency. R is the amplitude of the unmodulated traveling wave with phase factor $e^{i(kx+\omega_0 t)}$, which is time independent in the wave frame. The trajectories of the fluid response to the unmodulated wave are given by ψ_0 alone. The quantities δ and α are the amplitude and frequency parameters of the modulation, which go to zero at the secondary Hopf bifurcation at $R = 3.266$, as found both analytically and numerically (see Knobloch and Weiss, 1987, for details). For that value, the curves of constant stream function (an integrable Hamiltonian) are shown in Fig. 8.32a. Increasing R beyond this bifurcation value results in finite α and δ. For $\alpha = 0.195$, which was found experimentally (Heinrichs et al., 1987), and taking $\delta = 0.4$, Hamilton's equations (8.6.15) were integrated for the Hamiltonian of (8.6.16). Taking initial conditions corresponding to the central island and the ξ-spanning trajectories around it, they obtained the results shown in Fig. 8.32b. The usual divided phase space for time-dependent Hamiltonians is observed. In the core region, particles are trapped and carried along with the wave. In the outer region, particles are untrapped and left behind by the wave. Between the core and the outer region is the chaotic separatrix layer, where diffusion and mixing occur.

A quantity of some interest is the dispersion of trajectories along x. For particles within the separatrix layer, diffusion across the separatrix moves trajectories from regions near the wave-trapped core to regions where the

(a)

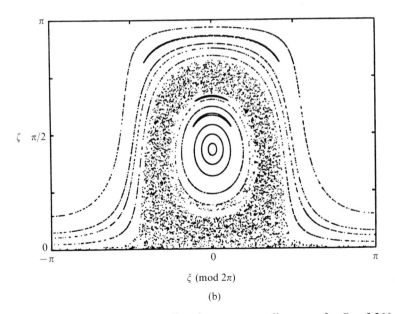

ξ (mod 2π)

(b)

Figure 8.32. (a) The co-moving stream lines in a pure traveling wave for $R = 3.266$, showing the heteroclinic orbit separating regions of trapped and untrapped particles. (b) Surface of section map for a modulated wave with $R = 3.266$, $\alpha = 0.195$, $\theta = 0$, and $\delta = 0.4$. Several initial conditions were started along the line $\xi = 0$.

8.6. The Fluid Limit

flow is streaming with respect to the wave. Removing the mod 2π from the ξ coordinate allows the resulting dispersion to be numerically determined for an infinite series of rolls along x; i.e., in the lab frame, one calculates

$$(\Delta x)^2(t) = \langle (x(t) - x(0))^2 \rangle - \langle x(t) - x(0) \rangle^2,$$

with the average taken over initial conditions. For diffusing particles $(\Delta x)^2 \propto t$, while for streaming particles $(\Delta x)^2 \propto t^2$. As with accelerator modes in action space (see Section 5.5d) we expect the motion near the KAM curves bounding the stochastic region to shadow the motion on the KAM curves, leading to streaming over long periods of time. Numerically integrating 10^3 initial conditions of an approximate separatrix map over $3.2 \cdot 10^6$ mapping iterations, Knobloch and Weiss (1989) obtain $(\Delta x)^2 \propto t^{1.93}$ indicating, as would be expected, the dominance of streaming. The long time correlations as the motion alternates between streaming and trapped trajectories can lead to $1/f^\alpha$-like power spectra (Geisel et al., 1987).

Boundary conditions require all physical systems to be three dimensional in space, and thus not completely described by the above picture. For example, with three spatial dimensions, Lagrangian turbulence can be obtained in a steady flow, which we have seen is not possible with two spatial dimensions. However, one might expect the phenomena described above to persist in systems with physical boundaries, provided the induced perturbations of the boundary are small.

APPENDIX A

Applications

In the Introductory Note, we described some applications of the methods treated in this monograph. Because we are most familiar with our own field of plasma confinement and heating, most of the examples used in the text are drawn from these areas. There are, however, a wealth of applications in other fields, such as planetary motion, accelerators and storage rings, solid state and molecular dynamics, chemical dynamics, fluid mechanics, ecology, etc. There are also closely related topics, such as application to quantum systems, that we have not treated. To partially redress this imbalance, we give an informal account of some of these applications. Since the areas are vast, no attempt at completeness is made.

Various conference proceedings and review articles are useful in directing the reader to this subject matter. For applications to solid state problems (not discussed here) the conference proceedings edited by Casati and Ford (1979) is useful. Dissipative systems and their application to a number of fields are described in a conference proceedings edited by Helleman (1980a). Extensive bibliographies in reviews by Chirikov (1979) and Helleman (1980) are also very helpful. Other proceedings and reviews are cited in the following sections.[1]

A.1. Planetary Motion

The early development of the subject matter of this monograph, as that of Hamiltonian mechanics, were inspired by the attempts to predict the motion of planetary systems over arbitrarily long periods of time. The most celebrated example is the *three-body problem*, and its simpler companion, the

---

[1] We have not updated this appendix from the first edition, since the literature has grown so rapidly. The references to applications, given here, are still of some historical interest.

A.1. Planetary Motion

restricted problem of three bodies. The first deals with the motion of three masses acted upon by their inverse square law gravitational attraction. In the simplified "restricted problem" one of the bodies is assumed massless, and its motion in the time-dependent field of the remaining two bodies is considered. In 1904, Whittaker, in presenting the developments up to that time, stated that the problem "has stimulated research to such an extent that since 1750 over 800 memoirs, many of them bearing the names of the greatest mathematicians, have been published on the subject."

It was shown, early, by Bruns (1887) and Poincaré (1892, p. 233), that the only global integrals of the motion that existed were those associated with the obviously conserved quantities (the so-called classical integrals), such as energy, and the vector momentum of the entire system. It was also determined during the same period that the restricted problem contained the essential complexity of the complete problem so that most effort was concentrated on the simpler system.

Perturbation methods (Poincaré, 1892; Von Zeipel, 1916) were originally developed to obtain approximate solutions to the three-body problem. These developments continued with new asymptotic expansion techniques based on the adelphic integral method of Whittaker (1964). They were developed and improved by Contopoulos and associates (e.g., Contopoulos, 1960, 1966) and, using Poisson brackets, by McNamara and Whiteman (1967). Another method achieving the same result was that of expansion in normal form by Gustavson (1966). The Poisson bracket method was developed in a systematic way, using Lie transform techniques, by Hori (1966), Garrido (1968), and Deprit (1969), again motivated by the problem of planetary motion.

A number of interesting mathematical results have been obtained, motivated, at least in part, by the three-body problem. We have already discussed the general theory of periodic orbits, associated with the work of Poincaré (1892) and Birkhoff (1927), and the development of the KAM theorem. Studies of stochasticity have also been motivated by an attempt to understand the chaotic motion of orbits near homoclinic points (see, for example, Siegel and Moser, 1971; Moser, 1973). Studies by Sitnikov (1960) and Alexeev (1970) showed for one form of the restricted three-body problem, near the separatrix of the motion of the light particle, that successive periods of the motion contained a random number of periods of the driving perturbation. Similar results of a more abstract topological nature have been obtained by Smale (1965).

Various techniques have been employed to understand the long-time behavior of astronomical motion, often by numerical and combinations of numerical and analytical techniques. The Hénon and Heiles (1964) potential (see Section 1.4) was analyzed numerically as an equivalent form to the restricted three-body problem. A review of the extensive literature related to this seminal example has been given by Churchill *et al.* (1979). Contopoulos and co-workers (Contopoulos, 1975), Ford and co-workers (Walker and Ford, 1969) and Kummer (1978), among others, examined simple systems of coupled oscillators, both analytically and numerically, to uncover the underlying structure of the three-body motion and its variants. As described in the previous chapters, this work, together with the mathematical results of the behavior at homoclinic points and the results of the KAM theorem, gives a complex picture of phase space motion for the restricted three-body problem, such as shown in Figs. 3.5 and 3.6.

The complete problem of planetary motion, in which the specific short-time orbits are determined from the perturbations of a number of other bodies, leads us into a specialized field beyond the scope of this monograph. Although it is generally thought that the results from the simpler problems give qualitative predictions of the behavior of the more complicated physical systems, many questions concerning the stability of the physical systems still exist. For general reviews of this large body of literature, monographs by Wintner (1947), Szebehely (1967), Siegel and Moser (1971), and Hagihara (1961, 1970–1976) can be studied. Some recent work is reviewed by Moser (1978) and Contopoulos (1979). Contopoulos and associates (Contopoulos and Papayannopoulos, 1980), have used these methods to treat other astronomical systems, such as galactic resonances.

A particularly interesting aspect of the long-time stability of the solar system concerns the effects resulting from higher phase space dimensionality, for which the KAM surfaces are nonisolating. The gaps in the rings of Saturn, near resonances between the rings and the inner satellites, may be due to this effect. Chirikov (1971) has studied this possibility for a related problem, that of the gaps in the asteroid belt near the resonances with Jupiter's orbital motion. His tentative conclusion is that the sizes of the gaps are consistent with the amplitudes of at least the main resonances, and the rate of Arnold diffusion would be sufficient to account for the emptying of these regions over the lifetime of the solar system. On the other hand, Contoupolos et al. (1978) have studied a model problem having three degrees of freedom, finding that Arnold diffusion, if it exists, is too weak to observe numerically in their case.

A.2. Accelerators and Beams

The study of beam dynamics in accelerators and storage rings has contributed substantially to the understanding of nonlinear coupling and adiabaticity and supplied a number of interesting examples. The developments have occurred principally since the introduction of the concept of strong focusing (Courant et al., 1952). Unfortunately much of the work has appeared in rather obscure or unpublished forms so that it is not widely known outside of the accelerator community.

Early treatments of the effect of resonances between the radial and vertical oscillations in strong focused synchrotrons appeared in CERN lecture notes

A.2. Accelerators and Beams

(Goward, 1953; Hine, 1953). Coupling to the synchrotron oscillations along the beam also introduces time-dependent driving terms, while the periodic nature of the focusing field introduces time-dependent modulation of the focusing term. These effects are usually treated in the linear approximation for which exact invariants of the motion can be found (e.g., Courant and Snyder, 1958; Lewis, 1967). However, linear instability, as treated in Section 3.3, can still occur. The increase in energy from the accelerating process also leads to an adiabatic change in the phase space trajectories, which is treated as in Section 2.3 (e.g., Slater, 1948).

Residual nonlinear effects are always present, which lead to the usual distortions of the phase plane near resonances of the harmonics of the principal frequencies. These nonlinearities become particularly important in Fixed Field Alternating Gradient (FFAG) Synchrotrons (Symon et al., 1956). An example of the treatment of such a resonance, using both the nonresonant and resonant perturbation theory of Sections 2.2 and 2.4, respectively, can be found in a review by Laslett (1967, p. 403). Other reviews and monographs that deal with some of these topics are in Green and Courant (1959), Kolomensky and Lebedev (1966), and Lichtenberg (1969).

Another aspect of the accelerator problems, indirectly related to the main subject of this monograph, is the transport of the phase space occupied by a collection of particles. Such systems are not quasi-periodic, and the primary problem is to understand the transformation of a set of initial conditions through a set of subsystems in which the transition from one subsystem to another is generally characterized by a step-function change in the parameters. The result of these step changes is *filamentation* of the phase space occupied by the particles. After coarse-graining the phase space, we find an increase in the effective phase space volume of the particles. Some aspects of this problem have been treated by Hereward et al. (1956) and Lichtenberg (1964). Related to it is the application of multiturn injection in FFAG synchrotrons (Symon and Sessler, 1956) and in storage rings. These applications, and their relation to the other types of beam transport, have been reviewed by Lichtenberg (1969).

Because storage rings must contain particles for extremely large numbers of revolutions, with beam lifetimes of hours or even days, they are very sensitive to resonant transport and may also be subject to Arnold diffusion. We have considered these mechanisms in Chapter 6. This effect may be particularly important in colliding beam experiments for which there is a periodic perturbation of the particles at least once per revolution, when the beams collide, leading to the so-called *beam–beam interaction*. Estimates of the rate of Arnold diffusion from this type of interaction have been made by Chirikov (1971), who found, particularly in the case where there was significant coupling to synchrotron oscillations, that beam lifetimes could be significantly shortened (see also Izraelev, 1980). Other mechanisms that have been investigated to explain shortened beam lifetimes include modulational diffusion (Tennyson, 1979) and resonance streaming (Tennyson, 1982).

Numerical studies of Arnold diffusion have often been motivated by the beam–beam interaction problem (e.g., Herrera *et al.*, 1979). Shortened beam lifetimes have been observed in experiments with colliding e^+–e^- beams (Wiedemann, 1979). However, no detailed comparison of theory and experiment has yet been made. Because of the large number of machine revolutions involved (generally greater than 10^8), the usual numerical computations, even in the mapping approximations, are impractical. A summary of these and other interesting results on the beam–beam interaction can be found in the conference proceedings edited by Month and Herrera (1979).

A.3. Charged Particle Confinement

We have already discussed the importance of studies of charged particle confinement in magnetic fields to the development of adiabatic invariants of nonlinear systems (see Section 2.3). In addition to the pioneering work of Alfvén (1950) and Northrop and Teller (1960), who obtained approximate solutions for the motion in the Earth's magnetic field, a number of other expansion procedures have been developed to approximate more closely the natural coordinate systems. Among these are the expansions of Lacina (1963, 1966) and Krilin (1967a, b) for axisymmetric mirror fields and Dragt (1965) for the axisymmetric dipole. The general procedures to obtain solutions to an arbitrary number of orders in the expansion parameter were developed by Bogoliubov and Mitropolski (1961) and by Kruskal (1957, 1962). A detailed second-order calculation was made by Northrop *et al.* (1966).

It was recognized, however, that all of these expansion procedures were divergent and that an exponentially small change in the magnetic moment occurred for each bounce in a magnetic mirror. Early calculations of this change were made by Hertweck and Schluter (1957) and by Chandrasekhar (1958). More complete calculations were made by Hastie *et al.* (1969) and by Howard (1971), who showed that the improper integrals that were averaged to zero by the method of averaging could be evaluated by contour integration in the complex plane and the method of steepest descent (stationary phase). Further developments and generalizations and comparison with numerically computed trajectories can be found in Cohen *et al.* (1978).

For an axisymmetric magnetic mirror the angular momentum is an isolating integral that reduces the degrees of freedom from three to two. Then, despite the jumps in the magnetic moment, it is still possible to have eternal confinement of large classes of particles. This is a consequence of the KAM invariants which, for small perturbations, can isolate the stochastic layers that arise from the nonlinear resonant interaction of the gyromotion with harmonics of the bounce motion (Chirikov, 1960). In terms of the jumps in magnetic moment, the bounce motion can also be considered as a mapping

A.3. Charged Particle Confinement

and the stability examined as in Sections 3.4 and 3.5. This approach has been taken by Chirikov (1971, 1979) and the limits of adiabaticity examined. In certain parameter ranges, we find periodic variations in the magnetic moment, at a period considerably slower than the bounce frequency. This phenomenon has been called *superadiabaticity* by Rosenbluth (1972). The loss of superadiabaticity, which can be physically associated with the loss of phase coherence from one passage through the midplane to the next, is equivalent to the destruction of KAM curves of the mapping.

The transition to global stochastic behavior, for which the particles can escape from the magnetic mirror, can be roughly determined numerically. Early numerical studies of this transition were made by Garren *et al.* (1958) for the axisymmetric mirror, by Dragt (1965) for the axisymmetric dipole, and by Siambis (1965) for the nonsymmetric mirror. However, computer calculations cannot unequivocally decide the question of the eternal stability of the mirror-confined particles. Experiments have been performed in axisymmetric fields at very low particle densities (negligible interparticle interaction) in order to shed further light on this question (e.g., Gibson *et al.*, 1963; Ponomarenko *et al.*, 1969). These experiments indicate that, although classical scattering appears to limit confinement time at high magnetic fields, there is a transition region at intermediate magnetic fields in which the confinement time is large but field-dependent and the particle loss cannot be accounted for. (For lower fields superadiabaticity fails and the particles are rapidly lost). Chirikov (1971, 1979) discussed these results, commenting that two hypotheses could be advanced to explain them: (1) a residual slow stochasticity exists when the perturbation is larger than that specified from a rigorous KAM criterion, but still considerably smaller than the numerically observed criteria of Chapter 4; or (2) small asymmetries or time variations of the magnetic field exist, which allow Arnold diffusion to occur. Chirikov estimated the magnitude of the Arnold diffusion and found it to be sufficient to account for the observed effect. However, no detailed parametric comparisons have been made. We agree with the conclusion of Chirikov that, although the question remains open, the latter hypothesis appears to be much more likely than the former.

A related problem is associated with the radial diffusion in tandem mirror systems that are stabilized by linked quadrupole windings. For axisymmetric systems, the constancy of angular momentum prevents radial diffusion that results from nonadiabatic magnetic moment changes (as distinguished from collisional diffusion). The asymmetry resulting from stabilizing windings breaks this invariance. For the tandem mirror, this effect is exacerbated by the long central cell that allows low-order resonances to exist between bounce and drift motion, leading to large radial excursions even in the absence of diffusion. The complete subject is quite complicated, and the reader is directed to the literature for further details (Ryutov and Stupakov, 1978; Cohen, 1979; Cohen and Rowlands, 1981).

A.4. Charged Particle Heating

A resonant interaction between two frequencies that modifies or destroys an invariant may be imposed on a system to accomplish some desired end, such as the heating of charged particles. In electron cyclotron resonance heating (ECRH), this is accomplished by resonance between the gyrofrequency and an electromagnetic wave. One method for calculating ECRH in a magnetic mirror is to obtain the change in energy in a single pass through resonance and then assume phase randomization between passes (Kuckes, 1968; Grawe, 1969; Eldridge, 1972). This approach is valid below a stochastic transition velocity and fails completely above an adiabatic barrier velocity, as described in Section 3.4. Seidl (1964) had earlier shown that a small resonant electric field led to superadiabatic oscillations, rather than stochastic heating. Following the approach of Seidl, and using the general notions of overlapping resonances to predict stochasticity, the transition from adiabatic to stochastic motion was investigated by Nekrasov (1970), Jaeger et al. (1972), and Lieberman and Lichtenberg (1973). A calculation including relativistic effects has been given by Bernstein and Baxter (1981). Numerical calculations by Sprott and Edmonds (1971) and experiments by Wyeth et al. (1975) and by Bardet et al. (1975) have given results in reasonable agreement with the theory. Calculations of the transition from superadiabaticity to stochasticity, similar to those for ECRH, have also been made for interaction of mirror confined particles with an r.f. field considered to arise from internally generated waves (Rosenbluth, 1972; Aamodt and Byers, 1972; Timofeev, 1974).

It is possible to have nonlinear interaction of the type described above, even in a uniform magnetic field. The nonlinearity arises from the finite gyroradius, which leads to a spectrum of harmonic frequencies in the motion. We have already considered this example in Section 2.4c to illustrate both accidental degeneracy (propagation of a wave at an angle to the field) and intrinsic degeneracy (propagation perpendicular to the field). However, the energies over which heating can occur are limited to those for which the interacting harmonics are large. This work has led to a detailed treatment of ion cyclotron heating (Karney, 1978, 1979).

Experimental observations (Lazar et al., 1982) of improved heating with two frequencies has stimulated theoretical work on this problem. Howard et al. (1982) and Rognlien (1982) have shown that the KAM barrier to heating can be increased by a factor of two in energy if two frequencies with the same total power are used. The physical mechanism appears to be due to the interspersal of resonant island structures, which allows their overlap at higher energies. The maximal effect is found for a low-order resonance of the difference frequency between the applied sources and the bounce frequency. The importance of the bounce resonance had already been noted by Smith et al. (1980). Another explanation of the improvement has been proposed by Samec et al. (1982). For single frequency heating, particles that pass through the resonance zone receive two heating kicks in quick succession that may cancel. A second frequency tends to wash out this effect, thus increasing the heating limit.

In the situations considered above, the self-consistent nature of the waves in a plasma has not been considered. For ECRH experiments, the r.f. field can generally be taken to be the vacuum field, provided the plasma frequency is less than the electron cyclotron frequency, a condition usually satisfied in experiments. For applications to ion heating or to situations in which the fields are self-generated, the lack of self-consistency may be a serious limitation to the results. In addition, the complicated spatial distribution of the vacuum fields or of the fields arising from the linear response of a plasma is also not generally considered.

The limitations described above point out one of the fundamental problems with the basic theoretical approach that requires that the perturbation terms in the system Hamiltonian be *a priori* given. Nevertheless, considerable success has been achieved in comparing the theory of stochastic ion cyclotron heating to experiments in tokamaks (Gormezano *et al.*, 1981).

There are a number of problems that are related to that of resonance heating that deserve mention. We have shown in Section 2.5c (for detailed treatments, see Motz and Watson, 1967; Cary and Kaufman, 1981) that a spatially varying r.f. field or an r.f. field in a spatially varying steady magnetic field can give rise to an average force (ponderomotive force). This can be used for r.f. confinement and has been considered for plugging the end loss due to scattering from a magnetic mirror. However, as we have seen, particles interacting in this manner are not necessarily adiabatic. This question has been examined by Lichtenberg and Berk (1975) who showed that the nonadiabaticity in magnetic mirrors serves as a limit to the useful confinement that can be achieved. A related calculation has been made for cusp fields by Hatori and Watanabe (1975). Although r.f. confinement does not appear to be practical for the main component of a mirror-confined fusion plasma, it may be useful to confine (and possibly heat) a warm plasma required for stabilization of a collective plasma mode. The self-consistent problem has also been considered in connection with r.f. confinement (Motz and Watson, 1967; Watari *et al.*, 1974) but not for the parameters for which stochastic effects are important.

A.5. Chemical Dynamics

In Section 8.6 we reviewed one of the main motivating examples for the study of dissipative systems, that of fluid turbulence. Another area of considerable interest is turbulence in chemical interactions. The mass action law governing the time evolution of a homogeneous chemical system leads to a set of nonlinear, first-order differential equations. There is one equation for each chemical species so that for M species we have an M-dimensional flow of the type considered in Section 7.1. We should not be surprised, then, to find the full range of motions described in Chapter 7, including sinks, limit cycles, and strange attractors.

In common experience most homogeneous chemical systems decay to a final stationary state (sink). In 1958, B. P. Belousov discovered periodic behavior in a simple laboratory reaction. Subsequent studies were undertaken by A. M. Zhabotinsky and collaborators (see Zaiken and Zhabotinsky, 1970); the reaction is now known as the Belousov–Zhabotinsky reaction. The basic constituents are bromide and bromate ions, an organic fuel, and an indicator dye in a water solution. When mixed in the proper concentrations, the system oscillates, changing from yellow to colorless to yellow twice a minute. The oscillations typically last for over an hour, until the organic fuel is exhausted. Many other oscillating chemical systems are now known (see Winfree, 1974, and Walker, 1978, for popular accounts). Although early models of the Belousov–Zhabotinsky system involved reactions among eleven chemical species, Field and Noyes (1974) have shown that the system can be approximated by a sequence of reactions involving only three chemical species, e.g., a three-dimensional flow in the chemical concentration phase space.

We have discussed in Chapters 7 and 8 the appearance of strange attractors for three-dimensional flows. Ruelle (1973) suggested that the Belousov–Zhabotinsky system, and chemical systems in general, could exhibit chaotic behavior of this type (usually called chemical turbulence). The theoretical and computational existence of chemical turbulence is now fairly well established (Rössler, 1976; Tyson, 1978; Timita and Tsuda, 1979). There have also been many experiments (Degn *et al.*, 1979; Schmits *et al.*, 1977; Wegmann and Rössler, 1978; Hudson *et al.*, 1979; Vidal *et al.*, 1980) that show strong evidence of chemically turbulent behavior. For example, by measuring the Fourier spectrum of the concentration of a chemical species as the flow rate is increased, Vidal and co-workers see a transition from singly periodic to doubly periodic behavior, followed by a transition to a chemically turbulent state.

Still more complicated and surprising effects occur in Belousov–Zhabotinsky systems that are nonhomogeneous. When left alone (unstirred) in a thin layer (about 2 mm thick), the fluid spontaneously forms moving colored patterns of great complexity, including spirals, arcs, and circles, which propagate along the layer and annihilate when they collide (see Zaiken and Zhabotinsky, 1970; Kopell and Howard, 1973; and Winfree, 1974, for pictures). There is no fluid motion in these systems, rather, the chemical species have nonuniform concentrations along the layer. The effects are produced by the reactions of the species and by their diffusion along the layer. These reaction–diffusion systems must be modeled by partial differential equations and are much more complicated to study than homogeneous chemical systems. Kopell (1980) shows analytically the existence of plane waves, shocks, and also time-periodic but spatially chaotic solutions for a simple model problem. Earlier (Kuramoto and Yamada, 1976), had shown numerically the existence of chaotic solutions in time. In their study the chaotic behavior is a consequence of the diffusion; i.e., the homogeneous system would exhibit only a periodic oscillation. Some experiments (Yamazaki et al., 1978, 1979) seem to verify this behavior (diffusion induced turbulence). It appears that the transition to turbulence is gradual; i.e., there is no sharp discontinuity between the ordered and turbulent regime in the experiment.

A.6. Quantum Systems

As mentioned in the Introduction, the invention of quantum mechanics gave fresh impetus to the formal development of classical perturbation theory (Born, 1927), Conversely, the recent advances in understanding the behavior of classical dynamics have stimulated a renewed interest in the behavior of quantum systems in their semiclassical limit $\hbar \to 0$. Of particular interest is the correspondence between the classical solutions (phase space trajectories) and the quantal solutions (wave solutions). In quantum mechanics a quantal formulation can be obtained from a classical formulation, but a quantal solution cannot be obtained from a classical solution. No general correspondence between the solutions is known except when the system is integrable. The motion then separates into that of N independent one-degree-of-freedom systems for both the quantal and the classical case. The quantization of the stationary classical system for this case is well understood (Keller and Rubinow, 1960; Percival, 1977). The method is known as EBK quantization and restricts the N classical actions to discrete values $I = (n + \alpha/4)\hbar$, where n is the vector of N quantum numbers and α is an integer vector whose components are the Maslov indices, which are determined from the topology of the invariant for each degree of freedom. The quantized energy levels $E_n = H(I_n)$ can coincide and cross each other as a system parameter is varied. The EBK wave function is similar to a WKB solution and exhibits turning points or *caustics* in configuration space, which correspond to those of the classical motion.

For near-integrable classical systems in which regular and stochastic motion coexist on the finest scales, the quantal analogs are not well understood. Insight has been developed by quantizing classical systems that have completely stochastic (K system) behavior. Some systems that have been studied are Sinai's billiard problem, Arnold's cat mapping (Berry, 1980), and the motion of a particle within a stadium having straight sides connected by semicircles (McDonald and Kaufman, 1979; Casati et al., 1980a). It was conjectured by Berry (1977) and Zaslavskii (1977) that the energy levels for a stochastic system should repel each other and should almost never coincide as a system parameter is varied. The energy level spacing should be peaked about a finite value rather than having its maximum at zero separation, which represents the clustering of eigenvalues characteristic of integrable Hamiltonians. These properties have been verified computationally for Sinai's billiard and for the stadium problem (McDonald and Kaufman, 1979; Berry, 1980, 1981; Casati et al., 1980a; see also Helleman, 1980) and are often taken as a definition of quantum stochasticity.

The situation for near-integrable systems is not so clear. It is thought that the classical hierarchy of very thin stochastic layers intermingled with KAM tori, and the classical hierarchy of islands within islands is wiped away in the limit of $\hbar$ small but finite. Thus the quantal phase space is "coarse-grained," and those classical regions (regular or stochastic) with a volume in the N-dimensional phase space much smaller than $(2\pi\hbar)^N$ can almost all be neglected (Percival, 1977). One might then obtain quantal behavior that appears regular even if the corresponding classical system appears stochastic. These observations appear to be confirmed in calculations by Casati et al. (1979) on the quantized standard mapping and by Marcus (1980) for the quantized, Hénon-Heiles problem. The latter shows explicitly that non-stochastic quantum mechanical states occur in energy regimes that are classically largely stochastic, and also reviews the applications of the techniques to molecular dynamics. Another consequence of coarse graining is the nonisolating character of the KAM tori for finite $\hbar$. For the quantized standard mapping, Ott et al. (1981) have demonstrated the diffraction of the wave function from a classically regular region to a stochastic region. Other accounts of some of the areas of research are given by Berman and Zaslavskii (1979), Chirikov et al. (1980a), Shepelyanski (1981) and in the volume edited by Casati and Ford (1979). The subject has been reviewed by Zaslavskii (1981).

Bibliography

Aamodt, R.E., and J. A. Byers (1972). *Phys. Rev. Lett.* **29**, 1305.
Aamodt, R.E., and A. F. Jaeger (1974). *Phys. Fluids* **17**, 1386.
Abarbanel, H.D.I. (1981). *Physica* **4D**, 89.
Abarbanel, H.D.I., R. Brown, and J.B. Kadtke (1989). Institute for Nonlinear Science, U.C. San Diego Preprint, INLS 1014.
Abarbanel, H.D.I., and J.D. Crawford (1981). *Phys. Lett.* **82A**, 378.
Ablowitz, M.J., A. Ramani, and H. Segur (1980). *J. Math. Phys.* **21**, 715.
Ahlers, G., and R. P. Behringer (1978). *Phys. Rev. Lett.* **40**, 712.
Alexeev, M.V. (1970). *Actes. Int. Cong. Math.* **2**, 893 (Gauthier-Villars, Paris, 1971).
Alfvén, H. (1950). *Cosmical Electrodynamics*, Oxford University Press, Oxford.
Anosov, D.V. (1962). *Soviet Math. Dokl.* **3**, 1068.
Anosov, D.V. (1963). *Soviet Math. Dokl.* **4**, 1153.
Anosov, D.V. (1967). *Proc. Steklov Inst. Math.* **90**.
Aref, H. (1984). *J. Fluid Mech.* **143**, 1.
Arnold, V.I. (1961). *Soviet Math. Dokl.* **2**, 501.
Arnold, V.I. (1962). *Soviet Math. Dokl.* **3**, 136.
Arnold, V.I. (1962b). *Soviet Math. Dokl.* **3**, 1008.
Arnold, V.I. (1964). *Russian Math. Surveys* **18**, 85.
Arnold, V.I. (1966). *Ann. Inst. Fourier* **16**, 319.
Arnold, V.I. (1979). *Mathematical Methods of Classical Mechanics*, Springer-Verlag, New York.
Arnold, V.I. (1983). *Geometrical Methods in the Theory of Ordinary Differential Equations*, Springer-Verlag, New York.
Arnold, V.I., and A. Avez (1968). *Ergodic Problems of Classical Mechanics*, Benjamin, New York.
Bak, P. (1986). *Phys. Today*, Dec., p. 39.
Bak, P., and C. Tang (1989). *J. Geophys. Res.* **94**, 15635.
Bak, P., T. Bohr, and M.H. Jensen (1988). In *Directions in Chaos*, Vol. 2, Hao Bai-lin (ed.), World Scientific, Singapore.
Bak, P., C. Tang, and K. Wiesenfeld (1988a). *Phys. Rev.* **A38**, 364.
Bardet, R., P. Briand, L. Dupas, C. Gormezano, and G. Melin (1975). *Nucl. Fusion* **15**, 865.
Belleman, R. (1964). *Perturbation Techniques in Mathematics, Physics, and Engineering*, Holt, Rinehart, and Winston, New York.
Belykh, V.N., F.N. Pedersen, and O.H. Sorensen (1977). *Phys. Rev.* **B16**, 4860.
Benettin, G., M. Casartelli, L. Galgani, A. Giorgilli, and J.M. Strelcyn (1978). *Nuovo Cimento* **44B**, 183; (1979), *Nuovo Cimento* **50B**, 211.

Benettin, G., and L. Galgani (1979). In G. Laval and D. Gressillon (eds.), *op. cit.*, p. 93.
Benettin, G., L. Galgani, and J.M. Strelcyn (1976). *Phys. Rev.* **A14**, 2338.
Benettin, G., L. Galgani, and A. Giorgilli (1987a). *Phys. Lett. A* **120**, 23.
Benettin, G., L. Galgani, and A. Giorgilli (1987b). *Commun. Math. Phys.* **113**, 87.
Benettin, G., L. Galgani, A. Giorgilli, and J.M. Strelcyn (1980), *Meccanica*, March, 21.
Ben-Mizrachi, A., I. Procaccia, N. Rosenberg, A. Schmidt, and H.G. Schuster (1985). *Phys. Rev.* **A31**, 1830.
Bensimon, D., and L.P. Kadanoff (1984). *Physica* **13D**, 82.
Bensimon, D., M.H. Jensen, and L.P. Kadanoff (1986). *Phys. Rev.* **A33**, 362.
Bergé, P., M. Dubois, P. Manneville, and Y. Pomeau (1980). *J. Phys. (Paris) Lett.* **41**, L-344.
Berman, G.P., and A.R. Kolovskii (1984). *Sov. Phys. JETP* **60**, 1116.
Berman, G.P., and G.M. Zaslavskii (1979). *Physica* **97A**, 367.
Berman, R.H. (1980). "Transition to Stochastic Behavior in a Deterministic System." Res. Lab. Elec., M.I.T. Report PRR 80/10.
Bernstein, G.M., and M.A. Lieberman (1990). *IEEE Trans. Circuits and Systems*, **37**, 1157.
Bernstein, G.M., M.A. Lieberman, and A.J. Lichtenberg (1989). *IEEE Trans. Commun.* **B7**, 1062.
Bernstein, I.B., and D.C. Baxter (1981). *Phys. Fluids* **24**, 108.
Berry, M.V. (1977). *Philos. Trans. Roy. Soc. (London)* **A287**, 237; *J. Phys.* **A10**, 2083.
Berry, M.V. (1978). In S. Jorna (ed.), *op. cit.*, p. 16.
Berry, M.V. (1980). In R.H.G. Helleman (ed.), *op. cit.*, p. 183.
Berry, M.V. (1981). *Ann. Phys.* **131**, 163.
Birkhoff, G.D. (1920). *Acta Math.* **43**, 1.
Birkhoff, G.D. (1927). *Dynamical Systems*, American Mathematical Society, New York.
Birkhoff, G.D. (1935). "Nouvelles recherches sur les systemes dynamiques," *Mem. Pont. Acad. Sci. Novi. Lyncaei*, **1**, 85.
Bivins, R.L., N. Metropolis, and J.R. Pasta (1973). *J. Comput. Phys.* **12**, 65.
Boghosian, B.M. (1987). "Covariant Lagrangian Methods of Relativistic Plasma Theory," Lawrence Berkeley Laboratory Report, LBL-23241, Berkeley.
Bogoliubov, N.N., and D.N. Zubarev (1955). *Ukr. Mat. Zh.* **7**, 5.
Bogoliubov, N.N., and Y.A. Mitropolsky (1961). *Asymptotic Methods in the Theory of Nonlinear Oscillations*, Gordon and Breach, New York.
Bohr, T., and G.H. Gunaraine (1985). *Phys. Lett.* **113A**, 55.
Born, M. (1927). *The Mechanics of the Atom*, Bell, London.
Bountis, T.C. (1978). "Nonlinear Models in Hamiltonian Mechanics and Statistical Mechanics," Ph.D. Thesis, University of Rochester, Rochester, New York.
Bountis, T.C. (1981). *Physica* **3D**, 577.
Bountis, T.C., H. Segur, and F. Vivaldi (1982). *Phys. Rev.* **A25**, 1257.
Bowen, R. (1975). *J. Differential Equations* **18**.
Brahic, A. (1971). *Astron. Astrophys.* **12**, 98.
Brambilla, M., and A.J. Lichtenberg (1973). *Nucl. Fusion* **13**, 517.
Bridges, R., and G. Rowlands (1977). *Phys. Lett.* **63A**, 189.
Brillouin, L. (1927). *Compt. Rend.* **183**, 24.
Brillouin, L. (1953). *Wave Propagation in Periodic Structures*, Dover, New York.
Brock, W.A. (1988). In *Differential Equations, Stability, and Chaos in Dynamic Economics*, W.A. Brock and A.G. Maliaris (eds.), North-Holland, New York, Chap. 10.
Broomhead, D.S., and G.P. King (1986). *Physica* **20D**, 217.
Broucke, R. (1969). *AIAA J.* **7**, 1003.
Brouwer, D., and Clemence, G.M. (1961). In *The Solar System III; Planets and Satellites*, G. Kuiper and B. Middlehurst (eds.), p. 31.

Bruhwiler, D.L., and J.R. Cary (1989). *Physica* **D40**, 265.
Brumer, P., and J.W. Duff (1976). *J. Chem. Phys.* **65**, 3566.
Bruns, H. (1887). Berichte de Kgl. Sächs. Ges. der Wiss. pp. 1, 55; *Acta. Math.* **xi**, 25.
Bunimovich, L.A., and Ya. G. Sinai (1980). *Commun. Math. Phys.* **78**, 247.
Callen, J.D. (1977). *Phys. Rev. Lett.* **39**, 1540.
Carlson, J.M., and J.S. Langer (1989a). *Phys. Rev. Lett* **62**, 2632.
Carlson, J.M., and J.S. Langer (1989b). *Phys. Rev.* **A40**, 6470.
Carreras, B., H.R. Hicks, and D.K. Lee (1981). *Phys. Fluids* **24**, 66.
Cary, J.R. (1978). "Lie Transforms and Their Use in Hamiltonian Perturbation Theory," Lawrence Berkeley Laboratory Report 6350; (1979), Ph.D. Thesis. University of California, Berkeley, California.
Cary, J.R. (1981). "Lie Transform Perturbation Theory for Hamiltonian Systems." *Phys. Rep.* **79**, 129.
Cary, J.R., and A. N. Kaufman (1977). *Phys. Rev. Lett.* **39**, 402; (1978), *Phys. Rev. Lett.* **40**, 1266; (1981), *Phys. Fluids* **24**, 1250.
Cary, J.R., and J.D. Meiss (1981). *Phys. Rev.* **A24**, 2664.
Cary, J.R., J.D. Meiss, and A. Bhattacharjee (1981). *Phys. Rev.* **A23**, 2744.
Cary, J.R. (1982). *Phys. Rev. Lett.* **49**, 276.
Cary, J.R., and J.D. Hanson (1986). *Phys. Fluids* **29**, 2464.
Cary, J.R., D.F. Escande, and J.L. Tennyson (1986). *Phys. Rev.* **A34**, 4256.
Cary, J.R., and R.T. Skodje (1988). *Phys. Rev. Lett.* **61**, 1795.
Casati, G. (1975). *Lett. Nuovo Cimento* **14**, 311.
Casati, G., B.V. Chirikov, F.M. Izrailev, and J. Ford (1979). In G. Casati and J. Ford (eds.), *op. cit.*, p. 334.
Casati, G., B.V. Chirikov, and J. Ford (1980). *Phys. Lett.* **77A**, 91.
Casati, G., G. Comparin, and I. Guarneri (1982). *Phys. Rev. A* **26**, 717.
Casati, G., and J. Ford (eds.) (1979). *Stochastic Behavior in Classical and Quantum Hamiltonian Systems*, Lecture Notes in Physics, Vol. 93, Springer-Verlag, New York.
Casati, G., I. Guarneri, and F. Valz-Gris (1980a). "On the Connection between Quantization of Nonintegrable Systems and the Statistical Theory of Spectra," preprint.
Casati, G., F. Valz-Gris, and I. Guarneri (1981). *Physica* **3D**, 644.
Chaiken, J., R. Chevray, M. Tabor, and Q.M. Chan (1986). *Proc. Royal Soc. Lond. A* **408**, 165.
Chaitin, G.J. (1975). *Sci. Amer.*, May.
Chandrasekhar, S. (1943). *Rev. Mod. Phys.* **15**, 1.
Chandrasekhar, S. (1958). In *The Plasma in a Magnetic Field*, R.K.M. Landshof (ed.), Stanford University Press, Palo Alto, California.
Chandrasekhar, S., and S.K. Trehan (1960). *Plasma Physics*, University of Chicago Press, Chicago.
Channel, P.J., and C. Scovel (1988). Los Alamos National Laboratory Report LA-UR-88-1828.
Chao, A.W. (1979). In *Nonlinear Dynamics and the Beam-Beam Interaction*, M. Month and J.C. Herrera, (eds.), AIP Conference Proceedings, Vol. 57, p. 42.
Chaté, H., and P. Manneville (1988). *Physica* **D32**, 409.
Chen, C., G. Gyorgyi, and G. Schmidt (1987). *Phys. Rev. A* **35**, 2660.
Chen, F. (1974). *Introduction to Plasma Physica*, Plenum Press, New York.
Chernikov, A.A., R.Z. Sagdeev, D.A. Usikov, and G.M. Zaslavskii (1987). *Phys. Lett.* **125**, 101.
Chhabra, A., and R.V. Jensen (1989). *Phys. Rev. Lett.* **62**, 1327.
Chillingworth, D.R.J. (1976). *Differential Topology with a View to Applications*, Pitman, London.
Chirikov, B.V. (1959). *At. Energ.* **6**, 630.
Chirikov, B.V. (1960). *Plasma Phys.* (J.N.E. Pt. C) **1**, 253.

Chirikov, B.V. (1971). "Research Concerning the Theory of Nonlinear Resonance and Stochasticity," CERN Trans. 71–40, Geneva.
Chirikov, B.V. (1978). *Soviet J. Plasma Phys.* **4**, 289.
Chirikov, B.V. (1979). *Phys. Rep.* **52**, 265.
Chirikov, B.V. (1979a). *Soviet. J. Plasma Phys.* **5**, 492.
Chirikov, B.V., J. Ford, and F. Vivaldi (1979). In M. Month and J.C. Herrera (eds.), *op. cit.*, p. 323.
Chirikov, B.V., and F.M. Izrailev (1976). *Colloques Internationaux du C.N.R.S.* No. 229, Paris.
Chirikov, B.V., and F.M. Izrailev (1981). *Physica* **2D**, 30.
Chirikov, B.V., F.M. Izrailev, and V.A. Tayursky (1973). *Comput. Phys. Commun.* **5**, 11.
Chirikov, B.V., F.M. Izrailev, and D.L. Shepelyanski (1980). "Dynamical Stochasticity in Classical Mechanics," preprint 80–209, Institute of Nuclear Physics, USSR Academy of Science, Novosibirsk; also with J. Ford and F. Vivaldi, preprint 81–70; (1981). *Soviet. Sci. Rev.* **C2**, 209.
Chirikov, B.V., F.M. Izrailev, and D.L. Shepelyanski (1980a). "Transient Stochasticity in Quantum Mechanics," preprint 80–210, Institute of Nuclear Physics, USSR Academy of Science, Novosibirsk; (1981). *Soviet. Sci. Rev.* **C2**, 209.
Chirikov, B.V., M.A. Lieberman, D. Shepelyanski, and F. Vivaldi (1982). *Physica* **14D**, 289.
Chirikov, B.V., and D.L. Shepelyansky (1984). *Physica* **D13**, 395.
Chirikov, B.V., and D.L. Shepelyansky (1986). Preprint 86–174, Institute of Nuclear Physics, Novosibirsk.
Churchill, R.C., G. Pecelli, and D.L. Rod (1979). In G. Casati and J. Ford (eds.), *op. cit.*, p. 76.
Cohen, R.H. (1979). *Nucl. Fusion* **19**, 1579.
Cohen, R.H., W.M. Nevins, and G. Rowlands (1981). *Phys. Fluids* **24**, 1584.
Cohen, R.H., and G. Rowlands (1981). *Phys. Fluids* **24**, 2295.
Cohen, R.H., G. Rowlands, and J.H. Foote (1978). *Phys. Fluids* **21**, 627.
Collet, P., and J.-P. Eckmann (1980). *Iterated Maps on the Interval as Dynamical Systems, Progress in Physics*, Vol. I, Birkhäuser Verlag, Basel.
Collet, P., J.-P. Eckmann, and O.E. Lanford (1980). *Commun. Math. Phys.* **76**, 211.
Collet, P., J.-P. Eckmann, and H. Koch (1981). *Physica* **3D**, 457.
Collet, P., J.-P. Eckmann, and H. Koch (1981a). *J. Stat. Phys.* **25**, 1.
Contopoulos, G. (1960). *Z. Astrophys.* **49**, 273.
Contopoulos, G. (1966). *J. Math. Phys.* **7**, 788.
Contopoulos, G. (1975). In *Dynamics of Stellar Systems*, A. Hayli, (ed.), Int. Astr. Union, p. 209.
Contopoulos, G. (1978). *Celestial Mech.* **17**, 167.
Contopoulos, G. (1979). In G. Casati and J. Ford (eds.), *op. cit.*, p. 1.
Contopoulos, G., and M. Montsoulas (1965). *Astron. J.* **70**, 817.
Contopoulos, G., and Th. Papayannopoulos (1980). *Astron. Astrophys.* **92**, 33.
Contopoulos, G., L. Galgani, and A. Giorgilli (1978). *Phys. Rev.* **A18**, 1183.
Courant, E.D., and H.S. Snyder (1958). *Ann. Phys.* **3**, 1.
Courant, E.D., M.S. Livingston, and H.S. Snyder (1952). *Phys. Rev.* **88**, 1190.
Crawford, J.D. (1989). "Introduction to Bifurcation Theory," Institute of Fusion Studies, University of Texas, Report ISFR 407–Review.
Cremers, J., and A. Hübler (1987). *Z. Naturforsch.* **42a**, 797.
Crutchfield, J.P., and B.A. Huberman (1980). *Phys. Lett.* **A77**, 407.
Crutchfield, J., D. Farmer, N. Packard, R. Shaw, G. Jones, and R.J. Donnelly (1980). *Phys. Lett.* **A76**, 1.
Crutchfield, J.P., and K. Kaneko (1987). In *Directions in Chaos*, Hao Bai-lin (ed.), World Scientific, Singapore.

Crutchfield, J.P., and B.S. McNamara (1987). *Complex Systems* **1**, 417.
Curry, J.H. (1978). *Commun. Math. Phys.* **60**, 193.
Curry, J.H. (1979). *Commun. Math. Phys.* **68**, 129.
Curry, J.H., and J.A. Yorke (1978). In *Lecture Notes in Mathematics 668*, p. 48, Springer-Verlag, New York.
Dangelmayr, G., and E. Knobloch (1987). *Philos. Trans. Royal Soc. Lond.* **322**, 243.
Degn, H., L.F. Olsen, and J.W. Perran (1979). *Ann. N.Y. Acad. Sci.* **316**, 625.
Deprit, A. (1969). *Celestial Mech.* **1**, 12.
Derrido, B., A. Gervois, and Y. Pomeau (1979). *J. Phys.* **A12**, 269.
de Vogelaere, R. (1958). In *Contributions to the Theory of Nonlinear Oscillations*, S. Lefschetz (ed.), **4**, 53, Princeton University Press, Princeton, New Jersey.
Dewar, R.L. (1976). *J. Phys.* **A9**, 2043.
Dorizzi, B., B. Gramaticos, A. Ramani, and P. Winternitz (1985). *J. Math. Phys.* **26**, 3071.
Dragt, A.J. (1965). *Rev. Geophys.* **3**, 255.
Dragt, A.J., and J.M. Finn (1976a). *J. Geophys. Res.* **81**, 2327.
Dragt, A.J., and J.M. Finn (1976b). *J. Math. Phys.* **17**, 2215.
Driscoll, C.F., and T.M. O'Neil (1976). *Phys. Rev. Lett.* **37**, 69.
Driscoll, C.F., and T.M. O'Neil (1978). *Rocky Mt. J. of Math.* **5**, 211.
Dubinina, A.N., and Yu. N. Yudin (1968). *Soviet Phys. JETP* **26**, 707.
Dubois, M., M.A. Rubio, and P. Bergé (1983). *Phys. Rev. Lett* **51**, 1446.
Dunnett, D.A., E.W. Laing, S.J. Roberts, and A.E. Robson (1965). *Plasma Phys.* (J.N.E. Pt. C) **7**, 359.
Dunnett, D.A., E.W. Laing, and J.B. Taylor (1968). *J. Math. Phys.* **9**, 1819.
Ecke, R.E., J.D. Farmer, and D.K. Umberger (1988). "Scaling of Arnold Tongues," Los Alamos National Laboratory, Report K764, Los Alamos.
Eckmann, J.-P. (1981). *Rev. Mod. Phys.* **53**, 643.
Eckmann, J.-P., H. Koch, and P. Wittwer (1981). "A computer assisted proof of universality for area-preserving maps," preprint UGVA-DPT-1981/04-345, Theoretical Physics, University de Genève.
Eckmann, J.P., and D. Ruelle (1985). *Rev. Mod. Phys.* **57**, 615.
Einstein, A. (1911). In *La Theorie du Rayonnement et les Quanta*. P. Langevin and M. DeBroglie (eds.) (Report on Meeting at Institute Solvay, Brussels, 1911). Gauthier-Villars, Paris, 1912, p. 450.
Eldridge, O. (1972). *Phys. Fluids* **15**, 676.
Eminhizer, C.R., R.H.G. Helleman, and E.W. Montroll (1976). *J. Math. Phys.* **17**, 121.
Escande, D.F. (1985). *Phys. Rep.* **121**, 166.
Escande, D.F., and F. Doveil (1981). *J. Stat. Phys.* **26**, 257; see also *Phys. Lett.* **83A**, 307; F. Doveil and D.F. Escande (1981). *Phys. Lett.* **84A**, 399.
Escande, D.F., F. Doveil, and A. Mehr (1982). In *Nonequilibrium Problems in Statistical Mechanics*, **2**, W. Horton, L. Reichl, and V. Szebehely (eds.), Wiley, New York.
Farmer, J.D. (1982). *Physica* **4D**, 336.
Farmer, J.D. (1982a). *Z. Naturforsch.* **37a**, 1304.
Farmer, J.D., J. Crutchfield, H. Froehling, N. Packard, and R. Shaw (1980). In R.H.G. Helleman (ed.), *op. cit.*, p. 453.
Farmer, J.D., and I.I. Satija (1985). *Phys. Rev.* **A31**, 3520.
Farmer, J.D., and J.J. Sidorowich (1987). *Phys. Rev. Lett.* **59**, 845.
Farmer, J.D., and J.J. Sidorowich (1988). In *Evolution, Learning, and Cognition*, Y.C. Lee (ed.), World Scientific, Singapore.
Feigenbaum, M.J. (1978). *J. Stat. Phys.* **19**, 25.
Feigenbaum, M.J. (1979). *J. Stat. Phys.* **21**, 669.
Feigenbaum, M.J. (1979a). *Phys. Lett* **74A**, 375.
Feigenbaum, M.J. (1980). *Los Alamos Science*, Summer, **4**.

Feigenbaum, M.J., L.P. Kadanoff, and S.J. Shenker (1982). *Physica* **5D**, 370.
Feit, S.D. (1978). *Commun. Math. Phys.* **61**, 249.
Fenstermacher, R., H.L. Swinney, and J.P. Gollub (1978). *J. Fluid Mech.* **94**, 103.
Fermi, E. (1949). *Phys. Rev.* **75**, 1169.
Fermi, E., J.R. Pasta, and S. Ulam (1955). "Studies of Nonlinear Problems," Los Alamos Rept. LA-1940; also in *Collected Works of Enrico Fermi*, **2**, 978. University of Chicago Press, Chicago (1965).
Field, R.J., and R.M. Noyes (1974). *J. Chem. Phys.* **60**, 1877.
Filonenko, N.N., R.Z. Sagdeev, and G.M. Zaslavskii (1967). *Nucl. Fusion* **7**, 253.
Finn, J.M. (1977). *Phys. Fluids* **20**, 1749.
Finn, J.M., and G.-Z. Sun (1987). *Comment. Plasma Phys. Control. Fusion* **2**, 7.
Foias, C., and G. Prodi (1967). *Rend. Sem. Mat. Univ. Padova* **39**, 1.
Foias, C., and Y. Treve (1981). *Phys. Lett.* **85A**, 35.
Fois, C., O. Manley, and T. Teman (1987). *Nonlinear Anal. Theory Meth. and Applications* **11**, 939.
Ford, J. (1975). In *Fundamental Problems in Statistical Mechanics*, Vol. 3, E.D.G. Cohen (ed.), North-Holland, Amsterdam.
Ford, J., and G.H. Lunsford (1970). *Phys. Rev.* **A1**, 59.
Ford, J., and J. Waters (1963). *J. Math. Phys.* **4**, 1293.
Ford, J., S.D. Stoddard, and J.S. Turner (1973). *Progr. Theoret. Phys.* **50**, 1547.
Forest, E., and M. Berz (1989). Lawrence Berkeley Laboratory Report LBL-25609, ESG-46.
Fraser, A.M. (1989). *Physica* **34D**, 391.
Fraser, A.M., and H.L. Swinney (1986). *Phys. Rev.* **A33**, 1134.
Fredrickson, P., J. Kaplan, E. Yorke, and J. Yorke (1983). *J. Differential Equations* **49**, 185.
Freis, R.P., C.W. Hartman, F.M. Hamzeh, and A.J. Lichtenberg (1973). *Nucl. Fusion* **13**, 533.
Frisch, U., and S.A. Orzag (1990). *Phys. Today*, Jan., 24.
Froehling, H., J.P. Crutchfield, D. Farmer, N.H. Packard, and R. Shaw (1981). *Physica* **3D**, 605.
Froeschlé, C. (1971). *Astrophys. Space Sci.* **14**, 110.
Froeschlé, C. (1972). *Astron. Astrophys.* **16**, 172.
Froeschlé, C., and J.-P. Scheidecker (1973). *Astrophys. Space Sci.* **25**, 373; *Astron. Astrophys.* **22**, 431; *Comput. Phys.* **11**, 423.
Froeschlé, C., and J.-P. Scheidecker (1975). *Phys. Rev.* **A12**, 2137.
Fukuyama, A., H. Mamota, and R. Itatani (1976). *Proceedings of the 3rd Symposium on Plasma Heating in Toroidal Devices*, E. Sindoni (ed.), Editrice Compositori, Bologna; see also (1977). *Phys. Rev. Lett*, **38**, 701.
Gadiyak, G.M., F.M. Izrailev, and B.V. Chirikov (1975). *Proceedings of the 7th International Conference on Nonlinear Oscillations*, Berlin **2**, 315.
Galeev, A.A., and R.Z. Sagdeev (1979). *Rev. Plasma Phys.*, M.A. Leontovich (ed.), **7**, 257.
Galgani, L., and G. LoVecchio (1979). *Nuovo Cimento* **B52**, 1.
Garren, A., R.J. Riddel, L. Smith, G. Bing, L.R. Henrich, T.G. Northrop, and J.E. Roberts (1958). *Proceedings of the 2nd UN Conference on Peaceful Uses of Atomic Energy* **31**, 65.
Garrido, L.M. (1968). *J. Math. Phys.* **10**, 1045.
Geisel, T., A. Zacherl, and G. Radons (1987). *Phys. Rev. Lett.* **59**, 2503.
Geisel, T., and J. Nierwetberg (1981). *Phys. Rev. Lett.* **47**, 1975.
Gell, Y., J. Harte, A.J. Lichtenberg, and W.M. Nevins (1975). *Phys. Rev. Lett.* **35**, 1642.
Giacaglia, G.E.O. (1972). *Perturbation Methods in Nonlinear Systems*, Applied Mathematical Sciences, No. 8, Springer-Verlag, New York.

Gibson, G., W. Jordan, and E. Lauer (1963). *Phys. Fluids* **6**, 116.
Giglio, G., S. Musazzi, and U. Perini (1981). *Phys. Rev. Lett.* **42**, 243.
Glass, L., and R. Perez (1982). *Phys. Rev. Lett.* **48**, 441.
Glazier, J.A., M.H. Jensen, A. Libchaber, and J. Stavans (1986). *Phys. Rev.* **A34**, 1621.
Goedde, C.G., A.J. Lichtenberg, and M.A. Lieberman (1990). *Physica D* **41**, 341.
Goldstein, H. (1951). *Classical Mechanics*, Addison-Wesley, Reading, Massachusetts.
Gollub, J.P., S.V. Benson, and J. Steinman (1980). In R.H.G. Helleman (ed.), *op. cit.*, p. 22.
Gorman, M., L.A. Reith, and H.L. Swinney (1980). In R.H.G. Helleman (ed.), *op. cit.*, p. 10.
Gormezano, C., W. Hess, G. Ichtchenko, R. Mogne, T.K. Nguyen, G. Tonan, G.W. Pacher, H.D. Pacher, F. Söldner, and J.G. Wegrone (1981). *Nucl. Fusion* **21**, 1047.
Goward, F.K. (1953). In *Lectures on the Theory and Design of an AG Proton Synchrotron*, p. 19, CERN, Geneva.
Gramaticos, B., B. Dorizzi, and A. Ramani (1983). *J. Math. Phys.* **24**, 2289.
Grassberger, P. (1985). In *Chaos in Astrophys*, J.R. Buchler (ed.), Reidel, Dordrecht, p. 193.
Grassberger, P., and I. Procaccia (1983). *Physica* **9D**, 189.
Grawe, H. (1969). *Plasma Phys.* **11**, 151.
Grebogi, C., S.W. McDonald, E. Ott, and J.A. Yorke (1983b). *Phys. Lett.* **A99**, 415.
Grebogi, C., E. Ott, and J.A. Yorke (1983). *Physica* **7D**, 181.
Grebogi, C., E. Ott, and J.A. Yorke (1985). *Ergodic Theory Dynamical Systems* **5**, 341.
Grebogi, C., E. Ott, and J.A. Yorke (1986). *Phys. Rev. Lett.* **56**, 1011.
Grebogi, C., E. Ott, and J.A. Yorke (1986a). *Phys. Rev. Lett.* **57**, 1284.
Grebogi, C., E. Ott, and J.A. Yorke (1987). *Physica* **24D**, 243.
Grebogi, C., E. Ott, and J.A. Yorke (1987a). *Science* **238**, 583.
Grebogi, C., E. Ott, J.A. Yorke, and H.E. Nasse (1987b). *Ann. N.Y. Acad. Sci.* **497**, 117.
Grebogi, C., E. Ott, and J.A. Yorke (1988). *Phys. Rev.* **A37**, 1711.
Green, G.K., and E.D. Courant (1959). *Handbuch der Physik* **44**, 300.
Greene, G. (1837). *Cambridge Philos. Trans.* **6**, 457.
Greene, J.M. (1968). *J. Math. Phys.* **9**, 760.
Greene, J.M. (1979a). *J. Math. Phys.* **20**, 1183.
Greene, J.M. (1979b). In M. Month and J.C. Herrera (eds.), *op. cit.*, p. 257.
Greene, J.M., R.S. MacKay, F. Vivaldi, and M.J. Feigenbaum (1981). *Physica* **3D**, 468.
Greene, J.M., R.S. MacKay, and J. Stark (1986). *Physica* **21D**, 267.
Greenspan, B., and P. Holmes (1981). In *Nonlinear Dynamics and Turbulence*, C. Berenblatt, G. Looss, and D.D. Joseph (eds.), Pitman, London.
Grieger, G., W. Ohlendorf, H.D. Packer, H. Wobig, and G.H. Wolf (1971). *IVth IAEA Conference on Plasma Physics and Controlled Fusion*, CN-28/H-3.
Grossman, S. and S. Thomae (1977). *Z. Naturforsch.* **32A**, 1353.
Guckenheimer, J., and P. Holmes (1983). *Nonlinear Oscillations, Dynamical Systems, and Bifurcations of Vector Fields*, Springer-Verlag, New York.
Gustavson, F. (1966). *Astron. J.* **21**, 670.
Gwinn, E.G., and R.M. Westervelt (1987). *Phys. Rev. Lett.* **59**, 157.
Hagihara, Y. (1961). In *The Solar System III; Planets and Satellites*, G. Kuiper and B. Middlehurst (eds.), p. 95.
Hagihara, Y. (1970–1976). *Celestial Mechanics*, MIT Press, Cambridge, Massachusetts.
Hall, L. (1981). "On the Existence of a Last Invariant of Conservative Motion," preprint UCID 18980, Lawrence Livermore Laboratory, Livermore, California.
Halsey, T.C., M.H. Jensen, L.P. Kadenoff, I. Procaccia, and B.I. Shraiman (1986). *Phys. Rev.* **A33**, 1141.
Hammel, S.M. (1990). "A Noise Reduction Method for Chaotic Systems," *Phys. Lett.* **148A**, 421.
Hastie, R.J., J.B. Taylor, and F.A. Hass (1967). *Ann. Phys.* **41**, 302.

Hastie, R.J., G.D. Hobbs, and J.B. Taylor (1969). In *Plasma Physics and Controlled Fusion Research*, IAEA, Vienna.
Hatori, T., and T. Watanabe (1975). *Nucl. Fusion* **15**, 143.
Hausdorff (1918). *Math. Annalen* **79**, 157.
Heinrichs, R., G. Ahlers, and D.S. Cannell (1987). *Phys. Rev. A* **35**, 2761.
Helleman, R.H.G. (1977). In *Statistical Mechanics and Statistical Methods*, U. Landman (ed.), Plenum Press, New York.
Helleman, R.H.G. (1978). In S. Jorna (ed.), *op. cit.*, p. 264.
Helleman, R.H.G. (1980). In *Fundamental Problems in Statistical Mechanics*, Vol. 5, E.G.D. Cohen (ed.), North-Holland, Amsterdam, p. 165.
Helleman, R.H.G. (ed.) (1980a). *Nonlinear Dynamics*, Ann. N.Y. Acad. Sci., Vol. 357, New York Academy of Sciences, New York.
Helleman, R.H.G. (1981). In *Nonequilibrium Problems in Statistical Mechanics*, Vol. 2, W. Horton, L. Reichl, and V. Szebehely (eds.), Wiley, New York.
Helleman, R.H.G., and T. Bountis (1979). In G. Casati and J. Ford (eds.), *op. cit.*, p. 353.
Helleman, R.H.G., and E.W. Montroll (1974). *Physica* **74**, 22.
Hénon, M. (1969). *Quart. Appl. Math.* **27**, 291.
Hénon, M. (1974). *Phys. Rev.* **B9**, 1925.
Hénon, M. (1976). *Commun. Math. Phys.* **50**, 69.
Hénon, M., and C. Heiles (1964). *Astron. J.* **69**, 73.
Hentschel, H.G.E., and I. Procaccia (1983). *Physica* **8D**, 435.
Hereward, H.G., K. Johnsen, and P. Lapastolle (1956). *Proceedings of the CERN Symposium on High-Energy Accelerators*, p. 179, CERN, Geneva.
Herman, M.R. (1985). "Sur les Courbes Invariantes par les Difféomorphismes de l'Anneau," Vol. 2, *Astérisque*.
Herrera, J.C., M. Month, and R.F. Peierls (1979). In M. Month and J.C. Herrera (eds.), *op. cit.*, p. 202.
Hertweck, F., and A. Schlüter (1957). *Z. Naturforsch.* **12a**, 844.
Hietarinta, J. (1983). *Phys. Rev. A* **28**, 3670.
Hietarinta, J. (1986). *Direct Methods for the Search of the Second Invariant*. FTL-8100, Turku Univ., Turku, Finland.
Hildebrand, F.B. (1948). *Advanced Calculus for Engineers*. Prentice-Hall, New York.
Hine, M.G.N. (1953). In *Lectures on the Theory and Design of an AG Proton Synchrotron*, p. 69, CERN, Geneva.
Hirsch, J.E., B.A. Huberman, and D.J. Scalapino (1981). *J. Phys.* **14A**, 3153.
Hirsch, J.E., M. Nauenberg, and D.J. Scalapino (1982). *Phys. Lett.* **87A**, 391.
Holm, D., J. Marsden, T. Ratiu, A. Weinstein (1983). *Phys. Rep.* **123**, 1.
Holmes, P.J. (1977). *Appl. Math. Modeling* **1**, 362.
Holmes, P.J. (1979). *Philos. Trans. Roy. Soc.* **292**, 41.
Holmes, P.J. (1980). *SIAM J. Appl. Math.* **38**, 65; see also subsequent ERRATA.
Holmes, P.J., and J. Marsden (1981). "Melnikov's Method and Arnold Diffusion for Perturbations of Integrable Hamiltonian Systems," preprint; (1982). *J. Math. Phys.* **23**, 669.
Holmes, P.J., and J.E. Marsden (1983). *Indiana Univ. Math. J.* **32**, 273.
Holt, C.R. (1981). "Construction of Integrable Hamiltonians," Ph.D. Thesis, University of Colorado, Boulder, Colorado.
Hori, G. (1966). *Astron. Soc. Japan* **18**, 287.
Howard, J.E. (1970). *Phys. Fluids* **13**, 2407.
Howard, J.E. (1971). *Phys. Fluids* **14**, 2373.
Howard, J.E., A.J. Lichtenberg, and M.A. Lieberman (1982). *Physica* **5D**, 243.
Howard, J.E. and S.M. Hohs (1984). *Phys. Rev.* **A29**, 418.
Howard, J.E., A.J. Lichtenberg, M.A. Lieberman, and R.H. Cohen (1986). *Physica* **20D**, 259.

Howard, J.E., and R.S. MacKay (1987). *J. Math. Phys.* **28**, 1036.
Howard, J.E., and R.S. MacKay (1987a). *Phys. Lett.* **A122**, 331.
Howland, R.A. (1977). *Celestial Mech.* **17**, 327.
Hu, B., and J. Rudnick (1982). *Phys. Rev. Lett.* **48**, 1645.
Huberman, B.A., and J.P. Crutchfield (1979). *Phys. Rev. Lett.* **43**, 1743.
Huberman, B.A., and J. Rudnick (1980). *Phys. Rev. Lett.* **45**, 154.
Huberman, B.A., and A.B. Zisook (1981). *Phys. Rev. Lett.* **46**, 626.
Hübler, A., R. Georgi, A. Muchler, W. Stelzl, E. Lüscher (1988). *Helv. Phys. Acta* **61**, 898.
Hudson, J.L., M. Hart, and D. Marinko (1979). *J. Chem. Phys.* **71**, 1601.
Ichikawa, Y.H., T. Kamimura, and T. Hatori (1987). *Physica* **29D**, 247.
Ichikawa, Y.H., T. Kamimura, T. Hatori, and S.Y. Kim (1989). Institute of Plasma Physics, Report IPPJ-910, Nagoya University, Nagoya.
Inozemtsev, V.I. (1984). *Physica Scripta* **29**, 518.
Izrailev, F.M. (1980). *Physica* **1D**, 243.
Izrailev, F.M., and B.V. Chirikov (1966). *Soviet. Phys. Dokl.* **11**, 30.
Izrailev, F.M., and T.A. Zhadanova (1974). "On Statistical Fermi Acceleration," preprint, Institute of Nuclear Physics, USSR Academy of Sciences, Novosibirsk.
Izrailev, F.M., M.I. Rabinovich, and A.D. Ugodnikov (1981). "Approximate Description of Three-Dimensional Dissipative Systems with Stochastic Behavior," preprint N17, Institute of Applied Physics, USSR Academy of Sciences, Gorky.
Izrailev, F.M., S.I. Mishnev, and G.M. Tunaikin (1977). "Numerical Studies of Stochasticity Limits in Colliding Beams," preprint 77-43, Institute of Nuclear Physics, USSR Academy of Sciences, Novosibirsk.
Izrailev, F.M., B. Timmerman, and W. Timmerman (1987). "Transient Chaos in a Generalized Hénon Map on the Torus," preprint 87-1, Institute of Nuclear Physics, Novosibirsk.
Jaeger, F., and A.J. Lichtenberg (1972). *Ann. Phys.* **71**, 319.
Jaeger, F., A.J. Lichtenberg, and M.A. Lieberman (1972). *Plasma Phys.* **14**, 1073.
Jaeger, H.M., C.-H. Liu, and S.R. Nagel (1989). *Phys. Rev. Lett.* **62**, 40.
Jeffries, C., and J. Perez (1982). *Phys. Rev.* **A26**, 2117.
Jensen, H.J., K. Christensen, H.C. Fogerty (1989). *Phys. Rev.* **B40**, 7425.
Jensen, M.H. (1987). In *Proc. ASI Geilo School*, R. Pynn and T. Riste (eds.), Plenum, New York; see also NORDITA preprint 87/45 S.
Jensen, M.H., P. Bak, and T. Bohr (1984). *Phys. Rev.* **A30**, 1960.
Jensen, M.H., and I. Procaccia (1985). *Phys. Rev.* **A32**, 1225.
Jensen, M.H., L.P. Kadanoff, A. Lichaber, I. Procaccia, and J. Stavans (1985). *Phys. Rev. Lett.* **55**, 2785.
Jensen, R.V. (1984). *Phys. Rev.* **A30**, 386.
Johnston, S., and A.N. Kaufman (1978). *Phys. Rev. Lett.* **40**, 1266.
Jorna, S. (ed.) (1978). *Topics in Nonlinear Dynamics*, Vol. 46, American Institute of Physics, New York.
Kadenoff, L.P., S.R. Nagel, L. Wu, and S.-M. Zhou (1989). *Phys. Rev.* **A15**, 6524.
Kadomtsev, B.B. (1976). *Soviet J. Plasma Phys* **1**, 389.
Kadomtsev, B.B., and O.P. Pogutse (1971). *Nucl. Fusion* **11**, 67.
Kaneko, K. (1984). *Progr. Theoret. Phys.* **72**, 1089.
Kaneko, K. (1989). *Physica* **34D**, 1.
Kaneko, K., and T. Konishi (1989). *Phys. Rev.* **A40**, 6130.
Kantz, H., and P. Grassberger (1985). *Physica* **17D**, 75.
Kaplan, J.L., and J.A. Yorke (1979). *Springer Lecture Notes in Mathematics*, Vol. 730, 204.
Kaplan, J.L., and J.A. Yorke (1979). *Comm. Math. Phys.* **67**, 93.
Karimabadi, H., and V. Agelapoulos (1989). *Phys. Rev. Lett.* **62**, 2342.

Karney, C.F.F. (1977). "Stochastic Heating of Ions in a Tokamak by RF Power," Ph.D. Thesis, MIT, Cambridge, Massachusetts.
Karney, C.F.F. (1978). *Phys. Fluids* **21**, 1584.
Karney, C.F.F. (1979). *Phys. Fluids* **22**, 2188.
Karney, C.F.F. (1983). *Physica* **8D**, 387.
Karney, C.F.F., and A. Bers (1977). *Phys. Rev. Lett.* **39**, 550.
Karney, C.F.F., A.B. Rechester, and R.B. White (1982). *Physica* **4D**, 425.
Kaufman, A.N. (1978). In S. Jorna (ed.), *op. cit.*, p. 286.
Kaufman, A.N., J. Cary, and N. Pereira (1978). *Proceedings of the Third Topical Conference on RF Plasma Heating*, Pasadena, California; (1979). *Phys. Fluids* **22**, 790.
Keeler, J.D. (1986). "Coupled Maps and Reaction–Diffusion Equations," U.C. San Diego, Institute for Nonlinear Science, preprint.
Keeler, J.D., and J.D. Farmer (1986). *Physica* **23D**, 413.
Keller, J., and S. Rubinow (1960). *Ann. Phys. N.Y.* **9**, 24.
Khinchin, A.Ya. (1964). *Continued Functions*, University of Chicago Press, Chicago.
Klein, A., and Li, C-T (1979). *J. Math. Phys* **20**, 572.
Knobloch, E. (1986). *Phys. Rev. A* **34**, 1538.
Knobloch, E., and J.B. Weiss (1987). *Phys. Rev. A* **36**, 1522.
Knobloch, E., and J.B. Weiss (1989). *Phys. Rev. A* **40**, 2579.
Kobayashi, T., N. Mori, H. Hata, T. Horita, T. Yoshida, and H. Mori (1989). *Progr. Theoret. Phys.* (*Lett.*) **82**, 1.
Kolmogorov, A.N. (1941). *Dokl. Akad. Nauk. SSSR* **30**, 299.
Kolmogorov, A.N. (1954). *Dokl. Akad. Nauk. SSSR* **98**, 527.
Kolmogorov, A.N. (1959). *Dokl. Akad. Nauk. SSSR* **124**, 754.
Kolmogorov, A.N. (1965). *Prob. Inform. Trans.* **1**, 3.
Kolomensky, A.A., and A.N. Lebedev (1966). *Theory of Cyclic Particle Accelerators*, North-Holland, Amsterdam.
Konishi, T. (1989). *Progr. Theoret. Phys., Suppl.* **98**, 19.
Konishi, T., and K. Kaneko (1990). "Diffusion in Hamiltonian Chaos and Its Size Dependence," Department of Physics, Nagoya University, Report DPNU-90-19.
Kopell, N. (1980). In R.H.G. Helleman (ed.), *op. cit.*, p. 397.
Kopell, N., and L.N. Howard (1973). *Stud. Appl. Math.* **52**, 291.
Kramers, H.A. (1926). *Z. Physik.* **39**, 828.
Krilin, L. (1967a). *Czech. J. Phys. B* **17**, 112.
Krilin, L. (1967b). *Czech, J. Phys. B* **17**, 124.
Krishnaprasad, P.S., and J.E. Marsden (1987). *Arch. Rational Mech. Anal.* **98**, 137.
Kruskal, M.D. (1957). *Proceedings of the Third Conference on Ionized Gases*, Venice.
Kruskal, M.D. (1962). *J. Math. Phys.* **3**, 806.
Kruskal, M.D., and N.J. Zabusky (1964). *J. Math. Phys.* **5**, 231.
Krylov, N. (1950). Studies on the Foundation of Statistical Physics, *Akad. Nauk. SSSR* M.-L.
Krylov, N., and N.N. Bogoliubov (1936). *Introduction to Nonlinear Mechanics*, Kiev; (1974), Princeton University Press, Princeton, New Jersey.
Kuckes, A.F. (1968). *Plasma Phys.* **10**, 367.
Kulsrud, R.M. (1957). *Phys. Rev.* **106**, 205.
Kummer, M. (1978). *Commun. Math. Phys.* **55**, 85.
Kuramoto, Y., and T. Yamada (1976). *Progr. Theoret. Phys.* **56**, 679.
Lacina, J. (1963). *Czech. J. Phys.* **B13**, 401.
Lacina, J. (1966). *Plasma Phys.* (J.N.E. Pt. C) **8**, 515.
Lahiri, A., and T. Nag (1989). *Phys. Rev. Lett.* **62**, 1933.
Lamb, G. (1980). *Elements of Soliton Theory*, Wiley, New York.
Landau, L.D. (1937). *Zh. Eksper. Theor. Fiz.* **7**, 203.
Landau, L.D. (1941). *C. R. Acad. Sci. USSR* **44**, 311; see also Landau, L.D., and E.M. Lifshitz (1959), *Fluid Mechanics*, Pergamon Press, London, Chapter 3.

Lanford, O. (1976). "Qualitative and Statistical Theory of Dissipative Systems," 1976 CIME School of Statistical Mechanics, Liguori Editore, Napoli.
Lanford, O. (1977). In *Turbulence Seminar*. P. Bernard and T. Ratiu (eds.), Springer Lecture Notes in Mathematics 615, p. 114, Springer-Verlag, New York.
Lanford, O.E. (1981). In *Hydrodynamic Instabilities and the Transition to Turbulence*, H.L. Swinney and J.P. Gollub (eds.), Springer-Verlag, New York, Chapter 2.
Laslett, L.J. (1967). In *Focusing of Charged Particles*, Vol. 2. A. Septier (ed.), Academic Press, New York, p. 355.
Laval, G., and D. Gresillon (eds.) (1979). *Intrinsic Stochasticity in Plasmas*, Les Editions de Physique, Courtaboeuf, Orsay, France.
Lazar, N., J. Barter, R. Dandl, W. DiVergilio, and R. Wuerker (1980). In N.A. Uckan (ed.), *op. cit.*
Lewis, H.R., Jr. (1967). *Phys. Rev. Lett.* **18**, 510.
Lewis, H.R., Jr. (1968). *J. Math. Phys.* **9**, 1976.
Lewis, H.R., and P.G.L. Leach (1980). "A Generalization of the Invariant for the Time-Dependent Linear Oscillator," preprint LA-UR 80-68, Los Alamos Scientific Laboratory, Los Alamos, New Mexico.
Li, T-Y., and J.A. Yorke (1975). *Amer. Math. Monthly* **82**, 985.
Liapunov. A.M. (1907). *Ann. Math. Studies* **17**, 1947.
Libchaber, A., and J. Maurer (1980). *J. Phys.* **41**, C3.
Lichtenberg, A.J. (1964). *Rev. Sci. Instr.* **54**, 1196.
Lichtenberg, A.J. (1969). *Phase Space Dynamics of Particles*, Wiley, New York.
Lichtenberg, A.J. (1979). In G. Laval and D. Gresillon (eds.), *op. cit.*, p. 13.
Lichtenberg, A.J. (1979a). In G. Casati and J. Ford (eds.), *op. cit.*, p. 18.
Lichtenberg, A.J. (1984). *Nucl. Fusion* **24**, 1277.
Lichtenberg, A.J. (1985). *Physica* **14D**, 387.
Lichtenberg, A.J., and H.L. Berk (1975). *Nucl. Fusion* **15**, 999.
Lichtenberg, A.J., and F.W. Jaeger (1970). *Phys. Fluids* **13**, 392.
Lichtenberg, A.J., and M.A. Lieberman (1976). *Nucl. Fusion* **16**, 532.
Lichtenberg, A.J., and G. Melin (1972). *Phys. Fluids* **16**, 1660.
Lichtenberg, A.J., M.A. Lieberman, and R.H. Cohen (1980). *Physica* **1D**, 291.
Lichtenberg, A.J., M.A. Lieberman, and N.W. Murray (1987). *Physica* **28D**, 371.
Lichtenberg, A.J., and B.P. Wood (1989). *Phys. Rev.* **A39**, 2153.
Lichtenberg, A.J., and B.P. Wood (1989a). *Phys. Rev. Lett.* **62**, 2213.
Lieberman, M.A. (1980). In R.H.G. Helleman (ed.), *op. cit.*, p. 119.
Lieberman, M.A., and A.J. Lichtenberg (1972). *Phys. Rev.* **A5**, 1852.
Lieberman, M.A., and A.J. Lichtenberg (1973). *Plasma Phys.* **15**, 125.
Lieberman, M.A., and J.L. Tennyson (1982). In *Nonequilibrium Problems in Statistical Mechanics*, Vol. 2, W. Horton, L. Reichl, and V. Szebehely (eds.), Wiley, New York.
Lieberman, M.A., and K. Tsang (1981). *Bull. Amer. Phys. Soc.* **26**, 1013.
Lindstedt, M. (1882). *Astrom. Nach.* **103**, 211.
Liouville, J. (1837). *J. Math. Pures Appl.* **2**, 16.
Littlejohn, R. (1978). Lawrence Berkeley Laboratory Report UCID-8091, Berkeley, California.
Littlejohn, R. (1979). *J. Math. Phys.* **20**, 2445.
Littlejohn, R.G. (1983). *J. Plasma Phys.* **29**, 111.
Littlewood, J.E. (1963). *Ann. Phys.* **21**, 233.
Livi, R., M. Pettini, S. Ruffo, M. Sparpaglioni, and A. Vulpiani (1985). *Phys. Rev. A* **31**, 1039.
Longcope, D.W., and R.N. Sudan (1987). *Phys. Rev. Lett.* **59**, 1500.
Lorenz, E.N. (1963). *J. Atmos. Sci.* **20**, 130.
Lorenz, E.N. (1980). In R.H.G. Helleman (ed.), *op. cit.*, p. 282.
Lozi, R. (1978). *J. Phys.* **39**, C5-9.
Lunsford, G.H., and J. Ford (1972). *J. Math. Phys.* **13**, 700.

McDonald, S.W., and A.N. Kaufman (1979). *Phys. Rev. Lett.* **42**, 1189.
MacKay, R.S. (1983). *Physica* **7D**, 283.
MacKay, R.S., J.D. Meiss, and I.C. Percival (1984). *Physica* **13D**, 55.
MacKay, R.S., J.D. Meiss, and I.C. Percival (1987). *Physica* **27D**, 1.
MacKay, R.S., and I.C. Percival (1985). *Comm. Math. Phys.* **98**, 469.
MacKay, R.S., and C. Tresser (1986). *Physica* **19D**, 206.
Mackey, M.C., and L. Glass (1977). *Science* **197**, 287.
McLaughlin, J.B. (1979). *Phys. Rev.* **A20**, 2114; see also (1981). *J. Stat. Phys.* **24**, 375.
McNamara, B. (1978). *J. Math. Phys.* **19**, 2154.
McNamara, B., and K. I. Whiteman (1966), "Invariants of Nearly Periodic Hamiltonian Systems," Culham Laboratory Report CLM P111, Culham, England.
McNamara, B., and K.J. Whiteman (1967). *J. Math. Phys.* **8**, 2029.
Mandelbrot, B.B. (1974). *J. Fluid Mech.* **62**, 331.
Manneville, P., and Y. Pomeau (1980). *Physica* **1D**, 219.
Marcus, R.A. (1980). In R.H.G. Helleman (ed.), *op. cit.*, p. 169.
Marsden, J., and T. Ratiu (1987). In *Seminar on New Results in Nonlinear Partial Differential Equations*, A.J. Tromba (ed.), F. Vieweg, Vol. E10, p. 101.
Marsden, J.E., and J.C. Simo (1990). In *Proceedings of 200th Anniversary of "Méchanique Analytique" of J. L. Lagrange*. A. Lichnerowicz (ed.).
Martin, S., and W. Martienssen (1986). *Phys. Rev. Lett.* **56**, 1522.
Maschke, E.K., and B. Saramito (1982). *Phys. Scripta* **T2/2**, 410.
Maselko, J., and H.L. Swinney (1985). *Phys. Scripta* **T9**, 35.
Mather, J.N. (1984). *Ergodic Theory Dynamical Systems* **2**, 301.
Mattheus, W.H., and D. Montgomery (1980). In R.H.G. Helleman (ed.), *op. cit.*, p. 203.
May, R.M. (1976). *Nature* **261**, 459.
May, R.M., and G.F. Oster (1976). *Amer. Natur.* **110**, 573.
Meiss, J.D. (1986). *Phys. Rev.* **A34**, 2375.
Meiss, J.D., J.R. Cary, C. Grebogi, J.P. Crawford, A.N. Kaufman, and H.D.I. Abarbanel (1983). *Physica* **6D**, 375.
Meiss, J.D., and E. Ott (1986). *Physica* **20D**, 387.
Melnikov, V.K. (1962). *Dokl. Akad. Nauk. SSSR*, **114**, 747; (1963). **148** 1257.
Melnikov, V.K. (1963). *Trans. Moscow Math. Soc.* **12**, 1.
Metropolis, N., M.L. Stein, and P.R. Stein (1973). *J. Combinatorial Theory* **15A**, 25.
Misiurewicz, M. (1980). In R.H.G. Helleman (ed.), *op. cit.*, p. 348.
Mo, K.C. (1972). *Physica* **57**, 445.
Molchanov, A.M. (1968). *Icarus* **8**, 203.
Month, M., and J.C. Herrera (eds.) (1979). *Nonlinear Dynamics and the Beam–Beam Interaction*, AIP Conference Proceedings No. 57, American Institute of Physics, New York.
Morosov, A.D. (1973). *USSR Comput. Math. Phys.* **13**, 45.
Morosov, A.D. (1976). *J. Differential Equations* **12**, 164.
Morosov, A.I., and L.S. Solov'ev (1966). *Rev. Plasma Phys.* **2**, 58.
Moser, J. (1962). *On Invariant Curves of Area-Preserving Mappings on an Annulus.* Nachr. Akad. Wiss. Göttingen, Math. Phys. Kl, p. 1.
Moser, J. (1966). *Ann. Acuola Normale Sup. Pisa Ser. III* **20**, 499.
Moser, J. (1973). *Stable and Random Motions in Dynamical Systems*, Annals of Mathematical Studies 77, Princeton University Press, Princeton, New Jersey.
Moser, J. (1978). *Math. Intelligencer* **1**, 65.
Moses, E., and V. Steinberg (1986). *Phys. Rev. A* **34**, 693.
Moses, E., and V. Steinberg (1988). *Phys. Rev. Lett.* **60**, 2030.
Motz, H., and C.J.H. Watson (1967). *Adv. Electron Phys.* **23**, 153.
Murakami, S., T. Sato, and A. Hasegawa (1988). *Physica* **32D**, 26.
Murray, N. (1991). *Physica* **52D**, 220.

Murray, N.W., M.A. Lieberman, and A.J. Lichtenberg (1985). *Phys. Rev.* **A32**, 2413.
Nauenberg, N., and J. Rudnick (1981). *Phys. Rev.* **B24**, 493.
Nayfeh, A. (1973). *Perturbation Methods*, Wiley, New York.
Nekhoroshev, N.N. (1977). *Usp. Mat. Nauk. USSR* **32**, 6.
Nekrasov, A.K. (1970). *Nucl. Fusion* **10**, 387.
Nevins, W.M., J. Harte, and Y. Gell (1979). *Phys. Fluids* **22**, 2108.
Newhouse, S., D. Ruelle, and F. Takens (1978). *Commun. Math. Phys.* **64**, 35.
Newhouse, S. (1980). *Publ. Math. IHES* **50**, 101.
Noid, D.W., M.I. Koszykowski, and R.A. Marens (1977). *J. Chem. Phys.* **67**, 404.
Northrop, T.G. (1963). *The Adiabatic Motion of Charged Particles*, Wiley, New York.
Northrop, T.G., and E. Teller (1960). *Phys. Rev.* **117**, 215.
Northrop, T.G., C.S. Liu, and M.D. Kruskal (1966). *Phys. Fluids* **9**, 1503.
Okuda, H., and J. Dawson (1973). *Phys. Fluids* **16**, 2336.
Oseledec, V.I. (1968). *Trans. Moscow Math. Soc.* **19**, 197.
Ott, E. (1981). *Rev. Mod. Phys.* **53**, 655.
Ott, E., J.D. Hanson, and T.M. Antonsen, Jr. (1981). *Bull. Amer. Phys. Soc.* **26**, 1013.
Packard, N.H., J.P. Crutchfield, J.D. Farmer, and R.S. Shaw (1980). *Phys. Rev. Lett.* **45**, 712.
Parker, T.S., and L.O. Chua (1989). *Practical Numerical Algorithms for Chaotic Systems*, Springer-Verlag, New York.
Penrose, O. (1970). *Foundations of Statistical Mechanics*, Pergamon, Oxford.
Percival, I.C. (1974). *J. Phys.* **A7**, 794.
Percival, I.C. (1977). *Adv. Chem. Phys.* **36**, 1.
Percival, I.C. (1979a). *J. Phys.* **A12**, L57.
Percival, I.C. (1979b). In M. Month and J.C. Herrera (eds.), *op. cit.*, p. 302.
Percival, I.C., and N. Pomphrey (1976). *Mol. Phys* **31**, 97.
Percival, I.C., and N. Pomphrey (1978). *Mol. Phys* **35**, 649.
Pesin, Ya. B. (1976). *Soviet Math. Dokl.* **17**, 196.
Pesin, Ya. B. (1977). *Russian Math. Surveys* **32**, 55.
Pettini, M., and M. Landolphi (1990). *Phys. Rev. A* **41**, 768.
Pogutse, O.P. (1972). *Nucl. Fusion* **12**, 39.
Poincaré, H. (1892). *Les Methodes Nouvelles de la Mechanique Celeste*, Gauthier-Villars, Paris.
Pomeau, Y., and P. Manneville (1980). *Commun. Math. Phys.* 77; see also Manneville and Pomeau (1980).
Pomeau, Y., J.C. Roux, A. Rossi, S. Bachelart, and C. Vidal (1981). *J. Phys. (Paris) Lett.* **42**, L-271.
Ponomarenko, V.G., L. Ya. Trajnin, V.I. Yurchenko, and A.N. Yasnetsky (1969). *Soviet Phys. JETP* **28**, 1.
Procaccia, L., and H. Schuster (1983). *Phys. Rev.* **A28**, 1210.
Pustylnikov, L.D. (1978). *Trans. Moscow Math. Soc.* **2**, 1.
Quispel, G.R.W., and J.A.G. Roberts (1988). *Phys. Lett.* **A132**, 161.
Rabinovich, M.I. (1978). *Soviet Phys. Usp.* **21**, 443.
Ram, A.K., K. Hizanidis, and A. Bers (1986). *Phys. Rev. Lett.* **56**, 147.
Rand, D., S. Ostlund, J. Sethna, and E. Siggia (1982). *Phys. Rev. Lett.* **49**, 132; see also *Physica* **6D**, 303 (1984).
Rannou, P. (1974). *Astron. Astrophys.* **31**, 289.
Rechester, A.B., and M.N. Rosenbluth (1978). *Phys. Rev. Lett.* **40**, 38.
Rechester, A.B., and T.H. Stix (1976). *Phys. Rev. Lett.* **36**, 587.
Rechester, A.B., and T.H. Stix (1979). *Phys. Rev.* **A19**, 1656.
Rechester, A.B., and R.B. White (1980). *Phys. Rev. Lett.* **44**, 1586.
Rechester, A.B., and M.N. Rosenbluth, and R.B. White (1981). *Phys. Rev.* **A23**, 2664.
Renyi, A. (1959). *Acta Mathematica* (Hungary) **10**, 193.

Roberts, J.A.G., T. Post, H.W. Capl, and G.R.W. Quispel (1991). In *Solitons and Chaos*, F. Lambert (ed.), Springer-Verlag, New York.
Rognlien, T.D. (1982). In N.A. Uckan (ed.), *op. cit.*, **2**, 545.
Rosenbluth. M.N. (1972). *Phys. Rev. Lett.* **29**, 408.
Rosenbluth, M.N., R.Z. Sagdeev, J.B. Taylor, and G.M. Zaslavskii (1966). *Nucl. Fusion* **6**, 207.
Rössler, O.E. (1976a). *Phys. Lett.* **57A**, 397.
Rössler, O.E. (1976b). *Z. Naturforsch.* **31a**, 259, 1168.
Roux, J.C., A. Rossi, S. Bochelart, and C. Vidal (1980). *Phys. Lett.* **77A**, 391.
Roux, J.C., R.H. Simoyi, and H.L. Swinney (1983). *Physica* **8D**, 257.
Ruelle, D. (1973). *Trans. N.Y. Acad. Sci.* **35**, 66.
Ruelle, D. (1980). *Math. Intelligencer* **2**, 126.
Ruelle, D., and F. Takens (1971). *Commun. Math. Phys.* **20**, 167.
Russell, D.A., J.D. Hanson, and E. Ott (1980). *Phys. Rev. Lett.* **45**, 1175.
Ryutov, D.D., and G.Z. Stupakov (1978). *Soviet Phys. Dokl.* **23**, 412.
Sagdeev, R.Z., and A.A. Galeev (1969). *Nonlinear Plasma Theory*, W.A. Benjamin, New York.
Saltzman, B. (1962). *J. Atmos. Sci.* **19**, 329.
Samec, T.K., B.L. Hauss, and G. Guest (1982). In N.A. Uckan (ed.), *op. cit.* **2**, 531.
Schmidt, G. (1962). *Phys. Fluids* **5**, 994.
Schmidt, G. (1979). *Physics of High Temperature Plasmas*, 2nd ed., Academic Press, New York.
Schmidt, G. (1980). *Phys. Rev.* **A22**, 2849.
Schmidt, G., and J. Bialek (1981). "Fractal Diagrams for Hamiltonian Stochasticity," preprint; (1982). *Physica* **5D**, 397.
Schmidt, G., and B.W. Wang (1985). *Phys. Rev. A* **32**, 2994.
Schmits, R.A., K.R. Graziani, and J.L. Hudson (1977). *J. Chem. Phys.* **67**, 3040.
Schuster, H.G. (1988), *Deterministic Chaos*, 2nd ed., VCH, Weinheim.
Segur, H. (1980). Lectures Given at the International School of Physics (Enrico Fermi), Varenna, Italy, unpublished.
Seidl, M. (1964). *Plasma Phys.* (J.N.E. Pt. C) **6**, 597.
Shaw, R. (1981). *Z. Naturforsch.* **36a**, 80.
Shenker, S.J. (1982). *Physica* **5D**, 405.
Shenker, S.J., and L.P. Kadanoff (1982). *J. Stat. Phys.* **27**, 631.
Shepelyanski, D.L. (1981). "Some Statistical Properties of Simple Quantum Systems Stochastic in the Classical Limit," preprint 81–55, Institute of Nuclear Physics, USSR Academy of Science, Novosibirsk.
Shimada, I., and T. Nagashima (1979). *Prog. Theor. Phys.* **61**, 1605.
Shraiman, B.I. (1984). *Phys. Rev.* **A31**, 3520.
Siambis, J. (1965). "Guiding Center Motion of Charged Particles in Combined Mirror Multipole Cusp Magnetic Fields," Ph.D. Thesis, University of California, Berkeley, California.
Siambis, J.G., and T.G. Northrop (1966). *Phys. Fluids* **6**, 2001.
Siegel, C.L., and J. Moser (1971). *Lectures on Celestial Mechanics*, Grundlehren der Mathematischen Wissenschaften, Vol. 187, Springer-Verlag, New York.
Simó, C. (1980). *J. Stat. Phys.* **21**, 465.
Sinai, Ya. G. (1962). *Proc. Int. Cong. Math.*; see also Sinai (1966).
Sinai, Ya. G. (1963). *Soviet Math. Dokl.* **4**, 1818.
Sinai, Ya. G. (1966). *Izv. Akad. Nauk. SSSR*, Mt. **30**, 15.
Sitnikov, K. (1960). *Soviet Phys. Dokl.* **5**, 647.
Slater, J.C. (1948). *Rev. Mod. Phys.* **20**, 473.
Smale, S. (1965). In *Differential and Combinatorial Topology*, S.S. Cairns, (ed.), Princeton University Press, Princeton, New Jersey.
Smale, S. (1967). *Bull. Amer. Math. Soc.* **73**, 747.

Smith, G.R. (1977). "Stochastic Acceleration by a Single Wave in a Magnetized Plasma," Ph.D. Thesis, University of California, Berkeley, California (Lawrence Berkeley Laboratory Report, LBL-6824).
Smith, G.R., J.A. Byers, and L.L. LoDestro (1980). *Phys. Fluids* **23**, 278.
Smith, G.R., and A.N. Kaufman (1975). *Phys. Rev. Lett.* **34**, 1613; see also (1978). *Phys. Fluids* **21**, 2230.
Smith, G.R., and N. Pereira (1978). *Phys. Fluids* **21**, 2253.
Solomonoff, R.J. (1964). *Information Control* **7**, 1.
Solov'ev, L.S., and V.D. Shafranov (1970). *Rev. Plasma Phys.* **5**, 1.
Sprott, J.C., and P.H. Edmonds (1971). *Phys. Fluids* **14**, 2703.
Stavans, J., F. Heslot, and A. Libchaber (1985). *Phys. Rev. Lett.* **55**, 596.
Stern, D. (1970a). *J. Math. Phys.* **11**, 2771.
Stern, D. (1970b). *J. Math. Phys.* **11**, 2776.
Stix, T.H. (1973). *Phys. Rev. Lett.* **30**, 833.
Stoddard, S.D., and J. Ford (1973). *Phys. Rev.* **A3**, 1504.
Stoker, J.J. (1950). *Nonlinear Vibrations*, Interscience, New York.
Swinney, H.L., and J.P. Gollub (1978). *Phys. Today* **31** (August), 41.
Symon, K.R. (1970). *J. Math. Phys.* **11**, 1320.
Symon, K.R., D.W. Kerst, L.W. Jones, L.J. Laslett, and K.M. Terwilliger (1956). *Phys. Rev.* **103**, 1837.
Symon, K.R., and A.M. Sessler (1956). *Proceedings of the CERN Symposium on High-Energy Accelerators*, p. 44, CERN, Geneva.
Szebehely, V. (1967). *Theory of Orbits*, Academic Press, New York.
Szepfalusy, P., and T. Tel (1986). *Phys. Rev.* **A34**, 2520.
Tabor, M. (1981). *Adv. Chem. Phys.* **46**, 73.
Tabor, M. (1989). *Chaos and Integrability in Nonlinear Dynamics*, Wiley, New York.
Takens, F. (1971). *Indag. Math.* **33**, 379.
Takens, F. (1981). In *Lecture Notes in Mathematics*, Vol. 898, D.A. Rand and L.S. Young (eds.), Springer-Verlag, New York, p. 366.
Taylor, J.B. (1974). *Phys. Rev. Lett.* **33**, 1139.
Taylor, J. B., and E.W. Laing (1975). *Phys. Rev. Lett.* **35**, 1306.
Tennyson, J.L. (1979). In M. Month and J.C. Herrera (eds.), *op. cit.*, p. 158.
Tennyson, J.L. (1982). *Physica* **5D**, 123.
Tennyson, J.L., M.A. Lieberman, and A.J. Lichtenberg (1979). In M. Month and J.C. Herrera (eds.), *op. cit.*, p. 272.
Timofeev, A.V. (1974). *Nucl. Fusion* **14**, 165.
Timofeev, A.V. (1978). *Soviet Phys. JETP* **48**, 656.
Toda, M. (1970). *Progr. Theoret. Phys. Suppl.* **45**, 174.
Toda, M. (1974). *Phys. Lett.* **48A**, 335.
Tomita, K., and I. Tsuda (1979). *Phys. Lett.* **A71**, 489.
Treve, Y. (1978). In S. Jorna (ed.), *op. cit.*, p. 147.
Treve, Y. (1981). *J. Comput. Phys.* **41**, 217.
Tsang, K.Y. (1986). *Phys. Rev. Lett.* **57**, 1390.
Tsang, K.Y., and M.A. Lieberman (1984). *Physica* **11D**, 147.
Tsang, K.Y., and M.A. Lieberman (1984a). *Phys. Lett.* **103A**, 175.
Tsang, K.Y., and M.A. Lieberman (1986). *Physica* **21D**, 401.
Tyson, J.J. (1978). *J. Math. Biol.* **5**, 351.
Uckan, N.A. (ed.) (1982). *Proceedings of the 2nd Workshop on Hot Electron Ring Physics*, Oak Ridge National Laboratory, Oak Ridge, Tennessee.
Ueda, Y. (1980). In *New Approaches to Nonlinear Problems in Dynamics*, P.J. Holmes (ed.), SIAM, Philadelphia, p. 311.
Ulam, S.M. (1961). *Proceedings of the 4th Berkeley Symposium on Mathematical Statistics and Probability*, University of California Press, Berkeley, Vol. 3, p. 315.

Ulam, S.M., and J. Von Neumann (1947). *Bull. Amer. Soc.* **53**, 1120.
Umberger, D.K., and D. Farmer (1985). *Phys. Rev. Lett.* **55**, 661.
van der Weele, J.P., T.P. Valkering, H.W. Capel, and T. Post (1988). *Physica* **153A**, 283.
van Zeyts, J.B.J. (1981). Private communication; see Helleman (1981).
Vidal, C., J-C. Roux, S. Bachelert, and A. Rossi (1980). In R.H.G. Helleman (ed.), *op. cit.*, p. 377.
Von Zeipel, H. (1916). *Ark. Astron. Mat. Fys.* **11**, No. 1.
Waddell, B.V., M.N. Rosenbluth, D.A. Monticello, and R.B. White (1976). *Nucl. Fusion* **16**, 528.
Waddell, B.V., B. Carreras, H.R. Hicks, J.A. Holmes, and D.K. Lee (1978). *Phys. Rev. Lett.* **41**, 1386.
Walker, J. (1978). *Sci. Amer.* **239**, No. 1, 152.
Walker, G.H., and J. Ford (1969). *Phys. Rev.* **188**, 416.
Wang, M.C., and G.E. Uhlenbeck (1945). *Rev. Mod. Phys.* **17**, 523.
Warnock, R.L., and R.D. Ruth (1987). *Physica* **26D**, 1.
Watari, T., S. Hiroe, T. Sato, and S. Ichimaru (1974). *Phys. Fluids* **17**, 2107.
Wegmann, K., and O.E. Rössler (1978). *Z. Naturforsch*, **33a**, 1179.
Wentzel, G. (1926). *Z. Phys.* **38**, 518.
White, R.B., D.A. Monticello, M.N. Rosenbluth, and B.V. Wadell (1977). *Plasma Phys. Contr. Nucl. Fusion*, Int. Atom Energy Agency, Vienna **1**, 569.
Whiteman, K.J., and B. McNamara (1968). *J. Math. Phys.* **9**, 1385.
Whittaker, E.T. (1964). *A Treatise on the Analytical Dynamics of Particles and Rigid Bodies*. Cambridge University Press, Cambridge.
Widom, M., D. Bensimon, L.P. Kadanoff, and S.J. Shenker (1983). *J. Stat. Phys.* **32**, 443.
Wiedemann, H. (1979). In M. Month and J.C. Herrera (eds.), *op. cit.*, p. 84.
Winfree, A. (1974). *Sci. Amer.* **230**, No. 6, 82.
Wintner, A. (1947). *Analytical Foundations of Celestial Mechanics*, Princeton University Press, Princeton, New Jersey.
Wolf, A., and J. Swift (1981). *Phys. Lett.* **83A**, 184.
Wolf, A., J.B. Swift, H.L. Swinney, and J.A. Vastano (1985). *Physica* **16D**, 285.
Wood, B., A.J. Lichtenberg, and M.A. Lieberman (1990). *Phys. Rev. A* **42**, 5885.
Wright, H.L., and B.N. Miller (1984). *Phys. Rev. A.* **29**, 1411.
Wyeth, N.C., A.J. Lichtenberg, and M.A. Lieberman (1975). *Plasma Phys.* 17, 679.
Yamada, T., and Y. Kuramoto (1976). *Progr. Theoret. Phys.* **56**, 681.
Yamazaki, H., Y. Oono, and K. Hirakawa (1978). *J. Phys. Soc. Japan* **44**, 335; (1979). **46**, 721.
Yeh, W.L., and Y.H. Kao (1983). *Appl. Phys. Lett.* **42**, 299.
Yoshida, H. (1984). *Celest. Mech.* **32**, 73.
Zabusky, N.J. (1962). *J. Math. Phys.* **3**, 1028.
Zabusky, N.J. (1967). In *Nonlinear Partial Differential Equations*, W.F. Ames (ed.), Academic, New York, p. 223.
Zabusky, N.J., and M.O. Kruskal (1965). *Phys. Rev. Lett.* **15**, 240.
Zaiken, A.N., and A.M. Zhabotinsky (1970). *Nature* **225**, 535.
Zaslavskii, G.M. (1977). *Soviet Phys. JETP* **46**, 1094.
Zaslavskii, G.M. (1978). *Phys. Lett.* **69A**, 145.
Zaslavskii, G.M. (1981). *Phys. Rep.* **80**, 157.
Zaslavskii, G.M., and B.V. Chirikov (1965). *Soviet Phys. Dokl.* **9**, 989.
Zaslavskii, G.M., and B.V. Chirikov (1972). *Soviet Phys. Usp.* **14**, 549.
Zaslavskii, G.M., and Kh.-R. Ya. Rachko (1979). *Soviet Phys. JETP* **49**, 1039.
Zaslavskii, G.M., M. Yu. Zakharov, R.Z. Sagdeev, D.A. Usikov, and A.A. Chernikov (1986). *Soviet Phys. JETP* **64**, 294; *JETP Lett.* **44**, 453.
Ziglin, S.L. (1984). *Func. Anal. Appl.* **16**, 181.
Zisook, A.B., and S.J. Shenker (1982). *Phys. Rev.* **A25**, 2824.

Author Index

Aamodt, R.E., 364, 650
Abarbanel, H.D.I., 328, 607
Ablowitz, M.J., 42
Ahlers, G., 635
Alexeev, M.V., 645
Alfvén, H., 4, 648
Angelopoulos, V., 240
Anosov, D.V., 305, 610
Antonsen, T.M., Jr., (654, Ott et al.)
Arnold, V.I., 3, 5, 14, 59, 60, 61, 66, 163, 175, 185, 188, 197, 198, 302, 306, 307, 373, 380, 528
Avez, A., 59, 163, 185, 188, 302, 306, 307

Bachelert, S., (652, Vidal et al.)
Bak, P., 527, 528, 530, 532, 534, 620, 621, 623, 624, 625, 628
Bardet, R., 650
Baxter, D.C., 650
Behringer, R.P., 635
Belykh, V.N., 534
Benettin, G., 245, 298, 299, 301, 305, 311, 315, 316, 317, 319, 438, 445
Ben-Misrachi, A., 538, 545
Bensimon, D., 172, 594
Benson, S.V., (635, 636, Gollub et al.)
Bergé, P., 547

Berk, H.L., 651
Berman, C.P., 449
Berman, G.P., 654
Berman, R.H., 255
Bernstein, G.M., 534, 535, 536, 537, 572, 573
Bernstein, I.B., 650
Berry, M.V., 150, 185, 193, 293, 303, 653, 654
Bers, A., 87, 126, (364, Ram et al.)
Berz, M., 173
Bhattacharee, A., (342, Cary et al.)
Bialek, J., 256, 279, 280
Birkhoff, G.D., 2, 93, 185, 290, 645
Bivins, R.L., 439, 443, 449
Bochelart, S., (638, Roux et al.)
Bogoliubov, N.N., 1, 93, 103, 648
Boghosian, B.M., 45
Boozer, A., 422
Born, M., 2, 72, 81, 134, 653
Bountis, T.C., 42, 154, 158, 159, 160, 201, 273, 513
Bowen, R., 610
Brahic, A., 216
Brambilla, M., 427
Bridges, R., 581
Brillouin, L., 224
Brock, W.A., 598
Broomhead, P.S., 599, 601, 602

Broucke, R., 209, 210
Bruhwiler, D.L., 364, 370, 371, 372
Bruns, H., 645
Bunimovich, L.A., 67, 342
Byers, J.A., 650, (650, Smith *et al.*)

Capel, H.W., (202, Roberts *et al.*), (219, van der Weele *et al.*)
Carlson, J.M., 620, 625, 626, 627, 629
Carreras, B., 436, 437, 438
Cary, J.R., 43, 134, (134, Kaufman *et al.*), 136, 151, 152, 342, 364, 369, 370, 371, 372, 425, 651
Casartelli, M., (311, Benettin *et al.*)
Casati, G., 312, 506, 644, 653, 654
Chaiken, J., 640
Chaitin, G.J., 309
Chandrasekhar, S., 293, 408, 648
Chao, A.W., 210
Chaté, H., 619
Chen, C., 523, 524, 525
Chernikov, A.A., 217
Chhabra, A., 597
Chillingworth, D.R.J., 561
Chirikov, B.V., 4, 29, 57, 62, 99, 109, 114, 151, 163, 169, 182, 183, 216, 219, 234, 236, 242, 244, 247, 253, 254, 256, 258, 264, 298, (312, Casati *et al.*), 313, 315, 334, 364, 369, (373, Gadiyak *et al.*), 374, 385, 392, 393, 395, 397, 398, 399, 403, 404, 405, 406, 412, 413, 418, 437, 438, 443, 444, 570, 644, 646, 647, 648, 649, (653, Casati *et al.*)
Chua, L.O., 458
Churchill, R.C., 645
Cohen, R.H., 94, (216, 226, 254, 328, Lichtenberg *et al.*), 406, 412, 413, 416, 648, 649
Collet, P., 458, 485, 504, 505, 509, 514
Contopoulos, G, 179, 326, 645, 646
Courant, E.D., 32, 224, 646, 647
Crawford, J.D., 328, 462, 466, 468
Cremers, J., 608, 610

Crutchfield, J., (68, Farmer *et al.*), 469, 470, (478, Packard *et al.*), 557, 605, 608, 614, 619, 638
Curry, J.H., 473, 633, 635

Dangelmayr, G., 640
Degn, H., 652
Deprit, A., 2, 134, 136, 645
de Vogeleare, R., 172, 201
Dewar, R.L., 134, 136
Dorizzi, B., 43
Doveil, F., 249, 274, 281, 284, 286, 287, 289
Dragt, A.J., 134, 168, 186, 187, 648, 649
Driscoll, C.F., 439, 448, 449
Dubois, M., 548
Dunnett, D.A., 72, 110, 128, 147

Ecke, R.E., 528, 529
Eckmann, J.-P., 458, 485, 504, 509, 513, 514, 598, 635, (635, Collet *et al.*)
Edmonds, P.H., 650
Einstein, A., 101
Eldridge, O., 650
Eminhizer, C.R., 152, 154, 195, 201
Escande, D.F., 250, 274, 279, 281, 284, 286, 287, 289, (371, Cary *et al.*)

Farmer, J.D., 68, 478, (478, Packard *et al.*), 528, 531, 602, 603, 605, 607, 610, 611, 615, 618
Feigenbaum, M.J., 6, 458, 485, 487, 491, 492, 494, 507, 531
Feit, S.D., 473
Fenstermacher, R., 635
Fermi, E., 2, 57, 191, 216, 439, 443
Field, R.J., 652
Filonenko, N.N., 423, 425
Finn, J.M., 45, 48, 134, 168, 186, 187, 210, 213, 423
Foias, C., 634
Foote, J.H., (94, 648, Cohen *et al.*)
Ford, J., 38, (62, Chirikov *et al.*), 128, 179, 216, 275, 281, 293, 298, 303, (312, Casati *et al.*), 315, (398,

Chirikov et al.), 441, 443, 444, (644, Casati et al.), 645, 654
Forest, E., 173
Fraser, A.M., 599, 600, 601, 602, 604
Frederickson, P., 602
Freis, R.P., 419, 421, 423, 424
Frisch, U., 630
Froehling, H., (68, Farmer), 478
Froeschlé, C., 298, 315, 373, 380, 381, 440
Fukuyama, A., 87, 126, 248, 270, 271

Gadiyak, G.M., 374
Galeev, A.A. 428, 435
Galgani, L., (213, 245, 298, Benettin et al.), 298, 299, (301, 305, 315, 316, 317, 318, 319, Benettin et al.), (326, Contopoulos et al.), (438, 441, 445, Benettin et al.)
Garren, A., 649
Garrido, L.M., 134, 645
Geisel, T., 497, 643
Gell, Y., 429, (429, 433, 434, 435, Nevins et al.)
Giacaglia, G.E.O., 81, 93, 134, 141
Gibson, G., 649
Giglio, G., 637
Giorgilli, A., (301, 305, Benettin et al.), (326, Contopoulos et al.), (438, 445, Benettin et al.)
Glass, L., 532, 598
Glazier, J.A., 594, 595
Goedde, C.G., 491, 585
Goldstein, H., 7, 36, 47
Gollub, J.P., 635, 636, 637
Gorman, M., 635
Gormezano, C., 651
Goward, F.K., 647
Gramaticos, B., 43
Grassberger, P., 478, 576, 577, 578, 589, 590, 591, 592, 603, 604, 606
Grawe, H., 650
Graziani, K.R., (652, Schmits et al.)
Grebogi, C., 549, 550–560, 571, 576, 579, 580, 590, 591, 597
Green, G.K., 647

Greene, J., 4, 169, 182, 200, 202, 248, 254, 257, 271, 273, 275, 276, 278, 311, 341, 514, 517, 518, 519
Greenspan, B., 561
Grossmann, S., 504, 506
Guarneri, I., (653, Casati et al.)
Guckenheimer, J., 468
Guest, G., (650, Samec et al.)
Gustavson, F., 54, 56, 645
Gwinn, E.G., 594
Gyorgyi, G., 523, 524, 525

Haas, F.A., (99, Hastie et al.)
Hall, L., 41, 43
Halsey, T.C., 478, 589, 590, 592, 593, 594
Hammel, S.M., 610, 612
Hamzeh, F.M., (419, 421, 424, Freis et al.)
Hanson, J.D., 43, 425, (476, 477, Russell et al.), (654, Ott et al.)
Hart, M., 429, (652, Hudson et al.)
Harte, J., (429, Gell et al.), (429, 433, 434, 439, Nevins et al.)
Hartman, C.W., (419, 421, 423, Freis et al.)
Hasegawa, A., 240
Hastie, R.J., 99, 648
Hatori, T., 348, 651
Hausdorff, 590
Hauss, B.L., (650, Samec et al.)
Heiles, C., 9, 38, 52, 294, 298, 645
Heinrichs, R., 640, 641
Helleman, R.H.G., 66, (152, 154, Eminhizer et al.), 154, 157, 158, 160, (195, 201, Eminhizer et al.), 201, 273, 293, 458, 485, 490, 504, 509, 513, 515, 516, 633, 654
Hénon, M., 4, 39, 52, 191, 192, 294, 298, 468, 472, 473, 645
Hentschel, H.G.E., 589
Hereward, H.G., 647
Herman, M.R., 289
Herrera, J.C., 648
Hertweck, F., 94, 648
Hicks, H.R., (436, 437, Carreras et al.)
Hietarinta, J., 43
Hine, M.G.N., 647

Hirakawa, K., (653, Yamazaki et al.)
Hiroe, S., (651, Watari et al.)
Hirsch, J.E., 540, 543, 545
Hizanidis, K., 364
Hobbs, G.D., (648, Hastie et al.)
Hohs, S., 219
Holm, D., 48, 213
Holmes, P.J., 48, 380, 458, 468, 561, 565, 567, 569
Holt, C.R., 41
Hori, G., 134, 645
Howard, J.E., 94, 209, 210, 211, 219, 292, 440, 648
Howard, L.N., 652
Howland, R.A., 134, 151
Hu, B., 538, 543
Huberman, B.A., 498, 506, 507, 557, 637
Hübler, A., 608, 610
Hudson, J.L., (652, Schmits et al.), 652

Ichikawa, Y.H., 348, 349
Ichimaru, S., (651, Watari et al.)
Itatani, R., (87, 126, 270, Fukuyama et al.)
Izrailev, F.M., 113, (313, Chirikov et al.), (374, Gadiyak et al.), 374, (392, 403, Chirikov et al.), 443, 444, 570, 582, 647, (653, Casati et al.), (654, Chirikov et al.)

Jaeger, F., 99, 113, 121, 248, 293, 364, 620, 625, 650
Jeffries, C., 548
Jensen, M.H., 528, 529, 531, 533, 594, 596, 597, 621, 625
Johnston, S., 134
Jordan, W., (649, Gibson et al.)

Kadanoff, L.P., 172, 249, 620, 624
Kamimura, T., 348, 349
Kaneko, I., 442, 533, 614, 616, 618, 619
Kantz, H., 576, 577, 578
Kao, Y.H., 548

Kaplan, J., 476, 557
Karimabadi, H., 240
Karney, C.F.F., 87, 90, 91, 92, 126, 128, 129, 342, 343, 650
Kaufman, A.N., 87, 126, 127, 134, 268, 651, 653, 654
Keeler, J.D., 613, 614, 615, 618
Keller, J., 653
Khinchin, A.Ya., 273
Kim, S.Y., 349
King, G.P., 99, 601, 602
Klein, A., 249
Knobloch, E., 640, 641, 643
Kobayashi, T., 545
Kolmogorov, A.N., 3, 71, 149, 175, 304, 309, 629
Kolomensky, A.A., 647
Kolovski, A.R., 449
Konishi, T., 442
Kopell, N., 652
Kramers, H.A., 1
Krilin, L., 648
Krishnaprasad, P.S., 48
Kruskal, M.D., 1, 77, 93, 103, (141, Northrop et al.), 439, 448, (457, Northrop et al.), 648
Krylov, N., 93, 103, 304
Kuckes, A.F., 650
Kulsrud, R.M., 77
Kummer, M., 645
Kuramoto, Y., 653

Lacina, J., 648
Lahiri, A., 541
Laing, E.W., 13, (72, 110, 131, Dunnett et al.), 132, (147, Dunnett et al.)
Lamb, G., 43, 439
Landau, L.D., 7, 322, 634
Landolfi, M., 445, 446, 447
Lanford, O., 66, 67, 461, 466, (509, Collet et al.)
Langer, J.S., 620, 625, 626, 627, 629
Laslett, L.J., 109, 647
Lazar, N., 650
Leach, P.G.L., 32
Lebedev, A.N., 647
Lee, D.K., (436, 437, Carreras et al.)

Lewis, H.R., Jr., 32, 647
Li, C-T., 249, 497
Liapunov, A.M., 298
Libchaber, A., 638
Lichtenberg, A.J., 14, 24, 58, (62, Tennyson et al.), 77, (99, Jaeger et al.), 99, 113, 121, 194, 195, 196, 216, 221, 222, 224, 226, 237, 238, 248, (248, Jaeger, et al.), 254, (210, 211, 219, 292, Howard et al.), 289, 290, 325, 328, 345, 348, 351, 352, 358, (374, 379, 382, 383, 384, 386, 390, 392, 394, Tennyson et al.), (419, 421, 423, Freis et al.), 427, (429, Gell et al.), (440, Howard et al.), (451, Goedde et al.), (647, 650, Jaeger et al.), 647, (650, Wyeth et al.), (650, Howard et al.), 650, 651
Lieberman, M.A., 58, 61, (62, Tennyson et al.), (99, Jaeger et al.), 99, 216, (216, Lichtenberg et al.), 221, 222, 224, (226, Lichtenberg et al.), (248, Jaeger et al.), 248, 289, 290, (210, 211, 219, 228, 292, Howard et al.), (324, Lichtenberg et al.), 325, 328, (344, Lichtenberg et al.), 351, 352, 368, 385, 386, (374, 379, 382, 383, 384, 386, 390, 392, Tennyson et al.), 393, 397, (404, 405, Chirikov et al.), (440, Howard et al.), (451, Goedde et al.), 534, 535, 536, 537, 650, (649, Jaeger et al.), (650, Wyeth et al.), (650, Howard et al.), 574, 575, 584, 585, 586, 587, 588
Lindstedt, M., 71, 74
Littlejohn, R., 45, 103, 134
Littlewood, J.E., 144
Liu, C.S., (141, 648, Northrop et al.)
Livi, R., 445
Livingston, M.S., (646, Courant et al.)
Lo Destro, L.L., (650, Smith et al.)
Longcope, D.W., 239
Lorenz, E.N., 6, 65, 66, 68, 69, 504, 631, 633
Lo Vecchio, G., 441
Lozi, R., 474
Lundsford, G.H., 128, 179, 275, 281

MacKay, R.S., 172, 209, 210, 249, 250, 289, 340, 341, (342, Greene et al.), 516, 532
Mackey, M.C., 598
Mamota, H., (87, 126, 270, Fukuyama et al.)
Mandelbrot, B., 478, 588
Manneville, P., 538, 539, 540, 619, 638
Marcus, R.A., 654
Marinko, D., (652, Hudson et al.)
Marsden, J., 380
Marsden, J.E., 45, 48, 213
Martin, S., 534
Maschke, E.K., 634
Maselko, J., 534
Mather, J.N., 172, 250, 289, 290
Matthaeus, W.H., 638
Maurer, J., 638
May, R.M., 458, 497
McDonald, S.W., 653, 654
McLaughlin, J.B., 561
McNamara, B., 53, 54, 55, 93, 128, 134, 135, 141, 144, 147, 148, 149, 151, 645
McNamara, B.S., 605, 608
Meiss, J.D., 172, 195, (339, Mackay et al.), 341, (342, Cary et al.), 343
Melnikov, V.K., 56, 186, 380
Metropolis, N., (443, Bivins et al.), 497
Miller, B.N., 441
Misiurewicz, M., 474
Mitropolsky, N., 1, 93, 103, 648
Montgomery, D., 638
Month, M., (648, Herrera et al.), 648
Monticello, D.A., (436, White et al.)
Montroll, E.W., (152, 154, 199, 201, Eminhizer et al.)
Morosov, A.D., 561
Morozov, A.I., 419
Moser, J., 3, 150, 163, 175, 181, 182, 183, 185, 222, 645, 646
Moses, E., 640
Motz, H., 651
Murakami, S., 240
Murray, N., 336, 338, 339, 342, (344, 348, 349, Lichtenberg et al.), 372
Musazzi, S., (637, Giglio et al.)

Nag, T., 541
Nagashima, T., 315
Nauenberg, M., 495
Nayfeh, A., 134
Nekhoroshev, N.N., 62, 398, 438
Nekrasov, A.K., 650
Nevins, W.M., (429, Gell et al.), 433, 434, 435
Newhouse, S., 553, 635
Nierwetberg, J., 497
Northrop, T.G., 141, 648
Noyes, R.M., 652

Olsen, L.F., (652, Degn et al.)
O'Neil, M.N., 439, 448, 449
Oono, Y., (653, Yamazaki et al.)
Orzag, S.A., 630
Oseledec, V.I., 298, 300, 301, 498
Oster, G.F., 497
Ott, E., 341, 342, 458, 474, 547, 550–560, 571, 576, 579, 580, 582, 590, 591, 597, 654

Packard, N., (68, Farmer et al.), 478, 598
Papayannopoulos, T., 646
Parker, T.S., 458
Pasta, J.R., 2, 443, (443, Bivins et al.)
Pecelli, G., (645, Churchill et al.)
Peierls, R.F., (648, Herrera et al.)
Penrose, O., 293
Percival, I.C., 172, 250, 339, 341, 654
Pereira, N., (134, Kaufman et al.), 268, 269
Perez, R., 532, 548
Perran, J.W., (652, Degn et al.)
Perrini, U., (637, Giglio et al.)
Pesin, Ya.B., 298, 304
Pettini, M., 445, 446, 447
Poincaré, H., 1, 14, 71, 75, 79, 93, 185, 645
Pomeau, Y., 538, 539, 540, 548, 638
Pomphrey, N., 250
Ponomarenko, V.G., 649
Post, H.W., (202, Roberts et al.), (219, van der Weele et al.)
Procaccia, L., 478, 533, 538, 545, 546, 547, 589, 591, 592, 603, 604, 606

Prodi, G., 634
Pustylnikov, L.D., 219

Quispel, G.R.W., 202

Rabinovich, M.I., (373, Izrailev et al.), 458
Ram, A.K., 364
Ramani, A., (42, Ablowitz et al.)
Rand, D., 531
Rannou, P., 311
Ratin, T., 45, 48, 213
Rechester, A.B., 27, 328, 334, 335, (346, Karney et al.), 353, 355, 423
Renyi, A., 589
Roberts, J.A.G., 202
Rod, D.L., (641, Churchhill et al.)
Rognlien, T.D., 650
Rosenbluth, M.N., 247, (328, 353, 354, 355, Rechester et al.), 419, 423, 425, (436, White et al.), 649, 650
Rossi, A., (634, Roux et al.), (652, Vidal et al.)
Rössler, O.E., 468, 652
Roux, J.C., 600, 634, 638, (652, Vidal et al.)
Rowlands, G., (94, Cohen et al.), 406, 412, 413, 416, 581, (648, Cohen et al.), 649
Rubinow, S., 653
Rudnick, J., 495, 498, 538, 543
Ruelle, D., 6, 458, 598, 635, (635, Newhouse et al.), 652
Ruffo, S., (445, Livi et al.)
Russell, D.A., 476, 477
Ruytov, D.D., 649

Sagdeev, R.Z., (237, Zarlavskii et al.), (423, Filonenko et al.), 429, 435, (247, 428, 435, Rosenbluth et al.)
Saltzman, B., 633
Samec, T.K., 650
Saramito, B., 634
Satija, I.I., 531
Sato, T., 240, 373, 381, (651, Watari et al.)

Scheidecker, J.-P., 298, 315, 440
Schlüter, A., 94, 648
Schmidt, G., 256, 259, 279, 280, 425, 519, 520, 522, (523, Chen et al.)
Schmits, R.A., 652
Schuster, H.G., 458, 491, 538, 541, 542, 543, 545, 546, 547
Segur, H., (42, Ablowitz et al.), 42, (42, Bountis et al.), 43
Seidl, M., 109, 650
Sessler, A., 219, 647
Shafranov, V.D., 420, 427
Shaw, R., 64, (68, Farmer et al.), 458, 468, 469, (478, Packard et al.), 497, 499, 500, 501, 502, 582
Shenker, S.J., 249, 530
Shepelyanski, D.L., 342, (392, 397, 403, 654, Chirikov et al.), 654
Shimada, I., 315
Shraiman, B.I., 531
Siambis, J., 649
Sidorowich, J.J., 605, 607, 610, 611
Siegel, C.L., 181, 645, 646
Simo, C., 213
Simó, C., 473, 474
Sinai, Ya.G., 60, 67, 304, 342
Sitnikov, K., 645
Slater, J.C., 647
Smale, S., 185, 560, 645
Smith, G.R., 27, 87, 126, 127, 265, 269, 650
Snyder, H.S., 32, 224, (646, Courant et al.), 647
Solomonoff, R.J., 309
Solov'ev, L.S., 419, 420, 427
Sparpaglioni, M., (445, Livi et al.)
Sprott, J.C., 650
Stark, J., (341, 342, Greene et al.)
Stavans, J., 636
Stein, M.L., (497, Metropolis et al.)
Stein, P.R., (497, Metropolis et al.)
Steinberg, V., 640
Steinman, J., (635, 637, Gollub et al.)
Stern, D., 93, 141, 144
Stix, T.H., 27, 423
Stoddard, S.D., (38, Ford et al.), 441
Strelcyn, J.M., (245, 298, 301, 305, 311, 315, 316, 317, 319, 320, Benettin et al.)

Stupakov, G.Z., 649
Sudan, R.N., 239
Sun, B.Z., 45, 48, 210, 213
Swift, J., 507, 508
Swinney, H.L., 449, 534, 599, 601, 635
Symon, K.R., 32, 219, 647
Szebehely, V., 646
Szepfalusy, P. 576

Tabor, M., 43, 250
Takens, F., 6, 182, 498, (499, Newhouse et al.), 598, 599, 635
Tang, C., 628
Taylor, J.B., (72, Dunnett et al.), (99, Hastie et al.), (110, 128, Dunnett et al.), 131, 132, (147, Dunnett et al.), 220, (247, 419, 423, Rosenbluth et al.), (648, Hastie et al.)
Tayursky, V.A., (313, Chirikov et al.)
Tel, T., 576
Teller, E., 648
Tennyson, J.L., 62, 179, 364, 365, 368, 374, 376, 377, 378, 382, 383, 384, 385, 386, 390, 392, 394, 406, 407, 409, 410, 411, 412, 647
Thomae, S., 504, 506
Timofeev, A.V., 364, 650
Toda, M., 36
Tomita, K., 652
Trajnin, L.Ya., (649, Ponomarenko et al.)
Tresser, C., 532
Treve, Y., 66, 446, 458, 474, 634
Tsang, K., 574, 575, 584, 585, 586, 587, 588, 590, 592
Tsuda, I., 652
Turner, J.S., (38, Ford et al.)
Tyson, J.J., 652

Ueda, Y., 565
Ugodnikov, A.D., (374, Izrailev et al.)
Uhlenbeck, G.E., 293, 321
Ulam, S., 2, 216, 443, 504
Umberger, D.K., 528
Usikov, D.A., (237, Zaslavskii et al.)

Valkering, T.P., (219, van der Weele et al.)
Valz-Gris, F., (653, Casati et al.)
van der Weele, J.P., 219
Van Zeyts, J.B.J., 490, 515
Vidal, C., (638, Roux et al.), 652
Vivaldi, F., (42, Bountis et al.), (62, 398, 404, Chirikov et al.)
Von Neumann, J., 504, 645
Von Zeipel, H., 1, 75, 79, 454
Vulpiani, A., (445, Livi et al.)

Waddell, B.V., (436, White et al.)
Walker, G.H., 247, 645
Walker, J., 652
Wang, M.C., 283, 321
Watari, T., 651
Watenabe, T., 651
Waters, J., 444
Watson, C.J.H., 651
Wegmann, K., 652
Weinstein, A., (48, 213, Holm et al.)
Weiss, J.B., 640, 641, 643
Wentzel, G., 1
Westervelt, R.M., 594
White, R.B., 328, (328, Rechester et al.), 334, 335, (346, Karney et al.), 353, (304, 354, Rechester et al.), 354, 436
Whiteman, K.J., 53, 54, 55, 93, 141, 144, 645
Whittaker, E.T., 7, 14, 16, 24, 39, 141, 645

Widom, M., 576
Wiedemann, H., 648
Winfree, A., 652
Wintner, A., 646
Wolf, A., 507, 508, 602, 603
Wong, B.W., 519, 520, 522
Wood, B.P., 237, 238, 242, 248, 358, 361, 363, 442
Wright, H.L., 441
Wyeth, N.C., 650

Yamada, T., 651
Yamazaki, H., 651
Yasnetsky, A.N., (649, Ponomarenko et al.)
Yeh, W.L., 548
Yorke, J., 476, 497, 549, 550–560, 571, 576, 579, 580, 590, 591, 597
Yoshida, H., 43
Yurchenko, V.I., (649, Ponomarenko et al.)

Zabusky, N.J., 439, 448
Zaiken, A.N., 652
Zakharov, M.Yu., (237, Zaslavskii et al.)
Zaslavskii, G.M., 216, 237, (247, Rosenbluth et al.), 298, (419, 423, Rosenbluth et al.), (422, Filonenko et al.), 519, 653, 654
Zhabotinsky, A.M., 652
Zisook, A.B., 506, 507
Zubarev, D.N., 93, 103

Subject Index

Accelerator modes
  effect on diffusion, 344, 345–349
  for standard mapping, 253
Action, 171–172
  change, for driven pendulum, 233–234
  change in resonance crossing, 369–372
Action-angle variables, 21ff
  for central force, 35
  for coupling resonance, 395–396
  for harmonic oscillator, 23
  for pendulum, 27, 82
  for two resonances, 282–283
Action integral, relation to integral invariant, 16–17
Action space, 375–380
  for billiards problem, 381–383
  for resonance streaming, 410, 417
Adiabatic barrier, *see* Barrier transition
Adiabatic invariant, 16, 24, 71, 93ff
  by canonical methods, 93ff
  of harmonic oscillator, *see* Harmonic oscillator
  from Lie transformations, 141ff
  by noncanonical methods, 102ff
Anosov-systems, *see* C-systems
Area-preserving transformation
  condition for perturbed twist map, 203
  definition, 19

Arnold diffusion, 60–62, 238, 373, 401, 646, 647, 649
  calculation of, 386ff
  for coupling resonance, 393ff
  examples of, 380ff
  for many resonances, 397ff
Arnold tongue, 528
Arnold web, 61, 379–383
Asteroid belt, 646
Asymptotic expansions, *see* Expansions, asymptotic
Attractors
  basin boundaries, 51ff, 558ff
  basins of, 461
  crises, *see* Crises
  definition of, 461ff
  semiattractor, 576–578
  simple, *see* Fixed point, Sink, Limit cycle
  strange, *see* Strange attractors
Autocorrelation, 342
  for reconstruction, 599
Autonomous Hamiltonian, *see* Hamiltonian, autonomous
Avalanches, *see* Self-organized criticality
Averaging
  applied to harmonic oscillator, 106–109

Averaging *(cont.)*
  Kruskal's method, 103–106
  method of, 71
  multiple time scale, 103
  over slower of fast phases, 111

Baker's transformation, 303, 607
Barrier transition, 246
  for electron cyclotron heating, 650
  for Fermi acceleration, 225
  for separatrix mapping, 237
  for standard mapping, 256
  summary of criteria for, 291ff
  for two resonances, 287–288
Basin of attraction, *see* Attractors
Beam–beam interaction, 647
Belousov–Zabotinsky reaction, 652–653
Bernoulli shifts, 307–309
Bifurcations, 461ff
  exchange of stability, 463–464, 484
  of Fermi mapping, 226–229
  Hopf, 463–464, 465, 541, 634–635
  for one-dimensional quadratic map, 482–485
  period-doubling, 209, 463; *see also* Period-doubling bifurcations
  pitchfork, 463–464, 484ff, 511, 540
  reverse, *see* Reverse bifurcations
  of standard mapping, 254–256
  tangent (saddle-node), 209, 462, 464, 496, 538
  for two-dimensional dissipative quadratic map, 510–512
  for two-dimensional Hamiltonian map, 512ff
  transcritical, 463
Bifurcation theory
  for Hamiltonian maps, 508ff
Billiards problem, 381ff
  coupling resonance, 385–386, 393–397
  thick layers, 382, 387–389
  thin layers, 382, 389–393

Canonical perturbation, *see* Perturbation theory, canonical

Canonical transformation
  general theory, 7ff
  to rotating coordinates, 111
Canonical variables, 8
Cantor set, 63, 374–375, 588
  two-scale, 588–589, 593–594
Cantorus, 195, 328
Capacity, *see* Fractal dimension
Casimir, 45ff, 213–215
Caustics, 653
Central force, 33–36
Characteristic exponents, *see* Liapunov exponents
Characteristic function, 343
Circle map, 525ff
Coarse-graining, 89, 647, 653
Complexity, 309
Conditionally periodic motion, 22
  for central force, 33
Conjugate variables, 8
Constant of the motion, 11
Continued fractions, 273–274, 286
Contraction of phase space volume, 459–466
Converse KAM theorem, 289–291
Coordinates
  delay, 598ff
  generalized, 8
  noncanonical, 44ff
  rotating, 109ff
Correlation dimension, *see* Fractal dimension
Correlation function, for quadratic map, 506–507
Correlations, 342ff, 506
Crises, 523, 548ff
  boundary crisis, 551ff
  exterior crisis, 550
  hysteresis, 555
  interior crisis, 550, 553
  in one-dimensional maps, 549–551
  transient chaos near, 579ff
C-systems, 305–307

Darboux's theorem, 45
Decomposable system, 295, 305

Subject Index   681

Degeneracy
  accidental, 113–114, 119, 122–123, 178
  intrinsic, 113–117, 178–179
  intrinsic, for Kepler problem, 36
  transition from accidental to intrinsic, 124–125
Delay coordinates, 578ff
Devil's staircase, 526
Differential delay equation, Mackey-Glass, 598, 603–606
Differential equations
  for dissipative flows, 458
  Floquet form for, 31
  linear, 29ff
  with periodic coefficients, 30–32
Diffusion, *see also* Fokker–Planck equation
  in action space, 320ff
  Arnold, *see* Arnold diffusion
  banana, 412, 427–428, 434
  in billiards problem, *see* Billiards problem
  coupling resonance, 393ff
  extrinsic, *see* Extrinsic diffusion
  Fourier path technique, 329ff
  higher order, 329ff
  many resonance, 397ff
  modulational, *see* Modulational diffusion
  neoclassical, 428
  oscillation center, *see* Resonance streaming
  Pfirsch–Schlüter, 428
  plateau, 427–428, 433–435
  pseudo-classical, 433
  quasilinear, *see* Quasilinear diffusion
  across resonance layers, 386–387
  along resonance layers, 373, 386, 399ff; *see also* Stochastic pump model
  self-consistent problem, 436ff
Diffusion coefficient, 322
  relation to friction coefficient, 322
Diffusion for many degrees of freedom, 437ff
  for attracting sheets, 440–441

Fermi–Pasta–Ulam system, 440
  for Lennard–Jones potential, 441–442
Diffusion in toroidal magnetic fields, 412, 418ff
  magnetic islands, 422–425
  magnetic surfaces, 420–422
  mapping for, 428ff
  numerical results, 433–435
  static fields, 425ff
  tearing and disruptions in tokamaks, 436–438
  time varying fields, 428ff
Diffusion of a parameter, 413ff
  mapping for, 413–415
  for oscillation center, 414–416
  relation to resonance streaming, 416–418
  in toroidal magnetic field, 428ff
Diffusion tensor, 408–411
Digital phase-locked loop
  circle map representation, 534–537
  transient chaos in, 572
Dimension, *see* Fractal dimension
Dipole, axisymmetric magnetic, 648
Dissipative systems, 62ff, 570ff
  resonance streaming in, 412
Duffing's equation, 565–567
  chaotic motion in, 567

Earthquake model, 625–628
Eigenvalues, 196ff
  of cat mapping, 306–307
  for Liapunov exponents, 300
  symmetry of, 197–199
  for two-dimensional mappings, 200ff
Eigenvectors, 196ff
  of cat mapping, 306–307
  for Liapunov exponents, 300
  for two-dimensional mappings, 200ff
Electron cyclotron resonance heating, 650–651
Elliptic orbits, for two-dimensional mapping, 205–206
Elliptic points, *see* Fixed point, elliptic
Energy–Casimir method, 213–215
Energy surface, 61, 376–380, 406–407

Entropy, 308–309, 446, 609; *see also* KS entropy
Equipartition of energy, 445–447, 452–453
  relaxation time, 447
Ergodic systems, 59–60, 295ff, 308
  baker's transformation, 59–60
  hard sphere gas, 59–60
  Lorenz attractor, 65ff
Errors
  numerical, *see* Numerical errors
  roundoff, *see* Roundoff errors
Euler equations, 214–215
Expansions, *see also* Perturbation theory
  asymptotic, 75, 93–95
  eliminate secularity, 74–75
  power series, 73ff
  superconvergent, 149ff
Extrinsic diffusion, 349ff, 360ff, 373, 406ff; *see also* Resonance streaming
  in Fermi acceleration, 351–352
  using Fourier paths, 353–354
  limit to global stochasticity determination, 351
  in the presence of resonances, 353ff
  from a simple calculation, 354–357

$f(\alpha)$, *see* Scaling index spectrum
Fermi acceleration, 57–59, 216ff
  absolute barrier, 225
  bifurcation phenomena, 226ff
  comparison with overlap criterion, 264
  conversion to standard mapping, 250–251
  dissipative mapping, 574
  exact mapping, 216–217
  fixed points, 223
  Fokker–Planck solutions for, 326ff
  Hamiltonian formulation, 229ff
  invariant distribution for dissipative, 583–588
  linear stability, 223–225
  models for, 216ff
  numerical results, 220ff
  sawtooth wall velocity, 221–222
  simplified mapping, 217–218
  stochastic transition velocity, 224–225
  transient chaos in, 574–576
  transport coefficients for, 324–326
  two-frequency, 292, 440
Fermi–Pasta–Ulam system, 443ff
Fibonacci numbers, 530
Filamentation, *see* Coarse-graining
Fixed points, 49
  bifurcations for Fermi mapping, 226–229
  elliptic, 42, 183–185, 205–206
  expanding about, 113, 119
  for Fermi mapping, 223–225
  hyperbolic, 184, 185–187, 206–207, 560ff
  linearizing about elliptic, 114
  for quadratic DeVogelaere map, 518
  for quadratic map, 482ff
  reflection and ordinary hyperbolic, 206–207
  for separatrix mapping, 236–237
  for standard mapping, 253–256
  for two-dimensional mappings, 200–201
  of twist mapping, 166
Fluids, chaotic behavior in, 628ff, 638ff
Fokker–Planck equation, 32ff
  canonical variables, 325–326
  diffusion and friction coefficients, 322
  for invariant distributions on strange attractors, 583–585
  steady-state solutions, 326–327, 337
  for transient chaos, 572–576
  transient solutions, 326–327
  transport coefficients, 324ff
  validity of, 323–324
Forecasting, *see* Reconstruction
Fourier mode expansions for fluid systems, 630ff
Fourier path technique, 329ff, 353–357
Fourier spectra, *see also* Power spectra of trajectories
Fractal, fat, 528
Fractal diagrams, 275, 279–281

Subject Index 683

Fractal dimension, 474–478
  capacity, 577, 588
  correlation dimension, 589, 591
  embedding, 591, 598ff
  generalized, 588ff
  Hausdorff, 551, 590
  information dimension, 576–578
  multifractals, 588ff
  for period-doubling, 595
  for quasiperiodicity, 596
  reconstruction from time series, 602ff
  relation to Liapunov exponents, 474–478, 591, 602
  scaling index spectrum, 588, 592ff
  for two-scale Cantor set, 593–594
Fractal generator, 591
Friction coefficient, 322
  relation to diffusion coefficient, 322

Galerkin approximation, *see* Fourier mode expansions
Generating function, 173
  Fourier series for, 80
  harmonic oscillator, 23
  Lie, 12, 135
  mixed variable, 8–9, 21
  near-identity, 97
  relation to phase space area, 171
  rotating coordinates, 110
  use in perturbation theory, 79
Golden mean, 249, 274, 277, 288, 291
Greene's method, 248–249, 271ff, 291
  continued fraction approximates, 273–276
  golden mean, 274, 277
  mean residue, 271–273, 276–277
  numerical procedure, 276
  numerical results, 276–278
Guiding center
  Hamiltonian, 102
  variables, 87–88

Hamiltonian
  accidentally degenerate, *see* Degeneracy, accidental

  action-angle form, 78
  autonomous, 10, 11, 14–15, 288
  average part, 80, 98
  canonical transformation of, 9, 79
  conversion to a mapping, 170–171
  definition of, 8
  formulation for Fermi mapping, 229ff
  formulation for standard mapping, 256–258
  Fourier series for, 80, 111–112
  for free particle, 375–378
  for harmonic oscillator, 23, 116
  higher order expansion, 81
  intrinsically degenerate, *see*, Degeneracy, intrinsic
  nonautonomous, 14–15
  oscillating part, 80, 98
  standard, 29, 113
  superconvergent transformation of, 152–154
  for two resonances, 281–282
Hamilton–Jacobi equation, 10, 21
  for central force, 35
Hamilton's characteristic function, 10
Hamilton's equations, 8
Hamilton's principal function, 10
Hard sphere gas, 60, 304, 442, 654
Harmonic oscillator, 23–24
  adiabatic invariant of, 100
  effect of resonance, 77
  Hamiltonian, 23, 116
  with slowly varying frequency, 75–77, 100–102, 106–109
Hausdorff dimension, 551, 590; *see also* Fractal dimension, Hausdorff
Heating, stochastic, 650–651; *see also* Fermi acceleration
  at electron cyclotron resonance, 650–651
  at ion cyclotron resonance, 650–651
  with two frequencies, 650–651
Hénon attractor, 472–474
  fractal dimension of, 477
  invariant distribution for, 581–582
  map in quadratic form, 510

Hénon attractor (*cont.*)
  noise reduction for, 611–613
  reconstruction of, 609
Hénon–Heiles system, 52–57, 645
  high-order fixed-point destabilization in, 281
  integrability conditions for, 42–43
  KS entropy for, 316–318
  periodic orbits in, 158–161
  quantized, 654
  relation to Toda lattice, 38
Heteroclinic points, 186–187
Homoclinic intersection, 560ff
  in Duffings equation, 565–567
  in mappings, 568–569
Homoclinic points, 186–187, 309, 604
Hyperbolic orbits, for two-dimensional mappings, 206–207
Hyperbolic point, *see* Fixed point, hyperbolic

Information dimension, 576–578
Integrable systems, 24ff
  central force, 33–36
  finding, 39–43
  harmonic oscillator, 23–24
  linear, 29–32
  Painlevé property, 43
  pendulum, 25–29
  soliton solutions, 43, 376
  Toda lattice, 36–39
  Whittaker method, 39
Integral
  isolating, 24, 33, 381, 648
  for Toda lattice, 39
Integration, symplectic, 172
Intermittency, 537ff
  in lattice maps, 617–619
Intrinsic degeneracy, *see* Degeneracy, intrinsic
Intrinsic diffusion, *see* Diffusion
Invariant
  adiabatic, 17, 24, 71, 93ff
  circle, 290
  curve, *see* KAM curve
  exponential variation of, 94
  global, 24
  hierarchy of, 99–100
  integral, 13–14
  relative integral, 15
  in rotating coordinates, 111
  for secondary resonances, 120
  for wave-particle interaction, 90
Invariant distribution, 498ff, 580ff
  calculation of, 580ff
  for dissipative Fermi map, 583–588
  from the Fokker–Planck, equation 583ff
  for a Hamiltonian system, 580
  for Hénon attractor, 581–582
  for logistic map, 500–501
  phase averaged, 583–585
  by reduction to a one-dimensional map, 581ff
  for tent map, 502
  for two-dimensional maps, 580ff
Invariant measure, *see* Invariant distribution
Invertible maps, 459
  chaotic behavior in, 460, 509
  one-dimensional, 478, 497
Involution, integrals in, 24
Involutions, product of, 169, 201–202
  calculation of Fermi mapping bifurcations, 227–229
Ion cyclotron heating, 650–651
Islands, *see also* Resonance
  amplitude of, 114, 121
  magnetic, 418ff
  primary, 51
  secondary, 51
  second-order, *see* Second-order islands
Islets of stability, 256; *see also* Accelerator modes

Jacobian matrix, 20, 203, 271, 459, 461, 509–510, 562–563
Jacobi's identity, 11

KAM surface, 46, 61, 378; *see also* KAM theory
  curve, 49

for Lagrangian turbulence in fluids, 643
KAM theory, 49, 174ff
  conditions for, 174
  converse, 289ff
  moderate nonlinearity for, 182–183
  smoothness condition, 180–182, 221–222
  sufficient irrationality for, 182–183
  sufficient nonlinearity for, 177–180
  tori, variational principal for, 249
Kaplan and Yorke map, 477
KdV equation, 443, 448
Kepler problem, 34
Kicked rotor, 219
Kolmogorov entropy, see KS entropy
Krein collision, 209
Krein's theorem, 209
KS entropy, 308
  determination of, 312ff
  for Hénon–Heiles system, 316–318
  relation to Liapunov exponents, 304–305
K-systems, 304ff, 654

Lagrangian, 7
  averaged, for closed orbits, 155–157
Landau model, see Transition to turbulence
Lattice maps, 613; see also Spatio-temporal chaos
Legendre transformation, 9, 593
Lennard–Jones system, 441–442
Liapunov exponent, 298ff
  determination of, 312ff
  for dissipative maps and flows, 460–461
  for high-dimensional systems, 447, 456
  higher order, 301–302
  for integrable systems, 312–313
  for logistic map, 504
  numerical calculation of, 315ff
  for one-dimensional map, 497, 499
  for quadratic map, 499
  relation between maps and flows, 302
  relation to fractal dimension, 376–377
  relation to KS entropy, 304–305
  symmetry of, 301
  for tent map, 502–504
  test for stochasticity, 315–317
Libration, 22
Lie transformations, 133ff
  adiabatic invariants, 141ff
  applied to pendulum, 138–140
  applied to pondermotive force, 144–147
  applied to slowly varying harmonic oscillator, 142–144
  comparison with mixed variables, 133–134
  elimination of secularity, 141–142
  evolution operator, 135
  general theory, 135ff
  generating function, 135
  Lie operator, 135
  perturbation series, 136ff
  resonant denominators, 138
  for second harmonic of standard mapping, 259–262
  second order invariants for wave-particle resonance, 147–148
  superconvergent series, 152ff
Limit cycle, 63, 462–466, 461
Linear stability, 196ff
  for Fermi mapping, 223–225
  in higher dimensions, 208–211
  for quadratic map, 483–485
  for separatrix mapping, 236–237
  for standard mapping, 253–256
  for two dimensional mapping, 202ff
Linearized motion, 195ff; see also Differential equations for a mapping
Liouville's theorem, 14, 459
Lipshitz cone, 290
Logistic map, 479ff, 614
Lorenz system, 65ff
  derivation of, 631–633
  extension to many modes, 633–634
  intermittent behavior in, 638

Mackey–Glass differential delay equation, *see* Differential delay equation
Magnetic confinement systems, 418ff
  axisymmetric mirrors, 648–649
  diffusion in, *see* Diffusion in toroidal magnetic fields
  levitrons, 361, 420–424
  nonsymmetric mirrors, 99–100, 649, 651
  tandem mirrors, 458
  tokamaks, 319–320, 423, 425, 430–438, 651
  toroidal, *see* Toroidal magnetic fields
  transport in, 418ff
Magnetic moment, 88, 99–100, 648–649
Mappings, 162ff; *see also* Quadratic mappings
  area-preserving, 52, 167–169
  canonical, 164ff
  conversion to, 170
  conversion to Hamiltonian form, 171
  cat, 305–306
  Chirikov–Taylor, *see* Standard mapping
  description of nonlinear, 188ff
  for diffusion in magnetic fields with time variations, 428–431
  for diffusion of a parameter, 413–415
  dissipative generalized standard, 510
  equivalence to Hamiltonian system, 162ff
  Fermi, *see* Fermi acceleration
  generalized standard, 168
  how obtained, 162
  integer, 311
  invertible, *see* Invertible maps
  involution products for, *see* Involutions
  Kaplan and Yorke, 376–377
  kicked rotor, 216, 219
  lattice; *see* Spatio-temporal chaos
  linearized, 195ff
  logistic, *see* Logistic map
  noninvertible, *see* Noninvertible maps
  one-dimensional, *see* One-dimensional noninvertible maps
  one-dimensional, for Fermi acceleration, 229
  one-dimensional, for Lorenz system, 69
  from perturbed linear oscillators, 237ff
  perturbed twist, 167, 200
  radial twist, 168
  relation between maps and flows, 302
  return, 458
  reversible, 202
  separatrix, *see* Separatrix mapping
  standard, *see* Standard mapping
  symplectic, 208
  tent, *see* Tent map
  twist, *see* Twist mapping
  two-dimensional dissipative, *see* Two-dimensional dissipative maps
  two-dimensional Hamiltonian, 200ff
  whisker, *see* Separatrix mapping
  Zaslavskii, 477
Markov process, 321
Maslov index, 653
Matrix
  eigenvalues and eigenvectors for, 196ff
  symplectic, 197–200
Mean residue, 271–273, 276–278
Mean, space and time, 295; *see also* Invariant
Melnikov–Arnold integral, 233–235, 391
Melnikov's method, 560ff
  application to Duffing's equation, 565–567
  application to mappings, 568–569
Method of averaging, *see* Averaging, method of
Mirror symmetry for quadratic map, 457, 481–482, 490, 504
Mixing, 302–304, 308
  for baker's transformation, 60
  for hard sphere gas, 60, 304
  for Lagrangian turbulence in fluids, 643

Subject Index

for logistic map, 504
for Lorenz attractor, 66
Mode-locking, 466, 526-528, 534
Modulation, of phase or frequency, 365-369
Modulational diffusion, 373, 399ff, 647
  multiplet layer formation, 364-369, 400-401
  thin layer, thick layer, and trapping regimes, 365-369
Momenta, generalized, 8
Multifractals, 588ff; *see also* Fractal dimension
Multiplet, *see* Modulational diffusion
Mutual information, 599

Navier-Stokes equation, 65, 629, 634
Near-integrable systems, 10, 48ff
  as area-preserving mappings, 167ff
  dissipative perturbation of, 571ff
  Fermi acceleration, 57-58
  Hénon and Heiles, 52-56
Nekhoroshev bound, 398
Newton's method
  relocated, 157
  for solving mapping equations, 168
  superconvergence of, 150-151
Noise, *see* Extrinsic diffusion, Numerical errors, Power spectrum
Noisy data, 595
Noise reduction, 610-613
Nonautonomous Hamiltonian, *see* Hamiltonian, nonautonomous
Noninvertible maps, 65, 532-533, 535-537; *see also* One-dimensional noninvertible maps
  chaotic behavior in, 497ff
Nonlinearity parameter, definition, 114
Numerical errors, 310-312
  in finding KAM curve, 275

One-dimensional noninvertible maps, 478ff
  logistic, *see* Logistic map
  quadratic, *see* One-dimensional quadratic map
  for Rössler attractor, 469

One-dimensional quadratic map, 479ff
  bifurcation phenomena, 482-485
  fixed points and stability, 481-485
  intermittent behavior in, 538ff, 638
  invariant distributions for, 498-502
  Liapunov exponent for, 497-499
  mirror symmetry in, 481-482
  periodic cycles for, 495-497
  power spectrum for, 493-495, 504-508
  renormalization procedures, 485-493
  reverse bifurcations of chaotic bands, 504
  trajectories for, 484-485
  transient chaos in, 576, 579
Orthogonal metrics, 418
Oscillation center diffusion, *see* Resonance streaming
Oscillator, harmonic, *see* Harmonic oscillator
  multidimensional, 374ff, 437ff
  nonlinear, 73
Overlap criterion, 247, 258ff, 291
  comparison with Fermi acceleration, 264
  for first and second harmonies, 259-262
  for many degrees of freedom, 440
  for multiplet, 364-369, 400
  including separatrix layer thickness, 262-264
  simple, 259

Particle-wave interaction, *see* Wave-particle interaction
Pendulum
  driven one-dimensional, 231ff
  driven, second-order islands for, 265ff, 268ff
  exact solution for, 27-29
  mean half-period for separatrix motion, 392-393
  phase or frequency modulated, 364-368
  second order solution by Lie transformations, 138-140
  solution by perturbation theory, 81-83

Period-doubling bifurcations, 65; *see also* Bifurcations, Reverse bifurcations
  in Hénon attractor, 472–474
  model for transition to turbulence, *see* Transition to turbulence
  in one-dimensional quadratic map, 482ff
  in Rössler attractor, 468–471
  in two-dimensional dissipative maps, 508ff
Periodic delta function, 171
  use in billiards problem, 387
  use in Fermi acceleration, 229
  use in standard mapping, 256
Periodic orbits, *see also* Fixed points
  in one-dimensional quadratic map, 482ff
  series solutions for, 154ff, 195
Perturbation theory
  adiabatic, 93ff
  applied to periodic orbits of the Hénon and Heiles system, 158–161
  canonical, 70ff
  canonical adiabatic, 96ff
  contrasting slow and small, 95–96
  with explicit time dependence, 85–87
  for Fermi acceleration, 229–230
  higher order, 72; *see also* Lie transformations
  using Lie transformations, *see* Lie transformations
  Newton's method, 150–151
  one degree of freedom, 78ff
  pendulum, 81–83
  power series expansion, 72–74
  for periodic orbits, 154ff
  secular, 72, 109ff
  small denominators, 75–78
  superconvergent, 71
  superconvergent methods, 149ff
  two or more degrees of freedom, 83ff
  variational method, 155–156
Phase space
  boundary, 13
  coarse graining, 85, 303, 647, 652
  contraction, 459–460

  for dissipative system, 459–460
  extended, 15–16
  incompressible flow in, 14
  in Lorenz system, 66
  motion, 12ff
  reduced, 16, 19
  tangent vector in, 264
  tiling, 238
  trajectory, 12
  transformation of, 171
  volume, 13
  volume contraction of, 459–460
  web, 239
Plateau diffusion, *see* Diffusion plateau
Poincaré–Bendixson theorem, 460
Poincaré–Birkhoff theorem, 183–185
Poincaré surface of section, *see* Surface of section
Poisson brackets, 10–11
  use with Lie transformations, 135
Ponderomotive force, 144–147, 460
Power spectrum, 634
  for fluid velocity, 636
  for heat transport, 636
  for Lorenz system, 67
  for one-dimensional quadratic map, 493–495, 504–508
  for Rössler attractor, 469–470
Prediction, short and long time, *see* Reconstruction
Probability density, *see also* Invariant distribution
  conditional, 329
  transition, 321

Quadratic form, invariance of, 199
Quadratic mappings
  De Vogelaere, 514
  Hamiltonian, 512ff
  Hénon's dissipative, *see* Hénon attractor
  Hénon's twist, 191ff
  logistic, *see* Logistic map
  one-dimensional noninvertible, *see* One-dimensional quadratic map
  two-dimensional dissipative, *see* Two-dimensional dissipative map

Subject Index 689

Quantum systems, 653–654
Quasiaccelerator mode, 345ff
Quasicrystal, 239
Quasimode, 444
Quasilinear diffusion, 329, 414
   corrections to, 334–335

Randomness, 309ff
   definition of, 309
   relation to complexity, 309–310
Random phase assumption, 323–324, 388
Random sequences, 309–310
Rayleigh–Bénard convection, 65, 630, 631–634
   in binary fluids, 640
Rayleigh numbers, 632
Reaction–diffusion systems, 613
Reconstruction, 597ff
   autocorrelation function for, 599
   delay coordinates for, 598ff
   embedding dimension for, 589ff
   forecasting, 606
   of fractal dimension, 602ff
   global, 608ff
   local, 604ff
   mutual information for, 599
   noise reduction, 610ff
   redundancy for, 600
   singular systems analysis for, 601
   statistical quantities, 602ff
Reduction, 44ff
Redundancy, 600
Renormalization, 485ff, 542ff
   functional, 491–493, 542ff
   for intermittancy, 542ff
   for one-dimensional quadratic map, 485ff
   for power spectra, 493–495, 504–508
   for resonance islands, 122
   for two-dimensional dissipative quadratic map, 509–512
   for two-dimensional Hamiltonian map, 512–519
   for two resonances, 249, 281ff, 291–292
   transformation, 286

Renormalization group, 491, 542ff
Repeller, chaotic, 576
Rescaling parameters, *see* Renormalization
Residue, *see* Mean residue
Resonance
   between unperturbed frequencies, 111
   coupling, 385–386, 393ff, 409–410
   effect of, 77–78
   global removal of, 128ff
   higher order, 117ff
   islands, 49, 70
   in multidimensional oscillations, 373ff
   overlap, *see* Overlap criterion
   passage through, 101
   primary, 51
   removal of, 110ff
   removal of secondary, 119–120
   secondary, 51, 109
   secondary in wave-particle interaction, 124–126
Resonance layer, 51, 61–62, 375–380, 383–384, 406–407
Resonance streaming, 406ff; *see also* Extrinsic diffusion
   diffusion calculation, 407–412
   geometric construction, 406–407
   numerical results, 410–412
   of oscillation center, 406–407, 414–416
   relation to diffusion of a parameter, 416–418
Resonance surface, *see* Resonance layer
Resonance vector, 374–380, 406–407
Resonant denominators, 85, 98
Reverse bifurcations
   in one-dimensional quadratic map, 505
   in Rössler attractor, 469–470
Reversibility, 202
Reynold's number, 629
Rigid body, 46–48, 214
Rössler attractor, 466–471
Rotation, 22
Roundoff errors, 310–312
Ruelle and Takens model, *see* Transition to turbulence

Sandpiles, *see* Self-organized criticality
Scaling index spectrum, 592ff
Schrödinger equation, nonlinear, 449
Second-order islands, 189, 194, 248, 264ff, 291
　near elliptic fixed points, 265ff
　near the separatrix, 268ff
　for wave-particle interaction, 268
Self-organized criticality, 620ff
　avalanches, 621
　coupled differential equations, 625ff
　coupled mappings, 620ff
　earthquake model, 625–628
　sandpiles, 620
　scaling spectra for, 622–625
Separatrix
　motion for driven pendulum, 231ff
　motion for Duffing's equation, 565–567
　for pendulum, 28
　stochastic motion near, 51, 560ff
　web, 237–239
Separatrix crossings, 363ff
Separatrix layer, *see* Resonance layer
Separatrix mapping, 236ff, 243
　conversion to standard mapping, 251
　use in overlap criterion, 262–264
Shadowing theorem, *see* Noise reduction
Sine-Gordon equation, 450
Singularity, *see* Fixed point
Singular systems analysis for reconstruction, 601
Sink, 63, 461–468, 570ff; *see also* Fixed point
Skew-symmetric product, 199
Small denominators, *see* Resonant denominators
Solitons, *see* Integrable systems, soliton solutions
Space average, *see* Mean, space
Spatio-temporal chaos, 613ff
　domain formation, 614
　intermittancy, 617–619
　lattice maps, 613, 620
　natural spatial wavelength, 617
Spectrum, *see* Power spectrum

Stability, 195ff
　formal, 212
　linear, 198–200, 208–212
　nonlinear (Liapunov), 213
　spectral, *see* linear
Stadium problem, 654
Standard mapping, 168, 250ff
　accelerator modes, 253
　barrier transition, 256
　bifurcations, 254–256
　conversion from Fermi to, 251
　conversion from separatrix to, 251
　corrections to quasilinear diffusion for, 334–338
　dissipative, 519, 573–574
　effect of noise on, 353ff
　fractal diagram for, 279–281
　Greene's method applied to, 271ff
　Hamiltonian form, 256–258
　higher order diffusion in, 328ff
　higher period fixed points, 254–256
　Liapunov exponent for, 314
　overlap criterion for, 258ff
　period 1 fixed points, 253
　quantized, 654
　summary of transition to global stochasticity for, 291–292
Stationary point, *see* Fixed point
Stochastic barrier, *see* Barrier transition
Stochasticity, *see also* Stochastic motion
　across a resonance layer, 62
　along a resonance layer, 61
　concepts of, 63, 302ff
　extrinsic, *see* Extrinsic diffusion
　global (or strong or connected), 51, 246, 291–292
　local (or weak or isolated), 246, 291–292
　parameter for standard mapping, 169, 250
　regions of, 51–52
　transition to global, 245ff
Stochastic layer, *see* Resonance layer
Stochastic motion, 293ff; *see also* Stochasticity, Diffusion
Stochastic pump model, 387ff
　diffusion calculation, 387ff, 396–397
　for modulational diffusion, 401–405

Subject Index

Stochastic web, 237–239
Strange attractors, 63ff, 457ff, 521ff, 652; *see also* Lorenz system, Rössler attractor, Hénon attractor, Lozi attractor, Fermi acceleration, Dissipative mapping
  calculation of invariant distributions for, 580ff
  geometric properties of, 474ff
  leaved structure of, 471, 586–588
  Roux attractor, 600–601
  topology of, 469–471
Stream function for fluids, 640–643
Strong focusing, 647
Structural stability, 305
Superadiabaticity, 650, 651
Superconvergence, *see* Perturbation theory, Lie transformations, KAM theory
Superstable orbits, 492, 531
Surface
  energy, *see* Energy surface
  KAM, *see* KAM surface
Surface of section, 17–20
  for billiards problem, 382–385
  for dissipative systems, 458
  for driven pendulum, 232
Symmetries, *see* Integrals, isolating
Symplectic integration, 172
Symplectic maps, 208
  in many dimensions, 442, 451
Symplectic matrix, 197–200
System
  decomposable, 295, 305
  dissipative, *see* Dissipative systems
  ergodic, *see* Ergodic systems
  integrable, *see* Integrable systems
  near-integrable, *see* Near-integrable systems
  quantum, 653–654
  reaction–diffusion, 652

Tangent map, 208
Tangent vector, in phase space, 298
Tent map, 502–504
Three-body problem, 1, 644–645
Three resonance theory for Arnold diffusion, *see* Stochastic pump model

Tiling, 237, 239
Time average, *see* Mean, time
Time-dependent Hamiltonian, *see* Hamiltonian, nonautonomous
Time-independent Hamiltonian, *see* Hamiltonian, autonomous
Toda lattice, 36–39
Topology, change of, 72
Toroidal magnetic fields, 418ff
  configurations, 418–419
  diffusion in, *see* Diffusion in toroidal magnetic fields
  drift islands in, 427
  drift surfaces, 425
  Hamiltonian form, 421–422
  magnetic islands, 418ff
  magnetic surfaces, 420–422
Torus
  in phase space, 164–165, 375
  translations on, *see* Twist mapping
Trajectories
  exponential divergence of, 298, 307
  of the Hénon mapping, 191–192
  mean exponential rate of divergence, 299
  on a phase space torus, 164–165
  regular, 20, 49–51
  stochastic, 51–52
Transformation
  area preserving, *see* Area preserving
  canonical, *see* Canonical transformation
  Lie, *see* Lie transformation
Transformation theory, 7ff
Transient chaos, 570ff
  chaotic repeller, 536
  chaotic semiattractor, 576
  in digital phase-locked loop, 572
  in dissipative Fermi map, 574–576
  Fokker–Planck equation for, 572
  in generalized Hénon map, 572
  local description, 576–580
  near a crisis, 579ff
  in quadratic map, 570, 576, 579
  in standard map, 573–574
  weak dissipation, 571–576
  in weakly dissipative twist maps, 571

Transition to turbulence, 634ff
  intermittent model, 638
  Landau model, 634–635
  period-doubling model, 636–637
  Ruelle and Takens model, 635–636
Turbulence
  chemical, 651–653
  fluid, 634ff
  Lagrangian, 640
  transition to, see Transition to turbulence
Turnstile, 345
Twist mapping, 166, 296–297
  Liapunov exponent of, 313
  transient chaos in, 571
Two-dimensional dissipative maps, 508ff
  period-doubling bifurcations in, 508ff
  quadratic, see Two-dimensional dissipative quadratic map
Two-dimensional dissipative quadratic map
  expansion of general map, 509–510
  renormalization of, 510–512
  standard form, 510
Two-thirds rule, 250, 287, 291–292

Variational principle, 8
Variational theory
  for closed orbits, 155–156
  for KAM tori, 249

Wave-particle interaction
  accidental degeneracy for oblique propagation, 122–123
  canonical perturbation theory for, 87ff
  global removal of resonances, 131–133
  Hamiltonian, 87
  intrinsic degeneracy for perpendicular propagation, 124
  level curves (invariant curves), 90–93
  numerical results, 126–129
  resonances, 89–90
  resonant, 122ff
  second-order islands for, 268
Wedge operator, 301, 564
Winding number, 526, 530
WKB solution, 76
Wronskian determinant, 30

Zaslavskii map, 477

Applied Mathematical Sciences

(continued from page ii)

52. *Chipot:* Variational Inequalities and Flow in Porous Media.
53. *Majda:* Compressible Fluid Flow and System of Conservation Laws in Several Space Variables.
54. *Wasow:* Linear Turning Point Theory.
55. *Yosida:* Operational Calculus: A Theory of Hyperfunctions.
56. *Chang/Howes:* Nonlinear Singular Perturbation Phenomena: Theory and Applications.
57. *Reinhardt:* Analysis of Approximation Methods for Differential and Integral Equations.
58. *Dwoyer/Hussaini/Voigt (eds):* Theoretical Approaches to Turbulence.
59. *Sanders/Verhulst:* Averaging Methods in Nonlinear Dynamical Systems.
60. *Ghil/Childress:* Topics in Geophysical Dynamics: Atmospheric Dynamics, Dynamo Theory and Climate Dynamics.
61. *Sattinger/Weaver:* Lie Groups and Algebras with Applications to Physics, Geometry, and Mechanics.
62. *LaSalle:* The Stability and Control of Discrete Processes.
63. *Grasman:* Asymptotic Methods of Relaxation Oscillations and Applications.
64. *Hsu:* Cell-to-Cell Mapping: A Method of Global Analysis for Nonlinear Systems.
65. *Rand/Armbruster:* Perturbation Methods, Bifurcation Theory and Computer Algebra.
66. *Hlaváček/Haslinger/Necasl/Lovísek:* Solution of Variational Inequalities in Mechanics.
67. *Cercignani:* The Boltzmann Equation and Its Applications.
68. *Temam:* Infinite Dimensional Dynamical Systems in Mechanics and Physics.
69. *Golubitsky/Stewart/Schaeffer:* Singularities and Groups in Bifurcation Theory, Vol. II.
70. *Constantin/Foias/Nicolaenko/Temam:* Integral Manifolds and Inertial Manifolds for Dissipative Partial Differential Equations.
71. *Catlin:* Estimation, Control, and the Discrete Kalman Filter.
72. *Lochak/Meunier:* Multiphase Averaging for Classical Systems.
73. *Wiggins:* Global Bifurcations and Chaos.
74. *Mawhin/Willem:* Critical Point Theory and Hamiltonian Systems.
75. *Abraham/Marsden/Ratiu:* Manifolds, Tensor Analysis, and Applications, 2nd ed.
76. *Lagerstrom:* Matched Asymptotic Expansions: Ideas and Techniques.
77. *Aldous:* Probability Approximations via the Poisson Clumping Heuristic.
78. *Dacorogna:* Direct Methods in the Calculus of Variations.
79. *Hernández-Lerma:* Adaptive Markov Processes.
80. *Lawden:* Elliptic Functions and Applications.
81. *Bluman/Kumei:* Symmetries and Differential Equations.
82. *Kress:* Linear Integral Equations.
83. *Bebernes/Eberly:* Mathematical Problems from Combustion Theory.
84. *Joseph:* Fluid Dynamics of Viscoelastic Fluids.
85. *Yang:* Wave Packets and Their Bifurcations in Geophysical Fluid Dynamics.
86. *Dendrinos/Sonis:* Chaos and Socio-Spatial Dynamics.
87. *Weder:* Spectral and Scattering Theory for Wave Propagation in Perturbed Stratified Media.
88. *Bogaevski/Povzner:* Algebraic Methods in Nonlinear Perturbation Theory.
89. *O'Malley:* Singular Perturbation Methods for Ordinary Differential Equations.
90. *Meyer/Hall:* Introduction to Hamiltonian Dynamical Systems and the N-body Problem
91. *Straughan:* The Energy Method, Stability, and Nonlinear Convection.
92. *Naber:* The Geometry of Minkowski Spacetime.
93. *Colton/Kress:* Inverse Acoustic and Electromagnetic Scattering Theory.
94. *Hoppensteadt:* Analysis and Simulation of Chaotic Systems.
95. *Hackbusch:* Iterative Solution of Large Sparse Systems of Equations.
96. *Marchioro/Pulvirenti:* Mathematical Theory of Incompressible Nonviscous Fluids.
97. *Lasota/Mackey:* Chaos, Fractals, and Noise: Stochastic Aspects of Dynamics, 2nd ed.
98. *de Boor/Höllig/Riemenschneider:* Box Splines.
99. *Hale/Lunel:* Introduction to Functional Differential Equations.
100. *Sirovich (ed):* Trends and Perspectives in Applied Mathematics.

(continued on next page)

101. *Nusse/Yorke:* Dynamics: Numerical Explorations.
102. *Chossat/Iooss:* The Couette-Taylor Problem.
103. *Chorin:* Vorticity and Turbulence.
104. *Farkas:* Periodic Motions.
105. *Wiggins:* Normally Hyperbolic Invariant Manifolds in Dynamical Systems.
106. *Cercignani/Illner/Pulvirenti:* The Mathematical Theory of Dilute Gases.